New Discovery and Exploration Methods of Porphyry and Epithermal Mineral Deposits

New Discovery and Exploration Methods of Porphyry and Epithermal Mineral Deposits

Guest Editors

Chao Yang
Yiwei Peng
Gangyang Zhang

MDPI

Basel • Beijing • Wuhan • Barcelona • Belgrade • Novi Sad • Cluj • Manchester

Guest Editors
Chao Yang
College of Earth and Planetary Sciences
Chengdu University of Technology
Chengdu
China

Yiwei Peng
College of Earth and Planetary Sciences
Chengdu University of Technology
Chengdu
China

Gangyang Zhang
College of Earth and Planetary Sciences
Chengdu University of Technology
Chengdu
China

Editorial Office
MDPI AG
Grosspeteranlage 5
4052 Basel, Switzerland

This is a reprint of the Special Issue, published open access by the journal *Minerals* (ISSN 2075-163X), freely accessible at: https://www.mdpi.com/journal/minerals/special_issues/ON73U03CAB.

For citation purposes, cite each article independently as indicated on the article page online and as indicated below:

Lastname, A.A.; Lastname, B.B. Article Title. *Journal Name* **Year**, *Volume Number*, Page Range.

ISBN 978-3-7258-8183-3 (Hbk)
ISBN 978-3-7258-8184-0 (PDF)
https://doi.org/10.3390/books978-3-7258-8184-0

Cover image courtesy of Chao Yang

Contents

minerals

Editorial

Editorial for Special Issue "New Discovery and Exploration Methods of Porphyry and Epithermal Mineral Deposits"

Chao Yang, Yiwei Peng * and Gangyang Zhang

College of Earth and Planetary Sciences, Chengdu University of Technology, Chengdu 610059, China; yangchao_21@126.com (C.Y.); zhanggangyang@163.com (G.Z.)
* Correspondence: yiwei_peng@163.com

1. Background and Scope of the Special Issue

Porphyry deposits and their genetically and spatially associated epithermal deposits constitute one of the most economically significant mineral systems, attracting extensive attention from both industrial practitioners and academic researchers. Widely targeted in mineral exploration and genetic studies, this deposit system ranks among the most thoroughly investigated ore deposit types. Nevertheless, new discoveries and novel advances in its metallogenic mechanisms and exploration techniques continue to emerge. These findings deepen our fundamental understanding of porphyry mineralization and facilitate the formulation of optimized exploration strategies for prospective mineral prospecting.

To disseminate cutting-edge achievements regarding the metallogeny and exploration methodologies of porphyry and epithermal deposits, and keep relevant professionals informed about the latest research developments, we launch this Special Issue to share updated research insights on these two major deposit types.

This collection of 16 studies included in the Special Issue systematically investigates the geology, geochemistry, mineralogy, geochronology, hydrothermal evolution, metallogenic mechanisms, exploration indicators, and magmatic origins of diverse porphyry, skarn, epithermal, and trans-porphyry ore deposits distributed across Xizang (Tibet, China), Xinjiang (China), northern China, Türkiye, Canada, Australia, Brazil, Siberia, Hungary and France. These papers apply multiple analytical methods, including zircon U-Pb and Hf isotopes, Re-Os dating, fluid inclusions, stable S/C/O isotopes, SWIR spectroscopy, whole-rock geochemistry, indicator mineral chemistry (zircon, apatite, titanite, and pyrite), sulfide microanalysis and statistical discrimination models; they reveal hydrothermal fluid evolution, magmatic redox conditions, fluid mixing/precipitation mechanisms, magmatic source properties, alteration zoning, granite fertility evaluation, and magma reservoir processes. Collectively, these papers provide a comprehensive understanding of the porphyry–skarn–epithermal metallogenic systems, caldera-related epithermal mineralization, Archean–Paleoproterozoic ancient porphyry metallogeny, regional exploration vectors, and a novel trans-porphyry Sn–Ta–Nb–gem deposit metallogenic model, thus fulfilling the Special Issue's objective of disseminating cutting-edge progress in the field.

2. Individual Contributions of Each Paper

Vigneresse (2026) (Contribution 1) proposes a novel trans-porphyry deposit (TPD) metallogenic theory that is distinct from traditional subduction porphyry deposits, revealing low-crustal melting mechanisms for Sn-Ta-Nb-gem mineralization during large-scale shear between cratons within continental plates, internal decoupling, and vertical motion. The

 https://doi.org/10.3390/min16060595

study differentiates the formation processes of NYF/LCT pegmatites and gem minerals, providing new insights into porphyry-related deposit diversity for the Special Issue.

Guan et al. (2025) (Contribution 2) identify four pyrite types that record successive hydrothermal stages of the Zhunuo porphyry Cu deposit in Xizang, using LA-ICP-MS pyrite mineral geochemical data to reveal trace-element coupling behaviors and oxygen fugacity fluctuations during hydrothermal evolution. The study also establishes pyrite trace-element proxies and PLS-DA discrimination indicators, which effectively distinguish porphyry–epithermal systems on the Tibetan Plateau and contribute to the Special Issue's focus on hydrothermal processes and exploration indicator minerals.

Cheng et al. (2025) (Contribution 3) determine the petrogenesis of the Late Carboniferous subduction-related granites in the Baijianshan skarn Zn-Cu deposit via zircon U-Pb-Hf isotopes, whole rock geochemistry, and fluid inclusion characteristics. They clarify that the granites were derived from the juvenile crystal and formed in a subduction-related back basin, and ore-forming fluids originated from the mixing of magmatic and meteoric water. The research also confirms that the deposit forms a unified skarn–epithermal metallogenic system with adjacent epithermal deposits, advancing understanding of the correlation between skarn and epithermal systems highlighted in the Special Issue.

Kocatürk et al. (2025) (Contribution 4) characterizes multistage magmatic redox evolution and the continuous porphyry–skarn transition in the Eocene Tavşanlı Belt of Türkiye using whole-rock geochemistry, amphibole–biotite chemistry, and molybdenite Re-Os dating. The study constructs a unified syn–post-subduction genetic model for spatially associated Cu-Mo-W-Au mineralized systems, enriching the Special Issue's content on the metallogeny of porphyry-related deposits.

Hong et al. (2025) (Contribution 5) classify two types of early Oligocene ultrapotassic lamprophyre dykes in the Bangbule area of Xizang, constraining their lithospheric mantle source metasomatized by Indian continental crust melts through Sr-Nd isotopic analysis. The research proposes an open magma chamber model featuring fractional crystallization and mafic magma recharge, providing key insights into magmatic processes relevant to porphyry deposit formation.

Zhou et al. (2025) (Contribution 6) conduct in situ geochemical analysis of garnet and cassiterite from the Gongjuelong skarn Sn deposit, distinguishing the physicochemical conditions of different skarn stages. The study demonstrates that proximal skarn Sn and distal hydrothermal Pb-Zn-Ag deposits form an integrated granite-related metallogenic system in the Yidun Terrane, related to the late Cretaceous highly fractionated granites. This research proposed hypothesis highlights the potential prospecting of Sn mineralization beneath the hydrothermal Pb–Zn–Ag veins, aligning with the Special Issue's focus on deposit genesis and new exploration prospects.

Katz et al. (2025) (Contribution 7) systematically analyze the alteration lithogeochemistry and mass balance changes in the world-class Archean Côté Au(-Cu) porphyry deposit in Canada, defining zoned metal element assemblages and hydrothermal alteration footprints. The research further verifies the porphyry-type genetic model for this Archean gold system, contributing to the Special Issue's exploration of ancient porphyry deposits.

Quan et al. (2025) (Contribution 8) obtain systematic S/C/O isotope data of sulfides and carbonates from the Permian Drake epithermal Au-Ag field in Australia, revealing multiple mixed sources of ore-forming fluids and sulfur. The study clarifies metallogenic mechanisms in collapsed volcanic caldera settings, optimizing epithermal exploration strategies and addressing the Special Issue's focus on epithermal deposit exploration.

Li et al. (2025) (Contribution 9) integrate SWIR spectroscopy, EPMA, and XRD to study hydrothermal alteration minerals in the Sinongduo low-sulfidation epithermal deposit in the Gangdese belt of Xizang, establishing spatial alteration mineral zoning and white

https://doi.org/10.3390/min16060595

mica SWIR exploration models. The research predicts high mineralization potential in the northern Woruo area, advancing spectroscopy-based exploration methods highlighted in the Special Issue.

Wang et al. (2025) (Contribution 10) analyze muscovite SWIR spectral parameters (Al-OH wavelength and crystallinity) of the Yixingzhai gold deposit in Shanxi, revealing distinct spectral differences among porphyry, quartz vein, and breccia mineralization types. The study proposes muscovite crystallinity as a key indicator for porphyry gold exploration, enriching the Special Issue's content on exploration indicators.

Lee et al. (2025) (Contribution 11) conduct detailed geochronology and petrology on the Late Cretaceous porphyry and epithermal deposit in Canada's Dawson Range gold belt, identifying two newly recognized causative magmatic suites linked to Au mineralization. The research reveals a tectonic transition from subduction to local extension that controls regional porphyry belt metallogeny, contributing to the Special Issue's focus on geochronology and regional metallogeny.

Crépon et al. (2024) (Contribution 12) confirm the Neoarchean Chibougamau pluton hosts typical porphyry-type Cu-Au mineralization via whole-rock geochemistry, zircon chemistry, and fluid inclusion data. The study clarifies that moderately oxidized intermediate magmas provide metalliferous fluids, establishing an Archean porphyry metallogenic genetic model and expanding the Special Issue's coverage of ancient porphyry systems.

Lopes et al. (2024) (Contribution 13) constrain the hydrothermal conditions, fluid sources, and gold precipitation mechanisms of the Paleoproterozoic Raimunda porphyry Au deposit in Brazil's Amazonian Craton using fluid inclusion and stable isotope data. The research provides important prospecting implications for ancient Amazonia cratonic porphyry gold systems, supporting the Special Issue's exploration focus.

Berzin et al. (2024) (Contribution 14) evaluate granite metallogenic fertility in Siberia's Taimyr Peninsula using zircon, apatite, and titanite indicator minerals plus whole-rock geochemistry. The study identifies oxidized water-saturated fertile granitoid intrusions favorable for Cu-Mo mineralization and grades regional exploration targets, addressing the Special Issue's focus on indicator minerals and exploration strategies.

Biró et al. (2024) (Contribution 15) perform a comparative sulfide compositional study on porphyry, skarn, and carbonate-replacement mineralization in Hungary's Recsk complex, establishing distinctive trace-element geochemical fingerprints for each mineralization type. The research proposes sulfide trace-element evolutionary trends as effective hydrothermal system vectoring tools, contributing to geochemical exploration methods in the Special Issue.

Gao et al. (2024) (Contribution 16) systematically summarize the geological, geochemical, and geophysical characteristics of the Luming porphyry Mo deposit in NE China, revealing that hydrothermal breccias control mineralization and form distinct metal primary halo zoning. The study confirms the deposit belongs to a complete porphyry-skarn-epithermal system, providing regional prospecting guidance and aligning with the Special Issue's exploration and metallogeny focus.

Acknowledgments: The Guest Editors thank the authors, reviewers, and the Editorial Board, as well as the staff from *Minerals*, for their contributions to this Special Issue.

Conflicts of Interest: The authors declare no conflicts of interest.

List of Contributions:

1. Vigneresse, J.-L. Specificity of Ore Generation (Tin, Pegmatites, and Gems) in Trans-Porphyry Deposits. *Minerals* **2026**, *16*, 157. https://doi.org/10.3390/min16020157.
2. Guan, H.; Luosang, J.; Gao, L.; Xie, F. Pyrite Trace-Element Signatures of Porphyry-Epithermal Systems in Xizang: Implications for Metallogenic Discrimination and Hydrothermal Evolution. *Minerals* **2025**, *15*, 1113. https://doi.org/10.3390/min15111113.

https://doi.org/10.3390/min16060595

3. Cheng, F.; Zhang, S.; Wu, J.; Huang, B.; Zhang, D. Genesis of the Baijianshan Skarn-Type Zn-Cu Polymetallic Deposit, Chinese Eastern Tianshan: Constraints from Geology, Geochronology and Geochemistry. *Minerals* **2025**, *15*, 1107. https://doi.org/10.3390/min15111107.
4. Kocatürk, H.; Kumral, M.; Sendir, H.; Kaya, M.; Creaser, R.A.; Abdelnasser, A. Magmatic Redox Evolution and Porphyry–Skarn Transition in Multiphase Cu-Mo-W-Au Systems of the Eocene Tavşanlı Belt, NW Türkiye. *Minerals* **2025**, *15*, 792. https://doi.org/10.3390/min15080792.
5. Hong, C.; Wang, Y.; Zunzhu, S.; He, C.; Wang, H.; Wang, Y.; Bai, Y.; Yan, P.; Xun, S.; Cao, R.; et al. Mineralogy and Petrology of Ultrapotassic Lamprophyre Dykes in the Bangbule Area, Xizang, China: Evidence for Open Magma Chamber Fractionation and Mafic Magma Recharge. *Minerals* **2025**, *15*, 332. https://doi.org/10.3390/min15040332.
6. Zhou, Y.; Peng, Y.; Liu, C.; Tian, J.; Wang, Z.; Song, M.; Zhang, Y. Genesis of the Gongjuelong Sn Polymetallic Deposit in the Yidun Terrane, China: Constraints from the In Situ Geochemistry of Garnet, Cassiterite, and Quartz. *Minerals* **2025**, *15*, 314. https://doi.org/10.3390/min15030314.
7. Katz, L.R.; Kontak, D.J.; Dubé, B. Alteration Lithogeochemistry of an Archean Porphyry-Type Au(-Cu) Setting: The World-Class Côté Gold Deposit, Canada. *Minerals* **2025**, *15*, 256. https://doi.org/10.3390/min15030256.
8. Quan, H.; Graham, I.; Worland, R.; Adler, L.; Dietz, C.; Madayag, E.; Wang, H.; French, D. Interpreting the Complexity of Sulfur, Carbon, and Oxygen Isotopes from Sulfides and Carbonates in a Precious Metal Epithermal Field: Insights from the Permian Drake Epithermal Au-Ag Field of Northern New South Wales, Australia. *Minerals* **2025**, *15*, 134. https://doi.org/10.3390/min15020134.
9. Li, X.; Guo, N.; Li, C.; Deng, S.; Zhou, W. Shortwave Infrared Spectroscopy and Geochemical Characteristics of White Mica-Group Minerals in the Sinongduo Low-Sulfide Epithermal Deposit, Tibet, China. *Minerals* **2025**, *15*, 104. https://doi.org/10.3390/min15020104.
10. Wang, L.; Wu, S.; Lai, X.; Yang, W.; Sun, R.; Liu, P.; Yang, Y.; Ren, Y. The Short-Wave Infrared (SWIR) Spectral Exploration Identification and Indicative Significance of the Yixingzhai Gold Deposit, Shanxi Province. *Minerals* **2025**, *15*, 83. https://doi.org/10.3390/min15010083.
11. Lee, W.-S.; Kontak, D.J.; Sack, P.J.; Crowley, J.L.; Creaser, R.A. Geology, Petrology and Geochronology of the Late Cretaceous Klaza Epithermal Deposit: A Window into the Petrogenesis of an Emerging Porphyry Belt in the Dawson Range, Yukon, Canada. *Minerals* **2025**, *15*, 38. https://doi.org/10.3390/min15010038.
12. Crépon, A.; Mathieu, L.; Kontak, D.J.; Marsh, J.; Hamilton, M.A. An Archean Porphyry-Type Deposit: Cu-Au Mineralization Associated with the Chibougamau Tonalite–Diorite Pluton, Abitibi Greenstone Belt, Canada. *Minerals* **2024**, *14*, 1293. https://doi.org/10.3390/min14121293.
13. Lopes, A.A.C.; Moura, M.A. The Paleoproterozoic Raimunda Porphyry-Type Gold Deposit, Juruena Mineral Province, Amazonian Craton (Brazil): Constraints Based on Petrological, Fluid Inclusion and Stable Isotope Data. *Minerals* **2024**, *14*, 1185. https://doi.org/10.3390/min14121185.
14. Berzin, S.V.; Konopelko, D.L.; Petrov, S.V.; Proskurnin, V.F.; Berzon, E.I.; Kurapov, M.Y.; Golovina, T.A.; Chernenko, N.Y.; Chervyakovskiy, V.S.; Palamarchuk, R.S.; et al. Evaluation of Granite Fertility Utilizing Porphyry Indicator Minerals (Zircon, Apatite, and Titanite) and Geochemical Data: A Case Study from an Emerging Metallogenic Province in the Taimyr Peninsula, Siberian High Arctic. *Minerals* **2024**, *14*, 1065. https://doi.org/10.3390/min14111065.
15. Biró, M.; Raith, J.G.; Feichter, M.; Hencz, M.; Kiss, G.B.; Virág, A.; Molnár, F. Comparative Study of Sulfides from Porphyry, Skarn, and Carbonate-Replacement Mineralization at the Recsk Porphyry-Mineralized Complex, Hungary. *Minerals* **2024**, *14*, 956. https://doi.org/10.3390/min14090956.
16. Gao, B.; Dong, M.; Xie, H.; Liu, Z.; Li, Y.; Zhou, T. Discovery and Exploration of the Luming Porphyry Mo Deposit, Northeastern China: Implications for Regional Prospecting. *Minerals* **2024**, *14*, 718. https://doi.org/10.3390/min14070718.

minerals

MDPI

Article

Specificity of Ore Generation (Tin, Pegmatites, and Gems) in Trans-Porphyry Deposits

Jean-Louis Vigneresse

GéoRessources, UMR 7539, Université de Lorraine, BP 23, 54501 Vandoeuvre-lès-Nancy Cédex, France; jean-louis.vigneresse@univ-lorraine.fr

Highlights

- ore (Sn, gems) formation in Trans-Deposits TPD
- pegmatites LCT/NYF
- S- and A-type granites
- chemical evolution of pegmatites (polarizability)

Abstract

During the magmatic stage, base and rarer metals segregate from silicate melts to form ore deposits. The usual case is the porphyry (PD) type (Cu, Mo, and W) above subduction zones. The metal grade increases from some ppb or ppm up to percent levels. A new type of trans-porphyry (TPD) deposits (Sn, Ta, Nb, and gems) results from large-scale shear between cratons within continental plates, internal decoupling, and vertical motion. The bulk ore generation process develops along three stages: from magma generation, emplacement, and the formation of an immiscible magmatic phase (MIP), fluids, and melt. However, in TPD, metals segregate from the crust during melting below 800 °C, biotites break down, and the melt remains below the critical point (731 °C). Fluid advection competes with chemical diffusion, yielding the required enrichment. The subcritical MIP splits into a silicate-rich and an aqueous-rich phase, which are both incompatible with each other. Granite, pegmatites, and greisen coexist in the magma chamber. Their respective extraction from a composite mush involves electron exchanges between charges, or orbitals, yielding metal oxides through chemical diffusion. In contrast, in metals (Nb and Ta) observed in pegmatites, and also in gems, electrons rearrange their electronic cloud through their polarizability. Lastly, gems independently grow under the influence of the extremely hard fluids (Li, Be, and B). Magma generation, involving the lower crust (garnet and pyroxene), results in melts that form the two observed pegmatite groups (NYF and LCT), with each being associated with alkaline (A-type) or continental (S-type) granitic melts.

Keywords: ore formation; diffusion/advection; trans-porphyry deposits; porphtyry; gemstones

1. Introduction

1.1. Magma and Ore

Ore deposits of economic interest traditionally relate to magmatic activity depending on the tectonic settings, e.g., [1]. Initially, metals are present in the continental crust in proportions ranging from ppb to ppm. They are transported and concentrated into magmas during crustal melting due to thermal anomalies. This initial episode is a phase change and thus pertains to energy exchanges between its three components: temperature (T) and entropy (S); stress (σ) and strain (ε); and chemical potential (μ) and amount (N). All materials are considered viscous.

The conditions of metal transportation during melt emplacement and late individualisation are still partly debated. Ultimately, metals can enter crystalline phases into minerals or segregate to form specific ore minerals. The common ore grade may reach the percent level in some deposits of economic interest. The formation of magmatic ore, such as porphyry deposits (PDs), is equivalent to solving the enrichment of a specific element by a factor of four to six orders of magnitude. Three major hypotheses have been explored [2], which include source enrichment, chemical evolution of the magma, and late hydrothermal activity.

The first case relies on the observation of linear belts of such porphyries, suggesting an anomalous content in metals, either in the crust [3] or the mantle [4,5]. Nevertheless, enrichment rarely exceeds a maximum of one order of magnitude, far from the expected values.

The second model involves magmatic evolution through differentiation with magma specialization towards one metal, with eventual magma re-intrusion in a former body. This has been the preferred model for a long time [6,7], linked with the selective diffusion of elements.

The last model implies ultimate hydrothermal activity, illustrated by fluid inclusions in the rocks and alterations in the surroundings. It manifests through the exsolution of an immiscible fluid phase at the magmatic stage (MIP), followed by the brecciation of the surrounding rocks, vein formation, and hydrated mineral deposition [8,9].

1.1.1. Ore Formation in PD and MA

As a good example, in the continental crust, base (e.g., Cu, W, and Mo) and rare metals (Au and Ag) link with intermediate felsic magmas. They develop in linear belts of porphyry deposits (PDs), the heat source of which is provided by an underlying oceanic plate subduction (Sillitoe, [10]). This represents an infinite heat source at the scale of a magma chamber. Within plates, efficient heat sources may have a deeper origin, giving way to mafic intrusions (MAs) and sulfide metal assemblages (Ni and Cu), including platinum group elements (PGEs) [11]. Other tectonic settings need to benefit from a perennial heat source to drive magmatic activity. Such magmatic bodies have been grouped into trans-porphyry deposits (TPDs) since they develop in oblique plate convergence, yielding trans-pression/tension [12] with specific thermal conditions. They have consequences for the physical parameters of the magma (temperature, viscosity, and fluid content) and metal attractivity (chemical potential and diffusivity), whereas the ambient stress pattern may control partial melt extraction and interactions with the mobile fluid phases.

Briefly, ore formation in PDs or MAs can be summarized as melting, followed by the segregation, transport, and concentration of metals under one of the three phases (solid, melt, and fluid). The physical conditions of segregation and transport are addressed through viscosity, diffusivity, permeability, and the amount of fluid [13]. Furthermore, the metal attractivity for each phase pertains to chemistry.

1.1.2. The Specific Case of TPDs

TPDs also develop in linear belts due to the intense shearing between older and stiffer cratons. The differential motion between the cratons and the sheared crust represents a moderate heat source that may induce melting and magma chamber construction. This may vary with time according to the plate convergence, introducing rheological decoupling within the crust and vertical motion [14]. Heterogeneities within the sheared crust also modify the near-field stress pattern, leading to the formation of magmas, which are either derived from the crust or from the adjacent lower portion of cratons. Such

 https://doi.org/10.3390/min16020157

melts liberate metals that are later incorporated into one of the three phases of presence (solid, melt or fluid), except that they all originate from the continental crust, and not from the mantle, as for PDs. Their segregation and transport develop as in the PD, with the restriction that the temperature remains moderate, below 731 °C, maintaining the fluids in a subcritical state [15]. Accordingly, the temperature does not enhance metal diffusion, nor does it promote fluidity in the melt.

As a result, the process of ore formation shifts from a chemical multiphase flow problem (T and m) in PD towards a mechanical problem (T, μ, and σ) in TPD.

Nevertheless, the problem increases in complexity since magma develops in a subcritical state [12]. The MIP splits into two incompatible phases (Figure 1), one silicate-rich and the other aqueous-rich [16]. In consequence, the bulk magma chamber incorporates three sub-chambers, which independently evolve. Ore formation results in common ore minerals in the form of pegmatites and greisen, which are coarse-grained, equivalent to granite, but located in individualized pockets, as well as gemstones that form from the MIP components.

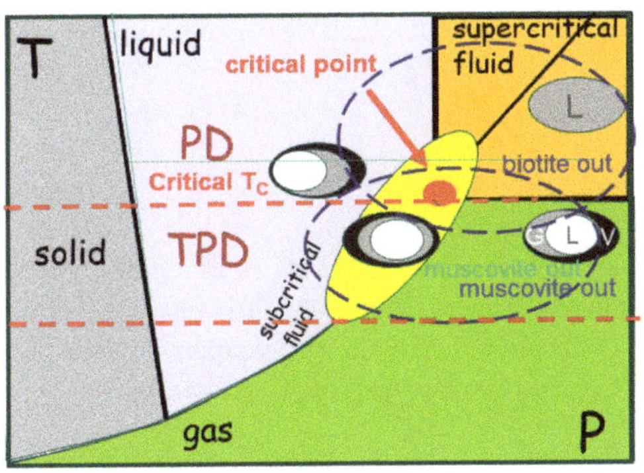

Figure 1. Schematic representation of phase change mapping under subcritical conditions. The system passes through the solid, liquid, and fluid conditions, whereas the gas phase is reached at high pressure. Melting is scaled as minerals (muscovite, biotite and amphibole) are involved. At 731 °C, the critical temperature, the fluid phase splits in two MIPs (liquid, vapour, glass), as illustrated by the melt inclusions (MI). The system cools from the supercritical state, where solubility is quasi-infinite. MPIs become subcritical, splitting into a silicate-rich and a fluid-rich phase, as observed in the melt inclusions.

1.1.3. Previous Models

Basically, two types of models have been designed to simulate the formation of ore deposits: the Eulerian and Lagrangian approaches [17] In the first case, an analytical approach describes the bulk system using an external frame, whereas a description of a discrete element attached to the system characterizes the second method.

The Eulerian description requires fixed system parameters, such as magma composition, temperature, melt viscosity, and near-field stress conditions. Chemistry rules are used to partition elements between the different phases, quantifying enrichment. This is the most debated, and most uncompleted, data base [18] The latter values are used to fix the enrichment factors for metals. However, the description remains imprecise in face of the randomly distributed different phases, leaving the interaction between them and their instabilities unexplored.

In contrast, techniques using Lagrangian descriptions, such as lattice Boltzmann models (LBM), examine the interactions between neighbouring particles (gas, metal, melt).

 https://doi.org/10.3390/min16020157

They provide estimates of shape evolution and metal enrichment, bracketed according to the ratio of partitioning between metals and MIP [19].

Both types of models provide solutions, which are unique in the direct problem and not too far from truth in the second but strongly dependent on the input conditions, that remain quite uncertain. See, for instance, the discussion about the amount and content of the additional elements (F, Li, Be) in the MIP, for which there is no direct constraint, since the MIP evaporates at the end of ore generation [11].

1.2. Energy Considerations

In consequence of the problems presented above, a different approach, including thermodynamics and energy considerations is suggested, which involves interactions and instabilities.

The origin of metal extraction, segregation, and transport before forming ore deposits is undoubtedly linked to magma production in the continental crust. Restricting their formation to particular continents does not modify these premises. However, addressing ore genesis from the perspective of a phase change supposes the energy budget is considered, and thus requires an examination of the different components of free energy, provided the system remains closed, avoiding the effects of dissipating components.

1.2.1. Energy and Melting

Magma melting is not simply an effect of temperature increases. Such an effect is primordial and produce mobile melting, but it also has consequences for the physico-chemical properties of the melt components. Temperature modifications generally enhance melt displacement. They also influence the exsolving of a composite fluid phase, which, in turn, has an effect on chemical attractivity for ionic metals. Finally, elements must adapt their structure, crystal lattice, or molecule formation, thus varying their entropy.

1.2.2. Energy Exchanges

A phase transition describes the energetic variations in a material involved in a reaction, for example, a transformation from solid to liquid [20]. Many textbooks on thermodynamics exist, mostly referencing solid, liquid, and gas states [21]. The reactions use pressure, temperature, and the amount of chemical element (T, P, N_i) in a function of time (reaction advance), (Figure 2a) with intensive variables such as entropy, volume, and chemical potential (S, V, μ_i)

$$dG = V\,dP - S\,dT + S_i\,m_i\,dN_i \tag{1}$$

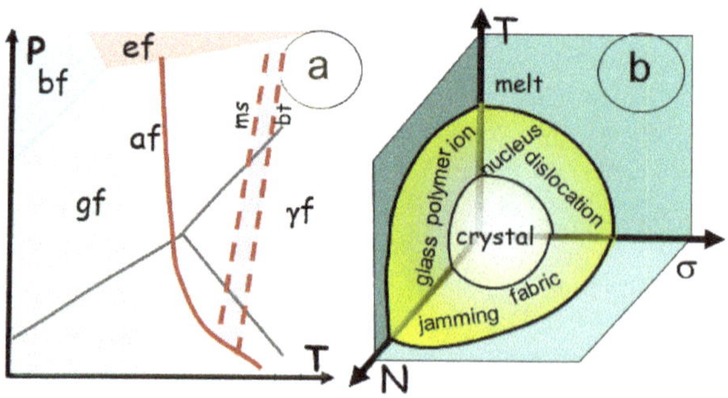

Figure 2. Phase diagram presented for different situations. (**a**). Traditional 2D diagram of pression temperature (P, T) for solid/liquid/gas transition with the binary boundaries including a triple point with the three phases in contact and a critical point above which the material enters a supercritical fluid,

https://doi.org/10.3390/min16020157

with disappearance of the meniscus between them. The triple and critical points are specified using their respective pressure and temperature. (**b**) A 3D phase diagram for crystal/melt modified from a former version of [22]. The usual variables of the energy are used, such as T, S, or temperature and entropy; N, μ, or density and chemical potential; and σ, ε or stress and strain. The binary transition between crystal and melt gives place to a transitional stage, where ions, glass, and polymers develop, creating a fabric, and jamming, and proceeds through nucleation and dislocations. During the phase transition, which commonly involves energy, exchanges occur between the components (S, μ, ε), allowing for the energy variation to be readjusted.

A simple PT diagram is no longer explicit enough, and the system must be considered within a higher-variance diagram (Figure 2b). In igneous rocks, the solid and the liquid phases are both viscous. In a first approach, a compressible gas phase with a restricted amount can be considered soluble in the melt. Therefore, in melt and solids, the energy changes can be expressed in terms of strain (ε), whereas stress (σ) is the applied pressure instead of V and P. The entropy (S) of the system represents its order, which increases from the crystalline network, polymers, and melt. It should evolve to a maximum value [23]. An average chemical potential (μ) and the amount N of chemical elements, taken as the mean value of their composition, may be used as a proxy for the melt and the solid rocks [17]. Consequently, the phase transition should be described with a higher-variance approach, including, at least, T, σ and N, and thus examining the variations in bulk energy along each coordinate and their complex interactions (Figure 2). For instance, enhanced melting or recrystallization (entropy changes) should affect, or be affected by, the melt flow in its components.

$$dG = \varepsilon\, d\sigma - S\, dT + \mu N \quad (2)$$

The bulk Gibbs free energy variation resulting from a reaction can be expressed by the difference between the sum of Gibbs free energy of the reactants and the products of the reaction.

1.2.3. The Case of Complex Systems

The evolution of phase transitions during melting (or crystallization) is commonly described in P-T diagrams indicating the physical entropy of the system (liquid, solid) (Figure 3a). For example, melting reactions for hydrous metapelites have been extensively described [24,25].

A thermodynamic review of metamorphic rocks' conditions has recently been issued [26]. It extends the simple binary system with no intermediate phases (Figure 3a) so the liquid and the solid coexist in a specific P-T-N range.

When the material incorporates multiple phases (e.g., binary with A and B), the transition goes through a high-variance domain delineated by the melting curves of each material, with a melting temperature (T_{mA} and T_{mB}) that may differ within each individual end-member (Figure 3b). Consequently, a 2D P-T diagram is not sufficient, and a T-N diagram can be introduced to describe the complex system (Figure 3b). A higher-variance domain develops in which a mixture of the end-members co-exists. Depending on the chemical system considered and the number of phases present, the transitional zone may be bivariant or multivariant with a eutectic point. The composition of the solid phase is a function of the lowest melting temperature of the mixture. In this space, energy varies as a function of the amount of each phase depending on P-T or σ-T. The structural effects, imposing order on the crystalline assemblage, are not considered. Energy variations are present and linked to interactions between the other components (S, ε, μ) and (T, σ, N) (Figure 2b). They develop as instabilities and form feedback loops, often minor, during direct modeling of phase transition.

https://doi.org/10.3390/min16020157

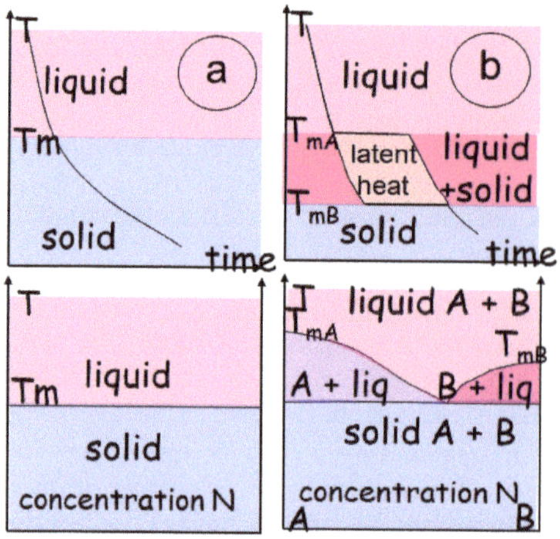

Figure 3. Two phase transitions, described as T versus time (t) and T versus concentration (N). (**a**) In a binary transition, a melting temperature (T_m) determines the abrupt change from a solid to a liquid; the concentration, in terms of elements, is similar in both states. (**b**) Conversely, when the material has several components, the transition adopts the path of the two melting curves for materials T_{mA} and T_{mB}. The result is a domain in which the composition follows the melting curves. The transitional zone may be a single domain, or several coexisting domains with a eutectic point and a eutectic composition determining the lowest melting temperature of the mixture.

1.3. Establishing the Problem

The complex problem of ore formation in magmatic settings does not start from scratch. It is based on field observations, laboratory observations, and experiments, as well as physico-chemical concepts, including the properties magma components (solid, melt, fluids). Nevertheless, addressing a phase change for a complex multiphase system requires consideration of the free energy balance between the components.

First, the free energy (G) for the melt and matrix relies on three major internal and external forces (Figure 2). These are controlled by temperature (T) and entropy (S), stress (σ) and strain (ε), chemical potential (μ_i), and the number of reactants (N_i). Their interactions drive the whole system to evolve toward a state of equilibrium characterized by a minimum value of Gibbs free energy. This requires describing a model of probability distributions by selecting the macrostate with the highest entropy [23]. However, system perturbations introduce instabilities, which are mostly irreversible. In addition, the processes of melting, extraction, and crystallization point to an open-system description and non-equilibrium thermodynamics. Since the phase transition does not allow for the system components to escape, a description using a Gibbs free energy may apply.

More commonly, the descriptions of geological processes involve materials with contrasted properties. Those present variations in state or ordering (from a crystal to a liquid), rheology (from solid rocks to viscous lavas), or chemistry (from felsic to mafic). The coexistence of such contrasted properties is a natural factor in instabilities. In addition, earth materials are not binary, e.g., with a weak and strong end-member, but encompass several intermediate phases governing each phase transition. Understanding the processes controlling transitional states and their intertwined mechanical, chemical, and thermodynamical properties should provide insights into their evolution towards equilibrium, providing insights into their instabilities and eventually contributing to predicting the evolution of the entire complex system.

The final aspects of ore formation are also strongly controlled by field observations.

 https://doi.org/10.3390/min16020157

1.3.1. TPD in the Field

The TPDs are characterized by localized intrusions within a transpressional zone of intense shearing [27]. Mechanical decoupling within the crust induces instability in the crust and vertical motion, in addition to the horizontal shear (Figure 4). The magmas are mainly derived from melting of the continental crust, with few traces of mantle-derived magmas and fluids (mafic melts and S compounds). Such magmatic bodies mostly originate during continental shear or very oblique collision.

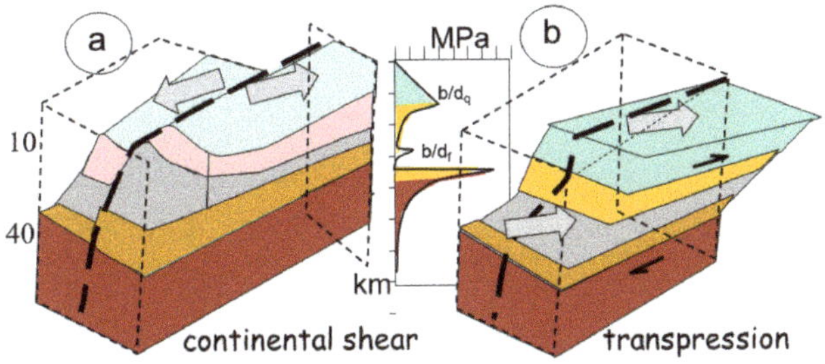

Figure 4. Regime of continental crust deformation (**a**) illustrating continental shear, manifested by the horizontal displacement of blocs, without vertical motion, and (**b**) transpressional dextral shear. Decoupling occurs at the brittle/ductile transition of quartz (bdq) and brittle/ductile transition of feldspar (bdf), above a brittle Moho. The regime of stress is also indicated (the so-called fir-tree), which reflects the maximum stress of the material.

The width of the shear zone acts as a focusing effect on the magma production. Though there are numerous worldwide examples, we present three cases to document the relation between shear zone and magma. When the sheared zone is restricted in width between two adjacent cratons, the resulting granitic massifs develop in small stocks along linear structures such as the Kivu Belt, or Kibarides, Central Africa, along 1500 × 150 km [28,29].

The continuing oblique continent–continent collision between the Australian and Pacific plates resulted in the formation of the Southern Alps, and Fiordland in southern New Zealand, with granite and pegmatites from the Mataketake Range, and the Stewart Island in the southwest of this orogen [30–32]. The Mesozoic Median Batholith, a S-type granite, is now dissected by the Alpine fault, though it intrudes into a 400 × 75 km shear zone.

In the Hercynian granitic belt [33–35] (Figure 5), S-type granitic intrusions run along 2500 × 600 km, from South Mountain Batholith (SMB), Nova Scotia [36]; through Land's End, Dartmoor (Cornwall), and Pontivy, Plouaret (Brittany); to Erzgebirge, Bohemia as Podlesi Cinovec (CI), Ehrensfriderdorf (EH) and Krusne Hory (KH) [37]. Globally, analyses of Bohemia have been issued in various specialized papers from the group of Rainer Thomas (GFZ, Potsdam) and Karel Breiter (Czech Geol Survey, Prague). These S-type granitic intrusions present root zones in the zone of local extension [38], but the bulk motion is shearing. In contrast, the banks of the shear zone give place to compression against the cratons, with subsequent relaxation of the crust. These occur in the place of so-called anorogenic pink alkali granites (A-type) or ferroan granites [39]. Examples (Figure 5) include Pleasant Ridge, New Brunswick [40]; Hensbarrow (HE) and St Austell, Cornwall [41]; northern Brittany (Aber-Ildut, Ploumanach (PL), Flamanville, Barfleur [42], and Vernéřov (VE) in Bohemia [43] (Figure 5). Transpressional zones are characterized by the coexistence of the two types of granitic magmas owing to heterogeneities in the crust. When the neighbouring craton is involved, the lower crust may undergo melting. Adopting a viscosity of 10^{17} Pa·s for the lower crust yields a delay of about 30 Ma between the S-type and the A-type granite emplacements.

https://doi.org/10.3390/min16020157

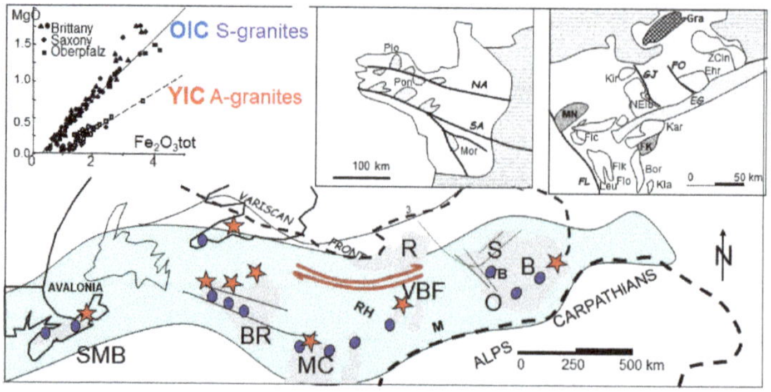

Figure 5. Reconstructed map of the Variscan deformation belt from eastern Canada to Bohemia. The belt intrudes on several granitic massifs, shown as dots (S-type) and stars (A-type). The whole belt is submitted to a dextral shear, presented by y shear zones. The magma categories (A- and S-types) are identified by MgO and Fe_2O_3 (inset).

1.3.2. Model of Continental Shear Zone

A continental shear zone develops in response to an oblique to nearly parallel convergence of cratons (Figure 4a). The two cratons present an older ticker and stiffer crust, which commonly occur with a granulitic dry lower crust, establishing its structural stability. The rheological response of the crust to stress is like that of a layered crust in 1D. The uppermost crust is controlled by quartz up to about 300 °C, when it becomes ductile (Figure 4). It corresponds to the seismic active zone, which manifests through rupture when the stress becomes too important. Below 20–22 km in depth, more abundant and stiffer feldspars reinforce the crust, first using subgrain rotations yielding ductile feldspathic brittle/plastic transition at about 450 °C. This is the place in which ductile shear zones anastomose gneissic shear zones that progressively flatten before encountering Moho at about 750 °C, where olivine takes control of a stiffer upper mantle that can undergo brittle deformation, as observed in seismic sections, in cases of large strain.

The general equations that govern mineral deformation (ε°) are a power law with stress (σ) through a factor (n), with a pre-exponential factor E_0, and an activation energy (Q) function of temperature [43].

$$e^{\circ} = E_0\ s^n\ d^p \exp(-Q/RT) \tag{3}$$

Depending on the minerals and experimental conditions, the grain size (d) may form an equation with a power law (p). The bulk trend is that reducing the grain size increases stiffness. The values of the most important minerals are summarized [44].

The bulk strength of the crust manifests in the so-called fir-tree aspect of the curve, with a linear strength increase up to ductile transition above a weaker layer (Figure 4b). Experimental deformation under varying ambient stress and a controlled shear rate have long been the basis of all models of crustal response to imposed stress field [43]. These are used to derive the values of the expected transitions in compression and simple shear experiments. In parallel, 2D and 3D numerical simulations of deformation reveal the importance of gravity in modifying the relief due to the compressional forces. However, they confirm the 1D layer model and can be used to investigate the non-linear aspects of the deformation behaviour in the case of rapid contrasts in rock properties or strain response to stress.

Such 1D models also serve to elucidate the geometric conditions internal to shear zones, identifying vorticity variations to take crustal heterogeneities into account [45].

 https://doi.org/10.3390/min16020157

The adopted geometry corresponds to a vertical shear zone within a weaker zone bracketed between adjacent cratons (Figure 4a). Owing to the 1D modelling, the vertical motion of the weaker central zone is not considered. The temperature profile with depth is controlled by the basal heat flow. Experimental conditions as well as field observations concluded that the basal friction induced by the heat flow remains unable to significantly alter the temperature profile with depth. Similarly, the friction of the shear planes, often manifest in local melt (tachylites), remains localized and has high strain rates, being rarely observed at usual tectonic rates (10^{-14} s^{-1}).

1.3.3. Continental Transpression

In the case of oblique convergence and melting, the deformation switches from a simple 2D geometry, shown by strike-slip along the fracture plane (Figure 4a), deviating from simple shear by adding a pure shear component across the zone, resulting in transpression (shortening) or transtension (extension) (Figure 4b). The result is a large spectrum of 3D deformation with its own complex strain reactions, modifying fabric rotation (the simple shear component) [44]. Owing to this perpendicular combination of a pure shear and a simple shear, the resulting deformation path becomes much more complex than for a simple shear zone.

The major difference with a continental shear zone results from the partitioning of the deformation. In the former case, the high-strain zone is localized in a subvertical plane to accommodate, through brittle fractures, the horizontal displacement between the two adjacent craton roots (Figure 4a). In contrast, during transpression, the upper part of the crust, controlled by quartz and feldspars, also reacts in a brittle mode, inducing fracture zones. However, the lower part plastically deforms under pure shear, flowing horizontally (Figure 4b) according to the basal frictional drag with the upper mantle. The conjunction of both patterns of deformation results in complex geometries of flow, resulting in mineral fabrics [46].

1.3.4. Implications of a Basal Friction

Another important character of transpression results from its driving force (Figure 4). A common fracture results from the accumulation of stresses induced by an ambient convergent stress pattern, usually noted $\sigma_1 > \sigma_2 > \sigma_3$. In a commonly stressed crust, when the major stress component is vertical, it gives place to extension. Depending on the relative intensity of the two horizontal components (i.e., when σ_1 is not the driving stress), deformation gives place to horizontal strike-slip or reverse faulting. In all cases, the stress intensity gradationally accumulates up to overcome the rock strength. The process is considered to be continuous and stationary during stress accumulation up to rupture.

Conversely, friction manifests in a discontinuous motion, known as stick-slip [47]. This manifests through alternate episodes of discrete slip, followed by an arrest, the stick phase. The interpretation of such transient effects relates to the simultaneous presence of a multiphase material that reacts individually to the applied stress. These are metal and air in the case of a common dry friction [48], or the melt and matrix in partially molten rocks [49].

A static coefficient of friction m_0 relates the vertical force (gravity) to the external horizontal force F pulling the system [50]. In the dynamic case, an external force applied to a two-phase material partitions between the two phases. A second term should be added with a state variable q function [51–53].

The general flow pattern, as experimentally determined for dry friction, becomes a competition between velocity and its variations (acceleration or slowing down), i.e., successive episodes of flow (velocity) constrained by enhanced (acceleration) or restricted flux

https://doi.org/10.3390/min16020157

(slowing down). The flow during dry friction consists of motion, interrupted by stops—the so-called stick-slip motion [49,50,53,54]. A spike (Dm), often called stiction, develops when the slip condition (acceleration) varies (Figure 6). The evolution that follows to readjust the friction coefficient takes the shape of a diffusive process, with a quantity D_c [51,52].

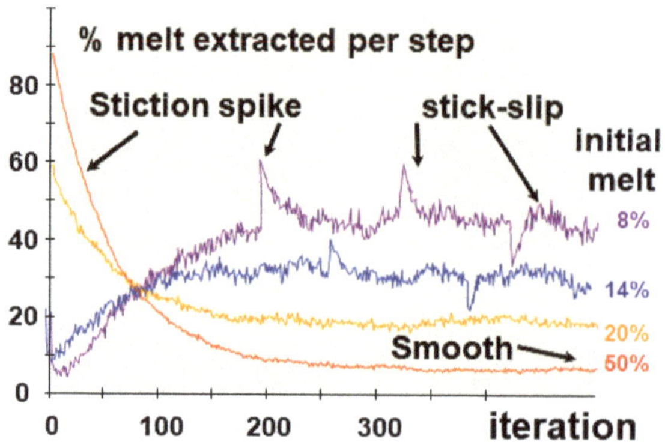

Figure 6. Simulation of melt extraction [49] through a cellular automaton. The amount of melt extracted by the step depends on the initial melt percentage (f_0), but produces a sporadic motion. When one parameter is changed during a run, a stiction spike appears, which rapidly returns to the stationary state. After normalization to the initial melt and the dimension of the cell network, values converge toward a coefficient of melt extraction (m_e). The spikes and the stick-slip like motion depend on the amount of flow (its content) and its accelerations. Two regimes, regular or stationary and transient, are observed depending on whether the initial conditions, especially permeability, vary during a run, inducing spikes (D_m) and stick-slip motion with amplitude (s_m).

The underlying state function is a combination of first and second derivatives, and the energy is evacuated by phonons, such as the musical tones of a violin string or the squeal of chalk on a chalkboard. This defines a general law linking stress and strain with V the slip velocity over an asperity.

$$\tau = \sigma\,[\mu_0 + \alpha \ln(V/V_0) + \theta] \qquad \text{with } d\theta/dt = 1 - V\theta/D_c \tag{4}$$

1.3.5. A State Variable for Friction

The frictional experiments allowed for a dynamic friction coefficient (m), as well as the motion timing, to be quantified. A state variable q is coined that would correspond to the expression of the energy with respect to the variance in the transitional zone [52,53].

$$q = q_0 + a\,\partial G(x)/\partial x + b\,\partial^2 G(x)/\partial x^2 \tag{5}$$

Coefficients a and b measure the velocity dependence of the friction [54]. Their difference (b-a) implies that velocity weakens when positive, i.e., when acceleration surpasses velocity. When the latter surpasses the former, instabilities develop. In other terms, advection takes over from diffusion. This eventually gives place to spinodal decomposition (Figure 7), thus partitioning the entropy between the components [55].

Similar effects are observed during simulations of melt extraction, i.e., the segregation of a mobile melt from a viscous matrix. Numerical simulations used cellular automata to reproduce melt motion [54], whereas in this interpretation, the variations were induced by varying the amount of the weak phase [53] In both situations, a stop and go motion is observed, with sudden bursts related to an increase in or shortening of the weak phase, followed by a diffusive return to the previous state. It remains difficult to calibrate time

https://doi.org/10.3390/min16020157

in such numerical experiments, but the stick-slip motion allows for their non-periodic intervals to be estimated as about 0.3 to 0.5 ky [55].

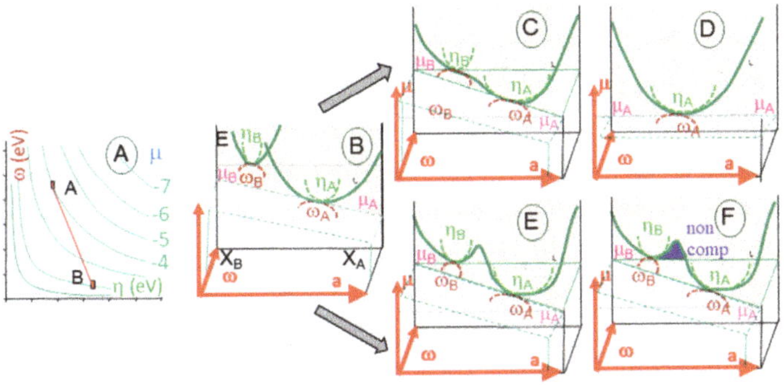

Figure 7. Chemical reaction between A and B elements, represented within a [w-h] diagram, with a being the content in one element (**A**). The descriptors illustrate the energy parabola (E), with its point of tangency to the chemical potential (m) and radius of curvature (h) and osculating circle (w) in a normal plane. The reaction between elements A and B supposes the two energy curves mingle into a single (**B**). Two situations may develop. The energy curve is distorted, with two minima (**C**), and returns to a stable state operates with a newly formed element, and the energy curve is again a parabola (**D**). When the energy gap is too large, positive energy remains (**E**) and the reactions yield two incompatible elements that do not mix (**F**). The incompatible zone is also marked by higher polarizability.

1.3.6. A Competition Between Advection and Diffusion

The stick-slip motion is commonly separated into two patterns according to the frequency of the alternate stop-and-go-motion. In a quasi-steady state, the frequency is as function of the velocity of the flow. Its duration is of the order of the second or lower; nevertheless, it is not periodical, as some sinusoidal functions may be (Figure 6). The amplitude varies continuously. When superimposed on this signal with a short period and small amplitude, some burst occurs, marked by a larger amplitude and a return to the previous steady state, as would occur for diffusion. Such bursts have a larger pseudo-period than the stop-and-go motion by about one order of magnitude. The common interpretation is an increase, or shortening, in the amount of the mobile phase, as suggested by the diffusive return to initial conditions.

2. Chemical Aspects

Chemical reactions are commonly interpreted as the re-equilibration of the chemical potential difference between elements [56]. In the presence of oxygen, metallic ions combine into metal oxide for a greater stability, thus lowering their possibility of inducing new reactions. During this process, there is no noticeable change in the structure of the nucleus of both elements.

2.1. Chemical Descriptors

Elements are considered as electronic clouds that could share other electrons via exporting or importing. For this purpose, elements are classified as acid or base depending on whether they accept or provide electrons, and as hard and soft depending on their ability to share electrons. Such quantities are called descriptors, namely electronegativity (η) and hardness (h). Electronegativity (χ) is the chemical potential (μ) changed of sign [57]. It is the first derivative of the energy when one electron is added to the electron cloud (N) under constant external potential v(r). Chemical hardness (η) is the second derivative of the energy under the same conditions [58] (Figure 9).

$$dG(N) = -N\,\chi + N^2\,h = N\,(\partial G/\partial N)_{v(r)} + N^2\,(\partial^2 G/\partial N^2)_{v(r)} \quad (6)$$

Under such conditions, the chemical potential is tangential to the energy curve, whereas hardness is akin to the radius of the curvature circle at this point. A third descriptor, electrophilicity (ϖ), is akin to the osculating circle [58], being equivalent to the normal tangent and related to the lateral deviation in the energy curve. This attempts to laterally deviate the energy curve (Figure 7).

Furthermore, polarizability (α) characterizes how the electronic cloud distorts in response to an external electric field [59]. It is defined as follows:

$$\alpha_{a,b} = -\left(\frac{\partial^2 E}{\partial F_a \partial F_b}\right); a, b = x, y, z \quad (7)$$

where F_a, F_b, and F_c the components of the applied electric field. Values are commonly reported in atomic units (1 a.u. = 27.211 eV), and polarizability is often inversely related to the cube of electronegativity [60], meaning it can serve as a proxy for unidirectional electronic resistance, and α, its average on geometric axes, as an estimate of the bulk polarizability.

2.2. Reaction Control

In ionic states, atoms fill their available orbitals (s, p, d, etc.), establishing electronic structures that define their reactivity (Figure 8). The frontier molecular orbitals, the highest occupied molecular orbital (HOMO) and the lowest unoccupied molecular orbital (LUMO), are particularly significant in this context, as their energies correspond to the ionization potential (IP) and electron affinity (EA), respectively. From these quantities, electronegativity and chemical hardness can be derived through finite difference approximations (Figure 8). These descriptors serve as fundamental indicators of an atom's chemical behaviour. Electronegativity, the negative of the chemical potential, governs an atom's ability to attract electron density, playing a central role in determining the direction and feasibility of chemical interactions. Hardness, in contrast, reflects the resistance of an electron cloud to deformation, influencing how easily a species can adapt to external perturbations.

Reactions involving small, highly charged ions (such as halogens) often proceed through charge-controlled (CC) mechanisms, where electron exchange occurs with minimal nuclear rearrangement. In such cases, the driving force of reactivity is dominated by the difference in chemical potential between species. Conversely, when polarizability is high, the electron cloud reorganizes, and orbital overlap and electrostatics start to dominate (Figure 8) ionic interactions (e.g., Cu^{2+}). The nature of the interaction–hard–hard or soft–soft—determines whether reactivity is primarily electrostatic or orbital-driven [61].

The energy levels of HOMO and LUMO correlate with fundamental electronic properties: the ionization potential and electron affinity. These, in turn, relate to electronegativity and hardness, which represent the first and second derivatives of the total energy with respect to the electron number. Electronegativity captures the tendency of a species to attract electron density, while hardness quantifies its resistance to charge deformation. Together, they help characterize how a system will respond during interactions with other chemical species. In reactions where the HOMO–LUMO energy gap is substantial, and the elements involved are less polarizable, the mechanism typically falls under the charge-controlled (CC) regime (Figure 8). Electron transfer dominates such reactions without requiring significant structural rearrangement. However, when soft species interact—those with diffuse orbitals and higher polarizability—orbital overlap and covalent character become more

 https://doi.org/10.3390/min16020157

influential, and frontier molecular orbital (FMO) mechanisms become prominent. In short, hard–hard interactions depend on CC, whereas soft–soft reactions are FMO (Figure 8).

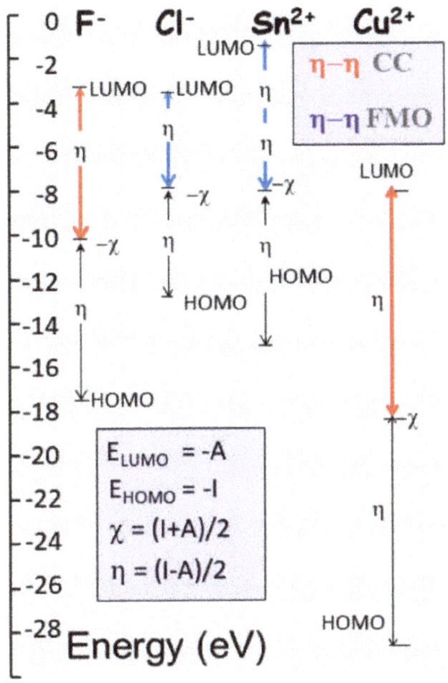

Figure 8. HOMO-LUMO definition and illustration for two fluids (F, Cl) in ionic state, which are respectively classified as hard and soft. Two metallic ions (Sn^{2+} and Cu^{2+}) are also represented, which are, respectively, soft and hard. From the energy level of the highest and lowest occupied/unoccupied molecular orbital (HOMO-LUMO), hardness and electronegativity are computed as the half sum and difference between ionization and electron affinity. Chemical hardness corresponds to the contrasted values of HOMO-LUMO, as evidenced for F^- and Cu^{2+}. The inset shows corresponding definition of the chemical descriptors, chemical potential (μ), and hardness (η), along with the major mechanisms, charge control (CC), and frontier molecular orbital (FMO) that control reactions.

A third mechanism may develop for larger molecules, which is directly related to the size of the electronic cloud and its polarizability [62]. The density polarization (DP) concept expresses the volumetric density of the electric dipole moments, describing how the material responds to an external electric field. Large molecules present large ionic radius, and often large charge. In the presence of an external shape, the electronic cloud deforms and polarizability enhances the formation of covalent bounds.

The three processes controlling chemical reactions determine the type of links between molecules, and thus their future ability to exchange with other elements, forming complex compounds.

2.3. Fitness Landscape Diagrams

Some simple principles rule these reactions. Three of them describe the bulk trend of the descriptors during a reaction. This represents a mapping of the chemical products.

First, the chemical potential should be similar in all components at equilibrium, justifying the use of geometrical averaging between descriptors for a molar amount of compounds or solvents. At equilibrium, the chemical potential tends to equalize among all constituents [63], supporting the use of geometric averaging for compounds:

$$\chi_{ABC} = (\chi_A{}^{a} + \chi_B{}^{b} + \chi_C{}^{c})^{1/(a+b+c)} \quad (8)$$

https://doi.org/10.3390/min16020157

Second, hard molecules arrange themselves to be as hard as possible [64], leading to the maximum hardness principle (MHP).

Third, a minimum electrophilicity principle (mEP) states that reactions should evolve toward minimum electrophilicity [65].

Finally, a minimum polarizability principle (mPP) favours the formation of stable molecules in regions of lower polarizability, typically associated with higher charge localization [66].

Consequently, given an initial product, characterized by its descriptors, it should be possible to predict the more probable path during a reaction through examining the variations in the descriptors. This represent a new mapping, or fitting landscape diagram, on which the potential wells should attract the reaction [67].

Such a transformation, which is fact is an isomorphism, can be used to plot the 11 common oxides used in magmatic petrochemistry to a 2D diagram using, for instance, [c-h] or [w-h] as new coordinates for one oxide 11. As an illustration (Figure 9), the 11 oxides plot on such diagram forms a triangle of descriptors of SiO_2, Fe_2O_3, and alkali (Na_2O) using poles. This diagram straightforwardly indicates the three major poles, i.e., silica, oxidation, and alkalis, within chemical analyses. Superimposed on this triangle, other parameter, such as polarizability (a), can be used as a third dimension, yielding a mapping with values of equal polarizability and their trends (Figure 9).

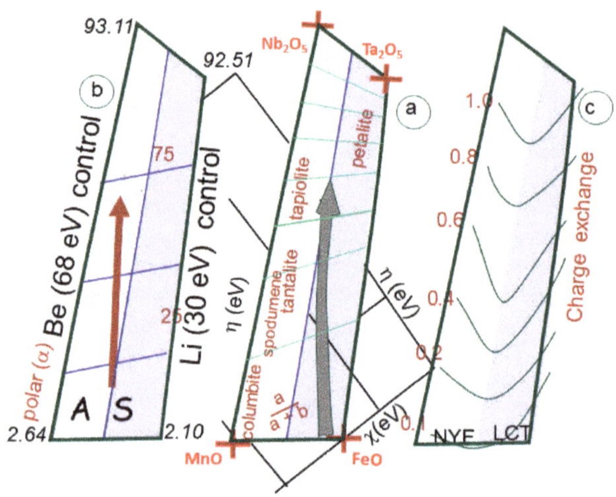

Figure 9. (**a**) The box is calibrated in a/(a + b), indicating the respective ratio of Fe and Mn compared to Ta and Nb. The fields of A and S granites, corresponding to NYF and LCT pegmatites, are indicated. The fields of columbite, tantalite, tapiolite and petalite, spodumene, and beryl are indicated. (**b**) The same box, but calibrated in polarizability, showing the large difference between oxides (about 2 to 100). (**c**) Charge exchange computed between the poles of the pegmatite box. The change is always low (below one unit), increasing from Fe-Mn to Ta-Nb oxides.

2.4. *Fitness Landscapes*

Using such descriptors, chemical reactions are considered analogous to electron flux, with a bulk velocity ruled by the chemical potential difference and its variations, akin to acceleration (or slowing down) through hardness (Equation (6)).

The chemical reactions are not limited to a unique category, e.g., the formation of oxides or sulfides (Figure 9) depending on the reactants that are present. Disequilibrium may occur if the chemical potential involves several successive reactions. For instance, silicate lava, e.g., basaltic, combines about 11 major oxides and fluids (Cl, S, OH, etc.).

 https://doi.org/10.3390/min16020157

During solidification, crystal precipitation modifies the composition of the melt, altering the chemical potential of the residual melt (Figure 10). This constitutes the so-called Bowen–Fenner trend [68,69]. Hence, the formation and precipitation of any phase is a disequilibrium process that follows a series of steps, also called the liquid line of descent, until a thermodynamically stable state, or equilibrium, is achieved [70]. The Bowen trend commonly evolves through successive steps (Figure 10), from ferro-magnesian minerals (olivine, pyroxene, amphiboles), to micas (biotite) and feldspars, and finally quartz. Known as fractional crystallization, it is a succession of forming crystals that are removed from the melt by setting according to successive potential lows.

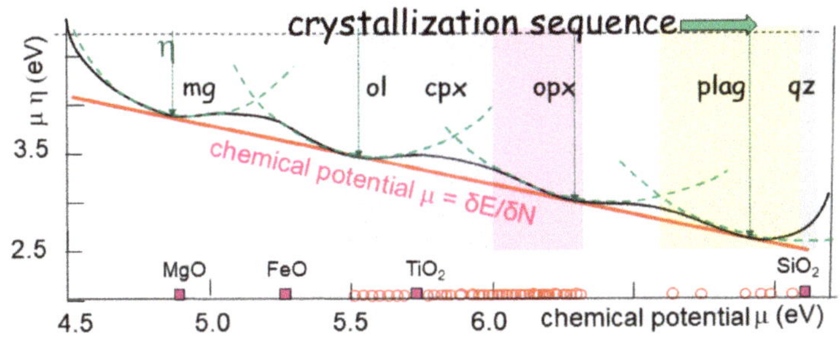

Figure 10. Sequence of crystallization from olivine to quartz, extended from MgO to SiO_2, according to Bowen's trend, but using the chemical descriptors showing a linear decrease in chemical potential. The descriptors are the first derivative, electronegativity (μ), which is the chemical potential change in the sign, plus the chemical hardness (η) and electrophilicity (ϖ), which are second derivatives that are tangential and normal, of the Gibbs energy. All values are measured with the same units (eV). The intermediate points are experimentally derived points of the line of descent from olivine to quartz [71]. The range of major minerals (olivine, pyroxenes, plagioclases, and quartz) is indicated. The evolution follows a line of descent on which the chemical potentials are tangent. The aperture of the curve is the chemical hardness, akin to the radius of curvature.

The chemical descriptors and the principles that rule the reactions offer an elegant mode of describing elements' evolution during chemical reactions [72]. In particular, descriptors pair allow for the reactions between multiple elements to be examined, namely, the 11 oxides used in geosciences within a 2D mapping, rather than a complex 11D space [11].

2.5. Cascading Energy

Reactions proceed through a charge exchange (*dN*), implying an energy exchange (*dE*).

$$dN = \frac{(\chi_A - \chi_B)}{(\eta_A + \eta_B)} = \frac{d\chi}{(2\eta_A + d\eta)} \tag{9}$$

and

$$dE = -\frac{(\chi_A - \chi_B)^2}{2(\eta_A + \eta_B)} = -\frac{d\chi^2}{2(2\eta_A + d\eta)} \tag{10}$$

The sign of *dN* indicates the direction of the electron flow, while *dE* is always negative, reflecting the energy stabilization upon charge redistribution.

The different crystallizing phases follow the order of their chemical potential, providing an analogy of a cascade of potential wells (Figure 11). Each mineral has a chemical potential value that progressively decreases (increasing the chemical electronegativity). The curve of the bulk free energy is tangential to each of these values, with a radius of curvature (akin to chemical hardness) that slightly increases (Figure 10). Since electrophilicity does not significantly vary, the lateral deviation (i.e., the radius of the osculating circle) does not vary, keeping the energy trend in a regular plane.

 https://doi.org/10.3390/min16020157

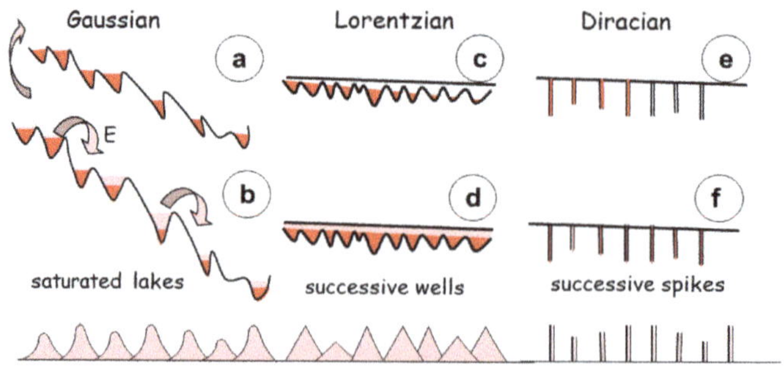

Figure 11. Cascading energy in saturated lakes. (**a**) In case of low global slope, few cascades occur, whereas the saturated lakes flow between them in case of strong slope. The change in instability style is progressive. (**b**) The process occurs in successive steps, releasing energy, between potential wells. Avalanches describe the transition since the local energy release adds to neighbouring changes. The global process is a Gaussian distribution of matter. (**c**) In case of successive potential wells, each energy variation is an individual process, that does not relate to the previous history of the flow. (**d**) It occurs through a Lorentzian distribution, triangular in shape. (**e**) in contrast, in a Diracian distribution, the filling occurs on a discrete point and the rest of the system is free. (**f**) in total the Dirac distribution is a suite of values at each successive point, mimicking the Dirac comb. The successive distributions are respectively caused by diffusion, chemical potential values, and conjugate of chemical hardness and polarizability.

The proximity of the potential wells and their amplitude imply a quasi-continuous trend toward a potential low. The trend runs from Fe oxides toward silica, as described by [68,69]. mimicking a liquid line of descent that was experimentally observed [73]. Immiscibility between liquids occurs when the difference in hardness, i.e., in the second derivative of Gibbs free energy, is too large to allow fir a smooth transition.

The cascading variation in the chemical potential represents a natural evolution of the chemical system towards a zone of lower energy (Figure 11). Although the reactions rapidly develop and the atoms are rearranged, the reaction is accompanied by an energy change as the new system is generated. This is obviously accompanied by an exchange in temperature. For most reactions, the reaction rate determines the proportion of the concentration of the products and reactants by time unit. It can be very slow, such as the natural oxidation of iron in contact with air, or quite instantaneous, scaling in $mol/m^3/s$.

Consequently, the chemical reactions behave identically to the stop-and-go motion of friction. Their rate depends on the available quantity of reactants, i.e., advection, and on the velocity at which new products are generated, i.e., diffusion.

2.6. Computational Aspects

We employed conceptual Density Functional Theory (cDFT) to calculate the chemical reactivity descriptors, in accordance with the foundational principles of the theory [74]. All computations were carried out using the code Multiwfn 3.8 [75,76], an open-source post-processing programme designed for the comprehensive analysis of molecular electronic structures.

The calculations are based on the B3LYP functional [77]. which incorporates the Lee–Yang–Parr correlation component. Geometry optimizations and property evaluations were performed at the B3LYP/6-311++G (2df,2p) level of theory. This level of theory was also used to calculate the external static potential (ESP), ensuring consistency across the dataset. The use of a uniform cDFT computational protocol allows for direct comparability of the derived descriptors across different systems and supports the assumption of data homogeneity.

https://doi.org/10.3390/min16020157

3. The Temperature Aspects

Temperature is a major controlling parameter during magma production. It controls the onset of melting, but it also has an influence on many other physical parameters (diffusion, viscosity). It links with energy through a quantity that describes the disorder of the system, increasing from a regular distribution (e.g., crystalline ordering) to a random distribution (e.g., a liquid or a gas). In information theory, the entropy is computed as the sum of all possible probabilities ($p(x) \log p(x)$) that can accommodate a given state [78,79]. In a chemical system, this probability is closely linked with the inverse of temperature.

3.1. Temperature and Entropy

The third component of the energy, temperature, could also be developed into a sum of the first and second derivative of the entropy, though it is less frequently determined as such. Hence, in the expression of the free energy dG (Equation (2)), the first derivative of the energy as a function of temperature is the entropy.

$$dG(T) = S_0 + T^{-1} (\partial G/\partial S) + T^{-2} (\partial^2 G/\partial S^2) \quad (11)$$

This mimics the development of the energy as formulated for stress/strain and chemical attractivity, emphasizing the relation between variations in temperature (e.g., melting under new magma batch) and back reactions due to restitic mineral formation (e.g., garnet). The idea is not new, but illustrates the concept of latent heat or specific heat capacity.

3.2. Physical Properties of the Magma

The fitness diagrams of the elements involved in the reactions (Figures 9 and 12) tell a complex story regarding the chemical reactions, quite different from a simple, diffusion-controlled evolution. Hence, the chemical evolution of the magma appears as a continuum of melting or crystallization, controlled by the successive wells of chemical potential (the so-called lines of descent, Figure 10). The trend, in terms of chemical descriptors, follows a bulk decrease in chemical potential through successive cascades (Figure 11). In contrast, the trend for the pegmatites, as indicated within the box of [Fe-Mn] and the [Nb-Ta] oxides, is at odds with the granitic magma trend. Another, different, trend determines the evolution of ionic tin towards its oxidation state (cassiterite) (Figure 9) and subsequent precipitation as ore [11].

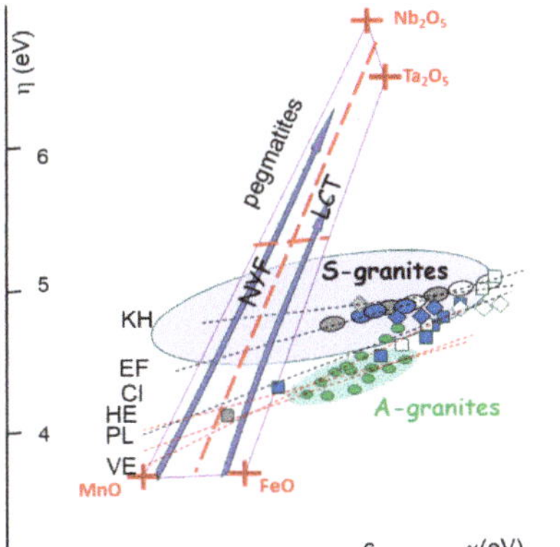

Figure 12. Chemical descriptors [h-c] for granitic magmas (plain), pegmatites (light grey on the soft side), and greisens (dark black on the hard side). Values plot on the lines indicate ore deposits within S-granites

https://doi.org/10.3390/min16020157

of Cinovec (CI), Ehrensfriderdorf (EF), and Krusne Hory (KH) in Bohemia. Also indicated are trends for similar rock types associated with A-type magmas (in green) with HE (Hensbarrow; St Austell, Cornwall), PL (Ploumanach), and Vernéřov (VE) in Bohemia (see location in Figure 5). Also plotted is the field for pegmatites, shown within a Mn-Fe-Ta-Nb box, indicating the separation of NYF and LCT ore formations between columbite and tapiolite. The fields of the S-granites (in blue), as well as the A-granites (green stars) are indicated. Both trends are at odds with the pegmatite box ranging between the [Mn-F] oxides and the [Nb-Ta] oxides. The trends for Sn[II] towards Sn[IV] compounds are shown in the inset.

3.3. Interactions Between the Energy Components

The energy partitioning into three components, along with the variability of their influence with time (velocity and acceleration), has drastic consequences regarding the rate of chemical reaction and the ultimate products of such reactions. The effects are more marked in TPD compared to MA and PD because the heat source is subject to more variations, essentially due to the convergence rate of the plates in motion. In PD, the heat coming from the mantle can be considered stationary in a first approximation.

3.4. MIP Development

The initial melting temperature has drastic consequences for the later evolution of the fluid–melt system [80]. During their evolution, magmas liberate a magmatic igneous phase (MIP), mostly aqueous, during two episodes. The first occurs during ascent and decompression, affecting fluid solubility in the melt. The second episode develops during crystallization by shrinking the amount of melt when crystals form [81]. The fluids result from the melting of hydrous minerals (Figure 1), such as muscovites above 700 °C [82], biotites over 800 °C [83], and amphiboles (900 °C) [84]. These hydrous minerals liberate water, which is immediately reincorporated into the melt and enhances melt production. The restitic mineral assemblage forms granulites, as observed from their seismic velocity in the lower crustal sequences.

The MIP attracts and carries metals that are under their ionic shape. Indeed, the viscosity of a gaseous MIP is about six orders of magnitude less than the melt. Metal transport shifts from diffusion to advection, accelerating by several orders of magnitude. The attractivity of MIP for metals would result in a similar amount of metal segregation. The metal immiscibility is controlled by the Coulombic properties, namely the electric charge (Z) and electronic radius (r). These both determine the formation of an electronic dipole moment when an electric charge, such as another ion, applies [85]. Diffusivity measures the influence along which a molecule interacts with the surroundings.

3.5. Subcriticality

The MIP's individualisation and its composition become more complex when considering temperature and pressure conditions. Indeed, at very high temperatures, the fluids are in a supercritical state. They may incorporate any quantity of matter. In intermediate felsic magmas, the temperature of the critical point is 731 $\pm$ 21 °C, which is valid for a pressure range from 1 to 5 kbar, and the critical water concentration is 26.5 $\pm$ 1.5% [86]. Below these conditions, the MIP splits into two incompatible phases (Figure 1): one silicate-rich and the second aqueous-rich [87]. In consequence, the bulk magma chamber incorporates three sub-chambers that independently evolve, yielding greisen, pegmatites, and granitic materials. Below the critical point, the system is subcritical, with properties of solubility.

Criticality controls the separation of the different phases within the magma. It traduces with an infinite solubility within the liquid [86], modifying the bulk density of the fluid phase (Figure 1). Below the temperature of the critical point, the solubility of the fluid is also enhanced, yielding a subcritical state (Figure 1).

https://doi.org/10.3390/min16020157

4. The Effects on the Magma Chamber

4.1. Magmas

The initial melting of the continental crust produces silicate-rich melts (>65% SiO_2). These hydrous minerals are mostly micas with volatiles (CO_2, Li, Be, B). The liberated water is immediately reincorporated into the melt. The restitic mineral assemblage forms granulites, as observed from their seismic velocity in the lower crustal sequences.

Conversely, at similar temperatures, the fluid-assisted partial melting of lower crustal minerals (garnet, pyroxene) results in alkaline magmas, which are richer in rare earth elements (REE).

4.2. Transport Properties (Viscosity, Diffusion)

During melting, metals are liberated from crystalline networks, reach an ionic state, and are free to move according to the diffusion in the melt (Figure 13). Unfortunately, the temperatures do not allow for much displacement. Hence, given an average lifetime of 1 Ma for the magma chamber, between 700 and 750 °C, the melt viscosity for a crustal melt computed from a rock composition [88] scales between 10^{-4} and 10^{12} Pa·s, without any possibility of differentiating between granite, pegmatites, and greisen [11]. At least 13% fluid is required to reach values allowing for lava flow. At the same temperature, diffusivity values for Sn range 10^{-17}–10^{-19} m^2/s, both equivalent to a characteristic length of less than 1 m for such a time interval [11].

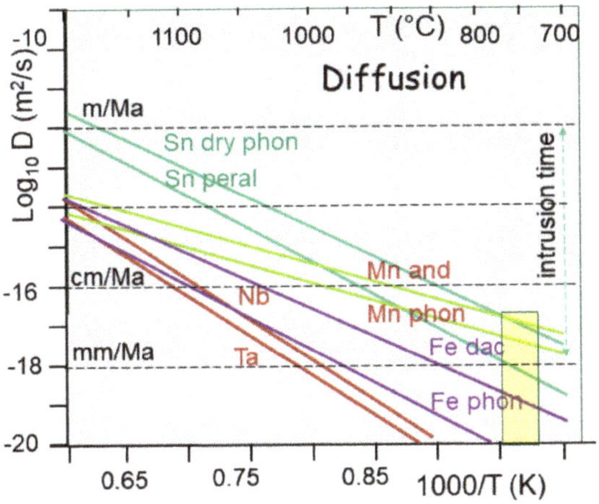

Figure 13. Diffusivity values (in Log10 and units in m^2/s) as a function of temperature curves [89] for Sn into wet trachyte (65% SiO_2) and dry phonolite (55% SiO_2). Complementing Sn, diffusivity values for Mn and Fe (dashes) in andesite and phonolite were extrapolated from experiments, since they are Arrhenian. Values for Nb and Ta are plotted on the same diagram. In an ore formation temperature range of around 750 °C, the diffusivity values for these metals range over four orders of magnitude. For comparison, the diffusivity of trace elements from Rb to Hf is indicated for wet trachyte [90], showing an inversion of values at about 1000 °C. The characteristic diffusion length that shows the diffusivity values in velocity is also indicated. The effect of this is quasi-ineffective at low temperatures, providing evidence that tin segregation is unrelated to simple chemical differentiation.

4.3. Ore Chemistry

The subcritical state of the MIP strongly affects the dissolution of granite into water from 300 °C to 450 °C [87]. Indeed, the supercritical water oxidation mechanism is commonly used to extract hazardous products from soils [91]. For pure water, the critical point is reached at 22.1 MPa and 374 °C, at which point ionic and polar species are extracted [92].

https://doi.org/10.3390/min16020157

In addition, the natural system is buffered along hematite–magnetite (HM), and can reach nickel–nickel oxide (NNO) buffers. As a result, other metal compounds (Sn Nb, Ta) are found as oxides.

A specific molecular dynamic (CPMD, [93] that is applied to supercritical water (scH_2O) shows that electrons cleave a cavity in the hydrogen bond network, and six H_2O molecules form the solvation shell (Figure 5 in [94]). At subcritical conditions, the H-bond network is not continuous, allowing electrons to localize in preexisting cavities or solvation shells formed by four water molecules. The progressive extent of the H-bonds is marked by a decreasing density. The electrochemical properties decrease alongside the dipole moment, modifying its polarizability.

4.4. Pegmatites and Greisen

The extraction of metals from TPD magmas is hampered by two factors, namely the paucity of metals and the poor transport properties within the melt.

The number of metals is not the real problem. In fact, there are enough metals in the lower continental crust compared to the oceanic crust. When melting, they are liberated from their network, become ionic compounds, and are free to move. However, metals' diffusivity towards the melt remains limited, not by their ionic radius (500–1100 pm) but by their poor activation energy (125–180 kmole/J). The low temperature imposed by the critical state (<730 °C) does not help. As a result, the characteristic diffusion velocities range from cm to dm per million years (Figure 13).

Their transport possibilities are linked with the low viscosity of the melt at such temperatures. The melts indicate viscosity values ranging from 10^7 to 10^{12} Pa·s, which should be compared to lavas flowing at viscosities of about 10^6 Pa·s. Even the 13% water and 3% F added during the computation result in such values [11].

In consequence, the magma splits into pegmatites and greisen, which are both associated with magmas from different sources. NYF pegmatites are associated with anorogenic A-type granites, rich in alkalis, Zr, Nb, Ga, and Y [94]. In contrast, LCT pegmatites link with peraluminous magmas, as observed in TPDs, which present a similar K/Rb ratio, but have a Ga content that is about one order of magnitude less than NYF [94]. A new classification is suggested that takes into account the importance of the initial melting temperature. Hence, TPDs are characterized by highly evolved magmas in the continental crust following the dehydration and melting of micas. They have a melting temperature of 700–800 °C.

The differentiation (K/Rb) plotted against the gallium (Ga) content in ppm shows a neat separation (Figure 14) that better illustrates the difference between the pegmatite types [94].

$$\ln(K/Rb) = -0.6\ \ln(Ga) + 2.1 \tag{12}$$

This represents an enrichment in Ga by a factor of log 50, which may be explained by the selective melting of pyroxenes or garnets in the lower crust at about 550 °C [95].

https://doi.org/10.3390/min16020157

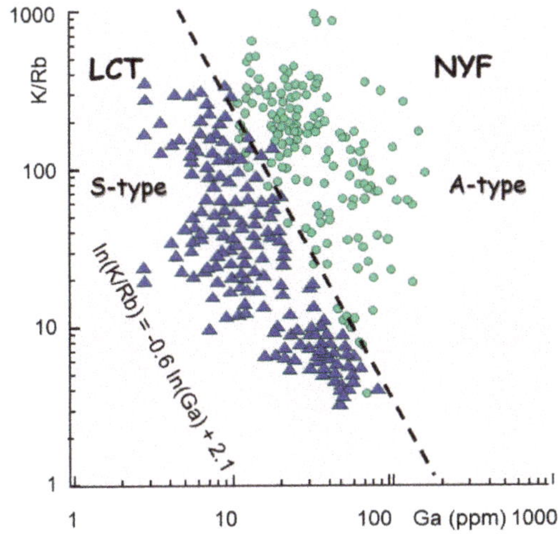

Figure 14. Commonly used pegmatite classification, in LCT and NYF, or S-type and A-type granites, represented in a diagram of K/Rb (differentiation) versus Ga content. Data are redrawn from [94]. They show a separation between the two groups that may be approximated by ln(K/Rb) = −0.6 ln (Ga) + 2.1. This indicates the important enrichment (ppm) in gallium (Ga), probably generated by garnet or pyroxene melting in the lower continental crust.

4.5. Usual Metals (Sn, Nb, Ta)

Tin (Sn) is the most common ore mineral mined in TPD systems. It is exploited as cassiterite, a tin dioxide (SnO_2). The electron configuration of tin orbitals is 5s2 4d10 5p2. This indicates that tin can lose two electrons from its 5p orbital to form a tin [II] ion (Sn^{2+}), and it can lose two additional electrons from the 5s orbital to form a tin [IV] ion (Sn^{4+}). Tin achieves a more stable electron configuration in both cases. Tin can form divalent compounds since its s subshell is more stable than its p subshell because of screening differences, and the s electrons do not bond. The tin [II] compounds are not as stable as tin [IV] compounds, in which all four outer electrons are involved in bonds. Consequently, Sn[II] acts as a reducing agent and may be converted into more stable Sn[IV] by losing two electrons.

Both tantalum (Ta) and niobium (Nb), first called columbium, are part of the refractor group of metals with a high melting point (close to 3000 °C). Their electronic configurations are 6s2 5d3 4f14 and 5s1 4d4, respectively. They both present several oxidation states, with the more common being +2 and +5, in which oxides such as TaO, and more often Ta_2O_5, are observed (Figure 9). Both are resistant to acid corrosion, although they dissolve in HF. As minerals, they are found close together in the high-oxidation state (V), as columbite (Fe, Mn)O Nb_2O_5) or columbite–tantalite (coltan) (Fe, Mn)O (Ta, Nb)$_2$O$_5$). As such, they are often represented in a box with poles [FeO—MnO] and [Ta_2O_5—Nb_2O_5], and alongside two trends (Figure 12), starting from the pole FeO-MnO, that are associated, respectively, with F-poor and F-rich fluids [96] (Table 1).

The bulk pegmatite-forming structure is commonly attributed to the progressive crystallization of a melt of granitic composition at a highly supersaturated level. During cooling, a boundary layer of the melt develops at the crystallization front. This may be the last liquid in the pegmatite body, accounting for the decreasing number of crystals and their increasing size. Mineral zonation manifests via plagioclase and a progressive increase in An content, from $An_{10–20}$ in the outer ones. This is followed by K-feldspars and quartz. The highly contrasting diffusivities of alkalis (Na, K) compared to those of other metals (Hf, Zr, Ti, Nb, and Ta) result in their differential migration, giving place to

 https://doi.org/10.3390/min16020157

boundary-layer pile-up with local saturation [97] his is not strange within a medium of high polarizability (Figure 9), extending from the positive ridge between FeO and MnO towards Ta and Nb oxides.

Table 1. Chemical descriptors (eV) for Fe, Mn, Ta, and Nb oxides.

Oxide	c (eV)	n (eV)	w (eV)	a (cm^3)
Nb_2O_5	6.00	6.89	2.61	93.11
Ta_2O_5	6.09	6.44	2.88	92.51
FeO	5.28	3.37	4.14	2.10
MnO	4.79	3.49	3.29	2.64

The decrease in chemical potential follows the habitual Bowen trend (Figure 10). With plagioclases' crystallization, the other incompatible elements in major minerals concentrate, giving the appearance of normally rare minerals in the sum of the first-formed and last-formed mineral assemblages of the rare-element pegmatites [98]. The final mineral assemblages contain an abundance of boron as tourmaline, lithium in spodumene, petalite, and lepidolite. They represent the standard mode of solidification by decreasing the chemical potential of the system through the successive energy cascade between potential wells (Figure 10).

Those crystals are irregularly zoned, especially in Ti, Li, Al, and Ge [99]. The variations observed are irregular, with an apparent spacing (wavelength) of about 50 mm, corresponding to about 0.5 sec in terms of crystallization time. The amplitude is more difficult to estimate, varying on a scale of 10 ppm for Ge, about 0.9 orders of magnitude for Li and Al. Ti is obviously the more mobile element, with its departure leading to a global increase, with a doubling of the concentration. When superimposed on such a large wavelength trend, smaller variations mimic a stick-slip or sporadic variation in the flux of elements (Figure 6). Spikes with larger amplitudes are also observed (Figure 6).

Both scales of zoning mimic those observed during melt extraction [11], indicating instability in short wavelengths (the strike slip) and some burst, or stiction spikes when back reactions increase or decrease the local melt.

4.6. Gems

Water-rich melt inclusions in some TPDs show a critical water concentration of about 31.6%, [99] which is huge and becomes uninterpreted when such MIs decrepitate, liberating the fluids. Hence, phase separation into glass, liquid, and vapour (G, L, V) does not occur during the supercritical state (Figure 1), obscuring their relative proportion and further interpretation. Notably, the origins of the volatile components (e.g., F, Be) remain poorly traceable.

When associated with such water-rich MI, hard, volatile chemical elements (Li, Be, B) appear to segregate and be transported in these fast-moving components, and then are freed. Owing to their chemical hardness, they concentrate and form zoned bodies with a border granitic zone and albite, with a central mantle core formed by Kfeldspar around the real core (beryl, tourmaline, and spodumene). The whole mineral assemblage indicates a water-rich origin with low temperatures (below 600 °C) within high-crystalline mush (about 10% restitic melt).

Such MIPs are enriched in volatiles, and the MI often documents two generations, one with moderate Be content (some 100s ppm), which occurs close to an abnormal MI with a high Be content (up to 10,000 ppm). Other metals occur in trace amounts (V, Cr), providing the final colours of the gemstones.

 https://doi.org/10.3390/min16020157

Nevertheless, tin ore has been mined in Bohemia for a long time, providing metals for wars and wealth for the lords. Such mining also obtained gemstones, such as garnet and, more frequently, emerald and beryl. The Habachtal deposit of emeralds, north of Saltzburg, Austria, was first mined by the Celts. It has been a large source of gemstones, able to favourably compete with the Egyptian mines–for instance, Mount Smaragdus or emerald Mountain, located in Wadi Sikait in Egypt's Eastern Desert, where mining probably began toward the end of the Ptolemaic period in the first century [100,101].

The emerald mineralization in the Habachtal is geologically and tectonically complex, and previous investigators identified fluid inclusion-based evidence for a hydrothermal/metamorphic origin for the emeralds. In this paper, we report the discovery of emeralds with a distinctly different inclusion population, including melt inclusions, which demonstrates that at least some, and probably most, of the emerald mineralization in the Habachtal occurred from an extremely fluid-rich pegmatite-like aluminosilicate melt under supercritical conditions, at high temperatures and under moderate pressures (~700 °C, 5 kbar).

5. Discussion

The present paper examines ore formation in a transpressional context, looking at the interactions between the energy components during the melting, segregation, and emplacement of a continental-derived magma, with inherently different viscous phases, including MIPs. The explanation for this is, apparently, at odds with the commonly formulated episodes of purely chemical evolution through differentiation. Hence, the latter is essentially controlled by diffusion.

Nevertheless, melting during transpression involves basal heat (strain imposed by the far-field external stress pattern). This indicates the first interaction between the energy components.

Second, magmatic evolution develops in a subcritical state, implying an oxidizing environment with drastic effects on the solubility of metals. This is an important aspect of the magmatism.

Third, the formed ores, which re often Sn derived from the continental crust, are observed in relation with pegmatites and greisens, specific water-rich magmas that remain within the magma chamber. This depends on the formation of volatile compounds, such as halogens or S-species.

Fourth, the evolution of such coarse-grain, aqueous magma, is specific and functions as a closed system.

Finally, the volatile elements (F, Li, Be, B) condensate as oxides, forming complex compounds of large molecules that aggregate to form gemstones. The most well-known formations are beryl and emerald. They segregate owing to their high polarizability and elevated chemical hardness, both serving as an antagonist to form complexes with other elements.

5.1. Caveat About Transpression

An important caveat should be noted for transpression and magmatism. All granitic intrusions (to be short) are not generated within transpressional zones. As an example, the magmas generated at the closure of a back arc are not considered to be purely derived from the shear of continental crust. They largely derive from mantle-derived sources, i.e., at the temperature of amphibole breakdown (900 °C).

Transpression is defined by the combination of horizontal convergence and vertical motion (Figure 5) due to the partitioning of the deformation [13].

https://doi.org/10.3390/min16020157

The existence of a vertical motion within the plate is a consequence of oblique convergence. Hence, the parallel motion of plates is continental shear, whereas orthogonal convergence is pure shear. This represents the first interaction between tectonics and magmatism (Figure 2).

A direct consequence of this vertical motion is the inherent decompression of rocks, which controls the solubility of fluids upon their first boiling. Indeed, the vertical motion is controlled by convergence velocity, depicted as cm/yr, equivalent to a rate of $3.2{\cdot}10^{-10}$ s^{-1}. This should be compared to the tectonic rate, which is about 10^{-14} s^{-1}. It also suggests that transpression itself is not sufficient to produce pegmatites and gems.

Depending on the convergence rate, the exsolution of MIPs is affected in the sense that the rapidly diffusing fluids (F, Li) are easily extracted from the melt, but other volatiles of a higher molar weight (Be, B) remain in the melt. This leads to the difference between TPD and the usual TP, in which it is the essentially the temperature that controls fluid exsolution.

5.2. Importance of a Subcritical State

In contrast with PDs, the heat source that generates, b, the plate subduction, and thus allows for the perennial melting of the mantle wedge and lower crust in TPDs strongly depends on the frictional stress generated by the oblique collision between continental plates. By nature, the oblique collision may vary in direction, thus modifying the transpressional character of the tectonics.

Nevertheless, the basal zone of friction remains close to the critical point, as attested by the lack of biotite breakdown involved, and the very low amount of mantle-derived components, especially S compounds. Consequently, the transport properties of the metals obtained through mobile MIP assist in the segregation of metals, in spite of their low diffusivity values (Figure 13).

The fact that the magma remains in a subcritical state also enhances metal segregation in the melt and MIPs [102]. The partition coefficients for metals into granitic melt and pegmatites is boosted by at least one order of magnitude.

5.3. Chemical Fate of the MIP Components

The fate of the magma mostly depends on the MIP conditions, since the silicate polymers that constitute melt are a function of the bulk temperature and metals are advected in such MIPs. However, the bulk evolution of these components, including oxidation, appears to be at odds with each other, for example, the melt, the fluids, ore formation, and pegmatites, as observed in the descriptors' diagrams (Figure 9).

The latter trend (oxidation) is the easiest to explain, since a medium under a critical state is also a powerful oxidation medium. It controls the evolution of tin that rapidly passes to cassiterite (SnO_2). In fact, recent studies demonstrate that Sn[IV] is already the dominant state in crustal rocks. Hence, Sn[II] compounds are not as stable as Sn[IV], which involves all four outer electrons in its bonding (see Section 4.5). The transition occurs just above the QFM buffer, with Sn[II] showing higher solubility than Sn[IV] (Figure 15). In consequence, two possibilities exist for tin ore formation, both under cassiterite. A fast cooling at a constant oxygen partial pressure (f_{O2}) reaches the HM buffer, yielding the precipitation of cassiterite (Figure 15). In contrast, the common trend is to follow the QFM buffer during a slow cooling process, also involving the precipitation of cassiterite.

https://doi.org/10.3390/min16020157

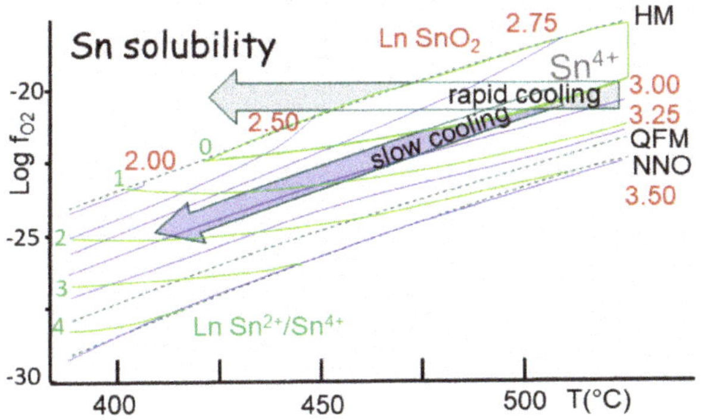

Figure 15. Solubility of tin (in log ppm), showing two aspects—Sn[II] and Sn[IV] (redrawn from [97]. Tin is essentially sensible to redox between the nickel–nickel oxide (NNO), quartz–fayalite–magnetite (QFM) and hematite–magnetite (HM) buffers—all indicated by dashes. The field of Sn[IV] dominant is indicated, close to the HM trend. Two modes of cooling are superimposed on the buffers. One is rapid at constant f_{O2}, and the second is slower, following the buffers; both are indicated from above 500 to about 420 °C.

5.3.1. The Origins of the MIP Components

This section focuses on the relationships between MIP and the genesis of pegmatites and gems.

The granitic melt by itself is the product of the melting of hydrous micas, muscovites, and biotites. The first are phyllosilicates with K and Al, whereas biotites are also phyllosilicates with Mg and Fe. These are the most important carriers of rare alkalis (Li, Rb, and Cs). Thus, mica breakdown liberates an important amount of Li, ranging from 5000 to 35,000 ppm.

The major elements enhancing the formation of pegmatites and gems are the volatile elements (F, Li, Be, B), which are also characterized by a large chemical hardness, making them unable to form compounds with other elements. They enter the composition of lepidolite and petalite (Li compounds), associated with A-granites. In contrast, S-granites link with B_3 and BO_3 compounds linked to six SiO_3 rings, forming beryls and tourmalines. It is surprising, and paradoxical, that the light lithophile elements (Li, Be, B) also present the highest molar weight (6.941, 9.012 and 10.80 g/mole) and a similar covalent radius (84 to 90 pm), but that their silicate oxide compounds present the highest polarizability values (from 540 to 900 units) (Figure 16). Consequently, their light weight identifies them as volatile elements, whereas their polarizability makes them incompatible with fluid solvent (F, Cl, S). The result is the formation and individualization of gemstones, but with a large crystallin network that accommodates the trace elements (Cr, Fe, V) that give them their colours. Among the rock-forming minerals, cordierite, calcic oligoclase, and muscovite control the fate of Be in this order owing to the compatibility of Be in these phases, illustrating its affiliation with alkaline earth elements. Boron's isotopic fractionation between modern biogenic carbonates and seawater was studied and found to be a function of pH. This reflects a major sink during the low-temperature alterations in the oceanic crust due to its adsorption on clay-rich sediments as the product of continental erosion rivers. Therefore, it seems normal to find volatiles that are associated with sediment-derived melt.

The reaction mechanisms that are involved are also in question. The chemical evolution of the magma components is usually ruled by electron exchanges between components (Figure 8). The chemical descriptors determine the reactions according to their respective hardness. They lead to the charge-controlled (CC) and frontier molecular orbital (FMO)

https://doi.org/10.3390/min16020157

control (see Section 2.2). Apparently, this is not the case for the highly polarized molecules in pegmatites and gems. The evolution of volatiles within the MIP is also problematic.

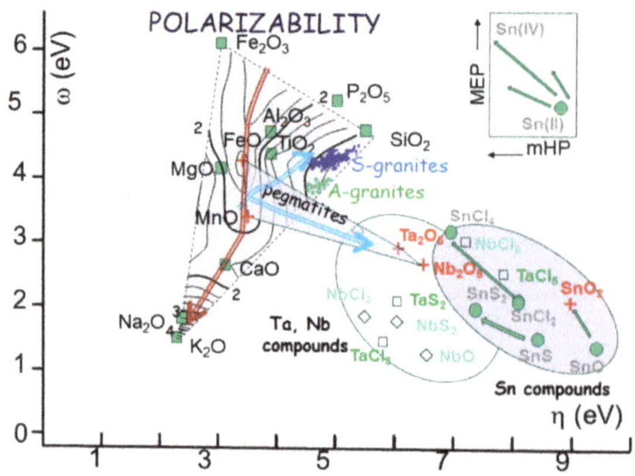

Figure 16. Field of the pegmatites within a [c-h] coordinates (rotated: (**a**) the change in chemical composition between A-type and S-type granites; (**b**) change in hardness; (**c**) charge exchange from one pole to the other). (**a**) Box of the pegmatites within [c-h] coordinates (rotated for convenience). (**a**) The box is calibrated in a/(a + b), indicating the respective ratio of Fe and Mn compared to Ta and Nb. The fields of A and S granites, corresponding to NYF and LCT pegmatites, are indicated. The fields of columbite, tantalite, tapiolite and petalite, spodumene, and beryl are indicated. (**b**) The same box, but calibrated in terms of polarizability, showing the large difference between oxides (about 2 to 100). (**c**) Charge exchange computed between the poles of the pegmatite box. The change is always low (below 1 unit), increasing from Fe-Mn to Ta-Nb oxides.

5.3.2. MIP in Pegmatites

For simplicity, pegmatites are classified as LCT and NFY, although a better definition would consider the host magma from S-type and A-granites. The discussion also involves the Fe-Mn and Nb-Ta oxides box (Figures 16 and 17). Within this box, trends are plotted reflecting the evolution of columbite, tantalite, and tapiolote on the side of A-granites, and petalite on the side of S-magmas (Figure 16). This is essentially based on the reactive content in each oxide.

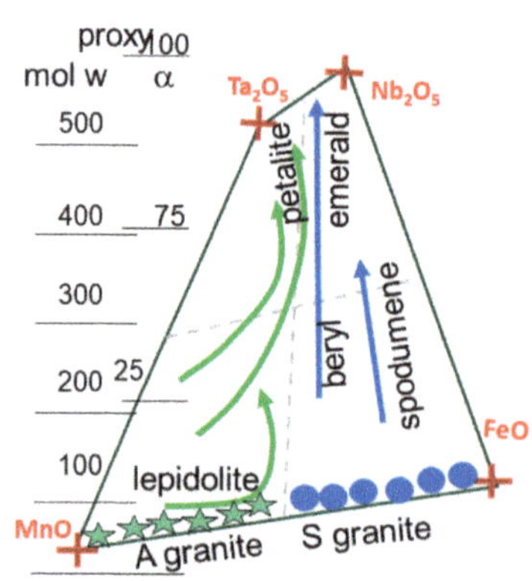

Figure 17. Box of the pegmatites within [c-h] coordinates (rotated for convenience and enlarged for readability). Within the box, the trends in gems are indicated: lepidolite and petalite for A-granites; beryls, emeralds, and spodumene for S-granites. As proxies, the molecular weight in g/mole and the value of polarizability (a) in cm^{-3} are indicated.

https://doi.org/10.3390/min16020157

Minerals yielding pegmatites present elevated polarizability (Figure 16b). For anions, polarizability is directly proportional to their charge and size. Polarizability increases as the volume occupied by electrons increases. This occurs because larger atoms have more loosely held electrons, in contrast to smaller atoms, with tightly bound electrons. Within halogens, F is the least polarizable, while Iodine is the most polarizable. Indeed, iodine has a larger and more diffuse electron cloud, allowing for electron movement within the electron cloud. Both Ta_2O_5 and Nb_2O_5 present heavy molecules, 442.9 and 265.8 g/Omole, respectively.

Of more interest, the charge exchange (Equation (9)) is computed from one pole (Fe O or MnO towards another place within the box (Figure 16c). The charges involved are not very large, hardly reaching 1 eV. This demonstrates that the evolution is nearly continuous between the elements constituting pegmatites, explaining the easy growth in crystals. No rate can be inferred from such diagrams.

The apparent symmetry between the sub-categories (NYF and LCT) also indicates that the process should occur via the same mechanism in both magma types. Nevertheless, the A-type seems controlled by Be, whereas Li should control the S-magma. This agrees with the deep origin within the crust of such fluids.

5.3.3. MIP in Gemstones

The situation becomes extreme when considering gemstones that are ubiquitously observed to be associated with LCT-type pegmatites (Figure 17). Such magmas present molar weight of around 50 g/mole with FeO and MnO, indicating 72 and 71 g/mole. In contrast, gems from the beryl ($Be_3Al_2(SiO_3)_6$ family show values between 276 and 538 g/mole, upward towards the extreme represented by emeralds ($Be_3Al_2(SiO_3)$, with values depending on the trace elements inducing the colours [14]. Gems from the boron series consist of tourmaline with the formula ($NaMg_3Al_6(BO_3)$ $(SiO_3)_6$ $(OH)_6$, and a molar mass of about 940 g/mole, correlated with polarizabilities. The latter values are difficult to estimate because of their low content in trace elements (V, Cr), providing their colour.

5.4. *Contrasting Behaviour of Metals, Pegmatites, and Gems*

The observations of melt inclusions in granites, ores, pegmatites, and gems tell a different story [11]. There, spatial distribution is non-random, corresponding to the different mechanisms ruling their intrinsic repartition.

Commonly, fluid and melt inclusions should be organized in a random or non-random manner. In the last case, three distributions are usually observed, revealing their mechanism of formation, following a Gauss, Lorentz, or Dirac pattern; each reflects the energy interaction between them. A randomly distributed pattern indicates no interaction, as expected. Conversely, a Gaussian distribution, in which each group of particles follows a normal shaped-bell distribution, indicates a diffusive interaction between particles. In a group, the density function of random variables presents an equal probability, with the expected value at the maximum of the curve, and its variance s^2 controls the width of the curve at mid-height (Figure 11a). In a second group, or Lorentzian distribution, the density function takes the shape of a triangle, the summit angle of which is the expected value, but has no variance (Figure 11b). At last, the triangular shape shrinks again, forming a Dirac distribution (Figure 11c), which is represented by a segment of line with the maximum at the expected value and nil elsewhere.

In a pioneering study, the distribution of melt inclusions in pegmatites was characterized as Gaussian, whereas tin deposits present a Lorentzian distribution, with sharp peaks, indicating the distribution in which both the expected value and its variance are undefined [11]. This corresponds to spikes in the sudden release of tin. In contrast, the rare melt inclusions in gems indicate a Dirac distribution, with only local spikes (Figure 11c).

https://doi.org/10.3390/min16020157

Their low formation temperature (400–450 °C) indicates a late crystallization, i.e., a highly evolved, fluid, quasi-mono mineral.

The interpretation of the three distributions becomes evident when considering the interactions between melt inclusions during their formation. In the case of tin ore, melt and ore present similar values of chemical hardness and polarizability. Their interactions are akin to diffusion, resulting in melt inclusions that are equally affected by the interactions during cascading energy (Figure 11). Conversely, in pegmatites, the increase in polarizability contrast turns the bell-shaped curve into a Lorentzian triangle. This effect is still reinforced for gems, with the double effect of chemical hardness and polarizability contrasts. Their low formation temperature (400–450 °C) indicates a late crystallization, i.e., a highly evolved, fluid, quasi-mono mineral, justifying the Dirac distribution.

6. Conclusions

When initiating this paper, the idea was to identify the formation of Sn ore deposits within continental magmatism free of mantle-derived heat sources. A first approach, dealing with the physical conditions of the magma (viscosity, metal diffusivity, and MIP), was soon revealed to be at odds with many acknowledged concepts. For instance, the problem of fluids in a subcritical state was generally largely underestimated. The coeval formation of pegmatites, greisen, and gemstones was treated as a continuous chemical evolution throughout diffusion. Again, the parameters (temperature, viscosity, diffusion, and fluid amount) were set without a real quantified amount and values. The energy interactions were totally ignored. The suggested model was too close to that of a former study [11], and we were surprised to see that rejections were successively presented, mostly based on formal grounds, without any comment on the ideas. This last point motivated the adoption of a totally different point of view, which is still mostly at odds with a gentle chemical differentiation.

Funding: This research received no external funding.

Data Availability Statement: The original contributions for this paper are included in the article. Further inquiries can be directed to the corresponding author.

Acknowledgments: This paper has benefited from studies on pegmatites and greisen by the groups of Rainer Thomas (GFZ, Potsdam) and Karel Breiter (Czech Geol Survey, Prague) for Erzgebirge, Bohemia; and Barrie Clarke (Dalhousie Univ) and Daniel Kontak (Laurentian Univ, Sudbury) for Nova Scotia. They are warmly acknowledged. J.L.V. was later very impressed with the experimental studies on supercritical fluids at the lab of Tsuchiya Noriyoshu, Tohokudai University Sendai, while on sabbatical at the AIST, Tsukuba, invited by Sunsho Ishihara. J.L.V. was invited to the IIT Kharagpur, India, where he was introduced to the HSAB concepts. The funding by CREGU and GeoRessources and Nancy is also acknowledged.

Conflicts of Interest: The author declares the non-existence of any financial interests/personal relationships.

References

1. Robb, L. *Introduction to Ore-Forming Processes;* John Wiley & Sons: Hoboken, NJ, USA, 2021.
2. Gardien, V.; Thompson, A.B.; Grujic, D.; Ulmer, P. Experimental melting of biotite+ plagioclase+ quartz±muscovite assemblages and implications for crustal melting. *J. Geophys. Res. Solid Earth* **1995**, *100*, 15581–15591. [CrossRef]
3. Arndt, N.T.; Lesher, C.M.; Czamanske, G.K. Mantle-derived magmas and magmatic Ni-Cu-(PGE) deposits. *Econ. Geol.* **2005**, 5–24. [CrossRef]
4. Richards, J.P. Porphyry copper deposit formation in arcs: What are the odds? *Geosphere* **2022**, *18*, 130–155. [CrossRef]
5. Frost, C.D.; Frost, B.R. On ferroan (A-type) granitoids: Their compositional variability and modes of origin. *J. Petrol.* **2011**, *52*, 39–53. [CrossRef]
6. Štemprok, M.; Novák, J.K.; David, J. The association between granites and tin-tungsten mineralization in the eastern Krušné hory (Erzgebirge), Czech Republic. *Monograph. Ser. Miner. Depos.* **1994**, *31*, 97–129.

 https://doi.org/10.3390/min16020157

7. Sillitoe, R.H. Porphyry copper systems. *Econ. Geol.* **2010**, *105*, 3–41. [CrossRef]
8. Williams-Jones, A.E.; Heinrich, C.A. 100th Anniversary special paper: Vapor transport of metals and the formation of magmatic-hydrothermal ore deposits. *Econ. Geol.* **2005**, *100*, 1287–1312. [CrossRef]
9. Barnes, S.J.; Staude, S.; Le Vaillant, M.; Piña, R.; Lightfoot, P.C. Sulfide-silicate textures in magmatic Ni-Cu-PGE sulfide ore deposits: Massive, semi-massive and sulfide-matrix breccia ores. *Ore Geol. Rev.* **2018**, *101*, 629–651. [CrossRef]
10. Spivack, A.J.; Palmer, M.R.; Edmond, J.M. The sedimentary cycle of the boron isotopes. *Geochim. Cosmochim. Acta* **1987**, *51*, 1939–1949. [CrossRef]
11. Vigneresse, J.L.; Poddar, A.; Chattaraj, P.K. Felsic trans-porphyry deposits and associated ore (Sn, Ta, Nb, pegmatites) viewed from physics, chemistry (cDFT) and geologic contexts. *Lithos* **2025**, 514–515, 108200. [CrossRef]
12. de Saint Blanquat, M.; Tikoff, B.; Teyssier, C.; Vigneresse, J.L. Transpressional kinematics and magmatic arcs. *Geol. Soc. Lond. Spec. Publ.* **1998**, *135*, 327–340. [CrossRef]
13. Thomas, R.; Davidson, P.; Appel, K. The enhanced element enrichment in the supercritical states of granite–pegmatite systems. *Acta Geochim.* **2019**, *38*, 335–349. [CrossRef]
14. Vigneresse, J.L.; Truche, L.; Richard, A. How do metals escape from magmas to form porphyry-type ore deposits? *Ore Geol. Rev.* **2019**, *105*, 310–336. [CrossRef]
15. Veksler, I.V.; Thomas, R.; Schmidt, C. Experimental evidence of three coexisting immiscible fluids in synthetic granitic pegmatite. *Am. Mineral.* **2002**, *87*, 775–779. [CrossRef]
16. Vigneresse, J.L.; Truche, L. Modeling ore generation in a magmatic context. *Ore Geol. Rev.* **2020**, *116*, 103223. [CrossRef]
17. Huber, C.; Bachmann, O.; Vigneresse, J.L.; Dufek, J.; Parmigiani, A. A physical model for metal extraction and transport in shallow magmatic systems. Cambridge University Press. *Geochem. Geophys. Geosyst.* **2012**, *13*, 8. [CrossRef]
18. Yakymchuk, C.; Brown, M. Consequences of open-system melting in tectonics. *J. Geol. Soc.* **2014**, *171*, 21–40. [CrossRef]
19. Ansermet, J.P.; Brechet, S.D. *Principles of Thermodynamics*; Cambridge University Press: Cambridge, UK, 2019.
20. Gasparik, T. *Phase Diagrams for Geoscientists*; Springer: Berlin/Heidelberg, Germany, 2014. [CrossRef]
21. Liu, A.J.; Nagel, S.R. Jamming is not just cool any more. *Nature* **1998**, *396*, 21–22. [CrossRef]
22. Hobbs, B.E.; Ord, A. The mechanics of granitoid systems and maximum entropy production rates. *Philos. Trans. R. Soc. A Math. Phys. Eng. Sci.* **2010**, *368*, 53–93. [CrossRef] [PubMed]
23. Vigneresse, J.L.; Truche, L. Chemical descriptors for describing physico-chemical properties with applications to geosciences. *J. Mol. Model.* **2018**, *24*, 231. [CrossRef]
24. White, A.J.R. Water, restite and granite mineralisation. *Aust. J. Earth Sci.* **2001**, *48*, 551–555. [CrossRef]
25. Vielzeuf, D.; Schmidt, M.W. Melting relations in hydrous systems revisited: Application to metapelites, metagreywackes and metabasalts. *Contrib. Mineral. Petrol.* **2001**, *141*, 251. [CrossRef]
26. Johnson, T.I.M.; Brown, M. Quantitative constraints on metamorphism in the Variscides of Southern Brittany—A complementary pseudosection approach. *J. Petrol.* **2004**, *45*, 1237–1259. [CrossRef]
27. Villeneuve, M.; Wazi, N.; Kalikone, C.; Gärtner, A. A review of the G4 "tin granites" and associated mineral occurrences in the Kivu Belt (Eastern Democratic Republic of the Congo) and their relationships with the last Kibaran tectono-thermal events. *Minerals* **2022**, *12*, 737. [CrossRef]
28. Wulff, K.; Bolhar, R. Geochemistry of alteration and host rock lithologies to the Bisie tin deposit, North Kivu Province, DR Congo. *Ore Geol. Rev.* **2025**, *178*, 106469. [CrossRef]
29. Batt, G.E.; Braun, J. The tectonic evolution of the Southern Alps, New Zealand: Insights from fully thermally coupled dynamical modelling. *Geophys. J. Int.* **1999**, *136*, 403–420. [CrossRef]
30. King, D.S.; Klepeis, K.A.; Goldstein, A.G.; Gehrels, G.E.; Clarke, G.L. The initiation and evolution of the transpressional Straight River shear zone, central Fiordland, New Zealand. *J. Struct. Geol.* **2008**, *30*, 410–430. [CrossRef]
31. Tulloch, A.J.; Ramezani, J.; Kimbrough, D.L.; Faure, K.; Allibone, A.H. U-Pb geochronology of mid-Paleozoic plutonism in western New Zealand: Implications for S-type granite generation and growth of the east Gondwana margin. *Geol. Soc. Am. Bull.* **2009**, *121*, 1236–1261. [CrossRef]
32. Lehmann, B. Tin granites, geochemical heritage, magmatic differentiation. *Geol. Rundsch.* **1987**, *76*, 177–185. [CrossRef]
33. Oncken, O. Transformation of a magmatic arc and an orogenic root during oblique collision and its consequences for the evolution of the European Variscides (Mid-German Crystalline Rise). *Geol. Rundsch.* **1997**, *86*, 2–20. [CrossRef]
34. Edel, J.B. Le socle varisque de l'Europe moyenne. Apports du magnétisme et de la gravimétrie. *Sci. Géologiques Bull. Mémoires* **1982**, *35*, 207–224. [CrossRef]
35. Kontak, D.J.; Ansdell, K.; Dostal, J.; Halter, W.; Martin, R.; Williams-Jones, A.E. The nature and origin of pegmatites in a fluorine-rich leucogranite, East Kemptville tin deposit, Nova Scotia, Canada. *Earth Environ. Sci. Trans. R. Soc. Edinb.* **2001**, *92*, 173–200. [CrossRef]
36. Breiter, K.; Förster, H.J.; Seltmann, R. Variscan silicic magmatism and related tin-tungsten mineralization in the Erzgebirge-Slavkovský les metallogenic province. *Miner. Depos.* **1999**, *34*, 505–521. [CrossRef]

 https://doi.org/10.3390/min16020157

37. Chauris, L. Pegmatites à allanite et molybdénite dans le granite de Ploumanac'h (Côtes-du-Nord). *Bull. Minéralogie* **1958**, *81*, 150–153. [CrossRef]
38. Edel, J.B.; Schulmann, K.; Rotstein, Y. The Variscan tectonic inheritance of the Upper Rhine Graben: Evidence of reactivations in the Lias, Late Eocene–Oligocene up to the recent. *Int. J. Earth Sci.* **2007**, *96*, 305–325. [CrossRef]
39. Ellis, S.; Fullsack, P.; Beaumont, C. Oblique convergence of the crust driven by basal forcing: Implications for length-scales of deformation and strain partitioning in orogens. *Geophys. J. Int.* **1995**, *120*, 24–44. [CrossRef]
40. Williamson, B.J.; Stanley, C.J.; Wilkinson, J.J. Implications from inclusions in topaz for greisenisation and mineralisation in the Hensbarrow topaz granite, Cornwall, England. *Contrib. Mineral. Petrol.* **1997**, *127*, 119–128. [CrossRef]
41. Breiter, K.; Müller, A.; Leichmann, J.; Gabašová, A. Textural and chemical evolution of a fractionated granitic system: The Podlesí stock, Czech Republic. *Lithos* **2005**, *80*, 323–345. [CrossRef]
42. Heslot, F.; Baumberger, T.; Perrin, B.; Caroli, B.; Caroli, C. Creep, stick-slip, and dry-friction dynamics: Experiments and a heuristic model. *Phys. Rev. E* **1994**, *49*, 4973. [CrossRef]
43. Bouchez, J.L.; Nicolas, A. *Principles of Rock Deformation and Tectonics*; Oxford University Press: Oxford, UK, 2021; ISBN 9780192843876.
44. Fossen, H.; Tikoff, B. Extended models of transpression and transtension, and application to tectonic settings. *Geol. Soc. Lond. Spec. Publ.* **1998**, *135*, 15–33. [CrossRef]
45. Fossen, H.; Tikoff, B.; Teyssier, C. Strain modeling of transpressional and transtensional deformation. *Nor. Geol. Tidsskr.* **1994**, *74*, 134–145.
46. Fossen, H.; Cavalcante, G.C.G. Shear zones—A review. *Earth-Sci. Rev.* **2017**, *171*, 434–455. [CrossRef]
47. Ruina, A.L. Unsteady motions between sliding surfaces. *Wear* **1986**, *113*, 83–86. [CrossRef]
48. Byerlee, J.D. Friction of rocks. *Pure Appl. Geophys.* **1978**, *116*, 615–626. [CrossRef]
49. Byerlee, J.D. The mechanics of stick-slip. *Tectonophysics* **1970**, *9*, 475–486. [CrossRef]
50. Vigneresse, J.L.; Cenki, B. Multiphase rheology as a cause for stick-slip like melt extraction. *Earth Planet. Sci. Lett.* **2022**, *577*, 117269. [CrossRef]
51. Ruina, A. Slip instability and state variable friction laws. *J. Geophys. Res.* **1983**, *88*, 10359–10370. [CrossRef]
52. Dieterich, J.H. Modeling of rock friction: 1. Experimental results and constitutive equations. *J. Geophys. Res.* **1979**, *84*, 2161–2168. [CrossRef]
53. Marone, C.; Cocco, M.; Richardson, E.; Tinti, E. The critical slip distance for seismic and aseismic fault zones of finite width. *Int. Geophys.* **2009**, *94*, 135–162. [CrossRef]
54. Voisin, C.; Renard, F.; Grasso, J.R. Long term friction: From stick-slip to stable sliding. *Geophys. Res. Lett.* **2007**, *34*, L13301. [CrossRef]
55. Wong, Y.Q.; Keller, T. A unified numerical model for two-phase porous, mush and suspension flow dynamics in magmatic systems. *Geophys. J. Int.* **2023**, *233*, 769–795. [CrossRef]
56. Vigneresse, J.L.; Burg, J.P. Continuous vs. discontinuous melt segregation in migmatites: Insights from a cellular automaton model. *Terra Nova* **2000**, *12*, 188–192. [CrossRef]
57. Myers, R. *The Basics of Chemistry*; Greenwood Publishing Group: Westport, CT, USA, 2009; p. 55; ISBN 978-0-313-31664-7.
58. Vigneresse, J.L. Revisiting immiscibility through DFT chemical descriptors. *Theor. Chem. Acc.* **2020**, *139*, 142. [CrossRef]
59. Ghanty, T.K.; Ghosh, S.K. Density functional study of the relationship between energy, hardness, and polarizability of molecules in nonequilibrium situations. *Int. J. Quantum Chem.* **1997**, *63*, 917–926. [CrossRef]
60. Nagle, J.K. Atomic polarizability and electronegativity. *J. Am. Chem. Soc.* **1990**, *112*, 4741–4747. [CrossRef]
61. Anderson, J.S.; Melin, J.; Ayers, P.W. Conceptual density-functional theory for general chemical reactions, including those that are neither charge-nor frontier-orbital-controlled. 1. Theory and derivation of a general-purpose reactivity indicator. *J. Chem. Theory Comput.* **2007**, *3*, 358–374. [CrossRef]
62. Guégan, F.; Abid-Charef, Y.; Hoffmann, G.; Chermette, H.; Morell, C. Finishing (off) the Klopman–Salem model: The importance of density polarization energy. *Theor. Chem. Acc.* **2023**, *142*, 104. [CrossRef]
63. Pearson, R.G. Absolute electronegativity and hardness: Application to inorganic chemistry. *Inorg. Chem.* **1988**, *27*, 734–740. [CrossRef]
64. Parr, R.G.; Chattaraj, P.K. Principle of maximum hardness. *J. Am. Chem. Soc.* **1991**, *113*, 1854–1855. [CrossRef]
65. Pan, S.; Sola, M.; Chattaraj, P.K. On the validity of the maximum hardness principle and the minimum electrophilicity principle during chemical reactions. *J. Phys. Chem. A* **2013**, *117*, 1843–1852. [CrossRef]
66. Chattaraj, P.K.; Fuentealba, P.; Jaque, P.; Toro-Labbé, A. Validity of the minimum polarizability principle in molecular vibrations and internal rotations: An ab initio SCF study. *J. Phys. Chem. A* **1999**, *103*, 9307–9312. [CrossRef]
67. Chattaraj, P.K.; Roy, D.R. Update 1 of: Electrophilicity index. *Chem. Rev.* **2007**, *107*, PR46–PR74. [CrossRef]
68. Bowen, N.L. The order of crystallization in igneous rocks. *J. Geol.* **1912**, *20*, 457–468. [CrossRef]
69. Fenner, C.N.; Piggot, C.S. The mass-spectrum of lead from bröggerite. *Nature* **1929**, *123*, 793–794. [CrossRef]

https://doi.org/10.3390/min16020157

70. Villiger, S.; Ulmer, P.; Müntener, O.; Thompson, A.B. The liquid line of descent of anhydrous, mantle-derived, tholeiitic liquids by fractional and equilibrium crystallization—An experimental study at 1· 0 GPa. *J. Petrol.* **2004**, *45*, 2369–2388. [CrossRef]
71. Faure, F.; Tissandier, L. Contrasted liquid lines of descent revealed by olivine-hosted melt inclusions and the external magma. *J. Petrol.* **2014**, *55*, 1779–1798. [CrossRef]
72. Lu, T.; Chen, F. Multiwfn: A multifunctional wavefunction analyzer. *J. Comput. Chem.* **2012**, *33*, 580–592. [CrossRef]
73. Lu, T.; Chen, Q. Realization of conceptual density functional theory and information-theoretic approach in Multiwfn Program. *Concept. Density Funct. Theory Towards A New Chem. React. Theory* **2022**, *2*, 631–647. [CrossRef]
74. Lee, A.L.; Lloyd, G.E.; Torvela, T.; Walker, A.M. Evolution of a shear zone before, during and after melting. *J. Geol. Soc.* **2020**, *177*, 738–751. [CrossRef]
75. Shannon, C.E. A Mathematical Theory of Communication. *Bell Syst. Tech. J.* **1948**, *27*, 379–423. [CrossRef]
76. Ellerman, D. An introduction to logical entropy and its relation to Shannon entropy. *Int. J. Semant. Comput.* **2013**, *7*, 121–145. [CrossRef]
77. Heřmanská, M.; Kleine, B.I.; Stefánsson, A. Supercritical fluid geochemistry in geothermal systems. *Geofluids* **2019**, *2019*, 6023534. [CrossRef]
78. Breton, N.L.; Thompson, A.B. Fluid-absent (dehydration) melting of biotite in metapelites in the early stages of crustal anatexis. *Contrib. Mineral. Petrol.* **1988**, *99*, 226–237. [CrossRef]
79. Thomas, R.; Davidson, P. Revisiting complete miscibility between silicate melts and hydrous fluids, and the extreme enrichment of some elements in the supercritical state—Consequences for the formation of pegmatites and ore deposits. *Ore Geol. Rev.* **2016**, *72*, 1088–1101. [CrossRef]
80. Tsuchiya, N.; Hirano, N. Chemical reaction diversity of geofluids revealed by hydrothermal experiments under sub-and supercritical states. *Isl. Arc* **2007**, *16*, 6–15. [CrossRef]
81. Boero, M.; Parrinello, M.; Terakura, K.; Ikeshoji, T.; Liew, C.C. First-principles molecular-dynamics simulations of a hydrated electron in normal and supercritical water. *Phys. Rev. Lett.* **2003**, *90*, 226403. [CrossRef]
82. London, D. Granitic pegmatites: An assessment of current concepts and directions for the future. *Lithos* **2005**, *80*, 281–303. [CrossRef]
83. Candela, P.A.; Piccoli, P.M. Magmatic processes in the development of porphyry-type ore systems. *Econ. Geol.* **2005**, 25–37. [CrossRef]
84. Barner, H.E.; Huang, C.Y.; Johnson, T.; Jacobs, G.; Martch, M.A.; Killilea, W.R. Supercritical water oxidation: An emerging technology. *J. Hazard. Mater.* **1992**, *31*, 1–17. [CrossRef]
85. Giordano, D.; Russell, J.K.; Dingwell, D.B. Viscosity of magmatic liquids: A model. *Earth Planet. Sci. Lett.* **2008**, *271*, 123–134. [CrossRef]
86. Brunner, G. Near and supercritical water. Part II: Oxidative processes. *J. Supercrit. Fluids* **2009**, *47*, 382–390. [CrossRef]
87. Zhang, Y.; Ni, H.; Chen, Y. Diffusion data in silicate melts. *Rev. Mineral. Geochem.* **2010**, *72*, 311–408. [CrossRef]
88. Kühne, T.D. Second generation Car–Parrinello molecular dynamics. *Wiley Interdiscip. Rev. Comput. Mol. Sci.* **2014**, *4*, 391–406. [CrossRef]
89. Breiter, K.; Hložková, M.; Korbelová, Z.; Galiová, M.V. Diversity of lithium mica compositions in mineralized granite–greisen system: Cínovec Li-Sn-W deposit, Erzgebirge. *Ore Geol. Rev.* **2019**, *106*, 12–27. [CrossRef]
90. Sanyal, A.; Bijma, J. A comparative study of the northwest Africa and eastern equatorial Pacific upwelling zones as sources of CO2 during glacial periods based on boron isotope paleo-pH estimation. *Paleoceanography* **1999**, *14*, 753–759. [CrossRef]
91. Handy, M.R.; Mulch, A.; Rosenau, M.; Rosenberg, C.L. The role of fault zones and melts as agents of weakening, hardening and differentiation of the continental crust: A synthesis. In *The Nature and Tectonic Significance of Fault Zone Weakening*; Geological Society of London: London, UK, 2001; Volume 186, p. 305. [CrossRef]
92. Phelps, P.R.; Lee, C.T.A.; Morton, D.M. Episodes of fast crystal growth in pegmatites. *Nat. Commun.* **2020**, *11*, 4986. [CrossRef]
93. Giuliani, G.; Groat, L.A.; Marshall, D.; Fallick, A.E.; Branquet, Y. Emerald deposits: A review and enhanced classification. *Minerals* **2019**, *9*, 105. [CrossRef]
94. Michaud, J.A.S.; Schmidt, C.; Naumova, M.A. Revisiting redox-driven pathways of tin cycle from source to economic deposit. *Sci. Rep.* **2025**, *15*, 34476. [CrossRef]
95. Gao, X.; Michaud, J.A.S.; Koch, L.; Zhou, Z.; Zhang, C.; Horn, I.; Allmeev, R.R.; Weer, S.; Holtz, F. Lithium isotope fractionation between mica, quartz, amphibole, feldspars, and granitic melt: Experimental approach and implications for natural granitic systems. *Geochim. Cosmochim. Acta* **2025**, *407*, 12–31. [CrossRef]
96. Simon, L.; Lécuyer, C.; Maréchal, C.; Coltice, N. Modelling the geochemical cycle of boron: Implications for the long-term δ11B evolution of seawater and oceanic crust. *Chem. Geol.* **2006**, *225*, 61–76. [CrossRef]
97. Liu, X.; Yu, P.; Xiao, C. Tin transport and cassiterite precipitation from hydrothermal fluids. *Geosci. Front.* **2023**, *14*, 101624. [CrossRef]

98. Lécuyer, C.; Grandjean, P.; Reynard, B.; Albarède, F.; Telouk, P. 11B/10B analysis of geological materials by ICP–MS Plasma 54: Application to the boron fractionation between brachiopod calcite and seawater. *Chem. Geol.* **2002**, *186*, 45–55. [CrossRef]
99. Černý, P.; Ercit, T.S. The classification of granitic pegmatites revisited. *Can. Mineral.* **2005**, *43*, 2005–2026. [CrossRef]
100. London, D. A petrologic assessment of internal zonation in granitic pegmatites. *Lithos* **2014**, *184*, 74–104. [CrossRef]
101. Giuliani, G.; Chaussidon, M.; Schubnel, H.J.; Piat, D.H.; Rollion-Bard, C.; France-Lanord, C.; Giard, D.; de Narvaaez, D.; Rondeau, B. Oxygen isotopes and emerald trade routes since antiquity. *Science* **2000**, *287*, 631–633. [CrossRef]
102. Thomas, R.; Rericha, A. Extreme element enrichment by the interaction of supercritical fluids from the mantle with crustal rocks. *Minerals* **2024**, *15*, 33. [CrossRef]

https://doi.org/10.3390/min16020157

minerals

Article

Pyrite Trace-Element Signatures of Porphyry-Epithermal Systems in Xizang: Implications for Metallogenic Discrimination and Hydrothermal Evolution

Hongzhong Guan [1], Jiancuo Luosang [2,3,*], Lutong Gao [2] and Fuwei Xie [2]

1 Tibet Zhongkai Holding Co., Ltd., Lhasa 850030, China; xizangreng@126.com
2 College of Earth and Planetary Sciences, Chengdu University of Technology, Chengdu 610059, China; 18663098035@163.com (L.G.); xiefuwei19@cdut.edu.cn (F.X.)
3 Tibet Xianglong Mining Co., Ltd., Xigaze 858500, China
* Correspondence: luosang8698@163.com

Abstract: The Zhunuo porphyry Cu deposit (2.9 Mt Cu @ 0.48%) in the Gangdese belt, southern Xizang, represents a key Miocene post-collisional system. This study integrates textural, major-, and trace-element analyses of pyrite from distinct alteration zones to unravel its hydrothermal evolution and metal precipitation mechanisms. Our study identifies four distinct pyrite types (Py1-Py4) that record sequential hydrothermal stages: main-stage Py2-Py3 formed at 354 ± 48 to 372 ± 43 °C (based on Se thermometry), corresponding to A and B vein formation, respectively, and late-stage Py4 crystallized at 231 ± 30 °C, coinciding with D-vein development. LA-ICP-MS data revealed pyrite contains diverse trace elements with concentrations mostly below 1000 ppm, showing distinct distribution patterns among different pyrite types (Py1-Py4). Elemental correlations revealed coupled behaviors (e.g., Au-As, Zn-Cd positive correlations; Mo-Sc negative correlation). Tellurium variability (7–82 ppm) records dynamic fO_2 fluctuations during system cooling. A comparative analysis of pyrite from the regional deposits (Xiongcun, Tiegelongnan, Bada, and Xiquheqiao) highlighted discriminative geochemical signatures: Zhunuo pyrite was enriched in Co-Bi-Ag-Pb (galena inclusions); Tiegelongnan exhibited the highest Cu but low Au-As; Xiquheqiao had the highest Au-As coupling; and Bada showed epithermal-type As enrichment. Partial Least Squares Discriminant Analysis (PLS-DA) identified Cu, As, and Bi as key discriminators for deposit types (VIP > 0.8), with post-collisional systems (Zhunuo and Xiquheqiao) showing intermediate Cu-Bi and elevated As versus arc-related deposits. This study establishes pyrite trace-element proxies (e.g., Se/Te, Co/Ni, and As-Bi-Pb) for reconstructing hydrothermal fluid evolution and proposes mineral-chemical indicators (Cu-As-Bi) to distinguish porphyry-epithermal systems in the Qinghai-Tibet Plateau. The results underscore pyrite's utility in decoding metallogenic processes and exploration targeting in collisional settings.

Keywords: pyrite; mineral chemistry; Zhunuo porphyry deposit; Xizang

1. Introduction

The giant Zhunuo porphyry deposit is located in the west part of the Gangdese belt in southern Xizang, containing 2.9 million tons of (Mt) Cu with an average grade of 0.48%. The discovery of this deposit significantly increases the exploration potential for porphyry deposit in the central-west part of the Gangdese belt [1,2]. As one of the most important

Miocene post-collisional porphyry Cu deposits in this belt, its hydrothermal evolution and related metal precipitation mechanism play an important role in understanding the formation of a giant porphyry deposit. Previous studies have discussed hydrothermal evolution by fluid inclusion and isotope studies, and conducted that Cu-Fe sulfides were precipitated during the chlorite ± muscovite alteration [3–6]. Meanwhile, in a porphyry system, the hydrothermal evolution is generally quite complex due to multiple mineralization pulses [7,8]. It is generally difficult to reveal the detailed hydrothermal evolution in a giant porphyry system using only fluid inclusion studies.

Recently, high spatial microanalysis of sulfides, using laser-ablation inductively coupled-plasma mass spectrometry (LA-ICP-MS), has been used to identify multiple fluid infiltration events [9,10]. Pyrite, the most widely distributed sulfide mineral [9], is a key indicator in diverse ore deposits. Its trace elements record ore-forming temperatures, environmental evolution, and denudation, aiding deep prospecting [11]. The enrichment of Au, Ag, and As preserves hydrothermal evolution and genetic signatures [11], making pyrite essential for studying fluid-to-metal enrichment processes. In porphyry systems, pyrite forms from magma to late mineralization stages, capturing the hydrothermal evolution [12–15]. In this contribution, we present textural results and major and trace element geochemistry of pyrite from different alteration zones to trace the hydrothermal evolution of the giant Zhunuo porphyry deposit in Xizang. In addition, we conducted a comparative study on the mineral chemistry of pyrite from different porphyry-epithermal deposits in the Qinghai-Tibet Plateau, establishing mineral chemical indicators for pyrite from different deposit types in this region.

2. Geologic Setting

The Gangdese belt is located in the south of the Lhasa terrane and hosts numerous porphyry-skarn Cu deposits, such as the Qulong [16], Jiama [17], Xiongcun [18], Bangpu [19], and Zhunuo deposits [6]. The Lhasa terrane can be subdivided into southern, central, and northern subterranes (Figure 1; [20]). The southern Lhasa subterrane, lying between the Luobadui-Milashan and Indus Yarlung fault systems, consists of extensive calc-alkaline magmatism, which has been interpreted to be related to the southward subduction of the Paleo-Tethys Ocean [21]. High positive $\varepsilon_{Hf}(t)$ of zircons from Jurassic magmatic rocks show the southern Lhasa subterrane is characterized by an oceanic basement and minor Late to Cretaceous volcano-sedimentary rocks [22,23]. The Gangdese batholith, which mainly comprises voluminous 243 to 8 Ma intrusions, dominates the southern Lhasa subterrane [24]. The Shiquan River-Nam Tso mélange zone separates the northern subterrane to the north and the central Lhasa subterrane to the south (Figure 1). The central Lhasa subterrane is characterized by a Precambrian crystalline basement and volcanic cover rocks [25]. The southward subduction of the Bangong-Nujiang ocean produced extensive calc-alkaline magmatism in the central Lhasa subterrane and intensive crustal growth that formed the northern Lhasa subterrnae (Figure 1; [25]).

The Zhunuo deposit is located near Angren, 200 km west of Xigaze (Figure 1). The detailed geology and geochronology of this deposit has been described by [5]. The Eocene volcanic rocks in the north of this deposit were components of the Pana Formation and were composed of rhyolite and andesite [5]. Three north-east trending reverse faults were distributed in the central of the Zhunuo deposit and cross-cut both Eocene volcanic rocks and intrusions. Furthermore, two north-west-trending inferred faults developed in the east and west of this deposit (Figure 2). Intrusions in this deposit included diorite porphyrite, dacite, lamprophyre, granodiorite, and quartz porphyry. Locally, breccias with outcrops less than 0.1 km^2 was observed within the quartz porphyry (Figure 2). Alteration on the surface was zoned and widely distributed in this deposit, including the potassic alteration in the southeast and the phyllic and propylitic alterations in the northwest (Figure 2). Potassic alteration is characterized by the mineral assemblage of biotite + K-feldspar + quartz + amphibole ± sericite ± calcite ± anhydrite ± chlorite ± rutile, with metallic minerals including chalcopyrite + pyrite + molybdenite ± magnetite. Phyllic alteration, a medium-to-low temperature hydrothermal alteration type, occurred in both shallow and deep sections of the mining area. Its mineral assemblage consisted of sericite + quartz ± K-feldspar ± muscovite ± chlorite, accompanied by metallic minerals such as chalcopyrite + pyrite + molybdenite ± covellite ± bornite. Propylitic alteration zones were identified by the diagnostic epidote + chlorite assemblage, with alteration minerals including epidote + chlorite + sericite ± calcite ± anhydrite. Mineralization was weak in these zones, with sporadic molybdenite veins and disseminated pyrite/chalcopyrite observed.

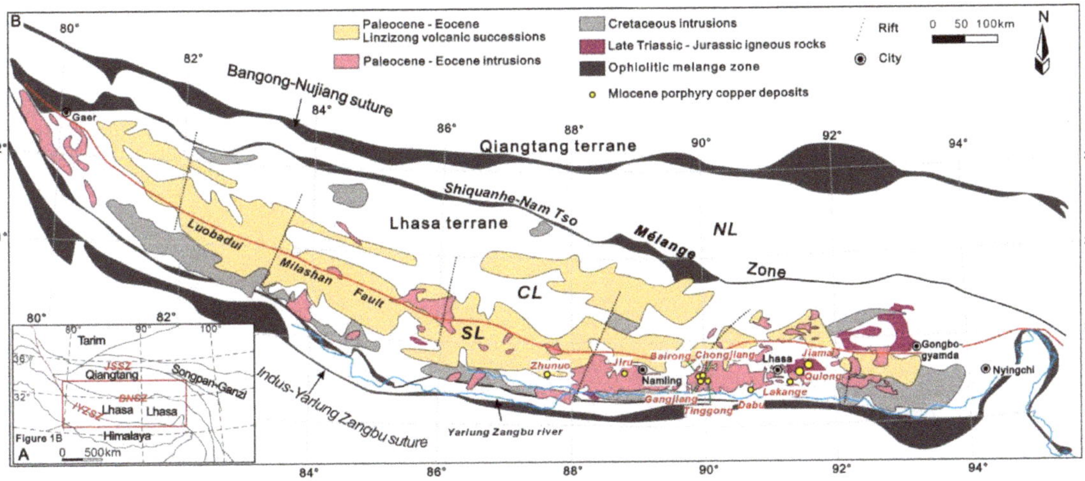

Figure 1. (**A**) Tectonic framework of the Tibetan Plateau (after [6]). (**B**) Geological map showing porphyry deposits in the Gandese Belt (after [6]). Abbreviations: CL: central Lhasa subterrane; SL: southern Lhasa subterrane; NL: northern Lhasa subterrane.

Veins were extensively developed in this deposit and could be divided into A, B, C, and D-type veins according to the nomenclature of [26]. The detailed vein characteristics have been described by [6]. Vein A was primarily associated with potassic-silicate alteration. Potassium feldspar occurred either as alteration halos around the vein or as an integral component of the vein itself. Quartz was predominantly granular, and the vein lacked symmetry. Vein B exhibited a symmetrical texture, with coarse-grained quartz and abundant sulfide mineralization (dominated by chalcopyrite and molybdenite, followed by pyrite). C veins were irregular and mainly composed of minerals such as chalcopyrite and chalcopyrite. Vein D appeared straight and narrow, displaying thin white sericite alteration halos. Sulfides within this vein were predominantly pyrite, with minimal chalcopyrite.

Orebodies in the Zhunuo deposit occurred with an irregular lens shape and were developed within the monzogranite porphyry. Spatially, orebodies developed within the potassic alteration zone and were overprinted by the phyllic alteration. Chalcopyrite and bornite dominated the economic Cu-Fe sulfides in this deposit. Molybdenite mostly occurred as flaky or scaly aggregates distributed in the B veins. Pyrite was also widely distributed in this deposit, especially within the D veins. Overall, the Cu grade had a positive correlation with the abundance of A and B veins. A decreasing trend of Cu grade with distance could be observed from the center of the monzogranite porphyry.

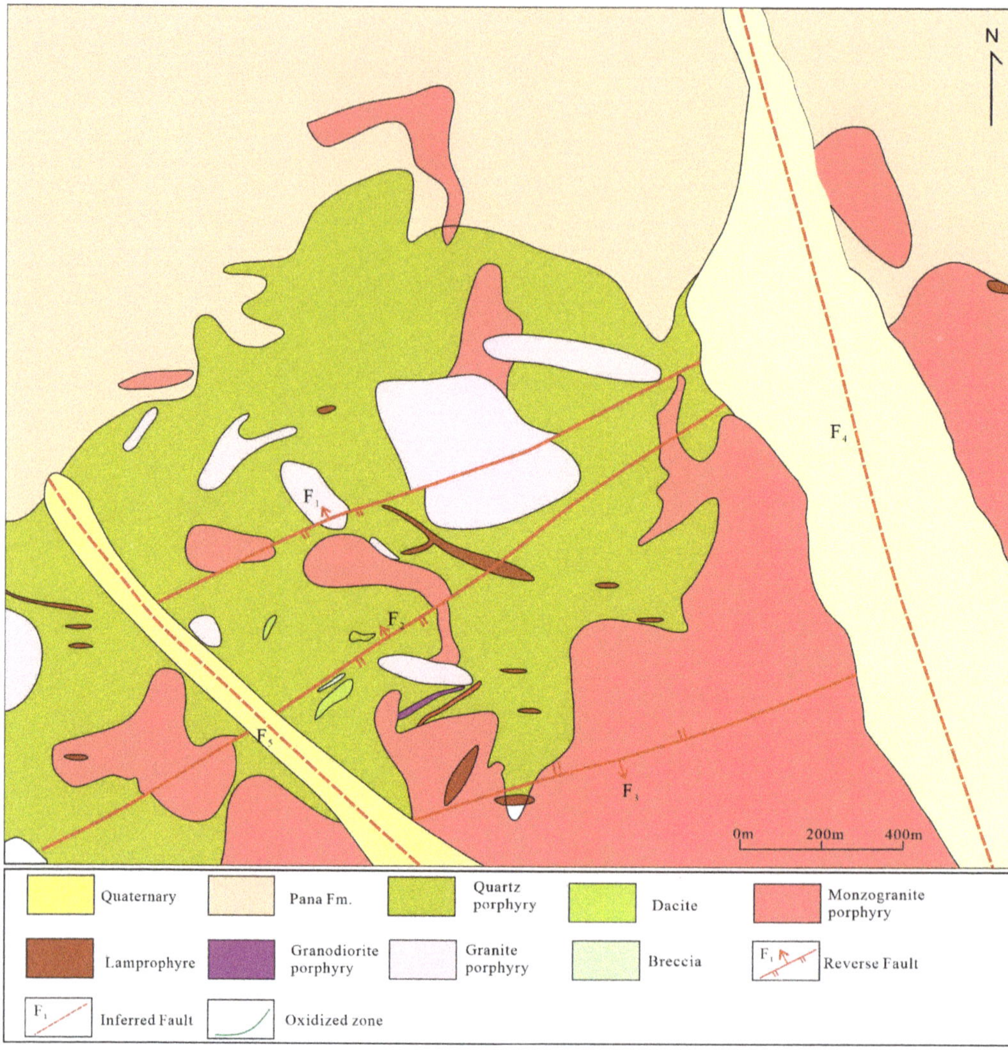

Figure 2. Geological map of the Zhuonuo deposit.

3. Sampling and Analytical Methods

Samples were collected from the Zhunuo deposit. Three types of pyrite were found in 13 samples (Table 1). The sampling borehole locations are shown in Figure 2. Thin sections were prepared for SEM characterization and LA-ICP-MS geochemical quantification. In addition, we collected pyrite mineral chemistry data from copper-gold deposits, including the Xiongcun porphyry deposit in an island arc setting [27], the Tiegelongnan porphyry-epithermal deposit in a continental margin arc setting [28], and the Bada epithermal deposit [29] and Xiquheqiao deposit [30] in a post-collisional setting.

Element analyses of pyrites in thin sections were conducted by LA-ICP-MS at Nanjing FocuMS Technology Co., Ltd. Teledyne Cetac Technologies Analyte Excite laser-ablation system (Bozeman, MT, USA) and Agilent Technologies 7700× quadrupole ICP-MS (Hachioji, Tokyo, Japan) were combined for the experiments. The 193 nm ArF excimer laser, homogenized by a set of beam delivery systems, was focused on the sulfide surface with fluence of 3.0 J/cm^2. Each acquisition incorporated 20 s background (gas blank), followed by spot diameter of 40 um at a 5 Hz repetition rate for 40 s. Helium(370 mL/min) was applied as carrier gas to efficiently transport aerosol out of the ablation cell, and was mixed with argon (~1.15 L/min) via the T-connector before entering the ICP torch. USGS polymetal sulfide-pressed pellet MASS-1 and synthetic basaltic glasses GSE-1G were combined for external calibration. Quantitative calculation of the element contents was performed using the "internal standard-free matrix normalization method". The measured isotope species included: ^{34}S, ^{57}Fe, ^{65}Cu, ^{66}Zn, ^{95}Mo, ^{23}Na, ^{25}Mg, ^{27}Al, ^{29}Si, ^{31}P, ^{39}K, ^{42}Ca, ^{45}Sc, ^{49}Ti, ^{51}V, ^{53}Cr, ^{55}Mn, ^{59}Co, ^{60}Ni, ^{71}Ga, ^{72}Ge, ^{75}As, ^{78}Se, ^{85}Rb, ^{88}Sr, ^{89}Y, ^{109}Ag, ^{111}Cd, ^{115}In, ^{118}Sn, ^{121}Sb, ^{125}Te, ^{133}Cs, ^{137}Ba, ^{183}W, ^{197}Au, ^{205}Tl, ^{209}Bi, ^{232}Th, $^{206/207/208}$Pb, and ^{238}U. The offline processing of analytical data was performed using ICPMSDataCal software 10.1. The analysis results along with their uncertainties and detection limits are presented in Tables S1 and S2.

Table 1. Classification of pyrite types in selected samples with their associated alteration/vein types and characteristics.

Types of Pyrite	Textural Occurrence	Associated Veins or Alterations	Alteration or Vein Mineral Assemblage	Formation Temperature Based on Inclusions	References
Py1	Dissemination	Phyllic alteration	Sericite, pyrite, quartz, K-feldspar, muscovite, and chlorite		
Py2	Dissemination	Phyllic and potassic alteration	Biotite, K-feldspar, quartz, amphibole, sericite, calcite, anhydrite, chlorite, rutile, and magnetite		
Py3	Predominantly vein-type with local dissemination	Phyllic alteration, A-vein, and C-vein	Quartz–pyrite–chalcopyrite veins, quartz–muscovite–pyrite–chalcopyrite veins, pyrite–chalcopyrite veins	340 °C to 400 °C	[6]
Py4	Vein	D-vein	Quartz-pyrite or pyrite veins	220 °C to 330 °C	[6]

Partial Least Squares Discriminant Analysis (PLS-DA) multivariate statistics: Pyrite trace element compositions were analyzed using Partial Least Squares Discriminant Analysis (PLS-DA), a supervised multivariate statistical method designed to classify labeled geochemical data [31,32]. PLS-DA has been widely applied in mineral geochemistry to distinguish various mineral phases, including iron oxides, tourmaline, native gold, rutile, chalcopyrite, and scheelite, thereby demonstrating its broad applicability and effectiveness [33–36].

Prior to analysis, the dataset was pre-processed to address censored values (i.e., those below detection limits). These were imputed using the robCompositions package in R 4.5.1, specifically employing the impKNNa function, which estimates missing values based on the Aitchison distance using k-nearest neighbors. Elements for which more than 40% of the data fell below detection limits were excluded from further analysis. The imputed data were then transformed using the centered log-ratio (clr) transformation, which provides a symmetric and orthonormal basis for compositional data [37].

PLS-DA score scatter plots (t-scores) were used to distinguish elemental associations within sample groups, while the corresponding loading plots (q_v*) revealed correlations among elements and their contribution to group separation. Elements plotting near one another in both loading and score plots exhibited positive correlations, whereas those in opposing quadrants were negatively correlated. Data points situated near the origin in both plots represent compositions close to the dataset mean and, therefore, are less effectively discriminated by the PLS-DA model.

Score contribution plots illustrate the relative enrichment or depletion of elements within specific sample groups, showing whether elemental concentrations are higher or lower compared to the overall mean. The Variable Importance in Projection (VIP) scores were also calculated to evaluate the influence of each element on group classification, with values greater than 0.8 considered significant for the model.

4. Pyrite Mineralogy and Textures

Pyrite is the most abundant sulfide mineral in the Zhunuo copper deposit, with an average concentration of approximately 0.64 wt.% in the ore. It predominantly occurs in two modes: vein-type (Figure 3A–C) and disseminated (Figure 3D–F). Pyrite-bearing veins include quartz–pyrite–chalcopyrite (A vein, Figure 3A), quartz–muscovite–pyrite–chalcopyrite (A vein, Figure 3B), and pyrite–chalcopyrite (C vein, Figure 3C) assemblages, with vein thicknesses ranging from 0.2 to 2 cm. Microscopically, disseminated pyrite is mainly characterized by anhedral granular and replacement textures, although euhedral granular, zoned, and cataclastic textures are also locally observed.

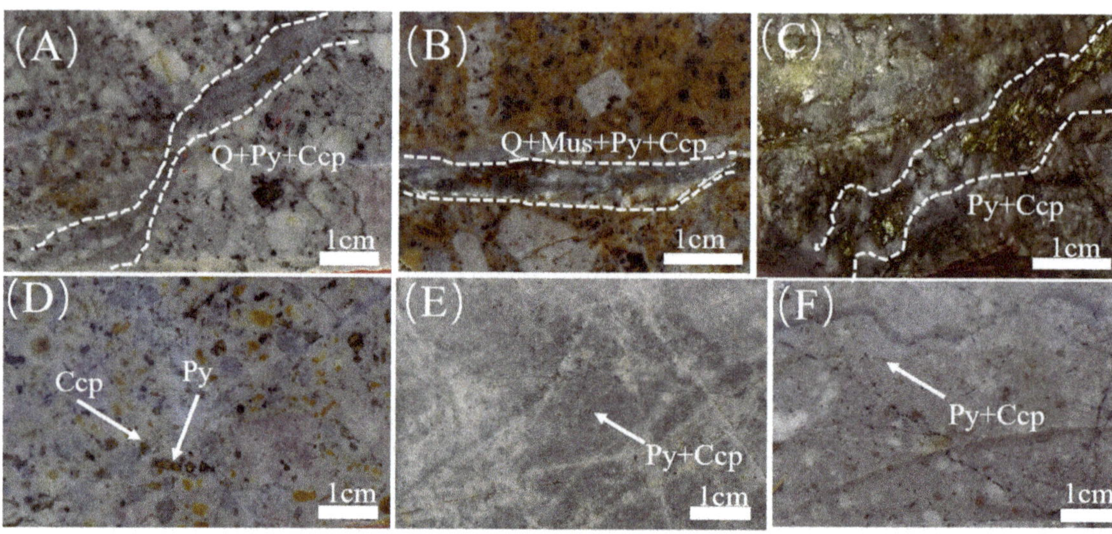

Figure 3. Macro-geological characteristics of the Zhunuo copper deposit. (**A**) Quartz-pyrite-chalcopyrite veins; (**B**) Chalcopyrite-pyrite veins; (**C**) Quartz-muscovite-chalcopyrite-pyrite veins; (**D**) Disseminated euhedral pyrite in weakly sericitized alteration zone; (**E**) Disseminated pyrite and chalcopyrite in strongly sericitized alteration; (**F**) Disseminated pyrite and chalcopyrite in silicified-weakly sericitized alteration zone.

Detailed core logging, petrographic analysis, and paragenetic relationships reveal the presence of three distinct types of pyrite in the Zhunuo deposit. Type 1 (Py1) represents subhedral pyrite disseminated in the matrix, with some pentagonal dodecahedral pyrite grains exhibiting features of brittle deformation (Figure 4A). Type 2 (Py2) occurs as disseminated pyrite within silicified zones exhibiting weak to intense sericitic alteration. This pyrite type is completely replaced by chalcopyrite (Figure 4B–D) and locally coexists with sphalerite (Figure 4D). Type 3 (Py3) occurs both as vein-hosted pyrite in sericitically altered zones with chalcopyrite, showing semi-euhedral to euhedral granular morphologies (Figure 4E,F), and as disseminated semi-euhedral to euhedral pyrite along biotite phenocryst margins or in the groundmass. Some Py3 grains contain Ti-rich sulfide or galena inclusions (Figure 4G,H), although most are partially to fully replaced by chalcopyrite (Figure 4H). Type 4 (Py4) consists exclusively of euhedral granular pyrite and is also replaced by chalcopyrite (Figure 4I).

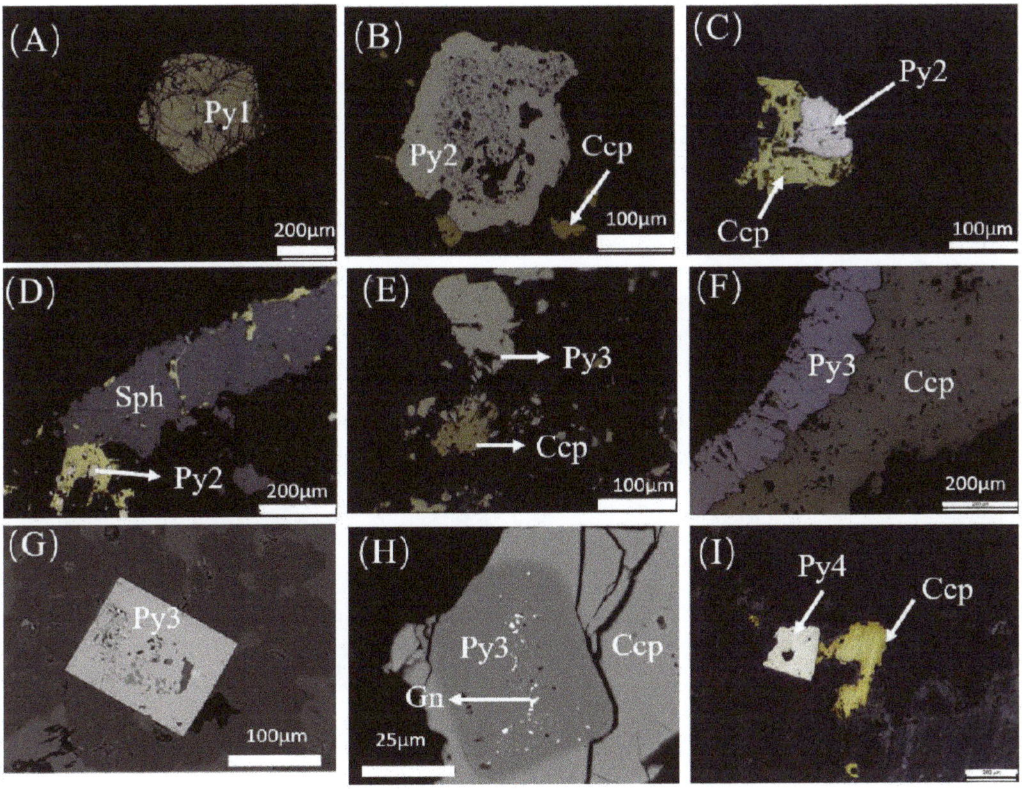

Figure 4. (**A**) Sample ZK404-138.7: Cataclastic euhedral Py1 in strong sericitic-altered zone. (**B**) Sample ZK2104-345: Coexisting Py2 (smooth edges, porous core) and chalcopyrite in a silicified-weak sericitic alteration. (**C**) Sample ZK7A08-237.35: Anhedral Py2 coexisting with chalcopyrite in weak sericitic alteration. (**D**) Sample ZK2307-195.2: Py3 replaced by chalcopyrite (coexisting with sphalerite) in a strong sericitic alteration. (**E**) Sample 7A04-223.4: Abundant Py3 with minor chalcopyrite in quartz-sulfide vein. (**F**) Sample ZK1106-505.4: Quartz-chalcopyrite + Py3 vein. (**G**) Sample ZK7A04-223.4: Disseminated euhedral Py3 in porphyry. (**H**) Sample 7A03-292.7: Matrix-hosted chalcopyrite coexisting with Py3 (containing galena inclusions) in a weak sericitic alteration. (**I**) Sample ZK7A03-292.7: Py4 associated with chalcopyrite.

5. Results

Pyrite hosts a large variety of trace elements, with the concentration of most elements being below 1000 ppm (Figure 5). Lead, Co and Ni have relatively higher concentration, with average contents slightly over 100 ppm. Concentration of Cu, As, Se, and Bi are lower with average values between 10 and 100 ppm. The average concentrations of Te, Cr, Zn, Ag and Ge are even lower ranging between 1 and 10 ppm. Other elements have low concentrations below 1 ppm (Figure 5), and many analyses are below detection limit (Table S1).

The concentration of elements varied among different groups of pyrite (Figure 6). Py3 had slightly higher Au contents compared to other groups in general, although the contents were still low, mostly less than 1 ppm (Figure 6A). Arsenic showed a similar trend to Au, but higher concentrations were obtained in Py2, with values close to 100 ppm (Figure 6B). Py4 had a more concentrated range of Au and As, whereas data for Py2 were more scattered with a wider box range for Au and As. Py4 had the highest content of Cu, Cd, and Ni (Figure 6A,E,F,H), compared with all of the other elements. Sc concentrations were also higher in Py2. Py3 had the highest contents of Bi and Mo relative to the other samples (Figure 6D). In contrast, Py1 had the lowest concentration of most elements, including As, Ag, Bi, Cu, Ni, Se, Sb, Te, and Zn (Figure 6B–E,H,K,M,N,P). Py3 had the highest concentrations of Sn, Sb, and Zn (Figure 6L,M) for all pyrite samples.

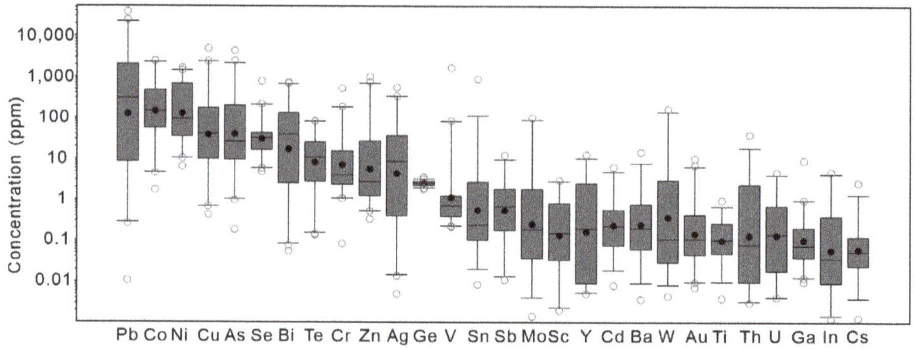

Figure 5. Trace element concentration of pyrite from the Zhunuo porphyry deposit. Boxes represent 25th to 75th percentiles; whiskers show 5% and 95% values; circles outside the boxes represent outliers (the top and bottom 5% of data); black circles are trimmed mean values (mean of values within top and bottom of whiskers); middle lines are median values.

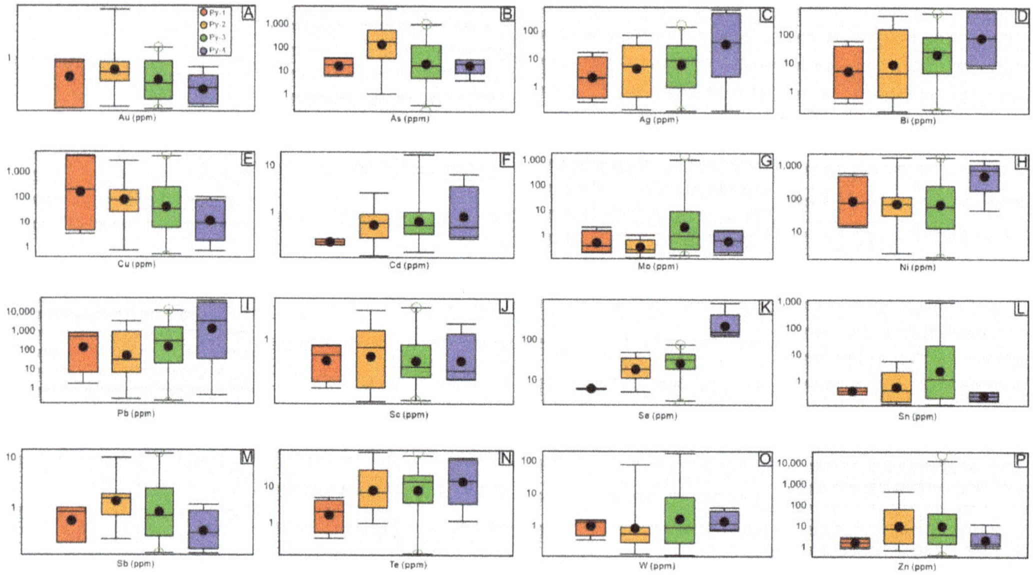

Figure 6. Box-whisker diagrams of key trace elements in various groups of pyrite from the Zhunuo deposit. Box-whisker information is the same as that in Figure 5.

Some elements showed a close correlation between one another. For example, Au had a positive correlation with As (Figure 7A), suggesting they might be incorporated into pyrite via the same process. Other coupled element groups included Zn and Cd (Figure 7B), Sb and Bi (Figure 7C), Te and Bi (Figure 7D), and Se and Bi (Figure 7E). However, for the Sb and Bi, and Te and Bi pairs, some Py2 plots did not show a positive correlation. Conversely, Mo and Sc displayed a negative correlation (Figure 7F), suggesting competitive substitution during their incorporation into the pyrite lattice.

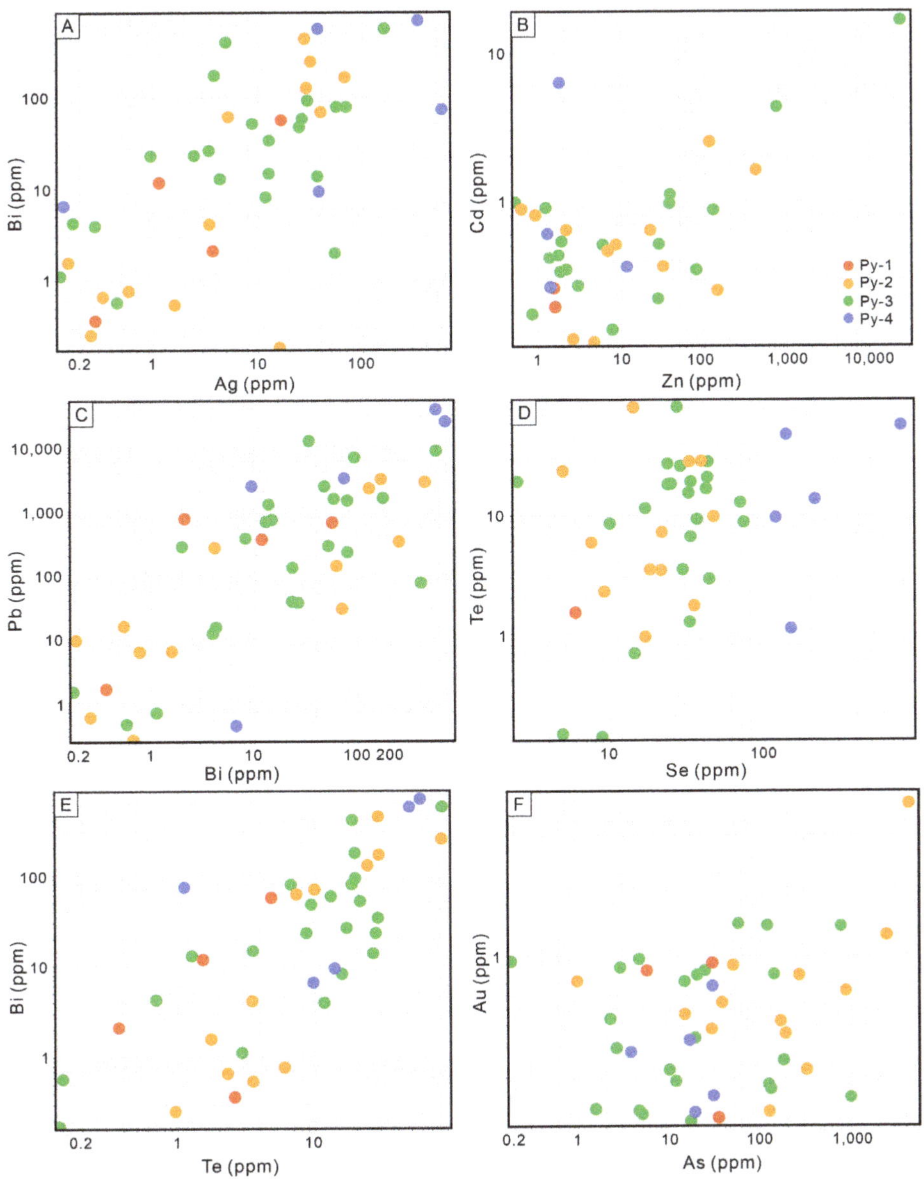

Figure 7. Biplot of key trace elements of pyrite from the Zhunuo deposit.

6. Discussion

6.1. Genetic Types of Pyrite and Occurrence States of Trace Elements

The major and trace elements of pyrite, such as Co-Ni-As compositions and Co/Ni ratios, can be used to identify its genetic type [38,39]. Sedimentary pyrite typically exhibits Co/Ni < 1, whereas magmatic-hydrothermal pyrite generally shows Co/Ni > 1, and stratabound pyrite related to volcanic activity or intermediate-mafic magmatic fluids displays variable Co/Ni ratios [38,40]. Ref. [11] proposed a typomorphic formula for pyrite-major elements: $\delta S = (S - 53.45)/53.45 \times 100\%$ and $\delta Fe = (Fe - 46.55)/46.55 \times 100\%$, where δS and δFe represent the deviation percentages of S and Fe contents (wt.%) from their theoretical values respectively. The δS and δFe values of the studied samples ranged from −0.14 to 0.56 and −0.15 to 1.81, respectively, clustering predominantly in the third quadrant (magmatic-hydrothermal pyrite field) and exhibiting Fe- and S-deficient characteristics.

When arsenic in pyrite is substituted for sulfur in the form of an As^{1-} anion, a negative correlation between As and S is observed [12]. However, no correlation existed between As and S in the pyrite from the Zhunuo copper deposit (Figure 8) coupled with highly heterogeneous As contents (0.18–4323.73 μg/g), suggesting that arsenic primarily occurred as mineral inclusions. The pyrite exhibited a positive Au-As correlation, although gold concentrations were generally low (<1 ppm), except for a few samples that reached 7–10 ppm. The predominance of intergranular gold further supported arsenic occurrence as As-bearing inclusions or nanoparticles.

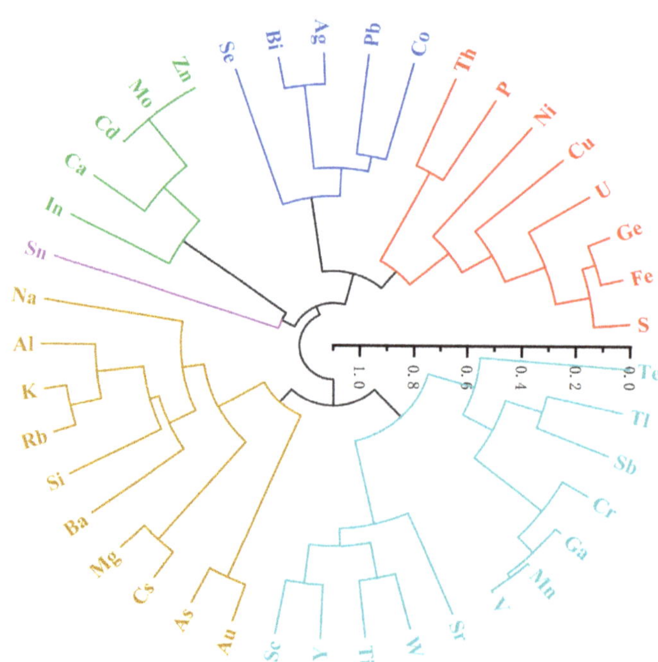

Figure 8. Circular dendrogram of a cluster analysis for pyrite from the Zhunuo copper deposit.

The absence of a Cu-Fe correlation in the Zhunuo pyrite precluded direct Cu^{2+} substitution for Fe^{2+} in the FeS_2-CuS_2 series. Copper concentrations exceeding 2000 ppm surpassed solubility limits during hydrothermal pyrite growth, indicating Cu incorporation via nano/micro-scale Cu-sulfide inclusions precipitated from Cu-saturated fluids at pyrite-fluid interfaces.

Elemental correlations (Ag-Pb-Bi, Zn-In; Figure 8) and microscopic observations of galena and sphalerite (Figure 4H) confirmed that these elements predominantly resided as mineral inclusions. The elevated Mo content in some pyrite samples was primarily attributed to the influence of molybdenite mineral inclusions during laser ablation.

6.2. Pyrite as a Record of Hydrothermal Fluid Properties

Selenium-rich pyrite typically crystallizes from low-temperature hydrothermal fluids [40,41]. A negative correlation exists between the average Se content in pyrite and the formation temperature across different deposit types. In the Zhunuo deposit, three main-stage pyrite types (Py2 and Py3) showed comparable Se concentrations (24 ± 14 ppm to 35 ± 22 ppm), while Py2 exhibited an order-of-magnitude increase (118–779 ppm). Applying the empirical thermometer $Se_{Py} = 5 \times 10^{13} \times T^{-4.82}$ (where T is in °C and Se_{Py} in ppm; [41]), we calculated formation temperatures of 372 ± 43 °C (Py2), 354 ± 48 °C (Py3), and 231 ± 30 °C (Py4). Py1 showed mostly undetected Se, but one sample reached a temperature of 480 °C. The undetected particles in the other samples could be due to the highly fragmented nature of pyrite. The Py2 and Py3 temperatures matched the fluid inclusion homogenization temperatures (340–399 °C) from quartz in the A and B veins [6]. These quartz varieties, characterized by low cathodoluminescence intensity and pore development, coexisted with Cu-Fe sulfides and showed reduced Cu/(Na + K) ratios, confirming their main mineralization-stage origin. In contrast, Py4 temperatures corresponded to D-vein quartz [6], indicating late-stage formation.

Unlike temperature-dependent Se, pyrite Te contents reflect oxygen fugacity (fO_2), with a higher Te indicating a lower fO_2 [42]. Porphyry deposits generally contain lower pyrite Te than Carlin, epithermal, or orogenic systems [41]. At Zhunuo, large Te variations (e.g., 81.87 ppm in Py2 cores vs. 7.28 ppm in rims) record dynamic fO_2 fluctuations during mineralization. In addition, the Te contents in pyrite showed significant variations across different alteration zones, indicating that Te enrichment may be independent of alteration types. Collectively, Se-Te systematics documented progressive cooling and oxidation of the Zhunuo hydrothermal system.

6.3. Pyrite Chemistry Plays a Significant Role in Porphyry-Epithermal Deposits in Xizang

Previous studies have shown that trace element contents (e.g., Te and Se) in pyrite vary significantly among different deposits [39–41]. In this study, we compared the trace element characteristics of pyrite from different deposit types in the Qinghai-Tibet Plateau. The available pyrite trace element data from four other porphyry-epithermal deposits in Tibet were compiled to further explore the implications for the genesis and exploration of porphyry-epithermal deposits. The four deposits included the Xiongcun porphyry Au-Cu deposit [27], Tiegelongnan porphyry-epithermal Cu (Au) deposit [28,43], Bada epithermal Cu-Au deposit [29], and the Xiquheqiao epithermal Cu-Au deposit [30].

In these deposits, Tiegelongnan showed the highest Cu content (Figure 9), but with a relatively low concentration of Au, As, Co, and Pb (Figure 9). This was attributed to the high-grade Cu mineralization of the Tiegelongnan deposit, which had a large part of over 0.8% Cu mineralization [43]. The Xiongcun deposit contained the lowest content of most of the studied elements, including Cu, Ni, Bi, Au, As, Ag, and Pb, except it had a relatively high Co content compared to the pyrite in other deposits (Figure 9). The pyrite from Xiongcun exhibited positive correlations between Au and Cu, Co, Ni, and As, indicating that the higher concentrations of Co, Ni, and As in pyrite corresponded to an elevated Au content [27]. However, the low abundance of these elements in Xiongcun pyrite suggest its gold-barren nature, as Au content shows a positive correlation with these elements [27], and gold is primarily partitioned into chalcopyrite instead [44]. The Bada deposit had a relative high content of Ni, Co, Bi, As, and Ag, and As was the highest compared with the other deposits (Figure 9). This could be due to the high As in the epithermal system [45,46]. The pyrite in the Xiquheqiao deposit had the highest Au content among these deposits, and relatively high As and Pb contents. In this deposit, As entered pyrite through isomorphic substitution, while Pb was incorporated into pyrite as nano- to micron-scale particulate minerals [30]. The migration, enrichment, and precipitation of Au were closely associated with As. The adsorption of Au by As-bearing pyrite and the high degree of As isomorphism (leading to destabilization of complexes) might be the primary factors controlling Au enrichment and precipitation in this deposit [30]. Compared to the data from the literature, the Zhunuo deposit had the highest Co, Bi, Ag, and Pb contents. These elements exhibited significant clustering (Figure 8). This phenomenon was associated with the frequent occurrence of galena mineral inclusions in pyrite from the Juno copper deposit (Figure 4D).

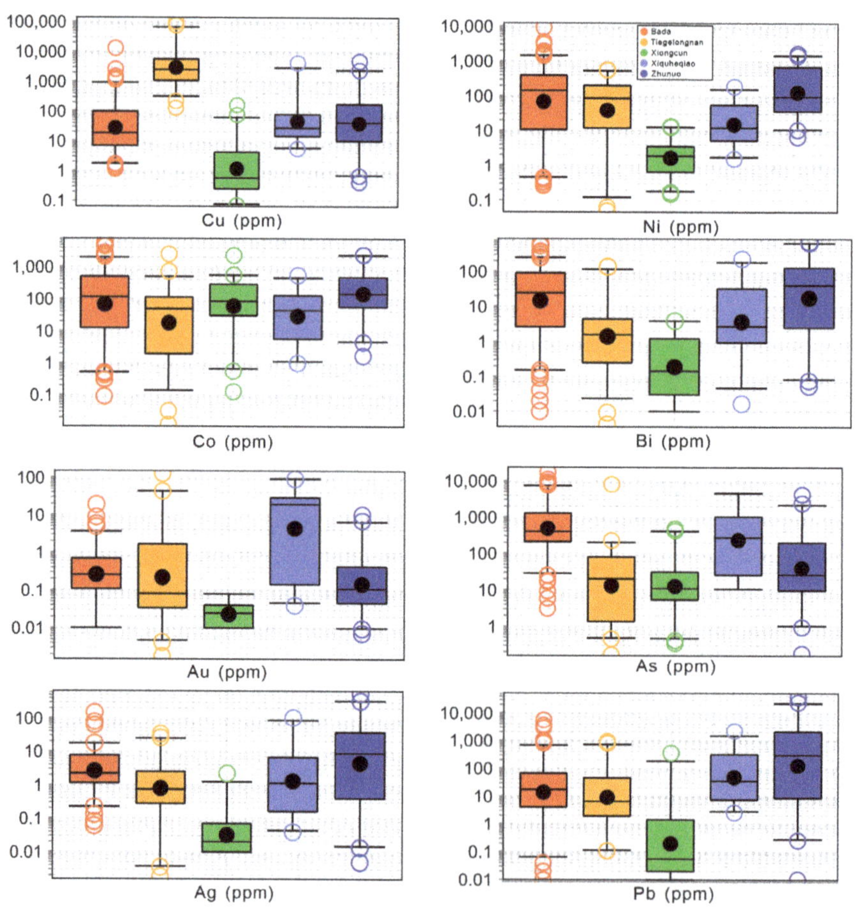

Figure 9. Box-whisker diagrams for a comparison of the pyrite trace elements from several porphyry deposits in Tibet. The box-whisker information is the same as in Figure 6.

In all these deposits, Au and As showed a positive correlation (Figure 10A), similar to that found at Zhunuo (Figure 7A), as well as in many other hydrothermal deposits [11,12,39,41]. In these deposits, Au also showed a positive correlation with Ag, although the correlation was not obvious in the samples from Xiquheqiao (Figure 10B). This might be attributed to the fact that deposits such as Xiongcun, Zhunuo, and Tiegelongnan are all associated with Ag, whereas the Xiquheqiao deposit lacks silver mineralization. Ag also shows a positive correlation with Pb and Bi as well (Figures 10C and 11D). The Xiongcun pyrite data are plotted at the lower concentration end, while Zhunuo pyrite data are located at the higher content end. Co and Ni mostly show a positive correlation in pyrite, but not for the Xiongcun pyrite, which instead had a negative correlation (Figure 10E).

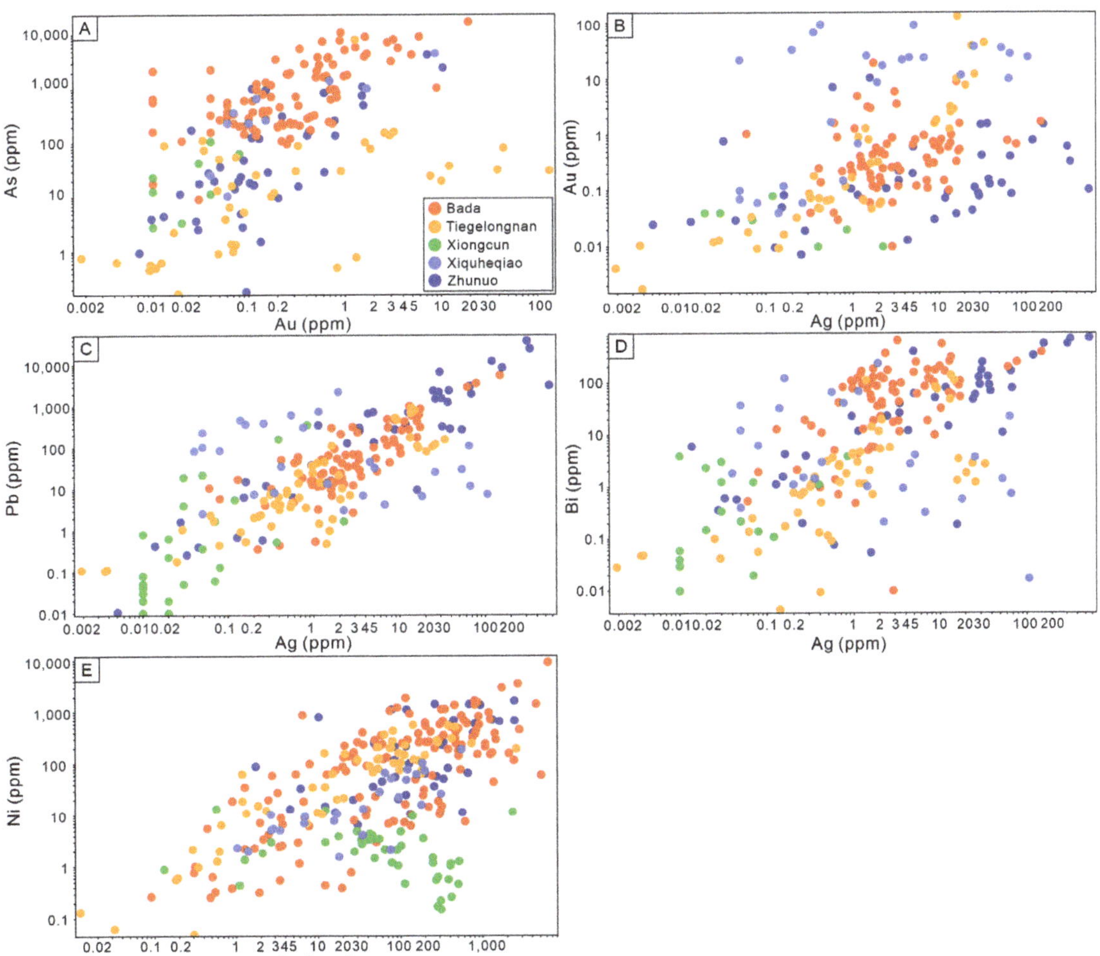

Figure 10. Biplot of key elements of pyrite from several different porphyry deposits in Tibet.

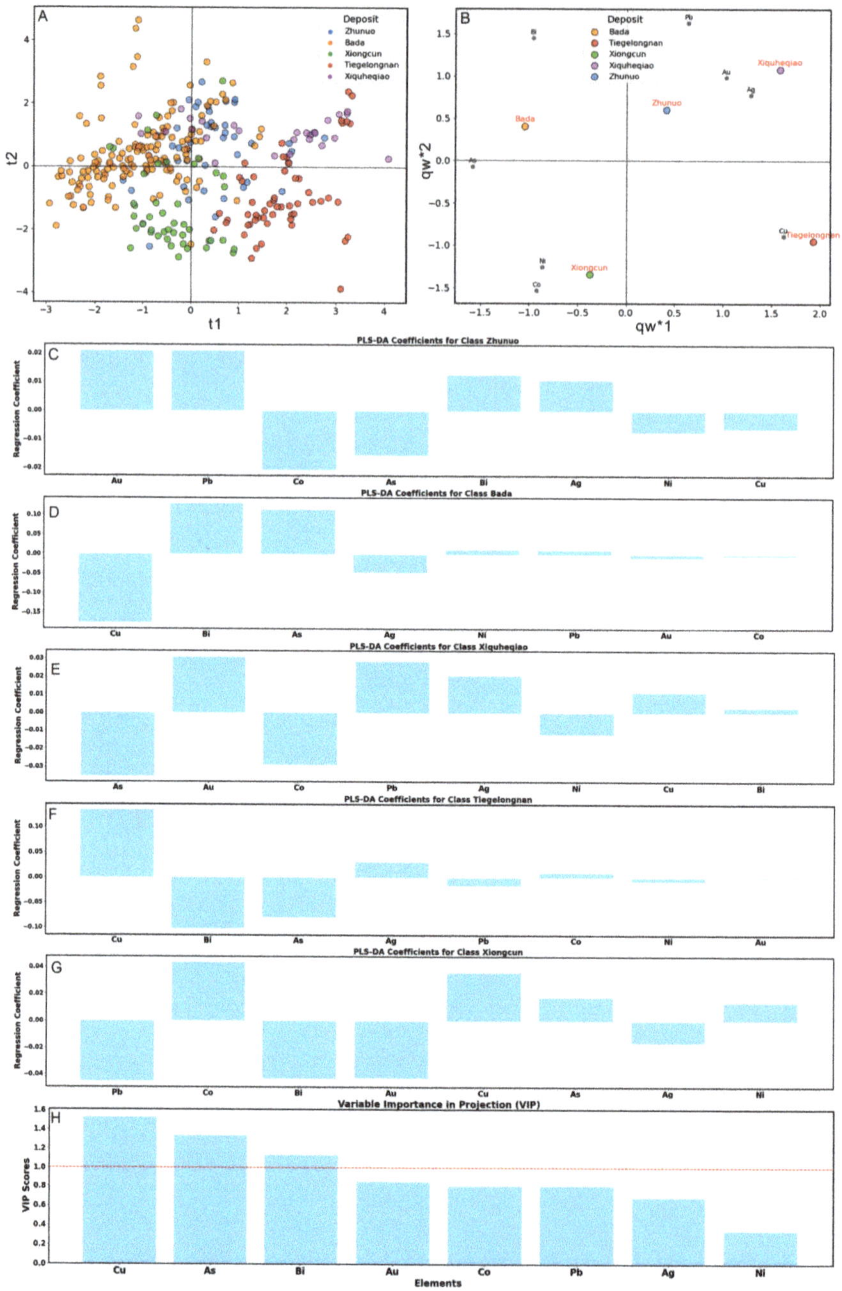

Figure 11. PLS-DA discrimination of pyrite from different porphyry deposits in Tibet. (**A**) The Score scatter plot of component t1 and t2 (first and second scores) plot showing the distribution of pyrite analysis from different deposit. (**B**) The qw*1 and qw*2 (first and second loading) plot showing correlations among elements and depositsLoading plot of component 1 and 2. (**C–G**) Regression coefficients of pyrite from each deposit, showing the contribution value of each element to the discrimination of each deposit. (**H**) Variable Importance in Projection (VIP) Score of selected trace elements, to quantify the importance of each element that contribute to the discrimination.

A biplot simply shows the relationship between two elements; however, the incorporation of elements into pyrite is usually more complicated and several elements may be incorporated simultaneously. As shown in Figure 10, Ag has a positive correlation with multiple elements including Au, Pb, and Bi. PLS-DA, as a multivariate statistic, can reveal the relationships between multiple elements. The PLS-DA diagram shows that it could discriminate pyrite from deposits well based on their various geochemical signatures (Figure 11). The Zhunuo pyrite data are plotted in the top-right quadrants of the score plot (Figure 11A), due to its close relationship to Bi, Pb, Au, and Ag (Figure 11B), which is consistent with the box-whisker plot suggesting Zhunuo pyrite has a relatively high content of Bi, Ag, Pb, and Au. The Xiquheqiao pyrite data were also plotted near the Zhunuo samples (Figure 11A), due to the close relationship with Pb, Au, Ag, and Bi. The Tiegelongnan deposit pyrite data were mostly plotted in the right and low quadrant (Figure 11A), due to the close relationship with Cu (Figure 11B), as suggested by the high Cu content in the box-whisker plot (Figure 11A). Pyrite from the Bada deposit were mostly plotted in the left quadrants (Figure 11A) because of the close correlation with As, Bi, Co, and Bi (Figure 11B), as there were high concentrations of these elements in this pyrite group. The Xiongcun pyrite data were mostly in the lower-left quadrant (Figure 11A), because of its high concentration of Co. Although Xiongcun data were also close to Ni (Figure 11), the VIP plot suggests that the Ni was low, thus it was not the most important discriminative element. Cu, As, and Bi had high VIP scores (Figure 11C), which suggest they are the most important elements for discriminating pyrite within these deposits. Therefore, the pyrite from the Xiongcun island-arc-type porphyry Cu-Au deposit in the Tibetan Plateau had the lowest Cu and Bi contents, while the pyrite from the Tiegelongnan continental margin arc-type porphyry-epithermal Cu-Au deposit exhibited the highest Cu content. In contrast, pyrite from collision-type porphyry and epithermal deposits showed intermediate amounts of Cu and Bi. The As content in pyrite from both island-arc and continental margin arc-type porphyry metallogenic systems was similar but lower than that found in deposits formed in post-collisional settings. Thus, As, Cu, and Bi can serve as discriminative indicators for different deposit types and metallogenic environments in the Tibetan Plateau. However, the compositional differences of pyrite in different deposits may also be controlled by the properties of the ore-forming fluids (e.g., temperature and composition) and their evolutionary processes, as well as the physicochemical characteristics of the pyrite types.

7. Conclusions

(1) The trace-element signatures and textural analysis of pyrite types in the Zhunuo deposit revealed a magmatic-hydrothermal origin with distinct mineralization stages. Main-stage pyrite (Py2 and Py3) formed at 354–372 °C, while late-stage Py4 reflected cooler D-vein conditions (231 °C). Dynamic fO_2 fluctuations were recorded by Te variability (7–82 ppm), indicating progressive oxidation during system cooling.

(2) Comparative analysis of pyrite from regional deposits (Xiongcun, Tiegelongnan, Bada, and Xiquheqiao) identified Cu-As-Bi as key discriminators (VIP > 0.8). Post-collisional systems (Zhunuo and Xiquheqiao) exhibited intermediate Cu-Bi and elevated As, contrasted with arc-related deposits. Zhunuo pyrite was uniquely enriched in Co-Bi-Ag-Pb, while Tiegelongnan showed high Cu but low Au-As, and Xiquheqiao displayed strong Au-As coupling.

(3) The study established pyrite trace-element proxies (e.g., As-Bi-Pb for Zhunuo, Se/Te for temperature, and Cu/As ratios) to reconstruct fluid evolution and distinguish porphyry-epithermal systems in collisional settings. These mineral-chemical indicators enhanced targeting strategies for Cu-Au mineralization in the Qinghai-Tibet Plateau.

Supplementary Materials: The following supporting information can be downloaded at: https://www.mdpi.com/article/10.3390/min15111113/s1, Table S1: The mineral—chemical characteristics of pyrite from different deposits in the Qinghai—Tibet Plateau; Table S2: The uncertainty of LA-ICP-MS in situ analysis of pyrite (1 Sigma) and The detection limit of LA-ICP-MS in situ analysis of pyrite.

Author Contributions: Conceptualization, H.G.; methodology, L.G.; software, L.G.; validation, H.G.; formal analysis, J.L.; investigation, H.G.; resources, J.L.; data curation, F.X.; writing—original draft preparation, H.G.; writing—review and editing, H.G.; visualization, F.X.; supervision, F.X.; project administration, F.X.; funding acquisition, J.L. All authors have read and agreed to the published version of the manuscript.

Funding: This research was supported by the Sichuan Provincial Foundation Youth Project (Grant number: 25QNJJ3968) and the Project of Tibet Xianglong Mining Co., Ltd. (Project No. 2508-ZC-2024-00096).

Data Availability Statement: Data are contained within the article.

Acknowledgments: Special thanks are given to the anonymous reviewers for their critical and constructive reviews.

Conflicts of Interest: Jiancuo Luosang is an employee of Tibet Zhongkai Holding Co., Ltd. Hongzhong Guan is an employee of Tibet Xianglong Mining Co., Ltd. The paper reflects the views of the scientists and not the company.

References

1. Sun, X.; Lu, Y.J.; McCuaig, T.C.; Zheng, Y.Y.; Chang, H.F.; Guo, F.; Xu, L.J. Miocene ultrapotassic, high-Mg dioritic, and adakite-like rocks from Zhunuo in Southern Tibet: Implications for mantle metasomatism and porphyry copper mineralization in collisional orogens. *J. Petrol.* **2018**, *59*, 341–386. [CrossRef]
2. Zheng, Y.Y.; Su, S.; Ci, Q.; Chen, S.Y.; Shi, S.D.; Chen, X.; Liu, P.; Li, J.Z. Metallogenic series and exploration breakthroughs of Zhunuo ore district in Xizang. *Miner. Depos.* **2024**, *43*, 1295–1315.
3. Li, M. Characteristics of Ore-Forming Fluid of the Zhunuo Porphyry Copper Deposit in the Gangdese Belt, Tibet. Master's Thesis, China University of Geoscience, Beijing, China, 2015.
4. Dai, J.; Chou, I.-M.; Pan, X.D.; Ding, J.; Huang, Y.; Yan, G.Q.; Shen, H.L. Fluid evolution of the Zhunuo porphyry Cu-Mo deposit in Gangdese Tract, Tibet, China: A fluid-inclusion investigation. *Ore Geol. Rev.* **2021**, *139*, 104484. [CrossRef]
5. Sun, X.; Hollings, P.; Lu, Y.J. Geology and origin of the Zhunuo porphyry copper deposit, Gangdese belt, southern Tibet. *Miner. Depos.* **2020**, *56*, 457–480. [CrossRef]
6. Sun, X.; Li, R.Y.; Si, X.B.; Xiao, K.; Deng, J. Timing and Mechanism of Ore Precipitation in Porphyry Cu Systems: Insight from LA-ICP-MS Analysis of Fluid Inclusions and In Situ Oxygen Isotope Analysis of Hydrothermal Quartz at Zhunuo Porphyry Cu Deposit, China. *Econ. Geol.* **2024**, *119*, 593–616. [CrossRef]
7. Mercer, C.N.; Reed, M.H.; Mercer, C.M. Time scales of porphyry Cu deposit formation: Insights from titanium diffusion in quartz. *Econ. Geol.* **2015**, *110*, 587–602. [CrossRef]
8. Li, Y.; Li, X.H.; Selby, D.; Li, J.W. Pulsed magmatic fluid release for the formation of porphyry deposits: Tracing fluid evolution in absolute time from the Tibetan Qulong Cu-Mo deposit. *Geology* **2018**, *46*, 7–10. [CrossRef]
9. Sykora, S.; Cooke, D.; Meffre, S.; Stephanov, A.S.; Gardner, K.; Scott, R.; Selley, D.; Harris, A. Evolution of Pyrite Trace Element Compositions from Porphyry-Style and Epithermal Conditions at the Lihir Gold Deposit: Implications for Ore Genesis and Mineral Processing. *Econ. Geol.* **2018**, *113*, 193–208. [CrossRef]
10. Li, H.G.; Zhang, Z.C.; Zhang, R.; Xie, Q.H.; Zhang, L.; Santosh, M. Geochemical discrimination of pyrite in diverse ore deposit types through statistical analysis and machine learning techniques. *Am. Mineral.* **2024**, *109*, 846–857. [CrossRef]
11. Yan, Y.T.; Li, S.R.; Jia, B.J.; Zhang, N.; Yan, L.N. Composition typomorphic characteristics and statistic analysis of pyrite in gold deposits of different genetic types. *Earth Sci. Front.* **2012**, *19*, 214–226.
12. Reich, M.; Deditius, A.; Chryssoulis, S.; Li, J.W.; Ma, C.Q.; Parada, M.A.; Barra, F.; Mittermayr, F. Pyrite as a record of hydrothermal fluid evolution in a porphyry copper system: A SIMS/EMPA trace element study. *Geochim. Cosmochim. Acta* **2013**, *104*, 42–62. [CrossRef]
13. Alford, L.; Gysi, A.; Hurtig, N.; Monecke, T.; Pfaff, K. Porphyry-related polymetallic Au-Ag vein deposit in the Central City district, Colorado: Mineral paragenesis and pyrite trace element chemistry. *Ore Geol. Rev.* **2020**, *119*, 103295. [CrossRef]

14. Velojic, M.; Erlandsson, V.B.; Melcher, F.; Onuk, P.; Jelenkovic, R.; Cvetkovic, V. Trace elements in pyrite from the Cukaru Peki porphyry Cu-high-sulfidation deposit, Serbia: Implications for ore evolution in a polyphase hydrothermal system. *Geol. Croat.* **2022**, *75*, 303–316. [CrossRef]
15. Xiao, X.; Zhou, T.F.; Hollings, P.; White, N.C.; Fan, Y.; Zhang, L.J.; Fu, B. Pyrite geochemistry in a porphyry-skarn Cu (Au) system and implications for ore formation and prospecting: Perspective from Xinqiao deposit, Eastern China. *Am. Mineral.* **2023**, *108*, 1132–1148. [CrossRef]
16. Yang, Z.M.; Lu, Y.J.; Hou, Z.Q.; Chang, Z.S. High-Mg diorite from Qulong in southern Tibet: Implications for the genesis of adakite-like intrusions and associated porphyry Cu deposits in collisional orogens. *J. Petrol.* **2015**, *56*, 227–254. [CrossRef]
17. Tang, J.X.; Deng, S.L.; Zheng, W.B.; Ying, L.J.; Wang, X.W.; Qin, Z.P.; Ding, F.; Li, F.J.; Tang, X.Q.; Zhong, Y.F.; et al. An exploration model for Jiama copper polymetallic deposit in Maizhokunggar County, Tibet. *Miner. Depos.* **2011**, *30*, 179–196.
18. Tang, J.X.; Lang, X.H.; Xie, F.W.; Gao, Y.M.; Li, Z.J.; Huang, Y.; Ding, F.; Yang, H.H.; Zhang, L.; Wang, Q.; et al. Geological characteristics and genesis of the Jurassic No. I porphyry Cu-Au deposit in the Xiongcun district, Gangdese porphyry copper belt, Tibet. *Ore Geol. Rev.* **2015**, *70*, 438–456. [CrossRef]
19. Zhao, X.Y.; Yang, Z.S.; Zheng, Y.C.; Liu, Y.C.; Tian, S.H.; Fu, Q. Geology and genesis of the post-collisional porphyry-skarn deposit at Bangpu, Tibet. *Ore Geol. Rev.* **2015**, *70*, 486–509. [CrossRef]
20. Zhu, D.C.; Zhao, Z.D.; Niu, Y.L.; Mo, X.X.; Chung, S.L.; Hou, Z.Q.; Wang, L.Q.; Wu, F.Y. The Lhasa Terrane: Record of a microcontinent and its histories of drift and growth. *Earth Planet. Sci. Lett.* **2011**, *301*, 241–255. [CrossRef]
21. Zhang, Z.M.; Dong, X.; Santosh, M.; Zhao, G.C. Metamorphism and tectonic evolution of the Lhasa terrane, central Tibet. *Gondwana Res.* **2014**, *25*, 170–189. [CrossRef]
22. Zhu, D.C.; Pan, G.T.; Chung, S.L.; Liao, Z.L.; Wang, L.Q.; Li, G.M. SHRIMP zircon age and geochemical constraints on the origin of Lower Jurassic volcanic rocks from the Yeba Formation, southern Gangdese, south Tibet. *Int. Geol. Rev.* **2008**, *50*, 442–471. [CrossRef]
23. Ji, W.Q.; Wu, F.Y.; Chung, S.L.; Li, J.X.; Liu, C.Z. Zircon U-Pb geochronology and Hf isotopic constraints on petrogenesis of the Gangdese batholith, southern Tibet. *Chem. Geol.* **2009**, *262*, 229–245. [CrossRef]
24. Xie, F.W.; Tang, J.X. The Late Triassic-Jurassic magmatic belt and its implications for the double subduction of the Neo-Tethys Ocean in the southern Lhasa subterrane Tibet. *Gondwana Res.* **2021**, *97*, 1–21. [CrossRef]
25. Zhu, D.C.; Li, S.M.; Cawood, P.A.; Wang, Q.; Zhao, Z.D.; Liu, S.A.; Wang, L.Q. Assembly of the Lhasa and Qiangtang terranes in central Tibet by divergent double subduction. *Lithos* **2016**, *245*, 7–17. [CrossRef]
26. Gustafson, L.B.; Hunt, J.P. The porphyry copper deposit at El Salvador, Chile. *Econ. Geol.* **1975**, *70*, 857–912. [CrossRef]
27. Wu, C.Y.; Lang, X.H.; Zhang, H.; Wang, X.H.; Zhang, P.; Deng, Y.L.; Lou, Y.M.; Jiang, K. Typomorphic characteristics and geological significance of pyrite from No. 1 deposit in the Xiongcun district, Tibet. *Bull. Mineral. Petrol. Geochem.* **2024**, *43*, 372–386. [CrossRef]
28. Zhang, Q.; Chen, W.; Yang, H.H.; Fu, X.L.; Wang, Q.; Li, Y.B.; Li, H.W.; Xin, T.; Liang, X.J.; Li, S.; et al. Characteristic and Significance of Pyrite at Tiegelongnan Cu (Au) Deposit, Duolong Ore District, Tibet. *Acta Geosci. Sin.* **2023**, *44*, 983–999.
29. Wu, H.; Zhu, X.P.; Jia, Q.; Zhang, B.; Li, Y.; Yang, L.C. Characteristics and environmental effects of in situ major and trace elements in pyrite from the Bada copper-gold deposit in Tibet. *Acta Geol. Sin.* **2023**, *97*, 2261–2280.
30. Bai, A.D.; Li, B.; Tang, G.; Wang, X.F.; Shu, H.W.; Yue, Y.; Li, Y.B. In situ trace element and sulfur isotopic compositions of pyrites from the Xiquheqiao copper-gold deposit, Mangkang County, Xizang, China. *Acta Mineral. Sin.* **2025**, *45*, 146–159. [CrossRef]
31. Makvandi, S.; Ghasemzadeh-Barvarz, M.; Beaudoin, G.; Grunsky, E.C.; McClenaghan, M.B.; Duchesne, C.; Boutroy, E. Partial least squares-discriminant analysis of trace element compositions of magnetite from various VMS deposit subtypes: Application to mineral exploration. *Ore Geol. Rev.* **2016**, *78*, 388–408. [CrossRef]
32. Huang, X.W.; Sappin, A.A.; Boutroy, E.; Beaudoin, G.; Makvandi, S. Trace Element Composition of Igneous and Hydrothermal Magnetite from Porphyry Deposits: Relationship to Deposit Subtypes and Magmatic Affinity. *Econ. Geol.* **2019**, *114*, 917–952. [CrossRef]
33. Makvandi, S.; Huang, X.W.; Beaudoin, G.; Quirt, D.; Ledru, P.; Fayek, M. Trace element signatures in hematite and goethite associated with the Kiggavik-Andrew Lake structural trend U deposits (Nunavut, Canada). *Miner. Depos.* **2021**, *56*, 509–535. [CrossRef]
34. Liu, H.M.; Beaudoin, G.; Makvandi, S.; Jackson, S.E.; Huang, X.W. Multivariate statistical analysis of trace element compositions of native gold from orogenic gold deposits: Implication for mineral exploration. *Ore Geol. Rev.* **2021**, *131*, 104061. [CrossRef]
35. Caraballo, E.; Dare, S.; Beaudoin, G. Variation of trace elements in chalcopyrite from worldwide Ni-Cu sulfide and Reef-type PGE deposits: Implications for mineral exploration. *Miner. Depos.* **2022**, *57*, 1293–1321. [CrossRef]
36. Miranda, A.C.R.; Beaudoin, G.; Rottier, B. Scheelite chemistry from skarn systems: Implications for ore-forming processes and mineral exploration. *Miner. Depos.* **2022**, *57*, 1469–1497. [CrossRef]

37. Egozcue, J.J.; Pawlowsky-Glahn, V.; Mateu-Figueras, G.; Barceló-Vidal, C. Isometric Logratio Transformations for Compositional Data Analysis. *Math. Geol.* **2003**, *35*, 279–300. [CrossRef]
38. Brill, B.A. Trace-element contents and partitioning of elements in ore minerals from the CSA Cu-Pb-Zn deposit, Australia. *Can. Mineral.* **1989**, *27*, 263–274.
39. Large, R.R.; Danyushevsky, L.V.; Hollit, C.; Maslennikov, V.; Meffre, S.; Gilbert, S.E.; Bull, S.; Scott, R.J.; Emsbo, P.; Thomas, H.; et al. Gold and trace element zonation in pyrite using a laser imaging technique: Implications of the timing of gold in orogenic and Carlin-style sediment-hosted deposits. *Econ. Geol.* **2009**, *104*, 635–668. [CrossRef]
40. Zhao, Z. *Principles of Trace Element Geochemistry*, 2nd ed.; Science Press: Beijing, China, 2016. (In Chinese)
41. Keith, M.; Haase, K.M.; Chivas, A.R.; Klemd, R. Phase separation and fluid mixing revealed by trace element signatures in pyrite from porphyry systems. *Geochim. Cosmochim. Acta* **2022**, *329*, 185–205. [CrossRef]
42. Gao, S.; Xu, H.; Li, S.X.; Santosh, M.; Zhang, D.; Yang, L.; Quan, S. Hydrothermal alteration and ore-forming fluids associated with gold-tellurium mineralization in the Dongping gold deposit, China. *Ore Geol. Rev.* **2017**, *80*, 166–184. [CrossRef]
43. Yang, C.; Beaudoin, G.; Tang, J.-X.; Song, Y.; Wang, L.-Q.; Huang, X.-W. Cu-sulfide mineralogy, texture, and geochemistry in the Tiegelongnan porphyry-epithermal copper system, Tibet, China. *Miner. Depos.* **2022**, *57*, 759–779. [CrossRef]
44. Simon, G.; Kesler, S.E.; Essense, E.J.; Chryssoulis, S.L. Gold in porphyry copper deposits: Experimental determination of the distribution of gold in the Cu-Fe-S system at 400° to 700 °C. *Econ. Geol.* **2000**, *95*, 259–270. [CrossRef]
45. White, N.C.; Hedenquist, J.W. Epithermal gold deposits: Styles, characteristics and exploration. *SEG Newsl.* **1995**, *23*, 9–13. [CrossRef]
46. Simmons, S.F.; Brown, K.L.; Tutolo, B.M. Hydrothermal transport of Ag, Au, Cu, Pb, Te, Zn, and other metals and metalloids in New Zealand geothermal systems: Spatial patterns, fluid-mineral equilibria, and implications for epithermal mineralization. *Econ. Geol.* **2016**, *111*, 589–618. [CrossRef]

minerals

MDPI

Article

Genesis of the Baijianshan Skarn-Type Zn-Cu Polymetallic Deposit, Chinese Eastern Tianshan: Constraints from Geology, Geochronology and Geochemistry

Fenwei Cheng, Shuai Zhang *, Jianxin Wu, Baofeng Huang and Di Zhang

Xinjiang Geological Bureau Hami Geological Brigade, Hami 839000, China
* Correspondence: zs748894242@163.com; Tel.: +86-18209026208

Abstract: The Baijianshan deposit is the sole skarn Zn-Cu polymetallic deposit in the Xiaoshitouquan ore field, Xinjiang, China. Its ore genesis remains controversial, which hinders understanding of the relationship between skarn-type Zn-Cu and adjacent epithermal Ag-Cu-Pb-Zn mineralization and consequently impedes further regional exploration. LA-ICP-MS U-Pb dating on zircons from the granite and granite porphyry from the mining area yielded ages of 311 ± 1.7 Ma and 312 ± 1.6 Ma, respectively. The corresponding zircon $\varepsilon_{Hf(t)}$ values and T_{DM} ages are 8.7–9.9 and 624–555 Ma for the granite, and 7.2–9.9 and 673–552 Ma for the granite porphyry. These granites are metaluminous, high-K calc-alkaline I-type granites, with high LREE/HREE ratios (4.92–9.03) and pronounced negative Eu anomalies. They are enriched in K, Th, U, Zr, and Hf, with significant depletions in Sr, P, and Ti. Combined geological and geochemical evidence indicate that these Late Carboniferous granites were derived from the juvenile crustal and formed in subduction-related back basin. Two-phase aqueous inclusions in the ore-bearing quartz and calcite have homogenization temperatures ranging from 117 to 207 °C and 112 to 160 °C, respectively, with the salinities in the ranges of 0.18~7.17 and 0.53~5.26 wt% NaCl eq. The S and Pb isotopic compositions of sulfides in the ores indicate that the ore-forming metals were sourced from the medium-acidic magmatite. The $\delta^{18}O_{H2O}$ and δD_{H2O} values of hydrothermal fluids range from −6.97% to −5.84% and −106.8% to −99.6%, respectively, suggesting that the ore-forming fluids originated from the mixing of magmatic and meteoric water. Fluid mixing and corresponding conductive cooling were identified as the principal mechanism triggering the metallic mineral precipitation. The Baijianshan skarn Zn-Cu polymetallic deposit shares contemporaneous magmatic-mineralization ages and analogous material sources with the epithermal polymetallic deposits in the Xiaoshitouquan ore field, collectively constituting a unified skarn-epithermal metallogenic system. This hypothesis indicates that the deep parts of the epithermal deposits within the Yamansu volcanic rocks possess potential for exploring the porphyry-skarn-type deposits.

Keywords: ore genesis; skarn Zn-Cu deposit; Baijianshan; harlik island arc; Chinese Eastern Tianshan

1. Introduction

In terms of the porphyry ore system, deposits of different end-members are the products of hydrothermal processes at different evolutionary stages and varying distances from the center of the causative intrusion [1–5]. These deposits comprise the porphyry-type mineralization formed by the hydrothermal fluids exsolved from the deep magma chambers, high- or intermediate-sulfidation epithermal deposits formed by the mixing of magmatic fluids and meteoric water, as well as low-sulfidation epithermal deposits formed by the magma-driven circulating meteoric water distal to the porphyries [1,5]. As another intrusion-related end-member in the porphyry ore system, the skarn deposits are defined as orebodies typically occurring within skarns and adjacent metasomatic rocks at the contact zones between aluminosilicate rocks (e.g., intrusions, volcanic rocks, and migmatites) and carbonates or other calcareous wall-rocks. These deposits form primarily through contact metasomatic reactions driven by the high-temperature hydrothermal fluids, generating both proximal Cu-Au skarns and distal Au/Zn-Pb skarn deposits [6–9]. Therefore, the mineralization processes of different deposit types within porphyry metallogenic systems and their temporal–spatial and genetic relationships can provide theoretical guidance for mineral exploration targeting in the mining districts [10–15].

The Xiaoshitouquan ore field is located in the eastern segment of the Harlik island arc belt (HIAB), Chinese Eastern Tianshan [16–18]. Since 1987, Team 704 of the Xinjiang Bureau of Geology and Mineral Resources has conducted mineral exploration in this district through geological reconnaissance, follow-up of geochemical anomalies, and deep drilling. These efforts have discovered numerous epithermal Ag-Cu-Pb-Zn polymetallic deposits (e.g., Qiongkuduke, Tongshan, Huangcaogou, and Xiaoshitouquan), as well as the Baijianshan skarn-type Zn-Cu polymetallic deposit [19–21]. All these epithermal deposits occur within the volcanic–volcaniclastic rocks of the Lower Carboniferous Yamansu Formation. Fluid inclusion and stable isotope analyses have been conducted on the epithermal deposits in the previous studies, preliminarily clarifying the ore genesis and mineralization process [22–24]. However, as a representative skarn deposit in the Xiaobaishitou ore field, the Baijianshan Zn-Cu polymetallic deposit has received insufficient attention. Current research on this deposit is limited to geological characteristics and exploration targeting [25–27]. As a consequence, magmatism-mineralization age, sources of ore-forming fluids and metals, as well as metal precipitation mechanism remain unclear. This knowledge gap constrains the understanding of its genesis and regional metallogeny, thus hindering the deep/peripheral and regional-scale exploration.

Thus, this study presents an integrated investigation on the petrography, whole-rock geochemistry, zircon U-Pb dating and Hf isotopes for the intrusive rocks from the Baijianshan deposit to constrain the timing of magmatism and mineralization, petrogenesis, and tectonic setting. Additionally, fluid inclusion and S-Pb-H-O isotopic analyses have been conducted in this study, aiming to constrain the sources of ore-forming fluids and metals, precipitation mechanism and ore genesis. These findings provide critical insights for further prospecting in the Baijianshan area and elsewhere in the region.

2. Regional Geology

The Tianshan orogenic belt, as a key component of the Central Asian Orogenic Belt, has an ongoing debate regarding its formation time. Some scholars propose that it started from the Early Permian [28], whereas others advocate for the Early Triassic [29]. Nevertheless, it is generally accepted that its formation was shaped by a combination of microcontinental collision and island arc accretion [30]. Based on the distribution of ophiolite belts, fault systems, and metamorphic rocks, the Tianshan region can be divided into three secondary units, including Northern, Central, and Southern Tianshan [29]. Northern Tianshan is predominantly characterized by the Devonian-Carboniferous calc-alkaline volcanic-sedimentary rocks, whereas Central and Southern Tianshan are distinguished by the pre-Cambrian crystalline basement and passive continental margin sediments. Eastern Tianshan refers to the eastern part of Central and Northern Tianshan [31]. It can be further subdivided into the Bogda–Harlik Belt, Danan Lake–Jueluotage Belt, and Central Tianshan Block (Figure 1a).

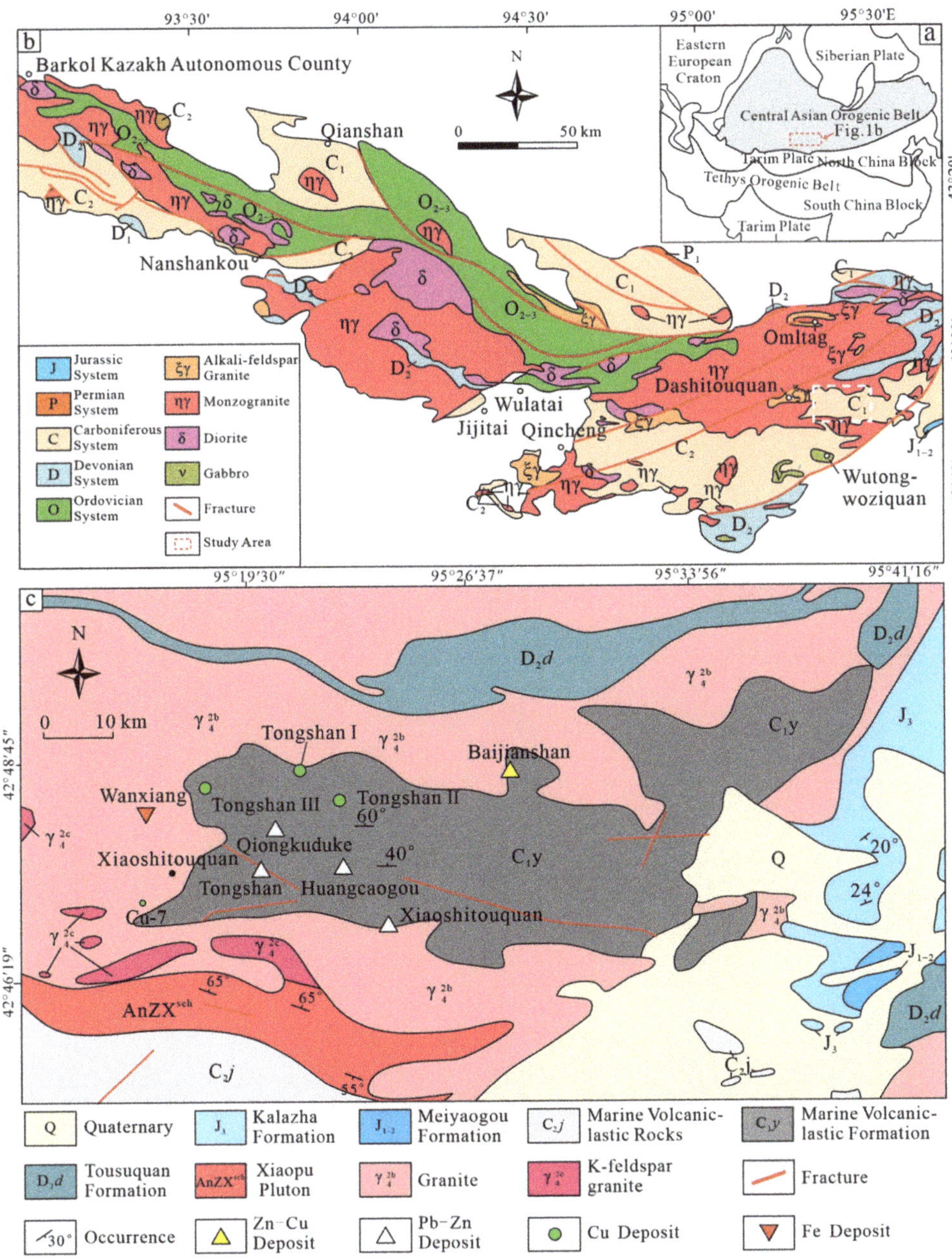

Figure 1. (**a**) Tectonic framework of the Central Asian Orogenic Belt [32], (**b**) simplified geological map of the HIAB (modified from [32]), and (**c**) geological map of the Xiaoshitouquan ore field (modified from [23]).

The Lower Carboniferous Yamansu Formation volcanic rocks are widespread in the Xiaoshitouquan ore field (Figure 1b). This stratigraphy represents a volcanic-sedimentary sequence of marine–continental transition, including tuff, tuffaceous sandstone, conglomerate, marble, and volcanic breccia. The thickness exceeds one kilometer, reflecting multi-phase faulting and central composite volcanic eruptions [20]. The regional structure generally exhibits a NW–SE trend. Secondary NW-trending faults, developed on both the northern and southern flanks, combine with NE-trending left-lateral strike-slip faults to form a conjugate shear zone [23]. The faults in the Xiaoshitouquan ore field are primarily oriented in a near-EW direction, with secondary faults trending near SN (Figure 1c). The stratigraphy as a whole exhibits near-SE trending compound anticline. The magmatic activity shows multiple phases [32–34]. Early Paleozoic (446–440 Ma) mafic-intermediate rocks (such as diorite and hornblende diorite) mark the initiation of subduction. Carboniferous (~350 Ma) diorite intrusions reflect the upwelling of mantle-derived melts during the subduction process. Large-scale Latest Carboniferous to Early Permian (299–296 Ma) felsic intrusions (biotite granite, potassium-feldspar granite) were widely emplaced and formed in a post-collisional extensional setting, covering more than 50% of the regional area.

3. Ore Deposit Geology

The exposed strata within the mining area are dominated by the second subgroup of the Lower Carboniferous Yamansu Formation (C_1y_2), consisting of marine–continental transitional facies volcanic clastic rocks, limestones, and marbles. Influenced by the magmatic intrusion, these strata have undergone varying degrees of thermal contact metamorphism, forming hornfels zones and marble + skarn zones with thicknesses of 30–50 m. The marble + skarn zones are closely associated with the Zn-Cu polymetallic orebodies (Figure 2). The structures within the mining area are dominated by two fault sets: an approximately east–west trending set and a northeast trending set. These faults are commonly observed within the gabbro in the northwestern part. Acid, intermediate-acid, and basic intrusive rocks are exposed within the mining area, intruding into the Lower Carboniferous Yamansu Formation. The rock types mainly include granite, granite porphyry, granodiorite, diorite, and gabbro (Figure 3).

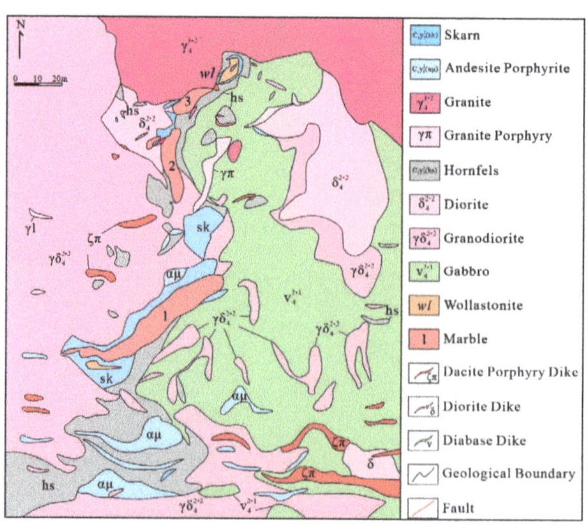

Figure 2. Geological map of the Baijianshan Zn-Cu polymetallic deposit [34].

Figure 3. Field geological photographs and hand specimen photomicrographs of intrusive rocks from the Baijianshan Zn-Cu polymetallic deposit. (**a**,**b**) Contact boundary between granite and skarn at Baijianshan; (**c**) flesh-red granite; (**d**,**f**) mineral composition predominantly quartz, K-feldspar, and plagioclase, with minor hornblende; (**e**) brownish-red granite porphyry.

Three zinc polymetallic orebodies (Zn-I, Zn-II, and Zn-III) and one copper polymetallic orebody (Cu-I) have been delineated in the mining area. The proven and probable reserves of Pb, Zn, and Cu are 302 t, 12,211 t, and 88 t, respectively. All the orebodies occur at depth and are controlled by the contact zone between the granitic intrusion and surrounding carbonates (Figure 4). As the largest Zn orebody, the Zn-I orebody has a length of 300 m and an inclined dip length of up to 178 m. The dip direction of this orebody ranges from 95° to 165°, with a dip angle of 38° to 63°. The Zn grade ranges from 0.56% to 35.44% (average 2.93%). The Cu-I orebody is located below the Zn-I orebody, with a length of approximately 75 m and a width of 50 m. The Cu grade ranges from 0.22% to 1.03% (average 0.49%).

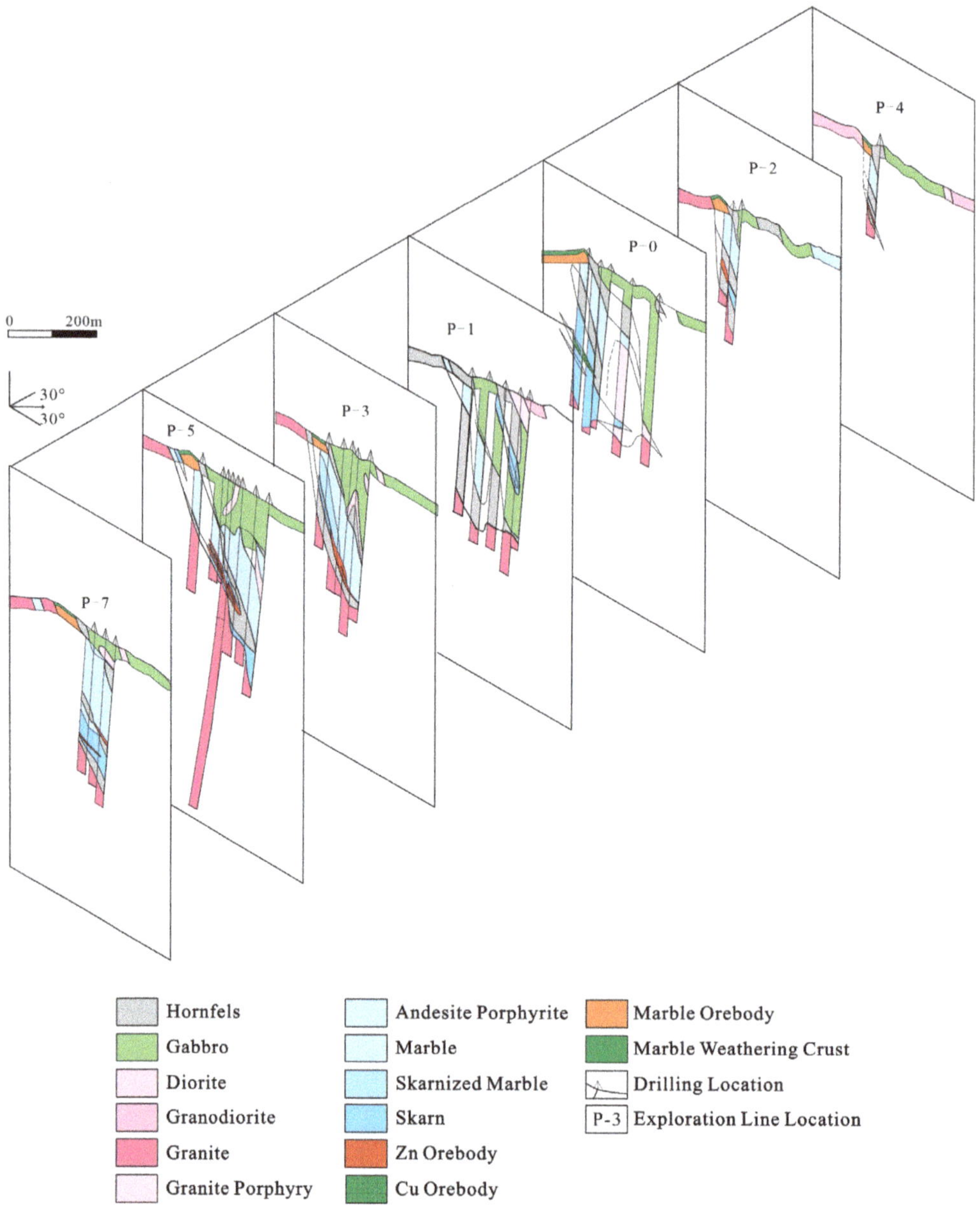

Figure 4. Composite section of the Baijianshan Zn-Cu polymetallic deposit [34].

The textures of ores mainly comprise fine-vein, net-vein, disseminated, nodular, and spotted (Figure 5). The primary metallic minerals include chalcopyrite, galena, sphalerite, and pyrite, with a small amount of ilmenite and magnetite. The non-metallic minerals mainly consist of garnet, pyroxene, wollastonite, quartz, and calcite (Figure 6). The hydrothermal alteration is characterized by skarnization, carbonation, and chloritization. Carbonation primarily occurs as light-colored calcite veins, accompanied by minor sulfides.

The orebodies are situated in the contact zones between intrusion and marble and are related to carbonation, skarnization, and chloritization.

Figure 5. Hand specimen photographs of representative ores from the Baijianshan Zn-Cu polymetallic deposit. (**a**) Sparse sphalerite in garnet matrix, (**b**) honey-colored wollastonite, (**c**) calcite replacing early-stage epidote, (**d**) paragenetic pyrite, chalcopyrite, sphalerite, and galena overprinted by later quartz-calcite veinlets, (**e**) honey-colored and milky calcite replacing epidote, (**f**) intergrown specularite and epidote replaced by quartz, (**g**) quartz-pyrite vein crosscutting pyritized wallrock, (**h**) oxidized pyrite exhibiting tarnish in wollastonite matrix, and (**i**) quartz-pyrite vein replacing early skarn minerals.

Figure 6. Photomicrographs of the mineral from the Baijianshan Zn-polymetallic deposit. (**a**) Zoned garnet exhibiting pale-yellow to white coloration, (**b**) zoned garnet under cross-polarized light, (**c**) the pyroxene was enclosed by calcite in the late stage of mineralization, (**d**) specularite partially altered to ilmenite, (**e**) sphalerite replacing chalcopyrite with euhedral pyrite grains, and (**f–i**) paragenetic association of sphalerite, chalcopyrite, pyrite, and galena; sphalerite and chalcopyrite occur as solid solution texture with minor Bi-bearing minerals in galena.

Figure 3c). The granite porphyry exhibits a brownish-red color with porphyritic texture and massive structure. Phenocrysts consist predominantly of feldspar (20%) and quartz (13%), with minor hornblende (10%) and pyroxene (7%). The matrix is composed of feldspar (25%) and quartz (25%, Figure 3d). Polished thin sections were made from both host rock samples and Zn-Cu ores of different hydrothermal stages. Subsequently, a total of sixteen samples with typical mineral assemblages were selected for further analysis of mineralogy, fluid inclusions, isotopes, and elemental geochemistry.

4.1. LA-ICP-MS Zircon U-Pb Dating

Zircons of the causative intrusions in the Baijianshan mining area were separated and selected for the U-Pb dating, including the granite and granite porphyry. In situ U–Pb dating of zircon was carried out using the LA-ICP-MS method at the Analytical Laboratory Beijing Research Institute of Uranium Geology. Analyses used a Coherent (Dieburg, Germany) Geolas 193 laser-ablation system coupled with Thermo Scientific (Bremen, Germany) HR-ICP-MS, with 24–32 μm spots, 6 Hz pulse rate, and 7 J/cm^2 energy density. Helium carrier gas was mixed with argon before entering the ICP-MS. Standard NIST SRM 610 [35] optimized the machine and calibrated trace elements. Zircon 91500 [36,37] served as the U-Th-Pb isotopic external standard, while Plešovice zircon [38] was the monitoring standard. Each analysis included 20 s gas-blank and 40 s sample data acquisition, with zircon 91500 repeated every six analyses for Pb/U fractionation and drift correction. Data were processed using ICPMSDataCal [39] for concordia diagrams.

4.2. Whole-Rock Major and Trace Elements Analyses

For the whole-rock major-trace element analyses, 7 samples of granite porphyry and 7 samples of granite were selected. Fresh whole-rock samples were crushed in a steel jaw crusher and powdered to <200 mesh using an agate mill. The major and trace elements analysis was completed by Chengdu Nanda Company, Chengdu, China. Major element compositions were determined using Rigaku (Tokyo, Japan) ZSX Primus III+ X-ray fluorescence spectrometers with accuracy better than 5%. Samples were placed in a $Li_2B_4O_7$ solution at a ratio of 1:10 and melted at a temperature of 1150–1250 °C. The melted sample was made into a glass disk. Trace elements were analyzed on a Perkin-Elmer (Singapore) NexION 1000G ICP-MS, with an analytical accuracy of better than 5%–8%. Sample powders were digested using an HF + HNO_3 mixture in high-pressure Teflon bombs at 185 ± 5 °C for 48 h. The USGS standard GSP-2 [40] were used for calibration of element concentrations of the unknowns.

4.3. In-Situ Zircon Hf Isotopes Analyses

In situ Hf isotope analysis of zircons was conducted at Analytical Laboratory Beijing Research Institute of Uranium Geology using a Coherent (Germany) Geolas 193 laser-ablation system coupled with Nu Plasma II MC-ICP-MS, with analytical parameters set at 44 μm spot size, 10 Hz pulse rate, and 10 J/cm^2 energy density. Helium carrier gas mixed with argon and trace nitrogen was applied to enhance Hf sensitivity. Zircon 91500 [41] served as the external standard, analyzed every ten samples alongside Plešovice zircon [37] as an unknown. The $n(^{176}Hf)/n(^{177}Hf)$ ratio for 91500 and Plešovice zircon was 0.2822779 and 0.2824571. Isobaric interference corrections utilized were $^{176}Lu/^{177}Hf$ via ^{175}Lu ($^{176}Lu/^{175}Lu$ = 0.02655 [42] and $^{176}Yb/^{177}Hf$ via ^{173}Yb ($^{176}Yb/^{173}Yb$ = 0.79381; [43]). Mass fractionation was corrected exponentially using $^{179}Hf/^{177}Hf$ = 0.7325 [42] and $^{173}Yb/^{171}Yb$ = 1.132685 [43].

4.4. Fluid Inclusion Microthermometry

Quartz and calcite from the main-ore stage were prepared to observe fluid inclusions. Microthermometry of fluid inclusions was conducted on a Linkam THMSG600 heating–freezing system at the College of Earth and Planetary Sciences, Chengdu University of Technology. Heating and freezing protocols followed Roedder (1984) [44]. The thermocouples were calibrated by measuring the melting temperatures of pure water (0.0 °C) and pure CO_2 (−56.6 °C) using synthetic fluid inclusions. The measurement accuracy of phase transitions in fluid inclusions was ±0.2 °C for temperatures below 0 °C, ±0.5 °C between 0 °C and 100 °C, and ±2 °C between 100 °C and 600 °C. Heating and cooling rates were controlled at 20 °C/min during heating or freezing processes most of the time, and were reduced to ±0.2 °C for ice-melting temperature and ±2 °C for homogenization (Th, L + V → L).

4.5. Stable Isotope Analysis

In situ sulfur isotope microanalysis by LA-MC-ICP-MS was conducted at Createch Testing (Tianjin) Technology Co., Ltd, Tianjin, China. Analyses employed a Thermo Scientific (Bremen, Germany) Neptune Plus multi-collector ICP-MS coupled with a RESOlution SE 193 nm solid-state laser system (Applied Spectra, West Sacramento, CA, USA). Target areas were selected based on post-scan imagery, with sulfide ablation performed using a spot diameter adjustable between 10 and 150 μm (30 μm for sphalerite; 20 μm for pyrite). Ablation parameters included an energy density of 3 J/cm^2 and a repetition rate of 5 Hz. Ablated aerosols were transported via high-purity He carrier gas to the MC-ICP-MS. ^{32}S and ^{34}S were measured using Faraday cups, with 0.131 s integration time per 200 data cycles (total acquisition time ≈ 27 s). Before analysis, instrument parameters were optimized using sulfide reference materials. To minimize matrix effects, matrix-matched sulfide standards were applied with a standard-sample-standard bracketing protocol for mass bias correction.

In situ lead isotope analyses of sulfides were conducted using a Thermo Fisher Scientific (Bremen, Germany) Neptune Plus MC-ICP-MS with a Coherent (Germany) Geolas HD excimer ArF laser-ablation system at Wuhan Sample Solution Analytical Technology Co., Ltd, Wuhan, China. Laser parameters were dynamically optimized: spot diameters (44–90 μm) adjusted in real time based on Pb signal intensity, pulse frequencies (4–10 Hz) modulated to maintain constant ablation rates, and fluence fixed at 5 J/cm^2. The gas flow protocol employed He as the primary transport medium, with post-ablation mixing of Ar to enhance plasma stability. Further details on the in situ Pb isotopic ratio analysis were provided by [45].

Hydrogen (δD) and oxygen ($\delta^{18}O$) isotope analyses of quartz separates from the quartz-sulfide veins were conducted at Createch Testing (Tianjin) Technology Co., Ltd., China, using a Finnigan MAT 253 mass spectrometer (Thermo Fisher Scientific, Germany). The δD values of fluid inclusions in minerals were analyzed using the Zn reduction method [46]. Oxygen gas was produced from the samples reacting with BrF_5 in externally heated nickel reaction vessels [47]. The H-O isotope results are reported relative to the Standard Mean Ocean Water (SMOW) with the precisions of ±1‰ for δD and ±0.2‰ for $\delta^{18}O$, respectively.

5. Results

5.1. Zircon U-Pb Ages

Zircon U-Pb dating results are listed in Supplementary Table S1. The analyzed zircons are colorless or light yellow, and are mainly short- to long-prismatic in shape, exhibiting high euhedrality and few fractures. Cathodoluminescence images show that most zircons are dark grayish-black and exhibit distinct magmatic oscillatory zoning, indicating typical magmatic zircons (Figure 8a,b).

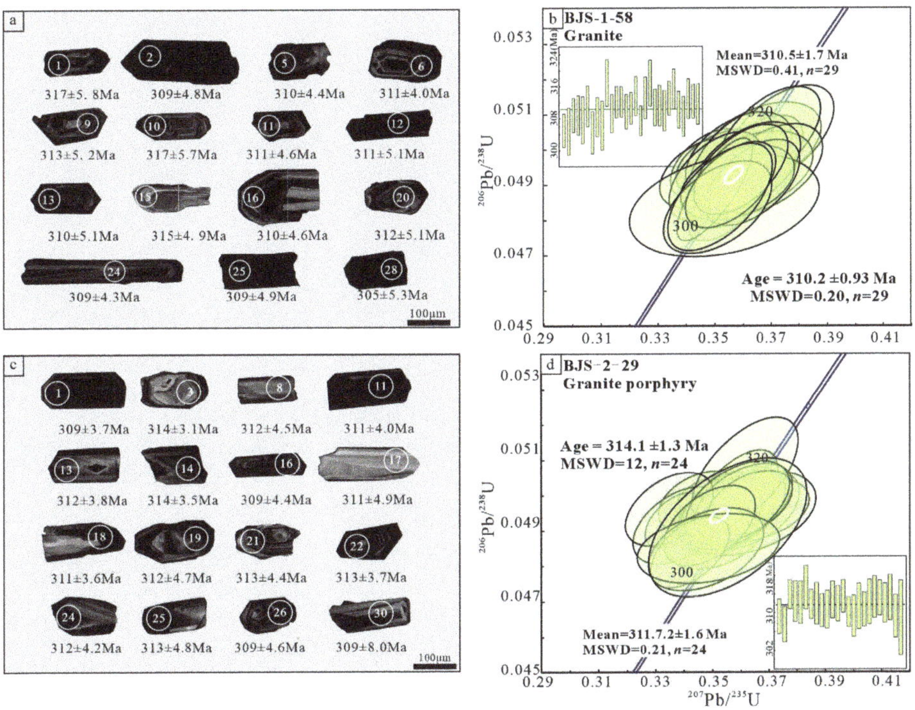

Figure 8. (**a**) representative cathodoluminescence images of zircon grains from the granite, (**b**) LA-ICP-MS U-Pb zircon concordia diagrams for the granite from Baijianshan, (**c**) representative cathodoluminescence images of zircon grains from the granite porphyry, (**d**) LA-ICP-MS U-Pb zircon concordia diagrams for the granite porphyry from Baijianshan.

Zircons of granite (BJS-1-58) exhibit grain sizes of 400–1200 μm with aspect ratios of 2:1–7:1 (Figure 8b). Analytical results show Th = (53.66–452.81) × 10^{-6}, U = (140–789) × 10^{-6}, and Th/U ratios of 0.30–0.63. Thirty spot analyses yielded 29 concordant data points, producing a weighted mean $^{206}Pb/^{238}U$ age of 311 ± 1.7 Ma (MSWD = 0.41) and a concordia age of 310 ± 0.9 Ma (MSWD = 0.20).

Zircons from granite porphyry (BJS-2-29) range from 350 to 700 μm in size with aspect ratios of 2:1–4:1 (Figure 8d). Measured compositions include Th = (31.73–553.22) × 10^{-6}, U = (90–726) × 10^{-6}, and Th/U = 0.29–0.76. Twenty-four zircon grains were analyzed, and these zircons yield a weighted mean $^{206}Pb/^{238}U$ age of 312 ± 1.6 Ma (MSWD = 0.21) and a concordia age of 314 ± 1.3 Ma (MSWD = 12).

5.2. *Whole-Rock Major and Trace Elements Compositions*

Major and trace element data of the two intrusive from the Baijianshan deposit are presented in Supplementary Table S2. The granite exhibits high silica content (SiO_2 = 73.79–76.86 wt%), potassium-rich (K_2O = 3.30–4.35 wt%), alkali-rich (7.25–8.27 wt%), calcium-poor (CaO = 0.91–1.97 wt%), magnesium-poor (MgO = 0.06–0.19 wt%), low Fe_2O_3 content (1.34–2.77 wt%), and low Al_2O_3 content (10.5–12.65 wt%). The K_2O/Na_2O ratio ranges from 0.76 to 1.73 (average 1.28). All sample points lie within the granite field in the TAS diagram (Figure 9a). The A/CNK and A/NK values range from 1.00 to 1.10 and 1.14 to 1.40, respectively, and all samples fall within the peraluminous series field and plot on the I-type granite side (Figure 9b). The samples exhibit low $Mg^{\#}$ values (19.47–28.56), and the Rittmann index (σ) are 1.65–2.18, classifying them as calc-alkaline rocks. In the w(K_2O) versus w(SiO_2) diagram, the samples mainly plot within the high-K calc-alkaline series field (Figure 9c). In the w(SiO_2) versus AR diagram, the samples fall into the alkaline-calc-alkaline series field (Figure 9d).

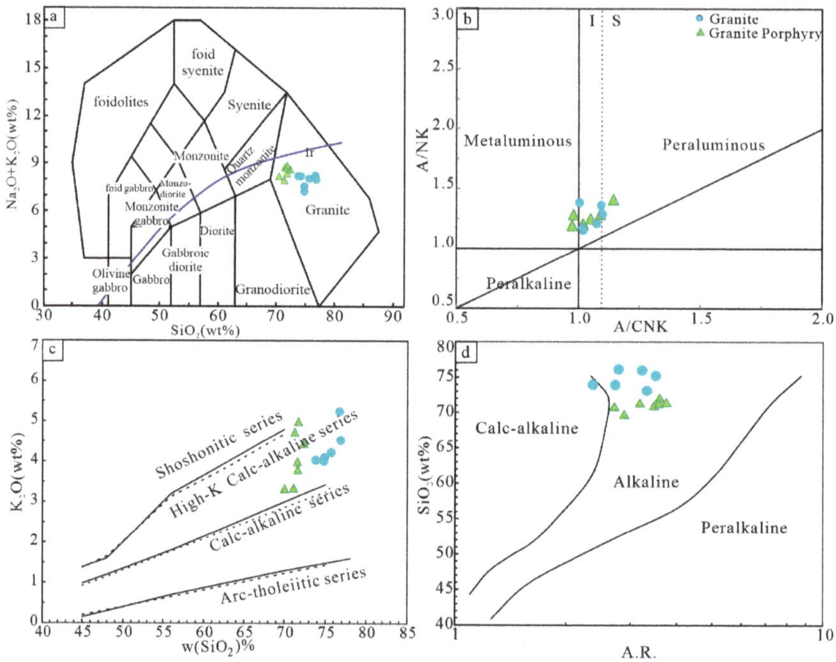

Figure 9. TAS classification and rock series discrimination diagrams of the typical intrusive rock from Baijianshan: (**a**) TAS classification [48]; (**b**) A/CNK vs. A/NK [49]; (**c**) SiO_2 vs. K_2O [50]; (**d**) SiO_2 vs. A.R. [51].

The granite porphyry samples exhibit relatively high SiO_2 content with a narrow range (70.26%–72.63%). Key compositional features include w(Al_2O_3) = 14.07–15.78 wt% (average 14.86), w(K_2O + Na_2O) = 7.95–9.14 wt% (average 8.66), w(Fe_2O_3) = 2.17–3.18 wt% (average 2.52), w(MgO) = 0.36–0.52 wt% (average 0.41), and w(TiO_2) = 0.22–0.37 wt% (average 0.37). The w(K_2O + Na_2O) ranges from 7.95% to 9.14% (average 8.64%), while the w(K_2O)/w(Na_2O) varies between 0.68 and 1.22 (average 0.92). These rocks show high total alkali content with comparable potassium and sodium levels, indicating slight sodium enrichment. The A/CNK and A/NK values range from 0.95 to 1.14 (average 1.03) and 1.17 to 1.41(average 1.26), respectively. The samples display low $Mg^{\#}$ values (20.19–25.48) and Rittmann index (σ) values ranging from 2.32 to 2.82, primarily classifying as calc-alkaline rocks. In TAS, w(K_2O)-w(SiO_2), A/NK-A/CNK, and w(SiO_2)-AR diagram, plots consistently classify the rocks as metaluminous high-K calc-alkaline series, plotting on the I-type granite field (Figure 9).

The total rare earth elements (REEs) in the Baijianshan granite are relatively low, with $\sum$REE = (133.94–189.15) $\times 10^{-6}$, $(La/Yb)_N$ = 3.51–8.79, and LREE/HREE = 4.92–9.03, showing a significant light–heavy rare earth element fractionation. The chondrite-normalized REE pattern shows LREE enrichment with a distinct negative Eu anomaly. (δEu = 0.23–0.41, Figure 10a). In the trace element spider diagram, it shows notable depletion in Ba, Nb, Sr, P, and Ti, with a slight enrichment in Zr and Hf, and clear enrichment in K, Th, and U (Figure 10b).

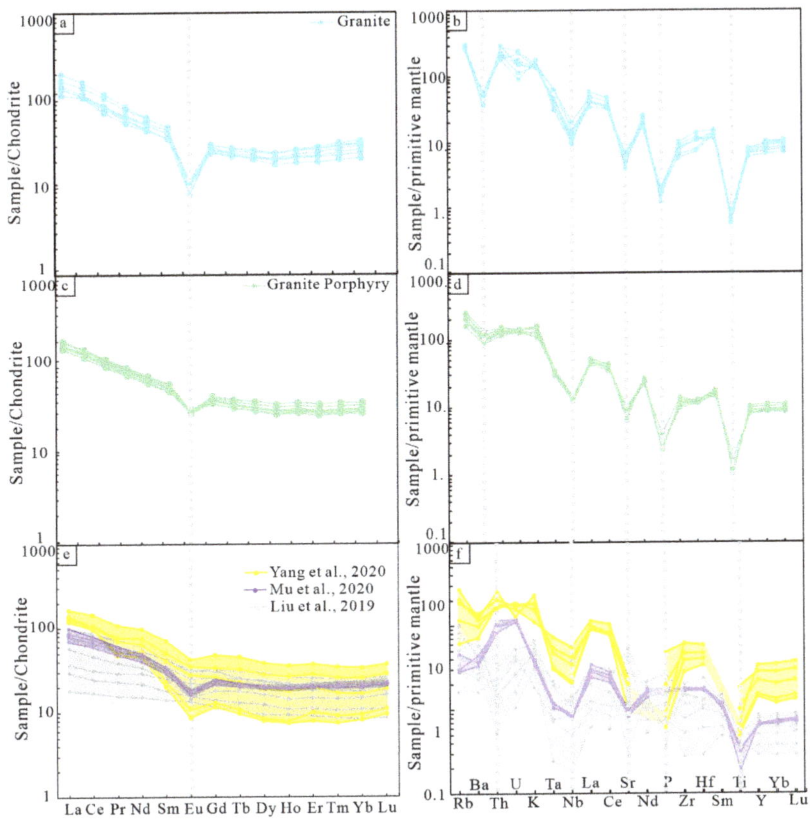

Figure 10. (**a**) chondrite-normalized REE patterns of granite from Baijianshan area; (**b**) primitive mantle-normalized trace element spider diagrams of granite from Baijianshan area; (**c**) chondrite-normalized REE patterns of granite porphyry from Baijianshan area; (**d**) primitive mantle-normalized trace element spider diagrams of granite porphyry from Baijianshan area; (**e**) chondrite-normalized REE patterns of the Yamansu volcanic rocks; (**f**) primitive mantle-normalized trace element spider diagrams of the Yamansu volcanic rocks. Normalizing values from [51]. Data are from [52–54].

The $\sum$REE in the granite porphyry is (160.13–198.25) $\times 10^{-6}$, $(La/Yb)_N$ = 4.79–5.62, and LREE/HREE = 5.31–5.88, with a noticeable light–heavy rare earth element fractionation. In the REE chondrite-normalized diagram, it also exhibits a right-leaning pattern (Figure 10c) with a negative europium anomaly (δEu = 0.56–0.75). In the trace element spider diagram, it shows strong depletion in Ba, Nb, Sr, P, and Ti, with slight enrichment in Zr and Hf, and noticeable enrichment in K, Th, and U (Figure 10d).

The trace element compositions of the two types of rocks show both differences and similarities. The granite exhibits a more pronounced negative europium anomaly compared to the granite porphyry. Both rocks are enriched in K, Th, U, Zr, and Hf and depleted in Sr, P, and Ti.

5.3. Zircon Hf Isotopes Compositions

This study conducted Lu-Hf isotopic analysis on geochronological samples from the Baijianshan granite (BJS-1-58) and granite porphyry (BJS-2-29) based on zircon U-Pb geochronology, with a total of 19 analysis points. The Hf isotope parameters were calculated using the weighted average age, and the data are presented in Supplementary Table S3. The $n(^{176}Hf)/n(^{177}Hf)$ ratio for the granite ranges from 0.282807 to 0.282857, with zircon $\varepsilon_{Hf(t)}$ values ranging from +8.7 to +9.9, corresponding to a two-stage model age (T_{DM2}) of approximately 624 to 555 Ma. The $n(^{176}Hf)/n(^{177}Hf)$ ratio for the granite porphyry ranges from 0.282781 to 0.282856, with zircon $\varepsilon_{Hf(t)}$ values of +7.2 to +9.9, corresponding to a two-stage model age (T_{DM2}) of approximately 673 to 552 Ma.

5.4. Fluid Inclusions

Based on petrographic observations, numerous fluid inclusions have been identified within quartz and calcite from the late quartz-sulfide stage of the Baijianshan Zn-polymetallic deposit. These inclusions are primarily classified by origin as primary and secondary inclusions. Based on phase state and composition at room temperature (25 °C), primary inclusions are classified into aqueous liquid-vapor inclusions and aqueous monophase inclusions (Figure 11). Quartz predominantly hosts aqueous liquid-vapor inclusions with sizes ranging from 1–5 μm (minor populations reaching ~10 μm), exhibiting vapor-to-liquid ratios (V/L) of 5%–10% and diverse morphologies including ellipsoidal, elongated, and irregular forms. In calcite, fluid inclusions occur exclusively as aqueous liquid-vapor types, typically distributed in clusters with subrounded, elongated, and irregular morphologies. These calcite-hosted inclusions display sizes concentrated within 1–5 μm and V/L ratios of 5%–30% (Figure 11).

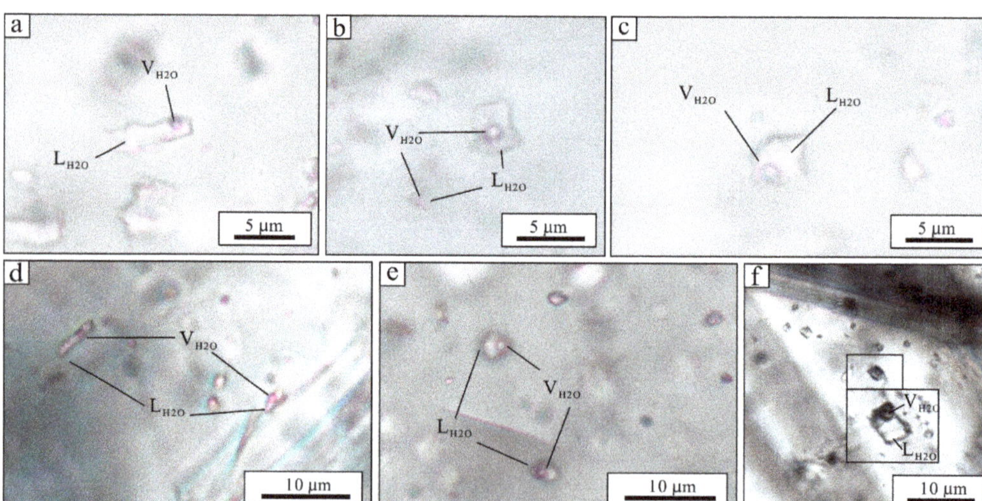

Figure 11. Photomicrographs of representative fluid inclusions from different stages in the he Baijianshan Zn-Cu polymetallic deposit. (**a–c**) Microscopic images of aqueous liquid-vapor fluid inclusions in quartz; (**d–f**) microscopic images of aqueous liquid-vapor fluid inclusions in calcite.

Homogenization temperatures were measured for large-diameter primary aqueous liquid-vapor inclusions in quartz and calcite. The microthermometry data of fluid inclusions are summarized in Supplementary Table S4 and shown in Figure 12. Quartz-hosted inclusions in late quartz-sulfide stage homogenize at temperatures from 117 to 207 °C (average 157 °C). The $T_{m,ice}$ vary from −4.5 to 0.5 °C (average −1.6 °C), corresponding to the salinities of 2.6–9.9 wt% NaCl eq (average 5.9 wt% NaCl eq). Microthermometric analysis of aqueous liquid-vapor inclusions in calcite yielded homogenization temperatures of 112 to 160 °C (average 143 °C), final ice-melting temperatures of −5.26 to −0.53 °C (average −1.8 °C), and calculated salinities of 0.53–5.26 wt% NaCl eq. (average 3.00 wt% NaCl eq).

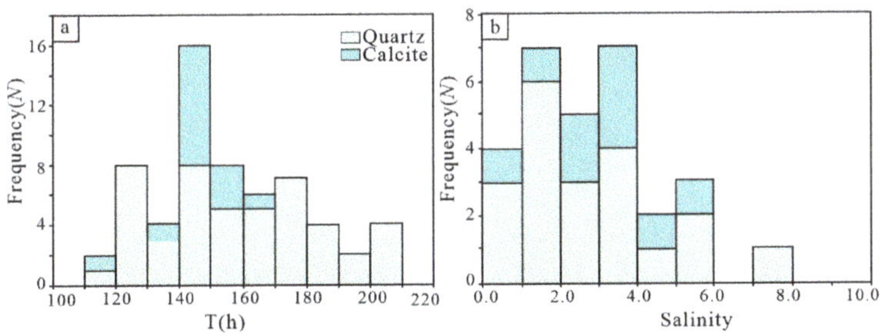

Figure 12. (**a**) Histograms of the homogenization temperatures; (**b**) histograms of the salinities of fluid inclusions in the ore-stage quartz and calcite from Baijianshan Zn-Cu polymetallic deposit.

5.5. S-Pb-H-O Isotopic Compositions

Results of in situ sulfur isotope analysis for metallic sulfides, including pyrite, chalcopyrite, sphalerite, and galena from the Baijianshan Zn-Cu deposit are presented in Supplementary Table S5. The $\delta^{34}S$ values of hydrothermal sulfides in ore range from −4.16‰ to 6.86‰ (average 3.90‰). Sulfur isotope compositions exhibit distinct ranges across sulfide species: Pyrite $\delta^{34}S$ values range from −4.16‰ to 1.71‰ (average −1.31‰), while sphalerite displays higher values between 4.76‰ and 6.25‰ (average 5.69‰). Chalcopyrite yields a narrow isotopic range of 6.10–6.39‰ (average 6.22‰), with galena showing intermediate values from 4.22‰ to 6.86‰ (average 5.01‰).

The in situ Pb isotopic data of hydrothermal sulfides from the Baijianshan deposits are listed in Supplementary Table S6. In situ lead isotopic measurements of galena from this deposit yielded characteristic ratios approximating $^{206}Pb/^{204}Pb$ = 18.119, $^{207}Pb/^{204}Pb$ = 15.541, and $^{208}Pb/^{204}Pb$ = 37.901.

The $\delta^{18}O_{quartz}$ and δD_{H2O} values of quartz-sulfide veins from the Baijianshan deposits are listed in Supplementary Table S7. The $\delta^{18}O_{quartz}$ values of quartz range from 8.68% to 9.81%. The $\delta^{18}O_{H2O}$ values obtained from this study were calculated using the quartz–water fractionation equation ($1000\ln\alpha Q\text{-}W = 3.38 \times 10^6/T^2 - 3.40$ [46]) and based on an average homogenization temperature of fluid inclusions of 153 °C. The calculated $\delta^{18}O_{H2O}$ values range from −6.97‰ to −5.84‰, and the measured δD_{H2O}-V_{SMOW} values range from −106.80‰ to −99.63‰.

6. Discussion

6.1. Timing of Magmatism and Mineralization

The oldest magmatic event in the HIAB occurred in the Middle Ordovician of the Early Paleozoic, related to the Early Paleozoic island arc evolution [16,55–57]. In the HIAB, the outcropped Middle Ordovician to Early Silurian amphibolite and medium-basic lava have formation ages concentrating between 469 and 447 Ma [16,26,33,56,58]. Magma activity in the Late Paleozoic was extremely intense within the study area, and many intrusions have now been dated with precise and reliable zircon ages (Supplementary Table S8). During the Carboniferous, large-scale intermediate-acid magmatic activity occurred in the HIAB, yielding U-Pb ages ranging from 352 to 330 Ma [34]. The zircon U-Pb ages of A-type granite and alkaline granite from the Nanshankou area are 352–351 Ma and 331 ± 2 Ma [59,60], respectively. The volcanic rocks are primarily andesites and rhyolites from the Yamansu Formation, with zircon ages ranging from 348 to 334 Ma [53,61]. Additionally, zircon ages of rhyolites and andesites from the Yamansu volcanic rocks in the southern part of Aqi Mountain are 324 ± 2 Ma and 319 ± 1 Ma, respectively [62]. The LA–ICP-MS zircon U–Pb age data of basalts and skarns from Yamansu skarn iron deposit yield ages of 324.4 ± 0.94 and 323.47 ± 0.95 Ma, respectively [62]. Previous studies have reported zircon ages for Late Carboniferous intrusions in the HIAB ranging from 320 to 307 Ma [21,32,63–66]. During the Early Permian, the region experienced widespread magmatic activity with zircon ages predominantly ranging from 298 to 288 Ma [67–71]. Recently, an amphibolite-peridotite dike was discovered in the Qincheng, Hami, with three sets of zircon U-Pb ages [72]: 298.4 ± 1.7 Ma, 297.7 ± 1.6 Ma, and 295.5 ± 1.6 Ma.

The weighted average zircon age obtained in this study for the granite in the Baijianshan mining area is 311 ± 1.7 Ma, which is closely consistent with that of the granite porphyry (312 ± 1.6 Ma). In the Xiaoshitouquan area, the zircon ages of granitic porphyry, monzogranite, and biotite granite from the Qiongkuduke Ag polymetallic deposit are 312 ± 1.7 Ma, 320 ± 1.5 Ma, and 314 ± 2.8 Ma, respectively [21]. These ages are consistent with the age range of the intrusions in the Baijianshan area. Currently, there are no reports on the metallogenic age of skarn-type and epithermal deposits in the study area. Therefore, the crystallization age of causative granite and the eruption age of hosting volcanic rocks are approximated as the metallogenic age. It is inferred that the metallogenic age of these deposits should be the same as or slightly later than the intrusive and eruptive ages. On the other hand, the metallogenic age of the Baijianshan skarn Zn-Cu deposit is roughly the same as that of the epithermal deposits in the Xiaoshitouquan area (e.g., Qiongkuduke and Tongshan).

6.2. Petrogenesis and Tectonic Setting of the Causative Granitoids

6.2.1. Petrogenetic Types

Granites are traditionally divided into the I-type, S-type, M-type, and A-type based on the nature of source rocks and geological setting [73,74]. Determining the rock type of granite requires a comprehensive analysis of petrographic evidence [73]. The Baijianshan granites exhibit no observed A-type granite diagnostic alkaline mafic minerals such as riebeckite-arfvedsonite, aegirine-aegirine-augite, or fayalite [74]. In terms of trace elements, they exhibit relatively low concentrations of high field strength elements such as Ga, Zr, and Nb with the values of Zr + Nb + Ce + Y all below the lower limit of A-type granites, leading to the conclusion that it does not belong to the A-type granite category [75].

The Baijianshan granites contain approximately 5% hornblende; thus, they are preliminarily classified as I-type granite based on petrography. According to the previous studies, typical S-type granites are strongly peraluminous rocks that develop minerals such as cordierite and muscovite, with an A/CNK value significantly greater than 1.1 [76]. The two types of granites in the Baijianshan area have A/CNK < 1.1, classifying them as sub-aluminous to weakly peraluminous rocks. The dark minerals are primarily biotite and amphibole, and there is a positive correlation between Rb and Th. The magma source area is likely to be weakly peraluminous or sub-aluminous, which is distinctly different from the characteristics of S-type granites.

Previous studies have further confirmed that P_2O_5 decreases progressively during the evolution of I-type granitoid magmas, whereas it increases during the evolution of S-type granitoid magmas [77,78]. In the SiO_2-P_2O_5 diagram, all the samples exhibit a clear negative correlation between SiO_2 and P_2O_5, displaying typical characteristics of I-type granites. The granites in the Baijianshan mining area are high-silica, sub-aluminous to peraluminous, and belong to the calc-alkaline to alkaline series. Both petrographic and geochemical characteristics indicate that these rocks are I-type granites.

6.2.2. Source of Magmas

In situ zircon Lu-Hf isotopes can provide detailed records of isotopic composition changes during magma mixing and differentiation, and serve as an effective isotopic tracer method for identifying mantle magma end-members in terms of discriminating the magma source region [79,80]. Generally, higher $\varepsilon_{Hf(t)}$ values and corresponding younger model ages suggest that the granite may originate from the mixing of mantle material within the continental crust or the recycling of newly formed crust [81,82]. In contrast, lower $\varepsilon_{Hf(t)}$ values and corresponding older model ages indicate that the granite may have originated from deep melting or re-melting of ancient crust. I-type granites can be formed by the addition of mantle-derived material during crustal remelting [83,84], or by partial melting of mafic igneous or metamorphic rocks within the crust [85].

In this study, the $\varepsilon_{Hf(t)}$ values of the Baijianshan granites range from 8.7 to 9.9 and 7.2 to 9.9, respectively, with corresponding two-stage model ages (T_{DM2}) being 808–693 Ma and 863–694 Ma. It is noteworthy that magmatic rocks in the Eastern Tianshan, regardless of lithology or age, consistently show positive Hf isotope values. For example, the $\varepsilon_{Hf(t)}$ values of the Xiaopu pluton and the Qincheng pluton (320–316 Ma) in the HIAB range from +8.0 to +13.8 and +10.8 to +16.7, respectively, with corresponding Hf model ages ranging from 822 to 450 Ma and 641 to 268 Ma, indicating that young crust served as the main source of the granite generation [66]. In addition, the $\varepsilon_{Hf(t)}$ values and T_{DM2} ages of the Omoertag alkaline granite (289 Ma) range from +8.7 to +12.5 and 600 to 430 Ma, respectively, suggesting that they may have been generated by partial melting of juvenile crust derived from depleted mantle material that had undergone crustal evolution [70]. Forming in the late stage of Early Carboniferous, alkali-feldspar granite from Nanshankou had $\varepsilon_{Nd(t)}$ values ranging from 5.66 to 6.12 and the Nd model ages are 600–620 Ma, which suggests that the magma may originate from the young crust [59]. Granitic intrusions at Balikun formed in the Early Permian (284–288 Ma) have variable Nd ($\varepsilon_{Nd(t)}$ = +1.92–+6.03) isotope compositions and young T_{DM} model ages (0.56–0.75 Ga) suggest that they were derived from juvenile crustal sources, probably earlier arc intrusion.

The Baijianshan granites exhibit positive $\varepsilon_{Hf(t)}$ values and their similar crustal model ages (863–693 Ma), suggesting the young crust served as the main source for the granite generation. Combined with the occurrence of mafic rocks in the region, these young materials may be genetically linked to neoteric mantle-derived mafic underplating magmatism.

The intrusive rocks from the Baijianshan mining area show depletions in P, Ti, Nb, and Ta, indicating characteristics of continental crust (Figure 10b,d). The Sr contents in the granites and granite porphyries from the Baijianshan mining area are relatively low (80.9–217 ppm; average 144 ppm), which lie within the range between the depleted mantle (20 ppm) and lower crust (290 ppm) [86]. It is significantly lower than that of the enriched mantle (~1100 ppm; [87]), suggesting that these granites cannot be the product of partial melting of enriched mantle. Experimental petrological studies have shown that the $Mg^{\#}$ of granites can be used to reflect the contribution of magma source region. $Mg^{\#}$ values less than 40 represent melts derived from mafic lower crust, while $Mg^{\#}$ values greater than 40 reflect melts predominantly sourced from the mantle [88]. The $Mg^{\#}$ values of the Baijianshan granites range from 14.8 to 35.3 (with an average of 23.4), indicating no contribution of mantle-derived magma. In conclusion, the formation of the Baijianshan granites is mainly related to the juvenile lower crust, accompanied by the contribution of some mantle materials.

Ba is an element that is highly enriched in the subduction zone fluids, whereas Th is relatively inert in aqueous fluids [89]. Therefore, the high Ba/Th ratios (>300) typically indicate a significant contribution from subduction zone fluids to the magma source region [89]. The Ba/Th ratios of the granite porphyries and granites in the Baijianshan area range from 59.6 to 74.6 (average 64.6) and 10.2 to 23.2 (average 17.2), respectively. The relatively low Ba/Th ratios suggest that the contribution of subduction zone fluids to the magma source region is not significant. Generally, if the magma is formed by partial melting of the mantle influenced by the subduction plate fluids, it typically has a higher Sr/Ce ratio (>20) [90]. However, the Baijianshan granites exhibit relatively low Sr/Ce ratios ranging from 2.7 to 23.5 (average 10.9), indicating that the contribution of subduction zone fluids to the source region magma is not significant. In the magmatic arc environment, the addition of subducted sediments and slab-derived fluids to the magma can be identified through the mass fraction of Th values and the (Th)/(Ce), (Th)/(Yb), (Sr)/(Nd), and (Sr)/(Th) ratios [91]. Magmas that incorporate sediments typically have high Th and Ce contents and (Th)/(Yb) values, whereas the magmas influenced by slab-derived fluids tend to have high Sr contents. As seen from the (Sr)/(Th)-(Th)/(Ce) and (Sr)/(Nd)-(Th)/(Yb) diagrams, it is evident that the Baijianshan granites resulted from the incorporation of subducted sediments (Figure 13).

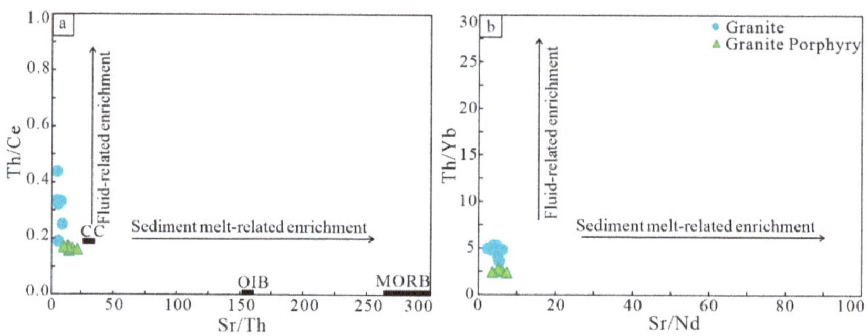

Figure 13. Diagrams of (**a**) Sr/Th vs. Th/Ce; (**b**) diagrams of Sr/Nd vs. Th/Yb [91].

6.2.3. Fractional Crystallization

As the SiO_2 content increases, the TiO_2, Al_2O_3, ^{T}FeO, MgO, and P_2O_5 contents of the Baijianshan granites show a systematic decrease (Figure 14), indicating that these variations may be related to the fractional crystallization of dark-colored minerals (such as pyroxene, amphibole), Fe-Ti oxides, feldspar, and apatite. The samples exhibit varying degrees of negative anomalies in Ba, Sr, and Eu, suggesting the fractional crystallization of plagioclase and potassium feldspar during the magma evolution [73,92]. The depletion of P and Ti may be associated with the limited fractional crystallization of apatite and Ti-oxide minerals, while the low MgO and FeO^{T} characteristics probably reflect the separation crystallization of magnesium-iron minerals (e.g., biotite).

6.2.4. Tectonic Setting

Currently, there is significant debate regarding the genesis and tectonic setting of the Late Paleozoic intrusive rocks in the HIAB, with four main viewpoints being proposed: (1) continental rift [58,93], (2) back-arc basin [59,63,94,95], (3) island arc [65,96,97], and (4) post-collision [98,99]. During the Silurian-Carboniferous period (440–310 Ma), large-scale intermediate to acidic island arc magmatic activities occurred in the Halrik Mountains [16,55,62,99]. The tectonic evolution of the Halrik Mountain subduction-related orogenic belt generally underwent four stages: (1) the oceanic basin contraction–back-arc extension stage (Carboniferous), (2) oceanic basin closure–arc (continent)-continent collision stage (Late Carboniferous), (3) post-collision–extension and compression stage (Early Permian to Middle Jurassic), and (4) intracontinental–compression and recompression stage (Late Middle Jurassic to Late Cenozoic) [64].

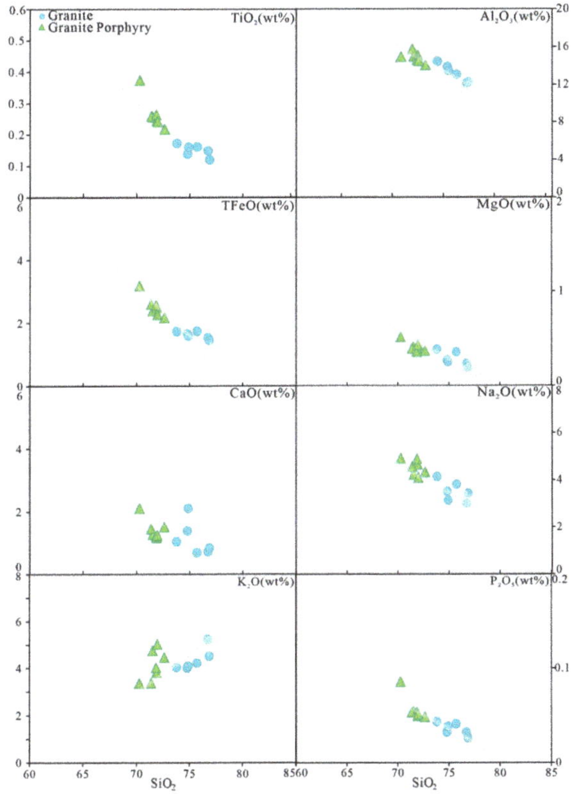

Figure 14. Harker diagrams of major elements of intrusive rock from Baijianshan.

The elemental geochemical characteristics of granite can reflect the tectonic environment in which it was formed [48,100,101]. The Baijianshan granites are enriched in large ion lithophile elements such as Rb, K, and Th, and depleted in high field strength elements such as Nb, Sr, and Ti, which is similar to the geochemical characteristics of typical island arc magmatic rocks. Moreover, Baijianshan granites are classified as I-type granite which is generally formed in the subduction setting. In the trace element Ta-Yb and Rb-(Y+Nb) tectonic discrimination diagrams, most samples fall into the volcanic arc granite area, with a small number falling into the post-collision granite and intraplate granite field (Figure 15a,b). The magma source region of Baijianshan granites may be strongly influenced by subduction processes.

The gneissoid-biotitic granite (311 ± 9 Ma) at Xiaopu formed in the extensional setting of back-arc [63]. The granite at Xiaopunan (312 ± 2 Ma) considered that the Omltag rock mass magma is formed in the island arc environment before the collision and is the product of the collision compression extension transition of the Hallik orogenic belt [102]. The granite from the Qiongkuduke Ag-Pb-Zn deposit (314–312 Ma) was formed in an island arc setting induced by the subduction of the Kanggurtag Ocean basin during the Late Carboniferous [21]. Previous studies suggest that the collision time in the Halrik area occurred approximately during the Late Carboniferous to Early Permian (310–289 Ma) [67,69,98,100]. The newly obtained zircon U-Pb ages of the Baijianshan granites in this study are 312–311 Ma, indicating that these granites formed in an extensional tectonic setting of back-arc basin during the subduction processes. In addition, the Baijianshan skarn deposit and epithermal deposits in the region have roughly contemporaneous mineralization ages [21,23]. Epithermal deposits generally form in subduction and back-arc extension tectonic settings rather than in a collisional setting, which also supports the formation of regional magmatic rocks in a subduction tectonic background.

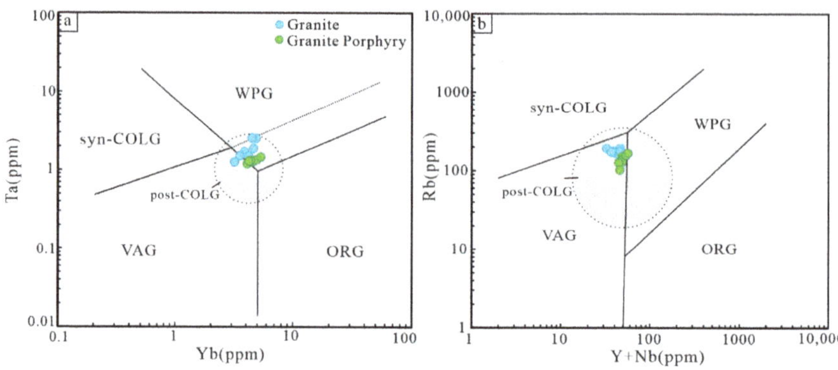

Figure 15. Tectonic discrimination diagrams for the Baijianshan granites: (**a**) Ta/Yb diagrams; (**b**) Rb/(Y+Nb) diagrams [48,103]. IAG: island arc granitoid; CAG: continental arc granitoid; CCG: continental collision granitoid; POG: post-orogenic granitoid; RRG: rift-related granitoid; CEUG: continental epeirogenic uplift-related granitoid; VAG: volcanic arc granitoid; syn—COLG: syn–collision granitoid; WPG: within-plate granitoid; ORG: ocean ridge granitoid; post—COLG: post-collision granitoid.

6.3. *Source of Ore-Forming Components*

The orebodies of the Baijianshan deposit are directly hosted in the skarns, which are in close contact with intermediate-acidic magmas. A large amount of garnet, wollastonite, and pyroxene are developed in the ores at early stage. In addition, the specularite and ilmenite in the oxide stage and quartz, calcite pyrite, and chalcopyrite in the quartz-sulfide stage are commonly observed. Such mineral assemblages show a typical alteration series of skarn-type deposits [6–9]. Therefore, the occurrence of orebodies and mineral assemblages indicate that the early ore-forming fluids may originate from the magmatic fluids, with progressive addition of meteoric water with time. Fluid inclusions are direct samples of ore-forming fluids preserved in minerals, which cannot only reflect the nature and evolution of ore-forming fluids, but also indicate the source of ore-forming fluids [44,104,105]. Abundant bi-phase aqueous inclusions are developed in the ore-stage quartz and calcite. The ore-forming fluids in the last stage belong to low-temperature (150–200 °C), low-salinity (0.18–7.17 wt% NaCl eq), and low-density (0.85–2.35 g/cm^3) H_2O-NaCl system, indicating the addition of meteoric water in the later stage. The H-O isotopic compositions of fluid inclusions can also effectively reflect the source of ore-forming fluids [46,106]. All the data points of H-O isotopic of ore-forming fluids in stage IV are located within the area between the meteoric water line and magmatic water in the H-O isotopic diagram (Figure 16), similar to that of typical skarn-type deposits in China, such as the Arqiale deposit ($\delta^{18}O_{H2O}$ = −5.2–−0.7‰, $\delta D_{V\text{-}SMOW}$ = −128–−109‰; [107]), the Xiaoyangchong iron deposit ($\delta^{18}O_{H2O}$ = −2.28–2.29‰, $\delta D_{V\text{-}SMOW}$ = −93.3–−74.4‰; [108]), and the Songyuan deposit ($\delta^{18}O_{H2O}$ = −6.39–1.41‰, $\delta D_{V\text{-}SMOW}$ = −85.6–−64.9‰; [108]). Accordingly, the observed H-O isotopic compositions for Baijianshan likely result from the mixing of magmatic fluids with meteoric water. Based on the geological, fluid inclusion, S-Pb isotopic compositions of sulfides and H-O isotopic compositions of quartz, it is believed that the early ore-forming fluids of the Baijianshan deposit were mainly magmatic fluids, and gradually mixed with meteoric water as the ore-forming process progressed, and eventually gradually transitioned to be dominated by meteoric water.

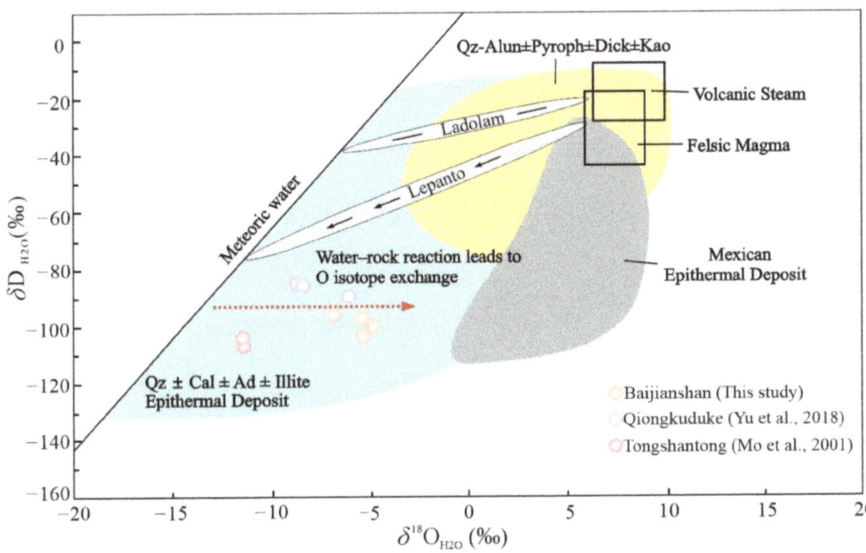

Figure 16. δD_{H2O} versus $\delta^{18}O_{H2O}$ plot of ore-forming fluids of the typical deposits in Xiaoshitouquan ore field [109]. The H-O isotopic data of Qiongkuduke and Tongshantong from [22,23].

Sulfur in the ore-forming fluids commonly have four sources: deep magmatic/mantle (generally $\delta^{34}S \approx -3‰$ to 3‰; [110]), seawater sulfate (generally $\delta^{34}S \approx 20 \pm 10‰$; [111]), sedimentary reduced sulfur (generally $\delta^{34}S$ as low as −40‰; [112]), and mixed sources (generally with a wide $\delta^{34}S$ range). The in situ $\delta^{34}S$ values of sulfides from the Baijianshan deposit exhibit a wide range from −4.16‰ to 6.86‰ (average 3.90‰). Given that the analyzed sulfides were not from the same sample, this variation may be attributed to differing contributions of heavy sulfur isotopes from the host strata, rather than disequilibrium sulfur isotope fractionation. Although the $\delta^{34}S$ values of sulfides vary greatly, it may still effectively indicate the source of sulfur. The ores in the Baijianshan deposit contain abundant sulfides and minor sulfosalt minerals, without sulfate minerals, indicating reducing ore-forming environment [113,114]. Under the reducing environment, the $\delta^{34}S$ values of sulfides are approximately equal to the total sulfur isotopic compositions ($\delta^{34}S_{\Sigma S}$) of the ore-forming fluids [113,114]. The in situ $\delta^{34}S$ values of sulfides from the Baijianshan deposit are close to those of magmatic sulfur and significantly differ from those of seawater and biogenic sulfur (Figure 17). Given the direct occurrence of Zn-Cu orebodies within the contact zone between skarn and granites, it is suggested that the sulfur in the ore-forming fluids primarily originated from the magmatic sources. In addition, the in situ Pb isotopic date of galena in the ores are located within the area between the mantle and orogenic evolution curve (Figure 18), indicating that the lead originated from a mixed source of the mantle and crust. Combined geology and S-Pb isotopic data indicate that the ore-forming metals at Baijianshan were mainly derived from medium-acidic magmatic rocks, with a minor contribution from the host rocks.

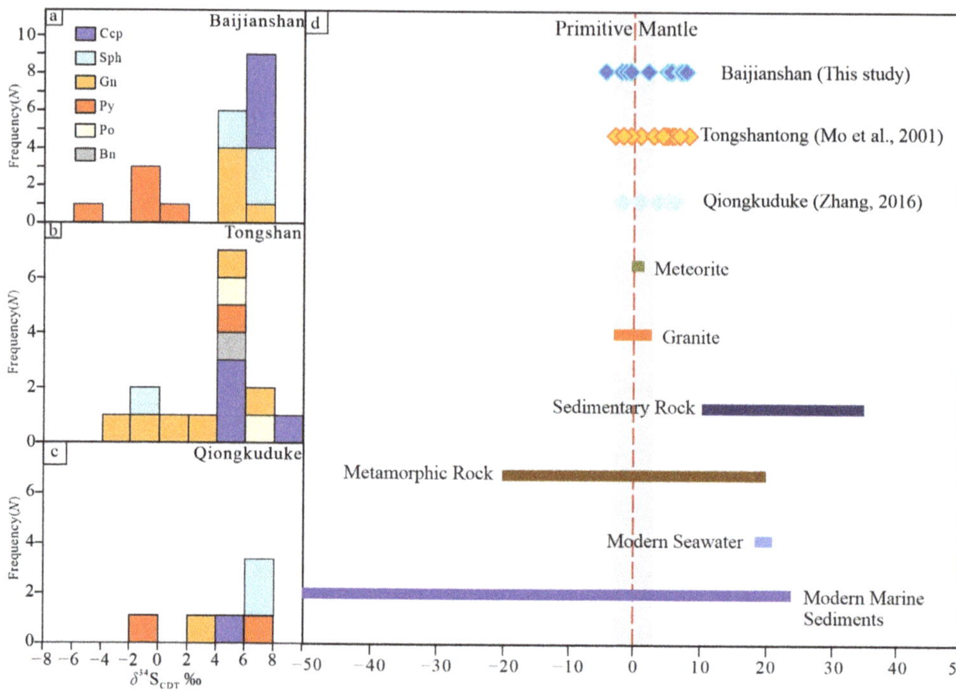

Figure 17. (**a**) Frequency histogram of values of $\delta^{34}S$ of the Baijianshan Zn-Cu polymetallic deposit; (**b**) Frequency histogram of values of $\delta^{34}S$ of the Tongshan Cu deposit; (**c**) Frequency histogram of values of $\delta^{34}S$ of the Qiongkuduke Ag deposit; (**d**) The Zn-Cu-Ag polymetallic deposit in Xiaoshitouquan ore field with those from different natural geological settings. The ranges for natural geological settings are taken from [112]. For data of Qiongkuduke and Tongshantong from [22,115].

6.4. Precipitation Mechanism

Fluid boiling, fluid mixing, and water–rock reactions commonly alter the physicochemical properties (e.g., T, pH and fO_2) of ore-forming fluids [116,117]. These processes are generally considered to be the principal mechanisms for the enrichment and precipitation of metallic minerals [118–120]. Regarding the skarn-type deposits, metal precipitation caused by fluid boiling and mixing is widely accepted [120,121]. For the skarn deposits containing low-temperature metallic minerals, the mixing between the magmatic fluids with high temperatures and salinities and groundwater/meteoric water with low temperatures and salinities is generally recognized as the key factor controlling the mineral precipitation [122–124]. The ore-forming elements of the Baijianshan deposit are dominated by Zn, accompanied by Cu, Ag, and Pb, which implies that the fluid mixing may have played a significant role in the precipitation of ore materials [122–124]. The ore-forming fluids during the main stage belong to low-temperature and low-salinity H_2O-NaCl system. This indicates that the influx of meteoric water into the mineralization system likely triggered metal precipitation. In addition, the H-O isotopic compositions of ore-forming fluids during the mineralization stage also indicates the involvement of meteoric water into the hydrothermal system. Therefore, integration of geological, fluid inclusion, and H-O isotopic compositions indicates that the mixing of magmatic-hydrothermal and meteoric water reduces the stability of metal complexes and causes the precipitation of ore-forming elements [125,126]. In fact, the mixing of magmatic fluids with meteoric water has been widely recognized as the critical precipitation mechanism in studies of numerous skarn deposits [107,127–129]. According to the results from fluid inclusion and H-O isotope, Altunbey and Kilic [127] demonstrated that the ore-forming fluids in the Cavuslu iron deposit in eastern Turkey formed by mixing magmatic fluids and meteoric water during the retrograde alteration stage. The importance of fluid mixing in the metal precipitation has also been demonstrated for the Jiawula large Pb-Zn-Ag deposit in the Great Hinggan Mountains of Inner Mongolia and the Mengya'a Pb-Zn Deposit in Tibet [128,129]. Moreover, similar observations have been reported for the Arqiale skarn deposit in Western Tianshan, where previous studies also attribute the primary cause of ore precipitation to the mixing of magmatic fluids and meteoric water [107].

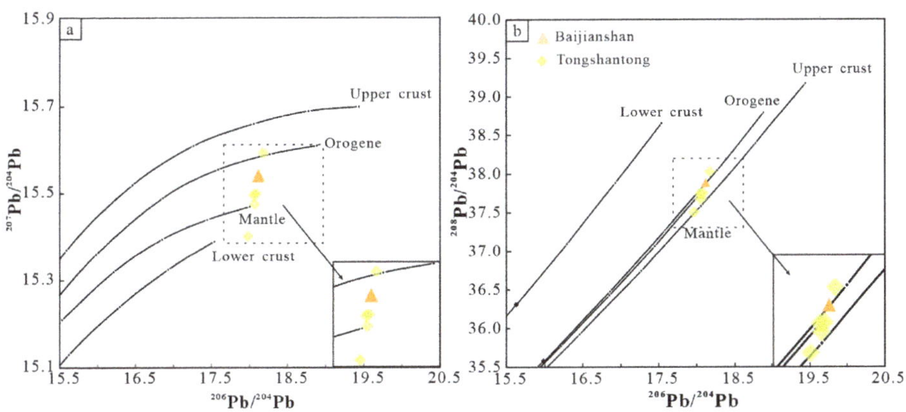

Figure 18. Plumbotectonic model for galen from the Baijianshan Zn-Cu polymetallic deposit and Tongshantong deposits [130]. (**a**) $^{207}Pb/^{204}Pb-^{206}Pb/^{20}4Pb$ model; (**b**) $^{208}Pb/^{204}Pb-^{206}Pb/^{204}Pb$ model.

6.5. Ore-Forming Process and Implications for Exploration

The HIAB is located in the eastern part of the Late Paleozoic Bogda-Halik Cu-Au-Ag polymetallic mineralization belt. It is a continental margin arc system formed during the subduction and consumption of the Paleo-Asian Ocean plate [28,29]. The Paleozoic tectonic evolution of the Halik area was accompanied by multiple episodes of magmatic activity [16,56,58,59,66,68–70]. The small-scale intrusive rocks were mainly formed during the Late Ordovician to Early Silurian, in an island arc environment, resulting from the northward subduction of the Kanggurtage Ocean basin plate [16,26,33,55]. The widely distributed granites and potassic granites formed in the Late Carboniferous (320–308 Ma) back-arc basin during subduction [32,63]. Intense magmatic activity occurred during the Early Permian (290–270 Ma) and formed in a post-collisional extensional tectonic environment, leading to the development of complex rock types (Figure 19a).

In the Late Carboniferous collisional–extensional stage, deep faults were activated, intruding residual magma from the chamber to form mineralized Mesozoic intermediate-acidic granites. The magmatic hydrothermal fluids exsolved from the crystallized granites migrated into the contact zone between the Baijianshan granite and carbonates, forming the prograde skarn assemblages (garnet–tremolite–silica) and later retrograde skarn assemblages. The ore-forming elements in the high-temperature and high-salinity magmatic fluids migrated as Cl^- complexes. with the hydrothermal system evolves, the meteoric water percolating along faults mixes with deep magmatic fluids, leading to major changes in the physical and chemical parameters of the ore-forming fluids. On the one hand, due to the sudden drop in temperature (quartz-sulfide stage homogenization temperature: 150–200 °C) and dilution of salinity (0.18–7.17 wt% NaCl eq), the stability of metal complexes was destroyed. On the other hand, the pH of ore-forming fluids increased due to the neutralization of acidic fluids by the carbonate host rocks, further promoting the precipitation of sulfides (Figure 19b).

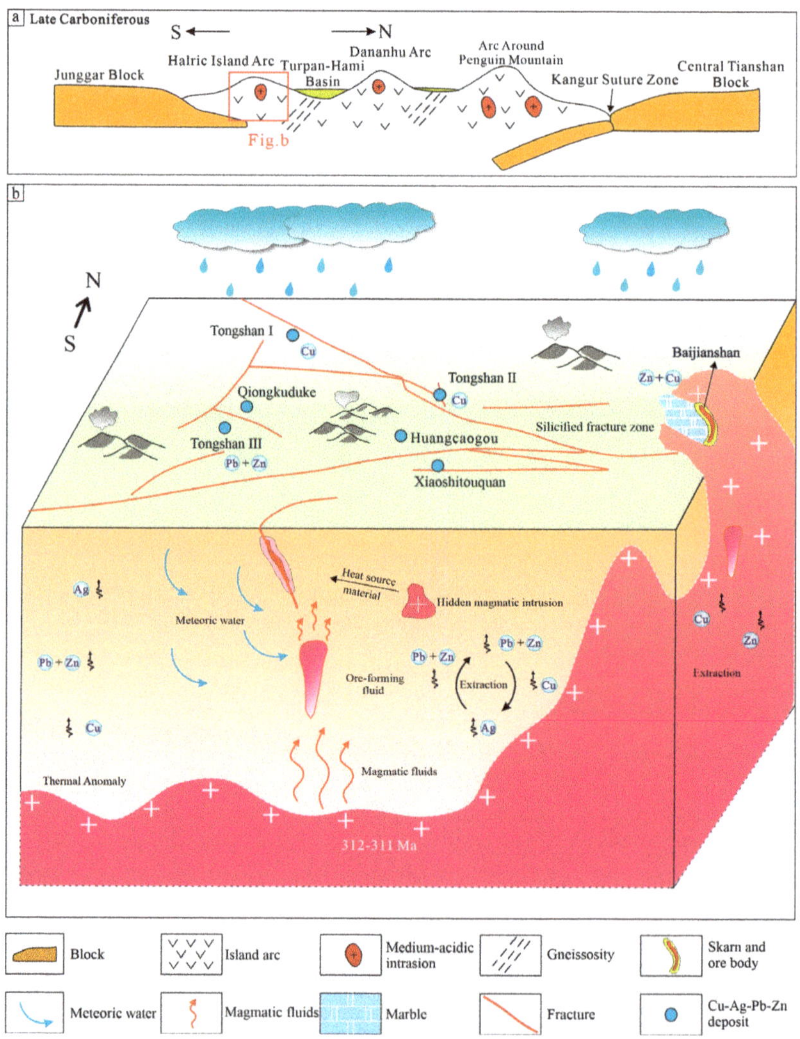

Figure 19. (**a**) A schematic diagram for the Late Carboniferous tectonic evolution of the Harlik, Eastern Tianshan, (**b**) plumbotectonic model for sulfides from the Baijianshan and Tongshantong deposits in Xiaoshitouquan ore field.

The Baijianshan deposit has a close genetic relationship with the regional intermediate-acidic magmatic rocks. This study reveals that the formation age of the regional magmatic rocks is Late Carboniferous (314–310 Ma), which can approximately represent the age of Zn-Cu mineralization. This age is consistent with those of epithermal polymetallic deposits in the Xiaoshitouquan ore field. In addition, the Baijianshan deposit has a similar ore-forming element combination (Cu-Pb-Zn-Ag) and consistent H-O-S-Pb isotopic compositions to the adjacent epithermal polymetallic deposits, indicating that these deposits have similar sources of ore-forming fluids and material. In summary, the Baijianshan skarn Zn-Cu deposit and numerous epithermal Ag-Cu-Pb-Zn polymetallic deposits (e.g., Qiongkuduke, Tongshantong, Huangcaogou, and Xiaobaishitou) in the Xiaoshitouquan ore field constitute a skarn-epithermal ore system (Figure 19b). The volcanic rocks in the Xiaoshitouquan ore field, located in the eastern part of the HIAB, are less eroded and has preserved a relatively complete epithermal system, during which silicification, sericite alteration, and vein-like Ag-Cu-Pb-Zn orebodies are widely developed (Figure 19b). To the northeast of this area, the large-scale exposure of the Halilik granite and occurrence of skarn mineral assemblages indicate a highen degree of erosion. Therefore, there is good potential to explore porphyry-skarn Cu polymetallic deposits at the deep part of the identified epithermal deposits within the Yamansu volcanic rocks. Furthermore, more attention should also be paid to the skarn-type Cu-Pb-Zn deposits in the contact zone between the Halilik granite and carbonates or calcareous volcanic rocks.

7. Conclusions

1. The granite and granite porphyry within the Baijianshan deposit yielded zircon U-Pb ages of 312–311 Ma. The granites at Baijianshan exhibit characteristic features of I-type granites. The magmas were primarily derived from the juvenile crust. These granites were formed in subduction-related back-arc basin.

2. The ore-forming fluids were dominated by the magmatic fluids at the early stage, with progressive input of meteoric water over time. The S-Pb isotopic compositions suggest that the ore-forming metals were sourced from medium-acidic magmatic rocks, with a minor contribution from the host rocks. Fluid mixing acts as the key mechanism triggering the precipitation of metallic minerals.

3. The Baijianshan skarn Zn-Cu polymetallic deposit and the adjacent epithermal polymetallic deposits in the Xiaobaishitou ore field show close relationship in time, space, and genesis, collectively constituting a skarn-epithermal ore system. This hypothesis indicates that the deep parts beneath the epithermal deposits within the Yamansu volcanic rocks possess good potential for exploring the porphyry-skarn deposits.

Supplementary Materials: The following supporting information can be downloaded at https://www.mdpi.com/article/10.3390/min15111107/s1, Table S1. LA-ICP-MS zircon U-Pb data for the studied intrusive rocks from the Baijianshan skarn deposit. Table S2. Whole-rock major (wt.%) and trace element (ppm) compositions of the intrusive rocks from the Baijianshan skarn deposit. Table S3. Zircon Hf isotopic data for the intrusive rocks from the Baijianshan skarn deposit. Table S4. Microthermometric and salinity data of fluid inclusions from the Pianyanzi Au deposit. Table S5. Sulfur isotope compositions of sulfides from the Baijianshan deposit. Table S6. Lead isotopic compositions of galena in Baijianshan deposit. Table S7. H-O isotopes of hydrothermal minerals from the Biajianshan deposit. Table S8. Zircon U-Pb ages of volcanic rocks in Harlik mountains, eastern of Xinjiang.

Author Contributions: Conceptualization, F.C. and S.Z.; data curation, F.C.; formal analysis, S.Z. and J.W.; funding acquisition, B.H.; project administration, S.Z.; validation, J.W. and B.H.; visualization,

D.Z. and B.H.; writing—original draft, F.C.; writing—review and editing, S.Z. All authors have read and agreed to the published version of the manuscript.

Funding: This research was funded by the Copper-Polymetallic Deposit Reconnaissance at Tongshan, Hami City, Xinjiang (Project No. K24-XJ015) and the Incremental Copper-Polymetallic Deposit Reconnaissance at Tongshan, Hami City, Xinjiang (Project No. K25-XJ071).

Data Availability Statement: The original contributions presented in this study are included in the article/Supplementary Material. Further inquiries can be directed to the corresponding author.

Acknowledgments: We express our gratitude to the Chengdu University of Technology for their guidance in the field.

Conflicts of Interest: The authors declare no conflicts of interest.

References

1. Sillitoe, R.H. Porphyry copper systems. *Econ. Geol.* **2010**, *150*, 3–41. [CrossRef]
2. Hedenquist, J.W.; Lowenstern, J.B. The role of magmas in the formation of hydrothermal ore deposits. *Nature* **1994**, *370*, 519–527. [CrossRef]
3. Richards, J.P.; Spell, T.; Rameh, E.; Razique, A.; Fletcher, T. High Sr/Y Magmas Reflect Arc Maturity, High Magmatic Water Content, and Porphyry Cu ± Mo ± Au Potential: Examples from the Tethyan Arcs of Central and Eastern Iran and Western Pakistan. *Econ. Geol.* **2012**, *107*, 295–332. [CrossRef]
4. Singer, D.A.; Berger, V.L.; Menzie, W.D.; Berger, B.R. Porphyry copper deposit density. *Econ. Geol.* **2005**, *100*, 491–514. [CrossRef]
5. Cooke, D.R.; Deyell, C.L.; Waters, P.J.; Gonzales, R.I.; Zaw, K. Evidence for magmatic-hydrothermal fluids and ore-forming processesin epithermal and porphyry deposits of the Baguio district, Philippines. *Econ. Geol.* **2011**, *106*, 1399–1424. [CrossRef]
6. Sangster, D.F. The Contact Metasomatic Magnetite Deposits of Southwestern British Columbia. Ph.D. Thesis, University of British Columbi, Vancouver, BC, Canada, 1964.
7. Shu, Q.H.; Chang, Z.S.; Mavrogenes, J. Fluid compositions reveal fluid nature, metal deposition mechanisms, and mineralization potential: An example at the Haobugao Zn-Pb skarn, China. *Geology* **2021**, *49*, 473–477. [CrossRef]
8. Niiranen, T.; Poutiainen, M.; Mänttäri, I. Geology, geochemistry, fluid inclusion characteristics, and U-Pb age studies on iron oxide-Cu-Au deposits in the Kolari region, northern Finland. *Ore Geol. Rev.* **2007**, *30*, 75–105. [CrossRef]
9. Zhao, Y.M.; Lin, W.W.; Bi, C.S.; Li, D.X. Basic Geological Characteristics of Skarn Deposits of China. *Bull. Chin. Acad. Geol. Sci.* **1986**, *14*, 59–87.
10. Jenner, F.E.; O'Neill, H.S.C.; Arculus, R.J.; Mavrogenes, J.A. The magnetite crisis in the evolution of arc-related magmas and the initial concentration of Au, Ag and Cu. *J. Petrol.* **2010**, *51*, 2445–2464. [CrossRef]
11. Sun, W.D.; Liang, H.Y.; Ling, M.X.; Zhan, M.Z.; Ding, X.; Zhang, H.; Yang, X.Y.; Li, Y.L.; Ireland, T.R.; Wei, Q.R.; et al. The link between reduced porphyry copper deposits and oxidized magmas. *Geochim. Cosmochim. Acta* **2013**, *103*, 263–275. [CrossRef]
12. Chiaradia, M. Copper enrichment in arc magmas controlled by overriding plate thickness. *Nat. Geosci.* **2014**, *7*, 43–46. [CrossRef]
13. Hou, Z.Q.; Yang, Z.M.; Lu, Y.; Kemp, A.; Zheng, Y.; Li, Q.; Tang, J.; Yang, Z.; Duan, L. A genetic linkage between subduction- and collision-related porphyry Cu deposits in continental collision zones. *Geology* **2015**, *43*, 247–250. [CrossRef]
14. Deng, J.; Wang, Q.; Sun, X.; Yang, L.; Groves, D.I.; Shu, Q.; Gao, L.; Yang, L.; Qiu, K.; Wang, C.; et al. Tibetan ore deposits: A conjunction of accretionary orogeny and continental collision. *Earth-Sci. Rev.* **2022**, *235*, 104245. [CrossRef]
15. Yang, Z.M.; Cao, K. Post-collisional porphyry copper deposits in Tibet: An overview. *Earth-Sci. Rev.* **2024**, *258*, 104954. [CrossRef]
16. Cao, F.G.; Tu, Q.J.; Zhang, X.M.; Ren, Y.; Li, S.L.; Dong, F.R. Preliminary determination of the Early Paleozoic magmatic arc in the Karlik Mountains, East Tianshan, Xinjiang, China-Evidence from zircon SHRIMP U-Pb dating of granite bodies in the Tashuihe area. *Geol. Bull. China* **2006**, *25*, 923–927.
17. Jin, L.Y.; Zhang, J.; Zhu, Z.X.; Zhao, T.Y.; Xu, S.Q.; Li, P.; Chen, B.X. The Geologic Character and Tectonic Significance of Paleozoic Volcanic Rocks in the Harlik Mountains, Xinjiang. *Xinjiang Geol.* **2013**, *31*, 173–179.
18. Wang, X.W.; Xu, X.Y.; Ma, Z.P.; Chen, J.L.; Cui, F.L.; Zhu, X.H.; Sun, J.M. Geochemistry and Tectonic Setting of the Early Carboniferous Volcanic Rocks in the Eastern Section of the Bogda Orogenic Belt in Xinjiang. *Geol. Explor.* **2015**, *51*, 108–122.
19. Fang, T.H.; Qing, K.Z.; Wang, S.L.; Wang, X.D.; Hui, W.D. Geological characteristics and exploration prospect of copper and Multi-Metal deposits in Xiaoshitouquan District, Eastern Xinjiang. *Xinjiang Geol.* **2002**, *20*, 371–374.
20. Wan, B.; Zhang, L.C.; Xu, X.W.; Sun, H. Geochemical characteristics of volcanic, sub-volcanic rocks in Xiaoshitouquan copper polymetallic deposit, eastern Tianshan, and its metallogenie setting. *Acta Petrol. Sin.* **2006**, *22*, 2711–2718.

21. Zhang, R.; Mao, Q.G.; Yu, M.J.; Fang, T.H.; Cheng, F.W. Geochemical Characteristics and Zircon U-Pb Ages of the Magmatite in the Qiongdukuke Ag-Pb-Zn Polymetallic Deposit of Xinjiang. *Geol. Explor.* **2017**, *53*, 270–282.
22. Mo, J.P.; Yi, S.K.; Wang, X.D.; Li, J.A. Study on geological and geochemical characteristics and genesis of Tongshan copper polymetallic deposit, Xingjiang. *Miner. Resour. Geol.* **2001**, *15*, 162–166.
23. Yu, M.J.; Wang, Y.W.; Mao, Q.G.; Wang, J.B.; Zhang, R.; Cheng, F.W.; Fu, W.W. Characteristics of Ore-Forming Fluids and Their Geological Significance of OiongkudukeAg-Polymetallic Deposit in the Xiaoshitouquan Area of Eastern Tianshan Mountains, Xinjiang. *Earth Sci.* **2018**, *43*, 3100–3111+3125.
24. Cheng, F.W.; Zhang, S.; Peng, Y.W.; Dong, S.Y.; Huang, B.F.; Zhang, D.; Song, M.W.; Liao, Z.Y.; Peng, L. Trace element characteristics of hydrothermal minerals and the genesis of the Tongshan copper deposit in the Xiaoshitouquan area, East Tianshan Mountains, Xinjiang. *Mineral. Petrol.* **2025**, *45*, 1–24.
25. Wei, J.Y.; San, J.Z.; Kang, F. Geological Characteristics and Preliminary Analysis of Prospecting Potential for the Baijianshan No. I Copper-Zinc Polymetallic Deposit, Hami City, Xinjiang Uygur Autonomous. *Xinjiang Non-Ferr. Met.* **2009**, *32*, 30–32.
26. Ma, X.H.; Chen, B.; Wang, C.; Yan, X.L. Early Paleozoic subduction of the Paleo-Asian Ocean: Zircon U-Pb geochronological, geochemical and Sr-Nd isotopic evidence from the Harlik pluton, Xinjiang. *Acta Petrol. Sin.* **2015**, *31*, 89–104.
27. Wei, J.Y. Geological Characteristics of the Baijianshan Zinc Polymetallic Deposit in Hami City, Xinjiang. *Xinjiang Non-Ferr. Met.* **2015**, *38*, 77–79+82.
28. Wang, B.; Faure, M.; Shu, L.S.; de Jong, K.; Charvet, J.; Cluzel, D.; Jahn, B.M.; Chen, Y.; Ruffet, G. Structural and Geochronological Study of High—Pressure Metamorphic Rocks in the Kekesu Section (Northwestern China): Implications for the Late Paleozoic Tectonics of the Southern Tianshan. *J. Geol.* **2010**, *118*, 59–77. [CrossRef]
29. Xiao, W.J.; Windley, B.F.; Hao, J.; Zhai, M.G. Accretion Leading to Collision and the Permian Solonker Suture, Inner Mongolia, China: Termination of the Central Asian orogenic belt. *Tectonics* **2003**, *22*, 1069. [CrossRef]
30. Jahn, B.M. The Central Asia orogenic Belt and Growth of the continental Crust in the Phanerozoic. *Geol. Soc. London Spec. Publ.* **2004**, *226*, 73–100. [CrossRef]
31. Shi, Y.; Wang, Y.W.; Wang, J.B.; Li, D.D. Mineralogical characteristics and petrogenetic significance of the amphibole in the Weiya V-Ti-magnetite deposit in the eastern Tianshan. *Miner. Explor.* **2016**, *7*, 45–52.
32. Zhu, X.H.; Zhu, T.; Zhang, X.; Xi, R.G.; Meng, Y.; Wang, K. Petrogenesis and Geological Implications of Late Carboniferous Leucogranites in Harlik Area, Eastern Tianshan. *Earth Sci.* **2018**, *43*, 4443–4458.
33. Liu, L.; He, X.F.; Li, J.T.; Yang, P.T.; Liang, B.; Su, H.; Yang, Y.D.; Liu, Y.Z.; Dai, Z.H. Petrogenesis and Tectonic Signficances of the Qincheng Tianshengquan Pluton in the Harlik Orogen of Eastern Xinjiang. *Bull. Geol. Sci. Technol.* **2017**, *36*, 86–96.
34. Cheng, F.W.; Huang, B.F.; Che, C.; Wu, L.N.; Cheng, J.P.; Chen, M.X. Detailed Investigation Report: Baijianshan Marble Mine and Zinc Polymetallic Mine in Hami City, Xinjiang. 2016, 1–133, *unpublished*.
35. Jochum, K.P.; Weis, U.; Stoll, B.; Kuzmin, D.; Yang, Q.C.; Raczek, I.; Jacob, D.E.; Stracke, A.; Birbaum, K.; Frick, D.A.; et al. Determination of Reference Values for NIST SRM 610–617 Glasses Following ISO Guidelines. *Geostand. Geoanal. Res.* **2011**, *35*, 397–429. [CrossRef]
36. Wiedenbeck, M.; Hanchar, J.M.; Peck, W.H.; Sylvester, P.; Valley, J.; Whitehouse, M.; Kronz, A.; Morishita, Y.; Nasdala, L.; Fiebig, J.; et al. Further characterisation of the 91500 zircon crystal. *Geostand. Geoanal. Res.* **2004**, *28*, 9–39. [CrossRef]
37. Wiedenbeck, M.; Alle, P.; Corfu, F.; Quadt, A.V.; Roddick, J.C.; Spiegel, W. Three natural zircon standards for U-Th-Pb, Lu-Hf, trace element and REE analyses. *Geostand. Newsl.* **1995**, *19*, 1–23. [CrossRef]
38. Sláma, J.; Košler, J.; Condon, D.J.; Crowley, J.L.; Gerdes, A.; Hanchar, J.M.; Horstwood, M.S.A.; Morris, G.A.; Nasdala, L.; Norberg, N.; et al. Plešovice zircon—A new natural reference material for U-Pb and Hf isotopic microanalysis. *Chem. Geol.* **2008**, *249*, 1–35. [CrossRef]
39. Liu, Y.S.; Hu, Z.C.; Zong, K.Q.; Gao, C.G.; Gao, S.; Xu, J.; Chen, H.H. Reappraisement and refinement of zircon U-Pb isotope and trace element analyses by LA-ICP-MS. *Chin. Sci. Bull.* **2010**, *55*, 1535–1546. [CrossRef]
40. Pretorius, W.; Weis, D.; Williams, G.; Hanano, D.; Kieffer, B.; Scoates, J. Complete Trace Elemental Characterisation of Granitoid (USGS G-2, GSP-2) Reference Materials by High Resolution Inductively Coupled Plasma-Mass Spectrometry. *Geostand. Geoanal. Res.* **2006**, *30*, 39–54. [CrossRef]
41. Blichert, T.J. The Hf isotopic composition of zircon reference material 91500. *Chem. Geol.* **2008**, *253*, 253–257. [CrossRef]
42. Blichert, T.J.; Chauvel, C.; Albarè, F. Separation of Hf and Lu for high-precision isotope analysis of rock samples by magnetic sector-multiple collector ICP-MS. *Contrib. Mineral. Petrol.* **1997**, *127*, 248–260. [CrossRef]
43. Fisher, C.M.; Vervoort, J.D.; Hanchar, J.M. Guidelines for reporting zircon Hf isotopic data by LA-MC-ICPMS and potential pitfalls in the interpretation of these data. *Chem. Geol.* **2014**, *363*, 125–133. [CrossRef]
44. Roedder, E. Fluid inclusions. *Rev. Mineral.* **1984**, *12*, 1–644.

45. Zhang, W.; Hu, Z.C.; Gunther, D.; Liu, Y.S.; Ling, W.L.; Zong, K.Q.; Chen, H.H.; Gao, S.; Xu, L. Direct lead isotope analysis in Hg-rich sulfides by LA-MC-ICP-MS with a gas exchange device and matrix-matched calibration. *Anal. Chim. Acta* **2016**, *948*, 9–18. [CrossRef]
46. Coleman, M.L.; Shepherd, T.J.; Durham, J.J.; Rouse, J.E.; Moore, G.R. Reduction of water with zinc for hydrogen isotope analysis. *Anal. Chem.* **1982**, *54*, 993–995. [CrossRef]
47. Clayton, R.N.; O'Neil, J.R.; Mayeda, T.K. Oxygen isotope exchange between quartz and water. *J. Geophys. Res.* **1972**, *77*, 3057–3067. [CrossRef]
48. Middlemost, E.A. Naming materials in the magma/igneous rock system. *Earth-Sci. Rev.* **1994**, *37*, 215–224. [CrossRef]
49. Maniar, P.D.; Piccoli, P.M. Tectonic discrimination of granitoids. *Geol. Soc. Am. Bull.* **1989**, *101*, 635–643. [CrossRef]
50. Peccerillo, R.; Taylor, S.R. Geochemistry of eocene calc-alkaline volcanic rocks from the Kastamonu area, Northern Turkey. *Contrib. Mineral. Petrol.* **1976**, *58*, 63–81. [CrossRef]
51. Wright, J.B. A simple alkalinity ratio and its application to questions of non-orogenic granite genesis. *Geol. Mag.* **1969**, *106*, 370–384. [CrossRef]
52. Sun, S.S.; McDonough, W.F. Chemical and isotopic systematics of oceanic basalts: Implications for mantle composition and processes. *Geol. Soc. Lond.* **1989**, *42*, 313–345. [CrossRef]
53. Yang, Y.S.; Chen, T.; Sang, M.S.; Zhang, Y.J. Geochemistry, zircon U-Pb age and geological significance of the Yamanu Formation volcanic rocks in the eastern segment of the East Tianshan. *West-China Explor. Eng.* **2020**, *32*, 93–96, 100.
54. Muzhapaer, M.; Gong, X.K.; Nijiati, A. Geochemical Characteristics and Tectonic Significance of Volcanic Rocks of the Lower Carboniferous Yamansu Formation in the Jueluotage Area. *Xinjiang Geol.* **2020**, *38*, 1–8.
55. Tian, J.; Liao, Q.A.; Fan, G.M.; Nie, X.M.; Wang, F.M. The discovery and tectonic implication of Early Carboniferous post-collisional I-type granites from the south of Karamaili in eastern Junngar. *Acta Petrol. Sin.* **2015**, *31*, 1471–1484.
56. Guo, H.C.; Zhong, L.; Li, L.Q. Zircon SHRIMP U-Pb dating of quartz diorite in the Koumenzi area, Karlik Mountains, East Tianshan, Xinjiang, China, and its geological significance. *Geol. Bull. China* **2006**, *25*, 928–931.
57. Li, J.Y.; Wang, K.Z.; Sun, G.H.; Mo, S.G.; Li, W.Q.; Yang, T.N.; Gao, L.M. Paleozoic active margin slices in the southern Turfan-Hami basin: Geological records of subduction of the Paleo-Asian Ocean platein central Asianregions. *Acta Petrol. Sin.* **2006**, *22*, 1087–1102.
58. Li, J.T.; He, X.F.; Liu, L.; Yang, P.T.; Liang, B.; Su, H.; Yang, Y.D.; Han, H.M.; Liu, Y.Z.; Dai, Z.H. Ordovician Tectonic Evolution of Harlik in Eastern Tianshan of Xinjiang: Constraints from LA-ICP-MS Zircon U-Pb Geochronology and Geochemistry of Volcanic Rocks. *Geoscience* **2017**, *31*, 460–473.
59. Wang, L.Y.; Liao, Q.A.; Xiao, D.; Luo, T.; Zhao, H.; Liu, H.F.; Wang, G.C. Petrogenesis and Tectonic Significance of Early Carboniferous A-Type Grainte in Harlik, Xinjiang. *J. Geomech.* **2016**, *22*, 1032–1048.
60. Liu, H.F.; Zhao, H.; Guo, R.L.; Wang, G.C.; Liao, Q.A. Geochronology, Geochemistry and Geological Implications of Early Carboniferous A-Type Granites in Harlik Area from the Eastern Tianshan. *Earth Sci.* **2022**, *47*, 2245–2263.
61. Luo, T.; Liao, Q.A.; Chen, J.P.; Zhang, X.H.; Guo, D.B.; Hu, Z.C. LA-ICP-MS Zircon U-Pb Dating of the Volcanic Rocks from Yamansu Formation in the Eastern Tianshan, and Its Geological Significance. *Earth Sci.* **2012**, *37*, 1338–1352.
62. Cui, C.; Yu, J.J.; Yang, W.Z.; Zhang, Y.H.; Cui, Y.C.; Yu, J.L. Geochronology, geochemistry and genesis of Yamansu Formation volcanic rocks of southern Aqi Mountain in Jueluotage tectonic belt, eastern Tianshan of Xinjiang. *Glob. Geol.* **2018**, *37*, 88–104.
63. Hou, T.; Zhang, Z.C.; Santosh, M.; Encarnacion, J.; Zhu, J.; Luo, W.J. Geochronology and geochemistry of submarine volcanic rocks in the Yamansu iron deposit, Eastern Tianshan Mountains, NW China: Constraints on the metallogenesis. *Ore Geol. Rev.* **2014**, *56*, 487–502. [CrossRef]
64. Sun, G.H. Dissertation submitted to Chinese Academy of Geological Science for Doctoral Degree. Ph.D. Thesis, Chinese Academy of Geological Sciences, Beijing, China, 2007.
65. Chen, X.J.; Shu, L.S.; Santosh, M.; Zhao, X.X. Island arc-type bimodal magmatism in the eastern Tianshan Belt, Northwest China: Geochemistry, zircon U–Pb geochronology and implications for the Paleozoic crustal evolution in Central Asia. *Lithos* **2013**, *168–169*, 48–66. [CrossRef]
66. Song, P.; Tong, Y.; Wang, T.; Huang, H.; Zhang, J.J.; Huang, W. Zircon U-Pb ages, genetic evolution and geological significance of Carboniferous granites in the Harlik Mountain, East Tianshan, Xinjiang. *Geol. Bull. China* **2018**, *37*, 790–804.
67. Zhao, M.; Shu, L.S.; Zhu, W.B.; Wang, C.Y.; Ma, R.S. Zircon U-Pb Dating of the Rocks from the Harlik Metamorphic Belt in Eastern Xinjiang and Its Geological Significance. *Acta Geol. Sin.* **2002**, *76*, 379–383.
68. Wang, C.S.; Zhang, Z.Z.; Wu, C.Z.; Tang, J.H.; Shao, Y.; Liu, S.H.; Feng, H.; Lei, R.X. Petrologic characteristics, zircon geochronology of dioritic enclaves in Early Permian Badashi monzogranite of East Tianshan Mountains and their geological implications. *Acta Petrol. Mineral.* **2009**, *28*, 299–315.

69. Chen, X.J.; Zhang, K.H.; Zhou, J. Geochronology and Geochemistry Characteristics of the Early Permian Monzogranite and Dioritic Enclaves of East Tianshan and Their Tectonic Implications. *Acta Geol. Sin.* **2016**, *90*, 2334–2354.
70. Chen, X.J.; Zhang, K.H.; Zhang, G.L.; Zhou, J. Characteristics, petrogenesis and tectonic implications of the Permian Omoertage alkaline granites in Harlik area, Xinjiang. *Acta Petrol. Mineral.* **2016**, *35*, 929–946.
71. Yao, Y.; Xing, W.W.; Jia, H.X.; Wang, Y.G.; Zhao, Z.N.; Ma, M.L.; Guo, S.E. Geochronology, Geochemistry and Tectonic Significance of Southern Xiamaya Granite in Harlik Area, Dongtianshan Mountain. *Mineral. Petrol.* **2020**, *40*, 30–43.
72. Du, X.F.; Wang, W.; Zhang, C.L.; Ma, H.D.; Zhu, B.Y.; Qiu, L. Zircon U-Pb age, geochemical characteristics, and tectonic implications of the early Permian ultrabasic dykes in the Harlik Mountain, east Tianshan, Xinjiang. *J. Geomech.* **2025**, *31*, 156–168.
73. Wu, F.Y.; Li, X.H.; Yang, J.H.; Zheng, Y.F. Discussions on the petrogenesis of granites. *Acta Petrol. Sin.* **2007**, *23*, 1217–1238.
74. Chappell, B.W. Aluminium saturation in I- and S-type granites and the characterization of fractionated haplogranites. *Lithos* **1999**, *46*, 535–551. [CrossRef]
75. Whalen, J.B.; Currie, K.L.; Chappell, B.W. A-type granites: Geochemical characteristics, discrimination and petrogenesis. *Contrib. Miner. Petrol.* **1987**, *95*, 407–419. [CrossRef]
76. Sylvester, P.J. Post-collisional strongly peraluminous granites. *Lithos* **1998**, *45*, 29–44. [CrossRef]
77. Pichavant, M.; Montel, J.M.; Richard, L.R. Apatite solubility in peraluminous liquids: Experimental data and an extension of the Harrison-Watson model. *Geochim. Cosmochim. Acta* **1992**, *56*, 3855–3861. [CrossRef]
78. Chappell, B.W.; White, A.J.R. I- and S-type granites in the Lachlan Fold Belt. *Trans. R. Soc. Edinburgh Earth Sci.* **1992**, *83*, 13–26.
79. Li, X.H.; Li, Z.X.; Li, W.X.; Liu, Y.; Yuan, C.; Wei, G.J.; Qi, C.S. U-Pb zircon, geochemical and Sr-Nd-Hf isotopic constraints on age and origin of Jurassic I- and A-type granites from central Guangdong, SE China: A major igneous event in response to foundering of subducted flat-slab? *Lithos* **2007**, *96*, 186–204. [CrossRef]
80. Wu, F.Y.; Li, X.H.; Zheng, Y.F.; Gao, S. Lu-Hf isotopic systematics and their applications in petrology. *Acta Petrol. Sin.* **2007**, *23*, 185–220.
81. Benbassat, M.; Carlson, R.W.; Puri, V.K.; Davenport, M.D.; Schriver, J.A.; Latif, M.; Smith, R.; Portigal, L.D.; Lipnick, E.H.; Weil, M.H. Pattern-based interactive diagnosis of multiple disorders: The MEDAS system. *IEEE Trans. Pattern Anal. Mach. Intell.* **1980**, *2*, 148–160. [CrossRef]
82. Mo, X.X.; Hou, Z.Q.; Niu, Y.L.; Dong, G.C.; Qu, X.M.; Zhao, Z.D.; Yang, Z.M. Mantle contributions to crustal thickening during continental collision: Evidence from Cenozoic igneous rocks in southern Tibet. *Lithos* **2007**, *96*, 225–242. [CrossRef]
83. Kemp, A.I.S.; Hawkesworth, C.J.; Foster, G.L.; Paterson, B.A.; Woodhead, J.D.; Hergt, J.M.; Gray, C.M.; Whitehouse, M.J. Magmatic and Crustal Differentiation History of Granitic Rocks from Hf-O Isotopes in Zircon. *Science* **2007**, *315*, 980–983. [CrossRef]
84. Collins, W.J.; Richards, S.W. Geodynamic significance of S-type granites in circum-Pacific orogens. *Geology.* **2008**, *36*, 559–562. [CrossRef]
85. Chappell, B.W.; Stephens, W.E. Origin of infracrustal (I-type) granite magmas. *Trans. R. Soc. Edinburgh Earth Sci.* **1988**, *79*, 71–86. [CrossRef]
86. Workman, R.K.; Hart, S.R. Major and trace element composition of the depleted MORB mantle (DMM). *Earth Planet. Sci. Lett.* **2005**, *231*, 53–72. [CrossRef]
87. Chen, B.; Zhai, M.G. Geochemistry of Late Mesozoic lamprophyre dykes from the Taihang Mountains, North China, and implications for the sub-continental lithospheric mantle. *Geol. Mag.* **2003**, *140*, 87–93. [CrossRef]
88. Zhang, D.Y.; Zhang, Z.C.; Encarnaci6n, J.; Xue, C.J.; Duan, S.G.; Zhao, Z.D.; Liu, J.L. Petrogenesis of the Kekesai composite intrusion, western Tianshan, NW China: Implications for tectonic evolution during Late Paleozoic time. *Lithos* **2012**, *146*, 65–79. [CrossRef]
89. Devine, J.D. Petrogenesis of the basalt-andesite-dacite association of Grenada, Lesser Antilles island arc, revisited. *J. Volcanol. Geotherm. Res.* **1995**, *69*, 1–33. [CrossRef]
90. Tatsumi, Y.; Eggins, S. *Subduction Zone Magmatism*; Blackwell: Cambridge, MA, USA, 1995.
91. Woodhead, J.D.; Eggins, S.M.; Johnson, R.W. Magma Genesis in the New Britain Island Arc: Further Insights into Melting and Mass Transfer Processes. *J. Petrol.* **1998**, *39*, 1641–1668. [CrossRef]
92. Liu, Z.C.; Liu, X.J.; Yu, L.J.; Yu, L.J.; Wang, J.G. Highly fractionated origin and magmtic-hydrothermal evolution of the Kampa leucogranites in the Tethyan Himalaya. *Nanjing Univ. Nat. Sci.* **2020**, *56*, 800–814.
93. Xia, L.Q.; Xia, Z.C.; Xue, X.Y.; Li, X.M.; Ma, Z.P.; Wang, L.S. Carboniferous Tianshan igneous megaprovince and mantle plume. *Geol. Bull. China* **2004**, *23*, 903–910.
94. Zhang, Y.Y.; Yuan, C.; Long, X.P.; Sun, M.; Huang, Z.Y.; Du, L.; Wang, X.Y. Carboniferous Bimodal Volcanic Rocks in the Eastern Tianshan, NW China: Evidence for Arc Rifting. *Gondwana Res.* **2017**, *43*, 92–106. [CrossRef]
95. Zhang, Y.Y.; Sun, M.; Yuan, C.; Long, X.P.; Jiang, Y.D.; Li, P.F.; Huang, Z.Y.; Du, L. Alternating Trench Advance and Retreat: Insights from Paleozoic Magmatism in the Eastern Tianshan, Central Asian Orogenic Belt. *Tectonics* **2018**, *37*, 2142–2164. [CrossRef]

96. Xiao, W.J.; Windley, B.F.; Badarch, G.; Sun, S.; Li, J.; Qin, K.; Wang, Z. Palaeozoic Accretionary and Convergent Tectonics of the Southern Altaids: Implications for the Growth of Central Asia. *J. Geol. Soc.* **2004**, *161*, 339–342. [CrossRef]
97. Li, L.; Xiao, W.; Windley, B.F.; Yang, H.; Jia, X.; Sang, M.; Abuduxun, N.; Liu, Y. Early Carboniferous Rifting of the Harlik Arc in the Eastern Tianshan (NW China): Response to Rollback in the Southern Altaids? *Am. J. Sci.* **2022**, *322*, 313–350. [CrossRef]
98. Chen, X.J.; Shu, L.S. Features of the post-colisional tectono-magmatism and geochronological evidence in the Harlik Mt., Xinjiang. *Acta Petrol. Sin.* **2010**, *26*, 3057–3064.
99. Yuan, C.; Sun, M.; Wilde, S.; Xiao, W.J.; Xu, Y.G.; Long, X.P.; Zhao, G.H. Post-Collisional Plutons in the Balikun Area, East Chinese Tianshan: Evolving Magmatism in Response to Extension and Slab Break-Off. *Lithos* **2010**, *119*, 269–288. [CrossRef]
100. Wang, C.S.; Gu, L.X.; Zhang, Z.Z.; Wu, C.Z.; Tang, J.H.; San, J.Z.; Li, G.G.; Li, Z.H. Petrogenesis and tectonic implications of the Permian alkaline granite and quartz-syenite assemblage in Harlik Mountains, XinJiang. *Acta Petrol. Sin.* **2009**, *25*, 3182–3196.
101. Defant, M.J.; Drummond, M.S. Derivation of some modern arc magmas by melting of young subducted lithosphere. *Nature.* **1990**, *347*, 662–665. [CrossRef]
102. Liu, L.; Li, J.T.; Ma, Y.C.; Li, S.H.; Wang, L.; Wang, F.J. Geochemical Characteristics and Zircon U-Pb Dating of the Omltag Massif in the Eastern Tianshan, Xinjiang. *Mineral. Petrol.* **2020**, *40*, 20–29.
103. Pearce, J.A.; Harris, N.B.W.; Tindle, A.G. Trace element discrimination diagrams for the tectonic interpretation of granitic rocks. *J. Petrol.* **1984**, *25*, 956–983. [CrossRef]
104. Chi, G.X.; Lu, H.Z. Validation and representation of fluid inclusion microthermometric data using the fluid inclusion assemblage (FIA) concept. *Acta Petrol. Sin.* **2008**, *24*, 1945–1953.
105. Chi, G.X.; Steel-Maclnnis, M. *Fluid Inclusions Studies*; Elsevier: Amsterdam, The Netherlands, 2024; pp. 1–412.
106. Goldfarb, R.J.; Grove, D.I. Orogenic gold: Common or evolving fluid and metal sources through time. *Lithos* **2015**, *233*, 2–26. [CrossRef]
107. Peng, Y.W.; Zou, H.; Bagas, L.; Shen, Y.F.; Shu, Z.P.; Su, J.; Liang, Q.D.; Wang, C.S.; Hu, Y.H.; Zhang, H. A newly identified Permian distal skarn deposit in the Western Tianshan, China: New evidence from geology, garnet U–Pb geochronology and S–Pb–C–H–O isotopes of the Arqiale Pb–Zn–Cu deposit. *Ore Geol. Rev.* **2022**, *143*, 104754. [CrossRef]
108. Zhang, S.S.; Li, W.; Wang, K.Y.; Yang, X.Y.; Han, C.S.; Yang, Y.L. Study of ore-forming fluid and ore forming age of skarn-type iron ore in the Fanchang area. *Acta Geol. Sin.* **2021**, *96*, 1297–1320.
109. Peng, Y.; Gu, X.; Su, J.; Zhang, Y.; Wang, J.; Wang, X.; Chen, X.; Song, M.; Shu, Z. Geology, geochemistry and genesis of tabei: A newly identified intermediate-sulphidation epithermal Pb-Zn deposit adjacent to low-sulphidation au deposit in the tulasu basin, Chinese western Tianshan. *Geol. J.* **2023**, *58*, 583–604. [CrossRef]
110. Chaussidon, M.; Lorand, J.P. Sulphur isotope composition of orogenic spinel lherzolite massifs from Ariege (North-eastern Pyrenees, France): Anion microprobe study. *Geochim. Cosmochim. Acta* **1990**, *54*, 2835–2846. [CrossRef]
111. Holser, W.T.; Kaplan, I.R. Isotope geochemistry of sedimentary sulfates. *Chem. Geol.* **1966**, *1*, 93–135. [CrossRef]
112. Hoefs, J. *Stable Isotope Geochemistry*, 6th ed.; Springer: Berlin, Germany, 2009; p. 285.
113. Ohmoto, H. Systematics of sulfur and carbon isotopes in hydrothermal ore deposits. *Econ. Geol.* **1972**, *67*, 551–578. [CrossRef]
114. Ohmoto, H.; Rye, R.O. Isotopes of sulfur and carbon. In *Geochemistry of Hydrothermal Ore Deposits*, 2nd ed.; Barnes, H.I., Ed.; Wiley: New York, NY, USA, 1979; pp. 509–567.
115. Zhang, R. The Discussion of the Metallogenic Characteristics and Genesis of the QiongKuduke Ag-Pb-Zn Deposit in Hami, Xinjiang, China. Master's Thesis, Chengdu University of Technology, Chengdu, China, 2016.
116. Drummond, S.E.; Ohmoto, H. Chemical evolution and mineral deposition in boiling hydrothermal systems. *Econ. Geol.* **1985**, *80*, 126–147. [CrossRef]
117. Williams-Jones, A.E.; Bowell, R.J.; Migdisov, A.A. Gold in solution. *Elements* **2009**, *5*, 281–287. [CrossRef]
118. Gu, X.X.; Liu, L.; Dong, S.Y.; Zhang, Y.M.; Li, K.; Li, B.H. Immiscibility during mineralization of Yinan Au-Cu-Fe deposit, Shandong Province: Evidence from fluid inclusions and H-O isotopes. *Miner. Depos.* **2010**, *29*, 43–57.
119. Lu, H.Z. Fluids immiscibility and fluid inclusions. *Acta Petrol. Sin.* **2011**, *27*, 1253–1261.
120. Li, W.; Xie, G.Q.; Mao, J.W.; Zhu, Q.Q.; Zheng, J.H. Mineralogy, Fluid Inclusion, and Stable Isotope Studies of the Chengchao Deposit, Hubei Province, Eastern China: Implications for the Formation of High-Grade Fe Skarn Deposits. *Econ. Geol.* **2019**, *114*, 325–352. [CrossRef]
121. Ni, P.; Fan, H.R.; Pan, J.Y.; Chi, Z.; Cui, J.M. Progress and Prospect of Fluid Inclusion Research in the Past Decade in China (2011–2020). *Bull. Mineral. Petrol. Geochem.* **2021**, *40*, 802–818.
122. Fang, J.; Chen, H.Y.; Zhang, L.; Zheng, Y.; Li, D.F.; Wang, C.M.; Shen, D.L. Ore genesis of the Weibao lead-zinc district, Eastern Kunlun Orogen, China: Constrains from ore geology, fluid inclusion and iso-tope geochemistry. *Int. J. Earth Sci.* **2015**, *104*, 1209–1233. [CrossRef]

123. Chen, F.C.; Deng, J.; Shu, Q.H.; Li, G.J.; Cui, X.L.; Zhao, F.; Wang, Q.F. Geology, fluid inclusion and stable isotopes (O, S) of the Hetaoping distal skarn Zn-Pb deposit, northern Baoshan block, SW China. *Ore Geol. Rev.* **2017**, *90*, 913–927. [CrossRef]
124. Shu, Q.H.; Chang, Z.S.; Hammerli, J.; Lai, Y.; Huizenga, J.M. Composition and Evolution of Fluids Forming the Baiyinnuo'er Zn-Pb Skarn Deposit, Northeastern China: Insights from Laser Ablation ICP-MS Study of Fluid Inclusions. *Econ. Geol.* **2017**, *112*, 1441–1460. [CrossRef]
125. Huang, H.L.; Tan, J.; Qiu, K.F.; Fu, J.M.; Cheng, X.H.; Li, F.; Yang, X.L. Fluid Inclusion Characteristics of Zhouwu Cu-Polymetal Deposit, Yingde, Guangdong Province. *Geol. Sci. Technol. Inf.* **2017**, *36*, 122–135.
126. Zhao, Y.Y.; Tan, J.; Liu, X.Y.; Zhang, M.; Chen, Y.; He, D. Inclusion features and geological significance of the Tonglüshan skarn-type copper-iron (gold) deposit in Daye, Hubei. *Bull. Geol. Sci. Technol.* **2020**, *39*, 64–74.
127. Altunbey, M.; Kilic, A.D. Fluid inclusion and oxygen isotope studies in garnets related to cavuslu Skarn iron mineralization, East Turkey. *J. Afr. Earth Sci.* **2019**, *149*, 465–473. [CrossRef]
128. Wang, L.Q.; Gu, X.X.; Tang, J.X.; Wang, H.; Cheng, W.B.; Jing, L.B.; Xiang, H.Y. Source and characteristics of ore-forming fluids in the Mengya'a Pb-Zn deposit, Tibet. *J. Chengdu Univ. Technol. (Sci. Technol. Ed.)* **2011**, *38*, 69–75.
129. Hu, D.L.; Jiang, S.Y.; Duan, D.F.; Xiong, S.F. Fluid origin and evolution of the Ruanjiawan W-Cu-(Mo) deposit from the Edong District in the Middle-Lower Yangtze River metallogenic belt of China: Constraints from fluid inclusions and H-O-C-S isotopes. *Ore Geol. Rev.* **2021**, *139*, 104428. [CrossRef]
130. Zartman, R.E.; Doe, B.R. Plumbotectonics-the model. *Tectonophysics* **1981**, *75*, 135–162. [CrossRef]

minerals

MDPI

Article

Magmatic Redox Evolution and Porphyry–Skarn Transition in Multiphase Cu-Mo-W-Au Systems of the Eocene Tavşanlı Belt, NW Türkiye

Hüseyin Kocatürk [1,2,*], Mustafa Kumral [1], Hüseyin Sendir [3], Mustafa Kaya [1], Robert A. Creaser [2] and Amr Abdelnasser [1,4]

1 Department of Geological Engineering, Faculty of Mines, Istanbul Technical University, Istanbul 34469, Türkiye; kumral@itu.edu.tr (M.K.); kayamusta@itu.edu.tr (M.K.); amrkhalil@itu.edu.tr (A.A.)
2 Department of Earth and Atmospheric Sciences, Faculty of Science, University of Alberta, Edmonton, AB T6G 2E3, Canada; rcreaser@ualberta.ca
3 Department of Geological Engineering, Faculty of Engineering and Architecture, Eskisehir Osmangazi University, Eskişehir 26040, Türkiye; hsendir@ogu.edu.tr
4 Department of Geology, Faculty of Science, Benha University, Benha 13511, Egypt; amr.khalil@fsc.bu.edu.eg
* Correspondence: kocaturkhu@itu.edu.tr

Abstract: This study explores the magmatic and hydrothermal evolution of porphyry–skarn–transitional Cu-Mo-W-Au systems within the Nilüfer Mineralization Complex (NMC), located in the westernmost segment of the Eocene Tavşanlı Metallogenic Belt, NW Türkiye. Through integration of field data, whole-rock geochemistry, Re–Os molybdenite dating, and amphibole–biotite mineral chemistry, the petrogenetic controls on mineralization across four spatially associated mineralized regions (Kirazgedik, Güneybudaklar, Kozbudaklar, and Delice) were examined. The earliest and thermally most distinct phase is represented by the Kirazgedik porphyry system, characterized by high temperature (~930 °C), oxidized quartz monzodioritic intrusions emplaced at ~2.7 kbar. Rising fO_2 and volatile enrichment during magma ascent facilitated structurally focused Cu-Mo mineralization. At Güneybudaklar, Re–Os geochronology yields an age of ~49.9 Ma, linking Mo- and W-rich mineralization to a transitional porphyry–skarn environment developed under moderately oxidized (ΔFMQ + 1.8 to +0.5) and hydrous (up to 7 wt.% H_2O) magmatic conditions. Kozbudaklar represents a more reduced, volatile-poor skarn system, leading to Mo-enriched scheelite mineralization typical of late-stage W-skarns. The Delice system, developed at the contact of felsic cupolas and carbonates, records the broadest range of redox and fluid compositions. Mixed oxidized–reduced fluid signatures and intense fluid–rock interaction reflect complex, multistage fluid evolution involving both magmatic and external inputs. Geochemical and mineralogical trends—from increasing silica and Rb to decreasing Sr and V—trace a systematic evolution from mantle-derived to felsic, volatile-rich magmas. Structurally, mineralization is controlled by oblique fault zones that localize magma emplacement and hydrothermal flow. These findings support a unified genetic model in which porphyry and skarn mineralization styles evolved continuously from multiphase magmatic systems during syn-to-post-subduction processes, offering implications for exploration models in the Western Tethyan domain.

Keywords: porphyry–skarn systems; Cu-Mo-W-Au mineralization; magmatic redox evolution; Re–Os geochronology; amphibole–mica chemistry; post-subduction magmatism

1. Introduction

Porphyry-related mineral systems are globally significant sources of critical metals such as copper, molybdenum, and, increasingly, gold. These deposits typically form within tectonically active orogenic belts and result from the interplay of magmatic and hydrothermal processes. A characteristic feature is a metal-enriched intrusive body that acts as the core of the system, surrounded by extensive zones of hydrothermal alteration and associated mineralization [1–6]. The evolution of these systems is strongly influenced by the behavior of magmatic fluids, which can exsolve from crystallizing magmas and interact with the surrounding host rocks. This process often leads to overlapping mineralization styles, including porphyry, skarn, greisen, and epithermal types—particularly in structurally complex settings or where there are strong lithological contrasts [7,8]. While tungsten enrichment is generally uncommon within central porphyry systems, it is frequently associated with peripheral skarn and vein-type systems. These tungsten-bearing systems typically form when metal-rich, reduced magmatic fluids interact with carbonate-rich host rocks, promoting tungsten deposition [7].

The Eocene Tavşanlı Porphyry Belt in northwest Türkiye, part of the Western Tethyan metallogenic domain, provides an ideal setting to investigate such hybrid mineral systems. This belt exemplifies the spatial and genetic relationships between porphyry and peripheral mineralization styles in a tectonically and magmatically complex setting [9] (Figure 1). Its eastern part hosts Cu-Mo porphyry deposits, including Muratdere and Kirazgedik, along with associated Cu-bearing skarns such as Gelemiç [10–12]. Conversely, the western segment is characterized predominantly by W-rich skarn systems, notably Kozbudaklar mineralization, which also exhibit Mo enrichment [13]. Within this western domain lies the Nilüfer Mineralization Complex (NMC), encompassing several centers such as Kozbudaklar, Güneybudaklar, and Delice. The NMC represents a hybrid mineral system having Cu, Mo, Au, and W, with mineralization styles comprising porphyry, skarn, and hydrothermal vein assemblages. This complex preserves a well-constrained temporal and spatial zoning of mineralization, serving as a representative example of hybrid magmatic-hydrothermal systems within the western portion of the Tavşanlı belt.

Previous studies in the Tavşanlı Zone of northwest Türkiye have primarily concentrated on isolated mineral deposits within the Eocene magmatic arc, notably the Kozbudaklar W-Mo skarn and the Kirazgedik Cu-Mo $\pm$ Au porphyry system. The Kozbudaklar skarn has been interpreted as a product of contact metasomatism between the Topuk Pluton and the Triassic İnönü Marble [13], while Kirazgedik has been linked to the South Tepeldağ Pluton and framed within a broader tectonomagmatic context of slab rollback during the waning stages of Neo-Tethyan subduction [12,14]. Although these studies provided foundational deposit-level insights, they did not explore the potential genetic and temporal relationships among the broader set of mineralized centers within the Nilüfer Mineralization Complex (NMC). As a result, the combination of porphyry, skarn, and vein-style mineralization across this area, and their integration into a coherent geodynamic and metallogenic framework, remained largely unresolved.

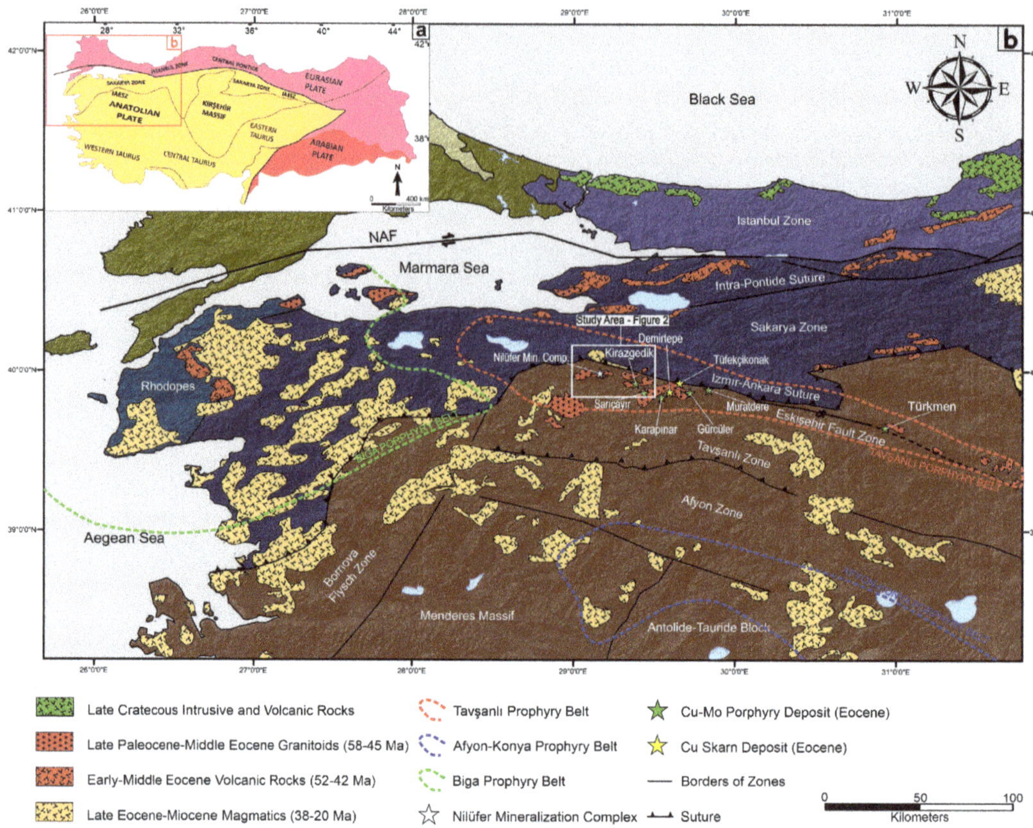

Figure 1. (**a**) Tectonic map of Turkey with major suture zones [15]; IAES: Izmir-Ankara-Erzincan suture, ITS: Inner Tauride suture. (**b**) Simplified geological map of Western Anatolia illustrating the major Eocene to Miocene porphyry belts and key tectonic features. NAF: North Anatolian Fault. Adapted from Kuşcu, et al. [16].

This study aims to evaluate whether the Nilüfer Mineralization Complex (NMC) represents a spatially zoned and evolutionarily continuous magmatic-hydrothermal system, rather than a collection of geologically isolated mineral deposits. We propose a comprehensive model of an integrated porphyry-skarn mineral system that includes four key sectors: Kirazgedik, Kozbudaklar, Güneybudaklar, and Delice (Figure 2). In this model, the mineralization associated with the Topuk Pluton, including the W-Mo skarns and related hydrothermal features, is hypothesized to be temporally and genetically linked to the Kirazgedik Cu-Mo porphyry system located to the east and hosted by the adjacent South Tepeldağ Pluton (Figure 2). To evaluate this hypothesis, a multidisciplinary methodological framework is applied. Field observations are combined with high-precision Re-Os dating of molybdenite. This is supported by electron microprobe analyses of amphibole and mica, which provide estimates of temperature, pressure, oxidation state, and volatile content (such as Cl and F), using established calibrations [17–19]. Whole-rock geochemical data and rare earth element patterns are used to evaluate the genesis and compositional evolution of all related intrusives associated with early porphyry to late-stage skarn systems [20–22].

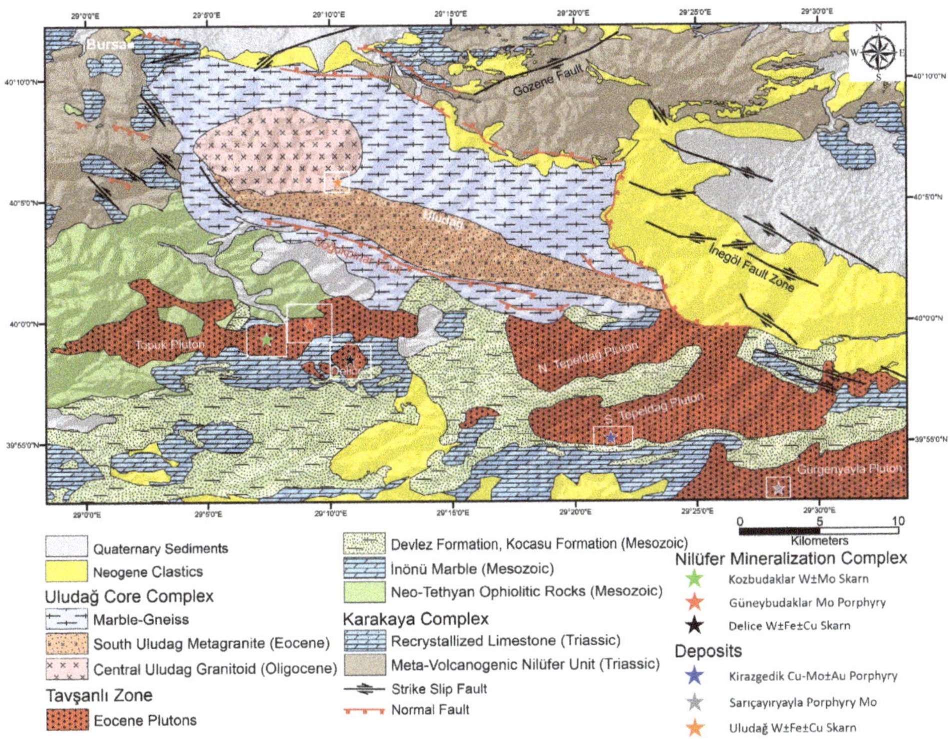

Figure 2. Schematic map of the Nilüfer Hybrid Mineralization Complex (NHMC), showing major intrusive bodies and the spatial distribution of porphyry, skarn, and vein-type mineralization styles. Modified after Kocatürk [9] and Kocatürk, et al. [14].

Consequently, this study investigates whether porphyry mineralizations of the Western Tavşanlı Metallogenic Belt and different skarn mineralizations listed under the Nilüfer Mineralization Complex represent isolated deposits or a spatially zoned and evolutionarily continuous magmatic-hydrothermal system. The presence of geochemical data, the alignment of mineralization with fault networks, and the temporal sequence among distinct deposit types indicate a response to syn- to post-subduction slab behavior, which influenced magma emplacement and ore fluid migration. As such, the study introduces a new exploration model for porphyry-skarn transitions in post-subduction settings, with broader implications for metallogenic interpretations in the Western Tethyan domain and similar orogenic belts. By integrating geochronological, petrological, and geochemical evidence, this work advances the understanding of how multi-style mineralization can evolve from a single, tectono-magmatically driven system.

2. Geological Framework

The Tavşanlı Zone (TZ), situated directly south of the İzmir-Ankara Suture Zone (IASZ), forms the northernmost margin of the Anatolide-Tauride Block and hosts the Tavşanlı Metallogenic Belt within the Bursa district (Figure 1). The zone preserves a complex tectonostratigraphic succession that records prolonged subduction-accretion processes and forms the crystalline basement upon which Eocene magmatism and associated ore-forming events are superimposed. The pre-Cenozoic basement of the TZ is composed of three primary lithostratigraphic packages. The lowermost unit is the Orhaneli Group, a high-pressure, low-temperature (HP-LT) metamorphic complex comprising quartz-rich greyschists (Kocasu Formation), massive marbles (İnönü Marble), and heterogeneous assemblages of metabasite, phyllite, and chert (Devlez Formation). These rocks exhibit mineral assemblages of glaucophane, lawsonite, phengite, and chlorite, which are characteristic of blueschist-facies metamorphism formed under subduction-related conditions [23,24].

Overlying this metamorphic basement is a Late Cretaceous accretionary complex, consisting of a serpentinite-matrix mélange that encloses blocks of basalt, radiolarian chert, pelagic limestone, and serpentinized ultramafic rocks—interpreted as fragments of the Neotethyan oceanic crust [14,25,26]. The uppermost stratigraphic component consists of an ophiolitic sequence, primarily composed of serpentinized harzburgite, dunite, and layered gabbroic cumulates, locally intruded by plagiogranites. These ophiolites display geochemical features consistent with a supra-subduction zone (SSZ) forearc origin and are interpreted as products of Late Cretaceous obduction processes, emplaced onto the passive margin of the Anatolide–Tauride Block [27,28].

The Eocene plutonism within the TZ is represented by several multiphase intrusions, notably the Topuk, Tepeldağ, and Gürgenyayla plutons, which are emplaced into this pre-Cenozoic basement (Figure 2). These intrusions differ in petrography, mineralogy, and emplacement age, as constrained by detailed field observations and U-Pb zircon geochronology [14,29]. The Gürgenyayla Pluton, situated in the eastern part of the study area (Figure 2), consists mainly of coarse-grained, equigranular granodiorite to monzogranite, with local syenitic facies and textural evidence of magma mingling, such as mafic microgranular enclaves. Its mineral assemblage includes plagioclase, quartz, K-feldspar, hornblende, and biotite, with zircon U-Pb crystallization ages indicating emplacement at approximately 51 Ma (51.02 ± 0.52 and 51.5 ± 0.1 Ma) [12,29] into structurally elevated basement blocks during a phase of post-subduction extension.

The Tepeldağ Pluton comprises two distinct intrusive phases (Figure 2). The South Tepeldağ phase is dominated by quartz monzodioritic to quartz monzonitic rocks that are rich in hornblende and biotite, displaying hypidiomorphic granular textures, and has been dated at 54.56 ± 0.21 Ma [14]. This phase is spatially associated with the Kirazgedik porphyry Cu-Mo system, suggesting a potential genetic relationship. The North Tepeldağ phase is more felsic, comprising quartz diorite to granodiorite with porphyritic textures, as an emplacement age of approximately 45.41 ± 0.34 Ma [29].

The Topuk Pluton, located in the central part of the study area (Figure 2), is a composite intrusion comprising mainly a granodioritic body and an evolved felsic cupola known as the Delice stock (Figure 2). The main phase consists of granodiorite, quartz diorite, quartz monzodiorite, and quartz monzonite, with porphyritic to hypidiomorphic textures. The Delice cupola includes tonalite, granodiorite, and alkali feldspar granite with both granular and porphyritic textures [14]. The primary mineral assemblages include plagioclase, quartz, K-feldspar, hornblende, and biotite, with muscovite present in more evolved phases. Zircon U–Pb dating constrains the crystallization ages of the main intrusive phases of the Topuk Pluton to 48.71 ± 0.44, while the Delice stock yields slightly younger ages, ranging from 47.3 ± 0.2 Ma to 45.9 ± 0.3 Ma [9,14,29].

Peripheral intrusive facies of the Topuk Pluton correspond spatially with the ore mineralization zones of the Nilüfer Mineralization Complex (NMC). The Kozbudaklar region hosts granodioritic rocks associated with W-Mo skarn mineralization. The Güneybudaklar region is characterized by quartz monzonitic and quartz dioritic apophyses exhibiting geochemical signatures consistent with late-stage magmatic evolution and metal enrichment.

The Delice region, marked by a felsic stock, corresponds to the structural cupola of the pluton and is associated with W-Fe-Cu skarn bodies (Figure 2).

Collectively, these Eocene intrusive bodies delineate a tectono-magmatic corridor within the Tavşanlı Zone (TZ) that reflects episodic emplacement and progressive geochemical differentiation. Despite the Topuk Pluton displaying structural and mineralogical continuity with the surrounding intrusions, field evidence—such as the presence of brittle faulting and polymictic hydrothermal breccias—suggests that the NMC developed under a distinct tectonic regime. This regime was likely influenced by a deep-seated oblique fault system, which controlled magma ascent, hydrothermal fluid migration, and the localization of mineralization. A detailed summary of petrographic characteristics, textures, and geochronological data for these plutonic units in the study area is presented in Table 1, after Altunkaynak et al. [29], Kocatürk [9], and Kocatürk et al. [14].

Table 1. Petrographic Summary of Eocene Plutons of Bursa District in the Tavşanlı Zone, after Altunkaynak, et al. [29], Kocatürk [9], and Kocatürk, et al. [14].

Pluton	Sub-Unit	Rock Type	Textures	Main Minerals	References
Topuk	Main Body	Grd, Qz-Di, Qz-Mzd, Qz-Mzn	Porphyritic, hypidiomorphic	Grd: Pl, Qz, Kfs, Hb, Bt; Qz-Mzd: Pl, Qz, Kfs, Hb, Bt; Qz-Mzn: Pl, Qz, Kfs, Bt, Hb	Altunkaynak, et al. [29]; Kocatürk, et al. [14]
	Delice	Ton, Grd, Afs-Gr	Granular, porphyritic	Ton: Pl ↑, Qz ↑, Hb, Bt, Kfs ↓; Grd: Pl, Qz, Kfs, Hb, Bt; Afs-Gr: Kfs ↑, Qz, Pl ↓, Bt, Ms	Kocatürk, et al. [14]
Tepeldağ	South Tepeldağ	Qz-Mzd to Qz-Mzn	Hypidiomorphic granular	Qz-Mzd: Pl, Qz, Kfs, Hb, Bt; Mzgr: Pl, Qz, Kfs ↑, Bt, Hbl	Kocatürk, et al. [14]
	North Tepeldağ	Qz-Di to Grd	Porphyritic	Qz-Di: Pl, Qz ↓, Hb, Bt, Kfs ↓; Grd: Pl, Qz, Kfs, Hb, Bt	Altunkaynak, et al. [29]
Gürgenyayla	Main Body	Grd to Mzgr	Hypidiomorphic granular	Grd: Pl, Qz, Kfs, Hb, Bt; Mzgr: Pl, Qz, Kfs ↑, Bt, Hb	Altunkaynak, et al. [29]

Abbreviations: Granodiorite (Grd), Quartz-Monzodiorite (Qz-Mzd), Quartz-Monzonite (Qz-Mzn), Tonalite (Ton), Alkali-Feldspar Granite (Afs-Gr), Quartz-Diorite (Qz-Di), Monzogranite (Mzgr). Pl (Plagioclase), Qz (Quartz), Kfs (K-feldspar), Hbl (Hornblende), Bt (Biotite), Ms (Muscovite), ↑ (higher abundance), ↓ (lower abundance).

3. Methodology

A total of 89 rock- and ore-related samples were systematically collected from four principal domains within the study area. These include (i) the Kirazgedik porphyry Cu-Mo system, spatially and genetically associated with the South Tepeldağ Pluton; and (ii) the Güneybudaklar, (iii) Kozbudaklar, and (iv) Delice regions, which collectively delineate the Nilüfer Mineralization Complex (NMC) and exhibit strong spatial and genetic associations with the Topuk Pluton. The sampling strategy was designed to encompass the full range of mineralization styles and associated lithological environments across these distinct mineralized centers. Targeted lithologies included multiphase intrusive rocks, hydrothermal alteration zones, skarn bodies, and ore-bearing structures such as veins, breccias, and mineralized intrusive contacts. Sampling was conducted along center-to-margin transects within each intrusive body to assess spatial variations in magmatic evolution and subsequent hydrothermal alteration. All collected samples were initially examined petrographically. Based on petrographic and textural criteria, a subset of 29 representative samples was selected for comprehensive analytical work. This included ore mineralogy, whole-rock geochemistry, electron probe microanalysis (EPMA) of amphibole and mica, and Re-Os geochronology of molybdenite-bearing samples.

The analytical data are summarized in the following data products:

- **Supplementary Table S1** details metadata for the 29 representative samples, including sample codes, lithological classifications, spatial information, and associated analytical datasets.
- **Supplementary Table S2** presents whole-rock geochemical data (major oxides and trace elements, including REEs) for the 29 representative samples.
- **Supplementary Table S3** reports ore element concentrations (Cu, Mo, W, Au, Ag, Bi, As, Te, Sb, Sn, Zn, Pb, Re) for the full set of 89 samples.
- **Supplementary Tables S4 and S5** provide quantitative EPMA results for mica and amphibole, respectively, from selected representative samples.
- **Supplementary Figures S1–S4** include representative back-scattered electron (BSE) images of amphibole and mica grains from across each study sector.
- **Table 2** presents the Re-Os isotopic data and corresponding model age for a molybdenite-bearing sample from the Güneybudaklar mineralization.

Table 2. Re–Os isotopic data and model age for molybdenite from the Güneybudaklar porphyry-skarn, with analytical uncertainties reported at the 2σ confidence level.

Re (ppm)	$\pm$ 2α	^{187}Re (ppb)	$\pm$2α	^{187}Os (ppb)	$\pm$2α	Model Age (Ma)	$\pm$2α with Decay Const (Ma)
34.65	0.10	21,780	61	18.125	0.011	49.93	0.21

3.1. Whole-Rock Geochemical Analysis (XRF & ICP-MS)

Whole-rock geochemical analyses were carried out on all collected samples to determine major, trace, and REE elements concentrations. Major oxides were analyzed by X-ray fluorescence (XRF) using fused bead methods, while trace and precious metals, including REEs, Au, and Ag, were quantified via inductively coupled plasma mass spectrometry (ICP-MS). Trace element determinations followed lithium metaborate fusion and nitric acid digestion. Analyses were conducted at the Geochemistry Research Laboratory (ITU/JAL) of Istanbul Technical University. Certified reference materials and internal standards were employed for calibration and quality control, ensuring high analytical precision and reproducibility.

3.2. Electron Probe Micro-Analysis (EPMA)

Mineral chemical compositions of amphibole and mica were determined through EPMA using a JEOL JXA-8900R electron microprobe (JEOL Ltd., Tokyo, Japan) at the University of Alberta's Electron Microprobe Laboratory. A total of 148 point analyses were conducted on 35 amphibole grains, and 130 point analyses were performed on 21 mica grains, with $\geq$3 analyses per grain to capture intra-grain compositional variation. Analyses were conducted on carbon-coated, polished thin sections under operating conditions of 15 kV accelerating voltage, 30 nA beam current, and a 2 µm beam diameter in spot mode. Major elements were analyzed with 20 = second peak and background counting times; fluorine (F) was analyzed with extended 40 s counting to enhance detection sensitivity. Calibration employed high-purity mineral standards: rutile (Ti), diopside (Mg), plagioclase (Si, Al, and Ca), fayalite (Fe), sanidine (K), tugtupite (Na, Cl), spessartine (Mn), magnesium fluoride (F), sanbornite (Ba), and chromium oxide (Cr). For F and Cl, relative uncertainties approached 5% due to their lower concentrations in mica. BSE imaging guided the selection of homogenous, unaltered grains for analysis.

3.3. Re-Os Geochronology

To constrain the timing of mineralization, a molybdenite-bearing sample was selected from a skarn zone at the northern contact of the Topuk Pluton, where quartz monzonitic intrusions intersect metagabbroic basement. Among all skarn-hosted domains in the NMC, the Güneybudaklar sector exhibited the highest molybdenum content (>4000 ppm Mo), making it the optimal candidate for Re-Os dating. Approximately 80 mg of molybdenite was isolated through metal-free mechanical separation techniques, including crushing, sieving, magnetic separation, and gravity concentration. Re-Os isotope analyses were conducted at the Crustal Re-Os Geochronology Laboratory (University of Alberta), following established procedures described in Selby and Creaser [30] and Markey et al. [31]. A mixed ^{185}Re–^{190}Os–^{188}Os double-Os spike was used for isotope dilution analyses. Molybdenite digestion, Re-Os separation, and purification were conducted using inverse aqua regia, solvent extraction, and ion exchange chromatography. Isotopic ratios were measured by negative thermal ionization mass spectrometry (NTIMS) using a ThermoScientific Triton instrument (Thermo Fisher Scientific Inc., Waltham, MA, USA). Procedural blanks were <3 pg for Re and <1 pg for Os and are considered negligible. Analytical accuracy was verified using the 8599 Henderson molybdenite standard (27.78 $\pm$ 0.07 Ma; Markey et al. [31]), yielding data consistent with the certified age. Age calculations used a ^{187}Re decay constant of 1.666×10^{-11} yr^{-1} [32]. The Re-Os geochronological result is presented in Table 2, summarizing the isotopic composition and model age.

4. Results

4.1. Ore Geology

The NMC comprises a variety of several ore deposit types, including porphyry, skarn, and vein-style mineralization, all situated within a structurally controlled intrusive pathway. In the current study, several mineralization types have been studied in detail, such as Kirazgedik Cu-Mo $\pm$ Au porphyry, Güneybudaklar Mo Porphyry-Skarn, Kozbudaklar W–Mo Skarn, and Delice W $\pm$ Fe $\pm$ Cu Skarn mineralization (Figure 2).

4.1.1. Kirazgedik Cu-Mo $\pm$ Au Porphyry Mineralization

The Kirazgedik Cu-Mo $\pm$ Au deposit is in northwest Turkey within the Tavşanlı Zone of western Anatolia, which forms part of the Western Tethyan metallogenic belt. This region experienced significant magmatic activity during the Eocene, represented by the South Tepeldağ pluton (U-Pb zircon ages of 54.56 $\pm$ 0.21 Ma), attributed to the closure of the Neo-Tethyan ocean and related slab dynamics [9,12,14]. The Eocene intrusions developed within a post-subduction extensional tectonic regime, associated with slab rollback of the subducting Tethyan slab rather than a sudden slab rollback. This tectonic setting marks a transition from subduction-related compression to transtensional deformation during the waning stages of Neo-Tethyan convergence [9]. As a result, a belt of granitic to dioritic stocks was emplaced throughout NW Anatolia, associated with numerous porphyry and skarn mineral deposits [5,6,12]. Kirazgedik mineralization is spatially associated with the South Tepeldağ Pluton, occurring near its eastern and southern margins (Figure 2). This porphyry Cu-Mo $\pm$ Au mineralization is hosted by the Eocene South Tepeldağ porphyritic quartz-monzonite to porphyritic quartz diorite intrusion and its surrounding older metamorphic wall rocks (Figures 2 and 3a,b).

Field observations at Kirazgedik reveal that intense hydrofracturing within the apical region of the South Tepeldağ porphyritic stock facilitated the development of a stockwork vein system and breccia zones, which host the principal Cu-Mo mineralization. This ore-bearing zone is centered on the intrusion, where pervasive fracturing and hydrothermal brecciation generated a dense network of quartz-sulfide veins and veinlets that extensively penetrate both the stock and the surrounding country rocks (Figure 3a,b).

Figure 3. Field, macroscopic, and microscopic investigations of the Kirazgedik Cu-Mo ± Au porphyry mineralization: (**a**,**b**) Pervasive fracturing and hydrothermal brecciation with network of quartz-sulfide veins and veinlets in the South Tepeldağ Pluton. (**c**) Macroscopic view of the porphyritic quartz-monzonite and porphyritic quartz diorite of the South Tepeldağ Pluton. (**d**) Plagioclase (pl) phenocrysts within a medium-grained microcrystalline matrix composed of plagioclase (pl) laths, hornblende (hb), biotite (bt), quartz (qz), and opaque minerals in the porphyritic quartz-monzonite. (**e**) Hornblende (hb) and plagioclase (pl) phenocrysts embedded in medium-grained microcrystalline matrix of plagioclase (pl), muscovite (ms), actinolite (act), quartz (qz) in the porphyritic quartz-diorite. (**f**,**g**) Quartz (qz), K-feldspar (kfels), secondary biotite (bt), sericite (ser), epidote (ep), and calcite veinlet (cal) are the alteration products in the Kirazgedik area. (**h**,**i**) Pyrite (py), chalcopyrite (cp), and molybdenite (Mo) are the main ore minerals that occurred in the Kirazgedik mineralization. The data are taken from Kocatürk [9] and Kocatürk, et al. [14].

Microscopically, the porphyritic quartz-monzonite consists predominantly of plagioclase phenocrysts, some of which are partially sericitized, set within a medium-grained microcrystalline matrix composed of plagioclase laths, hornblende, biotite, quartz, and opaque minerals. (Figure 3c,d). While the porphyritic quartz-diorite is mainly composed of hornblende and plagioclase phenocrysts embedded in a medium-grained microcrystalline matrix of plagioclase, muscovite, actinolite, quartz, and opaque minerals (Figure 3c,e). Gangue minerals include quartz (ubiquitous in all vein types) with the hydrothermal alteration minerals like K-feldspar and secondary biotite (within potassic alteration zones), sericite, and epidote with carbonates (locally in veins or late-stage fractures) (Figure 3f,g).

The dominant ore minerals are chalcopyrite and molybdenite, accompanied by abundant pyrite (Figure 3h,i). Chalcopyrite serves as the principal copper-bearing mineral, occurring in quartz veinlets and disseminations, whereas molybdenite appears as flakes and rosettes along late quartz veins or vein margins (Figure 3h,i). Geochemical analyses reveal that the Cu concentration in the altered sample reaches up to 1276 ppm, while Mo attains levels as high as 194.1 ppm (Supplementary Table S3). Although gold is not visually observable within the sample, geochemical data indicate trace amounts up to 1.1 ppm (Supplementary Table S3). This gold is likely present as submicroscopic inclusions or as an electrum intimately associated with pyrite and chalcopyrite minerals.

4.1.2. Güneybudaklar Mo-W ± Au Porphyry-Skarn Mineralization

The Güneybudaklar molybdenum mineralization is located within the Tavşanlı Zone of NW Turkey, part of a metallogenic belt characterized by Eocene magmatism that followed the Late Cretaceous collision between the Anatolide-Tauride platform and the Pontide (Sakarya) continent [9,14,33]. The deposit lies on the southern flank of Uludağ (Bursa province), where Eocene granitoid intrusions, notably the Topuk plutons, intruded Paleozoic-Mesozoic metamorphic sequences and Triassic carbonate units along the İzmir-Ankara suture zone [33,34]. The principal intrusive body associated with mineralization is the Topuk granite, which has U-Pb zircon ages of 48.71 ± 0.44 Ma [29]. This granite is an elliptical, east-west trending I-type fractured and moderately weathered granitoid that ranges in composition from granodiorite to granite (Figures 2 and 4a) and contains mafic microgranular enclaves, as well as peripheral margins of quartz monzodioritic to monzogabbroic affinity (Figure 4b); it is also crosscut by late-stage aplite dykes.

This pluton intruded a sequence dominated by Triassic İnönü Marble (recrystallized limestone) and other metamorphic lithologies (Figures 2 and 4c), establishing a critical juxtaposition of magmatic heat against reactive carbonate host rocks [34]. Granodiorite generally exhibits a hypidiomorphic granular texture composed of plagioclase feldspar (50%–65%, predominantly oligoclase), quartz (20%–30%), microperthite (<35%), and mafic minerals (5%–15%), primarily hornblende and biotite (Figure 4d). The monzogabbro enclave comprises approximately equal proportions of plagioclase (labradorite) and alkali feldspar (microperthite and orthoclase), in addition to mafic minerals, predominantly pyroxene (usually clinopyroxene) (Figure 4e).

The Mo-W ± Au mineralization at the Güneybudaklar mineralization is intimately linked both spatially and genetically to the metasomatic contact zone between the Topuk granodiorite intrusion and the surrounding Triassic İnönü Marble. This lithological boundary provides an ideal setting for skarn development due to the contrasting chemical reactivity and permeability between the silica-rich intrusive body and the carbonate host rock, resulting in the formation of both endoskarn within the intrusion and exoskarn in the adjacent marble (Figure 4c,f). Structural discontinuities, notably concentric and radial fractures related to pluton emplacement, acted as conduits for high-temperature magmatic-hydrothermal fluids. These structures, combined with lithological contacts, controlled fluid pathways and localized ore deposition (Figure 4a,c,f). A well-developed skarn system, characterized by a porphyry-style mineralized core having quartz-molybdenite veins in fractures, creating a proto-ore zone rich in molybdenum (and minor copper and gold) (Figure 4g), grading outward into a skarn shell hosted by the marble, forming garnet-rich (completely altered to epidote) and diopside-rich exoskarn zones with minor actinolite (Figure 4h,i). Ore mineralization aligns with this zonation pattern. Molybdenite occurs as veinlets and disseminations within the endoskarn and skarn, typical of the porphyry-style core (Figure 4g,j). Scheelite, often containing a powellite component due to Mo substitution, predominates in the outer skarn zones in which scheelite and powellite's characteristic bireflectance and internal reflections are found (Figure 4k). Retrograde alteration is evidenced by quartz-epidote (Figure 4h) and late-stage sulfide mineralization, including pyrite (Figure 4l). These late hydrothermal events introduced minor gold mineralization, typically hosted in quartz-sulfide veins crosscutting earlier skarn textures. In addition to the most extreme Mo-enriched sample (4032 ppm), W values as high as 866 ppm and the highest Au observed in the dataset (4.58 ppm), Te and Bi reach 97.8 ppm and 735 ppm, respectively, indicating a volatile-rich late-stage fluid signature (Supplementary Table S3).

Figure 4. Field, macroscopic, and microscopic investigations of the Güneybudaklar Mo-W ± Au porphyry-skarn mineralization: (**a**) Fractured and moderately weathered Topuk granitoid rock. (**b**) Mafic microgranular enclaves and peripheral margin of monzogabbro within the Topuk granitoid rock. (**c**) Metasomatic contact between Triassic İnönü Marble and Topuk granitoid rock. (**d**) Plagioclase (pl), quartz (qz), microperthite (mp), biotite (bt), and hornblende (hb) in the granodiorite of the Topuk pluton. (**e**) Plagioclase (pl) and microperthite (mp) with augit (aug) in the monzogabbro. (**f**) Diopside

zone in the skarn system along the metasomatic contact between Triassic İnönü Marble and Topuk granitoid rock. (**g**) Molybdenite (Mo)-rich ore with weathered iron-stained quartz matrix. (**h**) Garnet-rich skarn zone that completely altered to epidote (ep) associated with quartz (qz). (**i**) Diopside (di), quartz (qz), and actinolite (act) in the diopside-rich skarn zone. (**j**) Molybdenite (Mo) within diopside (di), typical of the porphyry-style core. (**k**) Molybdenite (Mo) mostly altered to powellite (po) associated with scheelite (sch). (**l**) Pyrite (py) occurred within quartz-epidote during the late-stage sulfide mineralization. The data are taken from Kocatürk [9] and Kocatürk, et al. [14].

4.1.3. Kozbudaklar W ± Mo Skarn Mineralization

The Kozbudaklar W ± Mo skarn deposit, located southeast of Bursa in northwestern Turkey, is hosted within the Triassic İnönü Marble of the Tavşanlı Zone, a metamorphic unit of the northern margin of the Anatolide block formed during Neo-Tethys closure [13,34]. During the Eocene, magmatism, interpreted as post-subduction in origin but commonly referred to as post-collisional in the regional literature, was triggered by slab rollback or progressive slab detachment following the cessation of Neo-Tethyan subduction. This magmatic activity led to the emplacement of the Topuk Pluton and the development of associated skarn systems (e.g., the Eğrigöz, Orhaneli, and Kestanbol plutons) [35,36]. The Topuk Pluton is represented by a medium- to coarse-grained, calc-alkaline, metaluminous I-type granodiorite [14]. Geochemical and petrologic evidence, including mafic microgranular enclaves of monzogabbro (Figure 4b), indicates magma mixing and mantle-crust interaction typical of volcanic-arc or collision settings. The pluton intruded the marble along a structurally weakened suture zone, causing contact metamorphism and extensive metasomatism that formed the Kozbudaklar skarn (Figure 5a,b). The skarn occurs along the pluton-carbonate contact, varying in thickness up to 850 m, and reflects a collision-related skarn system where magmatic heat and fluids metasomatized older carbonate strata, enabling W ± Mo mineralization [33,34].

The Kozbudaklar skarn W ± Mo mineralization developed both within the marble as exoskarn and within the granitoid as endoskarn adjacent to this contact (Figure 5a,b). The endoskarn is spatially restricted to the margin of the pluton and is characterized by metasomatic replacement of the granite, primarily by plagioclase and pyroxene minerals (Figure 5c). The exoskarn extends outward into the marble, displaying a clear mineralogical zoning with increasing distance from the intrusive contact. In the immediate contact zone, the skarn is dominated by clinopyroxene-rich assemblages composed mainly of diopsidic clinopyroxene with subordinate andradite garnet (Figure 5d,e). Moving further from the contact, the skarn transitions into garnet-rich lenses along the roof and walls of the pluton, composed predominantly of andradite garnet (partly altered to epidote) and calcite (Figure 5f,g). Scheelite, the primary ore mineral, is disseminated throughout these prograde zones during the initial stage of mineralization (Stage 1) (Figure 5h,i). In the subsequent late prograde stage (Stage 2), the skarn assemblage becomes more oxidized, indicative of elevated oxygen fugacity conditions. Scheelite continued to precipitate during this stage but became notably enriched in molybdenum, containing 7%–32% powellite component in solid solution, which imparts the W ± Mo character to the deposit [13]. Molybdenum is predominantly incorporated into scheelite as a solid solution rather than forming abundant discrete molybdenite minerals, which are scarce or absent in the deposit. Kozbudaklar displays the greatest W variability, ranging from 99 to 2558 ppm, at generally low Mo but locally elevated Au (such as 2.09 ppm), consistent with W-rich skarn mineralization (Supplementary Table S3).

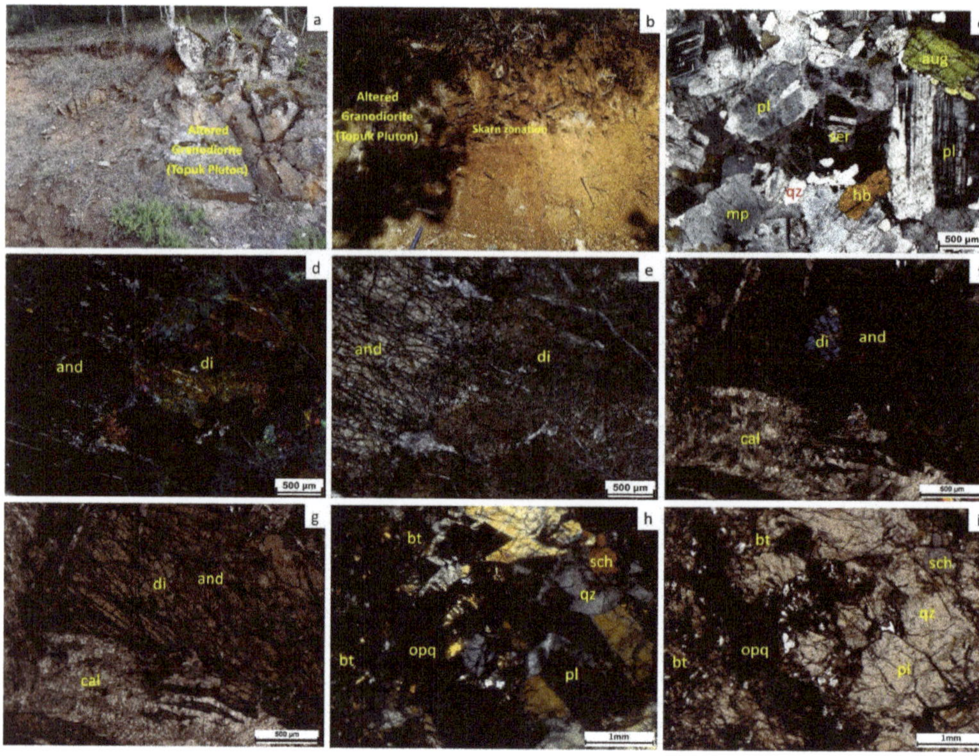

Figure 5. Field, macroscopic, and microscopic investigations of the Kozbudaklar W ± Mo skarn mineralization: (**a**) Altered granodiorite (Topuk pluton) at Kozbudaklar W ± Mo skarn deposit area. (**b**) Extensive contact metasomatism along the structurally weakened suture zone forming the Kozbudaklar skarn mineralization. (**c**) Endoskarn primarily formed from plagioclase (pl) and augite (aug) in the altered granodiorite having quartz (qz), microperthite (mp), sericite (ser), and hornblende (hb). (**d**,**e**) Diopside (di) and andradite (and) in the clinopyroxene-rich exoskarn zone. (**f**) and (**g**) Andradite (and) and calcite (cal) with little diopside (di) in the exoskarn garnet-rich zone. (**h**,**i**) disseminated scheelite (sch) throughout the prograde zones with the altered plagioclase (pl), biotite (bt), quartz (qz), and opaque minerals (opq).

4.1.4. Delice W ± Fe ± Cu Skarn Mineralization

The Delice W ± Fe ± Cu skarn deposit is located in the Tavşanlı Zone, a metallogenic belt characterized by Eocene magmatic activity, in northwest Turkey, south of Uludağ Mountain in Bursa Province [9]. The mineralization is genetically linked to Eocene magmatism (≈54 to 45 Ma), which occurred in a post-subduction extensional setting [9,14]. The mineralization formed at the contact between the Topuk Pluton, a zoned, calc-alkaline, I-type granitoid, and the Triassic İnönü Marble Formation (Figure 6a). The Topuk intrusion comprises tonalite-granodiorite to alkali-feldspar granite with mafic enclaves and late-stage granite-aplite dikes (Figure 6b). Tonalite and tonalite porphyry are the main stock of Topuk granite, having mainly plagioclase phenocrysts embedded in a microcrystalline matrix of plagioclase, quartz, biotite, and hornblende with minor microperthite and opaques (Figure 6c). U-Pb zircon ages date the pluton between 47.3 ± 0.2 Ma and 45.9 ± 0.3 Ma, placing mineralization firmly in the mid-Eocene [9,14]. Sr-Nd isotopes ($^{87}Sr/^{86}Sr \approx 0.706$–$0.707$, $\varepsilon Nd \approx -2$ to -4) confirm a mantle-derived magma with significant crustal interaction [9,13,14].

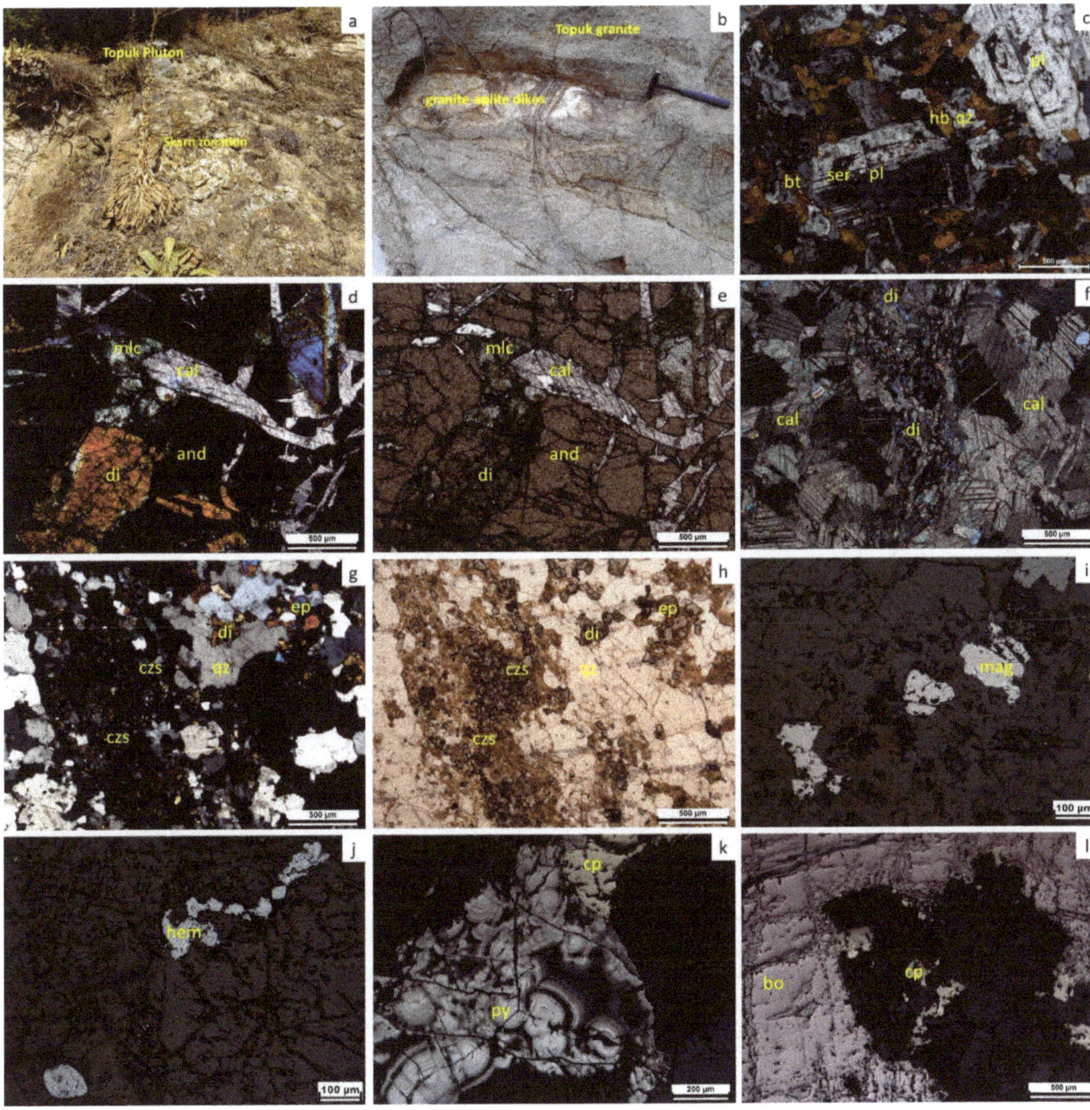

Figure 6. Field, macroscopic, and microscopic investigations of the Delice W ± Fe ± Cu skarn mineralization: (**a**) Metasomatic contact between the Topuk Pluton and Triassic İnönü Marble Formation at Delice. (**b**) Late-stage granite-aplite dikes cut through Topuk granite. (**c**) Plagioclase (pl) phenocrysts are embedded in a microcrystalline matrix of plagioclase (pl), quartz (qz), biotite (bt), and hornblende (hb) with minor sericite (ser) and opaques in tonalite porphyry. (**d**,**e**) Andradite (and) and diopside (di) with calcite (cal) in the prograde metasomatic alteration, forming calc-silicate skarn and latterly having malachite (mlc). (**f**) Diopside (di) formed within the calcite (cal) of the İnönü Marble Formation. (**g**,**h**) Epidote (ep), clinozoisite (czs), chlorite, and sericite within quartz (qz) occurring within the retrograde metasomatic alteration. (**i**) Idiomorphic to subhedral grains of magnetite (mag) formed in the early skarn stages. (**j**) Minor hematite (hem) occurs, reflecting locally variable oxidation states. (**k**) Colloform pyrite (py) occurred with chalcopyrite (cp). (**l**) Bornite (bo) occurred with chalcopyrite (cp).

The deposit exhibits a well-defined contact-metasomatic system. Skarn formation occurred as magmatic-hydrothermal fluids emanating from the Topuk Pluton penetrated the adjacent carbonate host along fractures, bedding planes, and fold hinges (Figure 6a). These fluids induced prograde metasomatic alteration, forming calc-silicate skarn composed of garnet (andradite) and pyroxene (diopside-hedenbergite) (Figure 6d–f). As the system cooled and fluids evolved, retrograde alteration overprinted the prograde skarn. This introduced amphiboles (tremolite-actinolite), epidote-clinozoisite, chlorite, and sericite, replacing earlier garnet and pyroxene, commonly associated with tungsten (scheelite) precipitation during fluid cooling and structural fracturing (Figure 6g,h). Mineralogically, scheelite is the principal tungsten ore mineral at Delice. It is disseminated in the skarn matrices and in cross-cutting veins. Importantly, relict wolframite is also found, primarily in the contact zone, partially replaced by scheelite, indicating an evolution from a more reduced, Fe-rich fluid to a more oxidized one. In terms of iron minerals, magnetite is abundant, especially in the early stage, where it can form idiomorphic to subhedral grains (Figure 6i). Minor hematite also occurs, and conversely some pyrrhotite is present, reflecting locally variable oxidation states (Figure 6j). Pyrite is the most widespread sulfide mineral, occurring throughout both the contact zone and the skarn (Figure 6k). However, other base-metal sulfides are more confined to the distal skarn lenses away from the contact, including chalcopyrite with bornite (Figure 6l). Malachite is represented by supergene minerals after Cu-bearing minerals (Figure 6d,e). Delice samples combine high W, peaking at 2391 ppm in D29, with notable Bi, 941 ppm in D28, and appreciable Au up to 1.22 ppm (Supplementary Table S3).

4.2. Whole-Rock Geochemistry

Whole-rock major and trace element geochemical data from key intrusive domains were compiled to elucidate the magmatic evolution of the NMC and its adjacent plutonic units. The dataset comprises representative samples from the Kirazgedik, Güneybudaklar, Kozbudaklar, and Delice sectors, each selected to capture distinct intrusive phases and stages of magmatic differentiation. Although the Kirazgedik samples originate from the South Tepeldağ Pluton and are not genetically associated with the NMC, they are included here as a regional compositional reference. All geochemical diagrams were generated using GCDkit software (version 6.0) [37].

Recent petrogenetic interpretations by Kocatürk et al. [14] reclassify the NMC intrusions as genetically linked phases of the Topuk Pluton. In contrast, the Kirazgedik samples represent a compositionally independent magmatic suite and serve as a baseline for evaluating regional geochemical variation. Major and trace element data for the four intrusive sectors were plotted using six geochemical discrimination diagrams (Figure 7a–f) and eight SiO_2-variation plots (Figure 8a–h). On the P–Q diagram (Figure 7a), Kirazgedik samples cluster within the quartz-monzodiorite to quartz-monzonite fields, whereas the Kozbudaklar samples span tonalite, quartz diorite, and granodiorite compositions. Delice samples are mostly confined to the tonalite–quartz diorite field, with a few plotting in the granite field, while Güneybudaklar is entirely represented within the granodiorite field. The SiO_2 versus FeO^t/MgO diagram (Figure 7b) reveals that most of the analyzed samples exhibit a calc-alkaline geochemical affinity, notably those from Kirazgedik, Kozbudaklar, Güneybudaklar, and Delice, with a subset of samples from Güneybudaklar and Delice straddles the transition zone toward tholeiitic compositions. Further insights are provided by the K_2O versus SiO_2 plot (Figure 7c), which classifies the majority of the samples as high-K calc-alkaline, with a few Delice samples extending into the shoshonitic field. Elevated K_2O contents are commonly associated with metasomatized lithospheric mantle sources or assimilation of crustal materials, both of which are frequent in post-collisional orogenic settings [38,39]. Supporting this interpretation, the $CaO–Al_2O_3/(FeO + MgO)–K_2O/Na_2O$ ternary diagram (Figure 7d) indicates source heterogeneity among the sample groups. Kirazgedik and Kozbudaklar samples tend to cluster within the high-K mafic, suggesting derivation from enriched mantle or lower crustal protoliths, whereas Delice samples show broader dispersion toward the low-K mafic rocks, implying significant crustal assimilation [40,41]. The plot of Dy/Yb versus SiO_2 (Figure 7e) offers further constraints on the depth and residual mineralogy of magma generation. Elevated Dy/Yb ratios in Kirazgedik samples suggest melting in the presence of residual garnet, consistent with a relatively deep lithospheric mantle source (>80 km), where garnet remains stable [42,43]. By contrast, lower Dy/Yb ratios in Güneybudaklar and Kozbudaklar point to melting under shallower conditions, likely involving amphibole as the dominant residual phase, which is typical for crustal or upper lithospheric levels. The Rb versus Y + Nb tectonic discrimination diagram (Figure 7f) places the majority of the samples, especially from Kirazgedik and Kozbudaklar, within the post-collisional granite (post-COLG) field. This tectonic setting is commonly associated with lithospheric thinning and slab rollback following continental collision [44,45]. Meanwhile, several Delice samples fall into the volcanic arc granite (VAG) and syn-collisional granite (syn-COLG) domains, reinforcing a complex magmatic history that includes both subduction-related and post-collisional processes.

Trace-element variation diagrams reveal coherent evolutionary trends. Compatible elements such as Ni and Mg decrease with increasing SiO_2 (Figure 8a,c), with Kirazgedik showing the highest concentrations and Güneybudaklar the lowest. Similarly, V and Sr contents decline with silica (Figure 8b,d); the most pronounced drop in Sr occurs in Güneybudaklar, where concentrations fall below 200 ppm. In contrast, incompatible elements such as Zr and Rb display increasing trends with SiO_2 (Figure 8e,f), attaining their highest levels in Güneybudaklar and, for Zr, also in Kozbudaklar. La exhibits a decrease with silica (Figure 8g), whereas K_2O/Na_2O ratios increase sharply, again culminating in Güneybudaklar (Figure 8h). Together, these diagrams delineate a systematic compositional progression from mafic to felsic lithologies and from low-K to high-K affinity across the four intrusive domains.

Chondrite-normalized REE patterns (Figure 9a–d) exhibit enrichment in light REEs (LREEs) and relatively flat heavy REE (HREE) distributions across all sectors. Kirazgedik samples display moderately fractionated REE patterns with weak negative Eu anomalies. In contrast, Güneybudaklar displays steeper LREE/HREE slopes and more pronounced negative Eu anomalies. Kozbudaklar exhibits moderately sloping REE trends, while Delice is characterized by flatter patterns and variable Eu anomalies, suggesting diverse degrees of plagioclase fractionation and crustal interaction.

Primitive mantle-normalized multi-element diagrams (Figure 9e,h) for all sectors exhibit similar overall patterns—namely enrichment in large-ion lithophile elements (e.g., Rb, Th, Ba) and depletion in high-field-strength elements (e.g., Nb, Ti). Although the general patterns are consistent across the suite, the magnitude of enrichment and depletion varies among the different intrusive sectors, reflecting subtle differences in source composition and magmatic evolution.

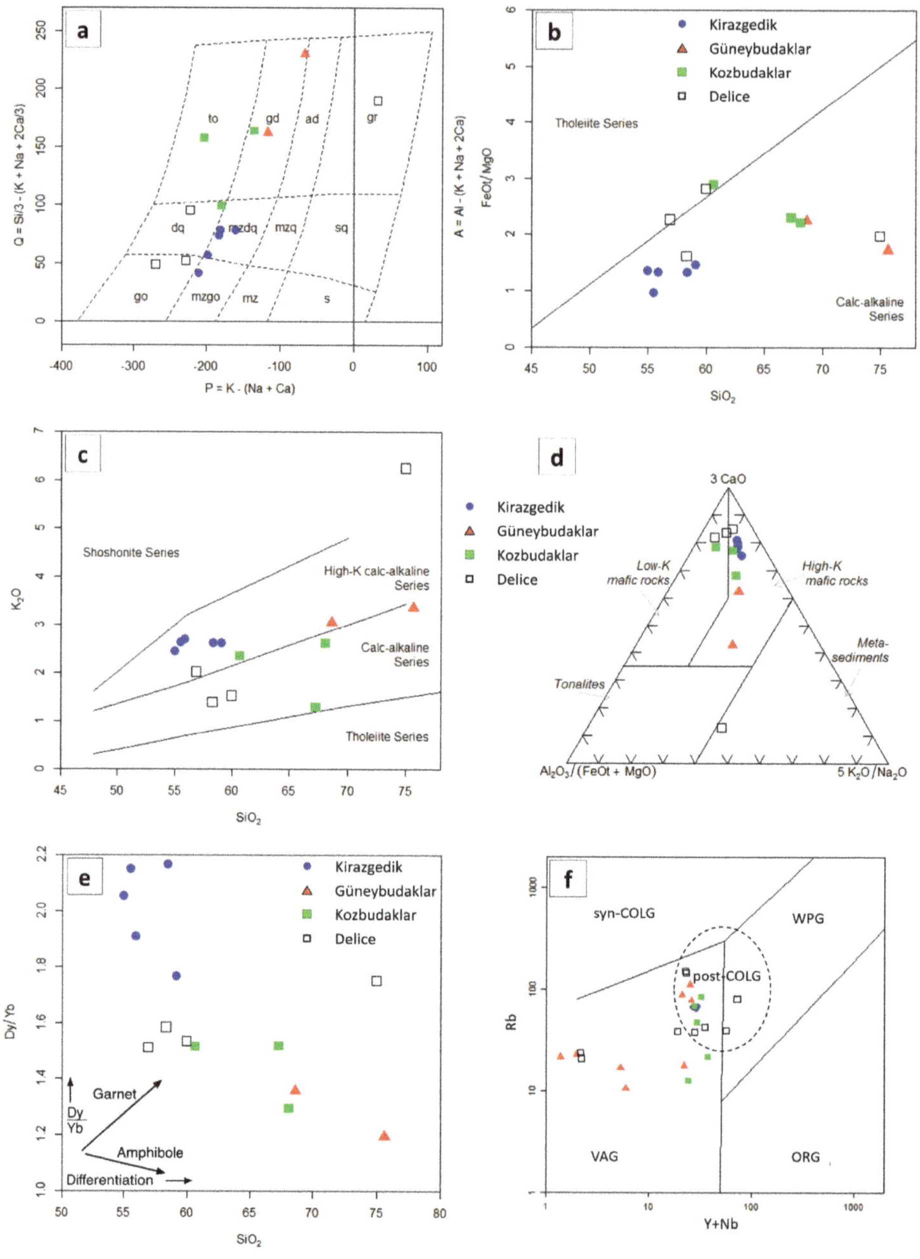

Figure 7. Major and trace element discrimination diagrams for intrusive samples. (**a**) P-Q diagram [46], abbreviations: adamellite (ad), gabbro or diorite (go/di), granite (gr), granodiorite (gd), quartz diorite (qd), quartz monzodiorite (qmzd), quartz monzonite (qmz), quartz syenite (qs), monzodiorite (mzd), monzonite (mz), syenite (s), tonalite (to); (**b**) SiO_2 versus FeO^t/MgO plot [47]; (**c**) K_2O vs. SiO_2 [48]; (**d**) $Al_2O_3/(FeO_t + MgO)$–$3CaO$–$5(K_2O/Na_2O)$ ternary plot for granitoids. The compositions of melts derived from several sources after Laurent, et al. [49]; (**e**) Dy/Yb vs. SiO_2 diagram [42]; (**f**) The Rb-(Y + Nb) diagram [50], Abbreviations: Ocean Ridge Granites (ORG), Within Plate Granites (WPG), Volcanic Arc Granites (VAG), and Syn-and Post-collision Granites (syn-COLG and post-COLG).

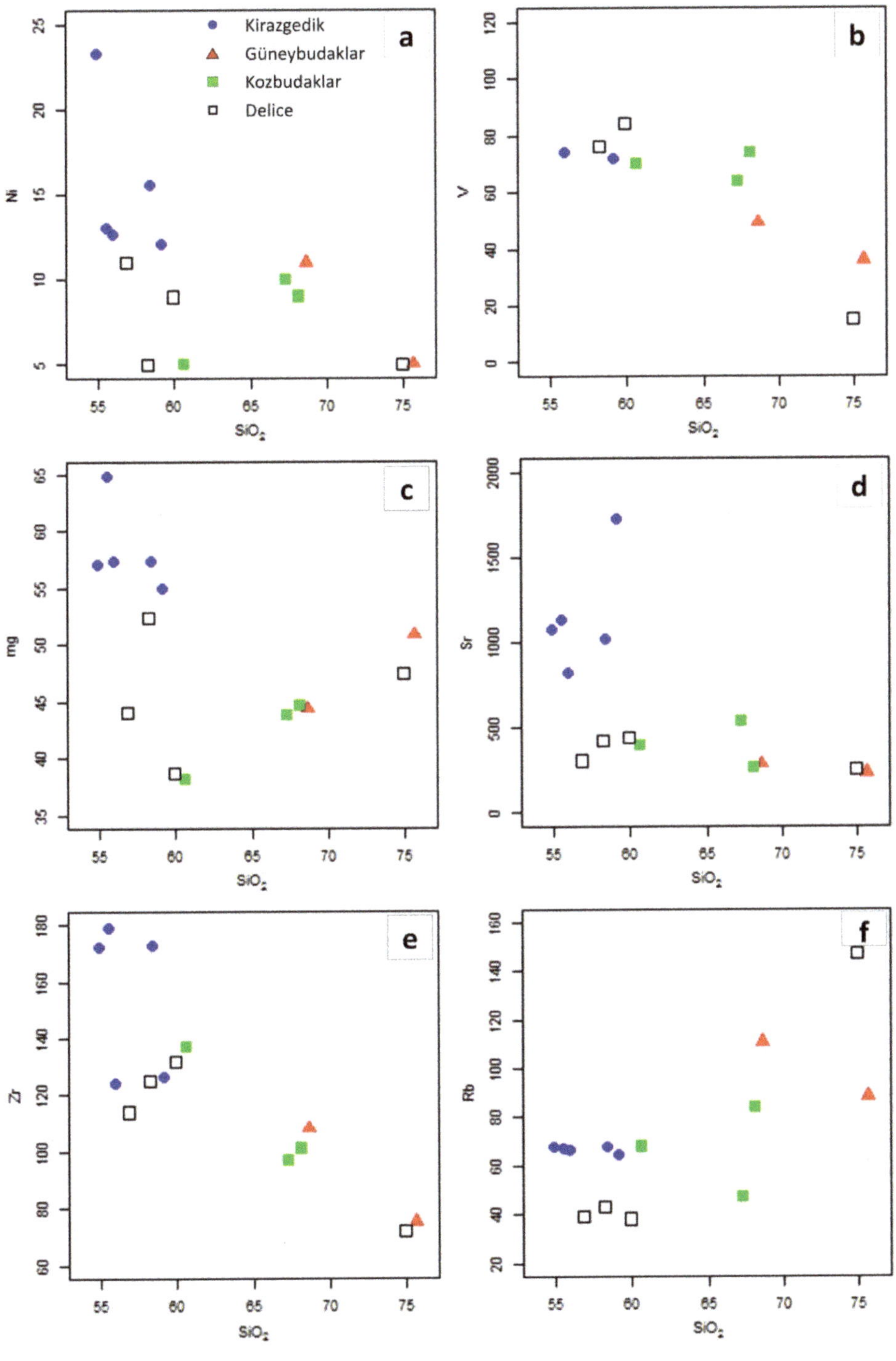

Figure 8. *Cont.*

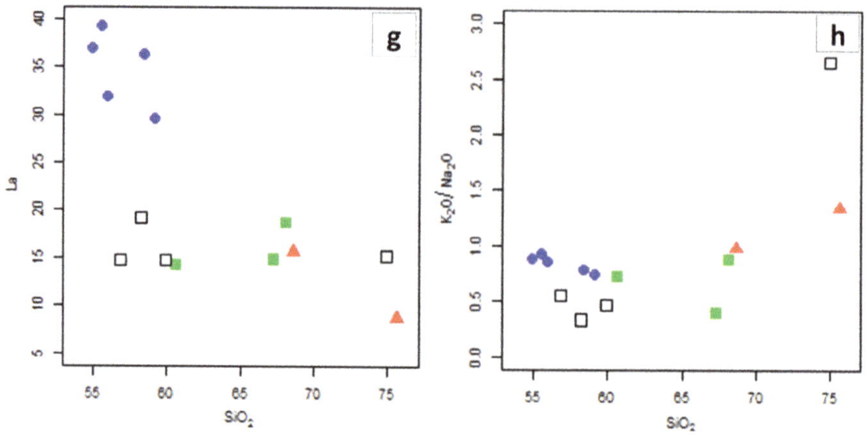

Figure 8. Selected trace element variation diagrams plotted against SiO_2 for samples from Kirazgedik and the three NMC sectors. SiO_2 versus Ni (**a**); V (**b**); Mg (**c**); Sr (**d**); Zr (**e**); Rb (**f**); La (**g**); and K_2O/Na_2O (**h**). Elements include Sr, Rb, Zr, V, and Ni, reflecting magma evolution trends and fractionation behavior. Interpretations follow the approaches of Gill [51], Pearce [52], and Rollinson [40].

4.3. Re–Os Geochronology

Re–Os isotopic dating of molybdenite from the Güneybudaklar skarn zone yields a model age of 49.93 ± 0.21 Ma (2σ). The analyzed sample displays moderate Re concentrations (34.65 ppm) (Table 2).

4.4. Mineral Geochemistry

4.4.1. Mica Geochemistry

Systematic compositional variations in mica across the four studied sectors reflect evolving magmatic conditions and varying degrees of fluid-rock interaction. Major element analyses of biotite display notable inter-sectoral differences. SiO_2 contents range from 33.7 to 40.0 wt.%, with the highest average observed in Kirazgedik (37.9 wt.%) and the lowest in Güneybudaklar (35.0 wt.%) (Table 3). TiO_2 contents peak at 4.3 wt.% in Kozbudaklar and Güneybudaklar, while Kirazgedik biotites show the lowest average (2.9 wt.%) (Table 3). A progressive increase in FeO is evident from Kirazgedik (avg. 12.5 wt.%) through Güneybudaklar (22.0 wt.%) and Kozbudaklar (22.9 wt.%), to a maximum in Delice (24.7 wt.%) (Table 3). Inversely, MgO decreases from 17.3 wt.% in Kirazgedik to 7.8 wt.% in Delice. The Mg number (XMg = Mg/[Mg + Fe], apfu) exhibits a declining trend from Kirazgedik (avg. 0.7) to Delice (avg. 0.4) (Table 3), implying advancing magmatic differentiation and oxidation.

Halogen systematics further discriminate the sector: Fluorine (F) concentrations are highest in Kirazgedik (1.3–1.5 wt.%, avg. 1.4 wt.%), followed by Kozbudaklar (0.4–0.5 wt.%, avg. 0.5 wt.%), Güneybudaklar (0.3–0.7 wt.%, avg. 0.6 wt.%), and the lowest in Delice (0.3–0.4 wt.%, avg. 0.4 wt.%) (Table 3). Corresponding atomic proportions (apfu) support these trends, ranging up to 0.6 apfu in Kirazgedik and from 0.2–0.3 apfu elsewhere. Chlorine (Cl) concentrations remain low and consistently low across all sites ($\leq$0.2 wt.%, $\leq$0.1 apfu), indicating minimal Cl enrichment (Table 3).

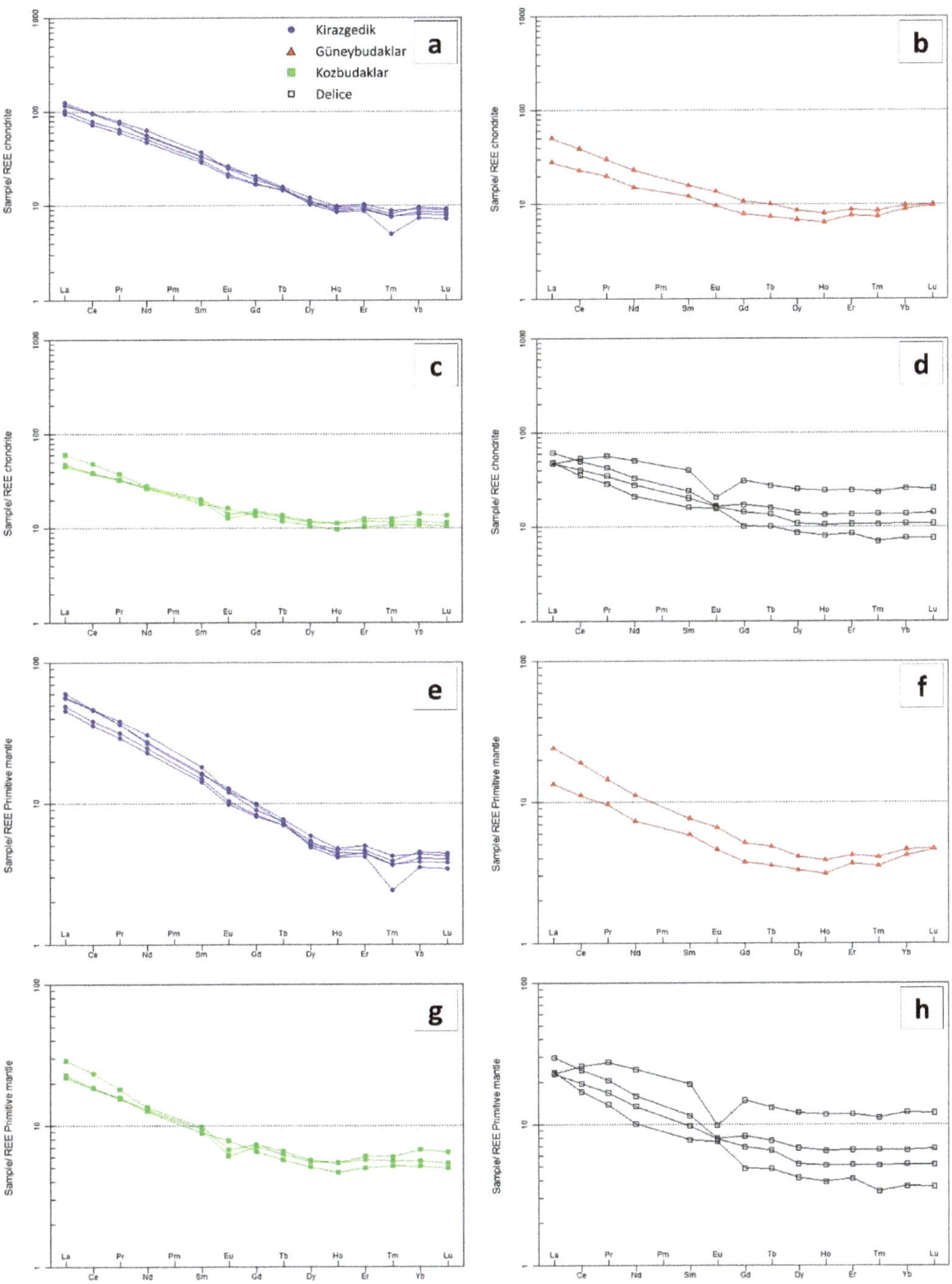

Figure 9. (**a–d**) Chondrite-normalized REE patterns for samples from (**a**) Kirazgedik, (**b**) Güneybudaklar, (**c**) Kozbudaklar, (**d**) Delice (normalization values from William and Boynton [53], (**e–h**) Primitive mantle-normalized multi-element diagrams for samples from (**e**) Kirazgedik, (**f**) Güneybudaklar, (**g**) Kozbudaklar, (**h**) Delice (normalization values from Sun and McDonough [54]).

Table 3. Summary of mica EPMA analytical results (wt.%).

DataSet	Kirazgedik (n = 11)			Güneybudaklar (n = 70)			Kozbudaklar (n = 15)			Delice (n = 34)		
	Min.	Max.	Av.	Min.	Max.	Av.	Min.	Max.	Av.	Min.	Max.	Av.
SiO_2	35.6	40	37.9	33.7	36.2	35	35	35.6	35.3	34.8	36.2	35.4
TiO_2	2.3	4.2	2.9	2.8	4.3	3.7	3.4	4.3	3.7	3.1	4.1	3.8
Al_2O_3	13.4	14.8	14.2	13.8	16.8	15.6	14	14.4	14.2	13.8	15.1	14.3
FeO	11.6	12.9	12.5	19.2	23	22	22	23.8	22.9	22.8	25.6	24.7
MnO	0.2	0.3	0.2	0.2	0.7	0.3	0.4	0.4	0.4	0.5	0.6	0.6
MgO	16.5	19.7	17.3	7.6	11.5	8.8	8.6	9.6	9.1	6.9	9.4	7.8
CaO	0	3.4	0.4	0	0.1	0	0	0.1	0	0	0	0
Na_2O	0.1	0.3	0.2	0	0.3	0.1	0.1	0.2	0.1	0	0.2	0.1
K_2O	5.6	10.4	9.5	9.4	10.4	9.9	9	10.2	9.9	9.6	10.2	10
Cr_2O_3	0	0.1	0	0	0.1	0	0	0.1	0	0	0.1	0
H_2O *	3.3	3.5	3.4	3.5	3.8	3.6	3.6	3.7	3.6	3.6	3.7	3.7
Li_2O *	0.7	1.9	1.3	0.1	0.8	0.5	0.5	0.7	0.6	0.4	0.9	0.6
F	1.3	1.5	1.4	0.3	0.7	0.6	0.4	0.5	0.5	0.3	0.4	0.4
Cl	0.1	0.2	0.2	0.1	0.2	0.1	0.1	0.2	0.1	0.1	0.2	0.1
Atoms per formula unit (apfu)												
Si	5.3	5.7	5.5	5.3	5.5	5.4	5.5	5.5	5.5	5.4	5.5	5.5
Al^{iv}	2.2	2.5	2.4	2.5	2.7	2.6	2.5	2.5	2.5	2.5	2.6	2.5
Al^{vi}	0	0	0	0	0.5	0.2	0	0.1	0.1	0	0.2	0.1
Ti	0.3	0.5	0.3	0.3	0.5	0.4	0.4	0.5	0.4	0.4	0.5	0.4
Cr	0	0	0	0	0	0	0	0	0	0	0	0
Fe	1.4	1.6	1.5	2.4	3	2.8	2.9	3.1	3	2.9	3.4	3.2
Mn	0	0	0	0	0.1	0	0	0.1	0.1	0.1	0.1	0.1
Mg	3.6	4.3	3.8	1.8	2.6	2	2	2.2	2.1	1.6	2.2	1.8
Ca	0	0.5	0.1	0	0	0	0	0	0	0	0	0
Na	0	0.1	0.1	0	0.1	0	0	0.1	0	0	0.1	0
K	1.1	1.9	1.8	1.9	2.1	2	1.8	2	2	1.9	2	2
F	0.6	0.7	0.6	0.2	0.4	0.3	0.2	0.2	0.2	0.1	0.2	0.2
Cl	0	0	0	0	0	0	0	0.1	0	0	0.1	0
OH *	3.3	3.4	3.3	3.6	3.8	3.7	3.7	3.8	3.7	3.7	3.8	3.8
Li *	0.4	1.1	0.8	0.1	0.5	0.3	0.3	0.4	0.4	0.3	0.5	0.4
TOTAL	19.7	20.4	20.2	19.8	20.1	19.9	19.8	20	20	19.9	20.1	20
Parameters												
Al^{total}	2.2	2.5	2.4	2.5	3.1	2.8	2.5	2.6	2.6	2.5	2.7	2.6
Mg-Li	2.7	3.6	3	1.5	2.1	1.7	1.7	1.8	1.7	1.2	1.7	1.4
Fe + Mn + Ti-Al^{vi}	1.8	1.9	1.9	2.7	3.2	3.1	3.2	3.5	3.4	3.4	3.8	3.6
X(Mg) = Mg/(Mg + Fe)	0.7	0.7	0.7	0.4	0.5	0.4	0.4	0.4	0.4	0.3	0.4	0.4
X(Fe) = Fe/(Fe + Mg)	0.3	0.3	0.3	0.5	0.6	0.6	0.6	0.6	0.6	0.6	0.7	0.6
Fe + Mn	1.5	1.6	1.6	2.5	3	2.9	2.9	3.1	3	3	3.4	3.3
Al^{iv} + Fe + Ti	1.7	1.9	1.8	2.9	3.8	3.5	3.4	3.6	3.5	3.4	3.9	3.7
Fe + Mn + Ti	1.8	1.9	1.9	3	3.5	3.3	3.3	3.6	3.5	3.4	3.9	3.7
FeO/(FeO + MgO)	0.4	0.4	0.4	0.6	0.7	0.7	0.7	0.7	0.7	0.7	0.8	0.8
$TiO_2 \times 10$	23.1	42.3	28.8	27.7	42.7	37.1	33.9	42.7	36.7	30.8	41	37.5
FeO + MnO	11.8	13.1	12.8	19.9	23.2	22.3	22.4	24.2	23.3	23.3	26.2	25.3
ln(Ti)	−1.4	−0.8	−1.2	−1.1	−0.7	−0.8	−0.9	−0.7	−0.9	−1	−0.7	−0.8
T (oC)	497.6	523.2	507.9	269.3	369	296.9	280.4	311	295.7	231.1	303.4	257.1

H_2O *, Li_2O *, OH *, and Li * values are estimated based on stoichiometric calculations.

In Figure 10a, all biotite compositions plot within the Mg–Fe mica compositional field. Kirazgedik biotites classify as Mg-biotite, while Delice, Güneybudaklar, and Kozbudaklar classify as Fe-biotite, consistent with late-stage magmatic evolution. In the $TiO_2 \times 10$–FeO + MnO–MgO ternary diagram (Figure 10b), Kirazgedik biotites fall within the "magmatic and newly formed biotite" field, while samples from other sectors align with the "primary magmatic" field, highlighting fluid-mediated overprinting during later magmatic stages. Ti-in-biotite thermometry indicates that biotites in Kirazgedik crystallized at the highest average temperatures (avg. ~508 °C), progressively decreasing in Kozbudaklar (~296 °C), Güneybudaklar (~297 °C), and Delice (~257 °C) (Table 3).

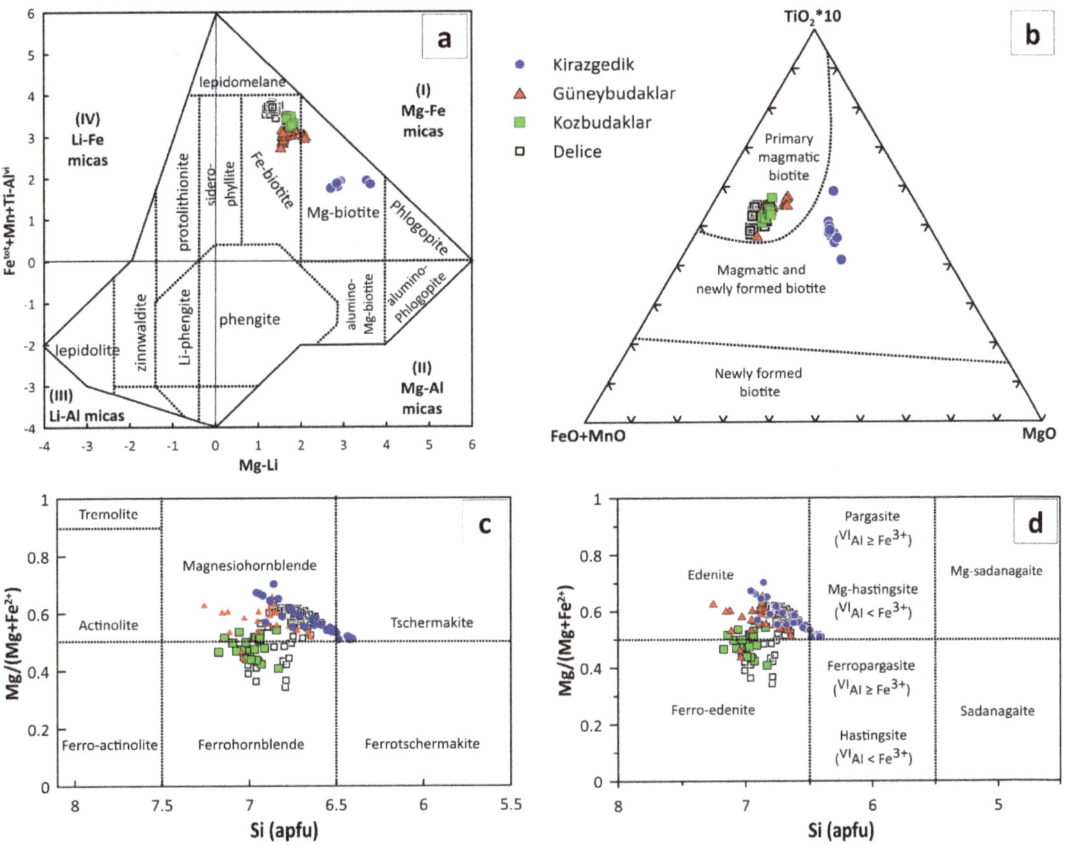

Figure 10. Mineral compositional classification and discrimination diagrams. (**a**) Mica classification in the Fe + Mn + Ti vs. Mg–Li field according to Tischendorf, et al. [55]; (**b**) Biotite typology in the $TiO_2 \times 10$-FeO + MnO-MgO ternary plotm, after Nachit, et al. [56]; (**c**,**d**) Amphibole classification using $Mg/(Mg + Fe^{2+})$ vs. Si (apfu), after Leake, et al. [57].

4.4.2. Amphibole Chemistry

Amphibole geochemistry similarly reflects inter-sectoral differentiation. SiO_2 content ranges from 42.1 to 49.2 wt.%, with the highest averages recorded in Kozbudaklar (46.3 wt.%) and Güneybudaklar (46.0 wt.%). TiO_2 reaches up to 1.9 wt.%, with Kirazgedik amphiboles averaging the highest values (1.3 wt.%) (Table 4). Al_2O_3 values range from 5.2 to 10.5 wt.% the highest averages in Delice and Kirazgedik (~9.0 wt.%) and the lowest in Kozbudaklar (7.4 wt.%) (Table 4). A systematic increase in FeO is evident from Kirazgedik (avg. 15.7 wt.%) to Delice (avg. 16.5 wt.%) and peaking in Kozbudaklar (avg. 19.3 wt.%) (Table 4). Conversely, MgO decreases from 12.2 wt.% in Kirazgedik to 11.5 wt.% in Delice. MnO values remain low (0.4–1.3 wt.%) across all sectors (Table 4).

Table 4. Summary of amphibole EPMA analytical results (wt.%).

	Kirazgedik (n = 28)			Güneybudaklar (n = 29)			Kozbudaklar (n = 25)			Delice (n = 66)		
Dataset	Min.	Max.	Av.	Min.	Max.	Av.	Min.	Max.	Av.	Min.	Max.	Av.
SiO_2	42.1	47.5	44.6	43.8	49.2	46.0	44.5	47.7	46.3	43.5	46.4	45.1
TiO_2	0.8	1.9	1.3	0.4	1.5	1.0	0.5	1.0	0.8	0.7	1.3	1.1
Al_2O_3	7.5	10.4	9.1	5.2	9.5	8.0	5.9	8.7	7.4	6.9	10.5	9.0
FeO	11.7	18.0	15.7	13.1	22.6	16.1	17.1	21.3	19.3	14.1	23.3	16.5
MnO	0.4	0.9	0.6	0.4	1.2	1.0	0.8	1.0	0.9	0.8	1.3	0.9
MgO	10.4	15.5	12.2	9.6	14.0	11.9	8.3	11.4	10.1	6.8	13.5	11.5
CaO	11.2	12.1	11.7	10.1	12.6	11.7	10.8	11.9	11.3	10.5	11.6	11.2
Na_2O	1.1	1.7	1.4	0.5	1.8	1.3	0.8	1.3	1.1	0.8	2.0	1.7
K_2O	0.4	1.6	1.0	0.4	1.0	0.7	0.4	0.9	0.6	0.4	0.8	0.6
					Atomic fraction							
Si	6.4	7.0	6.7	6.7	7.3	6.9	6.8	7.2	7.0	6.6	7.0	6.8
Ti	0.1	0.2	0.2	0.0	0.2	0.1	0.1	0.1	0.1	0.1	0.1	0.1
Al	1.3	1.9	1.6	0.9	1.7	1.4	1.0	1.5	1.3	1.3	1.9	1.6
Fe^{2+}	1.4	2.3	2.0	1.6	3.0	2.0	2.1	2.7	2.4	1.8	3.0	2.1
Mn	0.1	0.1	0.1	0.1	0.2	0.1	0.1	0.1	0.1	0.1	0.2	0.1
Mg	2.4	3.4	2.7	2.1	3.1	2.7	1.9	2.5	2.3	1.6	3.0	2.6
Ca	1.8	2.0	1.9	1.7	2.0	1.9	1.8	2.0	1.8	1.7	1.9	1.8
Na	0.3	0.5	0.4	0.2	0.5	0.4	0.2	0.4	0.3	0.2	0.6	0.5
K	0.1	0.3	0.2	0.1	0.2	0.1	0.1	0.2	0.1	0.1	0.2	0.1
H	2.0	2.0	2.0	2.0	2.0	2.0	2.0	2.0	2.0	2.0	2.0	2.0
$Mg/(Mg + Fe^{2+})$	0.5	0.7	0.6	0.4	0.7	0.6	0.4	0.5	0.5	0.3	0.6	0.6
Fe/(Fe + Mg)	0.3	0.5	0.4	0.3	0.6	0.4	0.5	0.6	0.5	0.4	0.7	0.5
Al IV	1.0	1.6	1.3	0.7	1.4	1.1	0.8	1.2	1.0	1.0	1.4	1.2
Na + K + Ca	2.2	2.7	2.5	2.0	2.6	2.4	2.2	2.4	2.3	2.2	2.5	2.4

In terms of atomic proportions, Si (apfu) ranges ftom 6.41–7.26, with the highest values in Kozbudaklar and Güneybudaklar. Fe^{2+} (apfu) increases from 1.97 in Kirazgedik to 2.07 in Delice, while Mg (apfu) decreases from 2.72 (Kirazgedik) to 2.56 (Delice) (Table 4). The Mg number (calculated as Mg#(Mg/[Mg + Fe^{2+}]) ranges from 0.34 to 0.70, with Kirazgedik amphiboles having the highest average (0.58), indicative of more primitive magmatic conditions. In the Mg/(Mg + Fe^{2+}) vs. Si diagram (Figure 10c), Kirazgedik amphiboles mostly plot within the magnesiohornblende field, with little representation in the tschermakite field, whereas Kozbudaklar and Delice extend from magnesiohornblende toward ferrohornblende, consistent with progressive Fe enrichment. Figure 10d reveals a systematic evolution from edenite and parasite/magnesio-hastingsite in Kirazgedik to ferro-edenite in Güneybudaklar, Kozbudaklar, and Delice, indicating increasingly oxidized, Fe-rich magmatic conditions.

Collectively, the geochemical evolution of both micas and amphiboles reflects a transition from high-temperature, Mg-rich mineral assemblages in Kirazgedik toward lower-temperature, Fe-enriched, and fluid-altered compositions in Güneybudaklar, Kozbudaklar, and Delice. These trends are coherent with whole-rock geochemistry and provide robust evidence for progressive magmatic differentiation, oxidation, and hydrothermal overprinting during the later stages of pluton emplacement.

5. Discussion

5.1. Tectono-Magmatic Framework and Temporal Constraints

Integrated geochronological, petrochemical, and mineralogical data delineate a complex, prolonged, and multi-phase magmatic evolution within the Nilüfer Mineralization Complex (NMC) and adjacent plutonic suites, extending over a temporal window of approximately 9 million years (Figures 7–10). High-precision U–Pb zircon and Re–Os molybdenite ages constrain the timing of successive magmatic episodes, beginning with the early crystallization of the South-Tepeldağ Pluton at 54.56 ± 0.21 Ma. This is followed by granodioritic intrusions in the Gürgenyayla Pluton, dated at 51.02 ± 0.52 Ma and 51.5 ± 0.1 Ma, respectively. Importantly, molybdenite mineralization in the Güneybudaklar sector yields a Re–Os age of 49.93 ± 0.21 Ma, preceding the principal emplacement of the Topuk Pluton (48.71 ± 0.44 Ma) and its associated bodies (47.35 ± 0.21 Ma and 45.90 ± 0.29 Ma) (Table 5).

Table 5. Summary of geochronological data for the Topuk, South-Tepeldağ, and Gürgenyayla plutons, along with their associated mineralized zones. The reported ages determined using U–Pb and Re–Os dating, as specified.

Pluton	Zone Description	Rock Type	Method	Age (Ma)	Reference
South-Tepeldağ	Porphyry Cu-Mo	Monzodiorite	LA-ICP-MS U-Pb	54.56 ± 0.21	Kocatürk, et al. [14]
Gürgenyayla	Porphyry Cu-Mo	Granodiorite	SHRIMP U-Pb	51.02 ± 0.52	Altunkaynak, et al. [29]
Gürgenyayla	Porphyry Cu-Mo	Granodiorite	CA-TIMS U-Pb	51.5 ± 0.1	Rabayrol, et al. [12]
Topuk	Güneybudaklar Mineralization	Molybdenite	Re-Os	49.93 ± 0.21	This Study
Topuk	Main Body	Granodiorite	SHRIMP U-Pb	48.71 ± 0.44	Altunkaynak, et al. [29]
Topuk	Delice	Tonalite	LA-ICP-MS U-Pb	47.35 ± 0.21	Kocatürk, et al. [14]
Topuk	Delice	AF-Granite	LA-ICP-MS U-Pb	45.90 ± 0.29	Kocatürk, et al. [14]

This temporally constrained magmatic evolution corresponds to a tectonic regime governed by continued Neo-Tethyan subduction, trench retreat, and progressive slab rollback, rather than abrupt slab rollback. The absence of pervasive thermal overprinting on high-pressure/low-temperature (HP/LT) metamorphic assemblages, as documented in regional studies [24,58], supports a slab-attached configuration during crustal reworking. Thermomechanical models [59] corroborate this scenario, predicting alternating phases of trench retreat, asthenospheric upwelling, and upper crustal extension—consistent with the spatiotemporal distribution and geochemistry of NMC magmatism. More recently, Kocatürk et al. [14] proposed that the Eocene magmatic activity in the Tavşanlı Zone reflects post-subduction rollback processes evolving from an initial transpressional regime to transtensional kinematics. This geodynamic framework is consistent with the syn- to post-subduction setting of the Tavşanlı Belt, which aligns with potassic arc environments described by Müller and Groves [60], particularly post-collisional systems influenced by rollback-related mantle melting. The high-K calc-alkaline signatures and moderate Mg# observed in the NMC intrusions are similar to post-subduction intrusions in Iran and the Balkans [50,61]. Additionally, the presence of transpressive fault systems controlling magma ascent supports the model of structurally focused potassic magmatism in extensional orogenic belts [60].

Whole-rock major and trace element geochemistry reveals a coherent calc-alkaline differentiation trend across the NMC intrusions, reflecting systematic compositional evolution from primitive to more evolved magmas. This geochemical continuum extends from low-silica, calc-alkaline composition in the Kirazgedik unit, through intermediate tonalitic–granodioritic phases observed in the Kozbudaklar and Delice units, and culminates in high-silica, high-K calc-alkaline granodioritic compositions in the Güneybudaklar sector (see Figure 7a–d). The majority of samples exhibit high-K calc-alkaline signatures, consistent with arc- or post-collisional magmatism. Notably, a subset of the Delice samples trends toward the shoshonitic field, suggesting more extreme potassium enrichment. This feature may reflect derivation from a metasomatically modified lithospheric mantle source, enriched in large ion lithophile elements (LILEs), or alternatively, significant crustal assimilation during magma ascent and evolution. Such geochemical signatures are commonly associated with post-orogenic tectonic environments, where lithospheric thinning and fluid fluxing facilitate the generation of potassic magmas [38,39]. The progressive depletion of compatible elements (e.g., Ni, V, Sr) and enrichment in incompatible elements (e.g., Rb, Zr, and K_2O/Na_2O) with increasing SiO_2 content strongly indicates a dominant control by fractional crystallization, particularly involving early removal of mafic (amphibole $\pm$ pyroxene) and feldspathic (plagioclase $\pm$ K-feldspar) phases. The Dy/Yb vs. SiO_2 diagram (see Figure 7e) provides additional constraints on REE partitioning during magma evolution. Kirazgedik samples, representing the earliest porphyry stage, are characterized by low SiO_2 and elevated Dy/Yb ratios, interpreted as a signature of garnet-dominated fractionation at deep crustal levels [42]. In contrast, the Kozbudaklar and Delice intrusions show intermediate Dy/Yb values and higher silica, reflecting amphibole-controlled differentiation at shallower depths. The most evolved Güneybudaklar samples display high SiO_2 and low Dy/Yb ratios, consistent with crystallization from highly fractionated, volatile-rich melts where garnet and amphibole are no longer major residual phases. Therefore, the geochemical and mineralogical trends from Kirazgedik to Güneybudaklar reflect a temporal-spatial progression from deep-seated garnet-influenced magmatism to shallow-level systems dominated by melt differentiation and volatile enrichment. This pattern is fully compatible with post-subduction slab rollback models, in which progressive decompression melting and increasing hydrous flux facilitate sustained magma ascent and crustal evolution in the Tavşanlı Belt. Moreover, REE and multi-element patterns (see Figure 9) further affirm a subduction-modified source, characterized by LREE enrichment and pronounced negative Nb, Ta, and Ti anomalies. Notably, the Güneybudaklar samples display steep REE slopes and the most pronounced negative Eu anomalies, implying advanced feldspar fractionation and residual melt evolution in the terminal phases of magmatism. Such geochemical signatures are consistent with protracted differentiation of hydrous, oxidized magmas at shallow crustal levels [62].

Biotite and amphibole chemistry across the studied units provides additional constraints on the magmatic and post-magmatic evolution, pointing to two distinct magmatic lineages. Kirazgedik samples are characterized by their calc-alkaline affinity and plot within mantle-derived, high-fO_2 fields (Figure 11a,b,d), which strongly suggest formation from hydrous, oxidized magmas originating in a subduction-influenced mantle source. The chemistry of amphiboles in these rocks further supports this interpretation, as most fall within the magmatic amphibole field and display signatures of crystallization under oxidizing conditions at relatively shallow crustal levels (Figure 11c,d). These findings are consistent with subduction-related magmatic systems where fluids released from the subducting slab enhance the oxidation state of the mantle wedge and promote amphibole stability during differentiation [63,64]. In contrast, the Güneybudaklar, Kozbudaklar, and Delice suites reveal a different evolutionary path. Their compositions trend toward more evolved, crustally influenced sources, plotting within the calc-alkaline orogenic suite with the peraluminous and anorogenic alkaline fields (Figure 11a), with Fe–Mg systematics suggesting a mix of crustal and mantle inputs (Figure 11b). Amphiboles from these units often exhibit lower oxidation states and, in some cases, show evidence of late- or post-magmatic re-equilibration (Figure 11c,d). This implies crystallization from magmas that were less hydrous and less oxidized, likely emplaced during post-collisional tectonic phases. Together, the contrasting characteristics of these magmatic suites highlight the shift from subduction-related to post-orogenic magmatism in NMC, reflecting broader crust–mantle interaction processes over time [18,65,66].

Taken together, these petrological, geochemical, and geochronological observations support a coherent model wherein hydrous, subduction-related magmas underwent polybaric differentiation and episodic emplacement into the upper crust under conditions shaped by slab rollback and transtensional deformation. The timing of molybdenite mineralization (~49.9 Ma) coincides with the inherited zircon domains (~50–52 Ma) in the earlier phase of the Topuk Pluton, indicating that hydrothermal fluid exsolution and mineralization were temporally linked with the early felsic intrusions. This temporal coupling supports the interpretation of a vertically integrated magmatic–hydrothermal system, in which progressive fractional crystallization, moderate levels of crustal contamination, and structurally focused fluid migration played key roles in concentrating metals [1,67,68]. The transition from transpressional to transtensional tectonics likely provided extensional conduits and permeability pathways, facilitating the ascent of magmas and hydrothermal fluids during the terminal stages of arc magmatism [1,69–72].

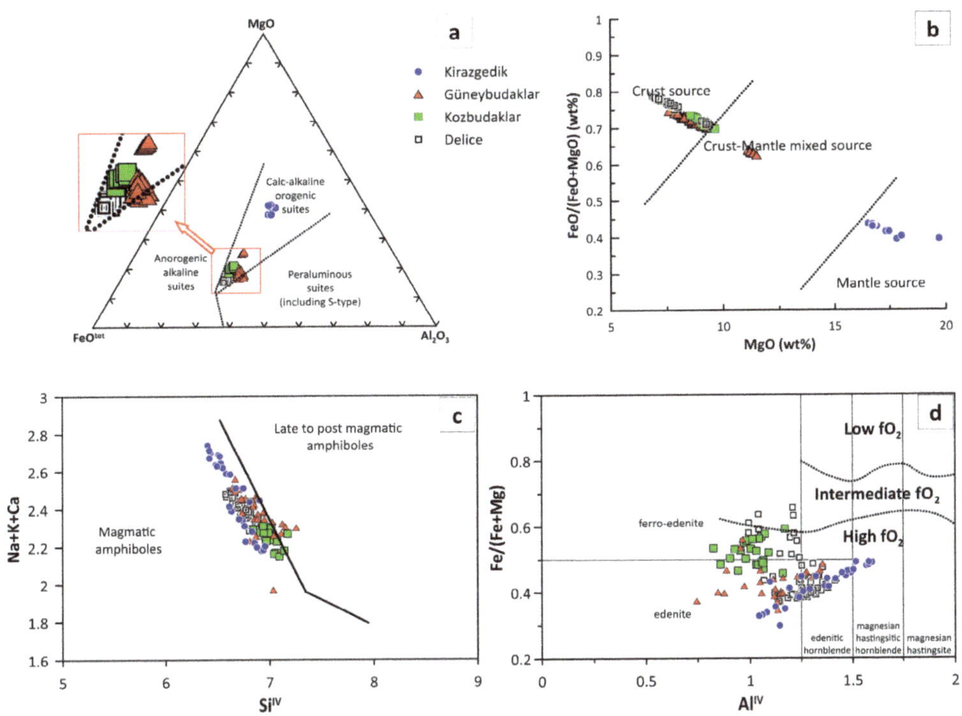

Figure 11. (**a**) MgO-FeOtot-Al_2O_3 discriminating diagrams for biotite, after Abdel-rahman [73]; (**b**) FeO/(FeO + MgO) vs. MgO (wt.%) binary diagram after Zhou [74]; (**c**) Na + K/Ca vs. Si^{IV} diagram distinguishing magmatic from late- to post-magmatic amphiboles, fields after Czamanske and Wones [75]; (**d**) Fe/(Fe + Mg) vs. Al^{IV} diagram for estimating relative oxygen fugacity, after Anderson and Smith [76].

5.2. Mineralogical Discriminants as Vectors in Exploration

Whole-rock and mineral-scale geochemistry across the NMC reveal pronounced spatial trends that integrate magmatic differentiation, volatile enrichment, and metal fertility. Among these, biotite halogen content—particularly fluorine—emerges as the most sensitive fluid-flux indicator. Fluorine concentrations in biotite decrease systematically eastward from ~1.4 wt.% at Kirazgedik to ~0.4 wt.% at Delice, whereas Cl remains uniformly low (<0.2 wt.%) (see Table 3). Notably, only biotite from the Güneybudaklar and Delice sectors simultaneously exhibits $X_{Mg} \leq 0.60$ and $F \geq 0.30$ wt.%, consistent with mineralization-fertile conditions (see Table 3; Figure 10a,b). Experimental petrology demonstrates that F-rich biotite crystallizes from H_2O–F-rich melts or fluids capable of transporting high-field-strength elements (HFSEs) such as W, Mo, and Au. Low Cl/F ratios further indicate acidic, sulfur-poor fluid conditions conducive to scheelite precipitation rather than chalcopyrite [77–79].

These trends are consistent with halogen systematics observed in amphibole compositions, which offer a complementary view of magmatic–hydrothermal evolution across the belt. Halogen concentrations in amphibole and biotite reveal a coherent spatial pattern that reflects progressive volatile depletion and fluid saturation. In the Kirazgedik porphyry system, both minerals are enriched in fluorine and chlorine (F > 0.4 wt.%, Cl > 0.3 wt.%), indicating an oxidized, volatile-rich magma capable of Cu-Mo transport. By contrast, the transitional Güneybudaklar samples show persistently high Cl (>0.3 wt.%) but decreased F (to ~0.1 wt.%), suggesting late-stage degassing or fluid unmixing during skarn evolution. This is consistent with evolved potassic magmas nearing volatile saturation, as observed in global analogues [60,80]. Similar halogen patterns in the MICA1 database also differentiate fertile from barren high-K suites [60]. Although post-emplacement degassing may reduce preserved F and Cl values [81,82], the persistence of Cl-rich, F-depleted signatures in Güneybudaklar and skarn-related intrusions implies continued fluid mobility and potential for ore deposition.

A similar spatial and compositional evolution is also observed in amphibole major oxide chemistry, reinforcing the halogen-based volatile trends. Samples from the Güneybudaklar and Delice plots in the ferro-edenite and ferro-pargasite fields (see Figure 10d), showing $X_{Mg} < 0.60$ and increased Al^{IV}, and containing up to 1.0 wt.% K_2O (see Table 4)—indicative of potassic, vapor-saturated melt conditions typical of cupola-stage degassing in a skarn system [18,83]. K-rich amphiboles are commonly recognized as a characteristic feature of porphyry–skarn systems [18,84].

Whole-rock major-oxide trends reinforce this evolving volatile scenario: SiO_2 increases from about 55 wt.% in Kirazgedik quartz-monzodiorites to >75 wt.% in Delice felsic dykes, whereas Fe_2O_3 reaches ~35 wt.% in skarn samples from Kozbudaklar and Delice, coincident with peak W concentrations (>2557 ppm). This enrichment is attributed to carbonate replacement and elevated oxygen fugacity—conditions favoring W and Bi metal transport via chloride–fluoride complexes [18,78].

Highly evolved apophyses at Güneybudaklar and Delice are further characterized by Rb/Sr > 1 and Zr > 250 ppm—classic characteristics of fractionated, volatile-rich granites [49,85]. Thermometry using Ti-in-biotite and amphibole thermometry indicates temperature decrease from ~500 °C at Kirazgedik to <300 °C at Delice, while fO_2 increases—marking the shift from Cu–Mo porphyry cores to W-rich skarn halos [18,78,86].

Field relationships corroborate this mineralogical vector. Chemically evolved intrusions in Güneybudaklar and Delice are structurally controlled by faults and breccia zones that acted as conduits for late-stage fluids. Kozbudaklar represents a transitional domain—Mo rich yet W laden (>2300 ppm) with moderate F in biotite—reflecting a Mo–W skarn system retaining porphyry characteristics. Meanwhile, Kirazgedik preserves primitive signatures (e.g., high X_{Mg} biotite, mantle-derived amphibole, Cu–Mo enrichment), typically of early compressional porphyry stages.

5.3. Thermo-Redox Evolution of Magmatic Pulses

Thermobarometric modeling based on amphibole compositions, using the empirical formulations of Ridolfi et al. [17] and magmatic fO_2 estimates from the Geo-fO_2 software (version 2.0) [87], provides critical insight into the temporal and spatial evolution of P–T–fO_2–H_2O conditions across the NMC (Supplementary Table S6). Each mineralized sector displays a distinctive and internally coherent thermodynamic trajectory, reflecting contrasting magmatic histories that align with their respective mineralization styles and geochemical signatures (Figure 12).

The Kirazgedik region is defined by amphibole compositions that cluster at high temperatures (mean ≈ 876 °C), with elevated pressures (up to 2.7 kbar), indicative of deeper magma storage (Figure 12a,b). A steep increase in ΔFMQ with decreasing temperature (up to 2.3) reflects the development of a pronounced redox gradient during rapid magma ascent, likely associated with progressive degassing and oxidation (Figure 12c). Despite these high pressures, amphibole H_2O contents remain relatively low (4.3–6.6 wt.%) (Figure 12e), suggesting that the parental magma was initially volatile-undersaturated and derived from a reduced mantle source within the sub-continental lithospheric mantle (SCLM), consistent with a compressional subduction setting [14,78]. The observed thermal–redox evolution—characterized by increasing oxidation with ascent—is compatible with the destabilization of early magnetite shells around the porphyry core and the influx of oxidized, slab-derived fluids [78,88].

In contrast, the Güneybudaklar region displays moderately lower crystallization temperatures (mean ≈ 826 °C) and a narrower pressure range (0.7–2.0 kbar) (Figure 12a,b), consistent with shallower magma emplacement. The ΔFMQ values in this region are relatively stable, with a median of approximately 1.4, suggesting that redox conditions were established early and remained constant throughout cooling (Figure 12c). Amphibole H_2O contents in this region are higher on average (4.7–7.3 wt.%) (Figure 12e), indicating a melt that was more hydrous and more oxidized than its Kirazgedik region. Although most amphiboles in Güneybudaklar display this oxidized and stable signature, a subset exhibits Kirazgedik-like trends in the ΔFMQ–H_2O relationship (Figure 12d), pointing to the possibility of early slab-fluid influx followed by overprinting by crustally buffered, oxidized fluids during later stages of skarn development [13,77,89]. This hybrid signature underscores the transitional nature of Güneybudaklar, straddling the boundary between porphyry-type magmatism and the onset of skarn-forming processes.

The Kozbudaklar region records the lowest average crystallization temperatures (mean ≈ 780 °C) and pressures (0.9–1.7 kbar) across the complex (Figure 12a,b). Its ΔFMQ–T trend is nearly flat (Figure 12c), centered around a ΔFMQ of ~1.1, reflecting little to no change in oxidation state during cooling. Amphibole H_2O contents are consistently high (6.2–7.4 wt.%) (Figure 12e), indicating that the magmatic system was water-rich and likely fluid-saturated through its evolution. The thermal and redox characteristics of Kozbudaklar suggest a shallow emplacement regime and limited magmatic recharge, consistent with the development of static, iron-rich skarn systems rather than dynamic porphyry-type roots.

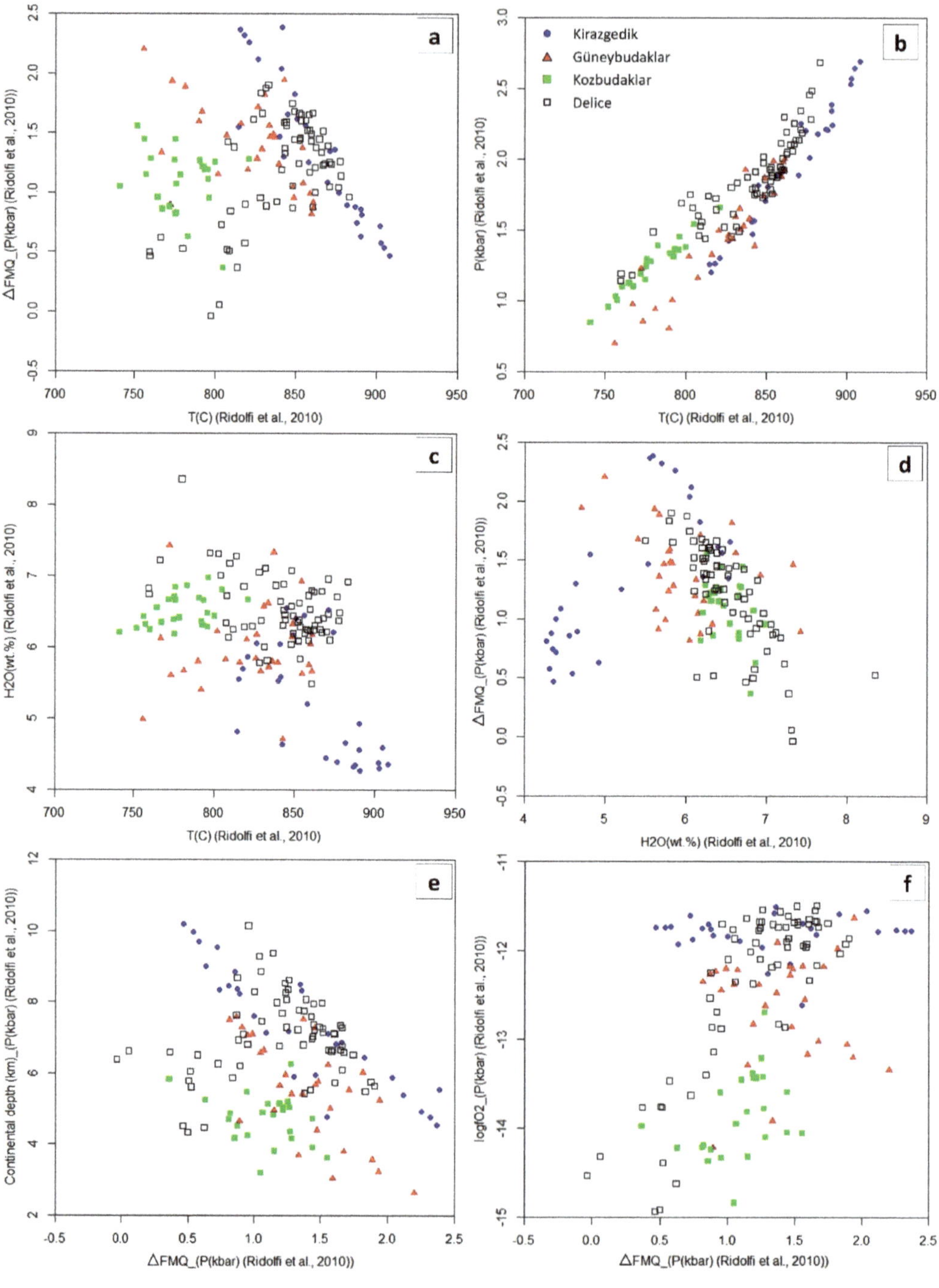

Figure 12. Thermobarometric and redox parameter estimations for amphiboles from the Kirazgedik, Güneybudaklar, Kozbudaklar, and Delice sectors [17,87]: (**a**) ΔFMQ (kbar) vs. temperature (°C); (**b**) Pressure (kbar) vs. temperature (°C); (**c**) H_2O content (wt.%) vs. temperature (°C); (**d**) ΔFMQ (kbar) vs. H_2O content (wt.%); (**e**) Continental crustal depth (km) vs. ΔFMQ (kbar); (**f**) $\log_{10}fO_2$ (kbar) vs. ΔFMQ (kbar).

The Delice region exhibits crystallization temperatures and pressures similar to those of Güneybudaklar (Figure 12a,b), but it is distinguished by a more complex and variable redox evolution. Amphibole ΔFMQ values at Delice range from 0.1 to 2.7, with an irregular decline as temperature decreases (Figure 12c). This oscillatory ΔFMQ behavior may reflect episodic magma recharge, heterogeneous wall–rock interactions, or differences in degassing efficiency during magmatic crystallization. Pressure estimates for Delice are broad (1.1–2.7 kbar), suggesting that intrusions were emplaced at multiple crustal levels (Figure 12b). Despite these fluctuations in fO_2, amphibole H_2O contents remain consistently high (up to 8.3 wt.%) (Figure 12e), indicating that the magmas were volatile-saturated systems and retain high fluid flux potential—ideal for the formation of W–Ce–Fe skarn assemblages [85,90].

A crucial distinction among sectors is observed in the relationship between ΔFMQ and calculated H_2O contents (Figure 12d). In Kirazgedik, a positive correlation between fO_2 and H_2O implies that the system became increasingly oxidized as juvenile, slab-derived fluids were fluxed. This trajectory is characteristic of early-stage magmas influenced by oxidizing volatiles such as SO_2 and Cl-bearing components released during slab dehydration [13,91,92]. These fluids are known to enhance both redox state and metal transport capacity in arc settings [77,93]. In contrast, the Güneybudaklar, Kozbudaklar, and Delice sectors exhibit either inverse or scattered ΔFMQ–H_2O relationships, suggesting decoupling of oxidation and volatile content. This pattern is interpreted as evidence of significant interaction with shallow-level or external fluids, including meteoric or metamorphic waters, which dilute primary magmatic signals and impose localized redox buffering through wall–rock reaction [85,94]. The divergence from the oxidized juvenile fluid signature of Kirazgedik highlights a fundamental shift in fluid provenance and redox control during the transition from compressional subduction to slab rollback.

The distribution of ΔFMQ values across sectors delineates a systematic redox structure within the magmatic plumbing system of the NMC (Figure 12f). Most samples cluster between ΔFMQ 1 and 1.5, consistent with moderately oxidized arc magmas [78,91]. However, amphiboles from Kirazgedik and parts of Güneybudaklar extend to higher ΔFMQ values, reflecting the influence of deeply sourced, oxidized slab-derived fluids. In contrast, the Kozbudaklar and Delice sectors, which crystallized at similar pressures, exhibit ΔFMQ values closer to the arc average, suggesting a progressive redox homogenization of the magmatic system over time. This convergence likely reflects long-term equilibration through magma recharge–mixing cycles, prolonged residence in upper crustal chambers, and overprinting by crustally derived or meteoric fluids [85,95]. Collectively, these trends support a vertically integrated, evolving magmatic–hydrothermal system that matured through both internal differentiation and external fluid interaction during the late stages of arc evolution in a rollback-dominated tectonic setting [13,92].

5.4. Litho-Structural Controls and Metallogenic Model

The coexistence of porphyry Cu-Mo and skarn W deposits is geologically uncommon due to their contrasting magmatic and redox requirements. Porphyry Cu-Mo systems are typically derived from oxidized, hydrous, calc-alkaline magmas, often associated with magnetite-series granitoids in convergent margin settings [1,96]. These magmas are characterized by high oxygen fugacity (fO_2), which facilitates the transport and deposition of chalcophile metals like Cu and Mo as sulfide complexes [97]. In contrast, W-bearing skarn systems generally form in association with reduced, ilmenite-series granitoids, often of S-type affinity, where lower fO_2 favors the stability of tungstate species (e.g., scheelite), and the involvement of crustal-derived magmas is more pronounced [7,98]. Nevertheless, spatial and temporal associations of both deposit types have been reported in several metallogenic provinces, challenging conventional redox-based models: the Yechangping Mo-porphyry/W-skarn system (East Qinling, China) and the giant Zhuxi W-Cu deposit (South China). These mining districts worldwide illustrate instances where porphyry copper-molybdenum (Cu-Mo) systems are spatially and temporally associated with tungsten (W) skarn mineralization. Although this coexistence is relatively rare, it underscores the role of specific geologic conditions—such as evolving magmatic oxidation states and mixed magmatic sources—in facilitating the simultaneous evolution of these mineralization styles.

In the Yechangping porphyry-skarn system (East Qinling, Luanchuan District, China), the Luanchuan metallogenic district hosts giant Mo deposits that grade into W-bearing skarns. For instance, the Yechangping deposit is a representative porphyry-skarn Mo-W system, featuring porphyry-style molybdenum mineralization accompanied by a tungsten-bearing skarn zone [99]. Other nearby intrusions (e.g., Nannihu–Sandaozhuang) similarly contain porphyry Mo orebodies with adjacent W-rich skarns [100]. These systems formed from Jurassic granitoids of intermediate oxidation state, and they illustrate how a single magmatic-hydrothermal system can generate Mo-sulfide ore in the cupola (porphyry) and scheelite in peripheral skarns [99]. Geological studies of Luanchuan deposits indicate that magma differentiation and fluid phase evolution led to varying fO_2 conditions—high fO_2 fluids precipitated Mo sulfides in the porphyry core, whereas relatively lower fO_2 at the limestone contacts favored scheelite deposition in skarns [99]. This explains the co-occurrence of Cu/Mo and W mineralization in what was once thought to be an exclusively "oxidized" porphyry environment. While in South China's Jiangnan ore belt (northern Nanling region), most granitoids are reduced ilmenite-series, typically yielding W–Sn skarns. However, recent discoveries in the Late Jurassic–Early Cretaceous Jiangnan belt include several giant W deposits with unusual Cu or Mo enrichment [101]. The Dahutang W–Cu deposit in Jiangxi is a prime example—it contains an estimated 1.1 million tonnes of WO_3 along with significant copper byproducts [101]. Such W–Cu skarn systems are interpreted to result from hybrid magmas or evolving redox conditions: an initially reduced granite generated the bulk of the tungsten (scheelite) mineralization, but pockets of more oxidized, sulfur-rich melt or magma mingling introduced chalcopyrite (Cu) mineralization in tandem [101,102]. In other cases, like the Yangchuling W–Mo deposit (Jiangnan belt), zircons record mixed Ce^{4+}/Ce^{3+} ratios, suggesting the magma fO_2 fluctuated during crystallization [102]. These Chinese examples show that W-skarn formation is not strictly limited to purely reduced systems—if a pluton's oxygen fugacity increases slightly (or a minor oxidized magma invades), it can carry chalcophile metals like Cu/Mo without fully suppressing scheelite deposition.

In our study, the NMC represents a vertically superimposed, laterally zoned magmatic–hydrothermal system developed during the late stages of subduction-related arc magmatism and post-subduction extension. The mineralization architecture observed—progressing from early Cu–Mo fractured porphyries to distal, relatively reduced W-dominant skarns—is controlled by a synergistic interplay between magma evolution, redox and volatile state, emplacement depth, and the inherited upper crustal structure. The temporal progression from Kirazgedik (~54.6 Ma) to Delice (~45.9 Ma) reflects an evolving magmatic system transitioning from mafic, mantle-derived, reduced melts to felsic, volatile-saturated, and oxidized intrusions. Early-stage Kirazgedik intrusions are interpreted as high-pressure, oxidant-deficient melts derived from a metasomatized sub-continental lithospheric mantle (SCLM), characterized by low H_2O and moderate fO_2, as inferred from amphibole-based thermobarometry and low Cl/F ratios in biotite [14,78] (Figure 13). The

Cu-Mo-Au-W association observed in both the Kirazgedik and Güneybudaklar systems closely resembles other porphyry-style deposits such as Skouries and Cadia Ridge-way, which are hosted in oxidized, high-K magmas [60,103]. Geochemical indicators—such as elevated fO_2 (~+1.5 ΔFMQ) and Cl-enriched amphiboles—collectively point to a fertile, volatile-saturated magmatic source. These features align well with the genetic models of porphyry ore formation proposed by Richards [8] and broadly applied in exploration frameworks [60].

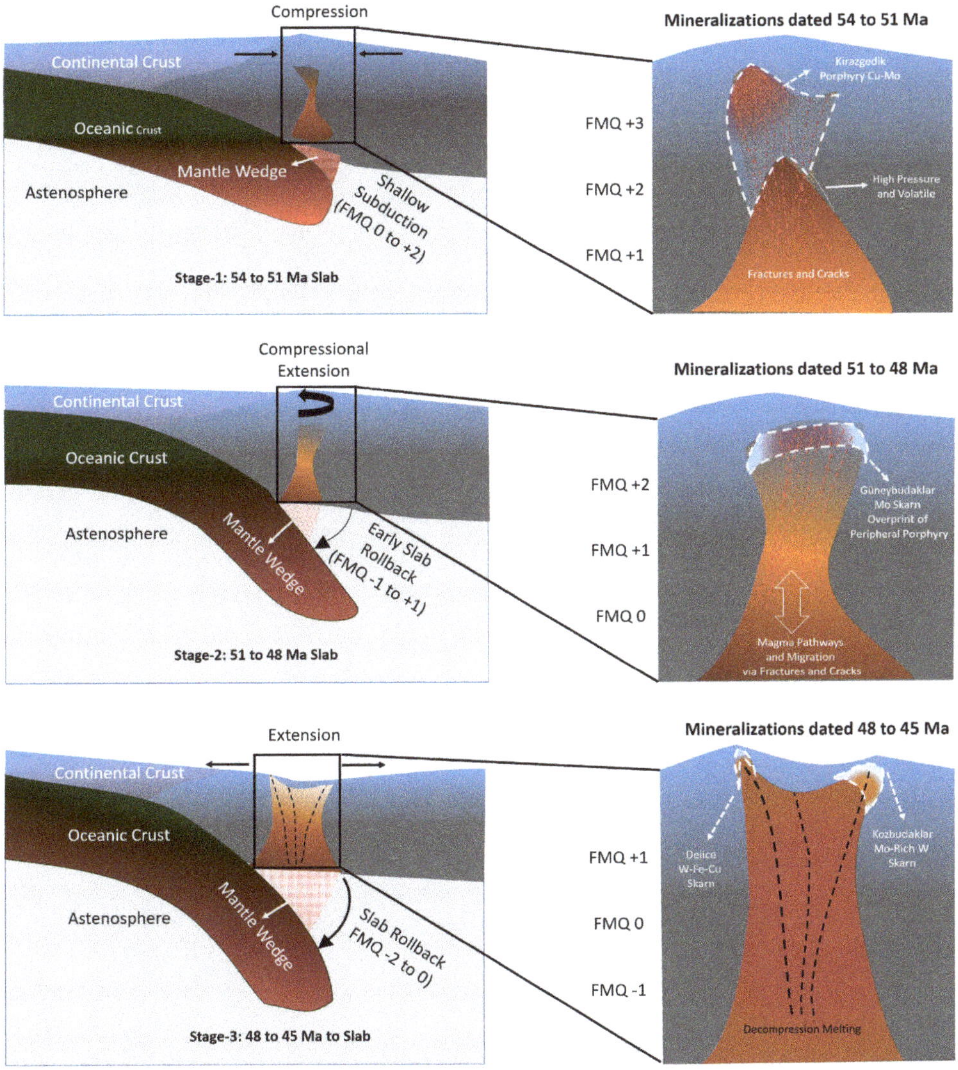

Figure 13. Tectonomagmatic evolution and associated mineralization stages in the NMC region from 54 to 45 Ma, illustrating slab dynamics, redox conditions (FMQ), and ore-forming processes.

As the system evolved, fractional crystallization, crustal assimilation, and the ingress of slab-derived oxidized fluids promoted higher H_2O content and elevated fO_2 conditions. These processes facilitated the formation of W, Mo, and Bi-rich mineralizing fluids, particularly in Güneybudaklar and Kozbudaklar. The elevated Al^{IV} and K_2O contents in amphiboles from these sectors, combined with high SiO_2 and Rb/Sr ratios in host rocks, indicate that these magmas had undergone extensive differentiation and Cl saturation at shallower crustal levels (~0.7–2.0 kbar), forming the cupola zones of evolving porphyry–skarn transitions [18,89]. Delice, the youngest and most evolved center, preserves felsic, fluid-charged intrusions with the highest H_2O contents and lowest crystallization temperatures (~<300 °C), suggestive of late-stage, low-pressure skarn-forming conditions. Amphibole ΔFMQ values at Delice fluctuate widely, possibly due to episodic magma pulses or interaction with external fluids (e.g., meteoric or metamorphic), reflecting an unstable redox environment during crystallization and late-stage hydrothermal fluid flow.

In addition to early-stage porphyry mineralization associated with oxidized high-K calc-alkaline magmas, the progressive rollback of the subducting slab likely triggered mantle wedge widening and enhanced contributions from carbonate platform melting (Figure 13). This evolution introduced reducing agents into the mantle wedge and facilitated variable redox states across the NMC intrusions. Similarly, the increased crustal input into high-K magmas—typical of post-collisional arc settings—is genetically linked to Cu-Mo-W-Au porphyry–skarn transitions [104,105].

Metal precipitation exhibited depth dependency; in deeper levels, magnetite and sulfide saturation promoted Cu and Mo deposition. Conversely, in shallower, acidic, and sulfur-poor conditions (e.g., Delice), scheelite and associated tungstate phases were stabilized by oxidized, F-bearing fluids [77,78]. Structural controls played a key role in channeling magmas and hydrothermal fluids. Major mineralized centers such as Güneybudaklar and Delice are aligned with oblique-slip fault zones and breccia corridors, interpreted as conduits formed during slab rollback-driven transtensional deformation [24,58]. These tectonic features enabled vertical superposition of porphyry, skarn, and epithermal mineralization styles, preserved as a continuous spectrum from deep-seated porphyry roots to shallow-level skarn caps and vein systems [106,107]. The absence of significant thermal overprinting on the surrounding HP–LT metamorphic basement is consistent with slab-attached rollback scenarios, wherein fluid and heat transport remained spatially restricted along structurally reactivated zones [85,91].

Integrating mineral chemistry, thermobarometry, redox proxies, and structural data, the NMC exemplifies a complete porphyry–skarn system evolving within a post-subduction tectonic context. Its metallogenic sequence—ranging from Cu–Mo (Kirazgedik) to Mo–W–Bi–Au (Güneybudaklar/Kozbudaklar), and finally W–Fe–Ce (Delice)—reflects both spatial zonation and temporal magmatic evolution from mantle-derived to crust-dominated sources. This progression parallels analogous systems in the Andes and the Central Asian Orogenic Belt [85,108], where oxidized, hydrous magmas and structurally complex settings control metal enrichment. The NMC thus serves as a representative hybrid porphyry–skarn system, shaped by concurrent changes in tectonics, magma chemistry, and lithospheric architecture.

6. Conclusions

The mineralized systems of the westernmost Eocene Tavşanlı Metallogenic Belt, exemplified by the Nilüfer Mineralization Complex (NMC) in NW Türkiye, record a systematic and spatially coherent evolution from high-temperature, oxidized porphyry-style Cu-Mo systems to more reduced, volatile-depleted W-Mo-bearing skarns. This metallogenic progression reflects a broader magmatic differentiation trend coupled with evolving emplacement depths and redox states during the post-subduction tectonic regime.

Kirazgedik represents the earliest and most thermodynamically distinct mineralization phase, formed from deep-seated, oxidized, hydrous magmas crystallizing at elevated temperatures (~930 °C) and pressures (~2.7 kbar). This porphyry Cu-Mo system is interpreted as the product of early post-subduction arc magmatism, marked by high magmatic fO_2 and water content. The transitional Güneybudaklar system, constrained by Re–Os molybdenite ages (~49.9 Ma), developed under moderately oxidized (ΔFMQ ~ 1.8–0.5) and water-rich conditions, facilitating Mo and W enrichment at the porphyry–skarn interface. Kozbudaklar, in contrast, reflects a more evolved and degassed magmatic state, where reduced conditions favored scheelite (W) mineralization in a distal skarn environment. Delice captures the most complex fluid history, characterized by felsic intrusions into carbonate rocks and recording a wide redox-volatility spectrum. The coexistence of both oxidized and reduced fluid signatures suggests prolonged, multi-stage fluid evolution involving both magmatic and externally derived components, as well as intense fluid–rock interaction during retrograde alteration. Collectively, these systems illustrate a continuous, genetically linked porphyry–skarn evolution controlled by progressive magmatic differentiation, redox evolution, and crustal emplacement depth, all modulated by a regional tectonic shift from transpression to transtension during slab rollback. Amphibole and mica compositions provide robust proxies for reconstructing magmatic conditions (P–T–fO_2–H_2O) and are shown to be effective tools for assessing magmatic fertility and fluid evolution in complex post-subduction systems. Integrated petrological, mineralogical, and geochemical analyses delineate the mineralization potential and refine exploration models in porphyry–skarn terranes of the Western Tethyan domain.

Supplementary Materials: The following supporting information can be downloaded at: https://www.mdpi.com/article/10.3390/min15080792/s1. Table S1. Sample Metadata; Table S2. Whole Rock All; Table S3. Ore All; Table S4. Mica EPMA; Table S5. Amphibole EPMA; Table S6. Amphibole Calculations; Figure S1. Representative BSE images of amphiboles and micas from Kirazgedik samples; Figure S2. Representative BSE images of amphiboles and micas from Güneybudaklar samples; Figure S3. Representative BSE images of amphiboles and micas from Kozbudaklar samples; Figure S4. Representative BSE images of amphiboles and micas from Delice samples.

Author Contributions: Conceptualization, H.K., M.K. (Mustafa Kumral), R.A.C. and A.A.; methodology, H.K., M.K. (Mustafa Kaya) and A.A.; software, H.K. and M.K. (Mustafa Kaya); validation, H.K., H.S., R.A.C. and A.A.; formal analysis, H.K., M.K. (Mustafa Kaya) and R.A.C.; investigation, H.K., M.K. (Mustafa Kumral), H.S., M.K. (Mustafa Kaya), R.A.C. and A.A.; data curation, H.K., M.K. (Mustafa Kumral), H.S., M.K. (Mustafa Kaya), R.A.C. and A.A.; writing—original draft preparation, H.K., M.K. (Mustafa Kaya) and A.A.; writing—review and editing, H.K., M.K. (Mustafa Kumral), R.A.C. and A.A.; visualization, H.K., M.K. (Mustafa Kumral), H.S., R.A.C. and A.A.; supervision, M.K. (Mustafa Kumral). All authors have read and agreed to the published version of the manuscript.

Funding: This research was funded by the Scientific and Technological Research Council of Türkiye (TÜBİTAK) through graduate student research grants awarded to Hüseyin Kocatürk and is derived from the corresponding author's PhD dissertation, completed at Istanbul Technical University (ITU, Turkey). Additionally, it was funded by the Scientific Research Project (BAP Project No: MHD-2025-46403) of Istanbul Technical University (ITU, Turkey).

Data Availability Statement: The data are available Tables 1–5.

Acknowledgments: Hüseyin Kocatürk gratefully acknowledges the financial support received from the Scientific and Technological Research Council of Türkiye (TÜBİTAK) through graduate student research grants, which contributed to the advancement of his doctoral studies and the production of this article as part of his PhD dissertation at Istanbul Technical University. The authors also acknowledge additional funding provided by Istanbul Technical University (ITU, Turkey) under the

Scientific Research Project (BAP Project No: MHD-2025-46403). They express their gratitude to the Geochemistry Research Laboratories (JAL) group of ITU for their valuable support in conducting analytical research, as well as to Beril Tanç Kaya. Additionally, the authors are much more grateful to the anonymous Editor and three reviewers for their constructive comments and helpful suggestions, which greatly improved the manuscript.

Conflicts of Interest: The authors declare no conflicts of interest.

References

1. Sillitoe, R.H. Porphyry copper systems. *Econ. Geol.* **2010**, *105*, 3–41. [CrossRef]
2. Cooke, D.R.; Hollings, P.; Walshe, J.L. Giant porphyry deposits: Characteristics, distribution, and tectonic controls. *Econ. Geol.* **2005**, *100*, 801–818. [CrossRef]
3. Kaya, M.; Kumral, M.; Yalçın, C.; Abdelnasser, A. Genesis and Evolution of the Yolindi Cu-Fe Skarn Deposit in the Biga Peninsula (NW Turkey): Insights from Genetic Relationships with Calc-Alkaline Magmatic Activity. *Minerals* **2023**, *13*, 1304. [CrossRef]
4. Kaya, M.; Kumral, M.; Yalçın, C.; Abdelnasser, A. Sulfur and Carbon–Oxygen Isotopic Geochemistry and Fluid Inclusion Characteristics of the Yolindi Cu-Fe Skarn Mineralization, Biga Peninsula, NW Turkey: Implications for the Source and Evolution of Hydrothermal Fluids. *Minerals* **2023**, *13*, 1542. [CrossRef]
5. Karaman, M.; Kumral, M.; Yildirim, D.K.; Doner, Z.; Afzal, P.; Abdelnasser, A. Delineation of the porphyry-skarn mineralized zones (NW Turkey) using concentration–volume fractal model. *Geochemistry* **2021**, *4*, 125802. [CrossRef]
6. Abdelnasser, A.; Kumral, M.; Zoheir, B.; Yilmaz, H. Evolution of the Tepeoba porphyry-skarn Cu-Mo-Au deposit, NW Turkey: New mineralogical and geochemical findings. *Ore Geol. Rev.* **2022**, *147*, 104967. [CrossRef]
7. Meinert, L.D.; Dipple, G.M.; Nicolescu, S.; Hedenquist, J.W.; Thompson, J.F.H.; Goldfarb, R.J.; Richards, J.P. World Skarn Deposits. In *Economic Geology 100th Anniversary Volume*; Society of Economic Geologists: Littleton, CO, USA, 2005; pp. 299–336.
8. Richards, J. Metallogeny of the Neo-Tethys arc in central Iran. In *Mineral Exploration and Sustainable Development*; Millpress: Rotterdam, The Netherlands, 2003; pp. 1237–1239.
9. Kocatürk, H. Genesis and Tectono-Metallogenic Evolution of Gold-Tungsten-Copper-Molybdenum (Au-W-Cu-Mo) Mineralizations Associated with Eocene Granitoids Around Uludağ (Bursa). Ph.D. Thesis, Istanbul Technical University, Istanbul, Turkiye, 2024.
10. McFall, K.; Roberts, S.; McDonald, I.; Boyce, A.J.; Naden, J.; Teagle, D. Rhenium enrichment in the Muratdere Cu-Mo (Au-Re) porphyry deposit, Turkey: Evidence from stable isotope analyses (δ34S, δ18O, δD) and laser ablation-inductively coupled plasma-mass spectrometry analysis of sulfides. *Econ. Geol.* **2019**, *114*, 1443–1466. [CrossRef]
11. Kuşcu, İ.; Tosdal, R.; Gençalioğlu-Kuşcu, G. Episodic Porphyry Cu (-Mo-Au) formation and associated magmatic evolution in Turkish Tethyan collage. *Ore Geol. Rev.* **2019**, *107*, 119–154. [CrossRef]
12. Rabayrol, F.; Hart, C.J.; Friedman, R.M.; Spikings, R.A. Diachronous Magmatic and Cu-Au-Mo Metallogenic Responses to Slab Roll-Back Initiation from Northwest Anatolia to the Balkans, Western Tethyan Eocene Magmatic Belt. In *Tectonomagmatic Influences on Metallogeny and Hydrothermal Ore Deposits: A Tribute to Jeremy P. Richards*; Sholeh, A., Wang, R., Eds.; Society of Economic Geologists, Inc.: Littleton, CO, USA, 2021; Volume 1, pp. 17–28.
13. Orhan, A. Evolution of the Mo-rich scheelite skarn mineralization at Kozbudaklar, Western Anatolia, Turkey: Evidence from mineral chemistry and fluid inclusions. *Ore Geol. Rev.* **2017**, *80*, 141–165. [CrossRef]
14. Kocatürk, H.; Kumral, M.; Creaser, R.A.; DuFrane, S.A.; Ünlüer, A.T.; Sendir, H.; Döner, Z.; Kaya, M.; Özdamar, Ş.; Abdelnasser, A. Magma nature and tectono-magmatic context of the Eocene Uludağ granitoids (NW-Türkiye): Insights into the Cenozoic geodynamics of the Tethyan Orogenic Belt. *Geochemistry* **2024**, *84*, 126170. [CrossRef]
15. Okay, A.I.; Tüysüz, O. Tethyan sutures of northern Turkey. *Geol. Soc. Lond. Spec. Publ.* **1999**, *156*, 475–515. [CrossRef]
16. Kuşcu, İ.; Tosdal, R.M.; Gençalioğlu-Kuşcu, G. Porphyry-Cu Deposits of Turkey. In *Mineral Resources of Turkey*; Springer: Berlin/Heidelberg, Germany, 2019; pp. 337–425.
17. Ridolfi, F.; Renzulli, A.; Puerini, M. Stability and chemical equilibrium of amphibole in calc-alkaline magmas: An overview, new thermobarometric formulations and application to subduction-related volcanoes. *Contrib. Mineral. Petrol.* **2010**, *160*, 45–66. [CrossRef]
18. Ridolfi, F.; Renzulli, A. Calcic amphiboles in calc-alkaline and alkaline magmas: Thermobarometric and chemometric empirical equations valid up to 1,130 °C and 2.2 GPa. *Contrib. Mineral. Petrol.* **2012**, *163*, 877–895. [CrossRef]
19. Li, X.; Zhang, C.; Almeev, R.R.; Zhang, X.-C.; Zhao, X.-F.; Wang, L.-X.; Koepke, J.; Holtz, F. Electron probe microanalysis of Fe2+/ΣFe ratios in calcic and sodic-calcic amphibole and biotite using the flank method. *Chem. Geol.* **2019**, *509*, 152–162. [CrossRef]

20. Shafaii Moghadam, H.; Xiao, W.; Karsli, O.; Aydin, F.; He, Y.; Chiaradia, M.; Griffin, W.L. Copper endowment of the magmatic rocks from Eastern Sakarya Zone (Türkiye): Insights from zircon and apatite geochemical evolution. *Contrib. Mineral. Petrol.* **2025**, *180*, 38. [CrossRef]
21. Duan, X.-X.; Ju, Y.-F.; Chen, B.; Wang, Z.-Q. Garnet geochemistry of reduced Skarn system: Implications for fluid evolution and Skarn formation of the Zhuxiling W (Mo) deposit, China. *Minerals* **2020**, *10*, 1024. [CrossRef]
22. Carter, L.C.; Tapster, S.R.; Williamson, B.J.; Buret, Y.; Selby, D.; Rollinson, G.K.; Millar, I.; Parvaz, D.B. A rapid change in magma plumbing taps porphyry copper deposit-forming magmas. *Sci. Rep.* **2022**, *12*, 17272. [CrossRef]
23. Plunder, A.; Agard, P.; Chopin, C.; Okay, A.I. Geodynamics of the Tavşanlı zone, western Turkey: Insights into subduction/obduction processes. *Tectonophysics* **2013**, *608*, 884–903. [CrossRef]
24. Okay, A.I.; Harris, N.B.; Kelley, S.P. Exhumation of blueschists along a Tethyan suture in northwest Turkey. *Tectonophysics* **1998**, *285*, 275–299. [CrossRef]
25. Plunder, A.; Agard, P.; Chopin, C.; Pourteau, A.; Okay, A.I. Accretion, underplating and exhumation along a subduction interface: From subduction initiation to continental subduction (Tavşanlı zone, W. Turkey). *Lithos* **2015**, *226*, 233–254. [CrossRef]
26. Okay, A.I. Tavşanli zone: The northern subducted margin of the Anatolide-Tauride block. *Bull. Miner. Res. Explor.* **2011**, *142*, 191–221.
27. Okay, A.I.; Satir, M. Geochronology of Eocene plutonism and metamorphism in northwest. *Geodin. Acta* **2006**, *19*, 251–266. [CrossRef]
28. Sarifakioğlu, E.; Özen, H.; Winchester, J.A. Whole rock and mineral chemistry of ultramafic-mafic cumulates from the Orhaneli (Bursa) ophiolite, NW Anatolia. *Turk. J. Earth Sci.* **2009**, *18*, 55–83. [CrossRef]
29. Altunkaynak, Ş.; Sunal, G.; Aldanmaz, E.; Genç, C.Ş.; Dilek, Y.; Furnes, H.; Foland, K.A.; Yang, J.; Yıldız, M. Eocene Granitic Magmatism in NW Anatolia (Turkey) revisited: New implications from comparative zircon SHRIMP U–Pb and 40Ar–39Ar geochronology and isotope geochemistry on magma genesis and emplacement. *Lithos* **2012**, *155*, 289–309. [CrossRef]
30. Selby, D.; Creaser, R.A. Macroscale NTIMS and microscale LA-MC-ICP-MS Re-Os isotopic analysis of molybdenite: Testing spatial restrictions for reliable Re-Os age determinations, and implications for the decoupling of Re and Os within molybdenite. *Geochim. Cosmochim. Acta* **2004**, *68*, 3897–3908. [CrossRef]
31. Markey, R.; Stein, H.J.; Hannah, J.L.; Zimmerman, A.; Selby, D.; Creaser, R.A. Standardizing Re–Os geochronology: A new molybdenite reference material (Henderson, USA) and the stoichiometry of Os salts. *Chem. Geol.* **2007**, *244*, 74–87. [CrossRef]
32. Smoliar, M.I.; Walker, R.J.; Morgan, J.W. Re-Os ages of group IIA, IIIA, IVA, and IVB iron meteorites. *Science* **1996**, *271*, 1099–1102. [CrossRef]
33. Orhan, A.; Demirbilek, M.; Mutlu, H. Mineral and whole-rock geochemistry of the Topuk Granitoid (Bursa, Western Anatolia, Turkey). In Proceedings of the EGU General Assembly Conference Abstracts, Vienna, Austria, 27 April–2 May 2014; p. 5103.
34. Orhan, A.; Mutlu, H. Geochemical characteristics and rare-earth element distributions of Kozbudaklar W-Skarn Deposit (Bursa, Western Anatolia). *Bull. Miner. Res. Explor.* **2017**, *155*, 115–130. [CrossRef]
35. Kumral, M.; Abdelnasser, A.; Budakoglu, M. Geochemistry of Hydrothermal Alteration Associated with Cenozoic Intrusion-Hosted Cu-Pb-Zn Mineralization at Tavşanlı Area, Kütahya, NW Turkey. *Minerals* **2016**, *6*, 13. [CrossRef]
36. Doner, Z.; Unluer, A.T.; Özdamar, Ş.; Sarıkaya, O.; Roden, M.F.; Kaya, M.; Kocaturk, H.; Kumral, M.; Esenli, F. The origin of alkali granites and Th-U $\pm$ REE enrichments in Kestanbol Magmatic complex (NW Anatolia) revisited: Evidences from bulk-rock geochemistry and isotopic data, zircon UPb, biotite Ar/Ar and apatite (UTh)/He geochronology. *Lithos* **2024**, *484–485*, 107751. [CrossRef]
37. Janoušek, V.; Farrow, C.M.; Erban, V. Interpretation of whole-rock geochemical data in igneous geochemistry: Introducing Geochemical Data Toolkit (GCDkit). *J. Petrol.* **2006**, *47*, 1255–1259. [CrossRef]
38. El Bouseily, A.; El Sokkary, A. The relation between Rb, Ba and Sr in granitic rocks. *Chem. Geol.* **1975**, *16*, 207–219. [CrossRef]
39. Foley, S.F.; Venturelli, G.; Green, D.H.; Toscani, L. The ultrapotassic rocks: Characteristics, classification, and constraints for petrogenetic models. *Earth-Sci. Rev.* **1987**, *24*, 81–134. [CrossRef]
40. Rollinson, H. *Using Geochemical Data: Evolution, Presentation, Interpretation*; Longman Scientific and Technical: London, UK, 1993; p. 352.
41. Le Maitre, R.W. *Igneous Rocks: A Classification and Glossary of Terms*; Cambridge University Press: Cambridge, UK, 2002; p. 237.
42. Davidson, J.; Turner, S.; Handley, H.; Macpherson, C.; Dosseto, A. Amphibole "sponge" in arc crust? *Geology* **2007**, *35*, 787–790. [CrossRef]
43. McKenzie, D.; O'nions, R. Partial Melt Distributions from Inversion of Rare Earth Element Concentrations. *J. Petrol.* **1991**, *32*, 1021–1091. [CrossRef]
44. Pearce, J.A.; Harris, N.B.; Tindle, A.G. Trace element discrimination diagrams for the tectonic interpretation of granitic rocks. *J. Petrol.* **1984**, *25*, 956–983. [CrossRef]

45. Bonin, B. Do coeval mafic and felsic magmas in post-collisional to within-plate regimes necessarily imply two contrasting, mantle and crustal, sources? A review. *Lithos* **2004**, *78*, 1–24. [CrossRef]
46. Debon, F.; Le Fort, P. A chemical–mineralogical classification of common plutonic rocks and associations. *Earth Environ. Sci. Trans. R. Soc. Edinb.* **1983**, *73*, 135–149. [CrossRef]
47. Miyashiro, A. Volcanic rock series in island arcs and active continental margins. *Am. J. Sci.* **1974**, *274*, 321–355. [CrossRef]
48. Peccerillo, A.; Taylor, S. Geochemistry of Eocene calc-alkaline volcanic rocks from the Kastamonu area, northern Turkey. *Contrib. Mineral. Petrol.* **1976**, *58*, 63–81. [CrossRef]
49. Laurent, A.; Janousek, V.; Magna, T.; Schulmann, K.; Mikova, J. Petrogenesis and geochronology of a post-orogenic calc-alkaline magmatic association: The Zulova Pluton, Bohemian Massif. *J. Geosci.* **2014**, *59*, 415–440. [CrossRef]
50. Pearce, J. Sources and settings of granitic rocks. *Epis. J. Int. Geosci.* **1996**, *19*, 120–125. [CrossRef]
51. Gill, J.B. *Orogenic Andesites and Plate Tectonics*; Springer Science & Business Media: Berlin/Heidelberg, Germany, 1981; Volume 16, p. 390.
52. Pearce, J.A. Trace element characteristics of lavas from destructive plate boundaries. In *Orogenic Andesites and Related Rocks*; Wiley: Hoboken, NJ, USA, 1982; pp. 528–548.
53. Boynton, W.V. Geochemistry of Rare Earth Elements: Meteorite Studies. In *Rare Earth Element Geochemistry*; Henderson, P., Ed.; Elsevier: New York, NY, USA, 1984; pp. 63–114. [CrossRef]
54. Sun, S.-S.; McDonough, W. Chemical and isotopic systematics of oceanic basalts: Implications for mantle composition and processes. *Geol. Soc. Lond. Spec. Publ.* **1989**, *42*, 313–345. [CrossRef]
55. Tischendorf, G.; Gottesmann, B.; Förster, H.-J.; Trumbull, R.B. On Li-bearing micas: Estimating Li from electron microprobe analyses and an improved diagram for graphical representation. *Mineral. Mag.* **1997**, *61*, 809–834. [CrossRef]
56. Nachit, H.; Ibhi, A.; Ohoud, M.B. Discrimination between primary magmatic biotites, reequilibrated biotites and neoformed biotites. *Comptes Rendus Géosci.* **2005**, *337*, 1415–1420. [CrossRef]
57. Leake, B.E.; Woolley, A.R.; Arps, C.E.; Birch, W.D.; Gilbert, M.C.; Grice, J.D.; Hawthorne, F.C.; Kato, A.; Kisch, H.J.; Krivovichev, V.G. Nomenclature of amphiboles; report of the subcommittee on amphiboles of the International Mineralogical Association, Commission on New Minerals and Mineral Names. *Can. Mineral.* **1997**, *35*, 219–246.
58. Pourteau, A.; Oberhänsli, R.; Candan, O.; Barrier, E.; Vrielynck, B. Neotethyan closure history of western Anatolia: A geodynamic discussion. *Int. J. Earth Sci.* **2016**, *105*, 203–224. [CrossRef]
59. Tirel, C.; Brun, J.-P.; Burov, E.; Wortel, M.; Lebedev, S. A plate tectonics oddity: Caterpillar-walk exhumation of subducted continental crust. *Geology* **2013**, *41*, 555–558. [CrossRef]
60. Müller, D.; Groves, D.I. *Potassic Igneous Rocks and Associated Gold-Copper Mineralization*, 4th ed.; Springer: Cham, Switzerland, 2016.
61. Müller, D.; Rock, N.; Groves, D. Geochemical Discrimination between Shoshonitic and Potassic Volcanic Rocks in Different Tectonic Settings: A Pilot Study. *Mineral. Petrol.* **1992**, *46*, 259–289. [CrossRef]
62. Park, J.-W.; Campbell, I.H.; Chiaradia, M.; Hao, H.; Lee, C.-T. Crustal magmatic controls on the formation of porphyry copper deposits. *Nat. Rev. Earth Environ.* **2021**, *2*, 542–557. [CrossRef]
63. Putirka, K. Special Collection: Rates and Depths of Magma Ascent on Earth: Amphibole thermometers and barometers for igneous systems and some implications for eruption mechanisms of felsic magmas at arc volcanoes. *Am. Mineral.* **2016**, *101*, 841–858. [CrossRef]
64. Luo, C.-H.; Wang, R.; Nebel, O.; Li, Q.-W. Amphibole fractionation as a key driver for oxidation of magmas in convergent margins. *Earth Planet. Sci. Lett.* **2024**, *641*, 118851. [CrossRef]
65. Frost, B.R.; Barnes, C.G.; Collins, W.J.; Arculus, R.J.; Ellis, D.J.; Frost, C.D. A geochemical classification for granitic rocks. *J. Petrol.* **2001**, *42*, 2033–2048. [CrossRef]
66. Moyen, J.-F. High Sr/Y and La/Yb ratios: The meaning of the "adakitic signature". *Lithos* **2009**, *112*, 556–574. [CrossRef]
67. Heinrich, C.A.; Candela, P.A. Fluids and ore formation in the Earth's crust. In *Treatise on Geochemistry*, 2nd ed.; Elsevier: Amsterdam, The Netherlands, 2014; Volume 13, pp. 1–28.
68. Codeço, M.S.; Weis, P.; Andersen, C. Numerical modeling of structurally controlled ore formation in magmatic-hydrothermal systems. *Geochem. Geophys. Geosystems* **2022**, *23*, e2021GC010302. [CrossRef]
69. Blanquat, M.D.S.; Tikoff, B.; Teyssier, C.; Vigneresse, J.L. Transpressional kinematics and magmatic arcs. *Geol. Soc. Lond. Spec. Publ.* **1998**, *135*, 327–340. [CrossRef]
70. Murphy, J.B. Igneous rock associations 7. Arc magmatism I: Relationship between subduction and magma genesis. *Geosci. Can.* **2006**, *33*, 145–168.

71. Oriolo, S.; Gómez, A.L.R.; Maffini, M.N.; Oyhantçabal, P.; Demarco, M.M.; Vargas Perucca, M.S.; Bastías Torres, M.V.; Rubinstein, N.A. Transtension, brittle-ductile shear zones and hydrothermal ore deposits: Towards quantitative structural and kinematic models. *J. Struct. Geol.* **2024**, *185*, 105173. [CrossRef]
72. Contreras-Reyes, E.; Díaz, D.; Bello-González, J.P.; Slezak, K.; Potin, B.; Comte, D.; Maksymowicz, A.; Ruiz, J.A.; Osses, A.; Ruiz, S. Subduction zone fluids and arc magmas conducted by lithospheric deformed regions beneath the central Andes. *Sci. Rep.* **2021**, *11*, 23078. [CrossRef]
73. Abdel-rahman, A.-F.M. Nature of Biotites from Alkaline, Calc-alkaline, and Peraluminous Magmas. *J. Petrol.* **1994**, *35*, 525–541. [CrossRef]
74. Zhou, Z. The origin of intrusive mass in Fengshandong, Hubei province. *Acta Petrol. Sin.* **1986**, *2*, 59–70.
75. Czamanske, G.K.; Wones, D.R. Oxidation during magmatic differentiation, Finnmarka complex, Oslo area, Norway: Part 2, the mafic silicates1. *J. Petrol.* **1973**, *14*, 349–380. [CrossRef]
76. Anderson, J.L.; Smith, D.R. The effects of temperature and fO2 on the Al-in-hornblende barometer. *Am. Mineral.* **1995**, *80*, 549–559. [CrossRef]
77. Audétat, A.; Günther, D.; Heinrich, C.A. Causes for large-scale metal zonation around mineralized plutons: Fluid inclusion LA-ICP-MS evidence from the Mole Granite, Australia. *Econ. Geol.* **2000**, *95*, 1563–1581. [CrossRef]
78. Loucks, R.R. Distinctive composition of copper-ore-forming arcmagmas. *Aust. J. Earth Sci.* **2014**, *61*, 5–16. [CrossRef]
79. Icenhower, J.P.; London, D. Partitioning of fluorine and chlorine between biotite and granitic melt: Experimental calibration at 200 MPa H_2O. *Contrib. Mineral. Petrol.* **1997**, *127*, 17–29. [CrossRef]
80. Reed, M.H.; Palandri, J. Sulfide mineral precipitation from hydrothermal fluids. *Rev. Mineral. Geochem.* **2006**, *61*, 609–631. [CrossRef]
81. Chelle-Michou, C.; Rottier, B.; Caricchi, L.; Simpson, G. Tempo of magma degassing and the genesis of porphyry copper deposits. *Sci. Rep.* **2017**, *7*, 40566. [CrossRef] [PubMed]
82. Chelle-Michou, C.; Chiaradia, M. Amphibole and apatite insights into the evolution and mass balance of Cl and S in magmas associated with porphyry copper deposits. *Contrib. Mineral. Petrol.* **2017**, *172*, 105. [CrossRef]
83. Cai, W.-y.; Wang, K.-y.; Li, J.; Fu, L.-j.; Li, S.-d.; Yang, H.; Konare, Y. Genesis of the Bagenheigeqier Pb–Zn skarn deposit in Inner Mongolia, NE China: Constraints from fluid inclusions, isotope systematics and geochronology. *Geol. Mag.* **2021**, *158*, 271–294. [CrossRef]
84. Zhao, X.; Yang, Z.; Zheng, Y.; Liu, Y.; Tian, S.; Fu, Q. Geology and genesis of the post-collisional porphyry–skarn deposit at Bangpu, Tibet. *Ore Geol. Rev.* **2015**, *70*, 486–509. [CrossRef]
85. Chiaradia, M. Adakite-like magmas from fractional crystallization and melting-assimilation of mafic lower crust (Eocene Macuchi arc, Western Cordillera, Ecuador). *Chem. Geol.* **2009**, *265*, 468–487. [CrossRef]
86. Song, G.; Cook, N.J.; Li, G.; Qin, K.; Ciobanu, C.L.; Yang, Y.; Xu, Y. Scheelite geochemistry in porphyry-skarn W-Mo systems: A case study from the Gaojiabang Deposit, East China. *Ore Geol. Rev.* **2019**, *113*, 103084. [CrossRef]
87. Li, W.; Cheng, Y.; Yang, Z. Geo-fO2: Integrated software for analysis of magmatic oxygen fugacity. *Geochem. Geophys. Geosyst.* **2019**, *20*, 2542–2555. [CrossRef]
88. Wilkinson, J.J. Triggers for the formation of porphyry ore deposits in magmatic arcs. *Nat. Geosci.* **2013**, *6*, 917–925. [CrossRef]
89. Dilles, J.; Barton, M.; Johnson, D.; Proffett, J.; Einaudi, M. *Part I. Contrasting Styles of Intrusion-Associated Hydrothermal Systems*; Guidebook Series; Society of Economic Geologists, Inc.: Littleton, CO, USA, 2000; Volume 32, pp. 1–162.
90. Krawczynski, M.J.; Grove, T.L.; Behrens, H. Amphibole stability in primitive arc magmas: Effects of temperature, H_2O content, and oxygen fugacity. *Contrib. Mineral. Petrol.* **2012**, *164*, 317–339. [CrossRef]
91. Richards, J.P. Magmatic to hydrothermal metal fluxes in convergent and collided margins. *Ore Geol. Rev.* **2011**, *40*, 1–26. [CrossRef]
92. Keppler, H. Fluids and trace element transport in subduction zones. *Am. Mineral.* **2017**, *102*, 5–20. [CrossRef]
93. Candela, P.A.; Piccoli, P.M. Magmatic processes in the development of porphyry-type ore systems. In *One Hundredth Anniversary Volume*; Hedenquist, J.W., Thompson, J.F.H., Goldfarb, R.J., Richards, J.P., Eds.; Society of Economic Geologists: Washington, DC, USA, 2005.
94. Heinrich, C.A. Fluid-fluid interactions in magmatic-hydrothermal ore formation. *Rev. Mineral. Geochem.* **2007**, *65*, 363–387. [CrossRef]
95. Dilles, J.H.; Kent, A.J.R.; Wooden, J.L.; Tosdal, R.M.; Koleszar, A.; Lee, R.G.; Farmer, L.P. Zircon compositional evidence for sulfur-degassing from ore-forming arc magmas*. *Econ. Geol.* **2015**, *110*, 241–251. [CrossRef]
96. Ishihara, S. The granitoid series and mineralization. In *Seventy-Fifth Anniversary Volume: 1905–1980*; Skinner, B.J., Ed.; Economic Geology Publishing Company: McLean, VA, USA, 1981.
97. Ballard, J.R.; Palin, J.M.; Campbell, I.H. Relative oxidation states of magmas inferred from Ce(IV)/Ce(III) in zircon: Application to porphyry copper deposits of northern Chile. *Contrib. Mineral. Petrol.* **2002**, *144*, 347–364. [CrossRef]

98. Candela, P.; Piccoli, P. Model ore-metal partitioning from melts into vapor and vapor/brine mixtures. In *Magma, Fluids and Ore Deposits*; Thompson, J., Ed.; Mineralogical Association of Canada Short Course: Québec, QC, Canada, 1995; Volume 23, pp. 101–127.
99. Huang, P.-C.; Yu, J.; Yao, J.-M.; Qi, N.; Qiu, Z.-W.; Chen, Y.-J. Case study of the large-scale Mo-W mineralization in Eastern Qinling, China: Geology and genesis of the Yechangping porphyry-skarn system. *Ore Geol. Rev.* **2024**, *165*, 105867. [CrossRef]
100. Zhan, Q.; Gao, X.-Y.; Meng, L.; Zhao, T.-P. Ore genesis and fluid evolution of the Sandaozhuang supergiant W-Mo skarn deposit, southern margin of the North China Craton: Insights from scheelite, garnet and clinopyroxene geochemistry. *Ore Geol. Rev.* **2021**, *139*, 104551. [CrossRef]
101. Ouyang, Y.; Hou, Z.; He, X.; Pan, X.; Wei, J.; Zeng, R.; Chen, Q. Geochronological, geochemical and isotopic constraints on the genesis of Cu mineralization in the giant Zhuxi W–Cu deposit, South China. *Ore Geol. Rev.* **2025**, *184*, 106767. [CrossRef]
102. Wu, S.; Sun, W.; Wang, X. A new model for porphyry W mineralization in a world-class tungsten metallogenic belt. *Ore Geol. Rev.* **2019**, *107*, 501–512. [CrossRef]
103. Sillitoe, R. Characteristics and controls of the largest porphyry copper-gold and epithermal gold deposits in the circum-Pacific region. *Aust. J. Earth Sci.* **1997**, *44*, 373–388. [CrossRef]
104. Soloviev, S.G.; Kryazhev, S.G.; Shapovalenko, V.N.; Dvurechenskaya, S.S.; Semenova, D.V.; Kalinin, Y.A.; Voskresensky, K.I.; Sidorova, N.V. The Glafirinskoe and related skarn Cu-Au-W-Mo deposits in the Northern Altai, SW Siberia, Russia: Geology, igneous geochemistry, zircon U-Pb geochronology, mineralization, and fluid inclusion characteristics. *Ore Geol. Rev.* **2021**, *138*, 104382. [CrossRef]
105. Soloviev, S.G. Late Palaeozoic high-K calc-alkaline to shoshonitic series magmatism and related W (–Mo–Cu–Au) metallogeny of the Kyrgyz Tien Shan, Central Asia. *Geol. Soc. Lond. Spec. Publ.* **2024**, *551*, 231–258. [CrossRef]
106. Kumral, M.; Kabiru, M.; Aydogan, M.S.; Kocaturk, H.; Unluer, A.T. The Soma polymetallic hydrothermal Pb–Zn-Cu(± Mo ± Ag ± Au) mineralizations in the south of Biga Peninsula, NW Turkey: Constraints from mineralogy, geochemistry, and fluid inclusion data. *Arab. J. Geosci.* **2022**, *15*, 1172. [CrossRef]
107. Kocaturk, H.; Sendir, H.; Kumral, M.; Unluer, A.; Budakoglu, M. Isotopic Features and Fluid Charecterization of Ore Forming Process Gurgenyayla and Bogazova Plutons (Kütahya/Bursa/Bilecik-Turkey). *Magmat. Earth Relat. Strateg. Met. Depos.* **2018**, 138–141.
108. Mao, J.; Pirajno, F.; Cook, N. Mesozoic metallogeny in East China and corresponding geodynamic settings—An introduction to the special issue. *Ore Geol. Rev.* **2011**, *43*, 1–7. [CrossRef]

minerals

MDPI

Article

Mineralogy and Petrology of Ultrapotassic Lamprophyre Dykes in the Bangbule Area, Xizang, China: Evidence for Open Magma Chamber Fractionation and Mafic Magma Recharge

Chenyu Hong [1], Yong Wang [1,2,*], Sangmu Zunzhu [1,3], Chuankai He [1], Haiyong Wang [3], Yongqiang Wang [1,4], Yun Bai [5], Penggang Yan [3], Suaijie Xun [1], Rui Cao [1] and Dengke Chen [1]

1 College of Earth and Planetary Sciences, Chengdu University of Technology, Chengdu 610059, China; shanhua22@163.com (C.H.); denk-chen888@outlook.com (S.Z.); 2023050025@stu.cdut.edu.cn (C.H.); 13830250347@163.com (Y.W.); 15528080498@163.com (S.X.); chendengke@stu.cdut.edu.cn (D.C.)

2 SinoProbe Laboratory, Chinese Academy of Geological Sciences, Beijing 100037, China

3 No. 6 Geological Party, Xizang Bureau of Geology and Mineral Exploration and Development, Lhasa 851400, China; zhengzhihao@stu.ctud.edu.cn (H.W.)

4 The Second Geological and Mineral Exploration Institute, Gansu Bureau of Geology and Mineral Exploration and Development, Lanzhou 730020, China

5 Sichuan Institute of Comprehensive Geological Survey, Chengdu 610081, China; 2023020183@stu.ctud.edu.cn

* Correspondence: wy_geology@163.com

Abstract: Ultrapotassic lamprophyre dykes in southern Xizang provide important insight into the deep processes and nature of the source region. In this study, the lamprophyre dykes in the Bangbule area of southern Xizang were investigated to understand their petrogenesis and source mantle characteristics. The lamprophyres, based on their mineral and chemical compositions, are classified into two types: type-1, mainly comprising biotite and clinopyroxene, and type-2, featuring annite, phlogopite, and clinopyroxene as phenocrysts. Mica ^{40}Ar-^{39}Ar dating indicates their formation in the early Oligocene (29.5 ± 0.2 Ma). Geochemically, both lamprophyre types show extremely high Sr [$(^{87}Sr/^{86}Sr)_i$ = 0.713936–0.716999] but low and consistent Nd [$(^{143}Nd/^{144}Nd)_i$ = 0.511991–0.512012] isotopic ratios. Type-1 lamprophyres are more enriched in SiO_2, Al_2O_3, and total alkalis ($Na_2O + K_2O$) but have lower $^{T}Fe_2O_3$, CaO, and MgO contents, as well as $Mg^{\#}$ values, compared to type-2, underscoring the differentiated nature of type-1. Geochemical evidence shows that the Bangbule lamprophyre originated from the partial melting of the lithospheric mantle metasomatized by subducted Indian continental crust-derived melts, followed by fractional crystallization. The phenocrysts of clinopyroxene, phlogopite, and K-feldspar exhibit disequilibrium textures and complex zoning between their cores and mantles, suggesting magma replenishment and mixing during their formation. The cores of the clinopyroxene and phlogopite phenocrysts exhibit low $Mg^{\#}$ values, while their mantles show high $Mg^{\#}$ values, indicating magma replenishment between later mafic and early evolved (ultrapotassic) magmas. Accordingly, an open magma chamber model that experienced fractional crystallization and replenishment was proposed for the generation of diverse lamprophyre dykes in Xizang.

Keywords: lamprophyre dyke; geochemistry; mineral chemistry; magma recharge

1. Introduction

As one of the most prominent continental collisional orogens, the Himalayan–Tibetan Plateau orogen has garnered significant attention from geologists [1]. The Lhasa terrane in the southern Tibetan Plateau, over the past 100 million years, has witnessed (1) Andean-type orogeny due to the subduction of the Neo-Tethys Ocean [1–5], (2) collisional orogeny between India and Eurasia during the Cenozoic [1–4,6–8], and (3) plateau uplift and intra-continental extension post-collision [1,2,4,6–8]. These significant geological events have led to the complex and diverse tectonic environment and multistage large-scale mineralization of the Tibetan Plateau [9,10], making the understanding of the dynamic evolution of this orogenic belt crucial for continental tectonics.

Ultrapotassic lamprophyres, distinct in their chemical composition from mafic dykes in the same region, usually carry important information about the mantle characteristics, processes such as partial melting, metasomatism, and post-collisional geodynamics related to magma upwelling [11–13]. For instance, lamprophyre dykes, hosting abundant phenocrysts, antecrysts, and xenocrysts with distinct zoning patterns, can reveal the intricate evolution of magmas during ascent [14–16]. Therefore, these complex zonation patterns present a prime case for investigating open-system magmatic evolution and for gaining a deeper understanding of the processes, depths, and timescales leading to volcanic eruptions [15,17–19].

Post-collisional ultrapotassic rocks are prevalent in the southern Lhasa terrane in the southern Tibetan Plateau [12,20] (Figure 1a). Extensive research has focused on the geochronology and whole-rock geochemistry of these rocks to elucidate their petrogenesis and to delineate the tectonic evolution of the region [12,21–27]. However, despite rigorous studies on the magma sources of these ultrapotassic rocks, the influence of magmatic systems on the evolution of ultrapotassic magmas, as well as the mixing and fractionation processes of ultrapotassic melts in crustal magma reservoirs, remains elusive.

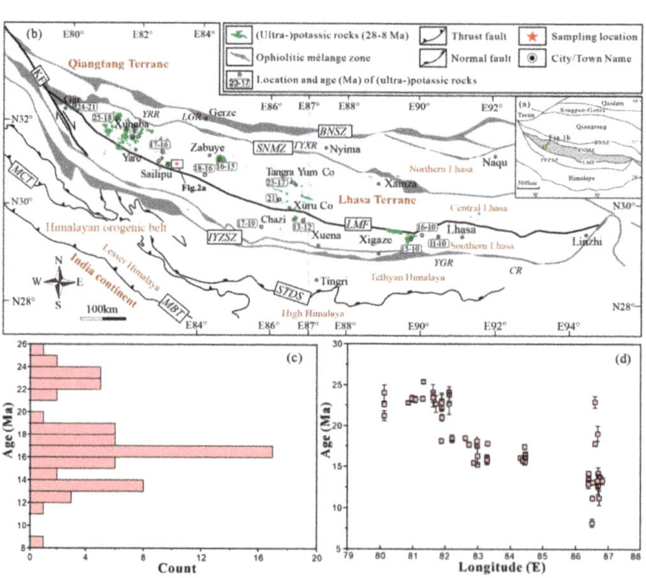

Figure 1. (**a**) Simplified tectonic map of the Tibetan Plateau. (**b**) Tectonic framework of the southern Tibetan plateau and spatial distribution of post-collisional ultrapotassic rocks [24,25]. (**c**) Age histogram of the ultrapotassic rocks. (**d**) Ages of the ultrapotassic rocks from west to east in Tibetan plateau. Age of ultrapotassic rocks [26]. BNSZ, Bangong–Nujiang suture zone; SNMZ, Shiquan River–Nam Tso mélange zone; LMF, Luobadui–Milashan fault; KF, Karakorum fault; IYZSZ, Indus–Yarlung Zangbo suture zone; STDS, Southern Tibetan detachment system; MCT, Main central thrust; MBT, Main boundary thrust. N-S trending rift: YRR, Yare rift; LGR, Lunggar rift; TYXR, Tangra Yum Co–Xuru Co rift; YGR, Yadong-Gulu rift; CR, Cuona rift.

The ultrapotassic lamprophyre dykes in the Bangbule area contain phenocrysts of clinopyroxene, phlogopite, and feldspar that exhibit diverse compositional zonation, presenting a prime case for examining open-system magmatic processes. (This study delves into the compositional variations of the phenocrysts of several minerals and proposes a post-collisional magma mixing model to account for the petrogenesis and diversity of the Bangbule lamprophyre in the southern Tibetan Plateau.)

2. Geological Setting

The Tibetan Plateau consists of several E-W trending terranes [21] from north to south, including the Songpan–Ganzi complex, Qiangtang terrane, Lhasa terrane, and Himalayan orogenic belt, separated by the Jinsha (JSSZ), Bangong-Nujiang (BNSZ), and Indus–Yarlung Zangbo (IYZSZ) suture zones [27] (Figure 1b). As a key tectonic unit of the Himalayan–Tibetan Plateau orogen, the Lhasa terrane is divided into northern (NL), central (CL), and southern (SL) subterranes, separated by the Shiquan River–Nam Tso Mélange Zone (SNMZ) and Luobadui–Milashan Fault (LMF), respectively [25,28] (Figure 1b).

The northern Lhasa terrane comprises Middle Triassic–Cretaceous sedimentary rocks, extensive Early Cretaceous volcanic rocks and volcano-sedimentary strata, and Cretaceous granite batholiths [25,28,29]. The central Lhasa terrane consists of the Precambrian crystalline basement, Cambrian–Permian sedimentary rocks, Late Jurassic–Early Cretaceous volcanic rock-bearing sedimentary strata [25,30], and localized Mesozoic and Cenozoic metamorphic rocks [31,32]. The southern Lhasa terrane is characterized by the juvenile crust and minor Precambrian crystalline basement, predominantly composed of Cretaceous-Tertiary Gangdese batholiths, Paleogene Linzizong volcanic succession, and some Triassic-Cretaceous volcanic-sedimentary rocks exposed in the eastern region [28], (Figure 1b).

Cenozoic ultrapotassic rocks in the western central Lhasa terrane are prominently exposed in the Gar, Xungba, Sailipu, Zhabuye, Tangra Yumco, Xuru Co, and Chazi areas (Figure 1b), forming a linear belt north of the IYZSZ [26]. Ultrapotassic mafic magmatism within the Lhasa terrane mainly occurs west of 87° E, while more evolved potassic magmatism is distributed throughout the terrane [26]. These ultrapotassic rocks are developed in the Miocene, which is consistent with the time of the post-collision extension (N-S rifting or graben) of the Lhasa terrane [24,33,34]. The age of the ultrapotassic magmatism ranges from 25 to 8 Ma and shows a decreasing trend from north to south and from west to east [26] (Figure 1c,d).

The ultrapotassic rocks comprise both extrusive and intrusive facies. The extrusive facies, in the form of lava, unconformably overlie Meso/Paleo-Cenozoic volcanic sedimentary strata [35]. The intrusive bodies emplaced into Gangdese granite batholiths, Linzizong volcanic succession, and underlying sedimentary formations as dykes. Mantle-derived xenocrysts and xenoliths composed of olivine, orthopyroxene, clinopyroxene, and spinel are found in the ultrapotassic volcanic fields of Xungba and Sailipu [24,35]. Additionally, crustal xenoliths are prevalent in the ultrapotassic volcanic rocks of the central Lhasa terrane [36].

The Bangbule Zn-Pb skarn deposit is located in the western portion of the N-S-trending Lunggar rift, near the Sailipu basin (Figure 1b). Lamprophyre samples in this study were all collected from drill cores in this deposit. Stratigraphic units exposed in this area include the Middle-Upper Devonian Chaguoluoma Formation, the Late Carboniferous-Early Permian Laga Formation, and the Quaternary (Figure 2a). The Chaguoluoma Formation in the Bangbule area primarily consists of limestone, which has been largely metamorphosed into marble. The Laga Formation is characterized by clastic rocks, including siltstone, slate, pebbled siltstone, and pebbled sandstone. Quartz porphyry, occurring as stocks or veins, typically intrudes into limestone and captures wall rocks [37]. The lamprophyre dykes were emplaced into quartz porphyry and carbonate rocks (Figures 2b and 3a), with contact zones frequently fragmented, indicating lamprophyre intrusion into the wall rocks (Figure 3a).

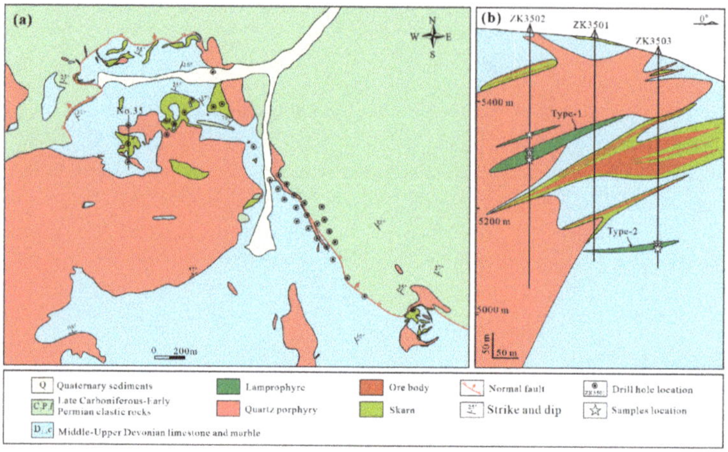

Figure 2. Geological map of the Bangbule deposit (**a**) and the No. 35 geological cross section of the deposit (**b**).

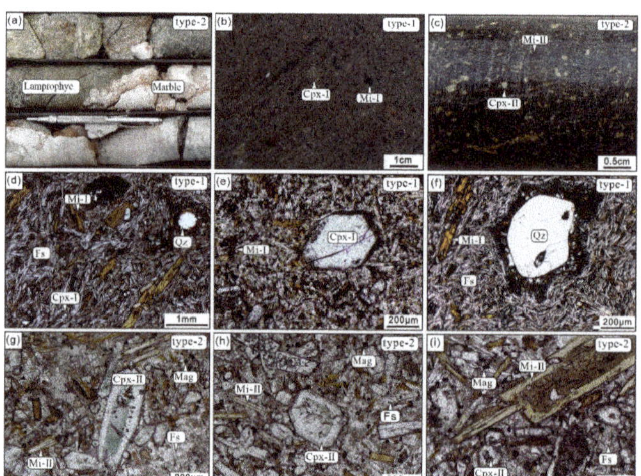

Figure 3. Photomicrographs showing the petrographic characteristics of the Bangbule lamprophyres. (**a**) Lamprophyre intruded into marble, from drillhole ZK3503 (351 m). (**b**) Hand-specimen photograph of type-1 lamprophyre from drillhole ZK3502 (185.5 m). (**c**) Hand-specimen photograph of type-2 lamprophyre from drillhole ZK3503 (347.5 m). (**d**) Euhedral platy biotite phenocrysts in lamprophyre with quartz xenocrysts, plane-polarized light. (**e**,**f**) Clinopyroxene and biotite phenocrysts in a groundmass of feldspar. Clinopyroxene has a dominating green core and a thin colorless rim, plane-polarized light. (**g**,**h**) Clinopyroxene and phlogopite phenocrysts in lamprophyre, clinopyroxene consist of a green core and a colorless surrounding mantle, plane-polarized light. (**i**) Mica shows a zonation characteristic in a plane-polarized light that consists of a darker core and a lighter rim. Abbreviations: Cpx, clinopyroxene; Mi, mica; Fs, feldspar; Mag, magnetite; Qz, quartz.

3. Petrography

The selected lamprophyres were fresh, unaltered, and unmetamorphosed. Based on the differences in host rocks and mineral compositions, the Bangbule lamprophyre can be divided into type-1 and type-2 (Figure 3b,c).

Type-1 is porphyritic, with phenocrysts of mica (biotite, Mi-I) and minor clinopyroxene (Cpx-I). The mica, the sole hydrous mineral phase, typically appears as euhedral platy phenocrysts or microphenocrysts, with some displaying zoning (Figure 3e). Some clinopyroxene crystals exhibit a dominant green core and a thin colorless rim, while others merely display an irregular green core. The groundmass of type-1 lamprophyres mainly consists of K-feldspar, alongside accessory apatite and Fe-Ti oxides (Figure 3d,e). Notably, quartz xenocrysts up to 2 mm in size were observed in this type and show typically melting corrosion structure (Figure 3f).

Type-2 lamprophyre has abundant lath-shaped phenocrysts of mica (phlogopite and annite, Mi-II) and clinopyroxene (Cpx-II). The clinopyroxene phenocrysts are euhedral to subhedral and comprise a green core and a colorless mantle rim under the microscope. The green cores often have irregular or embayed shapes (i.e., resorption/dissolution textures) and sharp boundaries with their mantles/rims (Figure 3g,h). The dark grey annite core is enclosed by the light brown phlogopite rim (Figure 3i). K-feldspar forms microcrystalline laths and constitutes the groundmass with sparse apatite and Fe-Ti oxides. In comparison, the abundance of K-feldspar as groundmass in type-2 lamprophyres is much lower than in type-1.

4. Analytical Methods

4.1. Whole-Rock Geochemical Analysis

Major and trace element analyses of the samples were conducted at the Beijing Geo-Analysis Co., Ltd., China. First, the samples were finely crushed to <200 mesh using an agate mill. After ultrasonic cleaning in distilled water, they were dried and ground into powders. Major elements were quantified using X-Ray fluorescence spectrometry (XRF), achieving an accuracy better than 3%. The compositions of trace elements, including rare earth elements (REEs), in the whole-rock samples were determined using an Agilent-7900 inductively coupled plasma mass spectrometer (ICP-MS) manufactured by Agilent Technologies, Santa Clara, CA, USA, with an estimated analytical accuracy of <10%. Loss on ignition (LOI) was measured as the weight loss of the samples after they were baked for one hour at a constant temperature of 1000 °C in a muffle furnace. Whole-rock fluorine (F) concentration was analyzed using the ion-selective electrode method detailed by He et al. [37].

4.2. Whole-Rock Sr–Nd Isotopic Analysis

Whole-rock Sr and Nd isotopic compositions were analyzed using a NEPTUNE Plus multi-collector inductively coupled plasma mass spectrometer (MC-ICP-MS) manufactured by Thermo Fisher Scientific, Waltham, MA, USA. The measurements were conducted at Beijing GeoAnalysis Co., Ltd., China, with the instrument operating in static mode. Before sample analysis, the parameters of the Neptune Plus MC-ICP-MS, including the plasma portion (e.g., the torque tube position and carrier gas flow rate) and ion lens settings, were optimized using NIST 987 (Sr 200 μg/L) and Thermo Fisher Tuning Fluid (Nd 200 μg/L) standard solutions to maximize the sensitivity. After chemical separation, the samples were introduced into the mass spectrum with 2% HNO_3 using the free atomization method, ensuring that the signal intensity for ^{88}Sr and ^{146}Nd was about 8 V (corresponding to Sr and Nd concentrations of approximately 200 μg/L). The injection system was flushed with

2% HNO_3 solution prior to analyzing subsequent samples. The raw data of isotopic ratios were corrected for mass fractionation by normalizing to $^{88}Sr/^{86}Sr$ = 8.375209 for Sr and $^{146}Nd/^{144}Nd$ = 0.7219 for Nd, following the exponential law. The initial $^{87}Sr/^{86}Sr$ and $^{143}Nd/^{144}Nd$ ratios were calculated based on the $^{40}Ar/^{39}Ar$ dating results.

4.3. Mica 40Ar-39Ar Dating

Mica (phlogopite) grains from type-2 lamprophyres were crushed and sieved to 40–80 μm and then meticulously handpicked under a binocular microscope for >99% purity at the Analytical Laboratory of the Hebei Institute of Regional Geology and Mineral Resources Survey, Langfang, China. Following ultrasonic cleaning, the samples were sealed in a quartz tube and wrapped with 1 mm thick cadmium skin for irradiation in an HFETR reactor for 24 h at the Chinese Institute of Atomic Energy, Beijing, China. After cooling to safe radiation levels, $^{40}Ar/^{39}Ar$ dating via stepwise heating was conducted at the Analytical Laboratory Beijing Research Institute of Uranium Geology, Beijing, China. During analysis, gas was released from samples through stepwise heating (800 to 1400 °C), and the gas released from each stage was purified using an NG PREP SYSTEM equipped with a Zr-Al pump (ThermoFisher Scientific, Waltham, MA, USA). The Ar isotopic composition was measured using an Argus VI noble gas mass spectrometer (ThermoFisher Scientific, Waltham, MA, USA). Argon isotopic interferences from K and Ca were assessed through irradiation of K_2SO_4 and CaF_2 pure salts, yielding correction factors of $(^{36}Ar/^{37}Ar)_{ca}$ = 0.0008057, $(^{39}Ar/^{37}Ar)_{ca}$ = 0.000713, and $(^{40}Ar/^{39}Ar)^{k}$ = 0.00213. The neutron flux gradient was determined using Fangshan biotite ZBH-25 (age: 132.7 ± 1.2 Ma, K content: 7.6%) in the quartz tube, with the irradiation constant of the sample calculated at J = 0.0005914. The mass discrimination factor of the mass spectrometer was derived from atmospheric argon's isotopic composition, with the ^{40}K decay constant set at $\lambda = 5.543 \times 10^{-10}\ a^{-1}$ [38]. Data reduction and plotting were performed using ArArCalc software© v2.4 [39]. Plateau ages distributed over at least 3 contiguous stages, with errors within 1σ.

4.4. Mineral Chemical Analysis and X-Ray Elemental Maps

Composition of pyroxene, mica, and feldspar were analyzed on polished thin sections using electron probe microanalysis (EPMA) technique at the Guangzhou Institute of Geochemistry, Chinese Academy of Sciences. The experiments were conducted using the JXA-8230 instrument manufactured by JEOL Ltd, Akishima, Japan. Backscattered electron images assisted in targeting specific analysis sites on mineral specimens. These microchemical analyses were conducted under an acceleration voltage of 15 kV and an electron beam current of 20 nA, with a beam diameter of 2 μm. Element counting lasted for 10–20 s at each peak and 5–10 s at each of two background positions. Inter-elemental effect corrections were applied using a ZAF (Z: atomic number; A: absorption; F: fluorescence) procedure.

High-resolution 2D elemental X-Ray mapping was generated for one mica and two clinopyroxene crystals and one feldspar glomerocryst. The mapping was carried out using the EPMA-1720 (manufactured by Shimadzu Corporation, Kyoto, Japan) at the Institute of Mineral Resources, Beijing General Research Institute of Mining and Metallurgy. The operational conditions included an acceleration voltage of 15 kV, a beam current of 20 nA, and a beam size of 1–5 μm. The sampling time was set at 12 ms per point. Na Ka, Mg Ka, and Al Ka were analyzed using a TAP crystal, while Ti Ka and Fe Ka were analyzed using a LIF crystal. K Ka, Ca Ka, Sr La, and Ba La were analyzed using a PET crystal, Shimadzu EPMA-1720.

5. Results

5.1. Major and Trace Elements

The data of the whole-rock major and trace elements of the studied lamprophyres are listed in Supplementary Table S1. Type-1 lamprophyres exhibit notably higher SiO_2 (58.80–59.80 wt.%), Al_2O_3 (14.51–14.67 wt.%), and $Na_2O + K_2O$ (10.08–10.22 wt.%) contents but lower $^{T}Fe_2O_3$ (4.74–5.15 wt.%), CaO (3.52–4.00 wt.%), and MgO (3.70–4.06 wt.%) contents and $Mg^{\#}$ values (59–63) than type-2 lamprophyres (SiO_2: 48.29–49.18 wt.%; Al_2O_3: 10.66–10.95 wt.%; $Na_2O + K_2O$: 6.90–7.41 wt.%; $^{T}Fe_2O_3$: 6.03–6.22 wt.%; CaO: 11.05–11.53 wt.%; MgO: 6.88–7.72 wt.%; $Mg^{\#}$: 69–71). The whole-rock geochemical data indicate that all the samples are ultrapotassic according to Foley's criteria (Figure 4). The LOI and CaO content of type-2 lamprophyres is significantly higher than that of type-1, consistent with the dominance of hydrous minerals (mica) and the influence of carbonate wall rocks. There are negative correlations on MgO vs. Al_2O_3 plots, but positive correlations for plots of MgO vs. $^{T}Fe_2O_3$, Ba vs. Sr, and MgO vs. Cr (Figure 5).

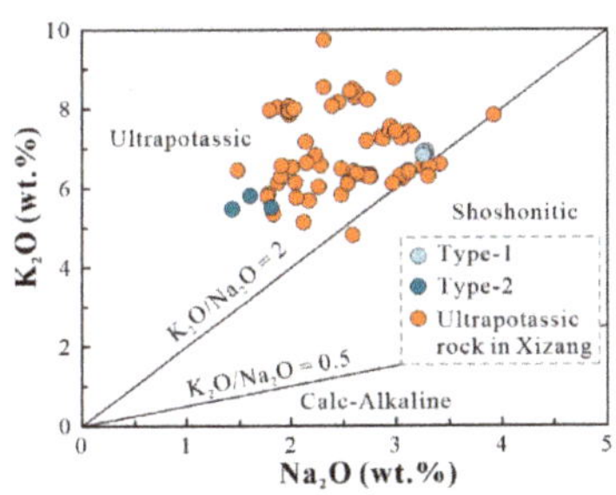

Figure 4. Na_2O (wt.%) vs. K_2O (wt.%) diagram of studied lamprophyres classifying them as ultrapotassic based on major oxide geochemistry. The fields demarcating ultrapotassic from shoshonitic and shoshonitic from calc-alkaline are based on a K_2O/Na_2O ratio of 2.0 and 0.5, respectively [40].

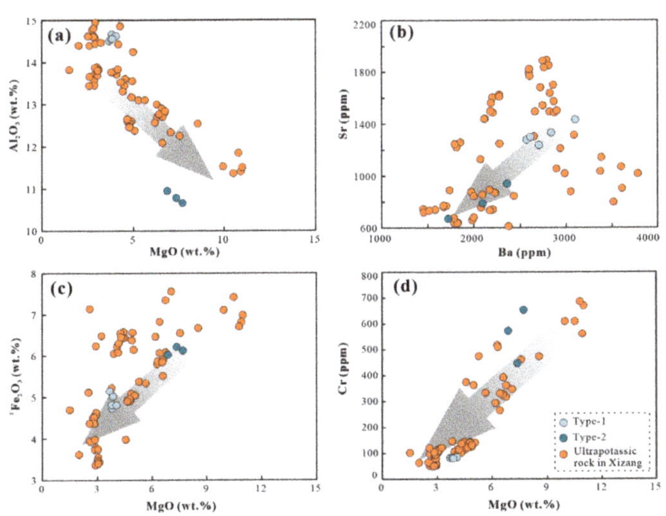

Figure 5. Bivariate plots (**a**) Al_2O_3 (wt.%), (**b**) Ba (ppm) vs. Sr (ppm), (**c**), MgO (wt.%) vs. $^{T}Fe_2O_3$ (wt.%), (**d**) MgO (wt.%) vs. Cr (ppm) for lamprophyres showing good fractionation trends. The arrows indicate the direction of fractional crystallization.

The two types of lamprophyres in Bangbule area differ markedly in Cr and Ni contents. Type-1 lamprophyres have much lower Cr and Ni contents (Cr content: 81–86 ppm, Ni content: 48–51 ppm) than type-2 lamprophyres (Cr content: 477–654 ppm, Ni content: 250–356 ppm). In the chondrite-normalized REE diagram (Figure 6a), both lamprophyre types show significant enrichments in LREEs relative to HREEs overall. Nevertheless, type-1 lamprophyres are more enriched in LREEs than type-2, with La_N/Yb_N ratios ranging from 36.82 to 38.85 and from 22.32 to 24.71, respectively. In the primitive mantle-normalized trace element pattern (Figure 6b), both types of lamprophyres are enriched in large-ion lithophile elements (LILEs) compared to high-field-strength elements (HFSEs), with variations in some trace elements. Notably, a significant positive Pb anomaly and negative Nb, Ta, and Ti anomalies can be observed.

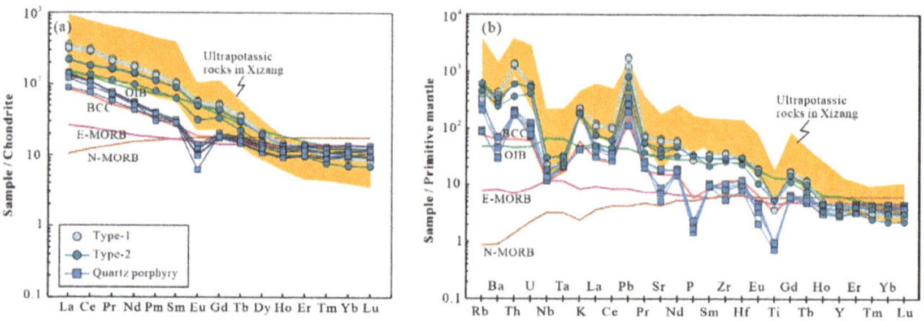

Figure 6. Chondrite-normalized REEs patterns (**a**) and primitive mantle-normalized trace-element patterns (**b**) of the lamprophyres from Bangbule. Primitive mantle and chondrite normalizing values and the OIB, E-MORB, and N-MORB [41]. Data of BCC (bulk continental crust) [42], quartz porphyry [37], and ultrapotassic rocks from southern Tibet [43] are shown for comparison.

5.2. ^{40}Ar-^{39}Ar Geochronology

A stepwise thermal release analysis of mica was performed in nine heating steps, ranging from 800 °C to 1400 °C. The analytical results are presented in Supplementary Table S2. The mica sample exhibits a roughly undisturbed age spectrum, yielding plateau ages of 29.5 ± 0.2 Ma (Figure 7), isochron ages of 29.5 ± 0.2 Ma with initial $^{40}Ar/^{36}Ar$ ratios of 270 ± 43 (Figure 7), and inverse isochron ages of 29.6 ± 0.2 Ma with initial $^{40}Ar/^{36}Ar$ ratios of 262 ± 38. Compared with other ultrapotassic rocks in Xizang, the Bangbule lamprophyre is slightly older (Figure 1c).

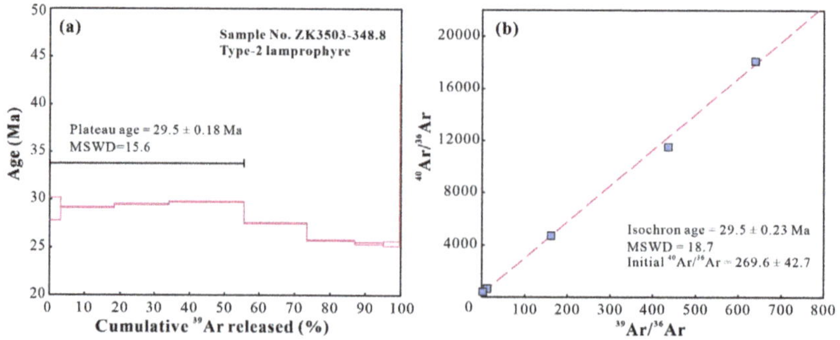

Figure 7. $^{40}Ar/^{39}Ar$ geochronological diagrams showing the plateau age (**a**) and isochron age (**b**) of the mica (phlogopite) from type-2 lamprophyre. The red solid line indicate yielding plateau ages, and the blue squares indicate the age data corresponding to the ^{39}Ar released during the heating process.

5.3. Whole-Rock Sr–Nd Isotopes

The initial Sr–Nd isotopic ratios of the lamprophyre samples were corrected to an emplacement age of 30 Ma. These lamprophyre samples have high initial Sr and low Nd isotopic ratios. The initial $^{87}Sr/^{86}Sr$ ratios varies from 0.713936 to 0.716999, and the consistent initial $^{143}Nd/^{144}Nd$ isotopic ratios (0.511991–0.512012, Figure 8, Supplementary Table S3). The remarkable consistency in the initial $^{143}Nd/^{144}Nd$ ratios suggests a similar source for all the samples. Depleted mantle Nd model ages (T_{DM}) ranging from 1.73 to 1.92 Ga, which is consistent with those of ultrapotassic rocks from the Lhasa terrane previously reported [24–26] (1.33–3.04 Ga).

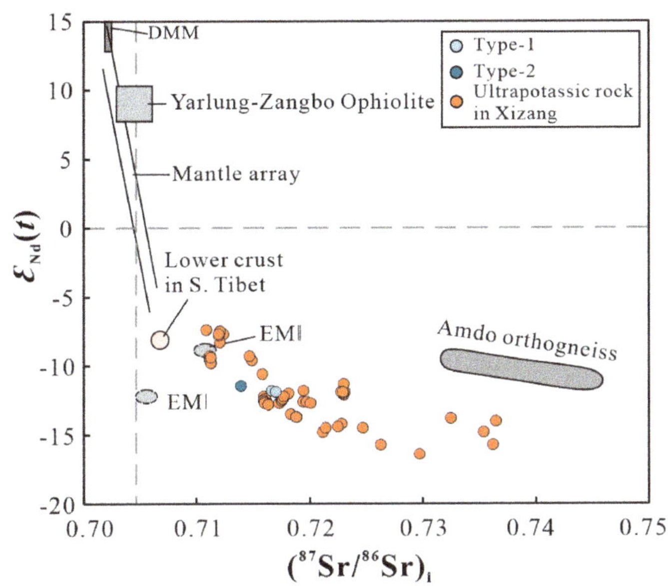

Figure 8. The diagram of $(^{87}Sr/^{86}Sr)_i$ vs. $\varepsilon_{Nd}(t)$ for the Bangbule lamprophyres and post-collisional ultrapotassic rocks in southern Tibet. Data sources: Mantle reservoirs, DMM, EMI, and EMII [44]. Yarlung Zangbo ophiolite [43–47]. Lower crust of south Tibet [45]. Amdo orthogneiss from (i.e., the upper continental crust of Tibet) [46]. Sr–Nd isotopes of ultrapotassic rocks [24–26].

5.4. Mineral Compositions

Clinopyroxene in the two types of lamprophyres differ in chemical compositions, as shown by electron probe microanalysis (Supplementary Table S4). Cpx-I has significantly lower $Mg^{\#}$ values (63–66) but higher Al_2O_3/TiO_2 ratios (8.33–18.34) compared to both zoned and unzoned Cpx-II ($Mg^{\#}$: 76–90, Al_2O_3/TiO_2: 1.80–6.78). They also show differences in other major elements like SiO_2, MnO, and Na_2O. The clinopyroxene are diopsidic, with $Wo_{42-43}En_{36-37}Fs_{20-22}$ for type-1 and $Wo_{41-45}En_{42-52}Fs_{6-14}$ for type-2 (Figure 9a). The green-core clinopyroxene exhibits reverse zoning, with cores characterized by higher Fe, Al, and Na contents but lower Mg content than their mantles, which is in contrast to the normal zoning of clinopyroxene (higher $Mg^{\#}$ in core), as indicated by elemental maps (Figure 10). Furthermore, Cpx-I's composition roughly matches Cpx-II's core composition.

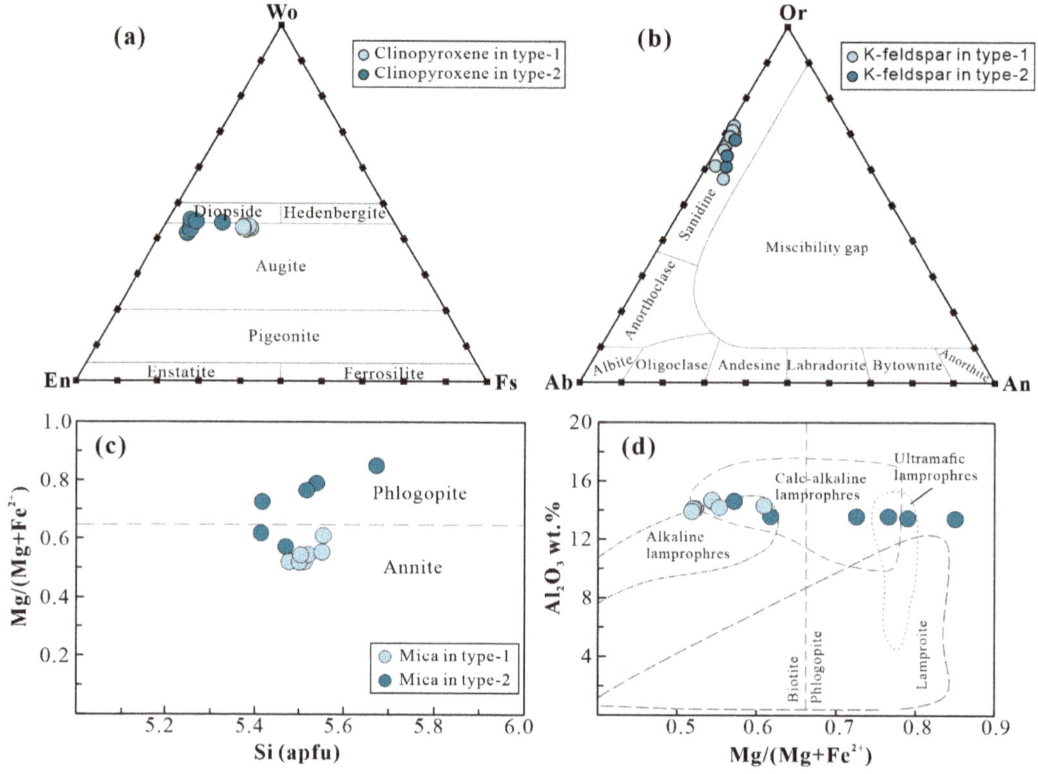

Figure 9. (**a**) En–Wo–Fs classification diagram for clinopyroxene of Bangbule lamprophyres [47]. (**b**) Triangular plot for classification of the feldspar of Bangbule lamprophyre [48]. (**c**) Mica of the studied samples on Si (apfu) vs. $Mg^{\#}$ diagram [49] and (**d**) $Mg^{\#}$ vs. Al_2O_3 diagram [50]. Samples are classified as annite and phlogopite, respectively, and showing affinity to both the global calc-alkaline and alkaline lamprophyres.

The two lamprophyre types have similar feldspar compositions. Feldspar phenocrysts of type-1 lamprophyre have Or, An, and Ab contents varies from 57.3 to 72.1, 1.1 to 5.9, and 26.8 to 37.0, respectively. The Or, An, and Ab contents of type-2 are 49.2–68.2, 3.3–31.3, and 19.5–34.4, respectively (Supplementary Table S5). Except for one analysis spot, all analysis spots are classified as orthoclase-sanidine on the An-Ab-Or diagram (Figure 9b), indicating a predominance of alkali-feldspar over plagioclase. Alkali-feldspars in type-1 lamprophyres show zonation, with cores enriched in K, Sr, and Ba but depleted in Na and Ca compared to their rims (Ba-rich K-feldspar, Supplementary Figure S1).

Mica from the two lamprophyre types also varies in chemical composition; III contains 11.52–13.31 wt.% MgO, 5.23–19.22 wt.% FeO, and 4.18–4.75 wt.% TiO_2 and has an $Mg^{\#}$ of 52–61 (average = 54). In contrast, IIII contains 12.94–21.2 wt.% MgO, 6.60–17.29 wt.% FeO, and 3.06–6.44 wt.% TiO_2 and has an $Mg^{\#}$ of 57–85 (average = 72) (Supplementary Table S6). The two mica types are classified as annite and phlogopite on the Si (apfu) vs. $Mg/(Mg + Fe^{2+})$ and $Mg/(Mg + Fe^{2+})$ vs. Al_2O_3 diagrams (Figure 9c,d), respectively. Mi-II's cores exhibit higher iron and titanium contents but lower magnesium contents than its rims (Supplementary Figure S2), with no notable differences in other major elements (e.g., Al and Na).

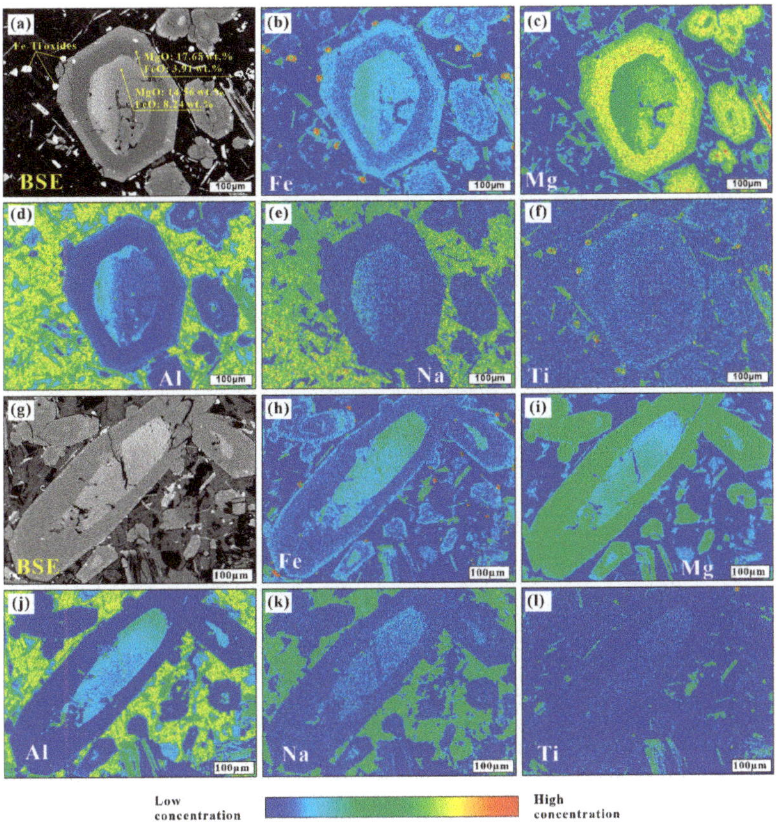

Figure 10. (**a**,**g**) Backscattered electron image of clinopyroxene phenocryst of type-2 lamprophyre. (**b**–**f**,**h**–**l**) EPMA elemental X-Ray maps of the studied Bangbule euhedral clinopyroxene phenocryst of type-2 lamprophyre. The grayscale and false-color compositional images illustrate the distinctive differences between the core and mantle in major elements.

5.5. T-P Conditions

In this study, we estimated the pre-intrusive temperatures and pressures of magma reservoirs beneath Bangbule using published clinopyroxene-only [51] and biotite-only thermobarometers [52]. P-T estimates were made using the clinopyroxene-only thermobarometer (Equations 32a and 32d) for lamprophyres. See Li and Zhang [52] for more details.

The calculated crystallization temperatures and pressures of Cpx-I are (1190–1220 °C) and 12.6–15.5 kbar (corresponding to a paleodepth of 47–58 km, avg. = 52; assuming 3.75 km/kbar as the pressure gradient; Table 1), whereas those of Cpx-II have ranges of (1220–1270 °C) and 4.4–14.1 kbar (16–53 km paleodepth). The pressure of the core of green-core clinopyroxene is high (14.1 kbar, 53 km palaeodepth), while other pressure values are relatively low and in a narrow range (4.4–9.5 kbar, 16–36 km palaeodepth, avg. = 24). These results are consistent with those of Li et al. [36]. Mi-I has temperatures of (820–850 °C) and pressures of 2.9–4.4 kbar (11–16 km palaeodepth, avg. = 15), and Mi-II has temperatures of (850–990 °C) 851–992 °C and pressures of 4.2–5.8 kbar (16–22 km palaeodepth, avg. = 18; Table 1). Overall, Cpx-I and Mi-I temperatures are lower than Cpx-II and Mi-II, respectively.

Table 1. Statistical Table of Clinopyroxene—Biotite Core—Rim Analysis Results.

Spot Number	Mineral		Crystal Number	Location	P/Kbar	Depth (km)	T/°C
ZK3502-185.45-C1-1	Clinopyroxene	Cpx-I	phenocryst 1	core	13.4	50	1207
ZK3502-185.45-C1-2			phenocryst 1	rim	14.6	55	1216
ZK3502-216.2-C2-01			phenocryst 2	core	15.5	58	1206
ZK3502-216.2-C3-02			phenocryst 3	core	14.4	54	1200
ZK3502-216.2-C3-03			phenocryst 3	core	13.2	50	1193
ZK3502-216.2-C3-04			phenocryst 3	core	13.3	50	1191
ZK3502-216.2-C1-01			phenocryst 4	rim	15.1	57	1203
ZK3502-216.2-C1-02			phenocryst 4	rim	13.2	50	1207
ZK3502-216.2-C1-03			phenocryst 4	rim	13	49	1201
ZK3502-216.2-C1-04			phenocryst 4	core	12.6	47	1195
ZK3503-347.2-C1-01		Cpx-II	phenocryst 1	core	14.1	53	1272
ZK3503-347.2-C1-02			phenocryst 1	rim	7.6	29	1244
ZK3503-347.2-C1-03			phenocryst 2	rim	7	26	1259
ZK3503-347.2-C1-04			phenocryst 2	core	9.5	36	1269
ZK3503-347.2-C2-03			phenocryst 3	core	5.3	20	1252
ZK3503-347.2-C2-04			phenocryst 3	rim	5.4	20	1248
ZK3503-347.2-C3-01			phenocryst 4	rim	6.1	23	1257
ZK3503-347.2-C3-02			phenocryst 4	rim	4.4	17	1220
ZK3502-185.45-2-01	Biotite	Mi-I	phenocryst 1	core	3.6	14	835
ZK3502-185.45-2-02			phenocryst 1	rim	4	15	841
ZK3502-185.45-3-01			phenocryst 2	core	3.7	14	839
ZK3502-185.45-3-02			phenocryst 2	rim	2.9	11	833
ZK3502-216.2-C2-02			phenocryst 3	core	4.4	17	832
ZK3502-216.2-C2-03			phenocryst 3	core	4.2	16	830
ZK3502-216.2-C2-04			phenocryst 3	core	4.2	16	832
ZK3502-216.2-C3-01			phenocryst 4	rim	4.3	16	830
ZK3503-347.2-C1-05		Mi-II	phenocryst 1	core	4.2	16	992
ZK3503-347.2-C1-06			phenocryst 2	core	4.3	16	965
ZK3503-347.2-C1-07			phenocryst 2	rim	5.4	20	963
ZK3503-347.2-C2-05			phenocryst 3	core	5.2	20	971
ZK3503-347.2-C3-03			phenocryst 4	rim	5.8	22	884
ZK3503-347.2-C3-04			phenocryst 5	rim	4.2	16	851

6. Discussion

6.1. Crustal Contamination

It is imperative to evaluate potential crustal contamination effects on melt compositions before discussion mantle source characteristics and petrogenetic processes. Critical petrogenetic evidence includes: (1) entrainment phenomena of quartz xenocrysts within type-1 lamprophyre dykes, demonstrating crustal interaction during magma ascent; and (2) anomalously elevated LOI in type-2, suggesting the influence of contamination by volatile-rich crustal materials. Therefore, the influence of crustal contamination cannot be neglected.

Firstly, the high SiO_2 content in Type-1 lamprophyre (~59 wt.%) is likely attributed to crustal contamination during magma intrusion into quartz porphyry. This conclusion is supported by petrographic evidence of quartz xenocrysts with resorbed textures (Figure 3f). In contrast, type-2 lamprophyres display significantly high CaO contents (~11 wt.%) compared to type-1 (~4 wt.%), potentially reflecting limestone contamination during emplacement. The limited variation of Nb/Y (0.78–1.11), Th/La (0.89–1.51) and Nb/Ta (13.8–18.9) and the enrichment of incompatible elements and light rare earth elements (LREEs) relative to both continental crust and quartz porphyry (Figure 6) indicate that crustal contamination would have a limited effect on the trace elements compositions of these rocks. Isotopic constraints further corroborate limited crustal influence, with both lamprophyre types exhibiting restricted $^{87}Sr/^{86}Sr$ (0.7136–0.7170) and $^{143}Nd/^{144}Nd$ (0.51201–0.51204) ranges that align with Tibetan ultrapotassic rocks (Figure 8). Collectively, these findings demonstrate that crustal contamination exerts substantial influence on major element geochemistry but shows negligible effects on trace element distributions (particularly high field strength elements) and radiogenic isotope systematics in Bangbule lamprophyres.

6.2. Nature of the Source Region

The MgO-Al_2O_3 negative covariation, Sr-Ba positive correlation, and week Eu anomalies (Eu/Eu* = 0.65–0.76) (Figure 5c,d and Figure 6a) suggest that the plagioclases were not a major residual phase in the magma source region of Bangbule lamprophyres. These lamprophyre samples exhibit diagnostic ultrapotassic signatures characterized by elevated K_2O contents and pronounced large-ion lithophile element (LILE) enrichment (the high Rb/Sr (0.16–0.47) but low Ba/Rb (5.63–13.47) ratios in both type-1 and type-2), indicative of derivation from a hydrous, volatile-bearing, K-rich mantle source containing mineral phases such as phlogopite or amphibole [53]. The high Rb/Sr (0.16–0.47) but low Ba/Rb (5.63–13.47) ratios in both type-1 and type-2 (Figure 11a) suggest the presence of phlogopite rather than amphibole in the melting source [54]. This is also demonstrated by the enrichment of K_2O relative to Na_2O in the sample, as partial melting of the phlogopite-bearing mantle generates ultrapotassic melts with $K_2O/Na_2O > 2$ [24].

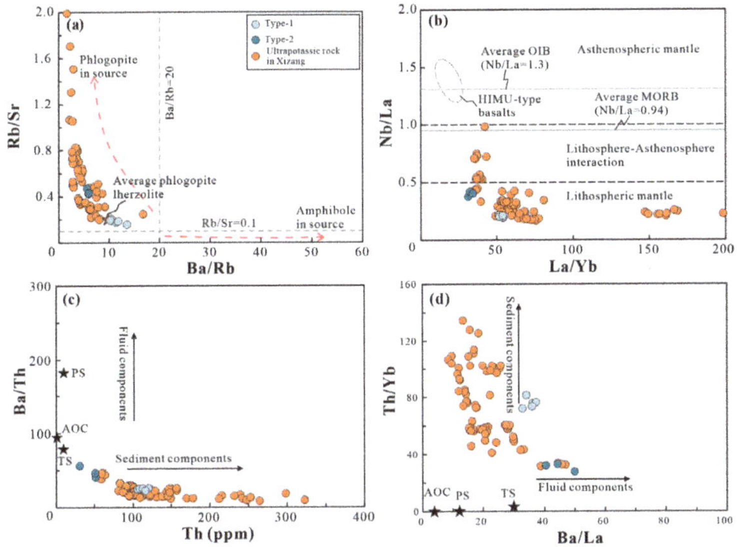

Figure 11. (**a**) Ba/Rb vs. Rb/Sr diagram [54]. Average phlogopite lherzolite [55]. The red dashed arrow indicates the source region (**b**) La/Yb vs. Nb/La variation in the studied lamprophyres showing their affinity to the lithospheric mantle [55]. The average Nb/La values for MORB and OIB [41]. (**c**) Th vs. Ba/Th, and (**d**) Ba/La vs. Th/Yb diagrams. The compositions of altered oceanic crust (AOC), terrestrial sediments (TS), and pelagic sediments (PS) [56].

The analyzed samples exhibit enrichment in incompatible elements such as LREEs and LILEs and depletion in HFSEs (especially Nb, Ta, and Ti) relative to neighboring trace elements (Figure 6). Generally, the decompressional melting of the asthenospheric mantle produces normal mid-oceanic ridge basalt (MORB) depleted in LREEs and LILEs [57] and positive Nb-Ta-Ti anomalies, contrasting with the characteristics of Bangbule lamprophyres (Figure 6). The low Nb/La (0.19–0.42) and La/Yb (31–54) ratios in the samples indicate an affinity for lithospheric mantle-derived melts [58] (Figure 11b), suggesting a lithospheric mantle origin of the Bangbule lamprophyres, which is supported by the high Ni

(250–356 ppm) and Cr (477–654 ppm) concentrations and high $Mg^{\#}$ (69–71) in type-2 lamprophyre. Both type-1 and type-2 lamprophyre have enriched Sr-Nd isotope composition with old T_{DM} ages (from 1.73 to 1.92 Ga), which are similar with other ultrapotassic rocks in Lhasa terrane that source from the metasomatic mantle during Indian continental subduction [24,59]. Given their high abundance of Ba, K, Pb, Sr, Li, and LREEs and the high Ba/La and Th/Yb ratios in the lamprophyre dykes (Figure 11c,d), it is more likely that the Bangbule lamprophyre originated from a metasomatic mantle with involvement of sediment melt-like components during Indian–Asia collision [60,61].

6.3. Petrogenesis

6.3.1. Fractional Crystallization

The studied lamprophyres exhibit geochemical co-variation trends with other ultrapotassic rocks in Lhasa terrane (Figure 5), supporting fractional crystallization as a key petrogenetic process [62]. Type-1 lamprophyres show significantly lower TiO_2 and P_2O_5 concentrations compared to type-2 counterparts, which can be attributed to early-stage fractionation of Fe-Ti oxides and apatite, respectively (Supplementary Table S1). The depleted Cr (81–86 ppm) and Ni (48–51 ppm) contents in type-1 lamprophyres relative to primitive mantle values suggest fractional crystallization of spinel and olivine during magma evolution. The positive correlations between MgO and $^{T}Fe_2O_3$ contents, as well as between Ni and Cr (Figure 5), suggest that the fractionation phases were dominated by mafic minerals such as olivine and clinopyroxene. This aligns with the abundance of clinopyroxene phenocrysts observed in both types of lamprophyres. Furthermore, the lower Mg# values (63–66) coupled with high Al_2O_3 concentrations of the Cpx-I phenocryst in type-1 lamprophyre (Supplementary Table S4) provide direct mineralogical evidence for advanced crystal fractionation in the parental magmas.

6.3.2. Magma Replenishment and Mixing

Mineral zoning typically correlates with magma composition, physicochemical condition changes, and magmatic processes like fractional crystallization, mixing, and recharge [63]. In this study, feldspar, clinopyroxene, and mica crystals exhibit clear compositional and textural zoning (Figure 10, Supplementary Figure S1). Pyroxene reverse zoning can be attributed to (1) magma mixing between co-genetic or distinct batches, (2) pulsed magma replenishment from the source, and (3) diffusion-related re-equilibration [63–65].

The following evidence indicates that Type-1 lamprophyre's magma was replenished with a more mafic co-genetic magma: (1) the consistent Sr-Nd isotopic characteristics of type-1 and type-2 lamprophyres suggest a common source region (Figure 8), (2) resorption of feldspar and clinopyroxene cores in Type-1 lamprophyres indicates entrapment by a high-temperature, primitive melt, (3) sharp contrasts in Ba, K, Na, Sr, and Ca contents between cores and rims (Supplementary Figure S1) rule out diffusion-related re-equilibration, which would cause gradual compositional changes, (4) reversely zoned clinopyroxene crystals with higher $Mg^{\#}$ in the rims indicate compositional similarity of the magmas [66,67], and (5) the occurrence of micro-ilmenite enclosed by magnetite (Figure 10b,f,h,l) suggests changes in melt composition [68].

Considering the distinct petrographic features and chemical variations in Fe, Mg, Cr, and Ni, the heterogeneity of type-1 and type-2 lamprophyric melts in the Bangbule area is likely caused by the replenishment of a mafic magma and associated magma mixing.

Therefore, mineral textures and geochemical evidence indicate that the Bangbule lamprophyre underwent magma replenishment and mixing during its evolution. Barometric calculations reveal that the rim of green-core clinopyroxene (Cpx-2) crystallized at a depth of ~29 km, while its core formed ~53 km below the paleo-surface, suggesting a two-stage magmatic evolution. Combined with prior discussions, we propose the following sequence: Initially, the early-stage Bangbule lamprophyre magma reservoir experienced crystallization differentiation of mafic minerals (e.g., pyroxene and olivine) at ~52 km depth before ascending to a shallow level. Subsequently, a more mafic magma with high $Mg^{\#}$ values intruded into the shallow magma reservoir at ~29 km depth, triggering magma mixing. During this process, early-formed minerals underwent resorption and developed mafic overgrowths [69], resulting in reverse-zoned phenocrysts. As the hybrid magmas continued to evolve, K-feldspar mantles or rims formed on the earlier resorbed, Ba-rich cores. In the shallow magma reservoir (the final emplacement site), mineral phases underwent gravitational sorting and accumulation, with interstices between cumulate minerals filled by late-stage K-feldspar and groundmass. Ultimately, the emplacement of this heterogeneous magma in the Bangbule area produced both type-1 and type-2 lamprophyres.

6.4. Implications for Ore Genesis of Porphyry Deposit in Gangdese

Previous studies have proposed that the fluid exsolution from the Miocene ultrapotassic magma during the magma mixing is a key process in the generation of post-collision porphyry Cu-Au deposits in Xizang [70,71]. Zhou et al. [40] proposed that the ultrapotassic magma may have high contents of Cl, which can be added into the porphyry system through magma mixing at shallow magma chamber. Furthermore, Wang et al. [41] also argued that the injection of water-enriched ultrapotassic magma into the felsic magma has played an important role in generation of fertile magmas with high water content and high oxygen fugacity in the Gangdese belt. Hence, the ultrapotassic magma played an important role in formation of giant porphyry deposit in the post-collision setting. From preceding discussion, we propose a magma recharge and mixing model for the petrogenesis of the ultrapotassic magmatism during its ascent (Figure 12). The late stage more mafic magma generally has distinct chemical compositions with the primary ultrapotassic magma [36]. Therefore, recharge of mafic magma is also important to form the fertile ultrapotassic magma and related porphyry deposit in Gangdese belt by changing its chemical composition, water contents, and temperature. Considering that the ultrapotassic magmatism in the Lhasa terrane has wide range of formation ages (~30 to 10 Ma) even in the same place, the later formed ultrapotassic rocks may have experienced more complex recharge and mixing processes, favoring formation of porphyry deposits.

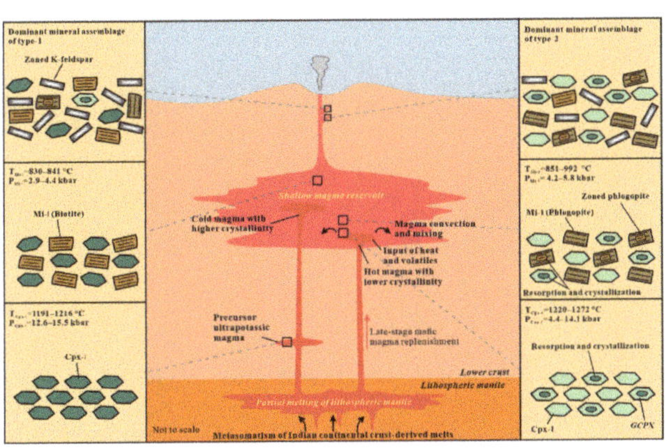

Figure 12. Schematic model of magma mixing and phenocryst evolution for the Bangbule ultrapotassic lamprophyre dykes. The black arrows indicate magma convection and mixing.

7. Conclusions

The Bangbule lamprophyre dykes can be classified into two types based on their mineralogy, mineral modes, and bulk-rock geochemistry. New mica ^{40}Ar-^{39}Ar dating of type-2 (basaltic) lamprophyres suggests their emplacement at ca. 30 Ma. The primitive ultrapotassic lamprophyre magmas originated from the partial melting of the lithospheric mantle in Tibet, which consists of phlogopite-bearing mantles. Both types likely evolved through varying degrees of fractional crystallization involving mafic mineral phases in the early stage of magma evolution. The overgrowth of mantles with high $Mg^{\#}$ values on clinopyroxene and biotite phenocrysts indicate a recharge of late-stage mafic magma. Clinopyroxene- and biotite-only thermobarometry calculations suggest a multi-level magma mixing model, in which the recharge of mafic magma changes the chemical composition, water content, and temperature of the fertile ultrapotassic magma, favoring the formation of giant porphyry deposits in the Gangdese Belt.

Supplementary Materials: The following supporting information can be downloaded at: https://www.mdpi.com/article/10.3390/min15040332/s1, Figure S1: Backscattered electron (BSE) image and EPMA elemental X-Ray maps of the studied Bangbule feldspar phenocryst of type-1 lamprophyre. The grayscale and false-color compositional images reveal the core and rim are obviously different in composition. Figure S2: Backscattered electron (BSE) image and EPMA elemental X-Ray maps of the studied mica phenocryst of Bangbule type-2 lamprophyre. The grayscale and false-color compositional images show compositional differences between the core and the rim; Table S1: Major and trace elements of the Bangbule lamprophyres. Table S2: $^{40}Ar/^{39}Ar$ stepwise heating analytical data of mica from the Type-2 lamprophyre. Table S3: Sr–Nd isotope composition of the Bangbule lamprophyres and ultrapotassic rocks from southern Tibet. Table S4: Major element compositions of the clinopyroxenes in lamprophyre. Table S5: Major element compositions of the feldspar in lamprophyre. Table S6: Major element compositions of the mica in lamprophyre. Ref. [72] is cited in Supplementary Materials.

Author Contributions: Conceptualization, Y.W. (Yong Wang) and C.H. (Chuankai He); methodology, Y.W. (Yong Wang) and C.H. (Chenyu Hong); software, C.H. (Chenyu Hong), S.X. and R.C.; validation, Y.W. (Yong Wang), Y.W. (Yongqiang Wang) and S.Z.; formal analysis, C.H. (Chenyu Hong); investigation, Y.B., P.Y. and D.C.; resources, S.Z. and H.W.; data curation, S.X. and R.C.; writing—original draft preparation, C.H. (Chenyu Hong); writing—review and editing, Y.W. (Yong Wang); visualization, C.H. (Chenyu Hong) and S.X.; supervision, Y.W. (Yong Wang) and H.W.; project administration, Y.W. (Yong Wang) and C.H. (Chuankai He). All authors have read and agreed to the published version of the manuscript.

Funding: This research was jointly supported by National Natural Science Foundation of China (Nos. 42302098, 42230813), the National Key Research and Development Project of China (No. 2022YFC2905002), and the Open Fund from SinoProbe Laboratory (Grant No.: SL202415).

Data Availability Statement: Data are contained within the article and Supplementary Materials.

Conflicts of Interest: The authors declare no conflicts of interest.

References

1. Mo, X.X.; Niu, Y.L.; Dong, G.C.; Zhao, Z.D.; Hou, Z.Q.; Zhou, S.; Ke, S. Contribution of syncollisional felsic magmatism to continental crust growth: A case study of the Paleogene Linzizong volcanic succession in southern Tibet. *Chem. Geol.* **2008**, *250*, 49–67.
2. Chen, J.S.; Huang, B.C.; Sun, L.S. New constraints to the onset of the India-Asia collision: Paleomagnetic reconnaissance on the Linzizong Group in the Lhasa Block, China. *Tectonophysics* **2010**, *489*, 189–209.
3. Gibbons, A.D.; Zahirovic, S.; Müller, R.D.; Whittaker, J.M.; Yatheesh, V. A tectonic model reconciling evidence for the collisions between India, Eurasia and intra-oceanic arcs of the central-eastern Tethys. *Gondwana Res.* **2015**, *28*, 451–492.

4. Westerweel, J.; Roperch, P.; Licht, A.; Dupont-Nivet, G.; Win, Z.; Poblete, F.; Ruffet, G.; Swe, H.H.; Thi, M.K.; Aung, D.W. Burma Terrane part of the Trans-Tethyan arc during collision with India according to palaeomagnetic data. *Nat. Geosci.* **2019**, *12*, 863.
5. Kapp, P.; DeCelles, P.G.; Gehrels, G.E.; Heizier, M.; Ding, L. Geological records of the Lhasa-Qiangtang and Indo-Asian collisions in the Nima area of central Tibet. *Geol. Soc. Am. Bull.* **2007**, *119*, 917–933.
6. Tan, X.D.; Gilder, S.; Kodama, K.P.; Jiang, W.; Han, Y.L.; Zhang, H.; Xu, H.H.; Zhou, D. New paleomagnetic results from the Lhasa block: Revised estimation of latitudinal shortening across Tibet and implications for dating the India-Asia collision. *Earth Planet. Sci. Lett.* **2010**, *293*, 396–404.
7. Hu, X.M.; Wang, J.G.; BouDagher-Fadel, M.; Garzanti, E.; An, W. New insights into the timing of the India-Asia collision from the Paleogene Quxia and Jialazi formations of the Xigaze forearc basin, South Tibet. *Gondwana Res.* **2016**, *32*, 76–92.
8. Owen, J.P. Geochemistry of lamprophyres from the Western Alps, Italy: Implications for the origin of an enriched isotopic component in the Italian mantle. *Contrib. Mineral. Petrol.* **2008**, *155*, 341–362.
9. Huang, F.; Chen, J.L.; Xu, J.F.; Wang, B.D.; Li, J. Os–Nd–Sr isotopes in Miocene ultrapotassic rocks of southern Tibet: Partial melting of a pyroxenite-bearing lithospheric mantle? *Geochim. Cosmochim. Acta* **2015**, *163*, 279–298. [CrossRef]
10. Pandey, A.; Rao, N.V.C.; Chakrabarti, R.; Pankaj, P.; Pandit, D.; Pandey, R.; Sahoo, S. Post-collisional talc-alkaline lamprophyres from the Kadiri greenstone belt: Evidence for the Neoarchean convergence-related evolution of the Eastern Dharwar Craton and its schist belts. *Lithos* **2018**, *320*, 105–117.
11. Ubide, T.; Mollo, S.; Zhao, J.X.; Nazzari, M.; Scarlato, P. Sector-zoned clinopyroxene as a recorder of magma history, eruption triggers, and ascent rates. *Geochim. Cosmochim. Acta* **2019**, *251*, 265–283.
12. Humphreys, M.C.S.; Blundy, J.D.; Sparks, R.S.J. Magma evolution and open system processes at Shiveluch Volcano: Insights from phenocryst zoning. *J. Petrol.* **2006**, *47*, 2303–2334.
13. Xing, C.M.; Wang, C.Y. Periodic mixing of magmas recorded by oscillatory zoning of the clinopyroxene macrocrysts from an ultrapotassic lamprophyre dyke. *J. Petrol.* **2020**, *61*, egaa103.
14. Nomade, S.; Renne, P.R.; Mo, X.X.; Zhao, Z.D.; Zhou, S. Miocene volcanism in the Lhasa block, Tibet: Spatial trends and geodynamic implications. Earth Planet. *Sci. Lett.* **2004**, *221*, 227–243.
15. Foley, S. Vein-plus-wall-rock melting mechanisms in the lithosphere and the origin of potassic alkaline magmas. *Lithos* **1992**, *28*, 435–453.
16. Yang, Z.M.; Goldfarb, R.; Chang, Z.S. *Generation of Postcollisional Porphyry Copper Deposits in Southern Tibet Triggered by Subduction of the Indian Continental Plate*; Special Publication 19; Society of Economic Geologists: Denver, CO, USA, 2016; pp. 279–300.
17. Zindler, A.; Hart, S.R. Chemical geodynamics. *Annu. Rev. Earth Plant. Sci.* **1986**, *14*, 493–571.
18. Miller, C.; Thöni, M.; Frank, W.; Schuster, R.; Melcher, F.; Meisel, T.; Zanetti, A. Geochemistry and tectonomagmatic affinity of the Yungbwa ophiolite, SW Tibet. *Lithos* **2003**, *66*, 155–172.
19. Xu, J.F.; Castillo, P.R. Geochemical and Nd–Pb isotopic characteristics of the Tethyan asthenosphere: Implications for the origin of the Indian Ocean mantle domain. *Tectonophysics* **2004**, *393*, 9–27.
20. Mahoney, J.J.; Frei, R.; Tejada, M.L.G.; Mo, X.X.; Leat, P.T.; Nagler, T.F. Tracing the Indian Ocean mantle domain through time: Isotopic results from old West Indian, East Tethyan, and South Pacific seafloor. *J. Petrol.* **2005**, *39*, 1285–1306.
21. Yin, A.; Harrison, T.M. Geologic evolution of the Himalayan-Tibetan Orogen. *Annu. Rev. Earth Planet. Sci.* **2000**, *28*, 211–280.
22. Hou, Z.Q.; Cook, N.J. Metallogenesis of the Tibetan collisional orogen: A review and introduction to the special issue. *Ore Geol. Rev.* **2009**, *36*, 2–24. [CrossRef]
23. Hou, Z.Q.; Yang, Z.M.; Qu, X.M.; Meng, X.G.; Li, Z.Q.; Beaudoin, G.; Rui, Z.Y.; Gao, Y.F.; Zaw, K. The Miocene Gangdese porphyry copper belt generated during post-collisional extension in the Tibetan Orogen. *Ore Geol. Rev.* **2009**, *36*, 25–31. [CrossRef]
24. Liu, D.; Zhao, Z.; Zhu, D.C.; Niu, Y.; DePaolo, D.J.; Harrison, T.M.; Mo, X.; Dong, G.; Zhou, S.; Sun, C.; et al. Postcollisional potassic and ultrapotassic rocks in southern Tibet: Mantle and crustal origins in response to India–Asia collision and convergence. *Geochim. Cosmochim. Acta* **2014**, *143*, 207–231.
25. Zhu, D.C.; Zhao, Z.D.; Niu, Y.; Mo, X.X.; Chung, S.-L.; Hou, Z.Q.; Wang, L.Q.; Wu, F.Y. The Lhasa Terrane: Record of a microcontinent and its histories of drift and growth. *Earth Planet. Sci. Lett.* **2011**, *301*, 241–255.
26. Guo, Z.F.; Wilson, M.; Zhang, M.L.; Cheng, Z.H.; Zhang, L.H. Post-collisional ultrapotassic mafic magmatism in south Tibet: Products of partial melting of pyroxenite in the mantle wedge induced by roll-back and delamination of the subducted Indian continental lithosphere slab. *J. Petrol.* **2015**, *56*, 1365–1406. [CrossRef]
27. Pan, G.T.; Wang, L.Q.; Li, R.S.; Yuan, S.H.; Ji, W.H.; Yin, F.G.; Zhang, W.P.; Wang, B.D. Tectonic evolution of the Qinghai-Tibet Plateau. *J. Asian Earth Sci.* **2012**, *53*, 3–14. [CrossRef]
28. Zhu, D.C.; Zhao, Z.D.; Niu, Y.; Dilek, Y.; Hou, Z.Q.; Mo, X.X. Origin and pre-Cenozoic evolution of the Tibetan Plateau. *Gondwana Res.* **2013**, *23*, 1429–1454.

29. Pan, G.T.; Ding, J.; Yao, D.S.; Wang, L.Q. *Guidebook of 1:1500000 Geologic Map of the Qinghai–Xizang (Tibet) Plateau and Adjacent Areas*; Chengdu Cartographic Publishing House: Chengdu, China, 2004; pp. 1–48.
30. Chen, Y.; Zhu, D.C.; Zhao, Z.D.; Meng, F.Y.; Wang, Q.; Santosh, M.; Wang, L.Q.; Dong, G.C.; Mo, X.X. Slab break off triggered ca. 113 Ma magmatism around Xainza area of the Lhasa Terrane, Tibet. *Gondwana Res.* **2014**, *26*, 449–463.
31. Kapp, P.; Yin, A.; Harrison, T.M.; Ding, L. Cretaceous–Tertiary shortening, basin development, and volcanism in central Tibet. *Geol. Soc. Am. Bull.* **2005**, *117*, 865–878. [CrossRef]
32. Dong, X.; Zhang, Z.M.; Liu, F.; Wang, W.; Yu, F.; Shen, K. Zircon U-Pb geochronology of the Nyainqêntanglha Group from the Lhasa terrane: New constraints on the Triassic orogeny of the south Tibet. *J. Asian Earth Sci.* **2011**, *42*, 732–739. [CrossRef]
33. Williams, H.M.; Turner, S.P.; Pearce, J.A.; Kelley, S.P.; Harris, N.B.W. Nature of the source regions for postcollisional, potassic magmatism in southern and northern Tibet from geochemical variations and inverse trace element modelling. *J. Petrol.* **2004**, *45*, 555–607. [CrossRef]
34. Williams, H.; Turner, S.; Kelley, S.; Harris, N. Age and composition of dikes in southern Tibet: New constraints on the timing of east–west extension and its relationship to postcollisional volcanism. *Geology* **2001**, *29*, 339–342. [CrossRef]
35. Zhao, Z.D.; Mo, X.X.; Dilek, Y.; Niu, Y.; DePaolo, D.J.; Robinson, P.; Zhu, D.C.; Sun, C.G.; Dong, G.C.; Zhou, S.; et al. Geochemical and Sr–Nd–Pb–O isotopic compositions of the post-collisional ultrapotassic magmatism in SW Tibet: Petrogenesis and implications for India intra-continental subduction beneath southern Tibet. *Lithos* **2009**, *113*, 190–212. [CrossRef]
36. Li, W.K.; Yang, Z.M.; Chiaradia, M.; Zhou, L.M.; Hou, Z.Q. Enrichment nature of ultrapotassic rocks in southern Tibet inherited from their mantle source. *J. Petrol.* **2021**, *62*, egab060. [CrossRef]
37. He, C.K.; Wang, Y.; Wang, H.Y.; Tang, J.X.; Yan, P.G.; Wang, Y.Q.; Fu, X.L.; Feng, Y.P. Geochronology, geochemistry and petrogenesis of the Bangbule quartz porphyry: Implications for the ore genesis. *Acta Geol. Sin. Engl. Ed.* **2023**, *97*, 744–758. [CrossRef]
38. Steiger, R.H.; Jager, E. Subcommission on geochronology: Convention on the use of decay constants in geo- and cosmochronology. *Earth Planet. Sci. Lett.* **1977**, *36*, 359–362. [CrossRef]
39. Koppers, A.A.P. ArArCALC—Software for 40Ar/39Ar age calculations. *Comput. Geosci.* **2002**, *28*, 605–619. [CrossRef]
40. Zhou, Y.W.; Yang, Z.M.; Zhou, L.M. Prospects of the origin and evolution of chlorine in collisional porphyry copper deposits. *Acta Petrol. Mineral.* **2023**, *42*, 417–441, (In Chinese with English abstract).
41. Wang, R.; Luo, C.H.; Sun, W.J.; Liu, Y.C.; Zhang, J.B. Progresses in the Study of High Magmatic Water and Oxidation State of Post-Collisional Magmas in the Gangdese Porphyry Deposit Belt. *Bull. Mineral. Petrol. Geochem.* **2021**, *40*, 1061–1077, (In Chinese with English abstract).
42. Rudnick, R.L.; Gao, S. Composition of the continental crust. *Treatise Geochem.* **2003**, *3*, 1–64.
43. Le Maitre, R.W. *Igneous Rocks: A Classification and Glossary of Terms. Recommendations of the International Union of Geological Sciences Subcommission on the Systematics of Igneous Rocks*; Cambridge University Press: Cambridge, UK, 2002; Volume 208.
44. Sun, S.S.; McDonough, W.F. Chemical and isotopic systematics of oceanic basalts, implications for mantle composition and processes. *Geo. Soc. Lond. Spec. Pub.* **1989**, *42*, 313–345. [CrossRef]
45. Zhang, S.Q.; Mahoney, J.J.; Mo, X.X.; Ghazi, A.M.; Milani, L.; Crawford, A.J.; Guo, T.Y.; Zhao, Z.D. Evidence for a widespread Tethyan upper mantle with Indian–Oceantype isotopic characteristics. *J. Petrol.* **2005**, *46*, 829–858. [CrossRef]
46. Niu, X.L.; Zhao, Z.D.; Mo, X.X.; DePaolo, D.J.; Dong, G.C.; Zhang, S.Q.; Zhu, D.C.; Guo, T.Y. Elemental and Sr–Nd–Pb isotopic geochemistry for basic rocks from Decun–Angren ophiolites in the Xigaze area, Tibet: Implications for the characteristics of the Tethyan upper mantle domain. *Acta Petrol. Sin.* **2006**, *22*, 2875–2888, (In Chinese with English abstract).
47. Miller, C.; Schuster, R.; Klötzli, U.; Frank, W.; Purtscheller, F. Post-collisional potassic and ultrapotassic magmatism in SW Tibet: Geochemical and Sr–Nd–Pb–O isotopic constraints for mantle source characteristics and petrogenesis. *J. Petrol.* **1999**, *40*, 1399–1424.
48. Harris, N.B.W.; Xu, R.; Lewis, C.L.; Hawkesworth, C.J.; Zhang, Y. Isotope geochemistry of the 1985 Tibet Geotraverse, Lhasa to Golmud. *Philos. Trans. R. Soc. Lond. Ser. A Math. Phys. Sci.* **1988**, *327*, 263–285.
49. Morimoto, N. Nomenclature of pyroxenes. *Mineral. J.* **1989**, *14*, 198–221. [CrossRef]
50. Smith, J.V. *Feldspar Minerals: Crystal Structure and Physical Properties*; Springer: Berlin/Heidelberg, Germany, 1974; Volume 1, p. 318.
51. Putirka, K.D. Thermometers and barometers for volcanic systems. Reviews in REV Mineral. *Geochem.* **2008**, *9*, 61–120.
52. Li, X.; Zhang, C. Machine learning thermobarometry for biotite-bearing magmas. *J. Geophys. Res. Solid Earth* **2022**, *127*, e2022JB024137.
53. Adam, J.; Green, T. Trace element partitioning between mica- and amphibole-bearing garnet lherzolite and hydrous basanitic melt: 1. Experimental results and the investigation of controls on partitioning behaviour. *Contrib. Mineral. Petrol.* **2006**, *152*, 1–17.
54. Furman, T.; Graham, D. Erosion of lithospheric mantle beneath the East African Rift system: Geochemical evidence from the Kivu volcanic province. *Dev. Geotecton.* **1999**, *24*, 237–262.

55. Reider, M.; Cavazzini, G.; D'yakonov, Y.S.; Frank-Kamenetskii, V.A.; Gottardi, G.; Guggenheim, S.; Koval, P.V.; Muller, G.; Neiva, A.M.R.; Radoslovich, E.W.; et al. Nomenclature of the micas. *Mineral. Mag.* **1999**, *63*, 267–279. [CrossRef]
56. Rock, N.M.S. *Lamprophyres*; Blackie: Glasgow, UK, 1991; 285p.
57. Salters, V.J.M.; Stracke, A. Composition of the depleted mantle. *Geochem. Geophys. Geosyst.* **2004**, *5*, Q05B07. [CrossRef]
58. Willbold, M.; Stracke, A. Trace element composition of mantle end-members: Implications for recycling of oceanic and upper and lower continental crust. *Geochem. Geophys. Geosyst.* **2006**, *7*, Q04004. [CrossRef]
59. Tian, S.H.; Yang, Z.S.; Hou, Z.Q.; Mo, X.X.; Hu, W.J.; Zhao, Y.; Zhao, X.Y. Subduction of the Indian lower crust beneath southern Tibet revealed by the post-collisional potassic and ultrapotassic rocks in SW Tibet. *Gondwana Res.* **2017**, *41*, 29–50. [CrossRef]
60. Zhao, L.; Guo, F.; Fan, W.M.; Huang, M.W. Roles of subducted pelagic and terrigenous sediments in Early Jurassic mafic magmatism in NE China: Constraints on the architecture of Paleo-Pacific subduction zone. *J. Geophys. Res. Solid Earth* **2019**, *124*, 2525–2550. [CrossRef]
61. Kirchenbaur, M.; Münker, C. The behaviour of the extended HFSE group (Nb, Ta, Zr, Hf, W, Mo) during the petrogenesis of mafic K-rich lavas: The Eastern Mediterranean case. *Geochim. Cosmochim. Acta* **2015**, *165*, 178–199. [CrossRef]
62. Price, R.C.; Stewart, R.B.; Woodhead, J.D.; Smith, I.E.M. Petrogenesis of high-K arc magmas: Evidence from Egmont volcano, North Island, New Zealand. *J. Petrol.* **1999**, *40*, 167–197. [CrossRef]
63. Ubide, T.; Gale, C.; Arranz, E.; Lago, M.; Larrea, P. Clinopyroxene and amphibole crystal populations in a lamprophyre sill from the Catalonian Coastal Ranges (NE Spain): A record of magma history and a window to mineral-melt partitioning. *Lithos* **2014**, *184*, 225–242. [CrossRef]
64. Sharma, A.; Pandey, R.; Rao, N.V.C.; Sahoo, S.; Belyatsky, B.; Dhote, P. Mineralogy and petrology of lamprophyre and dolerite dykes from the end-Cretaceous (~66 Ma) Phenaimata alkaline igneous complex, north-western India: Evidence for open magma chamber fractionation, mafic recharge, and disaggregation of crystal mush zone in a large igneous province. *Mineral. Petrol.* **2023**, *117*, 415–445.
65. Chen, L.; Zheng, Y.F.; Zhao, Z.F. Geochemical insights from clinopyroxene phenocrysts into the effect of magmatic processes on petrogenesis of intermediate volcanics. *Lithos* **2018**, *316*, 137–153. [CrossRef]
66. Streck, M.J. Mineral textures and zoning as evidence for open system processes. *Rev. Mineral. Geochem.* **2008**, *69*, 595–622. [CrossRef]
67. Tecchiato, V.; Gaeta, M.; Mollo, S.; Bachmann, O.; von Quadt, A.; Scarlato, P. Snapshots of primitive arc magma evolution recorded by clinopyroxene textural and compositional variations: The case of hybrid crystal-rich enclaves from Capo Marargiu Volcanic District (Sardinia, Italy). *Am. Mineral.* **2018**, *103*, 899–910. [CrossRef]
68. Zhang, Y.X.; Zeng, Z.G.; Gaetani, G.; Zhang, L.; Lai, Z.Q. Mineralogical constraints on the magma mixing beneath the Iheya Graben, an active back-arc spreading centre of the Okinawa trough. *J. Petrol.* **2020**, *61*, egaa098. [CrossRef]
69. Neave, D.A.; MacLennan, J. Clinopyroxene dissolution records rapid magma ascent. *Front. Earth Sci.* **2020**, *8*, 188. [CrossRef]
70. Yang, Z.M.; Lu, Y.J.; Hou, Z.Q.; Chang, Z.S. High-Mg Diorite from Qulong in Southern Tibet: Implications for the Genesis of Adakite-like Intrusions and Associated Porphyry Cu Deposits in Collisional Orogens. *J. Petrol.* **2015**, *56*, 2227–2254. [CrossRef]
71. Sun, X.; Lu, Y.J.; McCuaig, T.C.; Zheng, Y.Y.; Chang, H.F.; Guo, F.; Xu, L.J. Miocene ultrapotassic, high-Mg Dioritic, and adakite-like rocks from Zhunuo in Southern Tibet: Implications for mantle metasomatism and porphyry copper mineralization in collisional orogens. *J. Petrol.* **2018**, *59*, 341–386. [CrossRef]
72. Guo, Z.F.; Wilson, M.; Zhang, M.L.; Cheng, Z.H.; Zhang, L.H. Post-collisional, K-rich mafic magmatism in South Tibet: Constraints on Indian slab-to-wedge transport processes and plateau uplift. *Contrib. Mineral. Petrol.* **2013**, *165*, 1311–1340. [CrossRef]

minerals

MDPI

Article

Genesis of the Gongjuelong Sn Polymetallic Deposit in the Yidun Terrane, China: Constraints from the In Situ Geochemistry of Garnet, Cassiterite, and Quartz

Yuchang Zhou [1], Yiwei Peng [1,*], Chang Liu [2], Jianji Tian [2], Zhi Wang [1], Mingwei Song [1] and Yan Zhang [1]

1 College of Earth and Planetary Sciences, Chengdu University of Technology, Chengdu 610059, China; zyc2835145463@163.com (Y.Z.); 17828087322@163.com (Z.W.)

2 Division of Geology and Mineral Resources, Beijing Research Institute of Uranium Geology, Beijing 100029, China

* Correspondence: yiwei_peng@163.com; Tel.: +86-182-0046-6121

Abstract: Numerous skarn-type Sn and hydrothermal vein-type Pb–Zn–Ag deposits occur in the northern Yidun Terrane, China. The Gongjuelong skarn Sn polymetallic deposit, adjacent to the Haizishan granite, is situated in the central region of Yidun Terrane. The genesis of the Gongjuelong Sn deposit and its relationship with the adjacent Pb–Zn–Ag deposits remains controversial. The ore-forming process can be divided into three stages: the prograde stage (I), marked by the formation of garnet and pyroxene; the retrograde stage (II), which includes the epidote–actinolite sub-stage (II-1) and the quartz-cassiterite sub-stage (II-2); and the sulfide stage (III), consisting of the chalcopyrite–pyrrhotite sub-stage (III-1) and the arsenopyrite–sphalerite sub-stage (III-2). Two types of garnet (Grt-I and Grt-II) have been identified in stage I and both belong to the grossular–andradite solid solution. Grt-II ($Gro_{52-73}And_{25-45}Spe+Pyr+Alm_{2-3}$) contains slightly more Fe than Grt-I ($Gro_{64-76}And_{20-28}Spe+Pyr+Alm_{2-10}$). Grt-I is enriched in heavy rare-earth elements (HREEs) and depleted in light rare-earth elements (LREEs), whereas Grt-II is enriched in LREEs and depleted in HREEs. Grt-I has higher U contents and lower Th/U ratios than those of Grt II, indicating a lower oxygen fugacity for the earlier skarn alteration. In contrast to Grt-I, Grt-II shows a more significant negative Eu anomaly along with lower LREEs/HREEs. Therefore, Grt-I and Grt-II likely formed under mildly acidic and near-neutral conditions, respectively. The W (350–3015 ppm) and Fe (235–3740 ppm) contents and Zr/Hf ratios (18.7–49.4) of cassiterite from Gongjuelong are similar to those of cassiterite from the granite-related Sn deposits, as well as the Xiasai hydrothermal vein-type Pb–Zn–Ag deposit in the northern Yidun Terrane. The Ti/Ge ratio (0.06–1.13) and P contents (13.9–173 ppm) of quartz are also similar to those from the Xiasai Pb–Zn–Ag deposit, both of which resemble those of skarn-type deposits and Sn-associated quartz. Furthermore, the Ti/Zr ratio (average 33.2) of cassiterite at Gongjuelong are much higher than that of cassiterite at Xiasai (average 3.7), indicating that the Pb–Zn–Ag veins could represent the distal product of the "parent" granite. On the basis of combined evidence from geology, geochemistry, and published geochronology data, we propose that the proximal skarn-type Sn deposits and distal hydrothermal vein-type Pb–Zn–Ag±Sn deposits in the northern Yidun Terrane constitute an integrated ore system, which is genetically related to the late Cretaceous highly fractionated granites. This proposed hypothesis highlights the potential prospecting of Sn mineralization beneath the hydrothermal Pb–Zn–Ag veins, as well as the hydrothermal Pb–Zn–Ag veins controlled by faults/fractures within the strata around the Sn deposits and highly fractionated granites.

 https://doi.org/10.3390/min15030314

Keywords: garnet; cassiterite; quartz; Sn polymetallic deposit; Gongjuelong; Yidun Terrane

1. Introduction

Polymetallic vein systems associated with magmatic centers exhibit distinct element zoning patterns [1]. In the central area of the systems, metals such as Sn and Cu are typically enriched, with the minerals predominantly consisting of oxides (e.g., cassiterite) and sulfides, which precipitate under high-temperature, low-oxidation conditions. As the mineralizing fluids cool, the wall rocks adjacent to the veins gradually become enriched in Pb, Zn, and Ag through the precipitation of disseminated Pb–Zn sulfides [2,3]. Element zoning within the veins is more complex, resulting from the interplay between magmatic fluid evolution and subsequent mineralization processes. The spatial distribution relationship between Sn and Pb, Zn, and Ag suggests that these elements may originate from the same mineralizing system, with fluid evolution and magmatic processes controlling their distribution patterns [4–6]. The presence of Sn, along with Pb–Zn–Ag–bearing minerals, not only enhances the economic potential of the deposit but also supports the possibility of polymetallic mining. Understanding the relationship between Sn and Pb–Zn–Ag mineralization is essential for effective mineral exploration and economic evaluation.

Over the past 20 years, a number of skarn Sn and hydrothermal vein-type Ag–Pb–Zn±Sn deposits have been identified in the central Yidun Terrane, located in southwestern China [7–17] (Figure 1a,b). These mineral deposits form the Xiasai–Lianlong metallogenic belt, which has contributed over 4.6 million tons of Pb + Zn, 17,600 tons of Ag, and 0.27 million tons of Sn, making it one of the most important metallogenic regions in China for Ag, Pb–Zn, and Sn [17–21]. The skarn deposits of Sn–Ag (–Pb–Zn) include the Cuomolong, Lianlong and Gongjuelong deposit, whereas the hydrothermal (Sn–) Ag–Pb–Zn deposits are best represented by the Shaxi, Jiaogenma, and Xiasai sites. A genetic relationship between the mineralized systems and nearby granitic magmatic activity has been suggested through isotopic analyses. These analyses include Biotite ^{40}Ar-^{39}Ar dating of Haizishan and Queershan, zircon U-Pb dating of Rongyicuo, Xiasai, and Haizishan, as well as whole-rock Rb–Sr dating of Lianlong, Rongyicuo, and Haizishan. Geochronology suggests that ore-associated granites formed between 105 and 89 Ma [8,14,22,23]. These late Cretaceous granites are classified as A-type granites, which are characterized by high contents of SiO_2, Nb, Ta, Ga, and Y, and lower concentrations of Al_2O_3, CaO, MgO, Sr, and Eu [8,24].

Previous studies on the Sn–Pb–Zn–Ag polymetallic deposits in the central Yidun Terrane have primarily focused on the genesis of individual deposits [4]. However, whether the skarn Sn deposits and hydrothermal vein-type (Sn-) Pb–Zn–Ag deposits in this region formed as a result of the same metallogenic system remains a subject of debate. Li et al. [4,5] conducted geochronological studies using multiple methods: (1) LA-ICP-MS U-Pb dating of cassiterite from quartz-cassiterite veins yielded an age of 99.2 ± 0.8 Ma, (2) Rb–Sr dating of sphalerite from quartz–sulfide veins gave an age of 99 ± 3 Ma, and (3) zircon U-Pb dating of the mineralizing host rocks produced an age of 101 ± 1 Ma. Their studies confirmed a genetic link between Sn mineralization and Pb–Zn–Ag mineralization at the Xiasai deposit. However, research on their link to skarn Sn deposits remains insufficient.

The Gongjuelong Sn deposit is situated within the central Yidun Terrane. This study employs EMPA major element analysis on garnet and cassiterite from Gongjuelong, along with LA-ICP-MS trace element analysis on indicator minerals, including garnet, cassiterite, and quartz. The physicochemical conditions of mineralization remain unclear, and the genetic relationship between Sn mineralization and Pb–Zn–Ag mineralization has not yet been well defined. To address these questions, we will use geochemical data from garnet to discuss the metallogenic physicochemical conditions of the Gongjuelong Sn deposit. Additionally, we will analyze the geochemical data of cassiterite and quartz, combined with existing data from the Pb–Zn–Ag deposit in the central Yidun Terrane, to investigate the genetic relationship between Sn mineralization and Pb–Zn–Ag mineralization. This study aims to provide a theoretical foundation for future mineral exploration in the region.

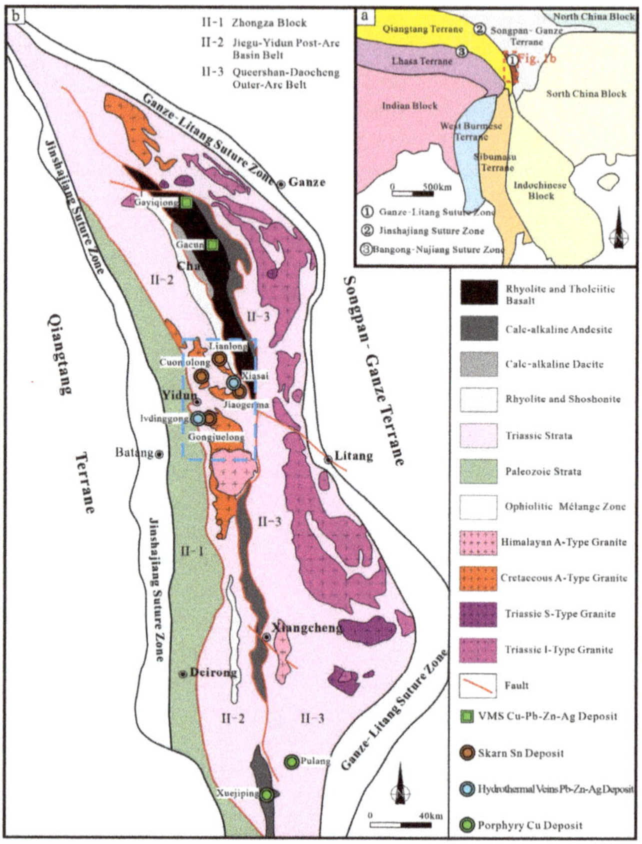

Figure 1. (**a**) Tectonic map of the Yidun Terrane (modified after [25]). (**b**) Regional geological map of the Chinese Yidun Terrane (modified after [24,26–28]).

2. Geological Setting

The Yidun Terrane is located in the eastern Tibetan Plateau, between the Qiangtang and Songpan–Garze terranes (Figure 1a). They are separated by two major oceanic sutures: the Ganze–Litang Suture and the Jinsha River Suture, which extend several thousand kilometers in the NNW direction [29–32]. The Xiangcheng–Geza Fault divides the Yidun Terrane into the Western Yidun Terrane (Zhongza Block) and the Eastern Yidun Terrane (Figure 1b). The Gongjuelong mining area lies on the western flank of the Yidun-Moyaba synclinorium, exhibiting diverse stratigraphy and lithology, with Permian, Triassic, and Quaternary sequences. The Permian strata (P) mainly consist of carbonate and clastic rocks, with minor volcanic intercalations, forming a belt along the Shamake-Namaqu Fault. The

Triassic strata are the most widely exposed, including the Tumugou, Qugasi, Dang'en, Lieyi, and Labaya Formations, which are primarily composed of metamorphic sandstone, slate, and limestone.

The Yidun Terrane forms a part of the Yangtze–Qiangtang tectonic domain. Its southeastern boundary adjoins the Yangtze Craton, whereas its western boundary is linked to the Qiangtang–Changdu Block. The regional tectonic framework includes the Yajiang Back-Arc Basin Fold Belt, the Ganza–Litang Ophiolitic Mélange Zone, the Yidun Arc Fold Belt, the Zhongza Block, and the Jinsha River Ophiolitic Mélange Zone (Figure 1b). The Yidun Terrane has undergone tectonic processes from the Indosinian to the Yanshanian and Himalayan periods, encompassing seafloor spreading, subduction, arc-continent collision, and intracontinental convergence [8,25,33]. The region is dominated by three primary structural orientations: NNW-SN, NW, and EW-NE.

Permian basalts and metamorphosed volcanic rocks are limited in exposure, primarily along the southern margins of the Zhongza Block and Yidun Terrane. The volcanic rocks of the Yidun Terrane display a north–south differentiation: the northern region is dominated by calc-alkaline andesites, rhyolites, and basalts [31], whereas the southern region features calc-alkaline andesites, trachytes, dacites, and minor alkaline basalts [26].

The Yidun Terrane underwent multiple magmatic events from the Triassic to Late Cretaceous, producing hundreds of intermediate to felsic plutonic intrusions. The distribution of these intrusions indicates the formation of several large-scale plutons during 230–204 Ma, whereas smaller, more dispersed intrusions and porphyries formed during 105–80 Ma [15,27,28,34,35]. The southern Yidun Terrane hosts Late Triassic oxidized and less-differentiated granitic porphyries and Late Cretaceous acidic porphyries. In the northern central part of the Yidun Terrane, the Changduoke–Genie Granite Belt extends for approximately 160 km and is composed of eight major granitic plutons, along with smaller stocks and dikes, including those at Hagala (Haizishan), Rongyicuo, Lianlong, and Ruoluolong. These NNW-trending plutons intrude Permian to Upper Triassic strata, creating alteration zones several hundred to over a thousand meters wide. The plutons are generally homogeneous, primarily composed of biotite monzogranite.

3. Ore Deposit Geology

The Gongjuelong Sn deposit is located in the eastern part of the Batang Deda Township, approximately at 99°27′10″ E and 30°14′10″ N, with an elevation of 4590 m. From west to east, the stratigraphy in the mining area includes the Upper Triassic Qugasi Formation (T_3q), comprising the third subunit (T_3q^3), composed of sandstone, slate, and quartzose conglomerates, and the fourth subunit (T_3q^4), consisting of crystalline limestone, bioclastic limestone, minor sandstone, and slate. The Tumugou Formation (T_3t^1) consists of quartzose conglomerates, pebbly sandstones, siltstones, and rhythmic sequences of slate (Figure 2a). These rocks show varying degrees of hornfelsic and marble alteration. The area is characterized by significant folding and faulting, with four major fault sets (NNW-, NW-, EW-, and NE-trending) intersecting, causing widespread deformation. NNW-trending faults dominate the structural framework, hosting NEE-trending ore-bearing secondary faults that intersect with EW-trending faults, reflecting compressional stress followed by extension. NW-trending faults, arranged en echelon, extend 2–5 km and show evidence of at least two periods of movement, playing a key role in ore control [36]. Two sets of joints (NW- and NE-trending) are also present: ore veins and dikes are located along the NW-trending joints, whereas NE-trending joints crosscut the ore veins. The fault activity sequence is NNW, followed by NW, EW, and NE, with the NNW- and NW-trending faults showing multiple phases of movement.

Magmatic rocks in the area belong to the Haizishan complex granite intrusion, which was formed during the first stage of the Late Yanshanian period. These rocks primarily consist of fine-grained biotite monzogranite, with common occurrences of granite stocks, apophyses, and dikes trending NNW and NW. The main ore-related pluton is a medium- to coarse-grained biotite monzogranite stock, measuring approximately 280 m in length and 50–70 m in width. It intrudes conformably into the fourth member of the Qugasi Formation, with a NW strike and NE dip.

Within an area of approximately 1 km^2, 11 ore bodies have been identified (Figure 2b,c). These deposits are mainly hosted in skarns, occurring along the contact zones between granite and marble or hornfels, as well as within secondary faults and interlayered slip zones. The ore deposits align in a NW direction along the eastern slope of Gongjuelong. The No. III orebody is situated along the contact zone between the granite stock and marble. Due to the NE dip of the contact planes of the stock, the SW section has marble as the footwall and granite as the hanging wall, whereas this relationship is reversed in the NE section. No. III orebody has five surface outcrops, with the longest being 163 m and a thickness ranging from 0.9 to 3.2 m, exhibiting vein-like or banded morphologies and local bead- or spindle-shaped structures. Other outcrops extend 1065 m, with a thickness of 0.24 m, showing vein-like and lenticular forms, all dipping NE at angles between 35° and 77°. Four additional ore bodies (I, II, IV, V) occur within minor faults or interlayered slip zones, showing vein-like and lenticular morphologies, with lengths of 132 m and thicknesses of 0.24 m. Some ore zones are concordant with the host rock structures, whereas others exhibit oblique relationships. The majority of the ore exhibits disseminated textures, with minor massive and vein-like ores. Alteration associated with mineralization is predominantly skarnization, along with greisenization, tourmalinization, fluorite alteration, silicification, and carbonate alteration.

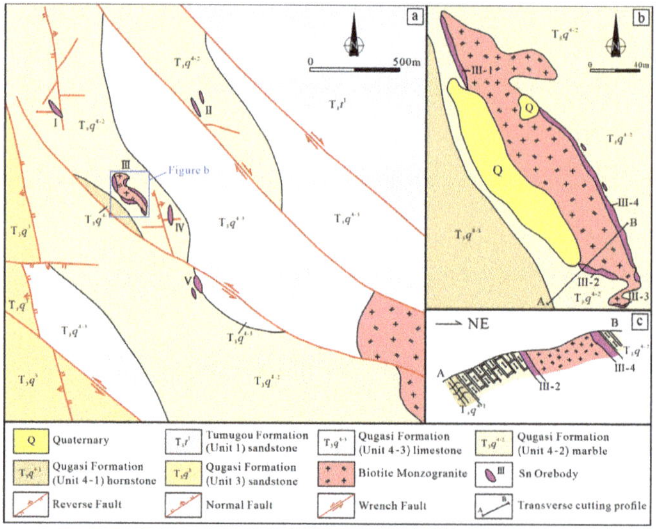

Figure 2. (**a**,**b**) Geological map, and (**c**) cross-section of the Gongjuelong deposit (modified after [36]).

The ore textures are mainly disseminated and massive, with minor vein and network-like structures. Ore minerals include cassiterite, chalcopyrite, pyrrhotite, and arsenopyrite, with minor sphalerite and pyrite. Gangue minerals consist primarily of quartz, calcite, garnet, diopside, epidote, and actinolite, with secondary and accessory minerals including hedenbergite, sillimanite, chlorite, apatite, and fluorite. Based on macroscopic observations and microscopic analysis, the mineralization of the Gongjuelong deposit can be divided into three stages: the prograde stage (I), represented by the garnet-pyroxene stage; the retrograde stage (II), which includes the epidote–actinolite sub-stage (II-1) and the quartz–cassiterite sub-stage (II-2); and the sulfide stage (III), consisting of the chalcopyrite–pyrrhotite sub-stage (III-1) and the arsenopyrite–sphalerite sub-stage (III-2).

The mineralization stages are described as follows (Figures 3 and 4): The skarn formation begins with the garnet–pyroxene stage (I), characterized by high-temperature, anhydrous silicate minerals. Hand specimens exhibit a distinct reddish hue, dominated by massive garnet and locally associated with diopside, while columnar apatite is occasionally enclosed by early garnet (Grt-I). As the system cools, the epidote–actinolite stage (II-1) initiates, marked by the formation of hydrous minerals such as coarse epidote (up to 15 cm) and radiating actinolite, accompanied by minor chlorite and sillimanite. The ore exhibits a dark green color, with quartz and calcite filling the interstitial spaces, while actinolite is partially altered to chlorite. The quartz–cassiterite stage (II-2), where cassiterite becomes predominant, occurring in disseminated textures with quartz and calcite. Cassiterite is brown, short-prismatic, and exhibits distinct zoning, occasionally intergrown with actinolite and sillimanite. Under the microscope, cassiterite displays well-developed euhedral forms, with some grains showing surface dissolution features and minor fractures. The chalcopyrite–pyrrhotite stage (III-1) follows, characterized by the precipitation of chalcopyrite and pyrrhotite in disseminated forms, closely associated with quartz and calcite. These sulfides replace skarn minerals, including epidote and actinolite, with minor arsenopyrite and sphalerite also present. Finally, the arsenopyrite–sphalerite stage (III-2) is marked by quartz–calcite–sulfide veins cross-cutting earlier disseminated ores. These veins contain large quartz and calcite crystals, with arsenopyrite and sphalerite as the main sulfides, accompanied by minor chalcopyrite, pyrrhotite, and fluorite. This sequence of stages reflects the evolution from high-temperature skarn formation to low-temperature hydrothermal alteration and sulfide mineralization.

Figure 3. (**a**–**c**) Hand-specimen photographs and (**d**–**f**) photomicrographs of ores from Gongjuelong showing the main ore types and mineral assemblages from the Gongjuelong Sn deposit. (**a**) Two types of garnet occur as granular and are associated with epidote in the dark green skarn, which exhibits an

emerald-like color due to the presence of epidote. (**b**) Self-formed cassiterite develops as vein-like structures within the quartz vein. (**c**) Sulfides, such as arsenopyrite, occur as massive and disseminated aggregates and are associated with quartz. (**d**) Grt-I exhibits rhythmic zoning and is replaced by Grt-II. (**e**) Garnet coexists with diopside (under plane-polarized light). (**f**) Grt-II coexists with diopside (under cross-polarized light). (**g**) Self-formed dark brown cassiterite crystals. (**h**) Cassiterite is brown and coexists with radial dark green actinolite. (**i**) Sulfides, such as pyrrhotite, chalcopyrite, and arsenopyrite, coexist within quartz. Ep, epidote; Grt, garnet; Cst, cassiterite; Qtz, quartz; Apy, arsenopyrite; Di Diopside.

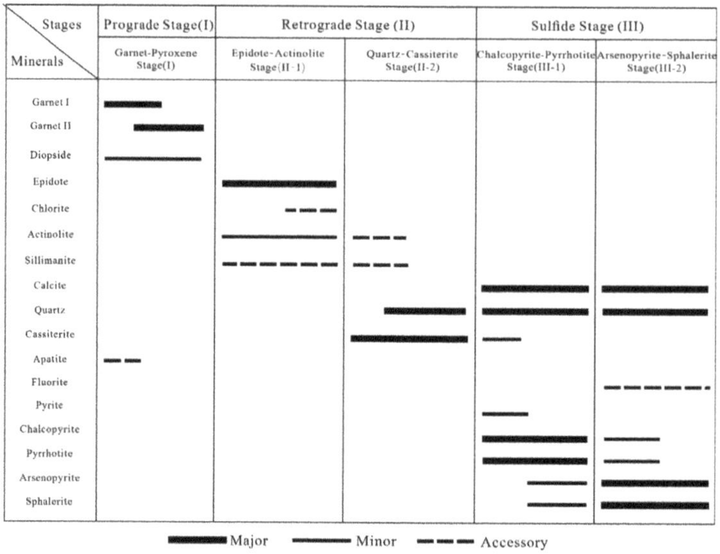

Figure 4. Paragenetic sequence of ore and gangue minerals in the Gongjuelong Sn deposit.

4. Sampling and Analytical Methods

In this study, major element analysis of garnet and cassiterite from the Gongjuelong deposit was conducted using electron probe microanalysis (EPMA). Additionally, trace element analysis was performed on two types of garnet from the garnet–skarn stage, cassiterite from the quartz–cassiterite stage, and quartz from the sulfide stage of the Gongjuelong deposit using LA-ICP-MS (Laser Ablation Inductively Coupled Plasma Mass Spectrometry).

4.1. Electron Probe Micro-Analysis (EPMA)

The composition of garnet was analyzed by Nanjing Hongchuang Geological Exploration Technology Service Co., Ltd., Nanjing, China. The electron probe microanalyzer used was a JEOL JXA-iSP100 (JEOL, Tokyo, Japan). The instrument was operated with an accelerating voltage of 15 kV, a beam current of 20 nA, and a beam diameter of 5 μm. For the elements Ti and Y, the peak counting time was set to 10 s, whereas for Si, Ca, K, P, Na, Mg, Al, Cr, Fe, and Mn, it was 5 s. The background counting time was set to half the peak time for both high- and low-background positions. All data were corrected for matrix effects using the ZAF correction method. The following standards were employed: 53-19 Chrome Diopside (Si), 53-23 Titanium Oxide (Ti), 53-19 Chrome Diopside (Ca), 53-40 Yttrium Al Garnet (Y), 53-13 Orthoclase (K), 53-20 Ap-5nA-20 um (P), 53-06 Albite (Na), 53-19 Chrome Diopside (Mg), 53-28 Almandine (Al), 53-24 Chromium Oxide (Cr), 53-26 Rhodonite (Mn), and 53-28 Almandine (Fe); the standard deviation (sigma) is typically 0.1–0.5 wt% for major elements (>1 wt%) and 0.01–0.1 wt% for trace elements (<1 wt%).

The composition of cassiterite was also analyzed by Nanjing Hongchuang Geological Exploration Technology Service Co., Ltd., Nanjing, China. The electron probe microanalyzer used was a JEOL JXA-iSP100 (JEOL, Tokyo, Japan). The accelerating voltage, beam current, and beam diameter were identical, set at 15 kV, 20 nA, and 5 μm, respectively. For Sn, the peak counting time was 5 s, whereas for Ti, Nb, Zr, Si, Al, W, Ta, Fe, and Mn, it was 10 s. Similar to the garnet analysis, the background counting time was half the peak time for both high- and low-background positions. The ZAF correction method was applied to all data to account for matrix effects. The following standards were used: 53-23 Titanium Oxide (Ti), 53-43 Cassiterite (Sn), 44-21 Niobium (Nb), 53-41 Cubic Zirconia (Zr), 53-19 Chrome Diopside (Si), 53-06 Albite (Al), 44-34 Tungsten (W), 44-33 Tantalum (Ta), 53-33 Hematite (Fe), and 53-26 Rhodonite (Mn).

4.2. Laser Ablation Inductively Coupled Plasma Mass Spectrometry (LA-ICP-MS)

Trace element analysis of cassiterite was performed at Guangzhou Tuoyan Analytical Technology Co., Ltd., Guangzhou, China. The laser ablation system used was a GeoLasPro 193 excimer laser solid sample system (Coherent, Santa Clara, CA, USA), coupled with an X Series 2 quadrupole inductively coupled plasma mass spectrometer (ICP-MS) from Thermo Fisher (Waltham, MA, USA). Helium (He) was employed as the carrier gas, whereas argon (Ar) served as the make-up gas. Analysis was carried out in single-point ablation mode with a laser spot size of 60 μm, frequency set to 6 Hz, and energy density of 7 J/cm^2. A 20 s background signal acquisition period was followed by 50 s for sample signal collection. NIST610 standard glass [37] was used as the external standard, and the theoretical Sn concentration in cassiterite was applied as the internal standard for trace element calibration. NIST612 was utilized as the monitoring standard to ensure data accuracy. Post-analysis data processing was conducted using software ICPMSDataCal 10.7; the standard deviation is generally 0.1–1 wt% for major elements and 0.01–0.1 wt% for trace elements.

For the trace element analysis of quartz, the study was performed at the Ore Deposit Geochemistry Microanalysis Laboratory, which is part of the State Key Laboratory of Geological Processes and Mineral Resources at China University of Geosciences, Beijing. The analysis was conducted using a GeoLasPro 193 nm excimer solid-state sampling system (Coherent, Santa Clara, CA, USA) coupled with an X Series 2 quadrupole ICP-MS (Thermo Fisher Scientific, Waltham, MA, USA). Helium (He) was used as the carrier gas, and argon (Ar) served as the compensating gas. The analysis was performed in single-spot ablation mode with a laser spot size of 60 μm, a frequency of 6 Hz, and an energy density of 12 J/cm^2. A 20 s period was used for background signal acquisition, followed by 50 s for the sample signal collection. NIST610 standard glass was employed as the external standard, and the theoretical silicon (Si) content in quartz was used as the internal standard for trace element calibration. A natural quartz standard, Q7, was utilized for monitoring purposes [38]. Data processing was conducted using software ICPMSDataCal 10.7 [39].

The trace element analysis of garnets was carried out at the Ore Deposit Geochemistry Microanalysis Laboratory at the State Key Laboratory of Geological Processes and Mineral Resources, China University of Geosciences, Beijing. The GeoLasPro-193 nm system from Coherent was used for laser sampling, and ion signal intensities were measured using Thermo Fisher's X-Series 2 ICP-MS. Helium was used as the carrier gas, with argon mixed as the make-up gas. The analysis was performed in single-spot ablation mode with a 32 μm spot size and a frequency of 6 Hz. Further details regarding the operating conditions and procedures are provided by Peng et al. [40].

5. Results

5.1. Major Elements of Garnet and Cassiterite

5.1.1. Garnet

The electron probe microanalysis (EPMA) results for garnet samples are presented in Table S1. The composition of garnet in Gongjuelong is presented in Figures 5 and 6.

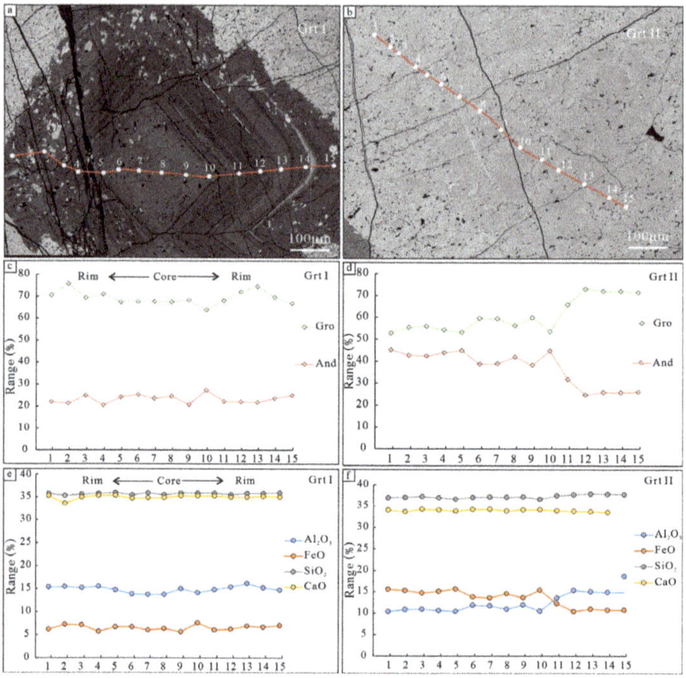

Figure 5. BSE images of Grt-I (**a**) and Grt-II (**b**), end member composition maps of Grt-I (**c**) and Grt-II (**d**), and the major element variation maps of Grt-I (**e**) and Grt-II (**f**) in Gongjuelong deposit.

Grt-I: The SiO_2 content ranges from 35.9% to 35.3%, with an average of 35.7%. The CaO content is 33.6–35.4%, averaging 35%. Al_2O_3 is relatively high, ranging from 13.8% to 16.1%, with an average of 14.9%. FeO content ranges from 5.7% to 7.6%, with an average of 6.6%. TiO_2, MgO, and MnO are lower (0.5%–3.1%, 0.4%–2.2%, and 0.2%–0.3%, respectively), and P_2O_5, Na_2O, and K_2O are all $\leq$0.01%. Based on the major elements, the garnet is primarily composed of grossular (Gro), with 64%–76%, followed by andradite (And) at 21%–28%, with small amounts of pyrope (Pyr) (2%–10%) and spessartine (Spe) (0.5%–0.8%). Grt-I belongs to the grossular (Gro)–andradite (And) series (Gro $_{64\text{-}76}$, And $_{20\text{-}28}$, Spe + Pyr + Alm $_{2\text{-}10}$, average = And_{70}, Gro_{24}, Spe + Pyr + Alm_6) (Figures 5 and 6, Table S1).

Grt-II: SiO_2 ranges from 36.7% to 37.9%, with an average of 37.3%. CaO ranges from 33.6% to 34.3%, with an average of 33.9%. The FeO content is higher, ranging from 10.4% to 15.7%, with an average of 15.5%, whereas Al_2O_3 content decreases (10.4%–15.4%, average 13.5%). The content of MnO is 0.5%–0.9%, and TiO_2 content is lower than Grt-I, ranging from 0.04% to 0.20%, with an average of 0.12%. Other elements (MgO, Na_2O, P_2O_5, K_2O) are all < 0.01%. Grt-II is mainly grossular (Gro) with some andradite (And), belonging to the Gro-And series ($Gro_{52\text{-}73}$, $And_{25\text{-}45}$, Spe + Pyr + $Alm_{1.5\text{-}2.4}$). Spe and Pyr range from 1.2% to 2.1% and 0.1% to 0.4%, respectively. The major element variations indicate an increase in FeO and a decrease in Al_2O_3 from Grt-I to Grt-II. Both Grt-I and Grt-II belong to the grossular–andradite solid solution series, but Grt-II contains a higher proportion of andradite and a lower proportion of grossular compared to Grt-I. The contents of other components, such as spessartine, pyrope, and almandine, show a significant decrease.

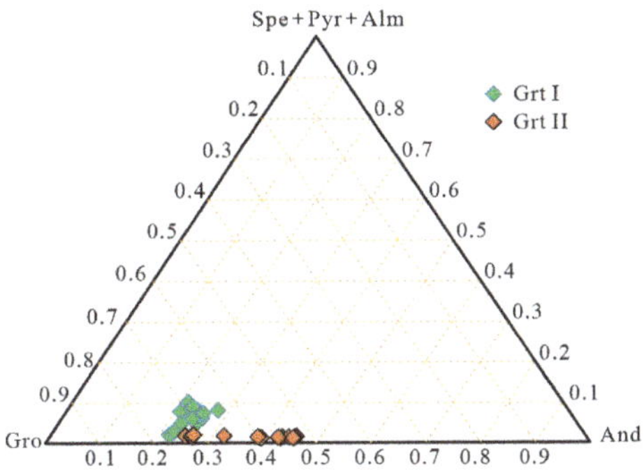

Figure 6. The end member composition diagram of garnets from the Gongjuelong Sn polymetallic deposit.

5.1.2. Cassiterite

The major element composition of cassiterite from the main mineralization stage of the Gongjuelong deposit was analyzed using electron probe microanalysis, and the results are presented in Table S2. The two samples (GJL-3-1 and GJL-3-3) reveal that the most abundant trace elements include Fe, Ti, Nb, Ta, and W, whereas the secondary trace elements are Zr, Al, Mn, and Si. Electron probe analysis detected small amounts of TiO_2 (<0.72 wt%), WO_3 (<0.35 wt%), FeO (<0.65 wt%), Nb_2O_5 (<0.07 wt%), and Ta_2O_5 (<0.07 wt%) in cassiterite from the Gongjuelong deposit. Additionally, trace amounts of ZrO_2 (<0.04 wt%), Al_2O_3 (<0.04 wt%), MnO (<0.02 wt%), and SiO_2 (<0.04 wt%) were also detected. In the binary diagrams, it is observed that the SnO_2 content decreases gradually from the core to the rim of the cassiterite, whereas the contents of TiO_2 and FeO increase. The contents of Ti, Fe, W, Nb, and Sn exhibit a negative correlation (Figure 7).

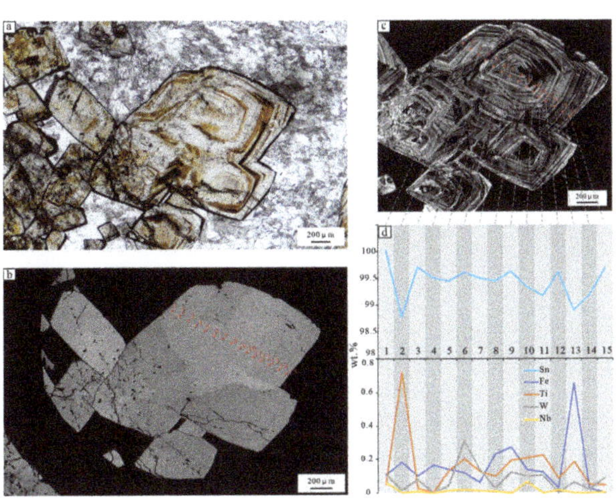

Figure 7. Photomicrograph (**a**), BSE image (**b**), CL (**c**), and compositional variations (**d**) of the cassiterite samples from the Gongjuelong deposit.

5.2. In Situ Trace Elements of Garnet, Cassiterite, and Quartz

5.2.1. Garnet

The LA-ICP-MS rare-earth element (REE) data for garnet (Table S3) reveal relatively high total REE contents, with ΣREE (excluding Y) ranging from 42.6 ppm to 203 ppm. For Grt-I, the total REE content is significantly higher, with ΣREE spanning from 165 ppm to 3973 ppm, averaging 1022 ppm. The content of LREEs ranges from 114 ppm to 3757 ppm, comprising 93% of the total rare-earth element (REE) content, whereas the HREEs range from 27.2 ppm to 216 ppm. The LREE/HREE ratio ranges from 2.3 to 32.8, with a mean of 11.6. The REE distribution pattern shows a pronounced enrichment in light REEs and a marked rightward inclination (Figure 8). A significant negative Eu anomaly (δEu = 0.1 to 0.7, average = 0.28) and a slight negative Ce anomaly (δCe = 0.8 to 1.0, average = 0.9) are observed. The $(La/Yb)_N$ ratios range from 1.8 to 97.3, indicating notable fractionation between light and heavy REEs, as well as between LREEs and HREEs.

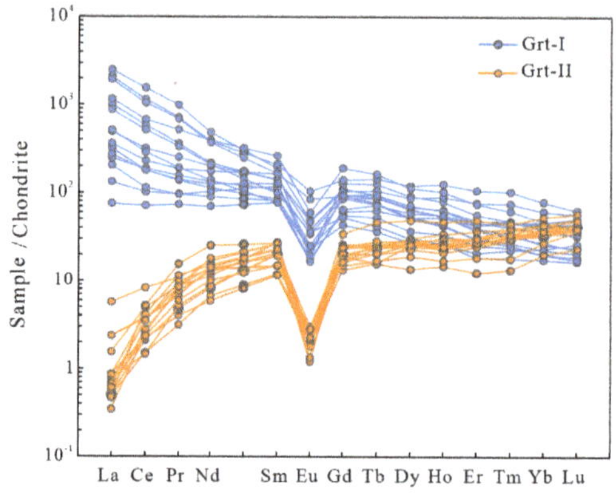

Figure 8. Chondrite-normalized REE patterns of the Gongjuelong garnet (normalizing values from [41]).

For Grt-II, the REE content is comparatively lower, with ΣREE values ranging from 31.2 ppm to 402 ppm, with an average of 62.9 ppm. The LREE content varies from 6.4 ppm to 332 ppm, whereas the HREE content ranges from 15.2 ppm to 70 ppm. The LREE/HREE ratio ranges from 0.2 to 4.7, with an average of 0.8. The REE distribution pattern shows enrichment in HREEs, with a moderate leftward slope. A moderate negative Eu anomaly (δEu = 0.04 to 0.2, average = 0.10) and a weak positive Ce anomaly (δCe = 0.7 to 1.9, average = 1.3) are noted. The $(La/Yb)_N$ ratios range from 0.01 to 5.4, indicating significant fractionation between LREEs and HREEs.

5.2.2. Cassiterite

The trace element content of cassiterite from the Gongjuelong deposit is summarized in Table S4. The results indicate that the trace element concentrations in cassiterite vary significantly (Figure 9). The most abundant trace elements are Fe, Ti, and W, with concentrations ranging from 235 to 3740 ppm, 350 to 3015 ppm, and 1.9 to 6364 ppm, respectively. All cassiterite grains analyzed contain detectable amounts of Nb and Zr, with Nb concentrations ranging from 1.6 to 694 ppm and Zr concentrations varying from 1 to 190 ppm. Additionally, other trace elements present in cassiterite include Mn (<23.7 ppm), Hf (0.1–5.8 ppm), Ta (<6.5 ppm), Sc (<14.8 ppm), Ga (0.2–1.7 ppm), Ge (<2.1 ppm), and U (0.7–138 ppm). The concentrations of REEs, Rb, and Sr in cassiterite are generally below detection limits. A positive correlation is observed between the contents of W, Mn, Fe, Hf, Zr, Nb, and Ta.

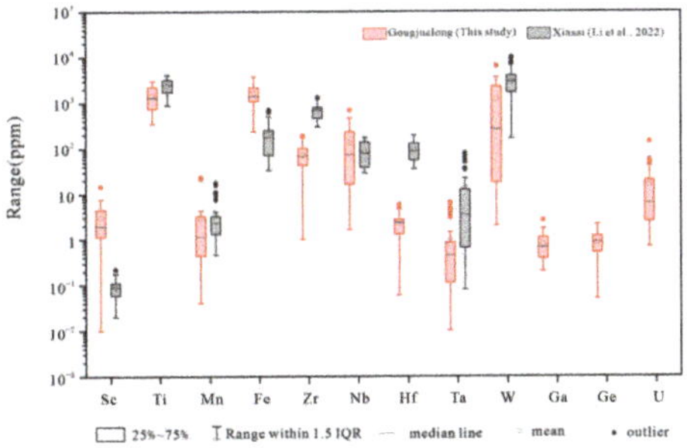

Figure 9. Box plots of selected trace elements in cassiterite from Gongjuelong (data of Xiasai are from [5]).

5.2.3. Quartz

The trace element content of quartz from the Gongjuelong deposit are provided in Table S5. and illustrated in Figure 10. The most abundant trace elements in quartz are Li, Al, Ti, Ge, and P. Among these, Li, Al, and Ti are primarily used for interpretation. The trace element concentrations of quartz in stage III of the Gongjuelong deposit are as follows: Li content ranges from 0.1 to 23.4 ppm, with an average of 10.0 ppm. Al concentrations vary between 35.8 and 185 ppm, with an average of 86.5 ppm. Ti ranges from 0.9 to 2.8 ppm, with an average of 1.0 ppm. Ge concentrations range from 1.6 to 5.5 ppm, with an average of 4.0 ppm. P content varies from 14.0 to 173 ppm, with an average of 64.3 ppm.

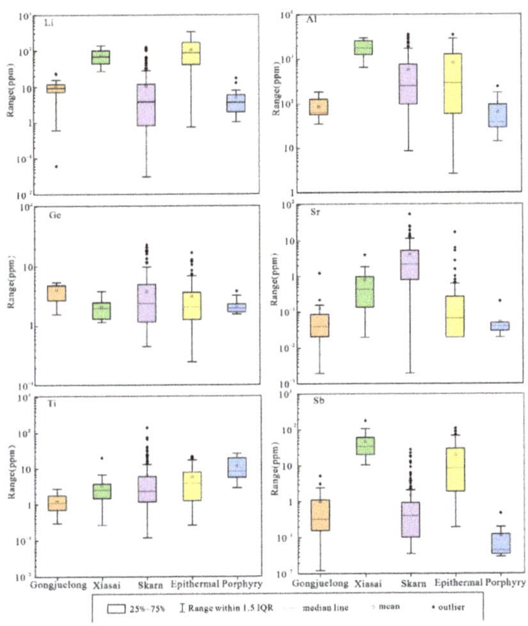

Figure 10. Box plots of selected trace elements in quartz from Gongjuelong, Xiasai and other different types of deposits [42]. Orange represents the Gongjuelong deposit, green represents the Xiasai deposit, purple represents skarn deposits, yellow represents epithermal deposits, and blue represents porphyry deposits.

6. Discussion

6.1. Substitution Mechanisms

6.1.1. Garnet

The REEs in garnet are valuable for tracing fluid sources and fluid–rock interactions, owing to their similar chemical properties, which allow for their fractionation under specific geological conditions [43]. The general formula of garnet is $X_3Y_2Z_3O_{12}$, where X denotes divalent cations, Y represents trivalent cations, and Z predominantly contains Si^{4+} in the silicon tetrahedral structure. The primary mechanisms of trace element incorporation into garnet crystals are surface adsorption, substitution, and isomorphic replacement [44]. U, REE^{3+}, and Y^{3+} commonly substitute for Ca^{2+} in the X site of the garnet lattice [45–47]. Based on the ionic radii and their effects on garnet's crystal structure, the likely substitution site for REEs is the octahedral X site (e.g., Ca^{2+}) [48]. Given that Eu^{2+} has an ionic radius of 1.12 Å, which is similar to that of Ca^{2+} (0.99 Å), and both ions share the same charge, Eu^{2+} is more likely to substitute for Ca^{2+} at the X site [45]. However, the substitution behavior of REE^{3+} and Eu^{2+} in garnet differs, which requires coupled substitution or the formation of vacancies to maintain charge balance [45,49]. Previous studies have proposed several mechanisms to maintain charge balance [45,48,50,51], which are summarized as follows:

$$[X^{2+}]VIII\text{-}2[X^{+}]VIII+[REE^{3+}]VIII \quad (1)$$

$$[X^{2+}]VIII\text{-}[REE^{3+}]VIII+[Si^{4+}]IV\text{-}[Z^{3+}]IV \quad (2)$$

$$[X^{2+}]VIII\text{-}[REE^{3+}]VIII+[Y^{3+}]IV\text{-}[Y^{2+}]IV \quad (3)$$

$$[X^{2+}]VIII\text{-}3[]VIII+[REE^{3+}]VIII \quad (4)$$

In these equations, X^{+} represents Na^{+}, X^{2+} mainly refers to Ca^{2+}, Z^{3+} represents Al^{3+} or Fe^{3+}, Y^{3+} is predominantly replaced by Al^{3+}, Y^{2+} corresponds to Mg^{2+} or Fe^{2+}, and [ ] denotes vacancies (Ca). When the Na^{+} content is high in garnet, Na^{+} can substitute into the lattice to maintain charge balance, thereby promoting the incorporation of REEs through the first mechanism [50,52]. However, the content of Na in garnet from Gongjuelong is predominantly below detection limits, indicating that the first mechanism is not the primary mechanism for REE incorporation. If REEs are incorporated into garnet through the second mechanism, charge compensation occurs by substituting trivalent cations into the tetrahedral sites [45,53]. In the element correlation diagrams for garnet samples from the Gongjuelong deposit, no clear correlation is observed between ΣREE contents and Al or Fe (Figure 11b,d), indicating that the second substitution mechanism is unlikely to be the dominant mechanism for REE incorporation. Furthermore, the low content of Mg in both garnet types (Table S1) further indicates that Mg is insufficient to facilitate this substitution. Consequently, the third substitution mechanism is unlikely to be the dominant process in this region. A certain degree of linear positive correlation with ΣREE is exhibited only by Eu. Therefore, the fourth substitution mechanism is likely the primary mechanism for REE incorporation into garnet. However, previous studies have demonstrated the substitution of REEs in garnet is not solely governed by crystal chemistry [54,55]; it is also significantly influenced by the physicochemical features of the fluid (e.g., pH, f_{O2}), the water/rock ratio, and the dynamics of mineral growth.

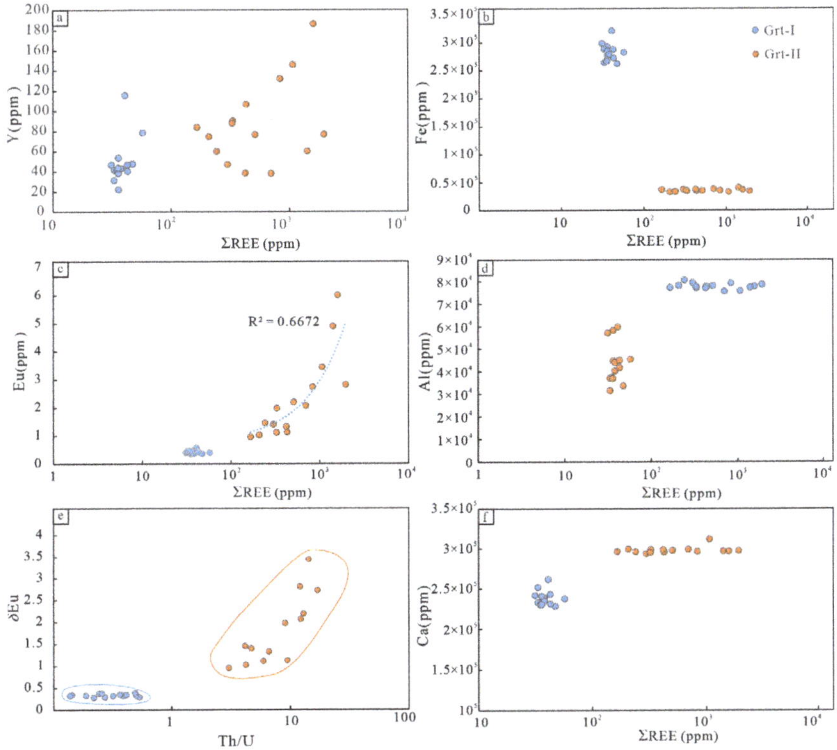

Figure 11. Diagram of (**a–d**,**f**) Y, Fe, Eu, Al, and Ca vs. ΣREE and (**e**) δEu vs. Th/U.

6.1.2. Cassiterite

The substitution mechanisms of elements in cassiterite primarily involve: (1) Fe^{3+} + (Nb, Ta)$^{5+}$ → 2Sn^{4+}; (2) 2(Nb, Ta)$^{5+}$ + (Fe, Mn)$^{2+}$ → 3Sn^{4+}; (3) Ti^{4+}, Zr^{4+}, U^{4+} → Sn^{4+}; (4) V^{5+} + Sc^{3+} → 2Sn^{4+} [55–58]. In magmatic cassiterite, such as that found in rare-element-rich granites and pegmatites, the dominant substitution mechanism is 2(Nb, Ta)$^{5+}$ + (Fe, Mn)$^{2+}$ → 3Sn^{4+} [59].

The Nb and Ta contents of cassiterite in the Gongjuelong Sn deposit are similar to those in the Xiasai deposit, whereas the Fe and Mn contents are significantly higher (Figure 12b). These values are comparable to those observed in cassiterite from tin-bearing granites and skarn ores in the Gejiu deposit [60]. The lack of clear correlation between Fe + Mn and Nb + Ta (Figure 12f) further indicates that none of these substitution mechanisms are the dominant processes in the Gongjuelong deposit. In cassiterite, the simplest charge balance mechanism is achieved by substituting two Sn^{4+} ions with a trivalent cation (e.g., Fe^{3+}, Al^{3+}, Sc^{3+}, Ga^{3+}) and a pentavalent cation (e.g., V^{5+}, Nb^{5+}, Ta^{5+}). Should this mechanism be dominant, a 1:1 charge balance between trivalent and pentavalent cations would be expected. However, the contents of trivalent cations (predominantly Fe^{3+}) are much greater than pentavalent cations (Table S4), and this saturation is also observed in the large tin deposits of Gejiu, Dachang, and Xiasai [5,60,61]. This suggests the need for an additional mechanism to accommodate the surplus trivalent cations. Therefore, we propose that the substitution mechanism Fe^{3+} + OH^- → Sn^{4+} + O^{2-} is the primary process [58,59].

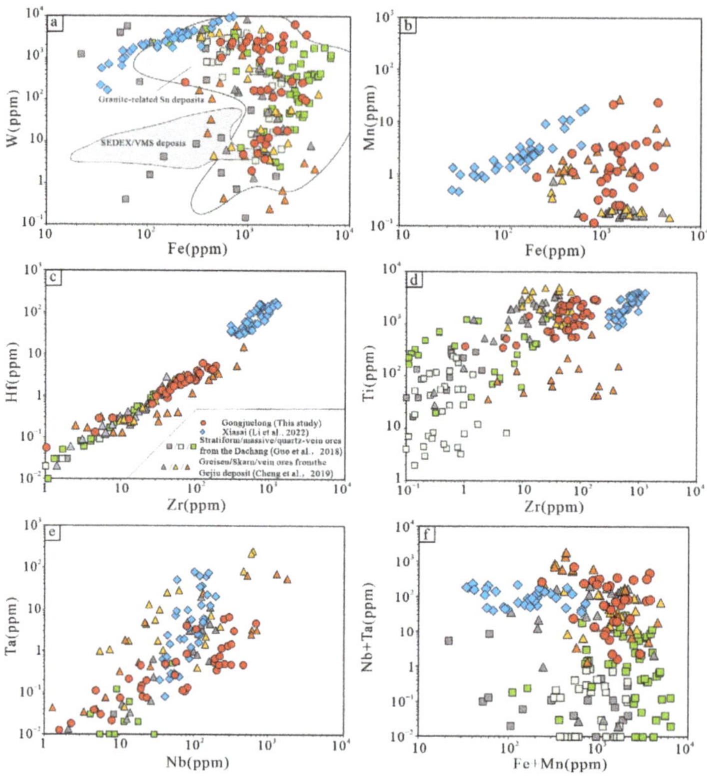

Figure 12. Trace element binary diagrams in cassiterite from the Gongjuelong deposit. Areas of granite-related and Sedex/VMS tin deposits are from [62]. Data from the Gejiu and Dachang Sn deposits are from [60,61], respectively. (**a**) W vs Fe (**b**) Mn vs Fe (**c**) Hf vs Zr (**d**)Ti vs Zr (**e**) Ta vs Nb (**f**) Nb+Ta vs Fe+Mn.

Moreover, cassiterite from the Gongjuelong deposit displays relatively high concentrations of W and Ti, which show a significant negative correlation with Sn (Figure 7d). These elements (W^{4+}, Ti^{4+}) can directly substitute for Sn^{4+} in the lattice.

6.1.3. Quartz

To interpret the incorporation and coupled substitution of major cations into the quartz crystal lattice, trace element concentrations were converted to atoms per formula unit (Table S5, Figure 13a–c). The content of Al^{3+} (190 apfu) is significantly higher than that of other trivalent cations, such as B^{3+} (73.5 apfu) and Sb^{3+} (0.5 apfu), in quartz from the Gongjuelong deposit. This is attributable to the fact that Al is commonly found in the Earth's crust, and its ionic radius is similar to that of Si^{4+} (r = 0.34 Å) and Al^{3+} (r = 0.47 Å). Consequently, the substitution of Al^{3+} for Si^{4+} in the quartz lattice, resulting in the formation of the $[AlO_4/M^+]^0$ structural center, is the primary mechanism for trace element incorporation in quartz [63]. The M+ ions can include H^+, Li^+, Na^+, K^+, Cu^+, and Ag^+ [64]. The trivalent cations are mainly dominated by Al^{3+}, followed by Sb^{3+} and B^{3+}. For monovalent cations, Li^+ is the most abundant, with Na^+ and K^+ following. The pentavalent cations are predominantly represented by P^{5+}. Overall, the incorporation of cations into the quartz crystal lattice is mainly governed by the coupled substitution processes: (Al^{3+}, Sb^{3+}, B^{3+}) + (Li^+, Na^+, K^+, H^+) $\leftrightarrow$ Si^{4+} and (Al^{3+}, Sb^{3+}, B^{3+}) + (P^{5+}) $\leftrightarrow$ $2Si^{4+}$ [64–67].

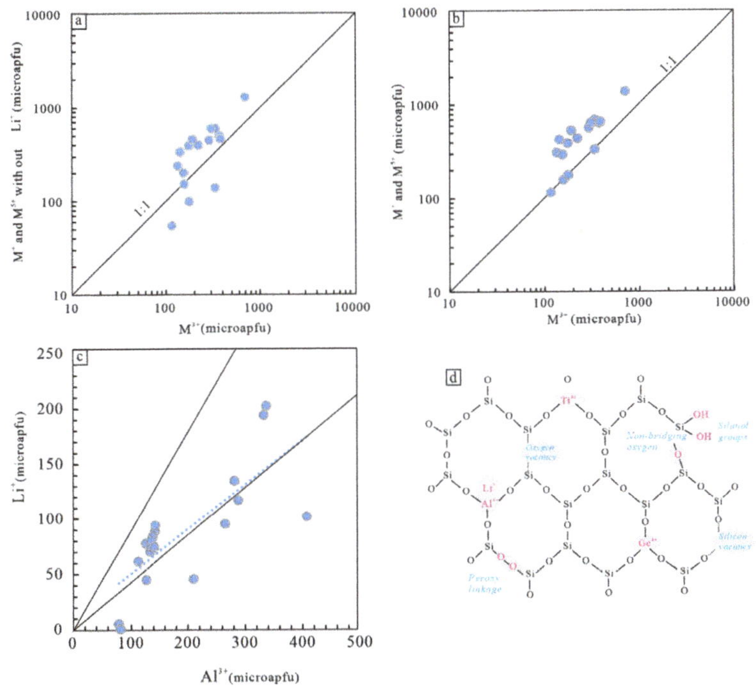

Figure 13. Plot of M^{3+} vs. M^{+} and M^{5+} without Li^{+} (**a**). Plot of M^{3+} vs. M^{+} and M^{5+} (**b**). Plot of Al^{3+} vs. Li+ (**c**) of quratz from Gongjuelong deposit. Schematic quartz structure showing the most common defect types (**d**) (modified after [67]).

Previous studies have shown that the substitution mechanism for trace elements entering quartz may involve $Sb^{3+} + Li^{+} \leftrightarrow Si^{4+}$, according to the Li^{+}/Al^{3+} ratio in quartz higher than 1:1. By contrast, the Li^{+}/Al^{3+} ratio is approximately 1:1, and the coupled substitution mechanism may be $Al^{3+} + Li^{+} \leftrightarrow Si^{4+}$. Conversely, when the Li^{+}/Al^{3+} ratio is less than 1:1, it suggests competition between Na^{+}, K^{+}, or P^{5+} with Li^{+}, which couples with Al^{3+} to replace Si^{4+} in the quartz lattice [63,68]. In the Gongjuelong deposit, the Li^{+}/Al^{3+} ratio (≈1:2) and $(M^{+}+M^{5+})/M^{3+}$ ratio (≈1:1) in quartz suggest two potential cation substitution mechanisms: (1) coupled substitution via $Al^{3+} + (Li^{+}/Na^{+}/K^{+}) \leftrightarrow Si^{4+}$ and (2) charge-balanced substitution through $Al^{3+} + P^{5+} \leftrightarrow 2Si^{4+}$.

6.2. Geochemistry Significance of the Garnets

Variations in the physicochemical conditions during the formation of garnet in skarn deposits affect the chemical composition of the garnet. Thus, the chemical compositions of garnet can serve as a proxy for the physicochemical conditions during its formation [53]. Previous studies have demonstrated that andradite typically forms in oxidizing environments, whereas grossular tends to form in reducing environments [68–70].

Grt-I in the Gongjuelong Sn deposit is primarily composed of grossular and minor amounts of andradite and trace amounts of manganese–aluminum garnet and magnesium–aluminum garnet ($Gro_{64\text{-}76}And_{20\text{-}28}Spe{+}Pyr{+}Alm_{2.39\text{-}10.33}$). It is similar to late-stage garnets from the skarn-type copper deposits in the Hongniu area of western Yunnan [71]. This indicates that Grt-I is enriched in Al and formed in a low-oxygen fugacity, moderately acidic environment. In terms of Grt-II, grossular remains dominant, and the contents of andradite increase. Grt-II is part of the solid solution series between grossular and andradite ($Gro_{52\text{-}73}And_{25\text{-}45}Spe+Pyr+Alm_{1\text{-}2}$), with the andradite component being slightly lower than that of grossular, suggesting that Grt-II formed in a weakly reducing, near-neutral environment. From Grt-I to Grt-II, a marked increase in FeO content is observed, whereas the Al_2O_3 content decreases slightly (Figure 5; Table S1). In Grt-II, the grossular component decreases, whereas the andradite component gradually increases. This suggests that the oxygen fugacity of garnet in the Gongjuelong deposit increased from Grt-I to Grt-II, with the ore-forming environment evolving from relatively reducing to relatively oxidizing conditions, reflecting the progressive changes in the ore-forming environment. However, the end-member composition remains dominated by grossular, indicating that both types of garnet in the Gongjuelong deposit formed under reducing conditions. Notable differences in the REE characteristics are observed between these two types (Figure 8). The REE distribution pattern of Grt-I exhibits a steep rightward skew, whereas Grt-II shows a more gradual leftward skew. The left-skewed REE distribution pattern in Grt-II is similar to that in garnet from the Shizhuyuan [72] and Jinchuantang [73] deposits in southern Hunan, China, whereas the right-skewed pattern in Grt-I is comparable to that in garnet from the Huangshaping deposit [74] in the same region. In addition to the differences in REE distribution patterns, the Eu anomalies in the two types of garnet also differ significantly with Grt-II showing a more pronounced negative Eu anomaly (Figure 8). Moreover, the LREE/HREE ratios differ between the two garnet types. Grt-I exhibits a higher LREE/HREE ratio (2.3 to 32.8, average 11.6). In contrast, Grt-II shows a lower ratio (0.2 to 4.7, average 0.8).

It has been proposed that pH is a key factor in REE fractionation [75]. In mildly acidic environments, the REE distribution pattern of garnet is more significantly influenced by Cl^-, which enhances the stability of soluble Eu^{2+} relative to REE^{3+}, leading to a pronounced positive Eu anomaly, LREE enrichment, and HREE depletion. On the other hand, near-neutral conditions lead to fluids that are enriched in HREEs and depleted in LREEs, typically showing negative or no Eu anomalies [75,76]. At Gongjuelong, Grt-I exhibits a clear negative Eu anomaly and a steep rightward skew in the REE distribution pattern, whereas Grt-II shows a more significant negative Eu anomaly than Grt-I, along with a gradual leftward skew in the REE distribution pattern. This suggests that Grt-I and Grt-II likely formed under mildly acidic and near-neutral conditions, respectively. The U content in garnet can act as a proxy for the redox conditions of hydrothermal fluid systems [41,59,72]. Due to differences in ionic radii, U^{4+} is more likely to substitute into garnet than U^{6+}. As a result, the lower U concentration and higher Th/U ratio in Grt-II suggest a more oxidizing environment (higher f_{O2}) compared to Grt-I (Figure 11e). Both Grt-I and Grt-II exhibit a strong linear correlation between Y and Ho (Figure 14a). In the Y/Ho-La/Ho diagram (Figure 14b), both Grt-I and Grt-II align along the same horizontal line, suggesting that they are products of the same ore-forming fluid. Wang [77] conducted micro-thermometry and Laser Raman Spectroscopy on fluid inclusions in garnet from the Gongjuelong deposit. The results indicate that the early ore-forming fluids were characterized by high homogenization temperatures (T_h = 435.2–500 °C) and high salinities (36.9–42.5 wt%), representing a $NaCl\text{-}H_2O\text{-}CH_4$ system. The presence of CH_4 indicates that the early ore-forming fluids were relatively reducing, consistent with the discussion of garnet.

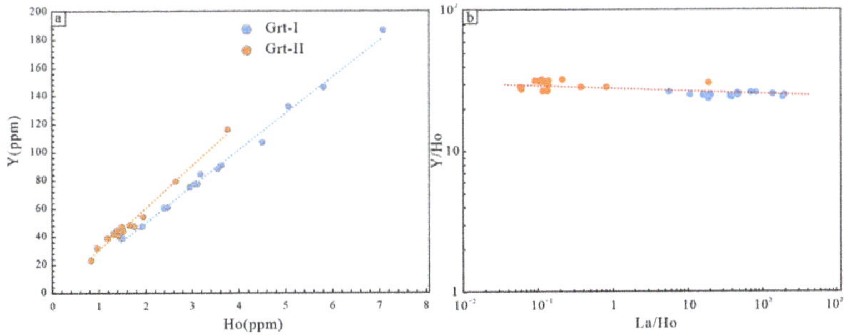

Figure 14. Diagrams of Y-Ho (**a**) and Y/Ho-La/Ho (**b**) of garnets in the Gongjuelong Sn polymetallic deposit.

6.3. Genetic Model and Implications for Exploration

To address the genetic relationship between Sn mineralization and Pb–Zn–Ag mineralization in the Yidun Terrane, this study systematically compares the geology, major/trace elements of cassiterite and quartz, isotopes (e.g., S, Pb, and H-O), fluid inclusions, and mineralization ages of the two deposit types using the data from this study and references (Table 1).

The Sn deposits and hydrothermal vein-type Pb–Zn–Ag±Sn deposits in the central Yidun Terrane are commonly located in the proximal and distal locations to the Late Cretaceous reduced and highly differentiated granite, respectively [78,79]. Most of the Sn deposits are situated within the contact zones between granite and Triassic carbonate rocks. Notable examples include the large-scale Cuomolong and Jiaogenma Sn deposit in the north of Rongyicuo pluton, as well as the small-scale Gongjuelong Sn deposit in the west of Haizishan pluton. In contrast, the hydrothermal vein-type Pb–Zn–Ag±Sn deposits generally occur within the metamorphosed sandstone and shale that are located distal to the causative granite intrusions. From the perspective of wall rock alteration, Sn±Ag deposits typically exhibit high-temperature alteration, including skarnization, greisenization, and tourmalinization. The primary ore minerals in these deposits include cassiterite, chalcopyrite, pyrite, and to a lesser extent, galena and sphalerite. Alternatively, Pb–Zn–Ag±Sn deposits are enriched in F, Si, and carbonates, with localized skarnization and greisenization. The primary ore minerals in these deposits include sphalerite and galena, with minor cassiterite and other minerals.

Based on the comparison of ion coordination and ionic radius with Sn^{4+}, several trace elements, such as Fe, Ga, V, Cr, Sc, Sb, W, U, Zr, Hf, Ti, Nb, and Ta, are considered to be compatible with cassiterite [56,60]. The fluid source, depositional circumstances, and kind of mineralization can all be inferred from the composition of cassiterite [60–62,80,81]. Cassiterite of the Gongjuelong Sn deposit is characterized by relatively high concentrations of W (1.9 to 6364 ppm) and Fe (235 to 3740 ppm), which are closely consistent with those of cassiterite from granite-related Sn deposits and Xiasai Pb–Zn–Ag deposit (Figure 12a). Cassiterite of the Gongjuelong deposit shows high contents of Zr and Hf, with Zr/Hf ratios ranging from 18.7 to 49.4. Cathodoluminescence (CL) imaging of cassiterite reveals alternately light and dark zoned bands, and no zircon inclusions are detected (Figure 7c). This suggests that hydrothermal alteration, rather than fluid exchange or zircon fractionation, is the main factor influencing the Zr/Hf ratio variations. Additionally, the presence of fluorite in the Gongjuelong deposit indicates that the mineralizing fluids are fluorine-rich. Fluorine-enhanced fluids are known to increase the mobility of Zr and preferentially extract Zr over Hf. The Zr/Hf ratios in Gongjuelong cassiterite are comparable to those in cassiterite from skarn ores (17.8–61.5, mean = 34.0) within the Gejiu deposit [60], aligning with characteristics typical of granite-related tin mineralization. Compared to the cassiterite of the Xiasai deposit, the cassiterite of the Gongjuelong deposit exhibits similar Ti concentrations (350 to 3015 ppm) and significantly lower Zr concentrations [5] (Figure 12c). The Ti/Zr ratios in Gongjuelong cassiterite vary by an order of magnitude, likely due to local element fractionation during crystal growth. However, the absence of Ti-rich or Zr-rich minerals in Gongjuelong suggests that crystallization has a minimal role in Ti/Zr fractionation. In granite-associated tin deposits, a declining Ti/Zr ratio in cassiterite typically reflects the gradual loss of Ti relative to Zr as the mineralizing fluids evolve and migrate away from the granite intrusion. This trend is consistent with the higher solubility of Zr compared to Ti in hydrothermal fluids [82]. Studies in the Gejiu polymetallic tin deposit have demonstrated that Ti/Zr ratios in cassiterite from various ore types (e.g., tin-bearing granite, greisenized ores, skarn ores, sub-oxidized ores, oxidized ores, and vein ores) steadily decrease with increasing distance from the host granite. A similar pattern is observed in the Bolivian tin metallogenic belt in Central South America [83]. The Fe/W and Zr/Hf ratios in cassiterite indicate that both belong to a granite-related magmatic–hydrothermal genesis. Moreover, the Ti/Zr ratio in cassiterite from Gongjuelong, located closer to the granite (average = 36.8), is markedly higher than that of the more distant Xiasai deposit (average = 3.7).

Under magmatic–hydrothermal ore-forming systems, hydrothermal quartz formed under low-temperature conditions (300 °C or below) contains high contents of Al and Li, whereas quartz formed under high-temperature conditions (approaching 400 °C) generally exhibits lower contents of Al and Li [84]. The Ge/Ti ratio serves as an effective indicator of magmatic fractionation due to the contrasting geochemical behaviors of Ge and Ti during magma evolution. Germanium (Ge) is a moderately incompatible element that tends to enrich in the melt as fractionation progresses, while titanium (Ti) is a compatible element primarily hosted in minerals such as ilmenite and rutile, causing its concentration to decrease with fractionation. As a result, the Ge/Ti ratio increases significantly with advancing magmatic fractionation, where higher Ge/Ti values indicate magmas with a higher degree of evolution [85]. Wang [42] collected trace element data from quartz in 1220 samples of various deposit types, including quartz from 196 pegmatites, 66 Carlin-type deposits, 52 porphyry deposits, 74 epithermal deposits, 203 skarn deposits, 565 granites, and 64 orogenic deposits. Based on machine learning, a novel discrimination diagram using the Ti/Ge ratio vs. P in quartz was proposed to classify deposit types. Therefore, the chemical composition of quartz can provide valuable insights into the hydrothermal environmental conditions and deposit types. The average concentrations of Li and Al of quartz in the Gongjuelong deposit are 10 ppm and 86.5 ppm, respectively. The quartz from the Xiasai deposit is enriched in both Li and Al, with concentrations approximately one order of magnitude higher than those in the Gongjuelong deposit. The variations in trace element compositions of quartz can effectively distinguish the genetic types of hydrothermal deposits with different styles of mineralization, which is due to the significant influence of temperature, pH, and pressure on the distribution of elements like Li, Al, P, Ti, and Ge in quartz [42,68,86,87]. The P (13.9 to 173.1 ppm; average 64.3 ppm), Al, and Li contents and Ti/Ge ratios (0.1 to 1.1; average 0.4) of stage III quartz from Gongjuelong are similar to those of sulfide-bearing quartz from the Xiasai Pb–Zn–Ag deposit (Figure 15a,b). Interestingly, both the quartz from Gongjuelong and Xiasai are consistent with those from the skarn-type deposits and reduced Sn-related Pb–Zn veins in trace elements (Figure 15). On the other hand, the Al and Ti contents of quartz from the Gongjuelong Sn deposit are closely consistent with those of skarn deposits and significantly distinct from those of epithermal and porphyry-type deposits (Figure 15c). In contrast to the Gongjuelong Sn deposit, the Xiasai Pb–Zn–Ag deposit has similar Ti content but higher Al content. Both of the date ranges of the two deposits are located between the fields of epithermal- and porphyry-type deposits, overlapping with the range of skarn deposits. However, the quartz from the Xiasai deposit shows higher Al content, positioning it closer to the range of epithermal-type deposits. The trace element contents in quartz and cassiterite from the Gongjuelong deposit exhibit certain similarities with those from the Xiasai deposit, both indicating a magmatic-related origin.

Recently, during our investigation of the Shaxi Pb–Zn–Ag deposit in the central Yidun Terrane, we identified high-temperature, halite-bearing tri-phase (solid–liquid–vapor) aqueous inclusions in quartz associated with sulfides. The vapor phase of these inclusions contains volatile components, including CH_4, N_2, and H_2S, further supporting a genetic link between Pb–Zn mineralization and magmatic–hydrothermal fluids in the region. On the other hand, the $\delta^{34}S$ values of sulfides from the Gongjuelong Sn deposit range from −6.74‰ to −3.56‰ [77], overlapping with those from the Xiasai Pb–Zn–Ag deposit (−10.1‰ to −1.2‰) [4]. Previous $\delta^{34}S$ measurements for the Lianlong pluton in the region yielded a value of −8.1‰ [7], similar to those from both the Gongjuelong Sn deposit and the Xiasai Pb–Zn–Ag deposit. This suggests that sulfur in both types of deposits possibly originates from the granitic magmas, with additional sulfur from wall-rock contamination. The Pb isotopic compositions of sulfides from the Gongjuelong deposit ($^{206}Pb/^{204}Pb$ = 18.182–18.826, $^{207}Pb/^{204}Pb$ = 15.575–15.704, $^{208}Pb/^{204}Pb$ = 38.444–39.083) are similar to those from the Xiasai deposit ($^{206}Pb/^{204}Pb$ = 18.702–18.731, $^{207}Pb/^{204}Pb$ = 15.680–15.718, $^{208}Pb/^{204}Pb$ = 39.000–39.128) [4,77]. These values are consistent with those found in orogenic belts and the upper crust, suggesting a shared crustal source for the lead in both deposits. The H-O isotopic compositions of quartz from the Gongjuelong deposit ($\delta D_{V\text{-}SMOW}$ = −117 to −98.4‰, $\delta^{18}O$ = 1.1 to 1.8‰) and the Xiasai deposit (substage II-2: $\delta D_{V\text{-}SMOW}$ = −104, δ18O = 8.4; substage II-3: $\delta D_{V\text{-}SMOW}$ = −105 to −138, $\delta^{18}O$ = −3.3 to −3.9) [4] both show characteristics of mixing between magmatic hydrothermal fluids and meteoric water. In conclusion, the ore-forming materials in both deposits primarily come from the granitic rocks, with sulfur and lead contamination from surrounding wall-rocks, while the ore-forming fluids are a combination of magmatic hydrothermal fluids and meteoric fluids. The above evidence indicates that the Pb–Zn mineralization seems to be influenced not only by a single magmatic hydrothermal process. The strata or surrounding rocks can also provide ore-forming materials. Therefore, during the infiltration of meteoric fluids or the intrusion of magmatic hydrothermal fluids, there is also a certain possibility that the Pb–Zn mineralization occurs due to the water–rock reaction with the surrounding wall-rocks. However, the magmatic hydrothermal process still plays an important role in the mineralization of Pb–Zn.

In the central part of the Yidun Terrane, the Cretaceous granitoids include the Rongyicuo, Ruolulong, Lianlong, Xiasai, and Haizishan plutons. Radiometric dating of whole-rock Rb–Sr isochrons has yielded ages ranging from 81 to 93 Ma for the Lianlong, Rongyicuo, and Haizishan plutons [8,14,34,88]. More precise zircon LA-ICP-MS U-Pb dating indicates that these granites of the central Yidun Terrane were emplaced between 94.4 and 108 Ma [4,78,89,90]. These findings confirm that the granitoids in this region primarily formed during the late Early Cretaceous to early Late Cretaceous periods. Previous studies on diagenetic and metallogenic ages have established that magmatic activity and metallogenesis in the central Yidun Terrane predominantly occurred during the Late Cretaceous. These processes were associated with an extensional environment that followed tectonic collision [77,91]. Two types of mineralization are identified in the central Yidun Terrane: skarn-type Sn and hydrothermal vein-type Pb–Zn–Ag. Both mineralization types are closely associated with the late Cretaceous A-type granites [4–8,17]. Furthermore, Li et al. [4] obtained an Rb–Sr isochron age of 99 ± 3 Ma for sphalerite in the quartz–chalcopyrite and sphalerite veins, which provides a close approximation of the Pb–Zn–Ag mineralization age. Regarding Sn mineralization, LA-ICP-MS U-Pb dating of cassiterite from the Xiasai quartz–cassiterite veins yielded a U-Pb age of 99.2 ± 0.8 Ma [5]. The cassiterite from the quartz–cassiterite vein of the Gongjuelong Sn deposit revealed a Tera–Wasserburg U-Pb lower-intercept age of 96.88 ± 1.36 Ma (unpublished data), which is closely consistent with the published ages of both the Pb–Zn–Ag mineralization (99 ± 3 Ma) and late Cretaceous granites (94.4–108 Ma).

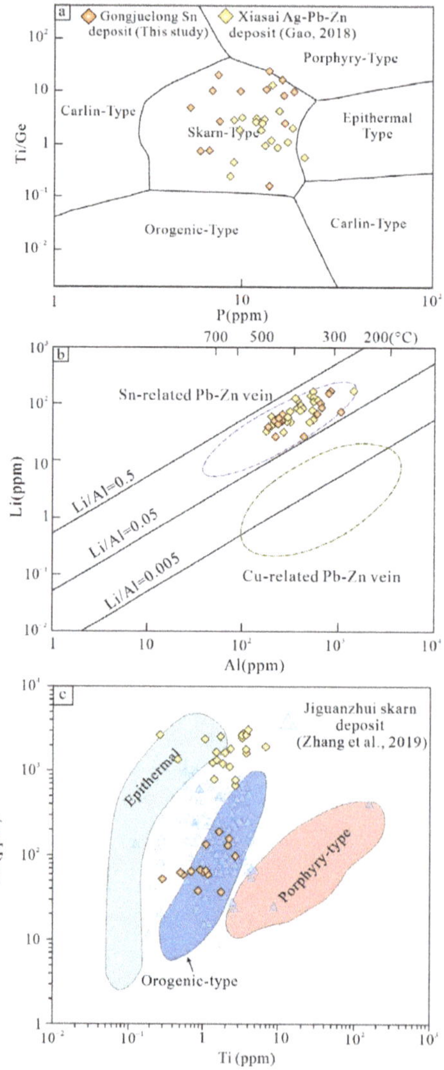

Figure 15. Discrimination diagrams of ore genesis based on quartz chemistry: (**a**,**b**) quartz Al-Ti and Ti/Ge-P diagrams to distinguish different deposit types; (**c**) quartz Li-Al diagram to distinguish vein-type Pb–Zn±Ag deposits related to Sn and Cu. These diagrams are modified after [42,86,88].

Table 1. Comparison of metallogenic factors between the Gongjuelong Sn deposit and the Xiasai Ag–Pb–Zn deposit.

Type	Location	Host Rock	Mineralization Age	Mineral Assemblage	Sulfur Isotopes (‰)	Lead Isotopes	Hydrogen and Oxygen Isotopes of Quartz (‰)	Fluid Inclusions T_h (°C)	Fluid Inclusions Salinities (wt%)	Reference
Gongjuelong skarn Sn deposit	Situated within the contact zones between granite and Triassic carbonate rocks.	Skarns, marble, or hornfels	Cassiterite U-Pb 97 ± 2 Ma	Ore minerals include cassiterite, chalcopyrite, pyrrhotite, and arsenopyrite, with minor sphalerite and pyrite.	$\delta^{34}S = (-6.74)–(-3.56)$	$^{206}Pb/^{204}Pb$ = 18.182–18.826 $^{207}Pb/^{204}Pb$ = 15.575–15.704 $^{208}Pb/^{204}Pb$ = 38.444–39.083	$\delta D_{V\text{-}SMOW}$ = (−117)–(−98.4) $\delta^{18}O$ = 1.1–1.8(unpublished)	Quartz: 172.6–232.1	5.2–13.6	[77]
Xiasai hydrothermal vein type Ag–Pb–Zn deposit	Occurs within the metamorphosed sandstone and shale that are located distal to the causative granite intrusions.	Sandstones, carbonate rock, and siliceous slate	sphalerite Rb–Sr 99 ± 3 Ma	Cassiterite, arsenopyrite, pyrrhotite, chalcopyrite, sphalerite, galena, Ag-bearing minerals, and native bismuth.	$\delta^{34}S = (-10.1)–(-1.2)$	$^{206}Pb/^{204}Pb$ = 18.711–18.748 $^{207}Pb/^{204}Pb$ = 15.711–15.752 $^{208}Pb/^{204}Pb$ = 39.104–39.225	Quartz from sub-stage II-2 $\delta D_{V\text{-}SMOW} = -104$, $\delta^{18}O = 8.4$ Quartz from sub-stage II-3 $\delta D_{V\text{-}SMOW}$ = (−105)–(−138), $\delta^{18}O = (-3.3)–(-0.9)$	Quartz from stage I: 423–481 Quartz from stage II-1: 285–386 Quartz from stage II-3: 158–242	14.9–19.0 3.5–8.0 3.4–5.7	[4]

These combined findings from the geology, geochronology, and geochemistry of hydrothermal minerals convince us to propose that the Sn mineralization is closely linked with the Pb–Zn–Ag mineralization in terms of time, space, and genesis. This magmatic–hydrothermal ore system consisting of proximal Sn and distal Pb–Zn–Ag mineralization was the product in response to the fluid exsolution of reduced and highly fractionated granites during the late Cretaceous. The ore metal zones are common in the magmatic–hydrothermal mineralization system, including the porphyry Cu-Mo-Au, greisen W-Sn, and skarn Fe-Cu/W-Sn ore systems [92–95]. Porphyry-type Cu-Mo-Au and skarn-type Fe-Cu mineralization are generally associated with oxidized, lowly differentiated and water-rich magmatic rocks, whereas skarn W-Sn mineralization is linked to reduced, highly differentiated and water-poor magmatic rocks [13,22,95]. Fe, Cu, and Au mineralization mainly come from partial melting of the mantle and magmatic crust, whereas W and Sn are primarily sourced from partial melting of the sedimentary crust [96]. Pb–Zn–Ag mineralization generally occurs in the peripheral strata around these systems. For instance, the Weilasituo Zn-Cu-Ag deposit in the southern Great Xing'an Range recently revealed W-Sn mineralization at depth [97,98]. Similarly, a series of hydrothermal Pb–Zn-Sb-Ag deposits (e.g., Zhaxikang, Keyue, and Suoyue) formed around the Cuonadong leucogranites, during which skarn-type W-Sn-Be and pegmatite-type Be-Nb-Ta mineralization were progressively discovered [99,100].

During exploration, areas exhibiting skarn alteration between the contact zones highly fractionated granite and carbonates should be prioritized for potential Sn mineralization, with Pb–Zn–Ag mineralization more likely to occur in the surrounding shallower strata. Recently, 16 rare metal veins (e.g., Be, Nb, Ta) were identified in the northern Haizishan granite [101]. Therefore, the region has potential not only for Sn–Pb–Zn–Ag mineralization but also for rare metal exploration.

7. Conclusions

1. The two types of garnet in the Gongjuelong Sn deposit, Grt-I and Grt-II, are part of the solid solution series of grossular and andradite and reflect an initially reducing ore-forming environment. The ore-forming environment shifted from a moderately acidic reducing to a weakly reducing neutral environment during the transition from Grt-I to Grt-II.
2. The trace element contents of garnet, cassiterite, and quartz from the Gongjuelong Sn deposit suggest that the Sn mineralization is favored by the reducing conditions developed by or around the magma intrusions. The quartz in this deposit formed at relatively high temperatures. The ore-forming fluid evolved from early-stage, high-temperature, reducing, moderately acidic fluids associated with garnet indicates that the Gongjuelong deposit is a proximal skarn deposit.
3. The magmatic–hydrothermal system in the central Yidun Terrane is centered on the Late Cretaceous reduced, highly differentiated granites. The system includes proximal skarn Sn deposits and distal hydrothermal vein-type Pb–Zn–Ag±Sn deposits. Further exploration should focus on identifying Pb–Zn–Ag mineralization and rare metal mineralization in distal, structurally controlled areas. The locations of Pb–Zn mineralization could serve as pathfinders for hindered Sn deposits at depth.

Supplementary Materials: The following supporting information can be downloaded at: https://www.mdpi.com/article/10.3390/min15030314/s1, Table S1: representative electron probe micro-analyzer (EPMA) results of garnets from the Gongjuelong deposit; Table S2: representative electron probe micro-analyzer (EPMA) results of cassiterites from the Gongjuelong deposit; Table S3: trace element analysis result (ppm) of the Gongjuelong garnet by laser ablation-inductively coupled plasma-mass spectrometry (LA-ICP-MS); Table S4: trace element analysis result (ppm) of the Gongjuelong Cassiterite by laser ablation-inductively coupled plasma-mass spectrometry (LA-ICP-MS); Table S5: trace element analysis result (ppm) of the Gongjuelong quartz by laser ablation-inductively coupled plasma-mass spectrometry (LA-ICP-MS).

Author Contributions: Conceptualization, Y.P.; Data curation, Y.Z. (Yuchang Zhou) and Y.Z. (Yan Zhang); Formal analysis, M.S.; Funding acquisition, C.L. and J.T.; Project administration, C.L. and J.T.; Validation, Y.Z. (Yuchang Zhou); Visualization, M.S. and Z.W.; Writing—original draft, Y.Z. (Yuchang Zhou); Writing—review and editing, Y.P. All authors have read and agreed to the published version of the manuscript.

Funding: This research was funded by the National Natural Science Foundation of China (No.42230311), the Mount Everest Scientific Research Plan of Chengdu University of Technology (No.80000-2024ZF11426), the Technology Support Project for the New Round of Mineral Exploration Breakthrough Strategic Action (No. ZKKJ202427), the Key Program of Sichuan Province Natural Science Foundation (No. 2024NSFSC1954), and the China National Nuclear Corporation (CNNC) (No. GT2201).

Data Availability Statement: The authors confirm that the data supporting the findings of this study are available within the article.

Acknowledgments: We express our gratitude to the Beijing Research Institute of Uranium Geology for their guidance in the field.

Conflicts of Interest: The authors declare no conflicts of interest.

References

1. Mulshaw, S.C.; Puig, C.; Spiro, B.; Buchanan, D.L. Genesis of epizonal Ag vein mineralization at San Bartolome in central Ecuador; textural evidence, fluid inclusions, and stable isotope geochemistry. *Econ. Geol.* **1997**, *92*, 210–227. [CrossRef]
2. Meinert, L.D.; Dipple, G.M.; Nicolescu, S. World Skarn Deposits. In *Economic Geology 100th Anniversary Volume*; Hedenquist, J.W., Thompson, J.F.H., Goldfarb, R.J., Richards, J.P., Eds.; Society of Economic Geologists: Littleton, CO, USA, 2005; pp. 299–336.
3. Welch, E.B.; Cooke, G.D. Internal phosphorus loading in shallow lakes: Importance and control. *Lake Reserv. Manag.* **2005**, *21*, 209–217. [CrossRef]
4. Li, Y.J.; Wei, J.H.; Tan, J.; Fu, L.B.; Li, H.; Ke, K.J. Albian–Cenomanian A-type granite-related Ag–Pb–Zn veins in the central Yidun Terrane, SW China: Constraints from the Xiasai deposit. *Miner. Depos.* **2019**, *55*, 1047–1070. [CrossRef]
5. Li, Y.J.; Zhang, R.Q.; Ji, H.; Wang, C.; Tan, J.; Wei, J.H. Chemistry and U–Pb geochronology of cassiterite in the Xiasai deposit, central Yidun Terrane (SW China): Link between Sn and Ag–Pb–Zn mineralisation. *Ore Geol. Rev.* **2022**, *149*, 105106. [CrossRef]
6. Chen, Y.-K.; Ni, P.; Pan, J.-Y.; Cui, J.-M. Tracing the magmatic-hydrothermal evolution of the Xianghualing tin-polymetallic skarn deposit, South China: Insights from LA-ICP-MS analysis of fluid inclusions. *Miner. Depos.* **2024**, *59*, 1553–1578. [CrossRef]
7. Qu, X.M. Metallogenic geological characteristics of the Lianlong skarn-type Sn-Ag polymetallic deposit in western Sichuan Province. *Acta Geol. Sin.* **2001**, *22*, 29–34.
8. Qu, X.M.; Hou, Z.Q.; Zhou, S.G. Geochemical and Nd, Sr isotopic study of the post-orogenic granites in the Yidun arc belt of northern Sanjiang region, southwestern China. *Resour. Geol.* **2002**, *52*, 163–172. [CrossRef]
9. Quan, L. Geological characteristics and genesis of Xiasai silver-polymetallic deposit in western Sichuan Province. *Miner. Depos.* **2003**, *22*, 121–128.
10. Lin, F.C.; Yang, J.R.; Chen, C.D.; He, X.G.; Lin, Q.; Lu, Y.X.; Liu, Y.C.; Xie, E.S. Potentiality and assessment of Cu–Ag–Pb–Zn–Sn mineral resources in the Yidun metallogenic belt. *Acta Geol. Sichuan* **2003**, *23*, 141–144.
11. Lin, Q. Geological features and prospecting potential for the Xialong Ag–Pb–Zn deposit in Batang, Sichuan. *Acta Geol. Sichuan* **2010**, *30*, 447–449.
12. Lin, Q. Geological features and prospecting criteria for the Jiaogenma Sn–Zn deposit in Litang, Sichuan. *Acta Geol. Sichuan* **2011**, *31*, 37–40.
13. He, X.G.; Li, S.H.; Xie, E.S. Geological features and prospecting potential for the Shaxi Ag–Pb–Zn deposit, Batang, Sichuan. *Acta Geol. Sichuan* **2004**, *24*, 77–81.

14. Ying, H.L.; Wang, D.H.; Fu, X.F. Timing and lead and sulfur isotope composition of Xiasai granite and silver polymetallic deposit in Batang, Sichuan Province. *Kuangchuang Dizhi (Miner. Depos.)* **2006**, *25*, 135–146.
15. Hou, Z.Q.; Zaw, K.; Pan, G.T.; Mo, X.X.; Xu, Q.; Hu, Y.Z.; Li, X.Z. Sanjiang Tethyan metallogenesis in SW China: Tectonic setting, metallogenic epochs and deposit types. *Ore Geol. Rev.* **2007**, *31*, 48–87. [CrossRef]
16. Zou, G.; Zheng, R.; Hu, S. Geological features of the Xiasai silver-polymetallic deposit in West Sichuan, China. *J.-Chengdu Univ. Technol.* **2008**, *35*, 93.
17. Fan, X. Geological features, ore-formation and mineralization zonation of the Hailong-Cuomolong skarn type Sn-polymetallic deposit in Batang, Sichuan. *Acta Geol. Sichuan* **2009**, *29*, 112–123.
18. Zhang, C.Q.; Wu, Y.; Wang, D.H.; Chen, Y.C.; Rui, Z.Y.; Lou, D.B.; Chen, Z.H. Brief introduction on metallogeny of Pb–Zn deposits in China. *Acta Geol. Sin.* **2014**, *88*, 2252–2268.
19. Zhang, D.Q.; Jiang, B.; Wang, D.H.; Wang, C.H.; Chen, Y.C.; Bai, G. A summary of resources characteristic and metallogenic regularity of silver deposits in China. *Acta Geol. Sin.* **2015**, *89*, 1008–1025.
20. Zhang, J.B.; Ding, J.H.; Nan, G.L. The characteristics and potential of tin resources in China. *Geol. China* **2015**, *42*, 839–852.
21. Chen, Z.H.; Wang, D.H.; Sheng, J.F.; Ying, L.J.; Liang, T.; Wang, C.H.; Liu, L.J.; Wang, Y.L. The metallogenic regularity of tin deposits in China. *Acta Geol. Sin.* **2015**, *89*, 1026–1037.
22. Reid, A.; Wilson, C.J.; Shun, L.; Pearson, N.; Belousova, E. Mesozoic plutons of the Yidun Arc, SW China: U/Pb geochronology and Hf isotopic signature. *Ore Geol. Rev.* **2007**, *31*, 88–106. [CrossRef]
23. Wang, Q.W.; Wang, K.M.; Kan, Z.Z.; Fu, X.F. *Granites and Associated Minerogenetic Series of Ore Deposits in Western Sichuan Province*; Geological Publishing House: Beijing, China, 2008.
24. Hou, Z.Q.; Qu, X.M.; Zhou, J.; Yang, Y.; Huang, D.; Lu, Q.; Tang, S.; Yu, J.; Wang, H.; Zhao, J. Collision-orogenic processes of the Yidun arc in the Sanjiang region: Record of granites. *Acta Geol. Sin. Chin. Ed.* **2001**, *75*, 497–505.
25. Deng, J.; Wang, Q.; Li, G.; Santosh, M. Cenozoic tectono-magmatic and metallogenic processes in the Sanjiang region, southwestern China. *Earth-Sci. Rev.* **2014**, *138*, 268–299. [CrossRef]
26. Chen, J.L.; Xu, J.F.; Ren, J.B.; Huang, X.X. Late Triassic E-MORB-like basalts associated with porphyry Cu-deposits in the southern Yidun continental arc, eastern Tibet: Evidence of slab-tear during subduction? *Ore Geol. Rev.* **2017**, *90*, 1054–1062. [CrossRef]
27. Zeng, P.S.; Li, W.C.; Wang, H.P.; Li, H. The Indosinian Pulang superlarge porphyry copper deposit in Yunnan, China: Petrology and chronology. *Acta Petrol. Sin.* **2006**, *22*, 989–1000.
28. Meng, J.Y.; Yang, L.Q.; Lu, L.; Gao, X.; Li, J.X.; Luo, Y.Z. Re-Os dating of molybdenite from the Hongshan Cu-Mo deposit in Northwest Yunnan and its implications for mineralization. *Acta Petrol. Sin.* **2013**, *29*, 1214–1222.
29. Peng, T.P.; Zhao, G.C.; Fan, W.M.; Peng, B.X.; Mao, Y.S. Zircon geochronology and Hf isotopes of Mesozoic intrusive rocks from the Yidun terrane, Eastern Tibetan Plateau: Petrogenesis and their bearings with Cu mineralization. *J. Asian Earth Sci.* **2014**, *80*, 18–33. [CrossRef]
30. Wang, B.Q.; Zhou, M.F.; Li, J.W.; Yan, D.P. Late Triassic porphyritic intrusions and associated volcanic rocks from the Shangri-La region, Yidun terrane, Eastern Tibetan Plateau: Adakitic magmatism and porphyry copper mineralization. *Lithos* **2011**, *127*, 24–38. [CrossRef]
31. Wang, B.Q.; Zhou, M.F.; Chen, W.T.; Gao, J.F.; Yan, D.P. Petrogenesis and tectonic implications of the Triassic volcanic rocks in the northern Yidun Terrane, Eastern Tibet. *Lithos* **2013**, *175*, 285–301. [CrossRef]
32. Wang, B.Q.; Wang, W.; Chen, W.T.; Gao, J.F.; Zhao, X.F.; Yan, D.P.; Zhou, M.F. Constraints of detrital zircon U–Pb ages and Hf isotopes on the provenance of the Triassic Yidun Group and tectonic evolution of the Yidun Terrane, Eastern Tibet. *Sediment. Geol.* **2013**, *289*, 74–98. [CrossRef]
33. Deng, J.; Wang, Q.F.; Li, G.J. Superimposed orogeny and composite metallogenic system: Case study from the Sanjiang Tethyan belt, SW China. *Acta Petrol. Sin.* **2016**, *32*, 2225–2247.
34. Hou, Z.Q.; Yang, Y.Q.; Qu, X.M.; Huang, D.; Lu, Q.; Wang, H.; Yu, J.; Tang, S. Tectonic evolution and mineralization systems of the Yidun arc orogen in Sanjiang region, China. *Acta Geol. Sin.-Chin. Ed.* **2004**, *78*, 109–120.
35. Li, W.C.; Pan, G.T.; Hou, Z.Q.; Mo, X.X.; Wang, L.Q.; Zhang, X.F. *Metallogenic Theory and Exploration Technology of Multi-Arc-Basin-Terrane Collision Orogeny in "Sanjiang" Region, Southwest China*; Springer Nature: Berlin/Heidelberg, Germany, 2023.
36. Li, Q.X.; Wang, X.M.; Zhang, Y.; Liu, Z.Q.; Zhao, L. *Cuola (Yidun) Sheet H-47-67-B Haizishan Sheet H-47-68-A 1/50,000 Regional Geological Survey Report: Mineral Part*; Sichuan Provincial Bureau of Geological and Mineral Exploration and Development, Regional Geological Survey Team: Chengdu, China, 1990.
37. Jochum, K.P.; Weis, U.; Stoll, B.; Kuzmin, D.; Yang, Q.; Raczek, I.; Jacob, D.E.; Stracke, A.; Birbaum, K.; Frick, D.A.; et al. Determination of reference values for NIST SRM 610–617 glasses following ISO guidelines. *Geostand. Geoanal. Res.* **2011**, *35*, 397–429. [CrossRef]

38. Audétat, A. Compositional evolution and formation conditions of magmas and fluids related to porphyry Mo mineralization at Climax, Colorado. *J. Petrol.* **2015**, *56*, 1519–1546. [CrossRef]
39. Liu, Y.S.; Hu, Z.C.; Gao, S.; Günther, D.; Xu, J.; Gao, C.G.; Chen, H.H. In situ analysis of major and trace elements of anhydrous minerals by LA-ICP-MS without applying an internal standard. *Chem. Geol.* **2008**, *257*, 34–43. [CrossRef]
40. Peng, Y.W.; Gu, X.X.; Lv, P.R.; Zhang, Y.M.; Cheng, W.B.; Wang, X.L. Genesis and tectonic setting of the Late Devonian Tawuerbieke gold deposit in the Tulasu ore cluster, western Tianshan, Xinjiang, China. *Int. Geol. Rev.* **2016**, *59*, 1344–1368. [CrossRef]
41. Irber, W. The lanthanide tetrad effect and its correlation with K/Rb, Eu/Eu∗, Sr/Eu, Y/Ho, and Zr/Hf of evolving peraluminous granite suites. *Geochim. Cosmochim. Acta* **1999**, *63*, 489–508. [CrossRef]
42. Wang, Y.; Qiu, K.F.; Hou, Z.L.; Yu, H.C. Quartz Ti/Ge-P discrimination diagram: A machine learning based approach for deposit classification. *Acta Petrol. Sin.* **2022**, *38*, 281–290.
43. Zhai, D.G.; Liu, J.J.; Zhang, H.Y.; Wang, J.P.; Su, L.; Yang, X.A.; Wu, S.H. Origin of oscillatory zoned garnets from the Xieertala Fe–Zn skarn deposit, northern China: In situ LA–ICP-MS evidence. *Lithos* **2014**, *190*, 279–291. [CrossRef]
44. Gaspar, M.; Knaack, C.; Meinert, L.D.; Moretti, R. REE in skarn systems: A LA-ICP-MS study of garnets from the Crown Jewel gold deposit. *Geochim. Cosmochim. Acta* **2008**, *72*, 185–205. [CrossRef]
45. Tian, Z.-D.; Leng, C.-B.; Zhang, X.-C.; Zafar, T.; Zhang, L.-J.; Hong, W.; Lai, C.-K. Chemical composition, genesis and exploration implication of garnet from the Hongshan Cu-Mo skarn deposit, SW China. *Ore Geol. Rev.* **2019**, *112*, 103016. [CrossRef]
46. Yu, J.; Li, N.; Zhang, B.; Ulrich, T.; Chen, X.; Hand, M.; Morrissey, L.J.; Payne, J.L.; Chen, Y.J. Perspective of Permian porphyry Cu-Au mineralization in Chinese Western Tianshan: Constraints from sulfide Re-Os dating and trace element study of the Kuruer deposit, Xinjiang. *Ore Geol. Rev.* **2022**, *142*, 104707. [CrossRef]
47. Carlson, W.D. Rates and mechanism of Y, REE, and Cr diffusion in garnet. *Am. Mineral.* **2012**, *97*, 1598–1618. [CrossRef]
48. Park, C.; Choi, W.; Kim, H.; Park, M.-H.; Kang, I.-M.; Lee, H.-S.; Song, Y. Oscillatory zoning in skarn garnet: Implications for tungsten ore exploration. *Ore Geol. Rev.* **2017**, *89*, 1006–1018. [CrossRef]
49. Enami, M.; Bolin, C.; Yoshida, T.; Kawabe, I. A mechanism for Na incorporation in garnet: An example from garnet in orthogneiss from the Su-Lu terrane, eastern China. *Am. Mineral.* **1995**, *80*, 475–482. [CrossRef]
50. Grew, E.S.; Marsh, J.H.; Yates, M.G.; Lazic, B.; Armbruster, T.; Locock, A.; Bell, S.W.; Dyar, M.D.; Bernhardt, H.-J.r.; Medenbach, O. Menzerite-(Y), a new species,{(Y, REE)(Ca, Fe2+) 2}[(Mg, Fe2+)(Fe3+, Al)](Si3) O12, from a felsic granulite, Parry Sound, Ontario, and a new garnet end-member,{Y2Ca}[Mg2](Si3) O12. *Can. Mineral.* **2010**, *48*, 1171–1193. [CrossRef]
51. Ding, T.; Ma, D.S.; Lu, J.J.; Zhang, R.Q. Garnet and scheelite as indicators of multi-stage tungsten mineralization in the Huangshaping deposit, southern Hunan province, China. *Ore Geol. Rev.* **2018**, *94*, 193–211. [CrossRef]
52. Xu, J.; Ciobanu, C.L.; Cook, N.J.; Zheng, Y.Y.; Sun, X.; Wade, B.P. Skarn formation and trace elements in garnet and associated minerals from Zhibula copper deposit, Gangdese Belt, southern Tibet. *Lithos* **2016**, *262*, 213–231. [CrossRef]
53. Jamtveit, B.; Wogelius, R.A.; Fraser, D.G. Zonation patterns of skarn garnets: Records of hydrothermal system evolution. *Geology* **1993**, *21*, 113–116. [CrossRef]
54. Bau, M.; Dulski, P. Distribution of yttrium and rare-earth elements in the Penge and Kuruman iron-formations, Transvaal Supergroup, South Africa. *Precambrian Res.* **1996**, *79*, 37–55. [CrossRef]
55. Neiva, A. Geochemistry of cassiterite and wolframite from tin and tungsten quartz veins in Portugal. *Ore Geol. Rev.* **2008**, *33*, 221–238. [CrossRef]
56. Murciego, A.; Sanchez, A.G.; Dusausoy, Y.; Pozas, J.M.; Ruck, R. Geochemistry and EPR of cassiterites from the Iberian Hercynian Massif. *Mineral. Mag.* **1997**, *61*, 357–365. [CrossRef]
57. Plimer, I.; Lu, J.; Kleeman, J. Trace and rare earth elements in cassiterite—Sources of components for the tin deposits of the Mole Granite, Australia. *Miner. Depos.* **1991**, *26*, 267–274. [CrossRef]
58. Möller, P.; Dulski, P.; Szacki, W.; Malow, G.; Riedel, E. Substitution of tin in cassiterite by tantalum, niobium, tungsten, iron and manganese. *Geochim. Cosmochim. Acta* **1988**, *52*, 1497–1503. [CrossRef]
59. Tindle, A.G.; Breaks, F.W. Oxide minerals of the Separation Rapids rare-element granitic pegmatite group, northwestern Ontario. *Can. Mineral.* **1998**, *36*, 609–635.
60. Cheng, Y.B.; Spandler, C.; Kemp, A.; Mao, J.W.; Rusk, B.; Hu, Y.; Blake, K. Controls on cassiterite (SnO_2) crystallization: Evidence from cathodoluminescence, trace-element chemistry, and geochronology at the Gejiu Tin District. *Am. Mineral.* **2019**, *104*, 118–129. [CrossRef]
61. Guo, J.; Zhang, R.Q.; Sun, W.D.; Ling, M.X.; Hu, Y.B.; Wu, K.; Luo, M.; Zhang, L.C. Genesis of tin-dominant polymetallic deposits in the Dachang district, South China: Insights from cassiterite U–Pb ages and trace element compositions. *Ore Geol. Rev.* **2018**, *95*, 863–879. [CrossRef]
62. Hennigh, Q.T. Cassiterite in Synsedimentary, Massive Sulfide Deposits: A Comparison to Cassiterite in "Granite-Affiliated" Tin Deposits Including Pegmatites. Master's Thesis, Colorado School of Mines, Golden, CO, USA, 1993; p. 129.

63. Fu, S.L.; Lan, Q.; Yan, J. Trace element chemistry of hydrothermal quartz and its genetic significance: A case study from the Xikuangshan and Woxi giant Sb deposits in southern China. *Ore Geol. Rev.* **2020**, *126*, 103732. [CrossRef]
64. Griffiths, J.; Owen, J.; Ward, I. Paramagnetic resonance in neutron-irradiated diamond and smoky quartz. *Nature* **1954**, *173*, 439–440. [CrossRef]
65. Müller, A.; Koch-Müller, M. Hydrogen speciation and trace element contents of igneous, hydrothermal and metamorphic quartz from Norway. *Mineral. Mag.* **2009**, *73*, 569–583. [CrossRef]
66. Breiter, K.; Badanina, E.; Ďurišová, J.; Dosbaba, M.; Syritso, L. Chemistry of quartz—A new insight into the origin of the Orlovka Ta-Li deposit, Eastern Transbaikalia, Russia. *Lithos* **2019**, *348*, 105206. [CrossRef]
67. Götze, J. Chemistry, textures and physical properties of quartz—Geological interpretation and technical application. *Mineral. Mag.* **2009**, *73*, 645–671. [CrossRef]
68. Misra, K. *Understanding Mineral Deposits*; Springer Science & Business Media: Berlin/Heidelberg, Germany, 2012.
69. Lu, H.-Z.; Liu, Y.; Wang, C.; Xu, Y.; Li, H. Mineralization and fluid inclusion study of the Shizhuyuan W-Sn-Bi-Mo-F skarn deposit, Hunan Province, China. *Econ. Geol.* **2003**, *98*, 955–974. [CrossRef]
70. Xie, G.Q.; Mao, J.W.; Zhu, Q.Q.; Yao, L.; Li, Y.H.; Li, W.; Zhao, H.J. Geochemical constraints on Cu–Fe and Fe skarn deposits in the Edong district, Middle–Lower Yangtze River metallogenic belt, China. *Ore Geol. Rev.* **2015**, *64*, 425–444. [CrossRef]
71. Gao, X.; Deng, J.; Meng, J.Y.; Yan, H.; Li, J.X.; Yang, C.H.; Sun, N.; Wei, C. Characteristics of garnet in the Hongniu skarn copper deposit, western Yunnan. *Acta Petrol. Sin.* **2014**, *30*, 2695–2708. [CrossRef]
72. Chen, J.; Halls, C.; Stanley, C. Tin-bearing skarns of South China: Geological setting and mineralogy. *Ore Geol. Rev.* **1992**, *7*, 225–248. [CrossRef]
73. Liu, X.F.; Yuan, S.D.; Shuang, Y.; Yuan, Y.B.; Mi, J.R.; Xuan, Y.S. In situ LA-ICP-MS REE analyses of the skarn garnets from the Jinchuantang tin-bismuth deposit in Hunan Province, and their significance. *Acta Petrol. Sin.* **2014**, *30*, 163–177.
74. Li, D.F.; Tan, C.Y.; Miao, F.Y.; Liu, Q.F.; Zhang, Y.; Sun, X.M. Initiation of Zn-Pb mineralization in the Pingbao Pb–Zn skarn district, South China: Constraints from U-Pb dating of grossular-rich garnet. *Ore Geol. Rev.* **2019**, *107*, 587–599. [CrossRef]
75. Zhang, Y.; Shao, Y.J.; Wu, C.D.; Chen, H.Y. LA-ICP-MS trace element geochemistry of garnets: Constraints on hydrothermal fluid evolution and genesis of the Xinqiao Cu–S–Fe–Au deposit, eastern China. *Ore Geol. Rev.* **2017**, *86*, 426–439. [CrossRef]
76. Allen, D.E.; Seyfried Jr, W. REE controls in ultramafic hosted MOR hydrothermal systems: An experimental study at elevated temperature and pressure. *Geochim. Cosmochim. Acta* **2005**, *69*, 675–683. [CrossRef]
77. Wang, Z. Geochemical Characteristics and Genesis of the Gongjuelong Tin Polymetallic Deposit, in the Central Yidun Terrane. Master's Thesis, Chengdu University of Technology, Chengdu, China, 2024.
78. Gao, X.; Yang, L.Q.; He, W.Y.; Groves, D. Redox conditions, compositional parameters, and indirect subduction-related source of Cretaceous Sn and Cu–Mo fertile post-subduction granites in the Yidun Terrane of eastern Tibet. *Ore Geol. Rev.* **2021**, *139*, 104506. [CrossRef]
79. Yang, F.C.; Li, W.C.; Jiang, X.J.; Li, C.; Wang, Z.Q.; Sun, H.Y.; Zhou, J.X. Late Cretaceous granitic intrusions and associated deposits in the Yidun Arc of the eastern Tibetan Plateau. *J. Asian Earth Sci.* **2020**, *192*, 104249. [CrossRef]
80. Chen, Y.-K.; Ni, P.; Pan, J.-Y.; Cui, J.-M.; Li, W.-S.; Fang, G.-J.; Zhao, Z.-H.; Xu, Y.-M.; Ding, J.-Y.; Han, L. Fluid evolution and metallogenic mechanism of the Xianghualing skarn-type Sn deposit, South China: Evidence from petrography, fluid inclusions and trace-element composition of cassiterite. *Ore Geol. Rev.* **2023**, *154*, 105351. [CrossRef]
81. Cai, J.; Su, W.; Shen, N.; Pan, J.; Zhao, H. In situ trace element compositions and UPb ages of cassiterite from tin-polymetallic deposits in the Dachang district, Guangxi, China: Implications for ore genesis and exploration. *J. Geochem. Explor.* **2023**, *247*, 107173. [CrossRef]
82. Taylor, R.G. *Geology of Tin Deposits*; Elsevier: Amsterdam, The Netherlands, 2014.
83. Gemmrich, L.; Torró, L.; Melgarejo, J.C.; Laurent, O.; Vallance, J.; Chelle-Michou, C.; Sempere, T.P. Trace element composition and U-Pb ages of cassiterite from the Bolivian tin belt. *Miner. Depos.* **2021**, *56*, 1491–1520. [CrossRef]
84. Jourdan, A.-L.; Vennemann, T.; Mullis, J.; Ramseyer, K. Oxygen isotope sector zoning in natural hydrothermal quartz. *Mineral. Mag.* **2009**, *73*, 615–632. [CrossRef]
85. Breiter, K.; Ďurišová, J.; Dosbaba, M. Quartz chemistry–A step to understanding magmatic-hydrothermal processes in ore-bearing granites: Cínovec/Zinnwald Sn-W-Li deposit, Central Europe. *Ore Geol. Rev.* **2017**, *90*, 25–35. [CrossRef]
86. Rusk, B. Cathodoluminescent textures and trace elements in hydrothermal quartz. In *Quartz: Deposits, Mineralogy and An-Alytics*; Springer Geology: Berlin/Heidelberg, Germany, 2012; pp. 307–329.
87. Gao, S.; Zou, X.; Hofstra, A.H.; Qin, K.; Marsh, E.E.; Bennett, M.M.; Li, G.; Jiang, J.; Su, S.; Zhao, J. Trace elements in quartz: Insights into source and fluid evolution in magmatic-hydrothermal systems. *Econ. Geol.* **2022**, *117*, 1415–1428. [CrossRef]
88. Li, Y.; Wei, J.; Chen, H.; Li, H.; Chen, C.; Hou, B. Petrogenesis of the Xiasai Early Cretaceous A-type granite from the Yidun Island Arc Belt, SW China: Constraints from zircon U-Pb age, geochemistry and Hf isotope. *Geotecton. Metallog.* **2014**, *38*, 939–953.

89. Zhang, F.Y.; Lai, S.C.; Qin, J.F. Petrogenesis and Geological Significance of the Late Cretaceous Haizishan Monzogranite from the Yidun Island Arc. *Geol. J. China Univ.* **2018**, *24*, 340.
90. Qu, L.F. Petrologic and Chronological Study of the Granite Belt Which Locate from Gaogong to the North Part of Genie in Yidun Island arc. Master's Thesis, Chengdu University of Technology, Chengdu, China, 2016.
91. Zhang, X.F.; Li, W.C.; Yang, Z.; Wang, Y.Q.; Yan, T.T.; Zhang, J.J. Temporal-spatial distribution and evolution implication of the composite intrusion in the Xiuwacu W-Mo deposit, SE Tibetan Plateau. *Sediment. Geol. Tethyan Geol.* **2022**, *42*, 105–121.
92. Li, Z.K.; Wang, C.; Wang, Z.Q.; Li, K.W.; Zhou, J.J.; Li, X.M.; Hu, J.; Zhang, J. A giant porphyry-skarn-epithermal Mo-W-Cu-Zn-Pb-Ag metallogenic system in the Nannihu ore field, southern North China Craton: Constraints from new geochronological and pyrite geochemical data. *Ore Geol. Rev.* **2023**, *163*, 105792. [CrossRef]
93. Peng, Y.W.; Gu, X.X.; Chi, G.X.; Xue, C.J.; Zhang, Y.M.; Yuan, P.; Wan, Y.; Wang, X.-L. Genesis of the Nailenggele Mo–Cu–Pb–Zn polymetallic orefield in the Boluokenu Metallogenic Belt, Western Tianshan, China: Constraints from geochronology, fluid inclusions and isotope geochemistry. *Ore Geol. Rev.* **2021**, *129*, 103940. [CrossRef]
94. Burt, D.M. Acidity-salinity diagrams; application to greisen and porphyry deposits. *Econ. Geol.* **1981**, *76*, 832–843. [CrossRef]
95. Gao, X. The Late Cretaceous Polymetallic Mineralization Related to Granitoid Systems in the Yidun Terrane, East Tibet. Ph.D. Thesis, China University of Geosciences, Beijing, China, 2018.
96. Sato, K. Sedimentary Crust and Metallogeny of Granitoid Affinity: Implications from the Geotectonic Histories of the Circum-J apan S ea Region, Central A ndes and Southeastern A ustralia. *Resour. Geol.* **2012**, *62*, 329–351. [CrossRef]
97. Gao, X.; Zhou, Z.H.; Breiter, K.; Ouyang, H.G.; Liu, J. Ore-formation mechanism of the Weilasituo tin–polymetallic deposit, NE China: Constraints from bulk-rock and mica chemistry, He–Ar isotopes, and Re–Os dating. *Ore Geol. Rev.* **2019**, *109*, 163–183. [CrossRef]
98. Han, L.; Pan, J.-Y.; Ni, P.; Chen, H. Cassiterite deposition induced by cooling of a single-phase magmatic fluid: Evidence from SEM-CL and fluid inclusion LA-ICP-MS analysis. *Geochim. Cosmochim. Acta* **2023**, *342*, 108–127. [CrossRef]
99. Cheng, W.B.; Peng, Y.W.; Su, J.; Chi, G.X.; Li, G.Q.; Lv, P.R. Geology, geochemistry and genesis of Keyue: A newly discovered Pb–Zn–Sb–Ag polymetallic deposit associated with magmatic center in Southern Tibet, China. *Ore Geol. Rev.* **2021**, *139*, 104546. [CrossRef]
100. Xie, L.; Tao, X.Y.; Wang, R.C.; Wu, F.Y.; Liu, C.; Liu, X.C.; Li, X.K.; Zhang, R.Q. Highly fractionated leucogranites in the eastern Himalayan Cuonadong dome and related magmatic Be–Nb–Ta and hydrothermal Be–W–Sn mineralization. *Lithos* **2020**, *354*, 105286. [CrossRef]
101. Fu, X.F.; HAO, X.F.; Ruan, L.S.; Liang, B.; Zhang, G.Q.; Hou, L.W.; Pan, M.; Tang, Y. Metallogenic characteristics and exploration prospecting of the 3R mineral resources in Sichuan, China. *Sediment. Geol. Tethyan Geol.* **2023**, *43*, 1–18.

minerals

MDPI

Article

Alteration Lithogeochemistry of an Archean Porphyry-Type Au(-Cu) Setting: The World-Class Côté Gold Deposit, Canada

Laura R. Katz [1], Daniel J. Kontak [1,*] and Benoit Dubé [2]

[1] Harquail School of Earth Sciences, Laurentian University, Sudbury, ON P3E 2C6, Canada; laura.r.katz@gmail.com
[2] Geological Survey of Canada, 490, Rue de la Couronne, Québec City, QC G1K 9A9, Canada
* Correspondence: dkontak@laurentian.ca

Abstract: Characterizing alteration and its geochemical signature provides critical information relevant to ore-deposit genesis and its related footprint; for porphyry-type deposits, zoned potassic-phyllic-propylitic alteration and metal enrichment are critical features. Here we integrate earlier lithological and mineralogical studies of the (10+ Moz Au) Archean Côté Gold porphyry-type Au(-Cu) deposit (Ontario, Canada) with identified alteration types to provide exploration vectors. The ca. 2740 tonalite-quartz diorite-diorite intrusive complex and co-temporal Au(-Cu) mineralization as disseminations, breccias and veins are co-spatial with ore-related alteration types (amphibole, biotite, muscovite). An early, locally developed amphibole event coring the deposit is followed by emplacement of a Au(-Cu) mineralized biotite-rich magmatic-hydrothermal breccia body and broad halo of disseminated biotite and quartz veining. These rocks record gains via mass balance calculations of K, Fe, Mg, LILE, and LREE with Au, Cu, Mo, Ag, Se and Bi. Later muscovite alteration is enriched in K, Rb, Cs, Ba, CO_2, and LOI with varied Au, Cu, Mo, Te, As, and Bi values. A strong albite overprint records extreme Na gains with the loss of most other elements, including ore metals (i.e., Au, Cu). Together these data define an Au-Cu-Mo-Ag-Te-Bi-Se core co-spatial with biotite breccia versus a peripheral stockwork and sheeted vein zone with a Te-Se-Zn-Pb-As association. These features further support the posited porphyry-type model for the Côté Gold Au(-Cu) deposit.

Keywords: Porphyry mineralization; Côté Gold; Archean; alteration; lithogeochemistry

1. Introduction

Alteration types and related geochemistry can define and distinguish different ore-deposit types, such as magmatic-hydrothermal versus metamorphic-hydrothermal types (e.g., [1–3]). That these parameters vary both among and within different deposits (e.g., porphyry Cu-Au-, Au-, and Mo-types [3,4]) are used to both constrain the nature and origin of deposits, but importantly also to define exploration vectors; the chemical signature of green-rock in porphyry settings being an excellent example of such application [5,6]. Additionally, elemental associations and their gains and losses are also used to constrain the nature and source of fluids. For example, these parameters have long been used to differentiate among Au deposits of metamorphic versus magmatic affinity (e.g., [2,7]). In this paper, we describe the alteration geochemistry of the recently discovered (2009/2010; [8]) world class ca. 2740 Ma Côté Gold Au(-Cu) deposit. It is a low-grade, large-tonnage, intrusion-related deposit located in the Swayze greenstone belt (SGB) of northern Ontario, Canada. The deposit has a total resource (i.e., measured and indicated) estimated at 365.5 Mt at 0.87 g/t Au (10.2 Moz Au) [9]. The mineralization, which occurs as veins, disseminations, and breccias, is hosted by a tonalite-quartz diorite-diorite intrusive complex and is temporally and spatially related to magmatic-hydrothermal alteration (amphibole, biotite, muscovite, albite) linked to the intrusive system. The well-documented features of the deposit, such as the oxidized hydrous nature of the causative magma, nature and types of mineralization and related alteration, strongly suggest it represents an Archean analogue to younger porphyry-type deposits dominant in the Phanerozoic [10–12]. Thus, henceforth the deposit is referred to as porphyry or porphyry-type. Importantly, the recognition of biotite-bearing veins and fractures overprinted by muscovite alteration coincident with Au mineralization at the deposit scale was instrumental in the exploration and discovery in 2019 of the geologically similar Gosselin Au(-Cu) deposit located adjacent and to the northeast of Côté Gold. The Gosselin Au(-Cu) deposit has a resource (measured and indicated) of 3.4 Moz (124.5 Mt @ 0.84 g/t; [9]). The Côté Gold and Gosselin deposits are important as they are the most significant Au deposits to date in the SGB, which is the southwestern extension of the exceptional Au endowed Abitibi greenstone belt (AGB; +300 Moz Au past production and current reserves/resources [13]. The Côté Gold deposit is also significant for defining a new metallogenic event in the Abitibi Subprovince [10,11,13]. Given the significance of this Au setting, here the focus is on characterizing the geochemical footprint of the alteration in order to constrain the deposit genesis and better define exploration parameters. Other relevant aspects of the nature of the fluids, as constrained from fluid inclusion and stable isotopes [11,14], which strongly favor a magmatic-hydrothermal origin.

Previous studies of this deposit setting focused on documenting the nature and timing of the host rocks, the hydrothermal alteration and related mineralization, age of mineralization, and a genetic model [10–12]. Although certain geochemical aspects were discussed by Kontak et al. [11], a detailed investigation of its alteration geochemistry and elemental mass balance were needed to improve the understanding of the ore system. Thus, to better characterize these aspects, a lithogeochemical study of 460 drill core samples from within the ore deposit envelope were collected for detail petrographic and lithogeochemical studies. These samples were used to expand on earlier work to provide a comprehensive documentation of the mineralogy, lithogeochemistry, and mass balance of the different styles of hydrothermal alteration in the Côté Gold deposit.

2. Regional Geological Setting

The Côté Gold deposit is located in the southeastern limb of the SGB, which is part of the Au-rich Abitibi Subprovince (Figure 1). The SGB is considered to be western extension of the AGB as it has a similar stratigraphy and structural history based on lithological and geochronological criteria [12,15,16]. Volcanic and plutonic rocks are diverse and include ultramafic to felsic types, as well as both chemical and clastic sedimentary rock types. These rock units range in age from 2750 to 2670 Ma [15,17] with six supracrustal groups recognized which, from oldest to youngest, are the Chester, Marion, Biscotasing, Trailbreaker, Swayze and Ridout [18]. The correlative assemblages within the southern AGB are respectively named the Pacaud, Deloro, Kidd-Munro, Tisdale, Blake River and Timiskaming assemblages [13,15,16].

The Côté Gold deposit is hosted by the ca. 2741 to 2739 Ma Chester Intrusive Complex (CIC [12]), a laccolith-shaped, subvolcanic, composite tonalite-quartz diorite-diorite intrusion. The CIC intruded the Chester Group (2750–2735 Ma; equivalent to the Pacaud episode in the AGB; Figure 2). The Arbutus Formation forms the lower formation in the Chester Group and earliest known rocks within the SGB. It is composed of mafic massive to pillowed flows ([18]) with sparse units of felsic tuff [19]. One U-Pb zircon age was obtained from the Arbutus Formation at 2748.2 ± 1.1 Ma [19].

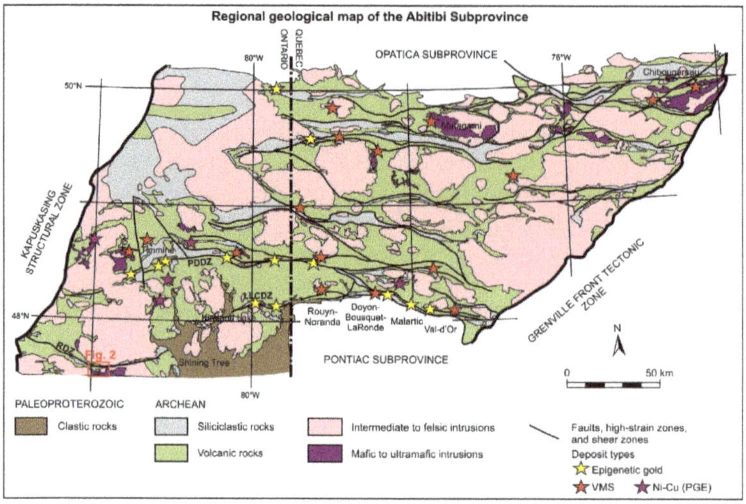

Figure 1. Regional geological map of the Abitibi Subprovince showing the distribution of major deposits and also the main structural features, in particular the Porcupine-Destor and Larder Lake-Cadillac deformation zones (PDDZ and LLCDZ, respectively; modified after [20]). The Ridout deformation zone (RDZ), located in the Swayze greenstone belt, occurs towards the bottom left of the map and the red outline represents the location of the district-scale map of the western part of the Chester Intrusive Complex where the Côté Gold deposit is located.

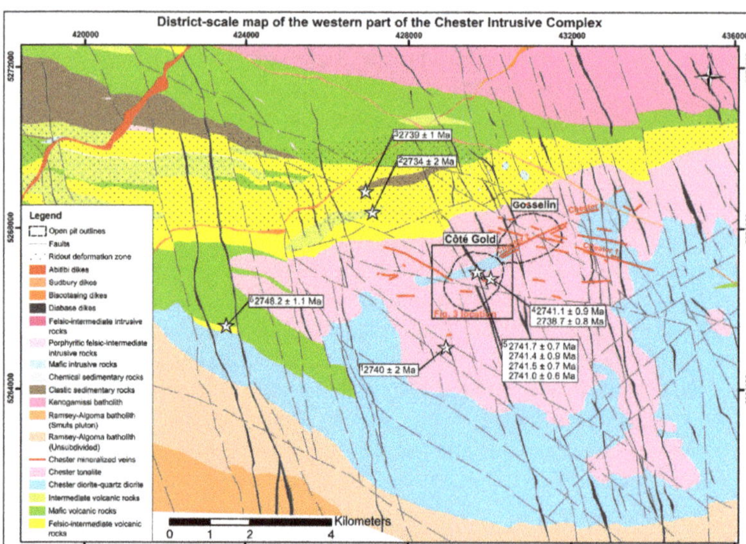

Figure 2. Simplified geological map of the western half of the Chester Intrusive Complex in the southeastern arm of the Swayze greenstone belt (modified after [19]). The white stars and related numbering show sample locations and ages for U-Pb zircon geochronology from [1] [21], [2] [22,23], [3] [16], [4] [11]), [5] [12] and [6] [19] and Figure 3 location is shown by the black solid outline.

The Yeo Formation (Figure 2), which is the upper formation of the Chester Group and overlies the CIC, is comprised of intercalated felsic and intermediate, massive to tuffaceous volcanic rocks, clastic sedimentary rocks and iron formations [18,19]. Two U-Pb zircon dates were obtained from felsic lapilli tuffs at 2734 ± 2 Ma [22,23] and 2739 ± 1 Ma [16], indicating this formation is broadly co-temporal with the CIC.

The SGB has a complex structural history of polyphase folding with development of multiple foliations, ductile high-strain zones, and late brittle faulting [16]. The preserved map pattern within the SGB formed during D_2 orogen-wide shortening (2696–2675 Ma; [16]), which contains several, regionally extensive, D_2 high-strain zones, such as the Ridout deformation zone (see RDZ in Figure 2) that are similar to the high-strain zones in the ABG (i.e., Larder Lake-Cadillac deformation zone) and constrained to <2670 Ma [13]. Importantly, as in the AGB, the D_2 event in the SGB is inferred to have been synchronous with the generation of orogenic-style gold mineralization [16,18,24,25].

In the past, the potential for Au gold mineralization in the SGB has been considered favorable based on both its proximity to and similar geology with the AGB [16,25], hence it is not surprising that many prospects and occurrences have been documented [26,27]. A few of these occurrences were developed into small past gold producers, such as the Jerome (56,897 oz Au) and Joburke (50,150 oz) deposits [28]. The Chester 1-Zone (Figure 2) or Chester Mine, located approximately 2 km east-northeast of the Côté Gold deposit, was exploited for high-grade, narrow, sheeted quartz-sulfide vein mineralization that was considered orogenic in origin, although this was reconsidered as intrusion-related by Kontak et al. [11]; these veins are noted for locally high Cu grades (i.e., 10 wt. % Cu). The Chester Mine was, however, eclipsed by the discovery of the Côté Gold deposit in 2009/2010 [8] and also, as noted above, by the more recent 2019 discovery of the Gosselin deposit, located near the former Chester Mine and just northeast of the Côté Gold deposit (Figure 2).

Metamorphic grade within the southern AGB ranges from sub-greenschist to greenschist grade; it is best seen as a pervasive saussuritization of calcic plagioclase ($An_{30–50}$) throughout the tonalite and diorite units of the CIC. Higher amphibolite grade conditions were locally attained next to large syn-volcanic and syn-tectonic intrusions. Peak metamorphism is estimated to have occurred from 2677 to 2643 Ma [29]. U-Pb dating of rutile in the Côté Gold deposit also helps constrains the age of metamorphism to ca. 2667 Ma [10].

3. Deposit Geology and Alteration Stages

3.1. Magmatic and Magmatic-Hydrothermal Phases of the Chester Intrusive Complex

The Côté Gold deposit, located in the northwestern part of the CIC, is hosted in a composite sill-like intrusive complex composed of tonalite, quartz diorite and diorite (Figure 3A). In the deposit, these rocks are intruded by breccias of both magmatic and magmatic-hydrothermal origin.

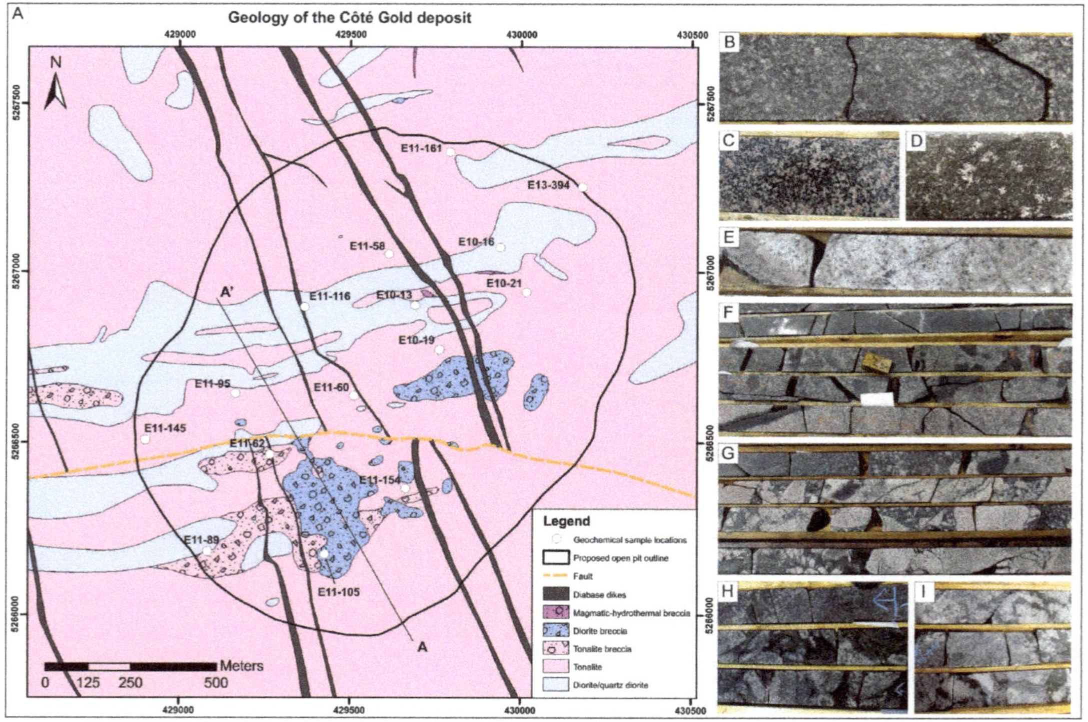

Figure 3. Geological map of the Côté Gold deposit (see location in Figure 2) and drill core photos (4.5 cm width) of host rocks. (**A**) Simplified geological map of Côté Gold deposit modified after Katz et al. ([10]) with the proposed open pit outline as of 2022 [9]. The deposit map was created using the first bedrock collared in 506 diamond drill holes and dozens of mapped outcrops in the deposit area. The white circles show the drill holes selected for geochemical analyses. (**B**) Melanocratic diorite with plagioclase phenocrysts. (**C**) Example of quartz diorite. (**D**) Melanocratic quartz porphyritic diorite. (**E**) Tonalite with minor fracturing and alteration. (**F**) Tonalite breccia along tonalite-diorite contact. (**G**) Diorite breccia with diorite matrix and biotite fracture-fill tonalite fragments. (**H**) Magmatic-hydrothermal biotite breccia with fine- to coarse-grained biotite-quartz-carbonate-chalcopyrite-pyrite matrix. (**I**) Magmatic-hydrothermal biotite breccia with in-situ texture that contains chalcopyrite-pyrite in the matrix and is overprinted by albite alteration.

At the deposit, as well as the surrounding region, intrusive phases of the CIC exhibit complex crosscutting relationships and textures suggesting their coeval emplacement [12,18,23,30,31]. These features include: (1) abundant fragments of diorite, quartz diorite, and tonalite of variable size (<1 cm to more than tens of meters) and texture (i.e., fine- to coarse-grain size) hosted in tonalite; (2) fragments of tonalite, and rarely diorite, hosted in diorite matrix; (3) tonalite bodies or dikes intruding diorite; (4) rounded, subrounded, or angular fragments of diorite with sharp to diffuse margins hosted in diorite; and (5) cuspate or lobate contacts (rare) between tonalite and diorite. Coeval emplacement of these felsic and mafic phases is supported by U-Pb ID TIMS dating of the dioritic (2741.5 $\pm$ 0.7 and 2741.0 $\pm$ 0.6 Ma) and tonalitic (2741.4 $\pm$ 0.9 and 2741.7 $\pm$ 0.7 Ma) bodies [12] and their inferred high-level setting is supported by Al-hornblende geobarometry ($\leq$1.3 $\pm$ 0.6 kbars; [12]). That dates for tonalite of 2740 $\pm$ 2 Ma [21] and 2741.1 $\pm$ 0.9 [11] and tonalite breccia at 2738.7 $\pm$ 0.8 Ma [11] overlap indicate their co-temporal emplacement. Details and photos of the magmatic and magmatic-hydrothermal phases were described in elsewhere [10,12] and below is a brief summary of the main magmatic and magmatic-hydrothermal phases.

Diorite and quartz diorite: In the deposit, several dioritic intrusions occur, including diorite (Figure 3B), quartz diorite (Figure 3C,D) and minor hornblende-plagioclase $\pm$ quartz pegmatite, all of which contain melanocratic to leucocratic phases. Diorite and quartz diorite bodies occur as few meters to <150 m thick intrusions. The hornblende-plagioclase $\pm$ quartz pegmatite can form either as pegmatitic segregations in diorite and quartz diorite or as distinct bodies that intrude diorite, quartz diorite and more rarely tonalite. The diorite is fine- to coarse-grained and dominated by subequal amounts of hornblende and plagioclase with minor titanite, ilmenite and magnetite with accessory apatite, zircon and tourmaline; rarely plagioclase phenocrysts occur. The quartz diorite, which is also amphibole-rich, is fine- to coarse-grained, commonly quartz and/or plagioclase porphyritic and contains the same minor and accessory phases as the diorite. These units are of tholeiitic to transitional nature [12].

Tonalite: The most common unit in the deposit forms numerous bodies of centimeter to >100 m in apparent width. It is a fine- to medium-grained leucocratic rock with equigranular quartz and plagioclase (Figure 3E) and rare quartz or plagioclase phenocrysts; accessory minerals include primary biotite, titanite, ilmenite, zircon, monazite and xenotime. The unit equates to a calc-alkaline to transitional, low-Al tonalite [32].

Tonalite breccia: This breccia occurs along the margins of tonalite and diorite and therefore is considered to be an intrusive breccia (Figure 3F). This unit contains angular to rounded fragments (<1 cm to several meters) of diorite or quartz diorite and rarely tonalite.

Diorite breccia: The second type of magmatic breccia is the result of a later injection of diorite that brecciates diorite, quartz diorite, tonalite and tonalite breccia, although tonalite is the most common fragment type (Figure 3G). The matrix is variable, occurring as medium- to coarse-grained melanocratic diorite or fine- to medium-grained quartz diorite or rarely a fine-grained evolved diorite. The unit occurs as an irregular-shaped body (500 m in length and 350 m in width; Figure 4) in the southern and central part of the deposit that dips moderately to steeply to the northwest [10].

Magmatic-hydrothermal breccias: Two types of magmatic-hydrothermal breccia are recognized based on their matrices: (1) magmatic-hydrothermal with an amphibole-rich matrix; and (2) magmatic-hydrothermal with a biotite-rich matrix (Figure 3H,I). As these bodies are rarely exposed on surface (Figure 3), much of their extent and features are based on our detailed core logging.

The magmatic-hydrothermal amphibole-rich breccia contains fragments of tonalite with rare diorite fragments. Its black to dark-green matrix is fine- to coarse-grained and consists of amphibole-quartz ± biotite ± carbonate ± pyrite ± chalcopyrite. It is the least abundant of the breccia types, forming meter-sized bodies, and appears to be restricted to the southern and central parts of the deposit.

The deposit is centered on several volumetrically significant breccia bodies, which include a magmatic-hydrothermal biotite breccia and the aforementioned diorite breccia. The magmatic-hydrothermal biotite breccia is the largest of the breccias, occurs in the central and northern parts of the deposit, is moderately-to-steeply northwest-dipping, and is up to approximately 600 m in length and 400 m in width at its largest [10]. It contains rounded to angular tonalite fragments and its matrix is subdivided into three assemblages based on the abundance of biotite, magnetite and carbonate: (1) fine-grained biotite-quartz ± epidote ± calcite ± pyrite ± chalcopyrite ± magnetite ± allanite ± titanite ± fluorite; (2) fine- to coarse-grained biotite-magnetite-quartz-ankerite-chalcopyrite-pyrite ± sphalerite ± allanite ± bastnaesite ± apatite ± titanite with rarely up to 50% magnetite; or (3) biotite-ankerite/calcite-quartz-pyrite ± allanite ± apatite ± magnetite ± chalcopyrite ± pyrrhotite with the coarse biotite hosted in finer-grained quartz and carbonate.

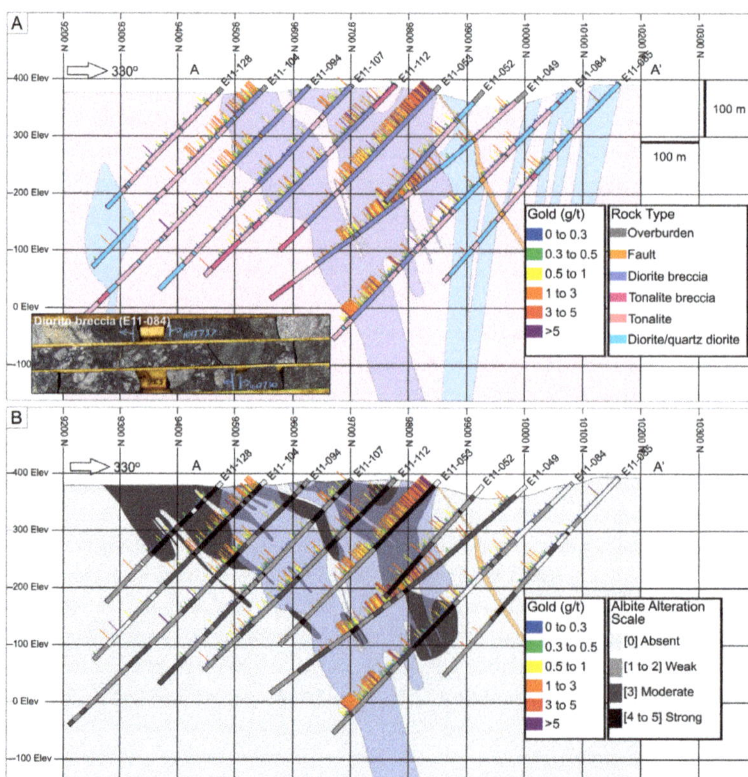

Figure 4. Vertical cross section of grid line 87+00 (A-A′) from the Côté Gold deposit modified after [12]. The location of the section is shown in Figure 3. (**A**) Simplified geological cross section showing the major rock types with gold values. (**B**) Simplified cross section showing the distribution and intensity of albite alteration with strong albite alteration depicted by dark grey shapes and the shape of diorite breccia (from **A**) depicted in blue.

3.2. Alteration Stages, Breccia Cement and Mineralization

The hydrothermal alteration and related mineralization developed in several stages with six alteration assemblages identified (see Table 1). Following on the nomenclature defined in Katz et al. [10], the alteration types from early to late are: amphibole (calcic), biotite (potassic), muscovite (phyllic), epidote (propylitic), albite (sodic) and chlorite. Representative photos of alteration assemblages are shown in Figure 5. The timing of intrusions, alteration stages, and mineralization all overlap at ca. 2740 Ma based on U-Pb (zircon, titanite) and Re-Os (molybdenite) dating [10–12]. A brief summary of the features of these different assemblages is shown in Figure 6 and discussed below from Katz et al. [10]. This summary is critical as it gives context to understanding the lithogeochemistry.

Amphibole vein and breccia stage: This occurs as rare amphibole-bearing veins and breccias of cm- to m width with a variable hornblende ± apatite ± titanite ± magnetite ± quartz ± albite ± pyrite ± chalcopyrite assemblage (Figure 5A,B; Table 1). Titanite of this stage was dated at 2745 ± 3 Ma [10].

These veins and breccias may cut barren diorite and tonalite and are typically Au-bearing (Figure 5A) but sulfide-poor. These veins and breccias appear to be localized to the central and southern parts of the deposit and their paucity, combined with a lack of continuity, precludes their significance in terms of the overall Au budget.

Biotite stage: The term biotite stage is used herein in contrast to the common terms of potassic or K-silicate alteration (e.g., [33]). The mineral assemblage varies with biotite ± quartz ± epidote ± calcite ± ankerite ± magnetite ± allanite ± pyrite ± chalcopyrite ± pyrrhotite ± apatite ± titanite ± ilmenite ± bastnaesite ± fluorite ± molybdenite ± sphalerite ± galena. It is the most widespread alteration and developed as disseminations, lining fractures (Figure 5C) and in veins (Table 1), in addition to cementing breccias (Figure 5D).

This stage also includes LREE-bearing minerals, such as allanite (Figure 5E,F) and bastnaesite, that occur as fine- to medium-grained, anhedral to euhedral crystals intergrown with biotite, epidote and titanite. They have been observed in disseminated biotite alteration, vein-controlled biotite, and the matrix of the magmatic-hydrothermal biotite breccia. These minerals were seen in 10 of 19 biotite-rich samples studied using SEM-EDS.

In drill core, black or brown spots of biotite occur as disseminations (typically 3–12 vol.%) replacing primary plagioclase and amphibole and also after secondary amphibole. Disseminated biotite occurs throughout the deposit, as well as outside the deposit. Veins and veinlets occur as stockwork (Figure 5C,G), discrete features, or subparallel sheeted sets with varied assemblages (Table 1); the sheeted veins cut and thus post-date the biotite breccia. The sheeted quartz ± carbonate ± biotite ± chalcopyrite ± pyrite ± pyrrhotite ± molybdenite veins form conjugate vein sets that consist of moderately to steeply dipping to the north, north-northeast, and northwest veins and moderately to steeply dipping to the south, southeast and southwest veins (Figure 2); these veins also occur outside the deposit area in the CIC and can be mineralized (e.g., Gosselin deposit). These veins define zones that are typically <5 to rarely >42 m in true width and the veins can be densely to widely spaced (i.e., few cm apart to >1 m spacing; [10]). They are typically overprinted by the muscovite stage with distinct muscovite haloes (Figure 5H).

Table 1. Summary of alteration assemblages at the Côté Gold deposit.

Timing	Early (Syn-Magmatism)		Transitional to Late (Post-Magmatism)			
Alteration type	Amphibole (*calcic*)	Biotite (*potassic*)	Muscovite (*phyllic*)	Epidote	Albite (*sodic*)	Chlorite
Alteration assemblage	**Hbl** ± Ap ± Ttn ± *Mag* ± Qz ± Ab ± *Bt* ± *Py* ± *Ccp*	**Bt** ± **Qz** ± Cal ± Ank ± Ep ± *Aln* ± Mag ± Ccp ± Py ± *Po* ± Mol ± *Ap* ± *Bst* ± *Ttn* ± *Ilm* ± *Fl* ± *Sp* ± *Gn*	**Ms** ± Qz ± Cal ± Py ± Ccp ± Po ± Chl ± *Rt* ± *Mol* ± *Tur*	**Ep** ± **Qz** ± Cal ± Chl	**Ab** ± Qz ± Cal ± *Ttn* ± *Ilm* ± *Chl*; **Ab** ± *Ttn* ± *Hem* ± *Rt* ± *Ap* ± *Chl (rare)*	**Chl** ± Qz ± Cal ± Ilm ± Rt ± Py ± Ccp
Vein types	**Hbl** ± **Ab**; **Hbl-Ap-Ttn** ± Mag ± *Py* ± *Ccp*; **Hbl-Qz-Ttn** ± Mag ± *Py* ± *Ccp*	**Qz** ± Mag ± Ccp ± Py; **Bt**; **Qz** ± **Bt** ± **Cb** ± **Ccp** ± **Py** ± **Po** ± Mol ± Ttn ± *Aln* ± *Bst* ± *Sp* ± *Gn*; **Bt-Qz-Ep** ± *Mag* ± *Ttn* ± Cb ± *Fl* ± *Py* ± *Ccp*	**Qz-Cb** ± **Py** ± **Ccp** ± **Po** ± Mol ± Chl ± Tur; **Qz** ± **Cal**; **Ms** ± Cal	**Ep**, **Qz-Ep**; **Ep** ± **Qz** ± Cal ± Chl ± *Py* ± *Ccp*	**Ab** ± **Qz** ± Cal; **Ab**-Ttn ± Cal; **Ttn**; **Qz-Cal-Ilm** ± Ttn; **Qz-Cal**-Ttn	**Chl** ± Qz ± Cal ± Py ± Ccp
Alteration style	Vein-controlled, breccias	Disseminations, vein-controlled, Breccias	Vein-controlled, Replacement	Disseminations, vein-controlled	Vein-controlled, replacement	Replacement

Note: Bold font indicates high abundance of minerals; regular font indiates moderate abundance; italics font indicates minor abundance. Mineral abbreviations: Ab = Albite, Aln = Allanite, Ank = Ankerite, Ap = Apatite, Au = Gold, Bt = Biotite, Bst = Bastnaesite, Cal = Calcite, Cb = Carbonate, Chl = Chlorite, Ccp = Chalcopyrite, Ep = Epidote, Fl = Fluorite, Gn = Galena, Hbl = Hornblende, Hem = Hematite, Ilm = Ilmenite, Mag = Magnetite, Mol = Molybdenite, Ms = Muscovite, Qz = Quartz, Rt = Rutile, Po = Pyrrhotite, Py = Pyrite, Rt = Rutile, Sp = Sphalerite, Ttn = Titanite, Tur = Tourmaline.

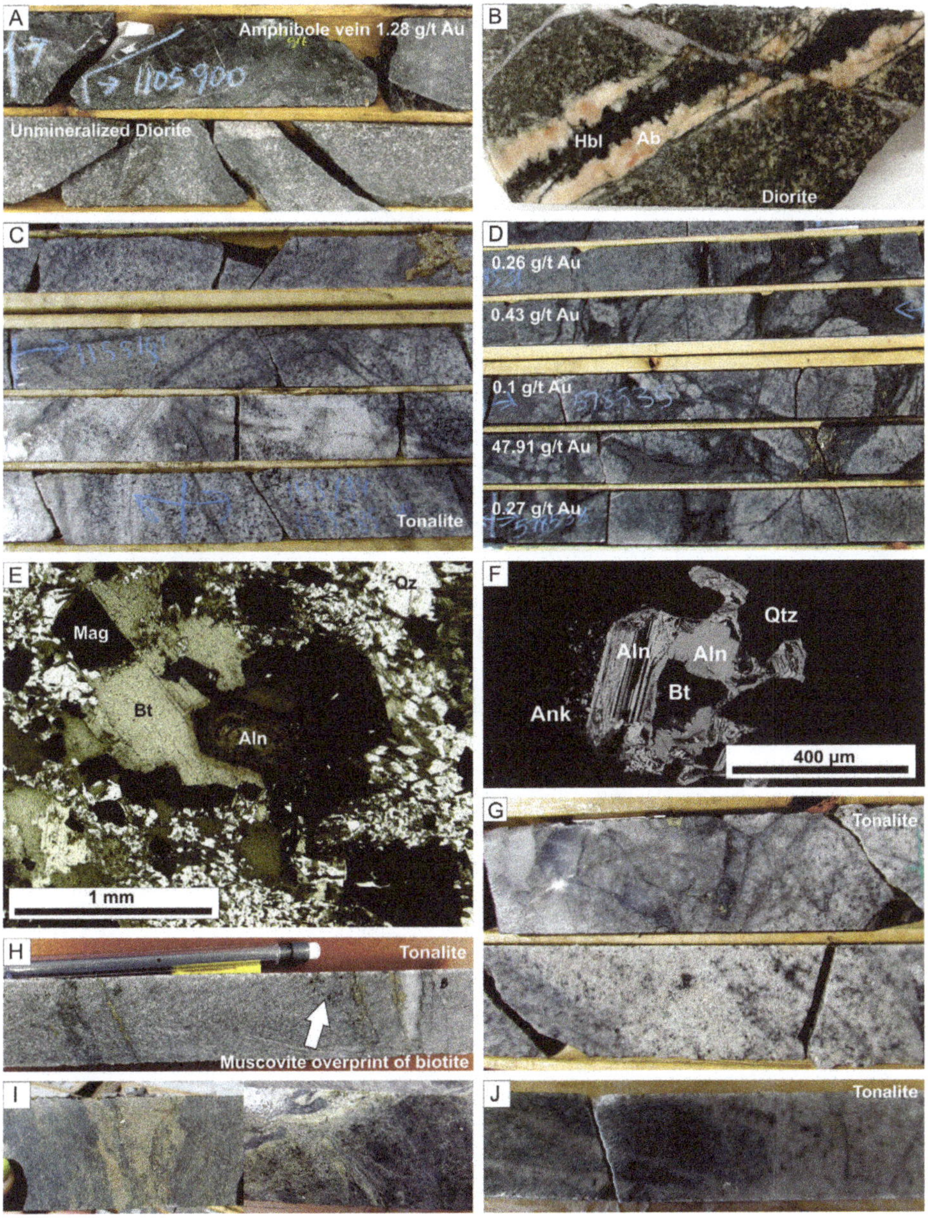

Figure 5. *Cont.*

Figure 5. Photographs showing representative examples of different alteration types at Côté Gold deposit. Note all drill core is 4.5 cm width. (**A**) Mineralized amphibole vein (1.28 g/t Au) cutting barren diorite in lower part of the photo. (**B**) Amphibole vein in diorite bordered by albite. (**C**) Disseminated and fracture-controlled biotite in tonalite with albite alteration exploiting the biotite stockwork. (**D**) Gold-bearing magmatic-hydrothermal biotite breccia hosting biotite and muscovite altered tonalite fragments. (**E**) Thin section in plane-polarized light (PPL) of matrix in magmatic-hydrothermal biotite breccia unit with biotite-quartz-magnetite-allanite assemblage. (**F**) Backscatter electron image of allanite and biotite with quartz and ankerite. (**G**) Muscovite alteration overprinting biotite-sulfide fractures in tonalite. (**H**) Pervasive muscovite alteration in tonalite overprinting sheeted quartz-biotite-pyrite veins and fractures. (**I**) Two examples of vein-controlled epidote alteration in diorite. (**J**) Pervasive muscovite altered tonalite overprinted by semi-pervasive albite alteration. (**K**) Thin section in crossed-polarized light of albite alteration with albite after primary calcic (An_{30}) plagioclase. (**L**) Pervasive albite alteration in diorite breccia with the fragment/matrix boundary still visible. Note some tonalite fragments are outlined by red-dashed lines and the abundance of secondary leucoxene in the matrix. (**M**) Chlorite alteration of magmatic-hydrothermal biotite breccia. Note presence of chlorite reaction rim around some fragments and presence of disseminated chlorite (after biotite) in tonalite fragments. (**N**) Thin section in PPL of magmatic-hydrothermal biotite breccia matrix with chlorite partially replacing biotite. Abbreviations: Ab = albite, Aln = allanite, Ank = ankerite, Bst = bastnaesite, Bt = biotite, Ccp = chalcopyrite, Chl = chlorite, Ep = epidote, Hbl = hornblende, Mag = magnetite, Py = pyrite, Qz = quartz, Ttn = titanite.

The magmatic-hydrothermal biotite breccia is closely associated with Au and Cu with an Au ± Cu ± Mo ± Ag ± Te association, which is also noted in the sheeted veins when proximal the breccia ([10] and see below). In general, mineralization is typically rare in disseminated biotite altered tonalite and diorite samples. However, gold mineralization is spatially associated with diorite breccia (Figure 4A and inset photo), which contains disseminated biotite alteration in the matrix and fragments. The main sulfides are pyrite and chalcopyrite with gold intergrown locally with the chalcopyrite. Magnetite may be abundant in magmatic-hydrothermal biotite breccia, particularly in assemblage 2 (up to 50 vol.%), but rare in disseminated and vein-controlled samples (<1%).

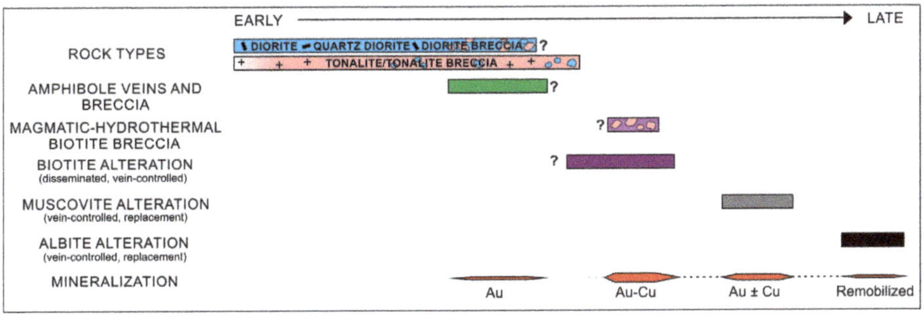

Figure 6. Simplified schematic paragenetic diagram showing the relative age relationships among the host rocks, hydrothermal alteration types, and mineralization at the Côté Gold deposit. Note that "?" refers to the unknown extent of the paragenesis.

Muscovite stage: This assemblage consists of muscovite ± quartz ± calcite ± pyrite ± chalcopyrite ± pyrrhotite ± chlorite ± rutile ± molybdenite ± tourmaline, thus analogous to phyllic alteration in porphyry deposits (e.g., [3,4]). It occurs as both fracture- and vein-controlled (Figure 5G,H) and replacement (e.g., plagioclase, amphibole), partially to wholly overprinting the biotite stage.

Muscovite alteration is most intense in the core of the deposit as a pervasive and fracture-controlled zone (~900 m long with an apparent width up to 400 m that dips shallowly to steeply to the northwest; [10]). The alteration zone is discordant to the two mineralized breccia bodies, with parts of biotite/chlorite altered diorite breccia in the south and magmatic-hydrothermal biotite breccia in the north of the deposit that contain little to no overprinting muscovite alteration.

This stage often spatially coincides with Au mineralization and has an Au ± Cu association, but Au grades are inconsistent. Sheeted veins with muscovite haloes are often mineralized (Figure 5H), however, as they have an original biotite assemblage some of the Au ± Cu ± Mo ± Ag may be inherited.

Epidote stage: Weakly developed patchy epidote occurs throughout the deposit that has been related to regional greenschist facies metamorphism. However, a zone of disseminated, vein-controlled and rarely strongly pervasive replacement epidote (Figure 5I) contains an epidote ± quartz ± calcite ± chlorite. Where veins occur, they often exhibit drusy textures.

This epidote zone is considered to be syn-intrusion in timing and occurs as relatively continuous alteration in a confined area (~900 m long and 150 m wide) found north-northeast of the magmatic-hydrothermal biotite breccia. It is cautiously interpreted to be analogous to epidosite alteration formed in high-temperature reaction zones due to the ascent of previous down-welling paleo-seawater heated by an underlying magmatic body [34]. This stage is not associated with Au mineralization, but one such epidote-quartz vein did contain visible gold.

Albite stage: This assemblage overprints and destroys all earlier alteration types (Figure 5C,J). The assemblage consists of albite ± quartz ± calcite ± titanite ± rutile ± chlorite. It is characterized by albite replacing calcic plagioclase (Figure 5K), destruction of magmatic textures, and strong white/grey bleaching (Figure 5L) that pervasively alters the rock. It commonly exploits preexisting fractures (Figure 5J) and occurs as haloes around veins. Where most intense (i.e., albitite; 10 to 11 wt.% Na_2O) is the development of light-pink to brown-red porous zones, typically <15 m wide. In such cases the secondary albite is accompanied by dissolution of quartz with the new porosity lined by a later generation of clear fluid-inclusion rich albite with trace titanite, hematite, rutile, apatite, zircon, and chlorite. This intense sodic alteration type is similar to episyenite rocks (i.e., sub-solidus quartz deficient alkali-feldspar rich rocks) widely recognized elsewhere and associated with a variety of mineralization types (e.g., U, Sn-W, Au [35]).

Near surface, the white/grey bleaching occurs discontinuously in the central and southeastern part of the deposit, whereas at depth it is strongest toward the center of the deposit and forms an alteration front of ~200 m wide that dips moderately to steeply to the northwest. Its timing is constrained by three U-Pb titanite ages averaging ca. 2740 Ma [10].

Although some zones of moderate to strong albite alteration contain gold, on the deposit-scale the gold is centered on the diorite breccia and magmatic-hydrothermal biotite breccia [10]. The association of Au with albite alteration is inconsistent and, thus, it is considered to overprint previously Au mineralized zones. Sulfides are generally not present in this stage where it is most strong and pervasive.

Chlorite stage: Chlorite is ubiquitous throughout the deposit area and occurs as disseminations, replacement and veins. Petrographically it is seen to partially to wholly replace primary plagioclase and amphibole, secondary amphibole, and particularly biotite (Figure 5M,N). Chlorite does not form its own alteration assemblage but may appear to when it wholly replaces biotite in areas. Thus, it is considered to relate to retrograde after biotite or regional greenschist facies metamorphism [10]. Where Au mineralization is spatially associated with chlorite, it is attributed to overprinting earlier higher-temperature hydrothermal biotite alteration.

In summary, there are several distinct alteration stages noted but the Au is only associated with specific stages in the evolution of the system: (1) matrix of the magmatic-hydrothermal biotite breccia; (2) in the matrix of biotite altered diorite breccia; and (3) areas with a higher density of biotite-bearing veins. Thus disseminated biotite altered tonalite or rocks lacking veining (i.e., diorite) are not going to be prospective for gold mineralization.

3.3. Post-Emplacement Dikes and Deformation

Several types of dike rocks are recognized in drill core and outcrops [12]. They have sharp contacts with all units of the CIC, vary in width (centimeters to several meters), and are distinguished based on macroscopic, petrographic and geochemical features. These intermediate to mafic dikes are typically Au barren and contain assemblages related to D_2 deformation and metamorphism.

Several small (<3 m wide), late-stage, east-west trending D_2-related deformation zones associated with the regional RDZ, which occurs approximately 3 km north of the Côté Gold deposit, overprint the major rock types, alteration types, and dike rocks. In addition, a west-striking, moderately to steeply dipping brittle deformation zone of about 55 m width cuts through the center of the deposit (Figure 3).

4. Methodology

4.1. Petrography

Petrographic analysis in both transmitted and reflect light of approximately 600 sections mainly from drill core (± outcrops) was used to identify alteration types and diagnostic hydrothermal minerals of each alteration stage in addition to the ore mineralogy and their textural relationships. Of these samples, 310 have complete whole-rock geochemical analyses.

4.2. *Laser Ablation-Inductively Coupled Plasma-Mass Spectrometry*

The trace element chemistry of chalcopyrite was determined using LA ICP-MS with results given in Table 2. It is noted that as the abundances of Co, Ni, Pd and Pt were at or below detection limits, these are not included in the data.

Table 2. Results of LA ICP-MS analyses of chalcopyrite.

		Fe	**Zn**	**Se**	**Ag**	**Cd**	**In**	**Sn**	**Sb**	**Te**	**Au**	**Pb**	**Bi**
Sample Number		**m.d.l. 5**	**m.d.l. 0.7**	**m.d.l. 1.8**	**m.d.l. 0.03**	**m.d.l. 0.12**	**m.d.l. 0.01**	**m.d.l. 0.07**	**m.d.l. 0.05**	**m.d.l. 0.24**	**m.d.l. 0.003**	**m.d.l. 0.05**	**m.d.l. 0.01**
E10-13-59	min.	281,800	79.2	167.3	2.67	0.61	14.96	6.72	0.04	n = 1 b.d.l.	0.041	2.87	3.68
Bx (n = 8)	max.	291,500	100	190.9	4.85	0.84	16.23	21.1	0.22	0.51	0.258	7.4	8.94
	avg.	286,200	87	179.3	3.49	0.71	15.58	17	0.11	0.17	0.142	5.1	6.3
	STD	3791	8	8.9	0.86	0.07	0.43	5	0.06	0.16	0.066	1.4	1.9
E11-161-	min.	284,900	790	136.2	138.6	21.1	19.55	48.5	0.03	0.24	0.006	1.98	4.35
337	max.	291,600	1900	141.3	204	47.9	20.23	114	0.05	0.41	0.023	2.36	7.09
Bx (n = 5)	avg.	288,220	1284	123.6	156	33.2	19.78	82	0.04	0.34	0.013	2.11	5.92
	STD	2895	495	25.4	24	11.8	0.25	26	0.01	0.06	0.006	0.14	0.96
E11-161-	min.	291,400	121	82.7	22.79	3.71	99.4	13.63	0.37	b.d.l.	0.004	9.46	1.7
340	max.	295,700	132.6	84.2	26.7	5.48	100.3	20.9	0.52	b.d.l.	0.017	14.1	5.4
Bx (n = 2)	avg.	293,550	127	83.5	24.7	4.60	99.9	17.3	0.44	b.d.l.	0.011	11.8	3.6
	STD	2150	6	0.8	2.0	0.88	0.4	3.6	0.08	b.d.l.	0.006	2.3	1.9
CL-10-03	min.	272,000	27.81	85.6	7.5	0.45	26.84	5.63	0.04	0.25	0.091	6	6.2
Vn (n = 10)	max.	288,900	37.1	98.6	15.73	0.83	29.45	70	0.21	2.27	0.973	32.4	36.7
	avg.	281,610	33.3	92	10.8	0.6	28.0	16	0.13	1.23	0.484	19	20.2
	STD	5299	2.6	4	2.6	0.1	0.7	20	0.06	0.66	0.291	9	9.8
CL-167	min.	296,100	954	100.9	130.1	15.19	6.6	18.48	0.21	0.67	0.119	3	2.33
SVn (n = 9)	max.	312,000	1760	157.9	259.2	29.2	8.32	26.54	0.60	1.48	0.293	6.07	4.72
	avg.	307,644	1267	123	193.0	20	7.3	21.4	0.32	0.95	0.2	4	3.10
	STD	5401	228	22	52.6	4	0.6	3.0	0.11	0.22	0.1	1	0.67
AuLk11	min.	270,800	29.2	91.6	0.13	0.46	2.94	0.72	0.07	n = 7 b.d.l.	0.107	2.4	1.91
Diss (n = 10)	max.	282,900	36.7	178	0.33	1.17	3.28	1.1	0.36	1.27	0.71	6.48	6.91
	avg.	276,490	32	152	0.21	0.84	3.1	0.9	0.20	0.74	0.3	4.5	3.82
	STD	3775	2	23	0.06	0.18	0.1	0.1	0.10	0.29	0.2	1.3	1.30
E10-44-56	min.	275,700	136	38.7	26.6	1.48	14.33	10.84	0.13	n = 4 b.d.l.	0.317	2.32	7.74
SVn (n = 10)	max.	291,000	960	61.9	142	8.3	27.47	14.2	0.81	4.1	17.2	120	119
	avg.	282,980	330	54	51	2.9	20.0	12.1	0.41	1.4	4	36	44
	STD	3989	278	7	36	2.0	4.9	1.0	0.25	1.2	5	38	41
E09-05-23	min.	280,300	131.5	195.8	13.31	1.65	16.8	28.81	0.05	1	0.112	1	1.94
Diss (n = 10)	max.	292,600	148.7	214.8	20.96	2.25	18.17	52.1	0.66	6.8	2.05	127	6.69
	avg.	288,000	140	205	16.7	2.04	17.7	40.8	0.14	4	0.44	14	3.4
	STD	3463	5	6	2.3	0.16	0.4	7.8	0.18	2	0.54	38	1.5
CL-155	min.	297,500	847	30.4	129.4	12.53	5.025	15.49	0.12	0.77	0.056	3.71	4.07
CVn (n = 6)	max.	317,200	1820	32.4	224	18.0	6.418	19.54	0.27	2.67	0.121	6.38	8.52
	avg.	309,183	1371	31.1	175	16	5.7	16.89	0.19	1.53	0.096	5.1	5.62
	STD	6397	391	0.8	33	2	0.5	1.48	0.05	0.64	0.021	0.9	1.52

Values in ppm; m.d.l. = mean detection limit and b.d.l. = below detection limit; samples < b.d.l. given value of detection limit in the average. Bx = breccia, Vn = vein, SVn = sheeted vein, CVn = Chester vein and Diss = disseminated.

Analyses were done at the Chemical Fingerprinting laboratory of Laurentian University (Sudbury, Canada) using a 20 ns pulse duration, 193 nm wavelength ArF excimer laser (Resonetics RESOlution M-50) employing a two-volume Laurin Technic sample cell ([36]) coupled to a Ar plasma quadrupole ICP-MS (Thermo X Series II). Ablation took place in He (650 mL/min) which was combined with Ar (800 mL/min) and N_2 for enhanced sensitivity (6 mL/min) prior to ionization. Traverse mode was used with a beam diameter (26–90 μm) and speed (10–35 μm/s) adjusted according to the grain size to provide adequate sampling time and maximum signal intensities; the one sample (CL-10-03) analyzed in spot mode used a beam diameter of 66 μm. For all analyses, laser repetition rate and fluence were 6 Hz and 5 J/cm^2, respectively, with each analysis typically consisting of 30 s of background followed by 30 s of ablation or longer (for traverses). The ICP-MS was operated with a forward power of 1450 W and an oxide production rate of <0.4% as determined by measuring ThO^+/Th^+ while ablating NIST SRM 612. Dwell times were set at 10 ms per analyte for all measurements. Data reduction was done with the Iolite software package (v. 2.5; [37]) using NIST SRM 610 (bracketing and between every 7 or fewer analyses) as an external reference and Cu (34.6%) as internal references for chalcopyrite. Mass 108 was used to determine Pd concentrations after being corrected for ^{108}Cd by the measured ^{111}Cd and assuming $^{108}Cd/^{111}Cd$ = 0.06953. Reference materials BHVO-2g and Po725 [38] were also analyzed periodically and typically return concentrations within 15% of their accepted values. Segments of the data displaying abnormal behavior (e.g., cracks or inclusions) were omitted during data reduction.

4.3. Lithogeochemistry

A total of 434 samples were collected from 15 drill holes throughout the deposit (see locations in Figure 2) with an additional 26 samples come from specifically selected drill intersections. Samples included all rock types in the deposit, including dike rocks. However, only least-altered and altered host rocks are discussed here. All of the geochemical data, in addition to location (drill collar, depth), a representative photo of the sample and a brief hand sample description can be found in a Geological Survey of Canada Open File Report 8040 ([39]). These data were used to assess down-hole changes in lithogeochemistry commensurate with the presence of the different alteration stages.

Analyses were done in two batches at Activation Laboratories in Ancaster, Ontario. Major elements were determined by lithium metaborate-tetraborate fusion and ICP-MS, whereas trace- and rare earth elements (REE) were determined by a combination of fusion ICP, fusion ICP-MS and total digestion ICP-MS. For As, Sb, Bi Se and Te, abundances were determined by a combination of aqua regia ICP-MS and nitric peroxide fusion ICP-MS, whereas Au, Pd and Pt were measured by fire assay. The FeO was determined using titration. For the anions, Cl was determined by instrumental neutron activation, B by prompt gamma neutron activation and F by ion selective method. Both S and CO_2 were determined by infrared analysis and Hg by cold vapor flow injection mercury technique.

4.4. Mineral Characterization

Numerous samples were studied using a JEOL 6400 scanning electron microscope (SEM) with a mounted Oxford Sight energy dispersive spectrometer (EDS) detector at the Mineral Analysis Centre, Laurentian University (Sudbury, Canada). Data was acquired using an accelerating voltage of 20 kV, 1.005 nA beam current, counting times of 5 to 10 s, and a working distance of 15 mm. Data were processed using Oxford Instrumentation AZTEC (v.4.0) software.

4.5. Mass Balance

To quantify changes in rock mass and elemental abundances in the different alteration types, the Grant [40] isocon method was performed on 115 samples collected along 14 logged drill holes and one outcrop sample. For these calculations, least-altered samples were first identified on the basis of petrographic and geochemical criteria and a group of five tonalite, four diorite and six quartz diorite samples selected from among the several hundred analyses to define the average compositions of these precursors; these data are given in Table 3 (tonalite) and Table 4 (diorite, quartz diorite). Using the latter averages, mass balance was done using representative samples for the specific alteration stages. In general, the elements used to define the isocons were Al_2O_3, TiO_2, Zr, and SiO_2. From this the average gains and losses for the specific units and alterations were calculated and are summarized in Table 5.

Table 3. Average composition of least-altered and altered tonalite samples from the Côté Gold deposit for mass balance calculations.

	Tonalite																	
	Least-Altered			**Disseminated/Stockwork Biotite**			**Fracture-ControlledSericite**			**Pervasive Sericite**			**Disseminated/Fracture-Controlled Epidote**			**Pervasive Albite**		
	Average	**STD**	**No.**	**Average**	**STD**	**No.**	**Average**	**STD**	**No.**	**Average**	**STD**	**No.**	**Average**	**STD**	**No.**	**Average**	**STD**	**No.**
SiO_2	77.052	2.0795	5	72.58	2.45	20	75.30	1.94	20	72.77	8.21	19	72.623	1.51	3	76.122	2.6468	9
Al_2O_3	12.206	0.58	5	13.38	0.89	20	12.30	0.69	20	12.68	1.17	19	13.407	0.75	3	13.042	1.6771	9
FeO	0.91	0.78	5	1.90	0.67	20	1.52	0.49	20	1.69	1.61	19	1.57	0.41	3	0.57	0.25	9
Fe_2O_3	1.32	0.90	5	1.89	0.67	20	0.45	0.38	20	1.31	3.15	19	1.37	0.24	3	0.24	0.20	9
FeOT	1.19	0.81	5	2.58	0.85	20	1.92	0.57	20	1.17	2.83	19	1.23	0.22	3	0.22	0.18	9
MnO	0.0138	0.0062	5	0.029	0.009	20	0.023	0.007	20	0.021	0.005	19	0.029	0.0073	3	0.0131	0.0042	9
MgO	0.33	0.23	5	0.79	0.33	20	0.43	0.11	20	0.70	0.41	19	0.89	0.44	3	0.24	0.09	9
CaO	1.48	0.50	5	2.70	0.77	20	1.90	0.49	20	1.90	0.80	19	3.76	0.49	3	1.37	0.62	9
Na_2O	5.44	0.51	5	4.57	0.52	20	4.54	0.53	20	2.44	1.45	19	2.93	0.92	3	5.63	1.18	9
K_2O	0.53	0.16	5	1.12	0.61	20	1.30	0.64	20	2.50	1.08	19	1.40	0.40	3	0.85	0.24	9
TiO_2	0.317	0.026	5	0.339	0.111	20	0.278	0.049	20	0.363	0.213	19	0.2963	0.0159	3	0.3164	0.103	9
P_2O_5	0.04	0.03	5	0.06	0.03	20	0.04	0.02	20	0.03	0.03	19	0.06	0.01	3	0.01	0.01	9
LOI	0.99	0.16	5	1.16	0.36	20	1.56	0.44	20	3.32	2.61	19	1.34	0.35	3	1.31	0.32	9
Total	99.71	0.56	5	99.57	0.73	20	99.79	0.98	20	99.89	1.03	19	99.90	0.52	3	99.73	0.68	9
Total S	0.10	0.06	5	0.07	0.13	20	0.07	0.14	20	1.17	3.95	19	0.07	0.07	3	0.09	0.13	9
CO_2	0.42	0.23	5	0.37	0.24	20	0.82	0.38	20	1.29	0.62	19	0.14	0.06	3	0.68	0.3319	9
F	0.0063	0.0022	5	0.009	0.007	20	0.007	0.005	20	0.034	0.029	19	0.030	0.0163	3	0.0117	0.0075	9
Au	98	110	5	48	82	20	371	1519	20	304	492	19	16	16	3	24	20	9
Cu	109.0	89.6	5	64.0	133.3	20	66.7	107.8	20	215.6	261.2	19	162.1	206.41	3	72.8	59.6	9
Mo	0.8	0.6	5	0.7	0.4	20	1.2	0.7	20	1.5	1.0	19	3.0	0.8	3	1.0	0.6	9
Ag	1.2	0.4	5	1.0	0.3	20	1.2	0.4	20	1.2	0.6	19	1.2	0.6377	3	1.2	0.5	9
Sb	0.4	0.3	5	0.4	0.2	20	0.3	0.2	20	0.3	0.3	19	0.2	0.1	3	0.3	0.3	9
Te	0.059	0.0528	5	0.059	0.046	20	0.048	0.039	20	0.073	0.059	19	0.027	0.024	3	0.0389	0.0443	9
Bi	0.17	0.17	5	0.09	0.16	20	0.05	0.06	20	0.30	0.59	19	0.24	0.09	3	0.07	0.06	9
Se	0.5	0.0	5	0.7	0.4	20	0.7	0.4	20	0.7	0.6	19	0.3	0.2	3	0.4	0.1	9
As	2.4	2.0	5	3.4	3.5	20	1.6	1.5	20	16.9	38.7	19	2.1	1.8	3	1.4	1.4	9
Pb	2	2	5	6	8	20	3	2	20	7	11	19	3	2	3	2	1	9
W	0.86	1.06	5	0.77	1.07	20	0.61	0.79	20	7.26	15.04	19	0.40	0.21	3	1.64	1.36	9
Zn	12.0	2.4	5	34.7	23.4	20	21.8	8.2	20	24.5	26.2	19	12.3	6.5	3	11.3	7.0	9
Li	5	3	5	11	5	20	7	2	20	10	5	19	7	3	3	4	2	9
Rb	14	6	5	37	20	20	36	16	20	85	41	19	56	24	3	31	10	9
Cs	0.1	0.1	5	0.6	0.7	20	0.4	0.3	20	0.8	0.5	19	0.6	0.1	3	0.3	0.2	9
Sr	121	60	5	176	65	20	107	34	20	73	36	19	385	135	3	86	32	9
Ba	139	69	5	260	133	20	314	164	20	405	213	19	342	47	3	172	93	9
V	14	5	5	32	15	20	17	6	20	22	11	19	31	10	3	13	3	9
Y	27.2	5.8	5	25.4	11.4	20	31.6	7.5	20	34.6	26.7	19	27.8	14.4	3	26.7	7.6	9
Zr	284	25	5	210	44	20	247	50	20	222	74	19	194	76	3	248	73	9
La	5.8	3.3	5	25.1	16.1	20	23.8	12.2	20	11.3	8.1	19	31.7	26.0	3	5.6	3.1	9
Ce	15.3	7.5	5	49.1	30.7	20	47.8	21.6	20	26.4	14.4	19	63.1	47.6	3	15.8	6.7	9
Dy	4.78	1.15	5	4.42	1.94	20	5.39	1.24	20	5.94	4.94	19	4.58	2.36	3	4.88	1.27	9
Yb	3.16	0.58	5	2.56	1.15	20	3.44	0.78	20	3.75	2.70	19	3.08	1.45	3	2.97	0.74	9
Lu	0.4434	0.0742	5	0.368	0.164	20	0.492	0.104	20	0.510	0.317	19	0.45	0.1821	3	0.4183	0.0958	9

Note: Major elements, total S, CO_2 and F are in wt.%, Au is in ppb and trace elements are in ppm.

Table 4. Average composition of least-altered and altered dioritic samples from the Côté Gold deposit for mass balance calculations.

	Quartz diorite															Diorite					
	Least-Altered			Disseminated Biotite			Fracture-Controlled/ Pervasive Sericite			Pervasive Albite			Disseminated Chlorite			Least-Altered			Pervasive Epidote		
	Average	STD	No.	Average	STD	No.	Average	STD	No.	Average	STD	No.	Average	STD	No.	Average	STD	No.	Average	STD	No.
SiO_2	55.97	2.14	6	57.74	3.90	6	62.76	5.33	5	65.22	5.16	9	55.02	1.39	6	50.33	1.36	4	49.52	2.97	3
Al_2O_3	14.44	0.51	6	15.14	0.30	6	15.46	2.81	5	18.07	3.33	9	14.32	0.96	6	15.91	1.43	4	16.33	0.05	3
FeO	6.80	0.99	6	5.93	0.87	6	2.70	1.39	5	0.83	0.61	9	6.32	1.30	6	7.80	1.04	4	4.93	1.96	3
Fe_2O_3	2.09	0.92	6	1.53	0.48	6	0.77	0.45	5	0.15	0.15	9	2.57	0.84	6	2.31	0.47	4	5.13	1.48	3
FeOT	1.88	0.83	6	1.37	0.43	6	0.69	0.40	5	0.14	0.14	9	2.31	0.76	6	2.07	0.42	4	9.55	1.01	3
MnO	0.151	0.022	6	0.1128	0.023	6	0.040	0.020	5	0.0239	0.0085	9	0.130	0.058	6	0.140	0.025	4	0.1207	0.0205	3
MgO	3.90	0.60	6	3.92	0.98	6	2.36	2.03	5	0.34	0.27	9	4.16	1.16	6	5.71	0.74	4	5.94	1.78	3
CaO	8.15	0.80	6	6.55	1.26	6	4.02	1.94	5	3.89	1.09	9	6.58	1.41	6	9.14	1.28	4	11.46	1.04	3
Na_2O	3.81	0.31	6	3.43	0.52	6	4.53	0.83	5	6.06	0.87	9	4.18	0.54	6	3.24	0.33	4	1.00	0.61	3
K_2O	0.46	0.09	6	1.25	0.52	6	1.54	0.57	5	1.67	0.59	9	0.57	0.27	6	1.11	0.67	4	0.98	0.07	3
TiO_2	1.741	0.535	6	0.9192	0.3483	6	1.1106	0.9215	5	0.969	0.4382	9	1.4067	0.1507	6	1.2853	0.2233	4	1.0883	0.1304	3
P_2O_5	0.22	0.12	6	0.11	0.06	6	0.12	0.08	5	0.03	0.04	9	0.31	0.13	6	0.21	0.11	4	0.25	0.21	3
LOI	1.87	0.63	6	2.51	1.43	6	3.79	0.69	5	2.77	0.75	9	3.76	1.47	6	2.31	0.64	4	3.39	0.42	3
Total	100.34	0.38	6	99.81	0.59	6	99.49	0.32	5	100.06	0.81	9	100.01	0.58	6	100.36	0.43	4	100.70	0.22	3
Total S	0.11	0.14	6	0.02	0.02	6	0.21	0.16	5	0.08	0.08	9	0.23	0.29	6	0.03	0.02	4	0.02	0.01	3
CO_2	0.59	0.31	6	0.88	0.95	6	1.89	0.96	5	1.72	0.68	9	1.76	1.10	6	0.59	0.39	4	0.21	0.11	3
F	0.082	0.020	6	0.0408	0.0383	6	0.043	0.020	5	0.0139	0.0084	9	0.0533	0.0125	6	0.1275	0.0327	4	0.1267	0.0386	3
Au	23	21	6	13	14	6	157	258	5	241	260	9	82	125	6	9	8	4	2	2	3
Cu	140.4	148.6	6	51.233	35.205	6	171.7	243.0	5	80.3	102.6	9	202.0	247.6	6	21.3	12.1	4	6.0	5.4	3
Mo	0.6	0.2	6	1.3	1.7	6	1.1	0.7	5	0.5	0.0	9	0.5	0.0	6	0.6	0.2	4	0.5	0	3
Ag	0.7	0.2	6	0.6	0.3	6	1.0	0.6	5	0.6	0.5	9	0.8	0.4	6	2.2	3.2	4	0.4	0.2	3
Sb	0.5	0.2	6	0.4	0.1	6	0.2	0.2	5	0.2	0.2	9	0.4	0.1	6	0.2	0.1	4	0.1	0.0	3
Te	0.075	0.057	6	0.1383	0.0773	6	0.214	0.2484	5	0.0244	0.033	9	0.1217	0.0279	6	0.0475	0.0225	4	0.01	0.00	3
Bi	0.07	0.05	6	0.03	0.01	6	0.29	0.44	5	0.07	0.04	9	0.05	0.03	6	0.05	0.02	4	0.11	0.02	3
Se	0.6	0.3	6	1.1	0.7	6	1.2	1.0	5	1.1	0.6	9	1.4	0.6	6	0.5	0.3	4	0.2	0.1	3
As	1.8	1.6	6	1.9	1.9	6	2.0	2.3	5	2.6	2.5	9	1.7	1.6	6	0.9	0.5	4	1.6	1.7	3
Pb	2	1	6	3	2	6	3	2	5	4	4	9	2	1	6	1	0	4	2	1	3
W	0.48	0.25	6	0.81	1.25	6	6.92	10.01	5	3.17	1.50	9	0.57	0.45	6	0.25	0	4	0.45	0.15	3
Zn	59.9	8.1	6	77.5	22.6	6	56.4	54.2	5	13.4	11.5	9	68.1	11.3	6	70.1	18.7	4	53.3	16.1	3
Li	13	3	6	27	12	6	29	26	5	6	3	9	26	13	6	19	7	4	28	8	3
Rb	12	3	6	49	22	6	55	13	5	56	22	9	15	7	6	46	36	4	52	29	3
Cs	0.3	0.2	6	1.0	0.6	6	0.5	0.2	5	0.6	0.3	9	0.3	0.1	6	1.1	1.4	4	0.9	0.8	3
Sr	262	85	6	246	35	6	161	46	5	168	88	9	198	54	6	278	74	4	465	304	3
Ba	82	26	6	255	125	6	280	175	5	240	152	9	125	107	6	177	132	4	104	1	3
V	194	19	6	125	33	6	74	38	5	33	17	9	144	46	6	226	20	4	180	20	3
Y	32.8	9.4	6	27.8	5.9	6	80.5	95.5	5	109.32	81.896	9	44.0	14.7	6	34.4	12.8	4	19.9	2.9	3
Zr	154	40	6	122	30	6	155	79	5	118	77	9	152	43	6	90	41	4	84	28	3
La	16.9	4.8	6	17.1	5.0326	6	15.4	10.5	5	12.8	6.7	9	22.4	4.9	6	18.8	10.0	4	21.7	19.4	3
Ce	46.2	16.9	6	39.2	14.3	6	60.1	59.4	5	53.0	34.1	9	55.7	18.8	6	56.2	26.5	4	53.2	41.0	3
Dy	6.02	1.53	6	5.03	1.26	6	14.46	17.74	5	20.90	17.55	9	8.11	3.01	6	6.57	2.57	4	3.76	0.61	3
Yb	3.50	0.95	6	2.92	0.76	6	8.83	10.59	5	11.28	7.83	9	4.45	1.78	6	3.44	1.18	4	2.05	0.30	3
Lu	0.5173	0.1444	6	0.416	0.1043	6	1.198	1.3835	5	1.4748	0.979	9	0.6408	0.2464	6	0.4805	0.1541	4	0.287	0.0407	3

Notes: Major elements, total S, CO_2 and F are in wt. %, Au is in ppb and trace elements are in ppm.

Table 5. Calculated gains and losses for different alteration types averaged for tonalite and diorite.

	Biotite		Muscovite		Epidote		Albite		Chlorite
	Tonalite	Quartz Diorite	Tonalite	Quartz Diorite	Tonalite	Diorite	Tonalite	Diorite	Diorite
SiO_2	**−0.4**	**−0.1**	**−0.3**	**10.2**	**−0.3**	**3.4**	**−5.8**	**28.6**	**−19.5**
Al_2O_3	**15.9**	**1.6**	**−5.8**	**5.2**	**12.6**	**8.0**	**−7.4**	**44.6**	**−20.1**
FeO	120.2	−15.5	−11.4	−61.0	6.5	−33.5	−69.8	−61.8	−10.0
FeO_T	129.6	−18.5	10.6	−73.8	54.1	1.6	−75.3	−64.6	−0.2
MnO	120.3	−27.5	−27.9	−73.8	31.7	−9.2	−43.3	−26.5	−3.0
MgO	151.1	−2.6	−11.5	−40.5	111.1	9.5	−69.8	−82.0	−10.3
CaO	92.0	−22.2	−30.0	−51.6	104.9	31.9	−35.2	19.9	−15.2
Na_2O	−11.2	−12.9	−47.0	16.8	−33.3	−67.4	108.3	65.3	2.9
K_2O	122.4	163.9	122.6	228.9	11.4	−7.7	−69.4	34.0	−61.5
TiO_2	**13.1**	**−48.8**	**6.7**	−37.3	**10.2**	**−11.0**	**−21.5**	**8.0**	29.3
P_2O_5	71.8	−48.7	−49.5	−48.1	60.4	25.3	−68.0	−70.1	134.6
LOI	24.6	30.4	184.3	99.3	−11.2	54.1	−64.5	−9.6	26.4
Total S	−27.0	−84.6	1580.4	100.2	−3.7	−38.3	−92.7	−51.8	1057.7
CO_2	−5.2	44.2	244.6	215.0	−82.4	−62.3	−52.6	12.2	69.5
F	56.5	−51.6	267.7	−48.3	342.9	4.4	−69.3	−60.0	10.4
Au	−48.3	−44.8	531.8	573.7	−95.6	−76.6	−92.8	90.8	431.9
Cu	−38.0	−64.7	235.1	20.1	151.1	−70.4	−69.6	−42.2	233.1
Mo	−4.2	107.6	102.1	85.2	169.6	−15.9	−38.9	−43.8	−66.2
Ag	−11.2	−16.3	22.3	38.3	5.1	−81.1	−13.2	−17.5	17.1
Sb	−7.5	−17.5	−14.8	−47.6	−40.6	−53.3	0.1	14.6	−22.8
Te	5.7	78.7	23.3	180.3	−42.0	−93.3	−52.1	−85.9	−25.7
Bi	−25.6	−60.6	249.9	296.1	423.1	143.4	−80.5	−70.7	42.4
Se	47.6	79.9	1.5	105.4	−50.1	−55.7	−45.1	9.3	10.5
As	48.4	2.2	390.5	10.3	41.8	81.8	−92.4	55.7	−26.5
Pb	146.7	29.2	20.6	57.2	−0.2	96.2	−76.4	54.7	−41.9
W	−5.6	64.9	837.6	1331.1	−32.2	89.2	−85.9	−43.3	−40.8
Sn	−40.0	−61.3	105.8	36.2	459.1	−60.6	−44.3	−7.5	101.9
Zn	205.9	25.3	−29.7	−7.5	−41.9	−20.1	−58.4	−70.6	−25.7
Li	149.4	102.5	−9.8	118.9	−6.9	57.0	−65.4	−72.1	−17.1
Rb	179.6	286.2	131.8	340.8	60.7	19.6	−67.1	27.2	−74.2
Cs	419.8	262.5	36.6	98.9	37.0	−9.8	−64.8	46.4	−70.9
Sr	53.7	−9.2	−58.7	−39.7	272.5	76.0	5.4	29.0	−31.8
Ba	98.0	202.3	54.9	236.6	12.4	−37.9	−61.6	6.1	−58.7
V	138.3	−37.6	−33.2	−62.8	96.2	−16.0	−44.4	−45.3	−2.9
Y	−1.3	−17.8	35.3	141.5	−9.1	−39.1	−30.4	68.0	33.8
Zr	−21.9	−20.6	**5.4**	**−1.4**	−18.6	**−1.3**	**0.6**	**−5.2**	**4.6**
La	357.8	−2.1	−55.1	−10.9	37.8	21.4	−55.4	3.1	10.6
e	240.1	−17.9	−46.6	27.7	36.3	−0.6	−46.1	9.2	20.1
Dy	−2.3	−19.0	33.6	136.1	−12.2	−39.8	−26.0	78.9	36.2
Yb	−14.2	−19.3	45.5	147.8	−7.3	−37.3	−28.7	58.1	28.9
Lu	−12.2	−22.1	37.8	127.5	−5.5	−37.2	−26.2	52.3	30.2

Note: Values for major elements, S, CO_2 and F are given in %, trace elements in ppm and Au in ppb. Bold font indicates elements used for isocon.

5. Results

In order to illustrate the distribution of the alteration and their diagnostic geochemical signature and use for discriminating the different stages, alteration geochemistry and mass balance are presented. Three down-hole geochemical profiles are presented in Figure 7 to show the diagnostic elements associated with a particular rock type or alteration. In addition, these geochemical profiles show which rock and/or alteration is spatially associated with the Au mineralization. Chondrite-normalized rare earth element (REE_{CN}) diagrams and several binary element plots are also used to geochemically fingerprint these stages. The trace-element chemistry of chalcopyrite is also used to see what possible insight it might provide in regard to ore-forming processes as until recently this has not been widely applied in porphyry settings (e.g., [41]). Lastly, mass balance calculations provide the elemental gains and losses and thus geochemical signature for fluids responsible for the different alteration stages (i.e., biotite, muscovite, albite, and chlorite) and related mineralization. Although the hydrothermal amphibole is spatially coincident with elevated Au, due to its overall general paucity it was not included as part of the mass balance, but the nature of its geochemistry is addressed.

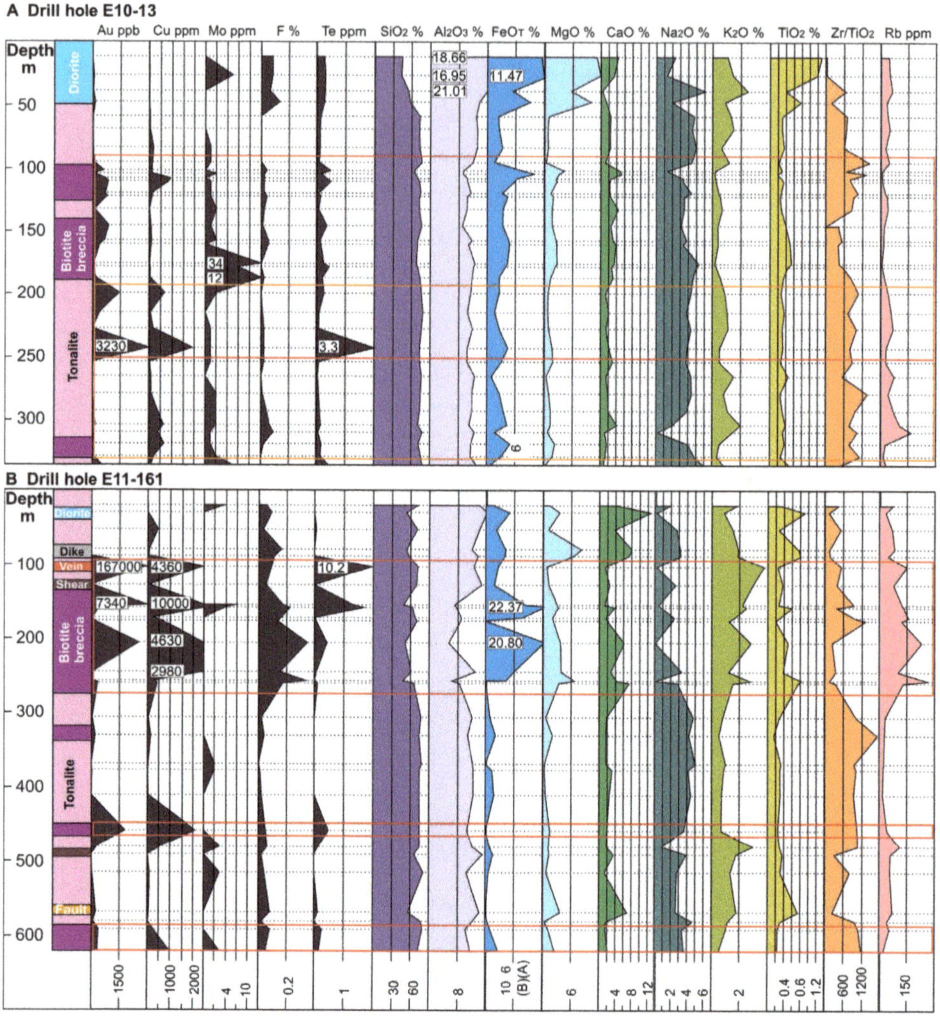

Figure 7. *Cont.*

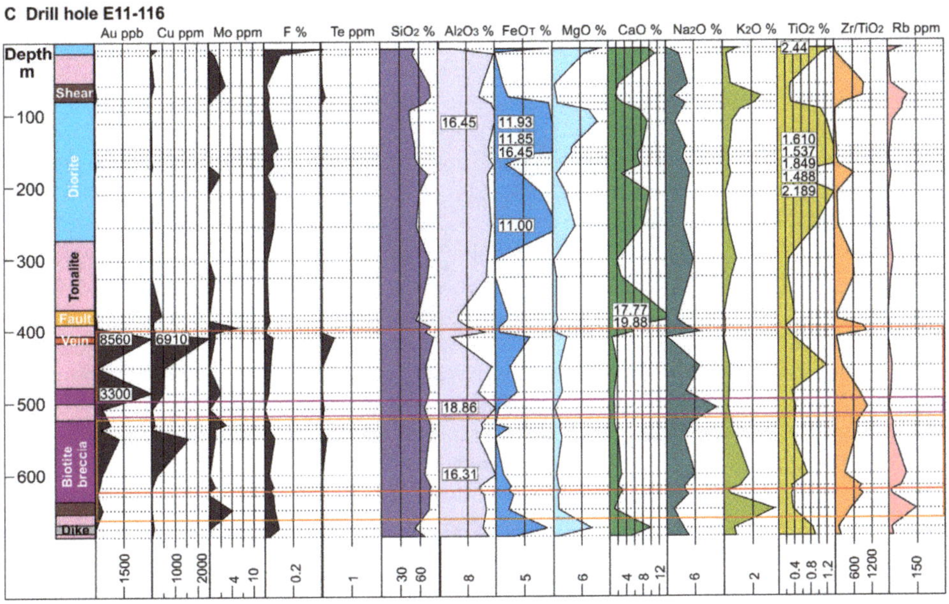

Figure 7. Lithological logs and corresponding geochemical profiles for selected elements from drill holes E10-13 (**A**), E11-161 (**B**) and E11-116 (**C**); see Figure 3 for locations. For each profile locations of analyzed samples are indicated by dashed horizontal lines with numbers in the profiles indicating off-scale values. Note scales for FeO and Na_2O (in wt. %) change between drill hole profiles. Red outline delineates the ore zone, the orange outline zones of muscovite alteration, and the purple outline zones of albite alteration.

5.1. Alteration Geochemistry and Elemental Behavior and Trends

Amphibole stage: The nature of the amphibole (i.e., veins and breccias) makes it difficult to geochemically characterize this stage using bulk methods, however, analyses of magmatic and hydrothermal amphibole provided in Katz et al. [10] are used to assess its signature. These data reveal that hydrothermal amphibole types contain higher SiO_2 and CaO than the magmatic equivalent, whereas TiO_2 is the reverse. In addition, although amphibole from the veins were found to contain similar Mg# (Mg values (100 × Mg/(Mg + Fe) in apfu) as the diorite and quartz diorite (Mg# 40 to 50) due to wallrock buffering, the breccia contained higher Mg# (i.e., 55 to 70), which coincided with higher F (0.2 to 0.5 wt. %) than in magmatic amphibole (F < 0.3 wt. %).

Biotite stage: The specific mineralogy of this alteration type provides some basis to geochemically fingerprint it with LREE- (e.g., allanite, bastnaesite) and F-rich phases (e.g., biotite, fluorite, apatite, titanite) being particularly useful. For the former, the presence of LREE-rich phases result in $La/Yb_N > 4$ for biotite altered tonalite (Figure 8A), whereas for the latter a strong correlation is manifested between F and K_2O due to the presence of biotite (Figure 8B with R = 0.93) with F most enriched in the magmatic-hydrothermal biotite breccia whereas the biotite altered tonalite and least-altered tonalite lack this F enrichment (Figure 8B). Biotite altered tonalite has a strong positive correlation between K_2O and Ba (r = 0.79; Figure 8C) and Rb (figure is not shown; r = 0.66), however the magmatic-hydrothermal biotite breccia shows a distinctly different trend between K_2O and Ba and poorer correlation (r = 0.45; Figure 8C). The latter observation may reflect changes in Ba partitioning in biotite through the paragenesis as it depends on the chemistry of the host rock (e.g., [42]), which here can be related to the change in Mg/(Mg + Fe) ratios of biotite (see [10]). Thus, the data suggest the increase La/Yb_N ratio and elevated K, Rb and Ba values can serve as proxies for the introduction of secondary biotite in the host rocks, whereas F is useful in identifying the magmatic-hydrothermal biotite breccia.

Geochemical profiles for drill holes E10-13, E11-161 and E11-116 (Figure 7A–C, respectively) demonstrate the spatial association of Au with magmatic-hydrothermal biotite breccia. Not only is there a strong association of Au with the breccia bodies, but these profiles also highlight a Au-Cu ± Te association. The latter is particularly evident in E11-161 which represents the least-altered of the profiles. This hole also shows clear increases in the abundances of K_2O, F, FeO_T, and Rb with the breccia (Figure 7B). Furthermore, a plot of Cu versus Au shows a positive correlation in the biotite breccia (r = 0.63) that is lacking in the biotite altered tonalite and diorite (red box area in Figure 8D).

Muscovite stage: This alteration type is relatively simple compared to the biotite stage in terms of its mineralogy and associated chemical signature (e.g., Rb, Ba, and F). For example, the profile for hole E11-116 (Figure 7C), which has a muscovite zone overprinting magmatic-hydrothermal biotite breccia and tonalite at 530 to 650 m depth, shows increases in K_2O and Rb. This latter feature is better illustrated in Figure 9A, a binary plot of Rb versus K_2O that shows a strong positive correlation (r = 0.89; Figure 9A) in addition to the muscovite-bearing samples being more Rb enriched. Additionally, the same zone shows weak to moderate positive correlations, respectively, between K_2O and F (r = 0.47; Figure 9B) and Ba (r = 0.60; Figure 9C). Again, we note that in some cases this alteration type locally overprints biotite alteration which may affect the correlations.

Although a correlation between Au and Cu is lacking in this alteration type, these zones are often coincident with Au ± Cu mineralization. For example, in hole E10-13 (Figure 7A) the Au-Cu mineralization continues from the footwall of the magmatic-hydrothermal biotite breccia into biotite and muscovite altered tonalite. Gold-only mineralization has been noted to be associated with zones having intense development of muscovite, however, such mineralization not as inconsistent and usually of lower grade [10,11] than the diorite breccia and magmatic-hydrothermal biotite breccia.

Since both the muscovite and biotite alteration equates to development of K-silicates, in addition to their respective mineralogical differences, some geochemical parameters are needed to further discriminate them. Thus, we note that whereas biotite altered tonalite lacks an apparent correlation between K_2O and F, a weak to moderate correlation is noted for sericite altered tonalite (Figure 9B). Additionally, in the Ba versus K_2O plot two trends are defined for biotite versus muscovite altered tonalites, which suggests Ba can be a discriminator (Figure 9C). Finally, the plot of Sr versus Rb (Figure 9D) shows Sr is relatively enriched in biotite versus muscovite altered tonalite.

Albite stage: This stage of alteration, recognized by its characteristic white bleaching and texturally destructive nature (Figure 10A–C), resulted in albite almost complete replacing primary minerals (e.g., plagioclase, hornblende, quartz), which thereby complicates the identification of the protolith (e.g., tonalite versus diorite). Thus, geochemistry can potentially provide the means to differentiate between the units. Although it is not uncommon for the high field strength element (HFSE), such as Zr, Ti, and Cr, to be conserved during hydrothermal alteration (e.g., [42,44,45]), this immobility must be first demonstrated. This immobility issue is addressed below.

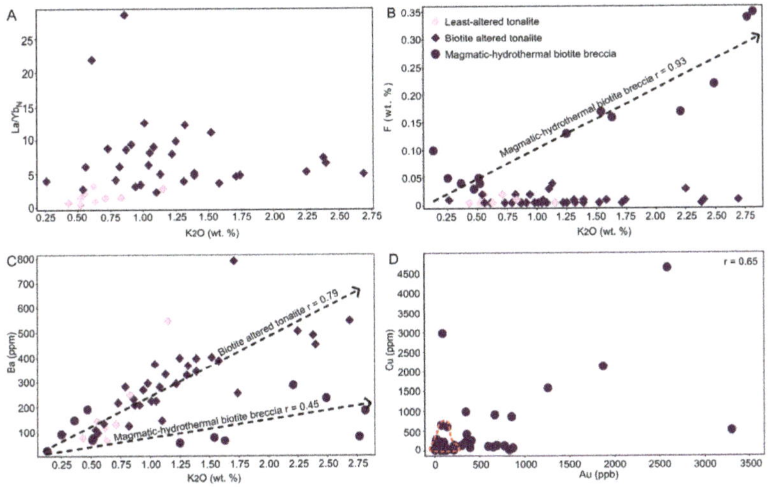

Figure 8. Binary element plots showing geochemical characterization of least-altered tonalite, biotite altered tonalite and magmatic-hydrothermal biotite breccia. (**A**) La/Yb_N versus K_2O showing increase in La/Yb_N for biotite altered tonalite (n = 32) compared to least-altered tonalite (n = 9). Chondrite normalizing values from [43]. (**B**) F versus K_2O showing strong positive correlation (r = 0.93) in magmatic-hydrothermal biotite breccia (n = 13), but no correlation for both least-altered and biotite altered tonalite samples. Note that two breccia samples are not included due to scale. (**C**) Ba versus K_2O showing strong positive correlation for biotite altered tonalite (r = 0.79) versus weaker correlation for magmatic-hydrothermal biotite breccia (r = 0.45). (**D**) Plot of Cu versus Au for the biotite breccia (n = 41), which shows a positive correlation (r = 0.65), and biotite altered tonalite that is outlined by the dashed red line in lower left. Note two samples with highest Au and Cu values plot off the diagram but are included in the regression.

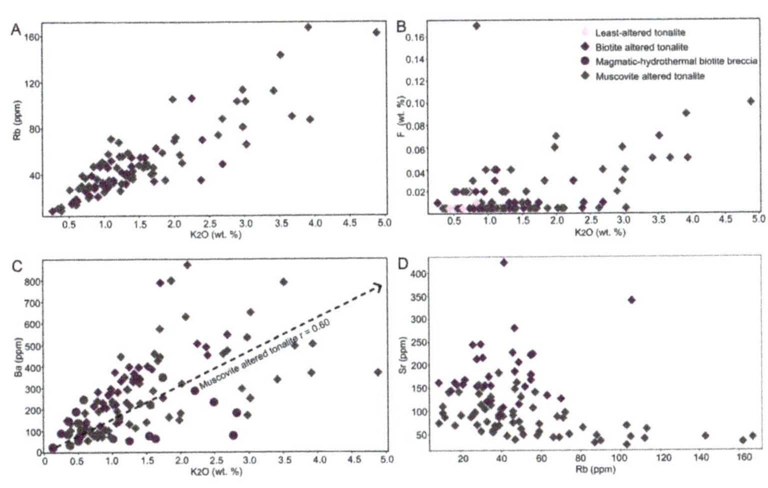

Figure 9. Binary element plots showing geochemical characterization of the muscovite altered tonalite. (**A**) Rb versus K_2O (n = 64) with r = 0.89 which compares to r = 0.66 for biotite altered tonalite (not shown). (**B**) F versus K_2O. (**C**) Ba versus K_2O showing different trends for biotite altered tonalite and magmatic-hydrothermal biotite breccia versus muscovite altered tonalite. (**D**) Sr versus Rb showing that biotite and muscovite altered tonalite define separate fields.

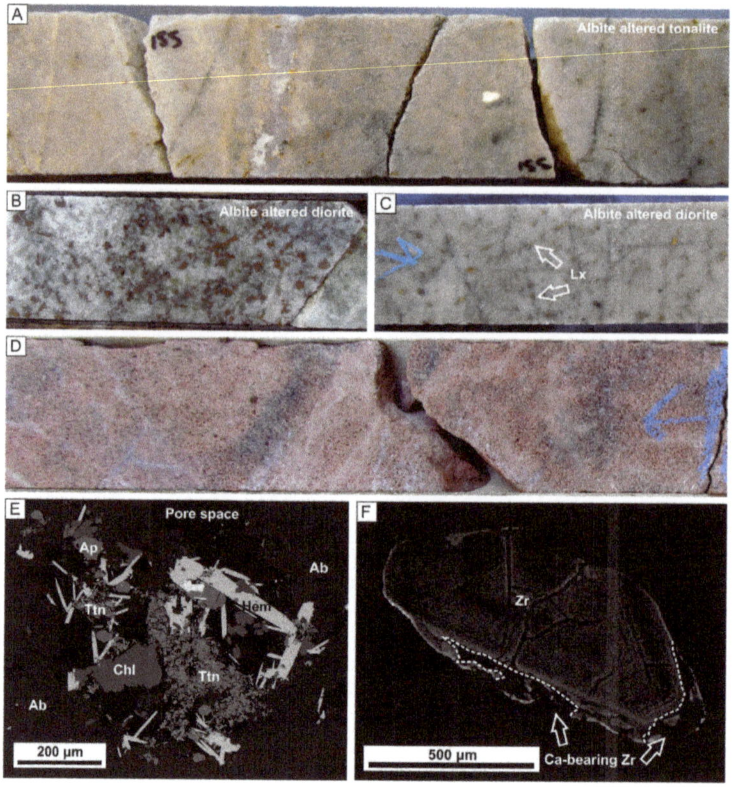

Figure 10. Drill core photos and backscatter electron (BSE) images used to illustrate the mobility of Ti and Zr during alteration. (**A**) Pervasive albite altered tonalite. (**B**) Pervasive albite altered diorite with albite replacing primary amphibole and calcic plagioclase with primary titanite or ilmenite pseudomorphed by brown rutile. (**C**) Strongly albite altered diorite with leucoxene after primary titanite or ilmenite; some of the leucoxene has been mobilized along the fractures. (**D**) An example of intense albite alteration (i.e., episyenite) with pore space created due to dissolution of quartz. (**E**) A BSE image of an episyenite where titanite, apatite, chlorite and hematite developed in pore space. (**F**) A BSE image of a zoned zircon in tonalite with dissolution-precipitation along rims, which have trace amounts of Ca.

For albite altered rocks, there is macro- and micro-scale evidence for variable mobility of the HFSE. Strong pervasive albite altered tonalite and dioritic rocks are usually distinguished by the presence/abundance of leucoxene; tonalite contains little to no leucoxene (Figure 10A) whereas diorite can have several percent (Figure 10B). Leucoxene, an alteration product of titanite and ilmenite, can be seen to be locally developed along fractures in albite altered diorite (Figure 10C). Similarly, in the rare occurrence of episyenite (Figure 10D) Ti-bearing minerals re-precipitate in the pore space developed from dissolution of primary quartz (Figure 10D,E). Zircon shows dissolution-precipitation textures along its margins and these areas contain trace Ca, P, Na, and F (Figure 10F); hydrothermal zircons have been documented to be enriched in Al, P, Y, Ca, REE, HFSE and F [46–50]. Although these textures suggest remobilization, it appears that dissolution and re-precipitation occurred only on a local scale with overall conservation of elements.

Based on the above, both Zr and Ti are considered to be relatively immobile and thus effective in distinguishing tonalite and diorite; this is illustrated in Figure 11A with a binary plot of SiO_2 versus Zr/TiO_2. In the latter it is apparent that $Zr/TiO_2 > 500$ separates the two lithologies using least altered samples, as indicated by the black dashed ellipses. In Figure 11B, different types of diorites are compared to tonalite with all having much lower Zr/TiO_2.

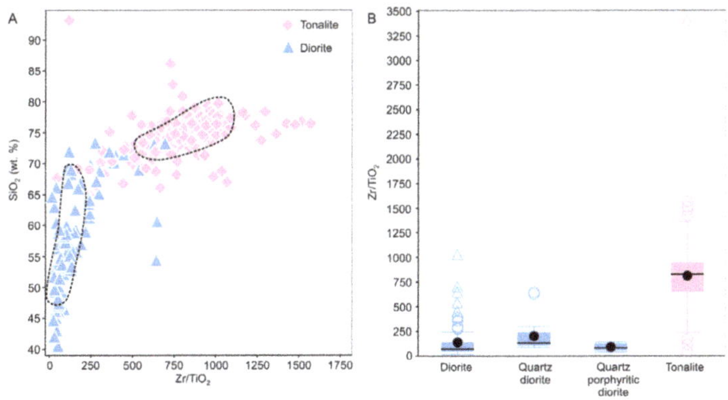

Figure 11. (**A**) Plot of SiO_2 versus Zr/TiO_2 used to discriminate between altered tonalite and diorite rocks. For comparison, the dashed black ellipses outline fields for least-altered tonalite (n = 9) and dioritic rocks (n = 16) and data falling outside these areas reflect scatter due to alteration. Note that a single outlier was removed from this diagram. (**B**) Box and whisker plots showing different Zr/TiO_2 ratios for tonalite (n = 209), diorite (n = 78), quartz diorite (n = 23) and quartz porphyritic diorite (n = 9). Black circle is the mean and central box the middle 50% of the data from Q1 to Q3. Circles are outliers and triangles far outliers.

Although tonalite and diorite are both altered by albite, their compositional difference results in different chemical changes. Tonalite is characterized by extreme enrichment in Na_2O accompanied by a decrease in SiO_2 (Figure 12A). Overall, it reflects a combination of the change in plagioclase composition (i.e., from $An_{30–50}$ to An_0) with variable loss of quartz. Thus for the tonalite a simple trend is defined in Figure 12A towards albite, whereas for the altered diorite a different trend is defined due to an increase in both Na_2O and SiO_2 owing to the partial to complete destruction of mafic minerals (e.g., hornblende) and change of plagioclase composition. Due to strong development of albite, the La/Yb_N ratio versus Na_2O can also be used to discriminate between biotite and albite altered rocks (Figure 12B).

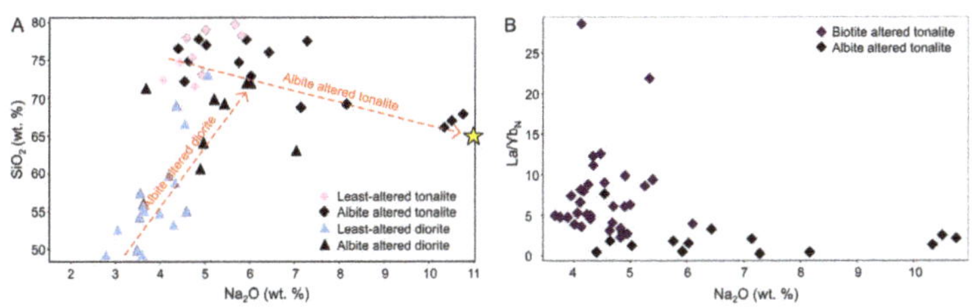

Figure 12. Geochemical characterization of the albite altered tonalite and diorite units. (**A**) Binary plot showing an increase in Na_2O and loss of silica for albite altered tonalite (n = 15) compared to an increase of Na_2O and SiO_2 silica in albite altered diorite (n = 9). Note that tonalitic samples plot towards pure albite (shown by the yellow star). The red dashed lines show a general trend of albite alteration inferred from the data. (**B**) Plot of La/Yb_N versus Na_2O that distinguishes albite altered samples from biotite altered samples. Chondrite normalizing values used are after [43].

Katz et al. [10] discussed the spatial relationship between Au mineralization and albite alteration as seen in both outcrop and drill core and noted its inconsistency. In addition, there is no correlation of Au with Cu in albite altered samples.

5.2. Rare Earth Element Geochemistry

The rare earth elements (REE) are typically found to be immobile in many hydrothermal ore systems, even when subjected to intense hydrothermal alteration, but exceptions occur (e.g., [45,51–59]). Thus, it is relevant to note here that in both the tonalite and diorite the REE concentrations, in particular the LREE, change dramatically in altered equivalents (Figure 13). Here we show that REE_{CN} patterns can be used to differentiate between some of the alteration stages.

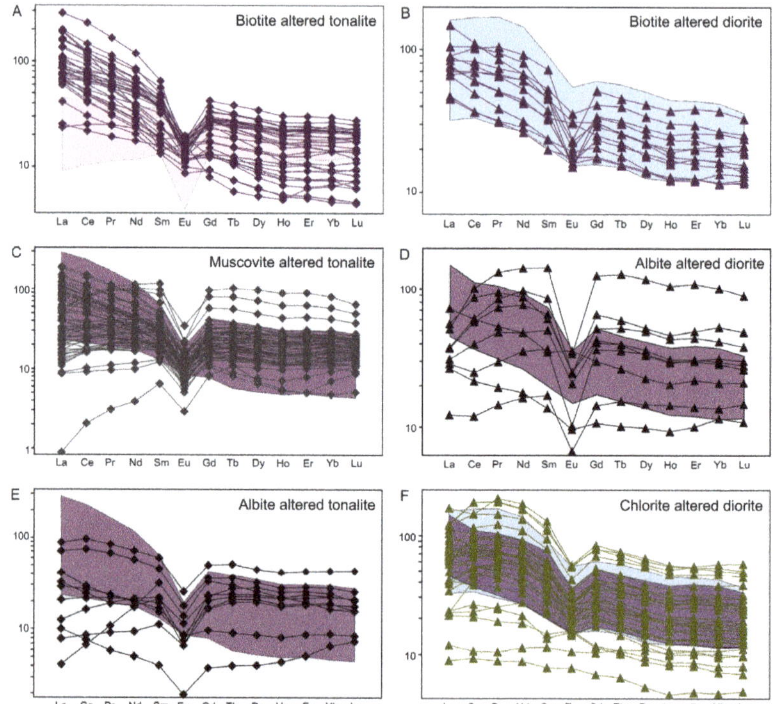

Figure 13. Chondrite-normalized REE diagrams (values from [43]) used to characterize different alteration stages. (**A**) Flat, unfractionated pattern typical of least-altered tonalite (n = 9) in shaded pink compared to LREE-enriched biotite altered tonalite (n = 32). (**B**) Comparison of least-altered diorite (n = 16) in the shaded blue with biotite altered diorite (n = 12). (**C**) Muscovite altered tonalite (n = 64) showing two REE patterns: (1) LREE-enriched with a negative slope (i.e., LREE > HREE); and (2) relatively flat slope with depletion from La to Pr, as seen in albite altered samples (see (**D**)). Note REE pattern for biotite altered tonalite (from (**A**)) is shown in purple shaded area. (**D**) Albite altered diorite (n = 9) showing depletion in LREE, especially La, Ce and Pr, compared to biotite altered diorite in purple shaded area (from B). (**E**) Selected samples of strong albite altered tonalite (n = 10) showing overall REE depletion, but particularly of LREE compared to biotite altered tonalite in the purple field (from (**A**)). (**F**) Chlorite altered diorite (n = 36) with concave-shaped LREE profile typical of the least-altered diorite, but also showing depletion of ∑REE. Shown for comparison are least-altered and biotite altered diorite in light-blue and purple shaded areas from (**B**), respectively.

Biotite altered samples have straight, negatively-sloped and strongly fractionated LREE patterns (i.e., La/Sm_N >> 1; Figure 13A) due to the presence of LREE-bearing minerals, such as allanite and bastnaesite (noted above, Figure 5E,F), which are LREE sinks (e.g., [60]). These patterns contrast markedly with the flatter REE patterns and lower ∑LREE exhibited by least-altered tonalite (pink shaded area in Figure 13A). The latter contrast for altered tonalite is less pronounced between least-altered and biotite altered diorite whereby a convex LREE pattern and negative slope characterizes the protolith (blue shaded area in Figure 13B) with the altered equivalent having similar ∑REE but straighter fractionation patterns. Thus, regardless of the LREE concentration it is apparent the straight negative-sloped LREE signature characterizes the biotite alteration and reflects the REE patterns of new mineral phases (e.g., bastnaesite, allanite).

Muscovite alteration is superimposed on biotite alteration zones, hence both muscovite altered tonalite and diorite are compared to biotite altered samples. Weak-to-moderate muscovite altered samples have negative-sloped LREE patterns similar to that in the biotite altered samples (shaded area in Figure 13C), whereas many of the moderate-to-intensely muscovite altered samples show flatter REE patterns with some distinctly LREE depleted. These contrasting patterns in the same alteration type suggest the elevated LREE signature was inherited from the earlier biotite alteration and survives where muscovite is weakly developed, but where more intense these minerals are destroyed, in particular with depletion of La and Ce. Muscovite altered samples are also characterized by larger negative Eu anomalies than for the biotite stage, which is consistent with destruction of plagioclase. In contrast, the HREE appear to be relatively unaffected by this alteration, except for a few samples (n = 3) with HREE enrichment.

The REE patterns for albite alteration also suggest mobilization of the REE (Figure 13D,E). Where biotite altered diorite is overprinted by albite alteration, there is a marked change in its patterns with a strong depletion of LREE and development of a positive slope from La to Sm, excepting for two samples that retain similar patterns to the diorite. There is also development of more negative Eu anomalies in most sample (Figure 13D), which is likely due to replacement of calcic plagioclase by albite. The REE patterns are consistent with the general abundance of secondary titanite in these albite altered rocks as hydrothermal titanite often has similar REE_{CN} patterns, in particular LREE enrichment with a convex profile (e.g., [61–63]). Similarly, albite altered tonalite typically has lower concentrations of LREE and flatter patterns compared to biotite altered tonalite (Figure 13E) but lacks the overall REE enrichment as in altered diorites due to a lack of titanite.

Chlorite altered diorite also shows some REE mobility compared to both least-altered and biotite altered diorite (Figure 13F). In general, the chlorite altered diorite samples display a similarly LREE enrichment as the least-altered diorite except for a few samples that contain the straight-sloped LREE pattern characteristic of biotite alteration, but the altered samples have the distinct convex patterns typical of titanite. In addition, a few samples with low ∑REE that have straight, slightly sloped to flat patterns with no Eu anomaly. In general, therefore, the REE abundance and patterns for chlorite altered diorite reflect that of the precursor diorite whether it be least-altered or biotite altered.

5.3. Whole-Rock Lithogeochemistry and Metal Associations

The summary for the whole rock data is presented in box and whisker plots in Figure 14, which highlight aspects already noted, such as: (1) enrichment of F in the magmatic-hydrothermal biotite breccia; (2) overall enrichment of all units in CO_2; (3) LREE enrichment (i.e., high La/Yb_N) in biotite altered tonalite; (4) Ba and Rb enrichment in both biotite and muscovite altered tonalite and diorite; and (5) higher Zr/TiO_2 values for the tonalite versus diorite.

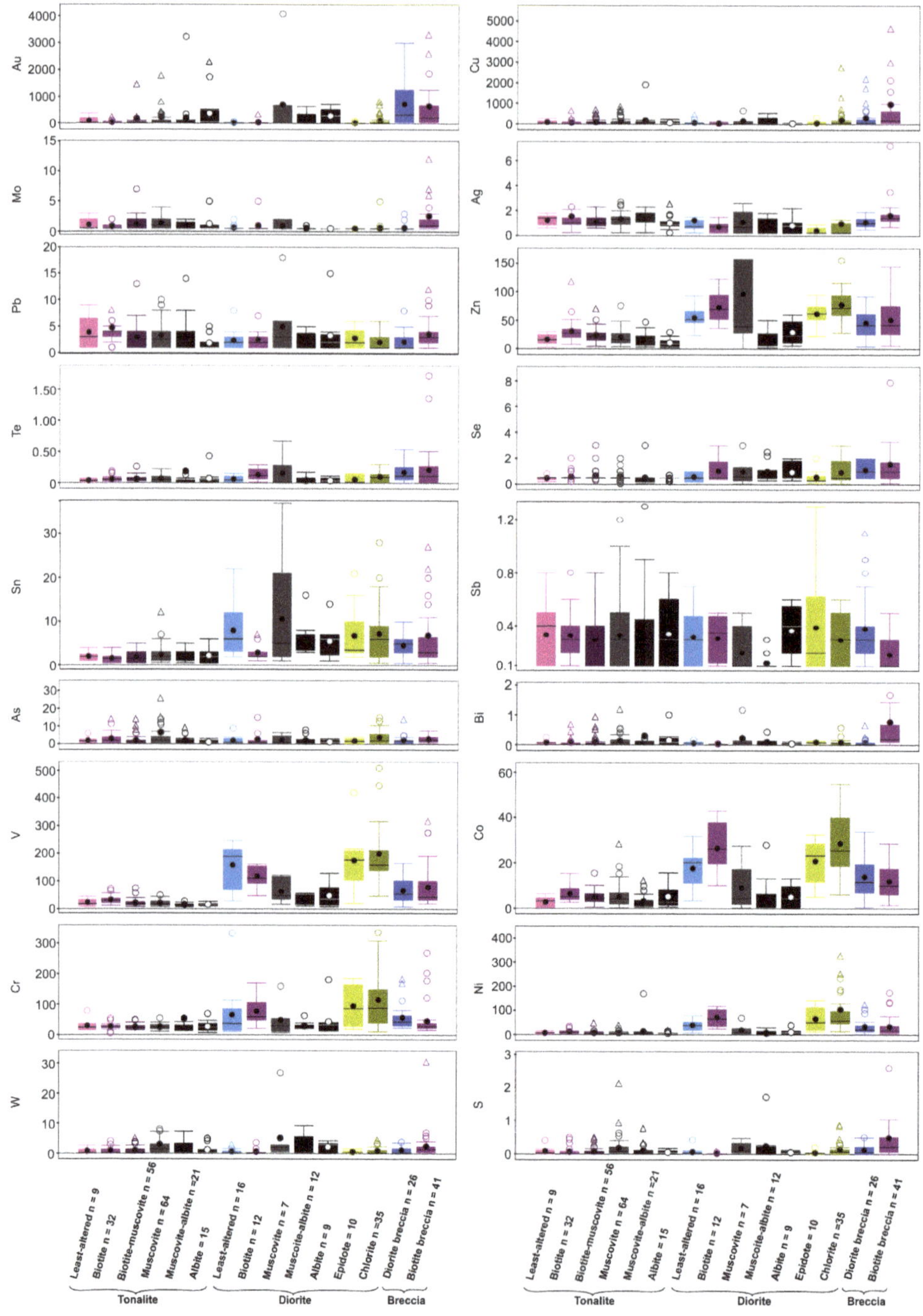

Figure 14. *Cont.*

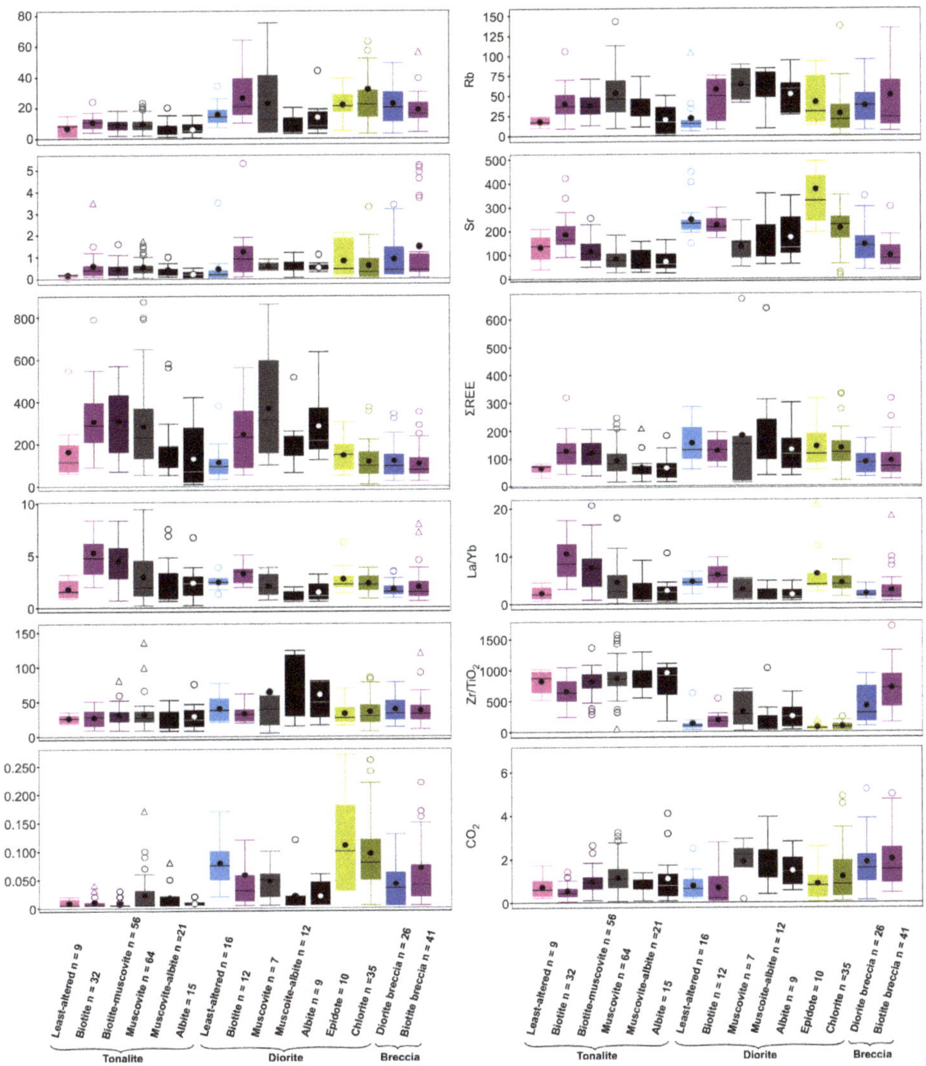

Figure 14. Box and whisker plots showing abundance of selected trace elements (ppm, but ppb for Au), volatiles (wt. %; F, S, and CO_2), and elemental ratios for different lithologies and alteration types. Note the use of different colours to discriminate the lithologies. The bottom of each box is Q1, the top is Q3 and the whiskers are the extreme values that are not outliers. An outlier (circle) is further than 1.5 × (Q3 − Q1), whereas a far outlier is further than 3 × (Q3 − Q1); outliers were omitted for several plots (Zn, Ag, Se, V, Cr, S, Rb, Sr, La, Lu, Cs, Y, Zr/TiO_2, La/Yb, F, La/Sm) and far outliers were also omitted for several plots (Au, Ag, Te, W, Cu, Ni, Pb, CO_2, Sn, Bi, Li, La/Yb, F) for clarity.

The whole-rock data are also used to assess different metal associations in the host and various altered equivalents. Overall, there is a lack of correlation of elemental enrichment in tonalite and diorite (Figure 14), but some generalizations are noted. The diorite unit is enriched compared to the tonalite in cafemic (i.e., Sr, Co, Cr, Ni, V, Zn) and lithophile (Li, Sn) elements, which reflects its abundance of primary amphibole and oxides versus secondary biotite and muscovite (e.g., [42]). For Au and Cu their enrichment is most evident in the two main breccias (e.g., diorite breccia and biotite breccia) and sheeted veins, as discussed below. Noted also is the overall depletion of all samples in As, Sb, Bi, and W.

Both the magmatic-hydrothermal biotite breccia and sheeted veins have an Au, Cu, Mo, Ag, Te, Bi, and Se signature (Figures 14 and 15). The geochemical profiles presented previously (Figure 7) also show the Au-Cu association in the breccia, along with variable Mo and Te. A well-developed ore shell and close correlation of Au and Cu grades suggests deposition of these metals synchronously with the magmatic-hydrothermal breccia event. In addition, a positive correlation between Bi and Te in the breccia is noted and consistent with the rare presence of tellurobismuthite (Bi_2Te_3). Ternary plots show the higher Te and Bi values correspond to higher Au grades in the breccia unit and sheeted veins (Figure 15A,B), but no such association is seen for the tonalite or dioritic rocks which indicates this is a hydrothermal signature. Higher Mo contents occur almost exclusively in the magmatic-hydrothermal biotite breccia versus diorite breccia, but in general the highest Mo does not equate to higher Au grades (Figure 15C).

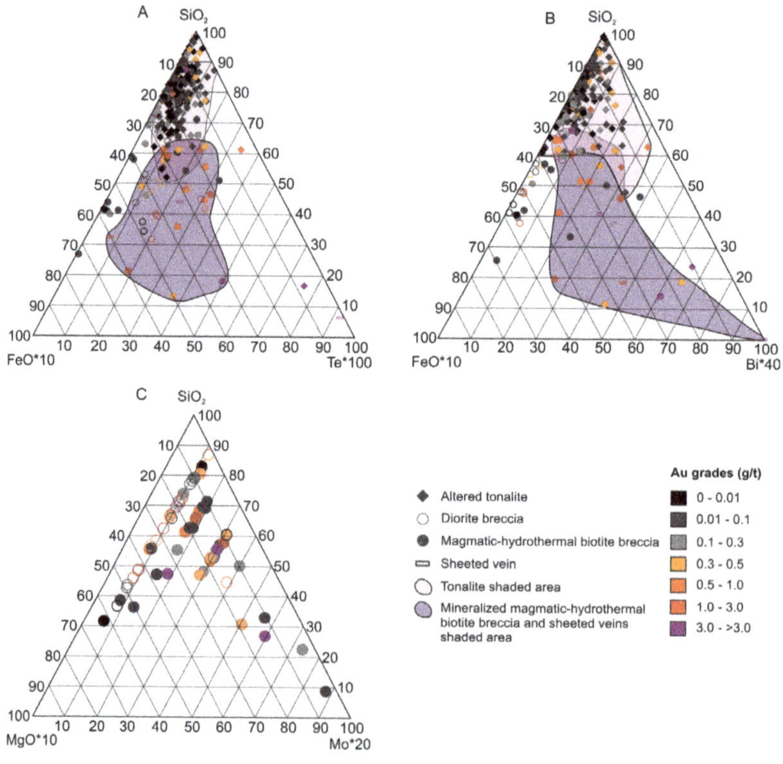

Figure 15. Ternary diagrams showing metal associations for diorite and magmatic-hydrothermal breccias. (**A**) SiO_2-FeO*10-Te*100 diagram showing enrichment of Te, as well as Fe, in both Au-bearing magmatic-hydrothermal biotite breccia and sheeted veins compared to tonalite. Note that diorite samples are not shown for clarity. (**B**) SiO_2-FeO*10-Bi*40 diagram showing Bi enrichment in Au-bearing magmatic-hydrothermal biotite breccia and sheeted veins. Some of the altered tonalite trends towards Bi and these samples are typically muscovite altered. (**C**) SiO_2-MgO*10-Mo*20 diagram showing regardless of Au grade the magmatic-hydrothermal breccia is enriched in Mo, whereas the diorite breccia is not.

5.4. Chalcopyrite Chemistry

A total of 70 analyses of chalcopyrite from nine samples are summarized in Table 2 and relevant binary plots in Figure 16. Eight samples representing the three styles of mineralization (i.e., breccia, disseminated, and vein) plus a high-grade Cu vein sample outside of the deposit envelope from the former Chester Mine were used. The data reveals highly variable amounts of Zn (28 to 1900 ppm), Ag (0.1 to 259.2 ppm), Se (30.36 to 214.8 ppm), Pb (1 to 127 ppm), Bi (1.7 to 119 ppm), Sn (0.72 to 114 ppm), In (5.025 to 100.3 ppm), Au (0.0041 to 17.2 ppm), Cd (0.45 to 47.9 ppm), Te (0 to 6.8 ppm) and Sb (0.04 to 0.81 ppm). Of these elements, Se, In, and Sb did not show correlations.

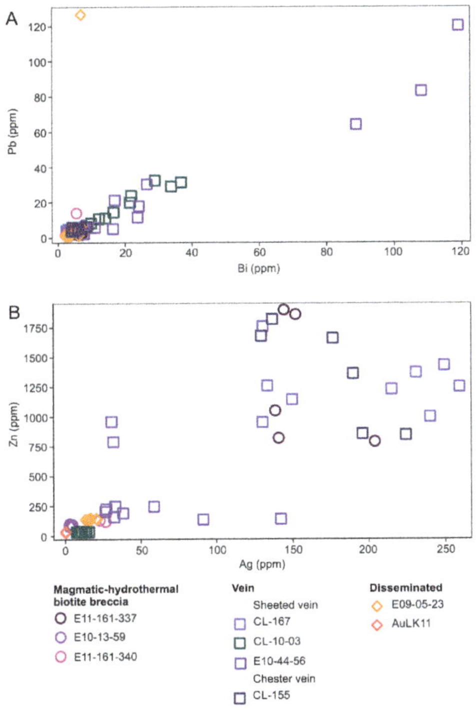

Figure 16. Bivariate plots showing abundances of selected trace elements in chalcopyrite for a variety of settings. (**A**) Pb versus Bi with a positive correlation (r = 0.77). (**B**) Zn versus Ag with a positive correlation with increases of Zn and Ag in magmatic-hydrothermal biotite breccia, sheeted veins, and Chester vein samples (r = 0.83).

Elevated Au in vein samples corresponds to enrichment of Bi and Pb with these two elements strongly correlated (r = 0.77; Figure 16A). Gold contents are low and variable within and among samples. Notably, chalcopyrite from the Au-bearing (above cut-off at >0.3 g/t Au) magmatic-hydrothermal biotite breccia samples have the lowest Au values (0.0041 to 0.258 ppm) suggesting it occurs either as micro-inclusions or as free gold. Likewise, Te is lowest in the breccia, which given its enrichment in this unit noted above (Figure 15A), suggests discrete Te-bearing phases. Interestingly, Te and Bi correlate moderately in vein and breccia material (R = 0.68 and 0.49, respectively), but no such correlation is noted in disseminated-type mineralization. Silver and Zn are strongly correlated (r = 0.83; Figure 16B) with the most elevated concentrations in chalcopyrite from breccia and vein material (i.e., sheeted vein and Chester vein), but not for disseminated chalcopyrite. Similarly, Cd is strongly correlated with Ag and Zn (note these plots are not included here).

5.5. Mass Balance

The geochemical data were used to quantify mass change and elemental gains and losses attending the different alteration types using the Grant method ([40]) Figure 17 shows that some combination of TiO_2, Zr, Al_2O_3 and SiO_2 defined the isocons for the various alteration types, thus these elements were conserved or immobile (i.e., black dots in Figure 17; see also the Zr/TiO_2 ratios in Figure 14). Additionally, as all the isocons have slopes of unity, except for chlorite altered samples, little if any mass or volume change occurred during alteration. Elemental changes (i.e., % gains/losses) for each alteration type per host rock are given in Table 5 and summarized in Figure 18. Note that for the isocon plots averages for each alteration type are used, but for the summaries of elemental changes presented in Figure 18 all samples are shown.

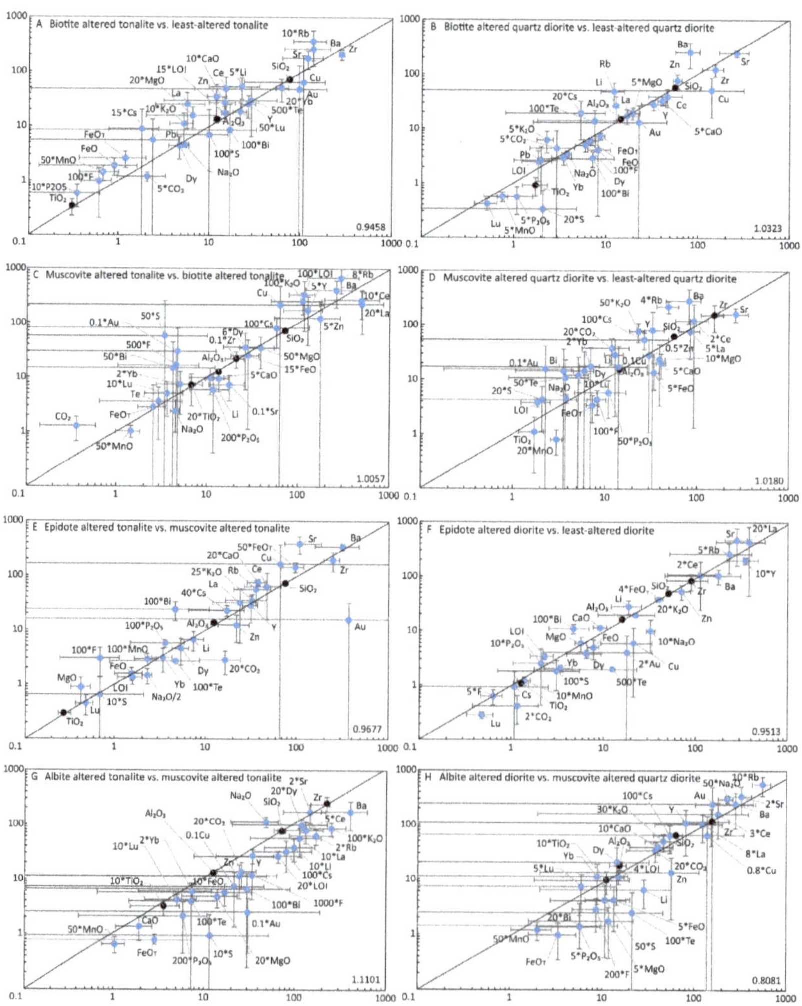

Figure 17. *Cont.*

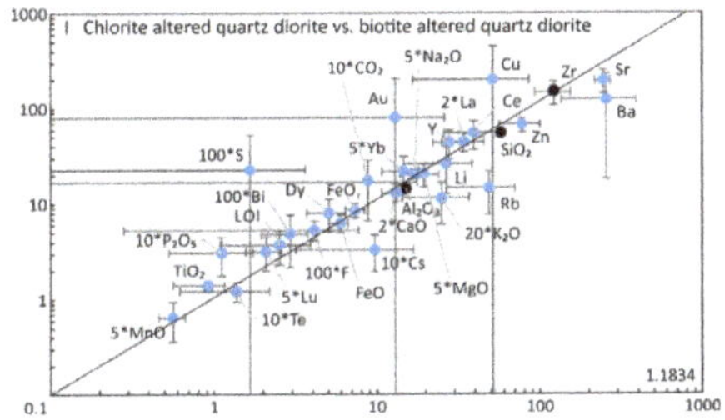

Figure 17. Summary of representative isocon diagrams using the most relevant elements for the different alteration types. Note data for various elements have been multiplied or divided to fit a common scale of the diagram. Black lines equate to the isocons, as defined by ratios of various immobile elements denoted by black circles, which were used for calculating the gains and losses (see Figure 18 and Table 5). The grey error bar lines represent the standard deviation.

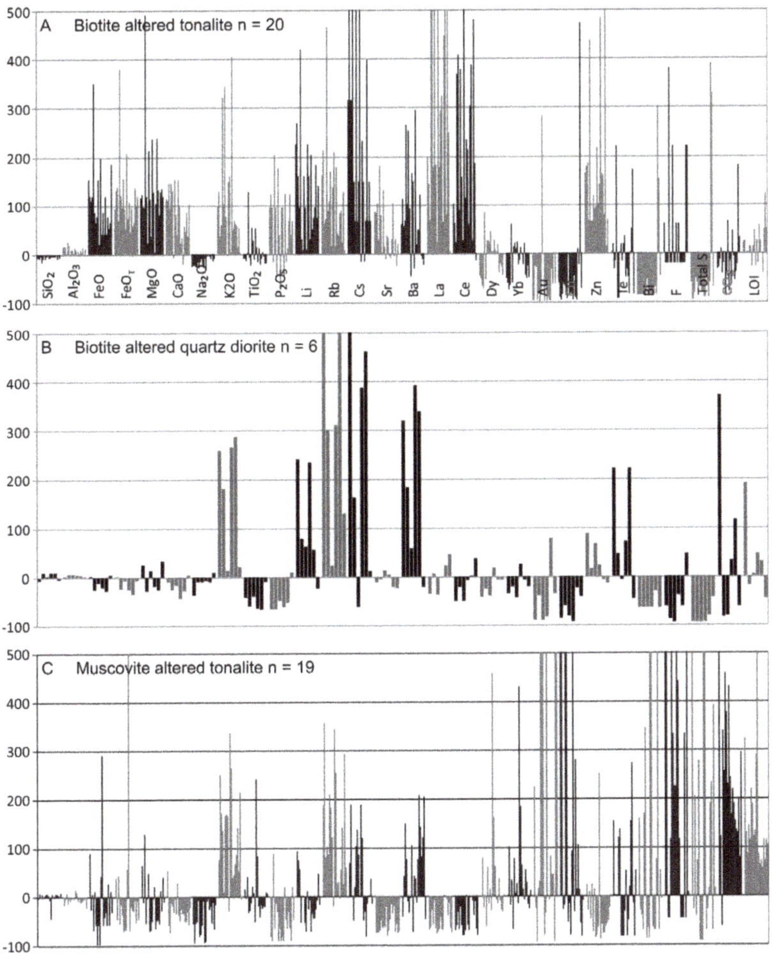

Figure 18. *Cont.*

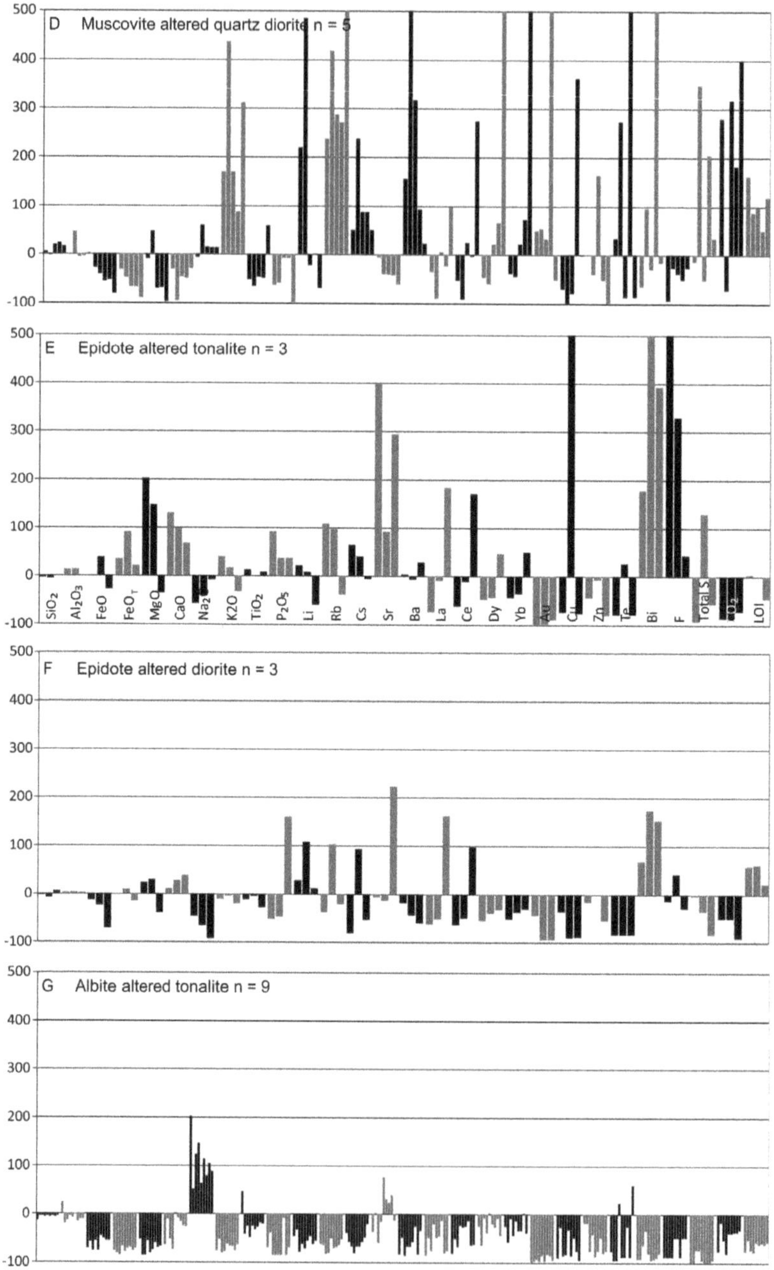

Figure 18. *Cont.*

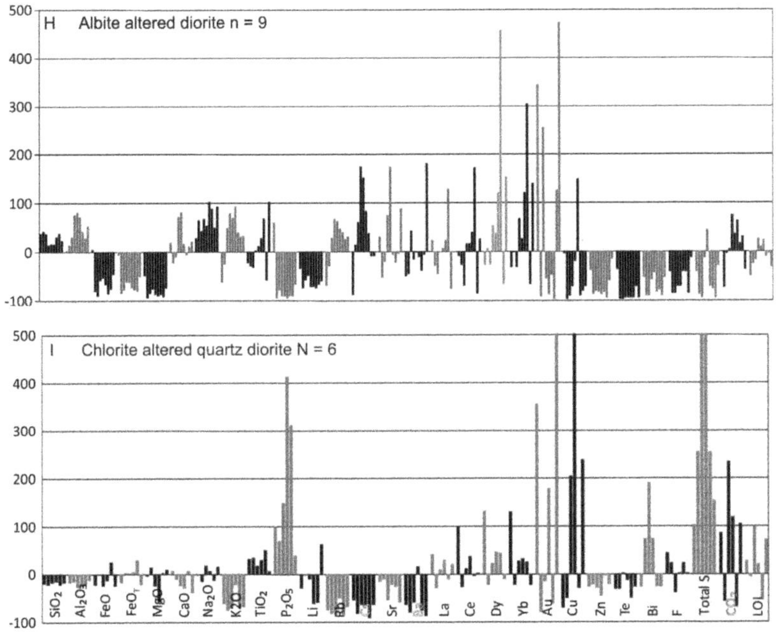

Figure 18. Summary diagrams for the gains and losses of major and trace elements in the main alteration types using the Grant-type isocon method (see Figure 17). Note that all gains are cut-off at 500%.

Biotite stage: Given that biotite alteration is early and widespread, unlike the rare and localized amphibole stage, altered samples are compared to least-altered tonalite and quartz diorite. Biotite altered tonalite samples include disseminated/stockwork types, whereas biotite altered quartz diorite samples used include only disseminated type.

For disseminated biotite altered tonalite (n = 20), most major elements (K_2O, FeO, FeO_T, MgO, MnO, CaO, P_2O_5) and LOI show gains, but for Na_2O there is minor loss (Figures 17A and 18A). The trace elements show moderate to strong gains for Li, Rb, Cs, Sr, Ba, La, Ce, and Zn, and F, with overall losses for Au, Cu, Bi, S, and CO_2. Several elements (e.g., Dy, Yb, Te, F, CO_2, and LOI) are variable with regard to gains and/or losses (Figure 18A).

For biotite altered quartz diorite (n = 6), there is consistent gain in K_2O, variable gains in CO_2 and LOI, overall weak losses for FeO, FeO_T, CaO, MnO, Na_2O with a pronounced loss for P_2O_5; MgO is variable (Figures 17B and 18B). For the trace elements, there are overall strong to moderate gains in Li, Rb, Cs, Ba, Zn, and Te, with strong to weak losses for Cu, Bi, and S (Figures 17B and 18B). Gold, Sr, Ce, Dy, Lu, and F display overall losses with individual samples displaying some gains, whereas La, CO_2 and LOI display overall gains, but individual samples may display losses (Figures 17B and 18B).

Both biotite altered tonalite and quartz diorite contain gains in K_2O, but the overall variable gains and losses for the other major oxides (i.e., FeO, FeO_T, MgO, CaO and P_2O_5) between the tonalite and quartz diorite are attributed to differences in their initial geochemistry. Thus, the lower initial abundances for most of these elements in tonalite is reflected in their substantial gains and the formation of secondary biotite, epidote, magnetite, apatite, and calcite. The noted and expected gains of K_2O, Li, Rb, Cs, Ba, Zn, F (tonalite only), and LOI in both units reflects the high partitioning of these elements in biotite (e.g., [42,64,65]). For the same reason, the gain of Sr in biotite altered tonalite is commensurate with gains of CaO and Fe_2O_3 (note this plot is not included) and reflects the addition of epidote, as well as some calcite, which is not seen for altered quartz diorite samples. Lesser and variable gains in Pb, Se, and As (Table 5) are also attributed to their partitioning into biotite [64].

The strong gains of La and Ce in biotite altered tonalite (also seen in REE plots; Figure 13) reflects the addition of LREE-bearing minerals (i.e., bastnaesite, allanite). In contrast, minor losses of LREE occur in biotite altered quartz diorite is attributed to initially higher abundances of LREE in the quartz diorite compared to tonalite.

Although individual samples show gains for Au, Cu, Bi, and S, there are overall losses of these elements for disseminated and stockwork biotite altered samples which relates to the general paucity of sulfides in the samples, as is also seen in the box and whisker plots (Figure 14). Lastly, gains in Te, similar to that seen in magmatic-hydrothermal biotite breccia, occur but not of the degree to constitute correlations with tonalite or dioritic rocks as noted above (Figure 15B).

Muscovite stage: The muscovite altered tonalite was compared to average biotite altered tonalite, its inferred precursor based on overprinting relationships (Figure 17C). In contrast, fracture-controlled and pervasive muscovite altered quartz diorite are compared to least-altered quartz diorite (Figure 17D).

The muscovite altered tonalite (n = 19) records overall addition of K_2O, F, CO_2, LOI, and variable S, whereas FeO, FeO_T, MnO, CaO, Na_2O, and P_2O_5 record losses (Figures 17C and 18C); however, both FeO and MgO are noted to vary with some individual samples recording gains. For the trace elements, strong to moderate overall gains are noted for Rb, Cs, and Ba, whereas losses occur for Sr, La, Ce, and Zn (Figures 17C and 18C). For several trace elements, both gains and losses are noted, as shown by Li, Dy, Yb, Au, Cu, Te, and Bi (Figure 18C).

The muscovite altered quartz diorite (n = 5) shows gains in K_2O, CO_2, LOI, and typically S and Na_2O, with overall losses of FeO, FeO_T, MgO, MnO, CaO, and P_2O_5 (Figures 17D and 18D). The trace elements show strong to moderate overall gains in Rb, Cs, and Ba, as does Au, but Sr, La, and F show overall losses (Figures 17D and 18D). Many trace elements variably show gains and losses, such as Li, Ce, Dy, Yb, Cu, Zn, Te, and Bi (Figure 17D).

As expected, K_2O, Rb, Cs, Ba, LOI, and CO_2 all show strong to weak gains in both muscovite altered tonalite and quartz diorite, which reflects addition of muscovite and carbonate during alteration, with loss of CaO and Sr due to muscovite replacing calcic plagioclase. In addition, loss of FeO, MgO, and P_2O_5 in muscovite altered tonalite is due to the breakdown of secondary biotite and apatite, whereas loss of FeO, MgO, CaO, TiO_2, P_2O_5, and F in muscovite altered quartz diorite is the result of replacement of the primary mineralogy (i.e., hornblende, plagioclase, titanite, ilmenite, and apatite) by muscovite. The overall strong gain of F in the altered tonalite reflects addition of muscovite, whereas overall loss of F in altered quartz diorite is due to its abundance in primary titanite and apatite. The inconsistency of Li (i.e., gains and losses) in both tonalite and quartz diorite rocks likely reflects it being inherited from early biotite alteration, which is supported by losses of Li concomitant with Mg. The consistent loss of both La and Ce in muscovite altered tonalite reflects destruction of LREE-bearing minerals originally present in the biotite altered samples. Conversely, the variability (i.e., gains and losses) of La and Ce in muscovite altered quartz diorite samples is attributed to some samples having inherited LREE-bearing minerals from earlier biotite alteration and destruction of these minerals in other samples.

The overall significant gains in Au and Cu associated with muscovite alteration is noted, but Au gains are not always commensurate with gains in Cu. In addition, this alteration stage is characterized by enrichments Mo, Ag, Te, Se, As, Pb, W and Sn. Gains of Fe may reflect the abundance of pyrite and corresponds to large gains in S in some samples.

Epidote stage: The averaged fracture-controlled and pervasive epidote altered tonalite were compared to average fracture-controlled muscovite altered tonalite as this is the inferred precursor (Figure 17E). In contrast, the averaged epidote altered diorite is compared to the average least-altered diorite (Figure 17F).

For the epidote altered tonalite (n = 3), the major elements show enrichments in FeO, FeO_T, MgO, MnO, CaO, K_2O, and P_2O_5, whereas weak to moderate losses for Na_2O (Figures 17E and 18E). Strong gains are noted for Sr, Bi, and F with inconsistent but overall gains in Rb, Cs, and Ba, whereas there are strong losses for Au and CO_2. Variable gains and losses are evident for Li, La, Ce, Dy, Yb, Cu, S and LOI (Figure 18E) depending on the sample.

Compared to the average least-altered diorite, the average epidote altered diorite (n = 3) shows overall gains in MgO, CaO, and P_2O_5, and LOI with minor to strong losses in FeO, Na_2O, K_2O, and CO_2 (Figures 17F and 18F); overall FeO_T is conserved. For the trace elements, there are typically weak to strong overall gains in Li, Rb, Sr, F, and Bi, whereas Ba, Dy, Yb, Au, Cu, Zn, and Te show losses. Cs, La, and Ce are variable, displaying both gains and losses (Figure 18F).

The inherent differences in geochemistry between the two host rocks partly accounts for notable differences in gains and losses (e.g., FeO_T, MgO, CaO). As expected, both epidote altered tonalite and diorite show enrichment CaO, however, FeO and FeO_T are on average conserved in altered diorite reflecting their higher original abundance compared to tonalite. Overall gains of FeO and MgO in tonalite samples may reflect accompaniment of chlorite with epidote. Addition of MgO and loss of K_2O in altered diorite samples reflects the addition of chlorite and the replacement of hornblende, respectively, whereas the gain of LOI in this unit relates to the abundance of epidote. The substantial gains of Sr in altered tonalite result from its substitution for Ca in epidote. In contrast, loss of Na_2O in both altered units reflects replacement of calcic plagioclase by epidote. The overall loss of LOI in tonalite samples is attributed to its LOI-rich precursor (muscovite altered tonalite) rock, whereas this is not the case for altered diorite. The general increase in Bi in both tonalite and diorite suggests an association of Bi in this alteration assemblage. The overall loss of Au and Cu reflects the general lack of mineralization in this alteration stage.

Albite stage: It is important to emphasize that the strong pervasive degree of this alteration in many cases precluded identification of the precursor unit, which was explored below with the precursor based on best fit according to field observations (see above and [10]).

For albite altered tonalite, it was compared to an inferred muscovite altered precursor based on crosscutting relationships. Samples were also compared to least-altered tonalite for comparison, but as results indicate gains in K_2O, Rb, Ba, LOI, CO_2, and Bi, which contradict petrographic observations, this supports use of the muscovite altered tonalite as the precursor unit. As for the diorite, muscovite altered quartz diorite is the preferred least-altered precursor even though it is not possible to choose between the quartz diorite or diorite. Despite large changes in several constituents, including losses and variability for most of the major oxides, particularly in the diorite, Al_2O_3, SiO_2, TiO_2, and Zr yield isocons with a slope of unity (Figure 17G,H) which validates use of the selected precursors.

In pervasive albite altered tonalite, there is strong enrichment in Na_2O and overall moderate to strong depletions in the other major oxides, including all the volatiles (F, S,

CO_2, LOI; Figures 17G and 18G). Notably, all the trace elements are also depleted except for Sr (Figure 18G).

In albite altered diorite, a different pattern of elemental change is noted. Whereas a weak to moderate enrichment of Na_2O is seen, strong losses in FeO, FeO_T, MgO, MnO, and P_2O_5 and all the volatiles (F, S, LOI) except CO_2 (Figure 17H and 18H) occur. Both gains and losses occur for CaO and K_2O and is sample dependent. Strong overall losses in trace elements include Li, Cu, Zn, Te, and Bi (Figures 17H and 18H), whereas Rb and Cs show overall gains. Many trace elements are variable and display gains or losses, such as Sr, Ba, La, Ce, Dy, Yb and Au (Figure 18H).

As expected, there is addition of Na_2O in the albite altered samples due to pervasive replacement of primary and secondary minerals, the latter from biotite and muscovite alteration with commensurate losses of FeO, FeO_T, MgO, MnO, P_2O_5, Li, Zn, Te, Bi, F, S and LOI. The overall loss of the REE is consistent with REE_{CN} patterns for albite altered samples. Gains of CaO and CO_2 in the altered diorite likely reflects carbonate, whereas enrichment in K_2O, Rb and Cs in the diorite reflects remnant muscovite and correlates with petrographic observations. In albite altered tonalite, CaO, K_2O, Rb, Cs, Ba, La, Ce, Dy, Yb, Au, Cu and CO_2 display consistent losses, whereas in albite altered diorite samples these elements display both gains and losses. The variability in the diorite compared to the consistent losses in the tonalite may relate to variability of the former samples as noted above or be inherited from earlier alteration. The strong decrease of S in both lithologies is due to the absence of sulfides in albite altered rocks. As for Au, its overall loss in the tonalite and gain in diorite reflects the inconsistent association of Au mineralization in albite altered rock samples, whereas the overall losses of most metals (i.e., Cu, Zn, Te, Bi, etc.) is due to the strong leaching capability of the fluids.

Chlorite stage: Only quartz diorite samples are included for this alteration. When compared to the precursor biotite altered quartz diorite, Al_2O_3, SiO_2 and Zr define an isocon indicating slight mass loss (Figure 17I). Based on the latter, samples show large losses for K_2O, large gains for P_2O_5 and S, and small variable changes for other major elements (FeO, FeO_T, MgO, CaO, Na_2O) and volatiles (LOI, CO_2) (Figures 17I and 18I). Losses for trace elements include Li, Rb, Cs, Sr, Ba, Zn, and Te. Several trace elements with variable changes but with overall gains are La, Ce, Dy, Yb, Au, Cu, and Bi (Figure 18I).

Overall losses of K_2O, Li, Rb, Cs and Ba reflects replacement of biotite by chlorite. Interestingly, gains in La and Ce in some samples occur and are probably due to remnant LREE-bearing minerals in chlorite altered samples inherited from earlier biotite alteration, but may also reflect in part newly formed phosphates, although not documented, as seen in strong addition of P_2O_5. This latter observation demonstrates that LREE-bearing minerals can be robust in chlorite altered samples (Figure 13F), which is consistent with petrographic observations [10]. Again, as expected gains in Au are not consistently associated with gains in Cu.

6. Discussion

The use of geochemistry in ore deposit studies is an important tool, not only for addressing ore genesis and related exploration (e.g., [66–74]), but also for ore deposit classification [7,69,70]. However, paramount in using geochemistry is identifying the true elemental gains and losses from mass balance, as noted above (e.g., [68]). In this particular case, the use of geochemistry can be helpful in discriminating between intrusion-related, which here we equate to porphyry-type for Côté Gold [10,11], versus orogenic deposit types.

It is well established that porphyry deposits have a spatial, temporal, and genetic link to fluids released from contemporaneous intermediate to felsic hydrous magmas [3,4,72,75]. Consequently, a zonal distribution of alteration types develop due to cooling and chemical equilibration of the fluids with the host intrusion and enveloping rocks [4]. This zonal alteration is typically manifested by a central potassic zone that gives way to a phyllic zone bordered by a narrow argillic zone; this is surrounded by a larger alteration halo of green rock, the propylitic zone that exceeds the limit of the orebody. In contrast, orogenic type deposits, generated during or immediately after the main phase of shortening and post-peak regional metamorphism, are characterized by a more focused and restricted alteration footprint, which lack the signature of magmatic-derived fluid [2,7,76–79]. In such orogenic deposits, the alteration (Fe-carbonatization, sericitization, and sulfidation) is focused along structural conduits (i.e., faults, shear zones) with a lateral and vertical zonation versus a zonation centered on the causative magmatic intrusive complex. Accordingly, these two deposit types have distinct characteristics that can be used to distinguish them, the porphyry-type being distinguished by: (1) an overlap of hydrothermal activity with magmatism; (2) a proven genetic association of the hydrothermal alteration with the causative intrusion; (3) distinct metal association and zonation; (4) fluid chemistry as reflected by lithogeochemistry, fluid inclusion types, and isotopic signatures; and (5) a contrasting chronology relative to the main phase of deformation (shortening). Even using these five parameters, the correct assignment of a deposit to an ore system, especially where overprinted by deformation and regional metamorphism, can prove difficult (cf. [80]). For example, this has been particularly challenging where others have argued in favor of Archean porphyry-type deposits, such as at Lac Troilus [81,82] and Malartic [83,84], as well as Phanerozoic reduced intrusion-related deposits [69,85,86].

6.1. Classification of the Côté Gold Deposit

The first two criteria noted above, which are partly based on geochronology, have been addressed previously [10–12]. Results for U-Pb dating of magmatic zircon and titanite and hydrothermal titanite in addition to Re-Os dating of molybdenite from auriferous veins and alteration zones all overlap at ca. 2740 Ma. Furthermore, these ages are significantly older than the known time (i.e., <2670 Ma) for orogenic-type mineralization in the AGB (see recent summary in [13]). This absolute timing of events is also supported by independent structural observations which have shown that regional main-phase of shortening deformation in the SGB (~2696 to 2675 Ma) overprints the veins and alteration related to the early magmatic-hydrothermal system at the Côté Gold deposit [87,88].

A porphyry-type model was suggested based on an inferred genetic link of the alteration types (i.e., amphibole, biotite, muscovite) to the causative dioritic intrusion [10]. The hydrothermal amphibole stage has been attributed to volatile exsolution of a causative dioritic magma based on a similar mineralogy (e.g., amphibole, apatite, titanite, magnetite, and quartz) and mineral chemistry for both magmatic and hydrothermal amphiboles. For the biotite stage, a genetic relationship is inferred based on the similar REE_{CN} patterns for the matrix of the magmatic-hydrothermal biotite breccia and diorite. In addition, in calcic systems derived from an original dioritic rock, epidote occurs as an integral component of the biotite alteration assemblage [4,89], as occurs at Côté Gold. Katz et al. [10] also attribute the muscovite and epidote alteration to the same processes as in porphyry systems, this being the result of mixing of acidic, magmatic sourced fluids with an external fluid of more neutral chemistry [3,4].

Although this is the first study to fully examine the alteration lithogeochemistry of the deposit, some aspects of the fluid chemistry were previously presented. Kontak et al. [11]

addressed the nature of the fluids using stable isotopes (O, C, S) for vein- and disseminated-style mineralization, as well as vein samples from the Chester vein system (i.e., former Chester Mine). The results for all samples are consistent with a dominantly magmatic source, but the data, in particular the calculated $\delta^{18}O_{H2O}$ values, also indicated input of a second fluid, either meteoric and/or seawater. More recently, Gourcerol et al. [14] showed that the fluid inclusions throughout the paragenesis is dominated by aqueous-carbonic fluids of variable XCO_2, salinity (7–35 wt. % eq. NaCl), and chemistry (i.e., Na- dominant with Ca, Fe, S, and F) with maximum homogenization values of 350 °C. These fluid properties were interpreted to indicate a magmatic origin and consistent with the summary of fluids associated gold deposit across the Abitibi greenstone belt [90,91]. Interestingly, Crépon et al. [92] have also documented a similar fluid chemistry (i.e., saline, aqueous-carbonic, Na-Ca-Fe-F) for a deposit setting in the Archean Chibougamau Cu-Au Camp located in the eastern part of the ABG interpreted to be of porphyry-type origin.

Proper classification of the Côté Gold deposit is critical, as its genesis has important implications for understanding the formation of and exploration for Au deposits in the Abitibi greenstone belt and Archean greenstone belts in general. Thus, further discussion below focuses on the nature of the fluids responsible for the observed hydrothermal alteration in terms of elemental gains and losses, REE mobility, and elemental associations. These data are also used to evaluate the porphyry-related model. The results from this study are fully examined and integrated with previous observations in order to establish a comprehensive genetic model for the deposit and geochemical criteria for exploration.

6.2. Chemical Signatures of Alteration Stages

Amphibole stage: Differences between the chemistry of magmatic and hydrothermal amphibole provides chemical information regarding the nature of the hydrothermal fluid. A fluctuating but general increase in fO_2 (i.e., higher Mg#) from magmatic to hydrothermal conditions has been noted in both amphibole and biotite from porphyry deposits [93,94] with amphibole noted to have higher Si, Mn, and Ca, but depletion in Fe, Ti, Na, K and Cl relative to less oxidized magmatic domains [94]. In Côté Gold, Katz et al. [10] showed hydrothermal amphibole in the breccia is zoned with Mg# increasing from core to rim which suggests relatively more oxidized conditions compared to the magmatic stage. In addition, these Mg-rich domains contain higher Si, Ca and F contents and lower Ti and Cl than magmatic amphiboles, which is also consistent with the change in fluid chemistry documented by Chivas ([94]). Thus, the documented change in mineral chemistry supports a genetic relationship from an early magmatic system which evolved to a hydrothermal-dominated system that records an increase in fO_2.

Biotite stage: Mass balance indicates a difference in the elemental enrichment of tonalite versus diorite, which likely reflects the different bulk chemistry of the host rocks. As the tonalite host is more sensitive to the changes, it best reflects the chemical signature of this alteration stage, which is enriched in Fe, Mg, Mn, Ca K, P, Li, Rb, Cs, Ba, LREE, Zn, Pb, Te, F, and LOI. That this signature is similar for different types of biotite assemblages (i.e., disseminated, veins, breccias) suggests a common fluid. Importantly, as the trace element signature of secondary biotite [10] shows appreciable quantities of Li (58 to 134 ppm), Rb (250 to 1408 ppm), Cs (7 to 70 ppm), Ba (22 to 3600 ppm), Mn (244 to 1594 ppm), and Zn (140 to 648 ppm), this validates the calculated gains noted above which are therefore related to this hydrothermal stage.

Both Rb and Cs readily substitute for K due to similar geochemical characteristics, thus such coupling has been noted in other alteration settings (e.g., [95]). Similarly, Li is known to substitute for Mg [95–97], which accounts for its enrichment in the Mg-rich biotite at Côté Gold (see [10]). In contrast, Ba is more enigmatic in potassic alteration zones in porphyry systems where both losses (e.g., Bajo de la Alumbrera [98] and Batu Hijau [99]) and gains (e.g., Sierrita [100]) are reported. Loss of Ba in some systems may reflect its compatible nature in regard to fractionation (e.g., [101]) or also relate to variable (i.e., lower) Kd values for hydrothermal versus magmatic biotite (e.g., [42]). Conversely, Ba may substitute for K in K-silicates [95,96]. Thus, the noted gain of Ba in both altered tonalite and diorite likely reflects a combination of Ba released from alteration of calcic plagioclase in the host rocks and addition from the fluids with subsequent sequestering in hydrothermal biotite and muscovite.

As with Ba, the geochemical behavior of Sr reflects the abundance of Ca [102], thus it should decrease with increasing fractionation. Consequently, Sr is usually depleted in potassic alteration zones of porphyry systems, and this is even more pronounced with extensive replacement of Ca-bearing minerals (i.e., plagioclase [102,103]). However, in this case the local gains of Sr noted (i.e., epidote altered diorite) likely reflects the coupled breakdown of calcic plagioclase in the biotite altered host and formation of epidote.

The noted addition of Zn is attributed not only to its incorporation into biotite, but also to sphalerite in the alteration assemblage. Enrichments in Zn and Pb in disseminated and stockwork-type biotite alteration outwards from the high-temperature breccia is consistent with base-metal zonation in porphyry systems (e.g., [4,104]). Zinc has been documented to be initially enriched in potassic alteration zones, but then subsequently remobilized outward (e.g., Panguna [93]); in the case of Côté Gold, Zn remains enriched in the biotite alteration stage relative to other alteration types but is a minor component of this stage.

Potassic alteration, as reflected by biotite and/or K-feldspar, is ubiquitous in porphyry systems and forms early in high-temperature core zones [3,4]. In contrast, these assemblages are not typical of orogenic gold deposits [2,76] and rarely documented for orogenic deposits hosted in granitoid rocks [7]. Additionally, that the biotite alteration is widespread and zoned (i.e., magmatic-hydrothermal biotite breccia core with peripheral veins and disseminations) at Côté Gold is thus wholly inconsistent with an orogenic setting. Furthermore, the Fe-Mg-K-Ca-bearing fluids, and also F, responsible for development of the biotite stage is more aligned with a magmatic fluid having exsolved from a co-spatial and contemporaneous diorite magma, albeit at depth. Lastly, the elemental gains associated with the biotite stage are similar to those widely documented in potassic alteration zones in porphyry deposits (e.g., Panguna [93], Sierrita [100], Bajo de la Alumbrera [98], and Batu Hijau [99]).

Muscovite stage: Mass balance indicates that for both tonalite and diorite host rocks this alteration stage has a similar geochemical signature with enrichment in K along with Rb, Cs, Ba, CO_2, and LOI with additional, albeit inconsistent, overall enrichment of Au, Cu, Bi, W, and S; moderate to weak enrichment is also noted for Te, As, Pb, Sn, Mo and Ag. Rubidium, Cs and Ba substitutes for K, whereas Sr is controlled by Ca [102,103], resulting in losses of both Ca and Sr due to replacement of calcic plagioclase by muscovite. No Sn- or W-bearing minerals have been noted, thus their gains are attributed to substitution into muscovite [97,105], which has been documented at Ann-Mason [106] and in many rare-metal granite settings (e.g., [107–109]).

Muscovite is common as part of alteration assemblages in both porphyry deposits, where it is represented by the phyllic zone [3,4], and also orogenic systems [2,7,76]. Gains and losses for the muscovite stage presented above are similar to those reported in porphyry systems (e.g., Bajo de la Alumbrera [98], Ann-Mason [106], and Darrehzar [110]), however, many of the elements are not discriminatory with respect to porphyry versus orogenic deposits. In fact, many orogenic deposits are also characterized by enrichments in

LILE (K, Rb, Ca, Li, Cs, Ti), As, S, and volatiles (H_2O, CO_2 [7]), thus not helpful in terms of discriminating a fluid source. The latter simply reflects the compatible behavior of these lithophile elements in muscovite (e.g., [95]). In contrast, the association of base metals (i.e., Cu, Mo, Ag, Bi, Te, Sn and W) is less characteristic of orogenic deposits [7,76,111,112]. Thus, this latter aspect, in addition to widespread formation of muscovite in the deposit versus being localized to faults and shears, and isotopic data [11] are together more indicative of a magmatic origin for these fluids.

Due to the typical addition of Rb and loss of Sr in phyllic alteration in porphyry systems, both these elements have been advocated as potential pathfinders for porphyry deposits enveloped by volcanic rocks, but inadequate where hosted in altered granites [102]. It is suggested that Rb and Sr are useful at Côté Gold as the loss of Sr in muscovite assemblages demonstrated above (Figure 9D) can be used to distinguish between it and biotite altered rocks.

Epidote stage: The most prominent geochemical features of this stage are addition of Ca accompanied by gains of FeO_T, Sr and Bi. Propylitic alteration in porphyry deposits is characterized not only by chlorite and epidote, both of which are sinks for Fe, but also of calcite and loss of albite [113]. Although chlorite and calcite accompany this alteration at Côté Gold, albite is not stable based on Na loss. The main chemical changes in propylitic alteration zones of porphyry systems is the addition of volatiles such as CO_2, $H_2O \pm S$ [3,33,113,114]. However, the lack of CO_2 addition likely reflects a lack cafemic elements in previously altered rocks due to their stripping out (i.e., see strong depletions of Fe and Ca in Figure 18C,D) that therefore precludes uptake CO_2 with alteration. Conversely, the noted gains in Bi is a feature not uncommon for porphyry or intrusion-related deposits [86,115,116], suggesting that magmatic fluids may have contributed to this signature [66]. The latter is further supported by the fact that Bi is the only element common to the biotite, muscovite, and epidote alteration zones, but this must be interpreted cautiously since inheritance may be a factor.

Albite stage: This alteration is characterized by gains in Na_2O and strong depletions in the other major oxides, trace elements, and metals. Thus in the progression from biotite to muscovite alteration into the albite stage there is a commensurate decrease in Ca, Fe, Mg, and the LREE that reflects albitization of the calcic plagioclase, mafic minerals, and destruction and remobilization of LREE-bearing minerals. Although albitization of precursor plagioclase is common in greenschist facies metamorphism and orogenic gold deposits [7], U-Pb dating of hydrothermal titanite at ca. 2740 Ma constrains this alteration stage to being syn-intrusion and mineralization.

Albite alteration in porphyry systems is attributed to low-temperature metasomatism, but its origin can be complex [3,4] owing to possible involvement of variable fluid types of either magmatic origin (e.g., [117]) or regionally sourced, non-magmatic, saline fluids [114,118]. For Côté Gold, Katz et al. [10] argued development of the albite stage related to Fe- and Mg-depleted, but Na-rich fluids originating from tonalitic melts co-temporal with the diorite. This Na-rich fluid is consistent with the Ca- and metal-deficient nature of the albite alteration.

Hydrothermal fluids responsible for sodic-calcic alteration in porphyry deposits are capable of leaching and transporting components from the precursor rock [3] or mobilizing what was added in earlier potassic alteration, as reflected by K, Fe, Cu, and S depletion in the Ann-Mason porphyry deposit [118]. The latter observation is thus consistent with the results of this study. Although drill core and outcrop assay results indicate that some strongly albite altered rocks may be Au-bearing [10], the overall loss of metals compared to the precursor rock (i.e., muscovite altered), as constrained from mass balance calculations, suggests metals were in fact leached and explains therefore the erratic nature of Au in this alteration type.

Chlorite stage: This stage of alteration was previously interpreted as an overprinting event given it inherits the mineral assemblage and sulfides of the biotite alteration, in addition to the Mg# of the precursor biotite [10]. The loss of K_2O, FeO, MgO, CaO, Li, Rb, Cs, Ba, Zn, and Te is consistent with chlorite overprinting and destroying the biotite assemblage. Further support for a replacement origin of the chlorite stage is suggested from: 1) similar REE patterns for chlorite altered diorite as its biotite altered precursor; and 2) the elemental signature of the chlorite altered rocks with gains in P_2O_5, Sn, S, Bi, CO_2, LOI, Au, and Cu, which is similar to that for muscovite altered rocks. That the presence of chlorite towards the core of the deposit, specifically in areas where albite alteration overprints earlier biotite and muscovite alteration, is further evidence of replacing earlier alteration assemblages.

6.3. *Implications of REE Enrichments and Depletions*

It was shown above that $\sum$REE and REE_{CN} profiles for the different alteration stages varied during ore system evolution. Mobility of trace elements, such as the REEs, during fluid-rock interaction is of interest because, in some cases, they are used to interpret hydrothermal processes. In this context, several studies have documented alteration related mobilization of the REE [45,56–59,119] in a variety of deposit types, including porphyry [51–55,120], greenstone-hosted orogenic gold [111,121], and hydrothermal U-REE [52,122,123]. The mobility of REE has been reported to be favored by several factors, such as low pH, changing temperature, high fluid/rock ratios, and importantly the abundance and type of ligand which stabilizes the REE complexes (CO_3^{2-}, F^-, Cl^-, $P_2O_5^{2-}$, SO_4^{2-}) in fluids [51,124–126].

Where alteration is related to an inferred magmatic fluid, such as the potassic zone of porphyries, the REE abundances and fluid signature relates to the magma chemistry [124,127]. Here biotite alteration resulted in apparent LREE enrichment for the tonalite, but not the diorite; this difference is likely due to the lower original $\sum$REE in tonalite compared to diorite and also fluid:rock exchange. In addition, the similarity of $\sum$REE in altered versus least-altered diorite supports its inherited magmatic origin for the biotite alteration.

In porphyry systems, phyllic alteration is accompanied by a decrease in temperature and lower pH with increasing fluid/rock ratios [3,128] with involvement of external fluids also occurring at this stage (e.g., seawater or meteoric fluids [129]. Under these conditions, the REE are variably noted to be progressively leached [51–53,55,130] or enriched [54,126] as alteration intensity increases. Depletion of LREE concomitant with alteration intensity has been documented in phyllic alteration zones of some porphyry systems (e.g., Bakircay and Ulutas prospects [51] and Murgul [55]), which is consistent with their mobility under intense fluid/rock interaction [125,131] and low pH conditions [51–53]. For Côté Gold, LREE depletion, in particular La, Ce and Pr, in muscovite altered samples is attributed to a similar change in fluid conditions with incursion of an external fluid, likely seawater [10,11,14] that occurred during this alteration stage.

Similar to the muscovite stage, strong depletion of LREE in the albite stage is favored by high fluid/rock ratios. In addition, the leaching nature of the fluid appears to have affected the stability of the secondary LREE-bearing phases (i.e., bastnaesite, allanite). Also relevant in the albite stage is the presence of secondary titanite, which has a low affinity for the LREE, and that when combined with high fluid/rock ratios attending this alteration resulted in the noted depletion of the LREE.

6.4. Metal Associations and Zoning

By virtue of the different processes in which porphyry and orogenic deposit types form, the nature of their fluids, and consequently, their metal associations differ. Thus, porphyry deposits are noted to typically have a Cu-Mo-Au-Ag-Pd-Te-Se-Bi-Zn-Pb signature [4] and also contain a distinct metal zoning with Cu ± Mo ± Au cores that give way outwards to enrichment in base metals such as Zn ± Pb ± Ag [4,104,132]. It is also noted, however, that both the metals present and their zoning can vary from system to system [3]. Conversely, the metallogenic geochemical signature of orogenic Au deposits has Au variably coupled to a broad range of metals (e.g., Ag, As, W, B, Sb, Te, Mo, and Bi) and notably with relatively low Cu, Zn and Pb contents; additionally, is the lack of a distinct deposit-scale zoning [7,76,77,111,112].

The magmatic-hydrothermal biotite breccia and sheeted veins are characterized by an Au-Cu-Mo-Ag-Te-Bi-Se association, whereas biotite disseminations/stockworks have a Te-Se-Zn-Pb-As association, although both Te and Se concentrations are generally on average lower than the breccia. In Au-rich porphyry deposits, magmatic-hydrothermal breccias that are generated early in the system are co-temporal with potassic alteration and have been documented to contain higher Au and Cu than the surrounding stockwork and disseminated zones (e.g., Panguna [133]; Endeavour 27 at Goonumbla [134]). Notably, the same zonation is documented at Côté Gold with an Au- and Cu-rich magmatic-hydrothermal biotite breccia and lesser well mineralized stockworks and disseminations extending outward and therefore, the zonation is consistent with Au-rich porphyry systems.

The muscovite alteration, which partially overprints the biotite alteration in the central part of the deposit and has an Au-Cu-Mo-Ag-Te-Bi-Se-Pb-As-W-Sn association similar to that of the breccia, but notably its Au, Cu, Mo, Ag, Te, and Bi concentrations are lower on average in the former compared to the latter. In porphyry deposit models, phyllic alteration is traditionally depicted as peripheral to the mineralized potassic alteration zone [4,135]. However, where this alteration is Au-bearing the zone has been documented to be within versus peripheral to the potassic alteration at several deposits (e.g., Granisle and Bell [136], Far Southeast [137], and Alumbrera [138]). The overall increase in Au ± Cu compared to disseminated/stockwork biotite altered samples, along with additional gains in S, Bi, W and Sn, suggest that some primary mineralization accompanied formation of the muscovite stage alteration.

Chalcopyrite trace element data is consistent with the metal associations noted above in both biotite and muscovite alteration stages. These data show an association of Te and Bi for chalcopyrite in both the magmatic-hydrothermal biotite breccia and sheeted veins but is lacking for the disseminated type. Elevated Ag and Zn values occur for some breccias and sheeted veins, and also the Chester vein sample. Thus, this association is not only just restricted to the deposit, but also extends distally outside the deposit area, which lends support to earlier interpretations that sheeted veins, both inside and outside the deposit, were part of the early hydrothermal system and associated with biotite alteration rather than later orogenic origin [11]. As highlighted before, the sheeted veins have muscovite alteration haloes that are characterized by elevated Bi and Pb and suggest the metal association is inherited from overprinting the earlier biotite alteration.

In summary, the metal associations and their zonation in the deposit are considered to more closely resembles those of porphyry systems rather than of orogenic gold deposits, which again supports a magmatic origin for the ore fluid versus a metamorphic one.

6.5. Genetic Model and Exploration Implications

The results and interpretations presented support Côté Gold having resulted from coupled magmatic-hydrothermal processes, consistent with our earlier interpretations [10,11]. Collectively these studies suggest the deposit shares attributes with Phanerozoic porphyry-type magmatic hydrothermal settings with respect to the nature of the host rocks (i.e., oxidized causative magma), overlap of magmatic and hydrothermal events, association of Au mineralization with a hydrous breccia body, distribution and nature of alteration types (i.e., potassic and phyllic), presence of stockwork and planar veins, and both temporal overlap and inferred genetic relationship of alteration with magmatism. Most recently, Meng et al. [139] have shown that the oxidation state of the causative intrusive complex at Côté Gold (i.e., CIC in Figure 2) overlaps with that of arc magmas (i.e., ΔFMQ = +1 to +2), whereas Mole et al. [140] document similar FMQ values characterize many Archean intrusive complexes across the southern Super Craton. Most recently, Crépon et al. [92] characterized a similar oxidized intermediate magma of dioritic nature related to the Archean (ca. 2720 Ma) Cu-Au mineralized Chibougamau Camp of the AGB, a plausible analogue to Côté Gold in regard to nature of mineralization, alteration, and fluid chemistry. Thus, an important conclusion from these few examples is that redox appears not to be a limiting factor in regard to forming porphyry-type mineralized intrusive settings in the Archean.

Importantly, the hydrothermal alteration and mineralization at Côté Gold is interpreted to result from magmatic fluids migrating upwards and outwards from the cooling and fractionating magma source at depth. In the deposit and surrounding area, the extent of hydrothermal biotite is widespread [10] with evidence for the area of infiltration of a similar magmatic fluid to be much larger. For example, a similar fluid chemistry is documented for sheeted veins in and around the deposit and in the former Chester Mine area based on metal associations and similar trace element abundances of chalcopyrite presented above, as well as isotopic signatures (δ^{34}S, δ^{18}O; [11]) and fluid inclusion thermometric and chemical data [14]. In addition, preliminary work on the Gosselin deposit, located just northeast from Côté Gold (Figure 2), indicates it shares similar host-rock types, alteration (e.g., biotite, muscovite), Au-Cu association, and fluid type (our unpublished observations; [9,141]). Thus, important for exploration is that the imprint of magmatic-sourced fluids extends well beyond the economic ore envelope of the Côté Gold deposit (i.e., open pit and resource shell) and strongly suggests the prolong extraction of mineralizing fluids from an evolving magma body at depth which is part of the generalized porphyry-deposit model [3].

The sequence of alteration types, their zonation in the deposit and their chemistry (i.e., gains and losses, REE patterns, metal associations, and isotopes) is consistent with the early hydrothermal stages (i.e., amphibole and biotite) having formed from high-temperature magmatic-derived fluids sourced at depth from an evolving hydrous dioritic magma with later alteration types (i.e., muscovite and epidote) reflecting dilution of the cooling magmatic fluids with down-welling oxidized surficial (i.e., seawater) fluids, as discussed by Katz et al. [10]. Our current of understanding of this world class Au deposit setting supports a magmatic origin (i.e., intrusion-related) specifically with an affinity to porphyry-type systems. This implies, therefore, the composition of the intrusions (e.g., diorites) and presence of and related geochemical indices of related alteration (e.g., biotite-bearing breccias, fracture-controlled biotite, and muscovite) are key exploration features.

7. Conclusions

This contribution follows on our earlier work that strongly suggests Côté Gold represents an Archean analogue to Phanerozoic oxidized intrusion-related ore (Au-Cu) deposits referred to as porphyry-type systems. Here we add to this by characterizing mineralogically and geochemically the distinct alteration types present in the deposit which overlap with those typical of porphyry-type deposit systems. Commencing with the early amphibole stage, its chemistry (e.g., Mg- and F-rich) reflects a hydrothermal versus magmatic origin. Using mass balance, the subsequent alteration stages were characterized. The biotite stage, manifest as magmatic-hydrothermal biotite breccia in addition to disseminated and vein-controlled biotite, is characterized by gains of Fe, Mg, Mn, Ca, K, P, Li, Rb, Cs, Ba, LREE, Zn, Pb, Te, F, and LOI, with additional enrichment of Au, Cu, Mo, Ag, Se and Bi noted in the magmatic-hydrothermal biotite breccia that, along with the diorite breccia, is the main host of the ore. The following muscovite stage shows gains in K, Rb, Cs, Ba, CO_2, and LOI with inconsistent but overall enrichments in Au, Cu, Bi, W, S, Te, As, Pb, Sn, Mo, and Ag. These gains represent both primary features of this alteration type, but some elemental inheritance due to earlier biotite alteration cannot be excluded. The locally developed epidote alteration is characterized by gains in Ca, FeO, Sr, and Bi, whereas the latest alteration stage involving strong development of albite is reflected in gains of Na with depletion of most other elements except for the HREE. Chlorite-bearing rocks are characterized by gains of P_2O_5, Sn, S, Bi, CO_2, LOI and Au and considered to reflect the overprinting the earlier biotite alteration assemblage. The metal associations and elemental ratios (i.e., La/Yb_N and Rb/Sr), along with the above geochemical criteria, provides vectors towards favorable alteration types that can be used in the exploration for similar intrusion-related deposits in Archean greenstone terranes.

The aforementioned alteration types are reflected in the behavior of the REE with their abundance and chondritic patterns reflecting the nature of the fluid. Thus, from an early high-temperature magmatic fluid to a later cooler fluid with lower pH (e.g., ingress of seawater) played an increasingly important role, which is supported by lithogeochemistry and published isotopic data. This study shows that the behavior of REE during different hydrothermal stages can effectively monitor changing fluid conditions and may be a useful tool to help constrain the genesis of hydrothermal ore deposits.

Significantly, all the geochemical data presented is consistent with our earlier and ongoing studies in which a porphyry-type deposit model for the Côté Gold deposit has and continues to be considered most plausible. The deposit is intrusion-related in the sense that Au mineralization is co-spatial and co-temporal to a magmatic center, in this case a dioritic intrusion, and that the hydrothermal system can be genetically related to this causative intrusion. Hence the nature and origin of the mineralization is considered to be dominantly magmatic and formed at ca. 2740 Ma more than 70 Ma before the main phase of shortening and regional metamorphism in the AGB.

Supplementary Materials: The lithogeochemical databased used in this study is found at Geological Survey of Canada, Open File 8040, p. 1–5 (https://geochem.nrcan.gc.ca/cdogs/content/pub/pub03684_e.htm, accesed on 15 February 2025).

Author Contributions: Conceptualization, L.R.K., D.J.K. and B.D.; methodology, L.R.K. and D.J.K.; software, L.R.K.; validation, L.R.K., D.J.K. and B.D.; formal analysis, L.R.K.; investigation, L.R.K. and D.J.K.; resources, L.R.K. and D.J.K.; data curation, L.R.K. and D.J.K.; writing—original draft preparation, L.R.K.; writing—review and editing, D.J.K. and B.D.; funding acquisition, D.J.K. and B.D. All authors have read and agreed to the published version of the manuscript.

Funding: This study was in part financed through the financial support of first Trelawney Mining and Exploration Inc. and then continued with IAMGOLD Corporation. Additional funding was also provided by the Geological Survey of Canada as part of the TGI-4 Lode Gold Project.

Data Availability Statement: Data are available as Supplementary material referred to in this contribution.

Acknowledgments: This contribution represents a portion the senior author's Ph.D. thesis at Laurentian University. The project was initiated by D. J. Kontak with Trelawney Mining and Exploration Inc. via D. Beilhartz and continued with IAMGOLD Corporation. Financial support was also provided by the Geological Survey of Canada as part of the TGI-4 Lode Gold Project. J. Petrus is thanked for his assistance in collecting the LA ICP-MS trace element data. Lastly, we thank the input and suggestions of reviewers which improved aspects of this contribution.

Conflicts of Interest: The authors declare no conflict of interest.

References

1. Černý, P.; Blevin, P.L.; Cuney, M.; London, D. Granite-Related Ore Deposits. In *One Hundredth Anniversary Volume*; Society of Economic Geologists: Littleon, CO, USA, 2005; pp. 337–370. ISBN 978-1-887483-01-8.
2. Goldfarb, R.J.; Baker, T.; Dubé, B.; Groves, D.I.; Hart, C.J.R.; Gosselin, P. *Distribution, Character, and Genesis of Gold Deposits in Metamorphic Terranes*; Society of Economic Geologists: Littleton, CO, USA, 2005; pp. 407–450.
3. Seedorff, E.; Dilles, J.H.; Proffett, J.M., Jr.; Einaudi, M.T.; Zurcher, L.; Stavast, W.J.A.; Johnson, D.A.; Barton, M.D. *Porphyry Deposits: Characteristics and Origin of Hypogene Features*; Society of Economic Geologists: Littleton, CO, USA, 2005.
4. Sillitoe, R.H. Porphyry Copper Systems. *Econ. Geol.* **2010**, *105*, 3–41. [CrossRef]
5. Baker, M.J.; Wilkinson, J.J.; Wilkinson, C.C.; Cooke, D.R.; Ireland, T. Epidote trace element chemistry as an exploration tool in the Collahuasi district, northern Chile. *Economic Geology* **2020**, *115*, 749–770. [CrossRef]
6. Wilkinson, J.J.; Baker, M.J.; Cooke, D.R.; Wilkinson, C.C. Exploration targeting in porphyry Cu systems using propylitic mineral chemistry: A case study of the El Teniente deposit, Chile. *Economic Geology* **2020**, *115*, 771–791. [CrossRef]
7. McCuaig, T.C.; Kerrich, R. P—T—T—Deformation—Fluid Characteristics of Lode Gold Deposits: Evidence from Alteration Systematics. *Ore Geol. Rev.* **1998**, *12*, 381–453. [CrossRef]
8. Rogers, J.R.; Beilhartz, D.; Kontak, D.J.; Katz, L.; Dubé, B.; McNicoll, V. The Côté Gold Deposit: Discovery of a New Generation Low-Grade, Multi-Million Ounce Gold Resource in the Archean Superior Province of Canada. In Proceedings of the NewGenGold Conference, Perth, Australia, 26–27 November 2013; pp. 159–174.
9. IAMGOLD. *Technical Report on the Côté Gold Project*; Report for NI 43-101; Iamgold: Toronto, ON, Canada, 2022.
10. Katz, L.R.; Kontak, D.J.; Dubé, B.; McNicoll, V.; Creaser, R.; Petrus, J.A. An Archean Porphyry-Type Gold Deposit: The Côté Gold Au(-Cu) Deposit, Swayze Greenstone Belt, Superior Province, Ontario, Canada. *Econ. Geol.* **2021**, *116*, 47–89. [CrossRef]
11. Kontak, D.J.; Creaser, R.A.; Hamilton, M.A. Geological and Geochemical Studies of the Côté Lake Au(-Cu) Deposit Area, Chester Township, Northern Ontario. In *Results from the Shining Tree, Chester Township and Matachewan Gold Projects and the Northern Cobalt Embayment Polymetallic Vein Project*; Ontario Geological Survey: Sudbury, ON, Canada, 2013; pp. 1–38.
12. Katz, L.R.; Kontak, D.J.; Dubé, B.; McNicoll, V. The Geology, Petrology, and Geochronology of the Archean Côté Gold Large-Tonnage, Low-Grade Intrusion-Related Au(–Cu) Deposit, Swayze Greenstone Belt, Ontario, Canada. *Can. J. Earth Sci.* **2017**, *54*, 173–202. [CrossRef]
13. Dubé, B.; Mercier-Langevin, P. Chapter 32: Gold Deposits of the Archean Abitibi Greenstone Belt, Canada. In *Geology of the World's Major Gold Deposits and Provinces*; Sillitoe, R.H., Goldfarb, R.J., Robert, F., Simmons, S.F., Eds.; Society of Economic Geologist: Littleton, CO, USA, 2020; pp. 669–708. ISBN 978-1-62949-312-1.
14. Gourcerol, B.; Kontak, D.J.; Thurston, P.C.; Tuba, G.; Katz, L. Reconstruction of the Pressure-Temperature-Composition Conditions of an Archean Porphyry-Type Deposit: The 2740 Ma Côté Gold Au(-Cu) Deposit, Canada. In Proceedings of the Program with Abstracts, Nancy, France, 23–29 June 2017.
15. Ayer, J.; Amelin, Y.; Corfu, F.; Kamo, S.; Ketchum, J.; Kwok, K.; Trowell, N. Evolution of the Southern Abitibi Greenstone Belt Based on U–Pb Geochronology: Autochthonous Volcanic Construction Followed by Plutonism, Regional Deformation and Sedimentation. *Precambrian Res.* **2002**, *115*, 63–95. [CrossRef]
16. van Breemen, O.; Heather, K.B.; Ayer, J.A. *U-Pb Geochronology of the Neoarchean Swayze Sector of the Southern Abitibi Greenstone Belt*; Geological Survey of Canada: Ottawa, ON, Canada, 2006.
17. Thurston, P.C.; Ayer, J.A.; Goutier, J.; Hamilton, M.A. Depositional Gaps in Abitibi Greenstone Belt Stratigraphy: A Key to Exploration for Syngenetic Mineralization. *Econ. Geol.* **2008**, *103*, 1097–1134. [CrossRef]
18. Heather, K.B. The Geological Evolution of the Archean Swayze Greenstone Belt, Superior Province, Canada. Ph.D. Thesis, University of Keele, Keele, UK, 2001.
19. Gemmell, T.P.; MacDonald, P.J. Precambrian Geology of the Yeo and Chester Townships Area. Chester Intrusive Complex, Southern Abitibi Greenstone Belt: Ontario Geological Survey, Preliminary Map. 2017. Available online: https://www.geologyontario.mines.gov.on.ca/publication/P3817 (accessed on 15 February 2025).

20. Monecke, T.; Mercier-Langevin, P.; Dubé, B.; Frieman, B.M. Geology of the Abitibi Greenstone Belt. In *Archean Base and Precious Metal Deposits, Southern Abitibi Greenstone Belt, Canada*; Monecke, T., Mercier-Langevin, P., Dubé, B., Eds.; Society of Economic Geologists: Littleton, CO, USA, 2017; pp. 7–49. ISBN 978-1-62949-220-9.
21. Heather, K.B.; van Breemen, O. An Interim Report on Geological, Structural, and Geochronological Investigations of Granitoid Rocks in the Vicinity of the Swayze Greenstone Belt, Southern Superior Province, Ontario. *Geol. Surv. Can. Pap.* **1994**, 259–268. [CrossRef]
22. Heather, K.B.; Shore, G.T. *Geology, Swayze Greenstone Belt, Ontario. Open File*; Ontario Geological Survey: Sudbury, ON, Canada, 1999.
23. Heather, K.B.; Shore, G.T. *Geology, Gogama, Swayze Greenstone Belt, Ontario. Open File*; Ontario Geological Survey: Sudbury, ON, Canada, 1999.
24. Hastie, E.C.G.; Kontak, D.J.; Lafrance, B. Gold remobilization: Insights from gold deposits in the Archean Swayze greenstone belt, Abitibi Subprovince, Canada. *Econ. Geol.* **2020**, *115*, 241–277. [CrossRef]
25. Hastie, E.C.G.; Kontak, D.J.; Lafrance, B.; Petrus, J.A.; Sharpe, R.; Fayek, M. Evaluating geochemical discriminants in Archean Au deposits: A Superior Province perspective with an emphasis on the Abitibi greenstone belt. *Econ. Geol.* **2023**, *118*, 123–155. [CrossRef]
26. Siragusa, G.M. *Geology, Geochemistry and Mineralization of the Southern Margin of the Swayze Belt*; Ontario Geological Survey: Sudbury, ON, Canada, 1993; pp. 1–144.
27. Fumerton, S.; Houle, K. *Mineral Showings, Occurrences, Deposits and Mines of the Swayze Greenstone Belt, Interim Report*; Open File Report 5871; Ontario Geological Survey: Sudbury, ON, Canada, 1993; 112p.
28. Heather, K.B.; Percival, J.A.; Moser, D.; Bleeker, W. *Tectonics and Metallogeny of Archean Crust in the Abitibi-Kapuskasing-Wawa Region*; Geological Survey of Canada: Ottawa, ON, Canada, 1995; pp. 1–148.
29. Powell, W.G.; Carmichael, D.M.; Hodgson, C.J. Conditions and Timing of Metamorphism in the Southern Abitibi Greenstone Belt, Quebec. *Can. J. Earth Sci.* **1995**, *32*, 787–805. [CrossRef]
30. Heather, K.B. *Regional Geology, Structure, and Mineral Deposits of the Archean Swayze Greenstone Belt, Southern Superior Province, Ontario*; Current Research, Part C; Geological Survey of Canada: Ottawa, ON, Canada, 1993; pp. 295–305.
31. Berger, B.R. Geological investigations south of Gogama. In *Summary of Field Work and Other Activities 2011*; Open File Report 6270; Ontario Geological Survey: Sudbury, ON, Canada, 2011; pp. 1–7.
32. Barker, F. *Trondhjemites, Dacites and Related Rocks*; Elsevier: New York, NY, USA, 1979.
33. Meyer, C.; Hemley, J.J. Wall Rock Alteration. In *Geochemistry of Hydrothermal Ore Deposits*; Holt: Winston, NY, USA, 1967; pp. 166–235.
34. Galley, A.G.; Hannington, M.D.; Jonasson, I.R. Volcanogenic Massive Sulphide Deposits. In *Mineral Deposits of Canada: A Synthesis of Major Deposit-Types, District Metallogeny, the Evolution of Geological Provinces, and Exploration Methods*; Special Publication; Geological Association of Canada: St. John's, NL, Canada, 2007; pp. 141–161.
35. Suikkanen, E.; Rämö, O.T. Episyenites—Characteristics, Genetic Constraints, and Mineral Potential. *Min. Metall. Explor.* **2019**, *36*, 861–878. [CrossRef]
36. Müller, W.; Shelley, M.; Miller, P.; Broude, S. Initial Performance Metrics of a New Custom-Designed ArF Excimer LA-ICPMS System Coupled to a Two-Volume Laser-Ablation Cell. *J. Anal. At. Spectrom.* **2009**, *24*, 209–214. [CrossRef]
37. Paton, C.; Hellstrom, J.; Paul, B.; Woodhead, J.; Hergt, J. Iolite: Freeware for the Visualisation and Processing of Mass Spectrometric Data. *J. Anal. At. Spectrom.* **2011**, *26*, 2508–2518. [CrossRef]
38. Sylvester, P.J.; Cabri, L.J.; Tubrett, M.N.; McMahon, G.; Laflamme, J.H.G.; Peregoedova, A. *Synthesis and Evaluation of a Fused Pyrrhotite Standard Reference Material for Platinum Group Element and Gold Analysis by Laser Ablation-ICPMS*; Geological Survey of Finland: Oulu, Finland, 2005; pp. 16–20.
39. Katz, L.R.; Kontak, D.J.; Dubé, B.; Mercier-Langevin, P.; Bécu, V.; Lauzière, K. *Whole-Rock Lithogeochemistry of the Archean Intrusion-Related Côté Gold Au(-Cu) Deposit, Ontario, Canada*; Geological Survey of Canada: Ottawa, ON, Canada, 2016; pp. 1–5.
40. Grant, J.A. The Isocon Diagram; a Simple Solution to Gresens' Equation for Metasomatic Alteration. *Econ. Geol.* **1986**, *81*, 1976–1982. [CrossRef]
41. Brodbeck, M.; McClenaghan, S.H.; Kamber, B.S.; Redmond, P.B. Metal(Loid) Deportment in Sulfides from the High-Grade Core of the Bingham Canyon Porphyry Cu-Mo-Au Deposit, Utah. *Econ. Geol.* **2022**, *117*, 1521–1542. [CrossRef]
42. Rollinson, H.R. *Using Geochemical Data: Evaluation, Presentation, Interpretation*; Longman Geochemistry Series; Taylor and Francis: Hoboken, NJ, USA, 2014; ISBN 978-0-582-06701-1.
43. Sun, S.-S.; McDonough, W.F. *Chemical and Isotopic Systematics of Oceanic Basalts: Implications for Mantle Composition and Processes*; Geological Society: London, UK, 1989; Volume 42, pp. 313–345. [CrossRef]

44. Petersen, M.D. The Use of the "Immobile" Elements Zr and Ti in Lithogeochemical Exploration for Massive Sulphide Deposits in the Precambrian Pecos Greenstone Belt of Northern New Mexico. *J. Geochem. Explor.* **1983**, *19*, 615–617. [CrossRef]
45. Petersson, J.; Eliasson, T. Mineral Evolution and Element Mobility during Episyenitization (Dequartzification) and Albitization in the Postkinematic Bohus Granite, Southwest Sweden. *Lithos* **1997**, *42*, 123–146. [CrossRef]
46. Corfu, F.; Hanchar, J.M.; Hoskin, P.W.O.; Kinny, P. Atlas of Zircon Textures. *Rev. Mineral. Geochem.* **2003**, *53*, 469–500. [CrossRef]
47. Putnis, A. Mineral replacement reactions: From macroscopic observations to microscopic mechanisms. *Mineral. Mag.* **2002**, *66*, 689–708. [CrossRef]
48. Hoskin, P.W.O.; Schaltegger, U. The composition of zircon and igneous and metamorphic petrogenesis. *Rev. Mineral. Geochem.* **2003**, *53*, 27–62. [CrossRef]
49. Schaltegger, U. Hydrothermal zircon. *Elements* **2007**, *3*, 51–79. [CrossRef]
50. Toscano, M.; Pascual, E.; Nesbitt, R.W.; Almodóvar, G.R.; Sáez, R.; Donaire, T. Geochemical discrimination of hydrothermal and igneous zircon in the Iberian Pyrite Belt, Spain. *Ore Geol. Rev.* **2014**, *56*, 301–311. [CrossRef]
51. Taylor, R.P.; Fryer, B.J. Multiple-Stage Hydrothermal Alteration in Porphyry Copper Systems in Northern Turkey: The Temporal Interplay of Potassic, Propylitic, and Phyllic Fluids. *Can. J. Earth Sci.* **1980**, *17*, 901–926. [CrossRef]
52. Taylor, R.P.; Fryer, B.J. Rare Earth Element Geochemistry as an Aid to Interpreting Hydrothermal Ore Deposits. In *Metallization Associated with Acid Magmatism*; John Wiley: New York, NY, USA, 1982; pp. 57–65.
53. Taylor, R.P.; Fryer, B.J. Rare Earth Element Lithogeochemistry of Granitoid Mineral Deposits. *Bull. Can. Inst. Mine Metall.* **1983**, *76*, 74–84.
54. Palacios, C.M.; Hein, U.F.; Dulski, P. Behaviour of Rare Earth Elements during Hydrothermal Alteration at the Buena Esperanza Copper-Silver Deposit, Northern Chile. *Earth Planet. Sci. Lett.* **1986**, *80*, 208–216. [CrossRef]
55. Schneider, H.J.; Oezguer, N.; Palacios, C.M. Relationship between Alteration, Rare Earth Element Distribution, and Mineralization of the Murgul Copper Deposit, Northeastern Turkey. *Econ. Geol.* **1988**, *83*, 1238–1246. [CrossRef]
56. Ward, C.D.; Mcarthur, J.M.; Walsh, J.N. Rare Earth Element Behaviour During Evolution and Alteration of the Dartmoor Granite, SW England. *J. Petrol.* **1992**, *33*, 785–815. [CrossRef]
57. Poitrasson, F.; Pin, C.; Duthou, J.-L. Hydrothermal Remobilization of Rare Earth Elements and Its Effect on Nd Isotopes in Rhyolite and Granite. *Earth Planet. Sci. Lett.* **1995**, *130*, 1–11. [CrossRef]
58. Fulignati, P.; Gioncada, A.; Sbrana, A. Rare-Earth Element (REE) Behaviour in the Alteration Facies of the Active Magmatic–Hydrothermal System of Vulcano (Aeolian Islands, Italy). *J. Volcanol. Geotherm. Res.* **1999**, *88*, 325–342. [CrossRef]
59. Parsapoor, A.; Khalili, M.; Mackizadeh, M.A. The Behaviour of Trace and Rare Earth Elements (REE) during Hydrothermal Alteration in the Rangan Area (Central Iran). *J. Asian Earth Sci.* **2009**, *34*, 123–134. [CrossRef]
60. Al-Ani, T.; Ahtola, T.; Kuusela, J.; Al-Ansari, N. Mineralogical and Petrographic Characteristics of Indium and REE-Bearing Accessory Phases in the Kymi Granite Stock, Southern Finland. *Nat. Resour.* **2018**, *9*, 23–41. [CrossRef]
61. Kontonikas-Charos, A.; Ciobanu, C.L.; Cook, N.J. Albitization and Redistribution of REE and Y in IOCG Systems: Insights from Moonta-Wallaroo, Yorke Peninsula, South Australia. *Lithos* **2014**, *208–209*, 178–201. [CrossRef]
62. Song, S.; Mao, J.; Xie, G.; Chen, L.; Santosh, M.; Chen, G.; Rao, J.; Ouyang, Y. In situ LA-ICP-MS U–Pb geochronology and trace element analysis of hydrothermal titanite from the giant Zhuxi W (Cu) skarn deposit, South China. *Miner. Depos.* **2019**, *54*, 569–590. [CrossRef]
63. Bruand, E.; Fowler, M.; Storey, C.; Laurent, O.; Antoine, C.; Guitreau, M.; Heilimo, E.; Nebel, O.; Bruand, E.; Fowler, M.; et al. Accessory mineral constraints on crustal evolution: Elemental fingerprints for magma discrimination. *Geochem. Perspect. Lett.* **2020**, *13*, 7–12. [CrossRef]
64. Johnson, C.A. Partitioning of Zinc among Common Ferromagnesian Minerals and Implications for Hydrothermal Mobilization. *Can. Miner.* **1994**, *32*, 121–132.
65. Foster, M.D. *Interpretation of the Composition of Trioctahedral Micas*; Professional Paper; US Geological Survey: Reston, VA, USA, 1960; Volume 354-B, pp. 1–146.
66. Prendergast, K. Application of Lithogeochemistry to Gold Exploration in the St Ives Goldfield, Western Australia. *Geochem. Explor. Environ. Anal.* **2007**, *7*, 99–108. [CrossRef]
67. Harris, J.R.; Wilkinson, L.; Grunsky, E.; Heather, K.; Ayer, J. Techniques for Analysis and Visualization of Lithogeochemical Data with Applications to the Swayze Greenstone Belt, Ontario. *J. Geochem. Explor.* **1999**, *67*, 301–334. [CrossRef]
68. Byrne, K.; Lesage, G.; Gleeson, S.A.; Piercey, S.J.; Lypaczewski, P.; Kyser, K. Linking Mineralogy to Lithogeochemistry in the Highland Valley Copper District: Implications for Porphyry Copper Footprints. *Econ. Geol.* **2020**, *115*, 871–901. [CrossRef]
69. Robert, F.; Poulsen, K.H.; Dubé, B. Gold deposits and their geological classification. In Proceedings of the Exploration 97: Fourth Decennial International Conference on Mineral Exploration, Toronto, ON, Canada, 14–18 September 1997; Gubins, A.G., Ed.; Exploration Geochemistry: Toronto, ON, Canada, 1997; pp. 209–220.

70. Carranza, E.J.M.; Sadeghi, M. Primary geochemical characteristics of mineral deposits – implications for exploration. *Ore Geol. Rev.* **2012**, *45*, 1–4. [CrossRef]
71. Cohen, D.R.; Bowell, R.J. Exploration geochemistry. In *Treatise on Geochemistry*, 2nd ed.; Holland, H.D., Turekian, K.K., Eds.; Elsevier: Oxford, UK, 2014; Volume 13, pp. 623–649. [CrossRef]
72. Halley, S.; Dilles, J.H.; Tosdal, R.M. Footprints: Hydrothermal alteration and geochemical dispersion around porphyry copper deposits. *Soc. Econ. Geol. Newsl.* **2015**, *No. 100*, 11–17. [CrossRef]
73. Kyser, K.; Barr, J.; Ihlenfeld, C. Applied geochemistry in mineral exploration and mining. *Elements* **2015**, *11*, 241–246. [CrossRef]
74. Gaillard, N.; Williams-Jones, A.E.; Clark, J.R.; Salvi, S.; Perrouty, S.; Linnen, R.L.; Olivo, G.R. The use of lithogeochemistry in delineating hydrothermal fluid pathways and vectoring gold mineralization in the Malartic district, Québec. *Ore Geol. Rev.* **2020**, *120*, 103351. [CrossRef]
75. Williams-Jones, A.E.; Heinrich, C.A. Vapor transport of metals and the formation of magmatic-hydrothermal ore deposits. *Econ. Geol.* **2005**, *100*, 1287–1312. [CrossRef]
76. Dubé, B.; Gosselin, P. Greenstone-Hosted Quartz-Carbonate Vein Deposits. In *Mineral Deposits of Canada: A Synthesis of Major Deposit-Types, District Metallogeny, the Evolution of Geological Provinces, and Exploration Methods*; Special Publication 5; Geological Association of Canada, Mineral Deposits Division: St. John's, NL, Canada, 2007; pp. 49–73.
77. Groves, D.I.; Goldfarb, R.J.; Robert, F.; Hart, C.J.R. Gold Deposits in Metamorphic Belts: Overview of Current Understanding, Outstanding Problems, Future Research, and Exploration Significance. *Econ. Geol.* **2003**, *98*, 1–29. [CrossRef]
78. Goldfarb, R.J.; Phillips, G.N.; Nokleberg, W.J. Tectonic setting of synorogenic gold deposits of the Pacific Rim. *Ore Geol. Rev.* **1998**, *13*, 185–218. [CrossRef]
79. Groves, D.I.; Goldfarb, R.J.; Gebre-Mariam, M.; Hagemann, S.G.; Robert, F. Orogenic gold deposits: A proposed classification in the context of their crustal distribution and relationship to other gold deposit types. *Ore Geol. Rev.* **1998**, *13*, 7–27. [CrossRef]
80. Goldfarb, R.J.; Pitcairn, I. Orogenic gold: Is a genetic association with magmatism realistic? *Miner. Depos.* **2022**, *58*, 5–35. [CrossRef]
81. Goodman, S.; Williams-Jones, A.E.; Carles, P. Structral Controls on the Archean Troilus Gold-Copper Deposit, Quebec, Canada. *Econ. Geol.* **2005**, *100*, 577–582. [CrossRef]
82. Fraser, R.J. The Lac Troilus Gold-Copper Deposit, Northwestern Quebec; a Possible Archean Porphyry System. *Econ. Geol.* **1993**, *88*, 1685–1699. [CrossRef]
83. Helt, K.M.; Williams-Jones, A.E.; Clark, J.R.; Wing, B.A.; Wares, R.P. Constraints on the Genesis of the Archean Oxidized, Intrusion-Related Canadian Malartic Gold Deposit, Quebec, Canada. *Econ. Geol.* **2014**, *109*, 713–735. [CrossRef]
84. De Souza, S.; Dubé, B.; Mercier-Langevin, P.; McNicoll, V.; Dupuis, C.; Kjarsgaard, I. Hydrothermal alteration mineralogy and geochemistry of the Archean world-class Canadian Malartic disseminated-stockwork gold deposit, southern Abitibi greenstone belt, Quebec, Canada. *Econ. Geol.* **2019**, *114*, 1057–1094. [CrossRef]
85. Sillitoe, R.H.; Thompson, J.F.H. Intrusion-Related Vein Gold Deposits: Types, Tectono-Magmatic Settings and Difficulties of Distinction from Orogenic Gold Deposits. *Resour. Geol.* **1998**, *48*, 237–250. [CrossRef]
86. Hart, C.J.R. Reduced Intrusion-Related Gold Systems. In *Mineral Deposits of Canada: A Synthesis of Major Deposit-Types, District Metallogeny, the Evolution of Geological Provinces, and Exploration Methods*; Special Publication No. 5; Geological Association of Canada, Mineral Deposits Division: St. John's, NL, Canada, 2007; pp. 95–112.
87. Smith, J. An Integrated Structural and Geochemical Study of Auriferous Sheeted Quartz Veins within the 2740 Ma Côté Gold Deposit, Swayze Greenstone Belt, Ontario. Unpublished Thesis, Laurentian University, Sudbury, ON, Canada, 2016.
88. Smith, J.; LaFrance, B.; Kontak, D.J. *A Comparative Study of the Deformation History of Auriferous Quartz Veins in the Archean Côté Gold Deposit and the Structural Evolution of the Spatially-Related Ridout Deformation Zone, Swayze Greenstone Belt, Northern Ontario*; Geological Association of Canada: Fredericton, NB, Canada, 2014.
89. Sillitoe, R.H. Gold-rich porphyry copper deposits: Geological model and exploration implications. In *Mineral Deposit Modeling*; Kirkham, R.V., Sinclair, W.D., Thorpe, R.I., Duke, J.M., Eds.; Geological Association of Canada: St. John's, NL, Canada, 1993.
90. Tuba, G.; Kontak, D.J.; Choquette, B.G.; Pfister, J.; Hastie, E.C.G.; van Hees, E.H.P. Fluid Diversity in the Gold-Endowed Archean Orogenic Systems of the Abitibi Greenstone Belt (Canada) I: Constraining the PTX of Prolonged Hydrothermal Systems. *Ore Geol. Rev.* **2021**, *135*, 1–31. [CrossRef]
91. Kontak, D.J.; Tuba, G. How Can Fluid Inclusion Studies Better Constrain Orogenic Gold Deposit Models: Case Studies from the Superior Province and Meguma Terrane, Canada? In Proceedings of the Society for Geology Applied to Mineral Deposits Annual Meeting, Quebec, QC, Canada, 5 February 2017.
92. Crépon, A.; Mathieu, L.; Kontak, D.J.; Marsh, J.; Hamilton, M.A. An Archean Porphyry-Type Deposit: Cu-Au Mineralization Associated with the Chibougamau Tonalite–Diorite Pluton, Abitibi Greenstone Belt, Canada. *Minerals* **2024**, *14*, 1293. [CrossRef]

93. Ford, J.H. A Chemical Study of Alteration at the Panguna Porphyry Copper Deposit, Bougainville, Papua New Guinea. *Econ. Geol.* **1978**, *73*, 703–720. [CrossRef]
94. Chivas, A.R. Geochemical Evidence for Magmatic Fluids in Porphyry Copper Mineralization. Part 1. Mafic Silicates from the Koloula Igneous Complex. *Contrib. Mineral. Petrol.* **1982**, *78*, 389–403. [CrossRef]
95. Kerrich, R.; Fryer, B.J. Lithophile-Element Systematics of Archean Greenstone Belt Au–Ag Vein Deposits: Implications for Source Processes. *Can. J. Earth Sci.* **1988**, *25*, 945–953. [CrossRef]
96. Rieder, M.; Cavazzini, G.; D'yakonov, Y.S.; Frank-Kamenetskii, V.A.; Gottardi, G.; Guggenheim, S.; Koval', P.V.; Müller, G.; Neiva, A.M.R.; Radoslovich, E.W.; et al. Nomenclature of the Micas. *Mineral. Mag.* **1999**, *63*, 267–279. [CrossRef]
97. Shannon, R.D. Revised Effective Ionic Radii and Systematic Studies of Interatomic Distances in Halides and Chalcogenides. *Acta Crystallogr.* **1976**, *32*, 751–767. [CrossRef]
98. Ulrich, T.; Heinrich, C.A. Geology and Alteration Geochemistry of the Porphyry Cu-Au Deposit at Bajo de La Alumbrera, Argentina. *Econ. Geol.* **2001**, *96*, 1719–1742. [CrossRef]
99. Idrus, A.; Kolb, J.; Meyer, F.M. Mineralogy, Lithogeochemistry and Elemental Mass Balance of the Hydrothermal Alteration Associated with the Gold-rich Batu Hijau Porphyry Copper Deposit, Sumbawa Island, Indonesia. *Resour. Geol.* **2009**, *59*, 215–230. [CrossRef]
100. Anthony, E.Y.; Titley, S.R. Patterns of Element Mobility during Hydrothermal Alteration of the Sierrita Porphyry Copper Deposit, Arizona. *Econ. Geol.* **1994**, *89*, 186–192. [CrossRef]
101. Jacobs, D.C.; Parry, W.T. Geochemistry of Biotite in the Santa Rita Porphyry Copper Deposit, New Mexico. *Econ. Geol.* **1979**, *74*, 860–887. [CrossRef]
102. Olade, M.A.; Fletcher, W.K. Primary Dispersion of Rubidium and Strontium around Porphyry Copper Deposits, Highland Valley, British Columbia. *Econ. Geol.* **1975**, *70*, 15–21. [CrossRef]
103. Armbrust, G.A.; Oyarzun, M.J.; Arias, J. Rubidium as a Guide to Ore in Chilean Porphyry Copper Deposits. *Econ. Geol.* **1977**, *72*, 1086–1100. [CrossRef]
104. Catchpole, H.; Kouzmanov, K.; Putlitz, B.; Seo, J.H.; Fontbote, L. Zoned Base Metal Mineralization in a Porphyry System: Origin and Evolution of Mineralizing Fluids in the Morococha District, Peru. *Econ. Geol.* **2015**, *110*, 39–71. [CrossRef]
105. Ahrens, L.H. The Use of Ionization Potentials Part 1. Ionic Radii of the Elements. *Geochim. Cosmochim. Acta* **1952**, *2*, 155–169. [CrossRef]
106. Cohen, J.F. Mineralogy and Geochemistry of Hydrothermal Alteration at the Ann-Mason Porphyry Copper Deposit, Nevada: Comparison of Large-Scale Ore Exploration Techniques to Mineral Chemistry. Unpublished Thesis, University of Oregon, Eugene, OR, USA, 2011.
107. Dostal, J.; Kontak, D.J.; Gerel, O.; Gregory Shellnutt, J.; Fayek, M. Cretaceous Ongonites (Topaz-Bearing Albite-Rich Microleucogranites) from Ongon Khairkhan, Central Mongolia: Products of Extreme Magmatic Fractionation and Pervasive Metasomatic Fluid: Rock Interaction. *Lithos* **2015**, *236–237*, 173–189. [CrossRef]
108. Kontak, D.J.; Dostal, J.; Petrus, J.A. Muscovite as a Monitor of Primary versus Secondary Rare-Metal Enrichment in Albite-Topaz Keratophyres from Ongon Khairkan, Mongolia. In Proceedings of the GAC-MAC-IAH Québec 2019, Québec City, QC, Canada, 12–15 May 2019.
109. Yin, R.; Han, L.; Huang, X.-L.; Li, J.; Li, W.-X.; Chen, L.-L. Textural and Chemical Variations of Micas as Indicators for Tungsten Mineralization: Evidence from Highly Evolved Granites in the Dahutang Tungsten Deposit, South China. *Am. Mineral.* **2019**, *104*, 949–965. [CrossRef]
110. Derakhshan, R.; Abdolzadeh, M. Geochemistry, Mineralization and Alteration Zones of Darrehzar Porphyry Copper Deposit, Kerman, Iran. *J. Appl. Sci.* **2009**, *9*, 1628–1646. [CrossRef]
111. Ludden, J.N.; Daigneault, R.; Robert, F.; Taylor, R.P. Trace Element Mobility in Alteration Zones Associated with Archean Au Lode Deposits. *Econ. Geol.* **1984**, *79*, 1131–1141. [CrossRef]
112. Gaboury, D. Parameters for the Formation of Orogenic Gold Deposits. *Appl. Earth Sci.* **2019**, *128*, 124–133. [CrossRef]
113. Djouka-Fonkwe, M.L.; Kyser, K.; Clark, A.H.; Urqueta, E.; Oates, C.J.; Ihlenfeld, C. Recognizing Propylitic Alteration Associated with Porphyry Cu-Mo Deposits in Lower Greenschist Facies Metamorphic Terrain of the Collahuasi District, Northern Chile--Implications of Petrographic and Carbon Isotope Relationships. *Econ. Geol.* **2012**, *107*, 1457–1478. [CrossRef]
114. Seedorff, E.; Barton, M.D.; Stavast, W.J.A.; Maher, D.J. Root Zones of Porphyry Systems: Extending the Porphyry Model to Depth. *Econ. Geol.* **2008**, *103*, 939–956. [CrossRef]
115. Sillitoe, R.H. Intrusion-Related Gold Deposits. In *Gold Metallogeny and Exploration*; Springer: Boston, MA, USA, 1991; pp. 165–209. ISBN 978-1-4612-7837-5.
116. Lang, J.R.; Baker, T. Intrusion-Related Gold Systems: The Present Level of Understanding. *Miner. Depos.* **2001**, *36*, 477–489. [CrossRef]

117. John, D.A. Geologic Setting, Depths of Emplacement, and Regional Distribution of Fluid Inclusions in Intrusions of the Central Wasatch Mountains, Utah. *Econ. Geol.* **1989**, *84*, 386–409. [CrossRef]
118. Dilles, J.H.; Einaudi, M.T. Wall-Rock Alteration and Hydrothermal Flow Paths about the Ann-Mason Porphyry Copper Deposit, Nevada; a 6-Km Vertical Reconstruction. *Econ. Geol.* **1992**, *87*, 1963–2001. [CrossRef]
119. Ludden, J.N.; Thompson, G. Behaviour of Rare Earth Elements during Submarine Weathering of Tholeiitic Basalt. *Nature* **1978**, *274*, 147–149. [CrossRef]
120. Norman, D.I.; Kyle, P.R.; Baron, C. Analysis of Trace Elements Including Rare Earth Elements in Fluid Inclusion Liquids. *Econ. Geol.* **1989**, *84*, 162–166. [CrossRef]
121. Kerrich, R.; Fryer, B.J. Archaean Precious-Metal Hydrothermal Systems, Dome Mine, Abitibi Greenstone Belt. II. REE and Oxygen Isotope Relations. *Can. J. Earth Sci.* **1979**, *16*, 440–458. [CrossRef]
122. McLellan, S.M.; Taylor, S.R. Rare Earth Element Mobility Associated with Uranium Mineralisation. *Nature* **1979**, *282*, 247–250. [CrossRef]
123. Metz, M.C.; Brookins, D.G.; Rosenberg, P.E.; Zartman, R.E. Geology and Geochemistry of the Snowbird Deposit, Mineral County, Montana. *Econ. Geol.* **1985**, *80*, 394–409. [CrossRef]
124. Lottermoser, B.G. Rare Earth Elements and Hydrothermal Ore Formation Processes. *Ore Geol. Rev.* **1992**, *7*, 25–41. [CrossRef]
125. Williams-Jones, A.E.; Migdisov, A.A.; Samson, I.M. Hydrothermal Mobilisation of the Rare Earth Elements—A Tale of "Ceria" and "Yttria". *Elements* **2012**, *8*, 355–360. [CrossRef]
126. Lottermoser, B.G. Rare-Earth Element and Heavy-Metal Behaviour Associated with the Epithermal Gold Deposit on Lihir Island, Papua New Guinea. *J. Volcanol. Geotherm. Res.* **1990**, *40*, 269–289. [CrossRef]
127. Flynn, R.T.; Wayne Burnham, C. An Experimental Determination of Rare Earth Partition Coefficients between a Chloride Containing Vapor Phase and Silicate Melts. *Geochim. Cosmochim. Acta* **1978**, *42*, 685–701. [CrossRef]
128. Gustafson, L.B.; Hunt, J.P. The Porphyry Copper Deposit at El Salvador, Chile. *Econ. Geol.* **1975**, *70*, 857–912. [CrossRef]
129. Pat Shanks, W.C. Stable Isotope Geochemistry of Mineral Deposits. In *Treatise on Geochemistry*; Elsevier: Amsterdam, The Netherlands, 2014; pp. 59–85, ISBN 978-0-08-098300-4.
130. Alderton, D.H.M.; Pearce, J.A.; Potts, P.J. Rare Earth Element Mobility during Granite Alteration: Evidence from Southwest England. *Earth Planet. Sci. Lett.* **1980**, *49*, 149–165. [CrossRef]
131. Kerrich, R.; Wyman, D.A. The Trace Element Systematics of Igneous Rocks in Mineral Exploration: An Overview. In *Trace Element Geochemistry of Volcanic Rocks: Applications for Massive Sulphide Exploration*; Short Course Notes; Geological Association of Canada: St. John's, NL, Canada, 1996; Volume 12, pp. 1–50.
132. Jones, B.K. Application of Metal Zoning to Gold Exploration in Porphyry Copper Systems. *J. Geochem. Explor.* **1992**, *43*, 127–155. [CrossRef]
133. Clark, G.H. Panguna Copper-Gold Deposit. *Aust. Inst. Min. Metall.* **1990**, *14*, 807–1816.
134. Jones, G.J. The Goonumbla Porphyry Copper Deposits, New South Wales. *Econ. Geol.* **1985**, *80*, 591–613. [CrossRef]
135. Lowell, J.D.; Guilbert, J.M. Lateral and Vertical Alteration-Mineralization Zoning in Porphyry Ore Deposits. *Econ. Geol.* **1970**, *65*, 373–408. [CrossRef]
136. Wilson, J.W.J.; Kesler, S.E.; Cloke, P.L.; Kelly, W.C. Fluid Inclusion Geochemistry of the Granisle and Bell Porphyry Copper Deposits, British Columbia. *Econ. Geol.* **1980**, *75*, 45–61. [CrossRef]
137. Hedenquist, J.W.; Arribas, A., Jr.; Reynolds, T.J. Evolution of an Intrusion-Centered Hydrothermal System; Far Southeast-Lepanto Porphyry and Epithermal Cu-Au Deposits, Philippines. *Econ. Geol.* **1998**, *93*, 373–404. [CrossRef]
138. Ulrich, T.; Günther, D.; Heinrich, C.A. The evolution of a porphyry Cu-Au deposit, based on LA-ICP-MS analysis of fluid inclusions: Bajo de la Alumbrera, Argentina. *Econ. Geol.* **2001**, *96*, 1743–1774. [CrossRef]
139. Meng, X.; Richards, J.P.; Kontak, D.J.; Simon, A.C.; Kleinsasser, J.M.; Marsh, J.H.; Stern, R.A.; Jugo, P.J. Variable Modes of Formation for Tonalite–Trondhjemite–Granodiorite–Diorite (TTG)-Related Porphyry-Type Cu $\pm$ Au Deposits in the Neoarchean Southern Abitibi Subprovince (Canada): Evidence from Petrochronology and Oxybarometry. *J. Petrol.* **2021**, *62*, 1–29. [CrossRef]
140. Mole, D.R.; Frieman, B.M.; Thurston, P.C.; Marsh, J.H.; Jørgensen, T.R.C.; Stern, R.A.; Martin, L.A.J.; Lu, Y.J.; Gibson, H.L. Crustal Architecture of the South-East Superior Craton and Controls on Mineral Systems. *Ore Geol. Rev.* **2022**, *148*, 105017. [CrossRef]
141. Dyer, J. Assessing the Nature of Breccia-Textured Rocks from a Au-Mineralized Porphyry-Type Setting in the Archean Chester Intrusive Complex, Northern Ontario. Unpublished Thesis, Laurentian University, Sudbury, ON, Canada, 2020.

minerals

MDPI

Article

Interpreting the Complexity of Sulfur, Carbon, and Oxygen Isotopes from Sulfides and Carbonates in a Precious Metal Epithermal Field: Insights from the Permian Drake Epithermal Au-Ag Field of Northern New South Wales, Australia

Hongyan Quan [1,2], Ian Graham [2,*], Rohan Worland [3,4], Lewis Adler [5], Christian Dietz [6], Emmanuel Madayag [2,7], Huixin Wang [8] and David French [2]

1 BGRIMM Technology Group, Beijing 100160, China; quanhongyan123@outlook.com
2 Earth and Sustainability Science Research Centre, School of Biological, Earth and Environmental Sciences, UNSW Sydney, Kensington, NSW 2052, Australia; elmadayag@gmail.com (E.M.); davidfrench474@gmail.com (D.F.)
3 White Rock Minerals Ltd., 12 Anderson Street, West Ballarat, VIC 3350, Australia; rworland@whiterockminerals.com.au
4 Cipango Limited, Suite 506 Level 5, 50 Clarence Street, Sydney, NSW 2000, Australia
5 Bioanalytical Mass Spectrometry Facility, UNSW Sydney, Kensington, NSW 2052, Australia; l.adler@unsw.edu.au
6 Central Science Laboratory, University of Tasmania, Hobart, TAS 7001, Australia; christian.dietz@utas.edu.au
7 Bowdens Silver Pty Ltd., 68 Maloneys Road, Lue, NSW 2850, Australia
8 Solid State & Elemental Analysis Unit, Mark Wainwright Analytical Centre, UNSW Sydney, Kensington, NSW 2052, Australia; huixin.wang@unsw.edu.au
* Correspondence: i.graham@unsw.edu.au

Abstract: The Drake Goldfield, also known as Mount Carrington, is located in north-eastern New South Wales, Australia. It contains a number of low–intermediate-sulfidation epithermal precious metal deposits with a current total resource of 724.51 metric tons of Ag and 10.95 metric tons of Au. These deposits occur exclusively within the Drake Volcanics, a 60 × 20 km NW-SE trending sequence of Late Permian volcanics and related epiclastics. Drilling of the Copper Deeps geochemical anomaly suggests that the volcanics are over 600 m thick. The Drake Volcanics are centered upon a geophysical anomaly called "the Drake Quiet Zone" (DQZ), interpreted to be a collapsed volcanic caldera structure. A total of 105 fresh carbonate samples were micro-drilled from diamond drillcores from across the field and at various depths. A pXRD analysis of these carbonates identified five types as follows: ankerite, calcite, dolomite, magnesite, and siderite. Except for three outlier values (i.e., −21.32, −19.48, and 1.42‰), the $\delta^{13}C_{VPDB}$ generally ranges from−15.06 to −5.00‰, which is less variable compared to the $\delta^{18}O_{VSMOW}$, which varies from −0.92 to 17.94‰. μ-XRF was used to analyze the elemental distribution, which indicated both syngenetic/epigenetic relationships between calcite and magnesite. In addition, a total of 53 sulfide samples (primarily sphalerite and pyrite) from diamond drillcores from across the Drake Goldfield were micro-drilled for S isotope analysis. Overall, these have a wide range in $\delta^{34}S_{CDT}$ values from −16.54 to 2.10‰. The carbon and oxygen isotope results indicate that the fluids responsible for the precipitation of carbonates from across the Drake Goldfield had complex origins, involving extensive mixing of hydrothermal fluids from several sources including those of magmatic origin, meteoric fluids and fluids associated with low-temperature alteration processes. Sulfur isotope ratios of sulfide minerals indicate that although the sulfur was most likely derived from at least two different sources; magmatic sulfur was the dominant source while sedimentary-derived sulfur was more significant for the deposits distal from the DQZ, with the relative importance of each varying from one deposit to another. Our findings contribute to a greater understanding of Au-Ag formation in epithermal environments, particularly in collapsed calderas, enhancing exploration strategies and models for ore deposition.

Keywords: stable isotopes; μ-XRF; epithermal deposit; Drake Volcanics; hydrothermal; carbonates; sulfides

1. Introduction

Epithermal deposits, which form at shallow depths and low temperatures, are significant sources of precious metals, such as Au and Ag [1]. Collapsed calderas are considered ideal locations for these types of deposits, as they provide the necessary conditions for the formation of mineralization from ascending thermal waters. These deposits typically occur in the shallow parts of magmatic systems, at depths ranging from ~50 to 1500 m, with ore minerals precipitating at temperatures between <150 °C and ~300 °C, often within associated volcano-sedimentary basins [2–5].

The Drake Goldfield, also known as Mount Carrington, is located in north-eastern New South Wales (NSW) Australia, ~5 km north-east of Drake Village, 44 km east of Tenterfield, and ~800 km north of Sydney (Figure 1). It contains a number of low–intermediate-sulfidation epithermal precious metal deposits (Figure 2), including Kylo, Strauss, Red Rock, Lady Hampden, Silver King, White Rock, and White Rock North [6], along with numerous other small deposits and prospects. Lady Hampden is unusual for the Drake Goldfield in being a Au-Ag deposit; All Nations, Gladstone Hill, and West Copper Deeps are all Cu prospects. The Strauss, Kylo, Silver King, and Lady Hampden deposits are collectively known as the Mount Carrington Group, as they are in close proximity to the historical Mount Carrington Au-Ag deposit. These deposits occur exclusively within the Drake Volcanics, which comprise a 60 × 20 km NW-SE trending sequence of Late Permian shallow volcanics and related epiclastics.

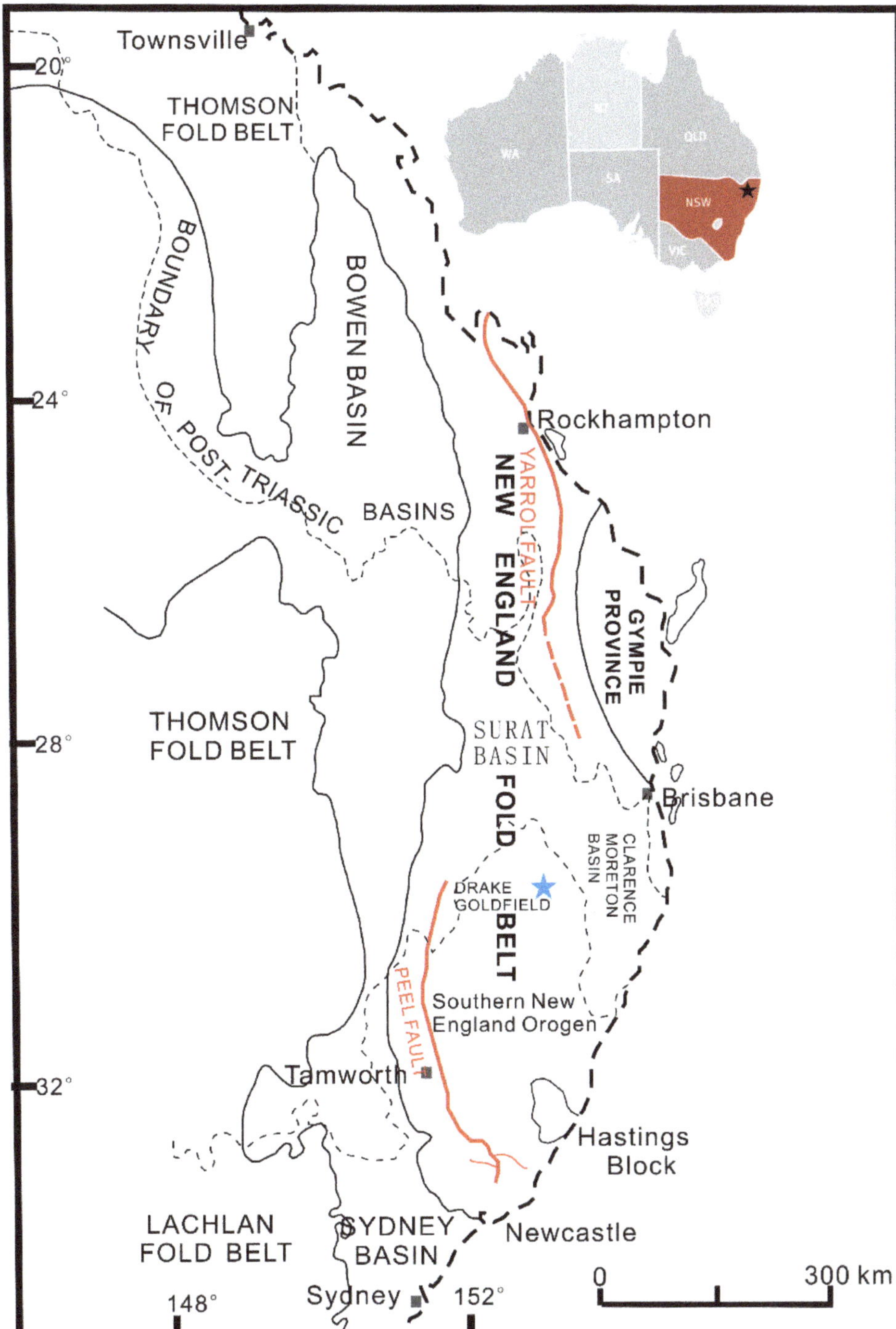

Figure 1. Simplified map showing the New England Orogen and the location of the Drake Goldfield (indicted by blue star; modified after Murray [7]).

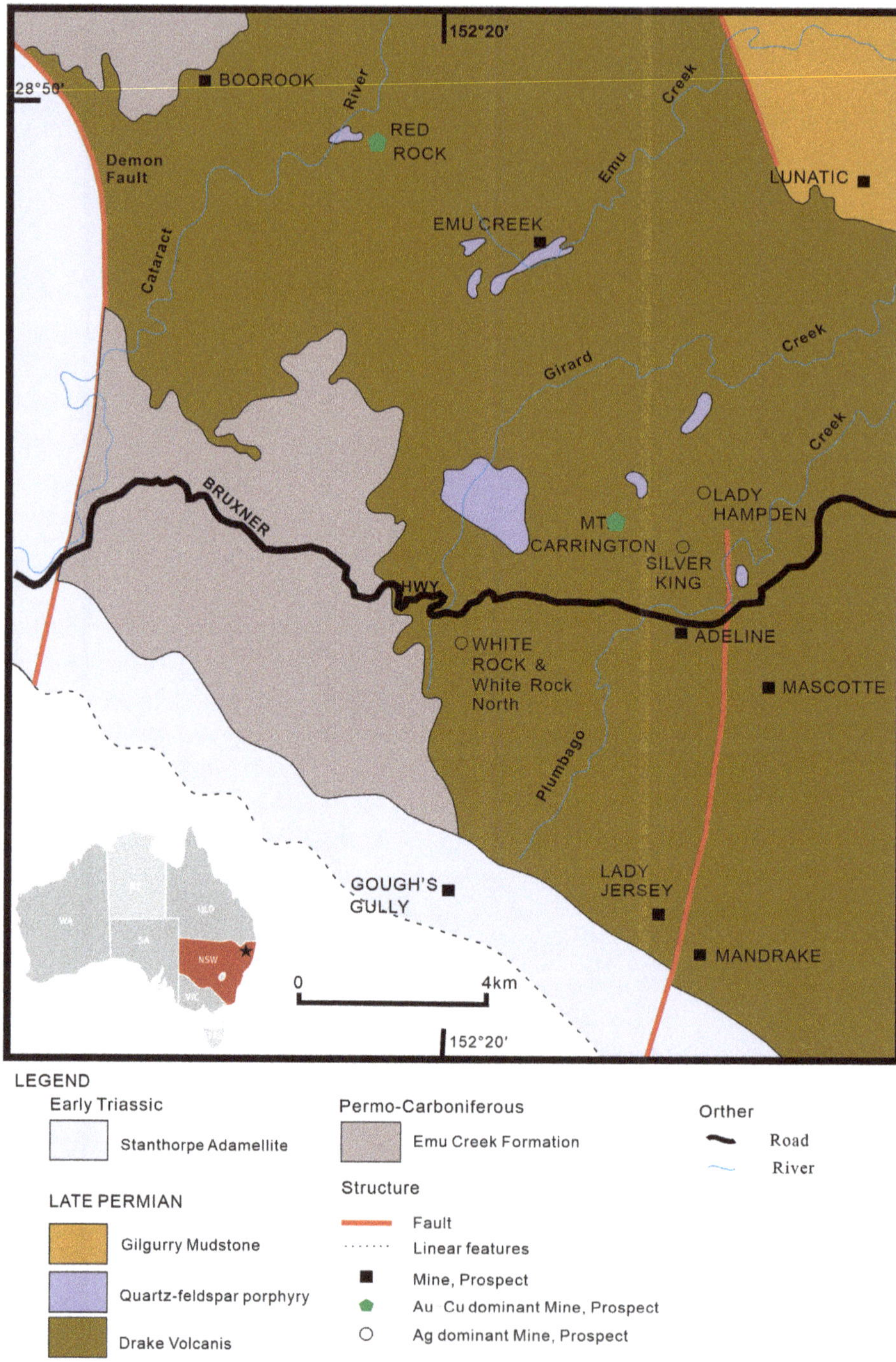

Figure 2. Simplified geological map of the Drake Goldfield showing the location of deposits and prospects within the Drake Volcanics (All Nations, Gladstone Hill, Strass, and Kylo are included in Mt Carrington, modified after White Rock Minerals [6] and Thomson [8]).

The first significant precious metal deposit discovered in the Drake Volcanics was the White Rock silver deposit which was discovered in 1886, and its geology was first described by Andrews [9]. Subsequently, many deposits of differing styles were discovered and mined. Although being known and mined for over 100 years, there has been no detailed study on the deposits associated with the Drake Volcanics as a whole, apart from that of Perkins [10] on the deposits of the Red Rock Field, which is one of the smallest of the fields within the larger Drake Goldfield. Recent work by White Rocks Minerals suggests that most of the economic precious metal deposits within the Drake Volcanics are centered upon a geophysical anomaly called "the Drake Quiet Zone" (DQZ), interpreted to be a collapsed volcanic caldera structure (Figure 3). The Drake Goldfield has been mined since the late 1800s for precious metals, particularly Au and Ag [6,9] with the most recent mining activity by Mt. Carrington Mines (1988 to 1990) producing 0.714 metric tons of Au and 434,870 oz of Ag [11]. Recent estimates suggest that the Drake Goldfield contains an Indicated Resource of 724.51 metric tons of Ag and 10.95 metric tons of Au [6]. The present study builds upon more recent unpublished studies on the relationships between mineralization, grade, and alteration assemblages/intensity on the White Rock and the Lady Hampden deposits [12,13]; the relationship between primary volcanic facies and mineralization and correlation between the White Rock and White Rock North deposits [14]; the structural controls and relationship between mineralization and alteration for the Strauss deposit [15]; and a detailed mineralogical and geochemical study on selected Ag-rich deposits within the Drake Goldfield [16].

As there were no comprehensive, published studies across the Drake Goldfield itself and on mineralization within the Drake Volcanics, the main objective of this study was to document the isotopic compositional variations of sulfides and carbonates that precipitated during different stages across the Drake Goldfield. A number of vein-like carbonates and sulfides from mineralized horizons from diamond drillcores from across the Drake Goldfield and at various depths were systematically studied for their carbon, oxygen, and sulfur isotopes in order to better understand the source of the mineralizing and alteration fluids and to provide information about the genesis of the Drake Goldfield.

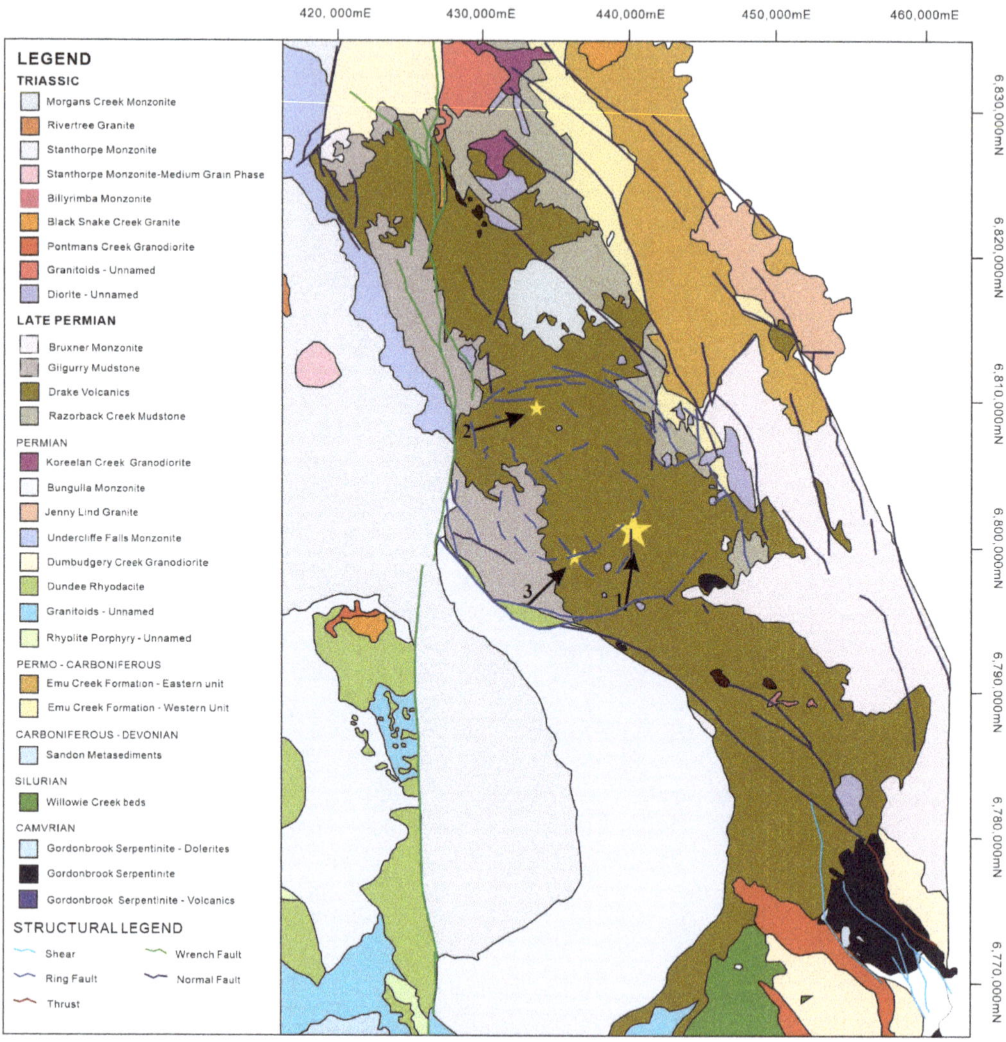

Figure 3. Interpreted lithostratigraphic and structural map of the Drake Volcanics and surrounding lithologies developed by Davies [17] from Geological Survey of New South Wales mapping (modified from Bottomer [18]) and interpretation of magnetic data, showing the location of the Drake Quiet Zone ("DQZ") (1—Mount Carrington Group, including All Nations, Gladstone Hill, Kylo, Strauss, West Copper Deeps, Lady Hampden, Guy Bell, Mozart and Silver King; 2—Red Rock; and 3—White Rock and White Rock North).

2. Geological Setting

2.1. The Southern New England Orogen

The Drake Goldfield is near the north-eastern boundary of the Southern New England Orogen (SNEO) or Southern New England Fold Belt [6,18] (Figure 1). The Southern New England Orogen is a sub-province of the New England Orogen (NEO), the youngest sub-province of the Tasmanides, whose formation is widely accepted to comprise two main cycles of compression and extension since the early Cambrian [7,19,20]. The NEO extends

~1600 km along the east coast of Australia from Townsville in northern Queensland to Newcastle in New South Wales (Figure 1). It is divided by the Clarence–Moreton Basin into northern and southern provinces, which are the Yarrol–Gympie Province and SNEO, respectively [19,21]. The SNEO is further divided into two regions, the Tamworth Belt and the Tablelands Complex, divided by the Peel-Manning Fault System [7,22]. The Tamworth Belt represents a relatively weakly deformed fore arc environment, while the Tablelands Complex comprises a series of accretionary wedge/subduction complexes [19,21]. During the uppermost Carboniferous to Lower/Middle Permian, the SNEO had been through an extensional cycle followed by widespread orogenic deformation, granitoid intrusion and volcanism [18,20]. The host rocks of the Drake Goldfield are mostly volcanics which formed during this period [6,23].

2.2. The Drake Volcanics

The Drake Volcanics were first briefly described by Andrews [9] and much later as a sequence of volcanic rocks some 20 km wide and 60 km long trending NNW, close to the north-eastern margin of the Tablelands Complex within the SNEO and bounded to the west by the Demon Fault (Figures 2 and 3) [7]. It has been shown through U-Pb zircon dating that the Drake Volcanics comprise middle Permian (~265 Ma [23]) acid to intermediate volcanics dominated by volcaniclastic andesitic units intruded by sub-volcanic andesite and rhyolite porphyries [18,24,25]. The Drake Volcanics have an estimated thickness of ~600 m in the Red Rock–White Rock–Lady Hampden area [18], while Thomson [8] estimated a maximum thickness of 900 m elsewhere.

The Drake Volcanics are calc-alkaline, conformably overlie the Razorback Creek Mudstone, and are conformably overlain by the Gilgurry Mudstone (Figures 2 and 3) [6,8,25–27]. The Carboniferous to Early Permian sedimentary Emu Creek Formations both underlie the Drake Volcanics and are fault-bounded on its eastern boundary, hosting the gold mineralization of the Tooloom and Lunatic Goldfields, while to the west, the Drake Volcanics are intruded by the Early Triassic Stanthorpe Monzonite pluton and structurally bound by the Demon Fault (Figures 2 and 3) [6]. The volcanics are relatively undeformed, with only slightly inclined bedding (~20°). Some researchers have suggested that the Drake Volcanics formed during rifting or graben formation [28,29], while others have suggested that they formed due to transform/strike slip fault movement [30].

Airborne magnetic imagery revealed a 250 km^2 roughly circular area of low magnetic response (the Drake Quiet Zone or DQZ) (Figure 3). This region with a diameter of approximately 20 km has been interpreted to represent a collapsed volcanic caldera structure with the circular shape and the quiet magnetic signature explained by the presence of a caldera filled with less magnetic volcanics than the surroundings [11,31]. This type of structure is a relatively common feature in a range of other epithermal Au-Ag deposits, including Creede (2488.28 metric tons Ag), Colorado, USA [32–34] and Round Mountain (513.21 metric tons Au), Nevada, USA [35–37]. As a collapsed caldera, the structure has been proposed to control the location of post-collapse intrusives, mineralization, and the extensive alteration within the region. Although Craighead and Gordon [11] believed that the flat to moderately dipping bedding planes acted as major fluid feeder structures, both field relationships and drill holes conclusively show that the mineralized veins are mostly subvertical in orientation, as expected, following epigenetic extensional faulting [37]. This is supported by the conclusions of Herbert [26], Bottomer [18], Perkins [27], and Houston [31] who all believed that structure played an important role in ore localization. Thick lobes and layers of dacitic–lithic and rhyolitic quartz–feldspar fiamme breccias typically observed throughout the area as a thick sequence have been interpreted as representing the caldera-forming eruption products [38].

3. Samples and Methodology

Prior to sample preparation for further analysis, samples were described, imaged, then made into thin sections or polished blocks, and analyzed using transmitted and reflected-light microscopy.

3.1. Sample Description

Carbonate deposition is mostly very late stage and is unrelated to the main mineralization event. The deposition was a multi-stage process involving a Ca-Mg carbonate stage, calcite forming cross-cutting veins, euhedral calcite lining cavities, late-stage calcite filling fractures, and siderite of meteoric origin on fracture surfaces of the Strauss open cut, comprising drusy euhedral crystals (to a few mm) of dark brown siderite lining fractures on altered andesite.

The carbonate samples were selected from carbonate veins from throughout the Drake Goldfield, covering a diverse range in terms of shape, thickness, and lateral continuity. The thickness of the veins ranged widely from <5 mm to >10 cm. The wide range in vein morphologies included one massive vein with straight and sharp boundaries (Figure 4b,e), stockwork veins (Figure 4a), breccia with carbonate-fill (Figure 4d) and sub-parallel veins (Figure 4b), vuggy infill with euhedral carbonate crystals (mostly calcite, Figure 4f), carbonate spots (Figure 4c), and zonation within individual veins occurred in some places (Figure 4e).

The principal occurrence of carbonates within the Drake Goldfield are as follows: (1) replacement/overprinting of earlier rock-forming minerals (e.g., plagioclase) in volcanic tuffs (Figure 5a,d); (2) vein infill as a main component of the carbonate $\pm$ quartz stage (Figure 5b,c,f). The vein carbonates are dominantly composed of fine to coarsely crystalline calcite or dolomite (Figure 5b,c,f). Three calcite generations were found within the veins: (1) fine-grained granular-textured calcite veinlets; (2) calcite replacing earlier coarse-grained dolomite; and (3) massive veins comprising coarse euhedral calcite crystals. Ankerite is rare, occurring as fine-grained veinlets, mostly replacing/overprinting calcite (Figure 5e). Dolomite occurs as massive veins and is locally banded with magnesite (Figure 4e) or calcite. Magnesite is more massive and coarsely crystalline, and interbanded with calcite/dolomite (Figure 4a,b). Siderite was observed as euhedral crystals on fracture surfaces within the Strauss open-cut or rimming/cross-cutting earlier magnesite in drill cores.

Sulfides were chosen from the vein and veinlet ores of the Drake Goldfield. Pyrite, sphalerite, and chalcopyrite are by far the most dominant sulfide species and were thus chosen for the sulfur isotope analyses, mainly occurring as sulfide $\pm$ quartz veins (Figure 6). Figure 6a–i show the main sulfide assemblages from the Drake Goldfield, including Ccp + Py + Qtz veins (Figure 6a), Py + Sph + Qtz veins (Figure 6b), Py veins, low-Fe Sph + Ccp + Qtz veins (Figure 6d,f), and low-Fe Sph + Ccp + Gn + Py + Qtz veins (Figure 6e).

Figure 4. Images of drill cores with carbonate veins from the Drake Goldfield: (**a**) sub-parallel calcite-magnesite veins in dacitic tuff from All Nations; (**b**) sub-parallel late-stage magnesite–calcite vein-cutting dacitic tuff from Kylo; (**c**) fine-grained calcite spots in dacitic tuff from Red Rock; (**d**) hydrothermal breccia with calcite infill from Silver King; (**e**) zoned dolomite and magnesite vein in dacitic tuff from Strauss; (**f**) euhedral calcite crystals in dacitic tuff from West Copper Deeps. Abbreviations: Dol—dolomite; Cal—calcite; Mgs—magnesite.

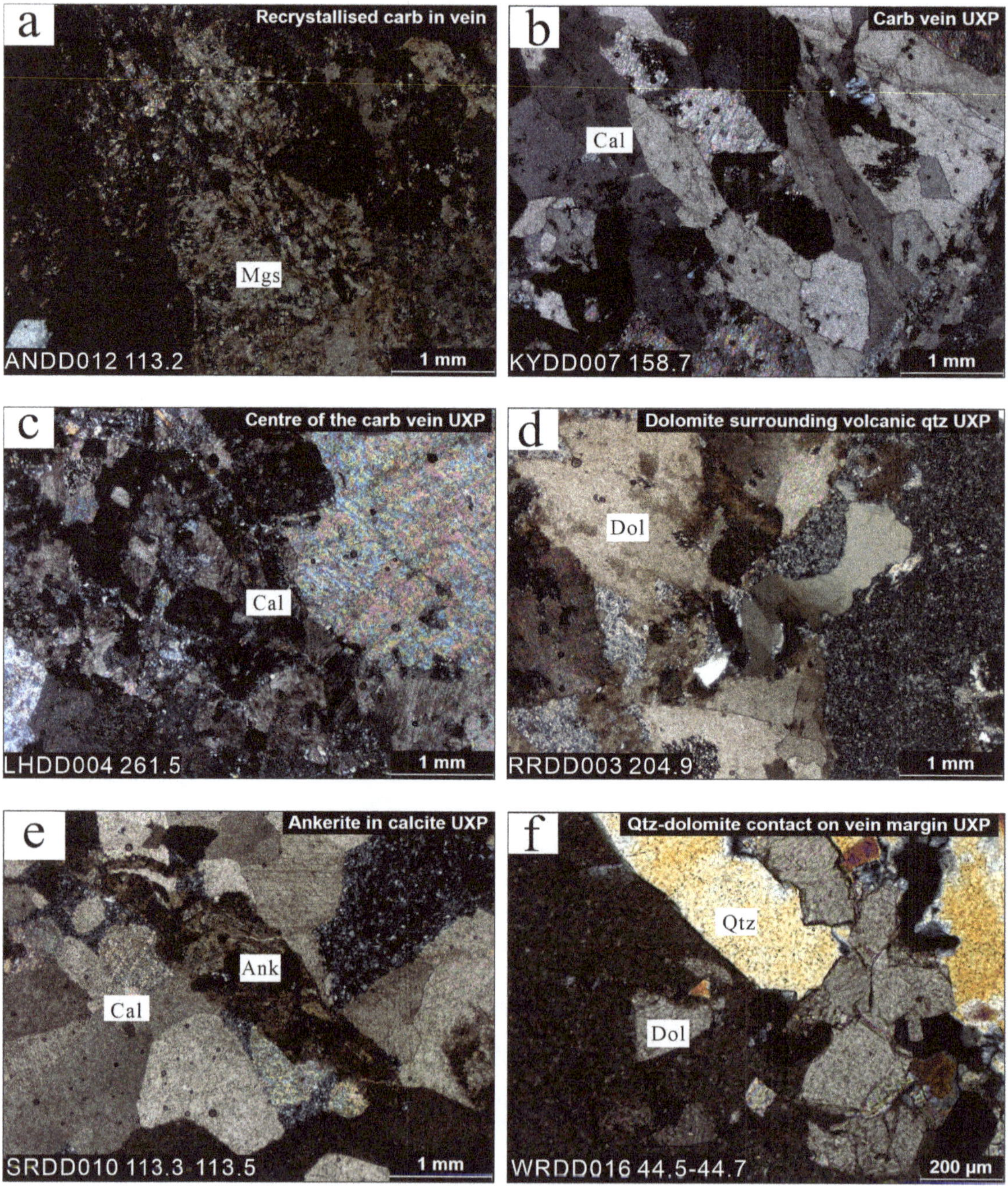

Figure 5. Typical textures and overprinting relationships of carbonates from across the Drake Goldfield: (**a**) recrystallized magnesite-overprinting dacitic tuff from All Nations; (**b**) calcite vein from Kylo; (**c**) calcite vein from Lady Hampden; (**d**) dolomite surrounding volcanic quartz from Red Rock; (**e**) ankerite overprint calcite from Silver King; (**f**) dolomite–quartz contact from White Rock North. Abbreviations: Dol—dolomite; Cal—calcite; Mgs—magnesite.

Figure 6. Images of drill core samples with vein sulfides and photomicrographs of sulfides and their associations and textures in reflected light (RL) from the Drake Goldfield: (**a**) chalcopyrite–pyrite–quartz vein within andesitic tuff from Gladstone Hill; (**b**) pyrite–sphalerite–quartz vein with open space in dacitic tuff from Kylo; (**c**) pyrite vein within dacitic tuff from Mozart; (**d**) coarse-grained low-Fe sphalerite–chalcopyrite–quartz vein from Red Rock; (**e**) low-Fe sphalerite–chalcopyrite–galena–pyrite–quartz vein within bleached dacitic tuff from Strauss; (**f**) low-Fe sphalerite–chalcopyrite vein within andesitic tuff from West Copper Deeps; (**g**) chalcopyrite and sphalerite replacing stage II pyrite from Kylo; (**h**) recrystallized pyrite within host andesite from Silver King; (**i**) sphalerite–galena intergrowth replacing pyrite with chalcopyrite exsolved in sphalerite from White Rock North. Abbreviations: Ccp—chalcopyrite; Py—pyrite; Sph—sphalerite; Gn—galena; Qtz—quartz.

The overall mineralogy for the deposits within the Drake Goldfield is relatively simple with the main ore comprising pyrite (some with micro to submicron inclusions of gold), with lesser sphalerite, chalcopyrite, and galena; minor electrum/gold and Ag minerals; along with supergene covellite and chalcocite–digenite. A variety of Ag minerals were identified, primarily of the tetrahedrite–tennantite group, along with minor occurrences of polybasite–pearceite species, and trace amounts of acanthite, native Ag, stephanite, pyrargyrite, and jalpaite. There are at least three generations of pyrite in the Drake Goldfield (Figure 6a–i): as small subhedral–euhedral disseminated pyrite in the host rocks (Figure 6c,h) (pyrite I); as euhedral coarse-grained pyrite (pyrite II, Figure 6c,g,i), and as framboidal-like or spheroidal textured pyrite (pyrite III, Figure 6h). Pyrite I formed before the main mineralization stage, while pyrites II and III formed during the main mineralization stage. Chalcopyrite occurs in two different generations, as exsolutions within sphalerite (Figure 6i) or as coarse grains rimming sphalerite or pyrite (Figure 6d,g). Sphalerite is often rimming pyrite (Figure 6g,i). Galena commonly exhibits mutually embayed grain boundaries with sphalerite suggesting co-crystallization (Figure 6i). For a detailed paragenetic sequence, please refer to the electronic Supplementary Material Figures S5–S7.

3.2. Sample Preparation

Fresh carbonate vein minerals were sampled from diamond drillcores covering a wide geographical and vertical distribution from across the Drake Volcanics and selected according to drill logs [6], with all either associated or spatially close to mineralization. A total of 105 samples were selected from 33 diamond drill holes from the Kylo, Lady Hampden, Silver King, All Nations, Mozart, Gladstone Hill, West Copper Deeps, Red Rock, Strauss, White Rock, and White Rock North deposits and one from the Strauss open cut. These deposits are distributed across the entire Drake Goldfield (Figure 2 and Figures S1–S4 in the Supplementary Material).

A further 53 sulfide samples from the main mineralization stage (mainly coarse-grained pyrite and sphalerite along with minor chalcopyrite and galena) were selected from across the Drake Goldfield (Figures S1–S4 in Supplementary Material) to investigate any spatial variation, and samples were also selected from various depths to investigate any vertical variation. Where possible, only coarse-grained and pure separates of sphalerite and pyrite were chosen.

The samples were then collected as ground powders using a tungsten-tipped micro-drill. They were then placed in small sample holders before being ground to <50 μm using an agate mortar and pestle to ensure compositional and grain-size homogeneity. After each sample was collected, the micro drill head was thoroughly cleaned with 85% acetone solution.

3.3. Portable X-Ray Diffraction

Carbonate species identification was made using an Olympus Terra 6400 portable X-ray diffraction (pXRD) (Olympus, Melburne, Australia) analyzer at the University of New South Wales, Sydney, Australia. In total, 15 mg of the sample with a uniform grain size <150 μm were loaded into the vibrating sample holder (VSH), which vibrated the sample without macroscopic movement of the holder [39]. This exposed crystallites in each sample to the X-ray beam at random orientations, thus helping to reduce orientation effects and allowing for superior analytical statistics [39]. The VSH was then introduced into the Olympus Terra pXRD, and the sample was subjected to XRD analysis using CoKα radiation, with a data collection range of 5 to 55° 2θ and an increment of 0.25°. Collected data were then analyzed and interpreted using HighScore Plus and Siroquant^{TM} (v.4) commercial interpretation software. Phase identification was conducted using HighScore Plus, while Siroquant^{TM} was used to estimate the quantitative abundance of mineral phases within the sample [40]. For full details on the use of pXRD for quantitative analysis of epithermal assemblages, see Burkett et al. [41].

3.4. Micro X-Ray Fluorescence

Micro X-ray fluorescence (μ-XRF) mapping of two carbonate veins, one from each of the All Nations and Kylo deposits, containing magnesite and calcite, was conducted using a Bruker M4 Tornado μ-XRF at the X-ray Fluorescence Laboratory, University of

New South Wales, Sydney, Australia. The instrument is equipped with a Rh tube with a maximum power of 30 W, a tube voltage of 45 kV, a current of 600 μA, and a 25 μm spot size in order to achieve a high spatial resolution. One silicon drift detector was used with an active area of 30 mm^2. For element-mapping analysis, spectra were acquired every 50 μm with an acquisition time set at 120 ms per pixel. All analyses were performed under vacuum (20 mbar) to facilitate the detection of light elements. Data analysis was performed in standardless mode using the Bruker M4 TORNADO built-in software FP MQuant.

3.5. Stable Isotope Measurements

Carbon and oxygen isotope analyses were carried out in the Bioanalytical Mass Spectrometry Facility, Mark Wainwright Analytical Centre, University of New South Wales, while sulfur isotopes were determined at the Isotope Ratio Mass Spectrometry facility, Central Science Laboratory, University of Tasmania.

For carbon and oxygen isotope analysis, 40 to 60 μg of carbonates were weighed out for each sample and then analyzed using a MAT 253 isotope ratio mass spectrometer equipped with a Kiel IV carbonate device (Thermo Fisher Scientific, Bremen, Germany). The carbonates were reacted with 2 drops of phosphoric acid at 70 °C to release CO_2. The reacted gasses were then subjected to cryogenic removal of water vapor and other non-condensable gasses, and the pure CO_2 was measured against a reference gas. For sulfur isotope determination, between 0.4 and 1.0 mg of finely powdered mineral sample was weighed in tin capsules and analyzed using flash combustion isotope ratio mass spectrometry on a varioPYRO cube coupled to an Isoprime100 mass spectrometer (Elementar Analysensysteme, Langenselbold, Germany). The SO_2 produced during combustion is separated from other reaction products in a trap, then fed into the mass spectrometer and measured against a reference gas.

Stable isotope abundances are reported in delta (δ) values as the deviations from conventional standards in parts per mil (‰) using the following equations:

$$\delta^{H}I\ (‰) = [(R_{sample}/R_{standard} - 1) \times 1000],$$

with $^{H}I = {}^{13}C$, ^{18}O, or ^{34}S and $R = {}^{13}C/{}^{12}C$, $^{18}O/{}^{16}O$, or $^{34}S/{}^{32}S$

$\delta^{13}C$ values are expressed as relative to VPDB (Vienna Pee Dee Belemnite). $\delta^{18}O$ can be standardized against either the VPDB or VSMOW (Vienna Standard Mean Ocean Water) scale. For the present work, the VPDB scale was adopted. The following equation was used for conversion [42] between VPDB and VSMOW scales:

$$\delta^{18}O_{VSMOW} = 1.3086\ \delta^{18}O_{VPDB} + 30.86$$

$\delta^{34}S$ values are reported relative to CDT (Canyon Diablo Troilite, a meteorite which created the Barringer crater, Arizona, USA).

International reference standards with known isotopic compositions, sulfur: IAEA-S-1 (−0.3‰), IAEA-S-2 (22.66‰), IAEA-S-3 (−32.3‰), IAEA-SO5 (0.49‰), NBS 123 (17.44 ‰), and NBS 127 (20.32‰); carbon and oxygen: NBS18 (−5.01 and −23.2‰ for carbon and oxygen, respectively) and NBS 19 (1.95 and 2.2‰ for carbon and oxygen, respectively), were measured for instrument calibration and quality assurance. The analytical performance of the instrumentation, drift correction, and linearity performance were calculated from the repetitive analysis of these standards. The analytical precision was 0.1‰ for both $\delta^{13}C$ and $\delta^{18}O$ and 0.2‰ for $\delta^{34}S$.

4. Results

4.1. Carbonate Species

The results of the pXRD analysis for the carbonates from the Drake Goldfield are presented in Supplementary Material Table S1. Five species of carbonates were identified, including ankerite, calcite, dolomite, magnesite, and siderite. Although commonly found in many epithermal deposits (e.g., Eroy et al. [43]), no manganese carbonates, such as kutnohorite or rhodochrosite, were found in the Drake Goldfield. The quantitative Siroquant results are also shown in Table S1.

4.2. μ-XRF

μ-XRF geochemical maps for two carbonate veins from the Drake Goldfield (Figure 7) show the relative variations in chemistry of the veins. There are relatively pure granular magnesite veins in ANDD012 123.55 (Figure 7a,b), which contain a thin prominent pure calcite vein enclosed within the magnesite (Figure 7b), indicating a clear epigenetic origin, and the calcite appears to be a late-stage cavity infill. In contrast, the carbonate veins in KYDD003 188 have a relatively complex texture (Figure 7c,d). The Mg and Ca distributions (Figure 7d) collectively show that there was a regularly changing Mg/Ca ratio during precipitation with coexisting magnesite, calcite, and magnesium-bearing calcite, indicating a syngenetic relationship. The pXRD confirmed the presence of magnesite, calcite, and magnesium-bearing calcite in these samples, consistent with the μ-XRF analysis.

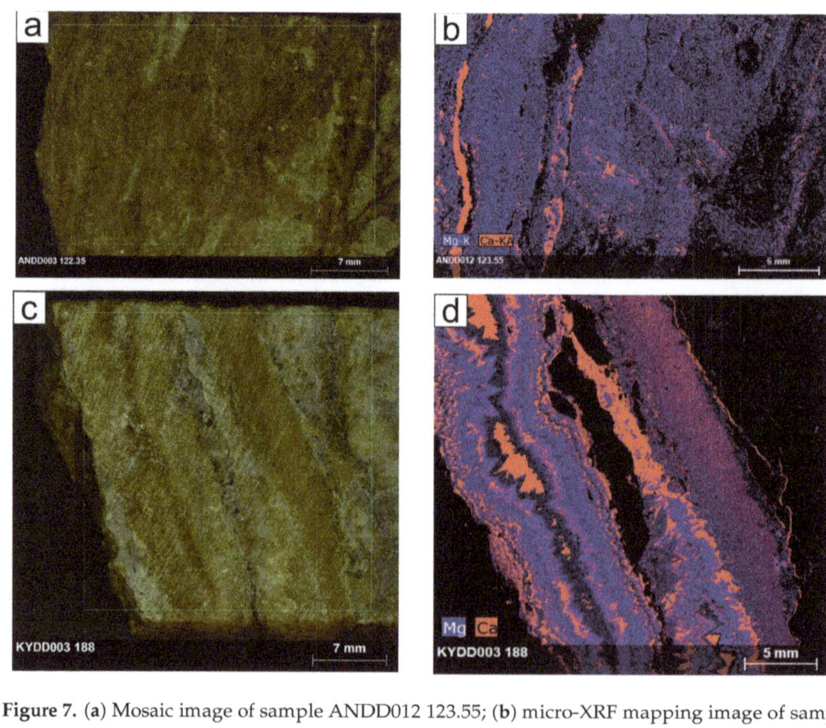

Figure 7. (**a**) Mosaic image of sample ANDD012 123.55; (**b**) micro-XRF mapping image of sample ANDD012 123.55 as a two-element overlay that includes Ca and Mg; (**c**) mosaic image of sample KYDD003 188; (**d**) micro-XRF mapping image of sample KYDD003 188 as a two-element overlay that includes Ca and Mg. Note: The red color depicts calcite, the purple color depicts magnesite; and red overlapping purple depicts magnesium-bearing calcite, with the intensity of each color representing the relative concentration.

4.3. *Stable Isotope Results*

4.3.1. Carbon and Oxygen Isotopes

The detailed carbon and oxygen isotope analyses from across the Drake Goldfield are presented in Supplementary Material Table S1. The range of C and O isotopic compositions for these carbonates are presented in Table 1.

Table 1. The range in carbon and oxygen isotopic compositions of carbonates from Drake Goldfield.

Mineral	Number of Samples	$\delta^{13}C_{VPDB}$ (‰)	Average Value for $\delta^{13}C_{VSMOW}$ (‰)	$\delta^{18}O_{VSMOW}$ (‰)	Average Value for $\delta^{18}O_{VSMOW}$ (‰)
Ankerite	4	−9.33–−5.92	−7.39	+5.81–+7.99	+6.63
Calcite	68	−21.32–+1.42	−9.13	−0.92–+17.94	+7.77
Dolomite	17	−13.02–−5.15	−8.47	+3.40–+10.59	+5.47
Magnesite	15	−10.44–−5.37	−7.99	+8.15–+15.84	+11.89
Siderite	1	−13.71	-	+17.11	-

4.3.2. Sulfur Isotopes

The results for sulfur isotopes are presented in Supplementary Material Table S2. A total of 53 sulfide samples (primarily sphalerite and pyrite) were analyzed, and these have a wide range of $\delta^{34}S_{CDT}$ values from −16.54 to 2.10‰. Specifically, $\delta^{34}S_{CDT}$ in pyrite ranges between −8.77 and +2.4‰ (mean −2.86‰—12 analyses) with one outlier of −16.33‰ from White Rock; in sphalerite between −11.50 and −0.53‰ (mean −3.62‰—23 analyses); and in a mixture of sulfides ranging from 8.38‰ to −0.64‰ (mean −4.03‰—16 analyses). To this, a single analysis each for chalcopyrite (−2.64‰) and galena (−11.57‰) are added.

5. Discussion

The study of stable isotopes can provide information regarding the diverse origins of fluids, metal sources, chemical conditions, as well as temperatures of ore formation (e.g., [Ohmoto et al. [44]; Huston [45]; Hoefs [46]). In this discussion, we compare the differences within the Drake Goldfield and attempt to explain the spatial distribution of the gold (copper) and silver-dominant deposits.

5.1. *Sources of Fluid for the Carbonates*

Late-stage carbonate samples from the Drake Goldfield have $\delta^{13}C_{VPDB}$ values ranging from −21.32 to 1.42‰, with 102 samples having a relatively limited variation between −15.06 to −5.00‰. This range for most of the analyses is consistent with carbonates from other epithermal deposits, such as the Kushikino Au-Ag deposit, Japan, and the Qiucun Au deposit, China [47–49], while the mean value of −8.65‰ is close to the mantle range (Figure 8) [50]. The $\delta^{13}C_{VPDB}$ value range for ankerite (−9.33 to −5.92‰), dolomite (13.02 to −5.15‰) and magnesite (−10.44 to −5.37‰) are relatively limited to the mantle range, while calcite (−21.32 to 1.42‰) has a much wider range, and siderite has a more negative value (−13.71‰). This suggests that carbon in ankerite, dolomite, and magnesite has a magmatic dominant source, while the carbon in calcite more likely results from significant mixing of fluids from a predominant magmatic source with those from a sedimentary organic carbon source. As the siderite occurs, lining fractures near the surface in the open cut, the negative value most likely indicates a groundwater source, which could be interpreted as meteoric water that has undergone interaction with the host andesites and organic matter as the waters percolated downwards via fracture networks from the surface. In the $\delta^{18}O$–$\delta^{13}C$ diagram (Figure 9 and Supplementary Material Figures S8–S10), the analyses collectively form an overall linear trend centered on an initial magmatic source trending to a more meteoric influence on one side and a low-temperature alteration influence on the other. Moreover, no data plot along any clear trend, indicating mixing of carbon derived from contaminated sedimentary rocks or high-temperature effects. The Razorback Creek Mudstone is an underlying sedimentary host rock; however, the effects of mixing with this potential carbon source are limited.

The carbon isotopes do not show any significant variation between the Au-Cu and Ag deposits, while the oxygen isotopes are slightly different (Figure 9). The $\delta^{18}O$ for carbonates from the Ag deposits trend to heavier values (concentrated in the range of 10 to 15‰), compared to those for the Au (Cu) deposits (concentrated in the range of 3 to 7‰). This difference also indicates that the carbon for the carbonates in the Au (Cu) deposits was dominantly from a magmatic source compared to that for the Ag deposits.

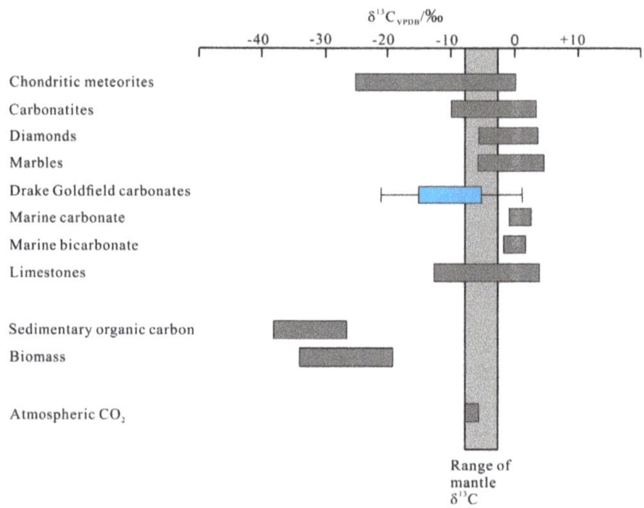

Figure 8. Comparative diagram of $\delta^{13}C_{VPDB}$ reservoir and carbonates (blue indicates range for most analyses) from the Drake Goldfield [50].

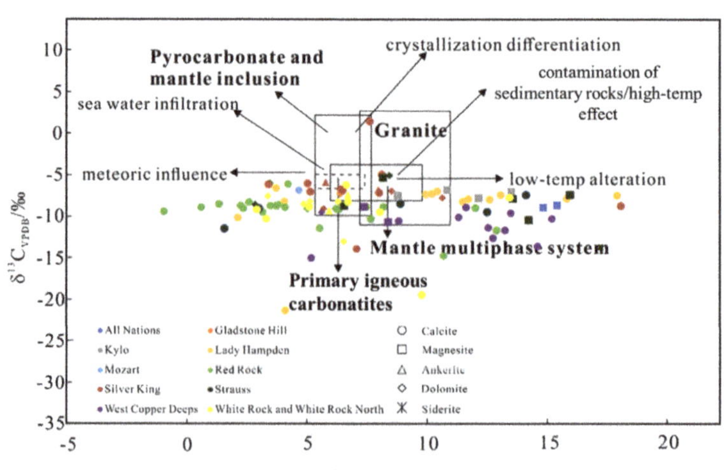

Figure 9. $\delta^{18}O$ versus $\delta^{13}C$ diagram for carbonates from the Drake Goldfield [51].

For these samples, the carbon and oxygen isotope trends are generally near horizontal in the $\delta^{18}O$–$\delta^{13}C$ diagram (Figure 9). This near-horizontal distribution of carbon and oxygen isotope values may be due to two processes: (1) degassing of CO_2; (2) reaction between the fluid and surrounding host rocks [52–54]. If the distribution of carbon and oxygen isotopes is caused by the degassing of CO_2, then the hydrothermal fluid is generally dominated by H_2O, and the degassing of CO_2 has no significant effects on the oxygen isotope composition but has a significant effect on the carbon isotopes within the fluid [55]. Though the number of samples is limited, the silver-rich deposits have a relatively wider range of $\delta^{13}C$, e.g., Lady Hampden (−21.32 to −6.16‰), Silver King (−13.72 to 1.4‰), and White Rock (North) (−19.48 to −6.21‰). This likely indicates that the carbonates in the silver-rich deposits have precipitated due to hydrothermal degassing. The observation of coexisting vapor-rich and liquid-rich fluid inclusions from Red Rock [27], and the presence of direct deposition of quartz (throughout the Drake Goldfield), adularia (mostly in the silver-rich deposits), and calcite (usually platy, present in White Rock and West Copper Deeps) into open spaces or colloform banded quartz (Red Rock and Kylo) could be the result of boiling in the Drake Goldfield [54,56]. However, the gold and copper-rich deposits have a narrower range of $\delta^{13}C$, e.g., Kylo (−7.83 to −6.79‰), Red Rock (−14.82 to −6.05‰), Strauss (−13.72 to −5.15‰), and West Copper Deeps (−15.06 to −8.84‰). Thus, the degassing of CO_2 might not be the only factor influencing the precipitation of carbonates across the Drake Goldfield. In hydrothermal fluids, the solubility of carbonates increases with temperature, decreases with pressure, and increases with decreasing solution pH [57,58]. Cooling in a closed system is unlikely to induce precipitation of carbonate from the hydrothermal fluid [51]. Although water–rock reactions involving magmatic-derived fluids and country rocks are common in epithermal deposits [59], they may have no significant effects on the carbon isotopes within the fluids [55], due to minor quantities of carbonates in the host Drake Volcanics. Therefore, the precipitation mechanism of carbonate in the gold- and copper-dominant deposits is most likely dominated by the boiling process with minor effects due to water–rock reactions.

On the $\delta^{18}O$ versus $\delta^{13}C$ diagram (Figure 9), data from the Gladstone Hill and Mozart deposits plot in the magma–mantle field. For the Silver King, White Rock, and White Rock North deposits, most of the points plot in the magma–mantle field with a slight shift to the negative $\delta^{18}O$ direction, suggesting that the fluids from which these carbonates precipitated from were partly meteoric in origin. For the All Nations, Kylo, and West Copper Deeps deposits, the data suggest that the carbon was derived from a magma–mantle source with a significant influence of low-temperature alteration fluids and also exhibits some characteristics indicative of recrystallization (Figure 5a). Only a few of the analyses for the Lady Hampden and Strauss deposits plot in the magma–mantle zone, with most trending towards higher $\delta^{18}O$ values and a few to lower values, presumably due to fluid reactions with the surrounding host rocks and minor localized mixing with meteoric water. For the Red Rock deposit, the carbon appears to have been mainly derived from a magma–mantle source but with a strong meteoric water influence.

The carbon in the carbonate veins throughout the Drake Goldfield likely originated from a magma–mantle source, subsequently mixed to varying extents with fluids linked to low-temperature hydrothermal alteration and meteoric water. Additionally, a few carbonates exhibit characteristics indicative of recrystallization. The spatial distribution of the Drake Goldfield deposits in the DQZ is shown in Figure 2, from which it can be seen that, except for the Red Rock field, all the gold- and copper-rich deposits (including Kylo, Strauss, Lady Hampden, All Nations, Gladstone Hill, and West Copper Deeps) are closer to the center of the DQZ than the silver-rich deposits (White Rock, White Rock North, Silver King, and Mozart). All the deposits in the Drake Goldfield are hosted within rhyolitic to andesitic volcanic tuffs. However, for the silver-rich deposits and the Red Rock deposit, mineralization is associated with intensive brecciation and adularia ± illite ± quartz ± pyrite alteration, characteristic of a typical rock-buffered low-sulfidation epithermal system [5,16]. In addition, the White Rock (North) deposits are hosted within a magmatic breccia, whereas the other deposits are hosted within various tuff and tuff–breccia units. On the other hand, the gold (copper)-rich deposits show characteristics of an intermediate-sulfidation epithermal system, having a common alteration assemblage comprising quartz ± kaolinite ± albite ± illite ± muscovite ± chlorite ± calcite ± siderite ± pyrite, associated with pyrite + chalcopyrite + low-Fe sphalerite + galena ± minor tennantite group minerals, tetrahedrite, and polybasite mineralization [15]. As we have discussed above, the fluids responsible for at least the carbonates in the gold and copper deposits (except for the Red Rock field) and Lady Hampden ($\delta^{18}O$ shift towards positive values in Figure 9) show an initial magmatic origin but were then more greatly affected by fluids responsible for low-temperature alteration. The carbonate-forming fluids for the silver-rich deposits show an initial magmatic fluid origin followed by later mixing with minor meteoric water during their evolution. This is supported by a $\delta^{18}O$ shift towards negative values, as illustrated in Figure 9 and ankerite overprinting calcite (Figure 5e). The metal zonation of the deposits within the central part of DQZ may relate to the localized distance from the fluid and/or heat source. For the Red Rock area, the carbonate-forming fluids also show a magmatic origin but have a much greater meteoric water influence. This may indicate a separate fluid and/or heat source for the Red Rock area. Although the carbon and oxygen isotope data for the carbonate veins from the White Rock and White Rock North deposits overlap with those of the Mt Carrington group silver-rich deposits, field relationships and other data (e.g., petrographic analysis, whole rock geochemistry) strongly suggest that these silver-rich deposits are from separate shallow level volcanic centers.

5.2. Sources of Sulfur

Sulfur in ore-forming fluids can originate from three main sources: (I) mantle-derived sulfur (0 ± 3‰ [60]), (II) seawater sulfur (>20‰), and (III) strongly reduced (sedimentary) sulfur (<0‰ [50]). The $\delta^{34}S$ values for the Drake Goldfield epithermal deposits are summarized in Figure 10a and Supplementary Material Table S2, from which it can be seen that the $\delta^{34}S$ values for the Drake Goldfield all cluster around −6 to 0‰. Pyrite is the most abundant sulfide and has a wider range in $\delta^{34}S$ values (−16.33 to 2.4‰), while sphalerite has a narrower range in composition (−6 to 0‰). Potential fractionation of $\delta^{34}S$ within the sulfides in the main stage is unlikely, as supported by the mutually embayed grain boundaries between galena and sphalerite (Figure 6i); similar $\delta^{34}S$ values were found for the galena and sphalerite in the same drill hole (WRDD019 in Supplementary Materials Table S2) and in the similar values for coexisting pyrite and sphalerite (SRDD020 10 and WRDD026 78.5 in Supplementary Materials Table S2).

The similar $\delta^{34}S$ values observed for both galena and sphalerite within the same drill hole (WRDD019, Supplementary Material Table S2), as well as for coexisting pyrite and sphalerite (SRDD020 10 and WRDD026 78.5 in Supplementary Material Table S2), suggest that isotopic reversals may have occurred. Sulfide minerals generally show isotopic fractionation in the following sequence: $\delta^{34}S$ pyrite > $\delta^{34}S$ sphalerite > $\delta^{34}S$ chalcopyrite > $\delta^{34}S$ galena [44]. The slight isotopic disequilibrium observed in our study is also supported by petrographic evidence, which shows early sulfides being replaced by later generations (Figure 6g,i).

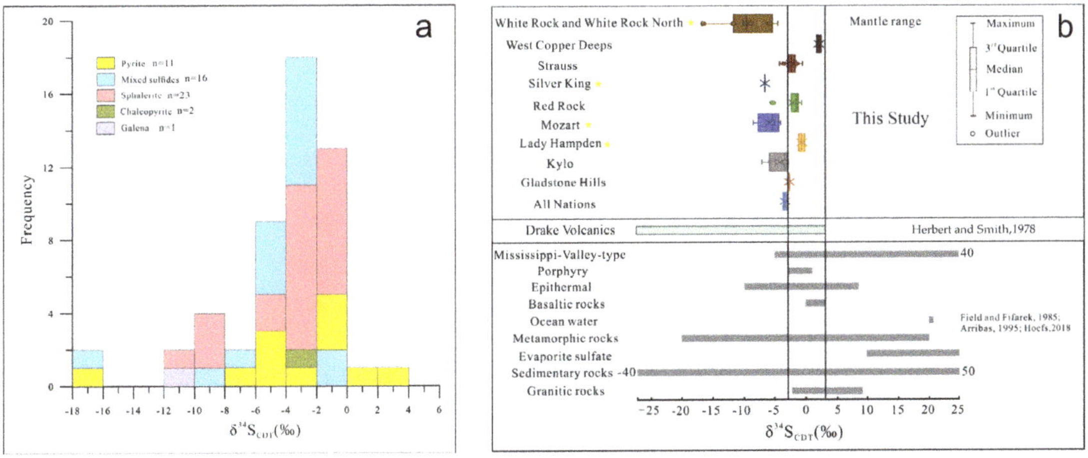

Figure 10. (**a**) Cumulative histogram of sulfur isotope compositions of sulfide minerals from the Drake Goldfield. (**b**) $\delta^{34}S$ values of sulfide minerals from the Drake Goldfield in comparison to common sources of sulfur (the ones marked with yellow stars are silver deposits, and the ones without any marks are Au deposits) [46,61–63].

Some deposits have a narrow compositional range and therefore suggest a single source of sulfur, while other deposits (e.g., White Rock) have a wide range that extends to more negative values. The broadening of the ranges may suggest mixing with other isotopic reservoirs or the operation of redox processes during epithermal mineralization [64]. However, there is no evidence for a significant change in oxidation state observed between the sulfides. The wide range of sulfur isotopes may in part be due to thermochemical sulfate reduction, as the reduced Razorback Creek Mudstone underlies the Drake Volcanics in the study area and could act as a reductant.

Herbert and Smith [63] reported $\delta^{34}S$ values for sulfides from 17 deposits (host rock included) from the Drake Volcanics ranging from −29.5 to 3‰ (Golden Age, Addisons, Red Rock 'K', Emu Creek, Lady Hampden, Mt Carrington, Kylo, Guy Bell, Pioneer, White Rock, Adeline, Mascotte, Lady Jersey, Golden Drake, Just In Time, Just In Time North, and the Drake Volcanics), most clustering between −4 and 2‰. They believed that this sulfur was derived from the local country rocks. Perkins [27] concluded that the sulfur for the Red Rock deposits was dominantly magmatic sulfur, while sulfur for the other deposits in the Drake Goldfield and the Drake Volcanics may be the result of seawater sulfur mixing with magmatic sulfur.

In this study, sulfides from the All Nations, Gladstone Hill, Kylo, Lady Hampden, Red Rock, Strauss, and West Copper Deeps have $\delta^{34}S$ values ranging from −6.96 to +2.4‰, with 89% of these close to 0 ± 4‰ (Figure 10a,b), similar to the range reported from other low-sulfidation epithermal deposits (−6 to 5‰ [61]). These sulfides have a relatively narrow range in $\delta^{34}S$ compositions, suggesting that they formed under stable physical and chemical conditions and were derived from a relatively homogeneous source [44]. The $\delta^{34}S$ values for these deposits are clustered near to 0‰, suggesting that the sulfur has a magmatic origin or has been leached from the host volcanic sequence (~1‰ [63]).

In contrast, the $\delta^{34}S$ data for Mozart, Silver King, White Rock, and White Rock North deposits show a wider and more negative range (Figure 10b) from −16.54 to −3.98‰, from which White Rock and White Rock North have a range from −16.54 to −4.4‰, similar to previously reported results (~ −15 to ~ −4‰ [63]). Variation in $\delta^{34}S$ values is common in epithermal deposits [45–47], and Richards [59] summarized that the $\delta^{34}S$ values in epithermal systems range widely from −15 to +8‰ (though mostly <0‰). Sulfur isotope values as high as −3.98‰ (close to 0 ± 3‰) suggest a magmatic sulfur source. The presence of a −16.54‰ value implies that a second source contributed minor amounts of sulfur to the Drake Goldfield and was similar to that of reduced sulfur in sedimentary rocks [50], suggesting that remobilized sulfur from a sedimentary source may have been incorporated into the hydrothermal fluids. Furthermore, remobilization of sulfur from a sedimentary source would be expected to produce a wider spread of $\delta^{34}S$ values [65]. Given that the Razorback Creek Mudstone underlies the Drake Volcanics in the study area, magmatic and sedimentary-derived sulfur mixing is certainly possible.

Furthermore, most of the deposits which have a magmatic dominant $\delta^{34}S$ source are gold- and copper-dominant, while the deposits that have a mixed $\delta^{34}S$ source are silver-dominant. Importantly, there is no noticeable $\delta^{34}S$ variation with depth either for individual deposits or across the Drake Goldfield as a whole.

6. Conclusions

Mineralization within the Drake Goldfield mainly occurs in the felsic intrusive units and andesitic/dacitic tuffs. There is a wide range of carbonate species within the Drake Goldfield, including ankerite, calcite, dolomite, magnesite, and siderite. Most of the carbonates having a vein or stockwork structure are of very late-stage origin and appear to have been the last phases to have precipitated from hydrothermal fluids across the Drake Goldfield. These carbonates are likely not related to the main mineralization event. The geochemistry and morphology show both syngenetic/epigenetic relationships between calcite, magnesium-bearing calcite, and magnesite, providing strong evidence for fluctuating Mg/Ca ratios in the fluids during carbonate precipitation and disequilibrium assemblages.

Apart from three outlier values, the carbonates are characterized by a wide range in $\delta^{18}O_{VSMOW}$ but a relatively narrow range of $\delta^{13}C_{VPDB}$. Carbon and oxygen isotopes of vein carbonates from the Drake Goldfield deposits indicate that the carbon had a mantle-derived source and that this then underwent limited mixing with meteoric waters and fluids associated with low-temperature alteration. Deposits dominated by gold and copper (except for the Red Rock field) show an initial magmatic origin for the late-stage fluid, which then mixed with fluids associated with low-temperature alteration processes during their evolution. In contrast, for the silver dominant deposits and the Red Rock deposit, although also having an initial magmatic origin, the fluids responsible for carbonate formation were more greatly affected by mixing with meteoric water.

Sulfur isotope ratios of sulfide minerals indicate that the sulfur was most likely derived from at least two different sources: a predominantly magmatic source (especially for the gold and copper dominant deposits) and a minor reduced sedimentary source (only for the silver-dominant deposits), the relative importance of each varying between deposits.

The combined carbon, oxygen, and sulfur isotopes clearly show that although the carbonate veins were derived from a wide range of fluids, these were different to those responsible for sulfide deposition. The different carbon, oxygen, and sulfur isotope characteristics from different deposits located across the Drake Goldfield collectively suggest a major heat and metal source below the center of the DQZ, which best explains the spatial distribution of the gold (copper) and silver-dominant deposits. The data also suggest that the Red Rock area may have had a second localized heat source, though this requires further work.

This study also has implications for further exploration for Ag-Au (Cu) and low–intermediate-sulfidation epithermal deposits within the Drake Goldfield. Au (Cu) intermediate-sulfidation deposits mainly occur within the center of the DQZ, while Ag and Au low-sulfidation epithermal deposits mainly occur within the periphery. In general, the Au (Cu) deposits within the Drake Goldfield show a limited variation in values for sulfur isotopes of the main mineralization stage sulfides (mostly close to 0‰ of $\delta^{34}S$) and oxygen isotopes of the carbonates, indicating magmatic dominant fluid sources. In contrast, the sulfur and oxygen isotopes for the Ag deposits have larger variations, suggesting substantial mixing processes during the fluid's evolution. The carbon and oxygen isotopes for the low-sulfidation epithermal deposits within the Drake Goldfield show an initial magmatic source which then mixed with meteoric fluids, while the intermediate-sulfidation epithermal deposits show an initial magmatic source, which then mixed with fluids associated with low-temperature alteration. In addition, the Ag low-sulfidation epithermal deposits show a greater variation in carbonate isotope values compared to the Au low-sulfidation epithermal deposits, reflecting increased fluid-mixing processes within the Ag low-sulfidation epithermal deposits.

Supplementary Materials: The following supporting information can be downloaded at: https://www.mdpi.com/article/10.3390/min15020134/s1, Figure S1: Location of diamond drill holes and open-cuts which were sampled for this study across the Drake Goldfield—overall distribution of the diamond drill holes in the Drake Goldfield. Note: For drill hole numbers please refer to Supplementary Tables S1 and S2 [66]; Figure S2: Location of diamond drill holes and open-cuts which were sampled for this study—zoom-in map of the White Rock and White Rock North deposits. Note: For drill hole numbers please refer to Tables S1 and S2 [66]; Figure S3: Location of diamond drill holes and open-cuts which were sampled for this study—zoom-in map of the Kylo, Strauss and All Nations deposits. Note: For drill hole numbers please refer to Tables S1 and S2 [66]; Figure S4: Location of diamond drill holes and open-cuts which were sampled for this study—zoom-in map of the Lady Hampden. Silver King and Mozart deposits. Note: For drill hole numbers please refer to Tables S1 and S2 [66]; Figure S5: Ore paragenesis for the Cu prospects within the Drake Goldfield; Figure S6: Ore paragenesis for the Au-dominant deposits within the Drake Goldfield; Figure S7: Ore paragenesis for the Ag-dominant deposits within the Drake Goldfield; Figure S8: $\delta^{18}O$ versus $\delta^{13}C$ diagram for carbonates from the Cu prospects [51]; Figure S9: $\delta^{18}O$ versus $\delta^{13}C$ diagram for carbonates from the Au-dominant deposits [51]; Figure S10: $\delta^{18}O$ versus $\delta^{13}C$ diagram for carbonates from the Ag-dominant deposits [51]; Table S1: Carbon and oxygen isotopic compositions of carbonates from across the Drake Goldfield. Note: Coordinates are documented using AMG AGD66 coordinate system. Abbreviations: Ank-ankerite; Dol-dolomite; Cal-calcite; Mgs-magnesite; Sd-siderite; Brt-barite; Ccp-chalcopyrite; Py-pyrite; Sp-sphalerite; Dkt- dickite; Ab-albite; Table S2: Sulfur isotope compositions of sulfides from the Drake Goldfield. Note: Coordinates are documented using AMG AGD66 coordinate system. Abbreviations: Ccp-chalcopyrite; Cv-covellite; Cc-chalcocite; Py-pyrite; Gn-galena; Sp-sphalerite.

Author Contributions: Writing—original draft preparation, H.Q.; writing—review and editing, I.G. and D.F.; formal analysis, H.Q., L.A., C.D. and H.W.; investigation, H.Q., E.M. and I.G.; supervision and writing input, R.W. All authors have read and agreed to the published version of the manuscript.

Funding: This research was funded by the PANGEA Research Centre, School of Biological, Earth and Environmental Sciences, UNSW Sydney, and White Rock Minerals Ltd. Hongyan Quan was supported in part by the China Scholarship Council (No. 201706170039).

Data Availability Statement: Data are contained within the article and Supplementary Materials.

Acknowledgments: We would like to thank Brett Nagel from White Rock Minerals for all his help during fieldwork and sample collection. Constructive and thorough reviews by Stuart Simmons greatly improved this manuscript. We would also like to thank all three anonymous reviewers for their insightful comments which greatly improved the revised manuscript.

Conflicts of Interest: The authors Hongyan Quan and Rohan Worland are respectively employees of BGRIMM Technology Group and White Rock Minerals Ltd. The paper reflects the views of the scientists and not the company.

References

1. Lindgren, W. *Mineral Deposits*; McGraw-Hill Book Company Inc.: New York, NY, USA, 1933; p. 930.
2. White, N.C.; Hedenquist, J. Epithermal gold deposits. Styles, characteristics and exploration. *SEG Newsl.* **1995**, *27*, 1–13. [CrossRef]
3. Simmons, S.F.; White, N.C.; John, D.A. Geological characteristics of epithermal precious and base metal deposits. In *100th Anniversary Volume*; Society of Economic Geologists: Littleton, CO, USA, 2005; pp. 485–522.
4. Sillitoe, R.H. Epithermal paleosurfaces. *Miner. Depos.* **2015**, *50*, 767–793. [CrossRef]
5. Hedenquist, J.W.; Arribas, A.; Gonzalez-Urien, E. Exploration for epithermal gold deposits. In *Reviews in Economic Geology*; Society of Economic Geologists: Littleton, CO, USA, 2000; Volume 13, pp. 245–277.
6. White Rock Minerals. Mt Carrington Project: Overview, Geological Setting, Resources, Development, Exploration Gold and Silver, and Exploration Copper. Available online: https://www.whiterockminerals.com.au/mt-carrington-overview (accessed on 7 August 2019).
7. Murray, C. Tectonic evolution and metallogenesis of the New England Orogen. In *New England Orogen-Tectonics and Metallogenesis*; Kleeman, J.D., Ed.; University of New England: Armidale, Australia, 1988; pp. 204–210.
8. Thomson, J. *Geology of the Drake 1: 100,000 Sheet*; 0724006699; Geological Survey of New South Wales, Department of Mines: Sydney, NSW, Australia, 1976; p. 202.
9. Andrews, E.C. *Report on the Drake Gold and Copper Field by EC Andrews*; Geological Survey; Department of Mines and Agriculture: Perth, WA, Australia, 1908; p. 46.
10. Perkins, C. *Mineralization in the Drake Volcanics. New England Orogen Tectonics and Metallogenesis*; University of New England: Armidale, NSW, Australia, 1988; p. 275.
11. Craighead, G.; Gordon, M. *White Rock Minerals (WRM): Low Cost Gold/Silver Start-Up Opportunity, Sydney*; Breakaway Investment Group: St. Augustine, FL, USA, 2016; pp. 6–19.
12. Chomiszak, G. Relationship Between Alteration Assemblages, Their Intensity and Mineralisation and Grade at the White Rock Epithermal Ag Deposit, Drake, Northeastern NSW. Unpublished Honours Thesis, University of New South Wales, Sydney, Australia, 2016.
13. Zhang, H. Relationship Between Alteration Assemblages, Alteration Intensity and Mineralisation and Grade for the Lady Hampden Epithermal Au-Ag Deposit, Drake, North-Eastern NSW. Unpublished Honours Thesis, University of New South Wales, Sydney, Australia, 2016.
14. White, V. Volcanic Facies and Its Relationship to Silver Mineralisation, White Rock and White Rock North Epithermal Deposits, Drake Goldfield, NE NSW. Unpublished Honours Thesis, University of New South Wales, Sydney, Australia, 2017.
15. Madayag, E. Controls on Mineralisation and Alteration in the Epithermal Strauss Au-Ag Deposit, Drake Goldfield, Northern NSW. Unpublished Honours Thesis, University of New South Wales, Sydney, Australia, 2020.
16. Lay, A. A Comparative Study of the Mineralogy and Geochemistry of Ore Minerals from Silver-Rich Polymetallic Deposits of the Lachlan and Southern New England Orogens, New South Wales, Australia. Unpublished Ph.D. Thesis, University of New South Wales, Sydney, Australia, 2019.
17. Davies, B. *Mt Carrington District: Structural Framework & Mineralisation*; Confidential company report prepared; Rex Minerals Ltd.: Pine Point, Australia, 2010; p. 44.
18. Bottomer, L. Epithermal silver-gold mineralization in the Drake area, northeastern New South Wales. *Aust. J. Earth Sci.* **1986**, *33*, 457–473. [CrossRef]
19. Leitch, E. The geological development of the southern part of the New England Fold Belt. *J. Geol. Soc. Aust.* **1974**, *21*, 133–156. [CrossRef]
20. Jessop, K.; Daczko, N.; Piazolo, S. Tectonic cycles of the New England Orogen, eastern Australia: A review. *Aust. J. Earth Sci.* **2019**, *66*, 459–496. [CrossRef]
21. Cawood, P.; Pisarevsky, S.A.; Leitch, E. Unraveling the New England orocline, east Gondwana accretionary margin. *Tectonics* **2011**, *30*, 1–15. [CrossRef]
22. Shaw, S.; Flood, R. The New England Batholith, eastern Australia: Geochemical variations in time and space. *J. Geophys. Res. Solid Earth* **1981**, *86*, 10530–10544. [CrossRef]
23. Waltenberg, K.; Blevin, P.; Bull, K.; Cronin, D.; Armistead, S. *New SHRIMP U–Pb zircon Ages from the Lachlan Orogen and the New England Orogen*; Geoscience Australia: Canberra, Australia, 2015; pp. 64–75.

24. Herbert, H. The Drake mineral field—An unique entity in eastern Australia. In Proceedings of the In Permian Geology of Queensland, Geological Society of Autralia, Queesland Diveision, Brisbane, Australia, 14–16 July 1982; pp. 367–377.
25. Cumming, G. *Geochemistry of the Drake Volcanics, Drake Region, Northern NSW: Preliminary Data Analysis*; unpublished report; White Rock Minerals Ltd.: Ballarat, Australia, 2011; p. 25.
26. Herbert, H.K. Gold-silver mineralisation within the Drake Volcanics of northeastern N.S.W. In Proceedings of the Permian Geology fo Queensland, Geological Society of Autralia, Queesland Diveision, Brisbane, Australia, 14–16 July 1982; pp. 401–412.
27. Perkins, C. The red rock deposit: A late Permian submarine epithermal precious metal system in Northastern New South Wales. In Proceedings of the Pacific Rim 87. International Congress on the Geology, Structure, Mineralisation and Economics of Pacific Rim, Gold Coast, Australia, 26–29 August 1987; pp. 895–898.
28. Brownlow, J. Diapirism and the development of four thermal provinces in north-eastern New South Wales. In Proceedings of the New England Geology, Voisey Symposium, Armidale, Australia, 7 October–8 November 1982; University of New England: Armidale, NSW, Australia, 1982; pp. 229–237.
29. Korsch, R. Early Permian Tectonic Events in the New England Orogen. New England Geology. In *Voisey Symposium*; University of New England: Armidale, Australia, 1982; pp. 35–42.
30. Flood, P.; Fergusson, C. The geological development of the northern New England Province of the New England Fold Belt. In Proceedings of the 1984 Field Conference. Volcanics, Granites and Mineralisation of the Stan Thorpe-Emmaville-Drake Region. Geological Society of Australia, Queensland Division, Brisbane, Australia, 6–10 August 1984; pp. 1–19.
31. Houston, M. The geology and mineralisation of the Drake mine area, northern NSW, Australia. In Proceedings of the New England Orogen, Eastern Australia, NEO 93, University of New England, Armidale, NSW, Australia, 2–4 February 1993; pp. 337–348.
32. Steven, T.A.; Eaton, G.P. Environment of ore deposition in the Creede mining district, San Juan Mountains, Colorado; I, Geologic, hydrologic, and geophysical setting. *Econ. Geol.* **1975**, *70*, 1023–1037. [CrossRef]
33. Barton, P.; Bethke, P.M.; Roedder, E. Environment of ore deposition in the Creede mining district, San Juan Mountains, Colorado; Part III, Progress toward interpretation of the chemistry of the ore-forming fluid for the OH Vein. *Econ. Geol.* **1977**, *72*, 1–24. [CrossRef]
34. Hayba, D.O. Environment of ore deposition in the Creede mining district, San Juan Mountains, Colorado; Part V, Epithermal mineralization from fluid mixing in the OH Vein. *Econ. Geol.* **1997**, *92*, 29–44. [CrossRef]
35. Sander, M.V.; Einaudi, M.T. Epithermal deposition of gold during transition from propylitic to potassic alteration at Round Mountain, Nevada. *Econ. Geol.* **1990**, *85*, 285–311. [CrossRef]
36. Henry, C.D.; Elson, H.B.; McIntosh, W.C.; Heizler, M.T.; Castor, S.B. Brief duration of hydrothermal activity at Round Mountain, Nevada, determined from Ar 40/Ar 39 geochronology. *Econ. Geol.* **1997**, *92*, 807–826. [CrossRef]
37. Rhys, D.A.; Lewis, P.; Rowland, J. Structural controls on ore localization in epithermal gold-silver deposits: A mineral systems approach. In *Applied Structural Geology of Ore-Forming Hydrothermal Systems, SEG Reviews*; Society of Economic Geologists: Littleton, CO, USA, 2020; Volume 21, pp. 83–145.
38. Cumming, G. *A Revised Stratigraphic Framework for Red Rock, Mozart, and White Rock, Drake Volcanics, Northern New South Wales*; Unpublished report; White Rock Minerals Ltd.: Ballarat, Australia, 2013; p. 18.
39. Sarrazin, P.; Chipera, S.; Bish, D.; Blake, D.; Vaniman, D. Vibrating sample holder for XRD analysis with minimal sample preparation. *Int. Cent. Diffr. Data Adv. X-Ray Anal.* **2005**, *48*, 156–164.
40. Taylor, J. Computer programs for standardless quantitative analysis of minerals using the full powder diffraction profile. *Powder Diffr.* **1991**, *6*, 2–9. [CrossRef]
41. Burkett, D.A.; Graham, I.T.; Ward, C.R. The application of portable X-ray diffraction to quantitative mineralogical analysis of hydrothermal systems. *Can. Mineral.* **2015**, *53*, 429–454. [CrossRef]
42. Friedman, I.; O'Neil, J.R. Compilation of stable isotope fractionation factors of geochemical interest. *Prof. Pap.* **1977**, *440*, 11. [CrossRef]
43. Leroy, J.L.; Hubé, D.; Marcoux, E. Episodic deposition of Mn minerals in cockade breccia structures in three low-sulfidation epithermal deposits: A mineral stratigraphy and fluid-inclusion approach. *Can. Mineral.* **2000**, *38*, 1125–1136. [CrossRef]
44. Ohmoto, H.; Rye, R.; Barnes, H. *Geochemistry of Hydrothermal Ore Deposits*; John Wiley & Sons: New York, NY, USA, 1979; pp. 517–611.
45. Huston, D. Stable isotopes and their signifi cance for understanding the genesis of volcanic-associated massive sulfide deposits: A review. In *Volcanic-Associated Massive Sulfide Deposits: Processes and Examples in Modern and Ancient Settings*; Society of Economic Geologists: Littleton, CO, USA, 1999; pp. 157–179.
46. Hoefs, J. Isotope fractionation processes of selected elements. In *Stable Isotope Geochemistry*; Springer: Berlin/Heidelberg, Germany, 2018; pp. 53–227.
47. Rye, R.O.; Ohmoto, H. Sulfur and carbon isotopes and ore genesis: A review. *Econ. Geol.* **1974**, *69*, 826–842. [CrossRef]

48. Matsuhisa, Y.; Morishita, Y.; Sato, T. Oxygen and carbon isotope variations in gold-bearing hydrothermal veins in the Kushikino mining area, southern Kyushu, Japan. *Econ. Geol.* **1985**, *80*, 283–293. [CrossRef]
49. Ma, Y.; Jiang, S.-Y.; Frimmel, H.E. Metallogeny of the Late Jurassic Qiucun epithermal gold deposit in southeastern China: Constraints from geochronology, fluid inclusions, and HOC-Pb isotopes. *Ore Geol. Rev.* **2022**, *142*, 104688. [CrossRef]
50. Rollinson, H.R. *Using Geochemical Data: Evaluation, Presentation, Interpretation*; Routledge: London, UK, 2014.
51. Liu, J. Basin fluid genetic model of sediment-hosted micro-disseminated gold deposits in the gold-triangle area between Guizhou, Guangxi and Yunnan. *Acta Miner. Sin.* **1997**, *17*, 448–456.
52. Zheng, Y.-F. Carbon-oxygen isotopic covariation in hydrothermal calcite during degassing of CO_2. *Miner. Depos.* **1990**, *25*, 246–250. [CrossRef]
53. Zheng, Y.-F.; Hoefs, J. Carbon and oxygen isotopic covariations in hydrothermal calcites. *Miner. Depos.* **1993**, *28*, 79–89. [CrossRef]
54. Simmons, S.F.; Christenson, B.W. Origins of calcite in a boiling geothermal system. *Am. J. Sci.* **1994**, *294*, 361–400. [CrossRef]
55. Zheng, Y. The modeling of stable isotopic system and application for ore deposit geochemistry. *Miner. Depos.* **2001**, *20*, 246–250.
56. Simmons, S.; Browne, P. Mineralogical indicators of boiling in two low sulfidation epithermal environments: The Broadlands-Ohaaki and Waiotapu geothermal systems. In *Geology and Ore Deposits*; Geological Society of Nevada: Reno, NV, USA, 2000; pp. 683–690.
57. Pokrovsky, O.S.; Golubev, S.V.; Schott, J.; Castillo, A. Calcite, dolomite and magnesite dissolution kinetics in aqueous solutions at acid to circumneutral pH, 25 to 150 C and 1 to 55 atm pCO_2: New constraints on CO_2 sequestration in sedimentary basins. *Chem. Geol.* **2009**, *265*, 20–32. [CrossRef]
58. Barnes, H.L. *Geochemistry of Hydrothermal Ore Deposits*; John Wiley & Sons: Hoboken, NJ, USA, 1997.
59. Shanks, W.P., III. *13.3—Stable Isotope Geochemistry of Mineral Deposits*; Elsevie: Amsterdam, The Netherlands, 2014; pp. 59–85.
60. Chaussidon, M.; Lorand, J.-P. Sulphur isotope composition of orogenic spinel lherzolite massifs from Ariege (North-Eastern Pyrenees, France): An ion microprobe study. *Geochim. Cosmochim. Acta* **1990**, *54*, 2835–2846. [CrossRef]
61. Field, C.; Fifarek, R. *Light-Stable Isotope Systematics in Epithermal Systems: Reviews in Economic Geology*; Society of Economic Geologists: Littleton, CO, USA, 1985; Volume 2, pp. 99–128.
62. Arribas, A., Jr. Characteristics of high-sulfidation epithermal deposits, and their relation to magmatic fluid. *Mineral. Assoc. Can. Short Course* **1995**, *23*, 419–454.
63. Herbert, H.; Smith, J. Sulfur isotopes and origin of some sulfide deposits, New England, Australia. *Miner. Depos.* **1978**, *13*, 51–63. [CrossRef]
64. Richards, J.P. Alkalic-Type Gold Deposits: A Review. In *MAGMAS, Fluids and Ore Deposits*; Short Course; Mineralogical of Association of Canada: Victoria, BC, Canada, 1995; Volume 23, pp. 367–400.
65. Zhou, J.; Huang, Z.; Bao, G.; Gao, J. Sources and thermo-chemical sulfate reduction for reduced sulfur in the hydrothermal fluids, southeastern SYG Pb-Zn metallogenic province, SW China. *J. Earth Sci.* **2013**, *24*, 759–771. [CrossRef]
66. Google Earth Pro, 2021. V7.3.3. 7786. Drake, NSW Australia. Lat -28.886327°, Lon 152.382921°, 531E. Retrieved from June 2021. Available online: https://earth.google.com/web/ (accessed on 22 January 2025).

minerals

MDPI

Article

Shortwave Infrared Spectroscopy and Geochemical Characteristics of White Mica-Group Minerals in the Sinongduo Low-Sulfide Epithermal Deposit, Tibet, China

Xuerui Li [1], Na Guo [1,*], Chunhao Li [1], Siyuan Deng [1] and Weirui Zhou [2]

[1] College of Earth and Planetary Sciences, Chengdu University of Technology, Chengdu 610059, China; lixuerui1026@126.com (X.L.); lchwilly001218@gmail.com (C.L.); dengsiy@126.com (S.D.)
[2] College of Engineering, Tibet University, Lhasa 850000, China; 18800435309@126.com
* Correspondence: cdut_guona@126.com; Tel.: +86-13540372037

Abstract: Sinongduo is the first low-sulfidation epithermal deposit to be found in the Gangdese metallogenic belt, Xizang, China. Through the integration of Shortwave Infrared (SWIR) spectroscopy, an electron probe microanalysis (EPMA), X-ray diffraction (XRD), and three-dimensional modeling, the following findings were obtained: (1) The alteration minerals in the belt predominantly belong to the white mica group, with mineral assemblages of muscovite → muscovite + illite → muscovite + illite + montmorillonite → muscovite + paragonite + montmorillonite → muscovite + paragonite + illite + montmorillonite from the outer edges to the ore zones. (2) On the basis of the number of interlayer cations, the white mica-group minerals were classified into three distinct types. (3) The SWIR spectral exploration model indicates the spatial distribution of the mineral combinations and their spectral characteristics in the low-sulfidation epithermal deposit. On the basis of these results, we predict that Woruo, an area located north of Sinongduo, has immense potential for future mineral exploration.

Keywords: low-sulfidation epithermal deposit; shortwave infrared; white mica-group mineral; Sinongduo; Woruo

1. Introduction

The Ag–Pb–Zn deposit in Sinongduo, the first low-sulfide epithermal deposit found in the Gangdese metallogenic belt, is located in the continental volcanic regions of the Tethys Himalayan orogenic belt [1,2]. It includes cryptoexplosive breccia-type Ag–Pb–Zn ore bodies, hydrothermal vein-type Pb–Zn–Ag ore bodies, and independent Ag ore bodies, which have been logged according to the characteristics of the host rock and the ore-forming elements [3]. The belt is estimated to have more than 300 thousand tons of Pb and Zn and more than 400 tons of Ag available for future mining prospects [4].

Alteration minerals are important for the classification of epithermal deposits, Hedenquist and Lowenstern concluded that the alteration minerals in low-sulfidation epithermal deposits were mainly adularia, muscovite, illite, bladed calcite, and ferromanganese carbonate (including siderite and rhodochrosite) [5]. We, using shortwave infrared (SWIR) spectroscopy, identified abundant white mica-group minerals in the cores, which had a close relationship with Ag–Pb–Zn ore bodies in the Sinongduo deposit [2,6].

Mica minerals are layered silicates with the same type of crystal structures. Variations in the nomenclature arise from the specific boundaries within a continuous solid-solution series [7,8]. Muscovite is an end member of the muscovite–paragonite solvus [9]. Paragonite is the ideal end member of the muscovite–paragonite series, with Na^+ instead of K^+ as an interlayer cation. It can undergo the substitution of Fe^{3+}, Fe^{2+}, or Mg^{2+} on the octahedral sheet, similar to other mica minerals, such as muscovite, phlogopite, and biotite [10]. Phengite is stable at high pressures and may play an important role in the mass transfer of K^+ [11,12]. Therefore, muscovite–paragonite series can be used as a geothermometer and geobarometer for exploration [13,14].

Muscovite is only a field term that includes fine-grained white micas with intermediate mineralogical compositions [15]. At temperatures below 300 °C, it forms a fine-grained potassium-deficient clay mineral called illite ($K_{0.6–0.8}Al_2(Al_{0.6–0.8}Si_{3.2–3.4})O_{10}(OH)_2$) [8,16].

Muscovite's structure features an interlayer cationic plane containing at least 0.6 cations per formula unit [17], which separates the adjacent interlayer planes. Minerals with an interlayer cation content greater than 0.95 are categorized within the muscovite–phengite series. If the cation content falls between 0.6 and 0.85, the muscovite is identified as interlayer-deficient mica. When the interlayer cation content approaches 0.65, the mineral resembles illite, while levels near 0.85 indicate it is closer to muscovite [18].

Variations in the stacking patterns of the structural unit layers result in different polytypes in the muscovite group, with $1M_d$, 1M, and $2M_1$ being the most common forms. The $1M_d$ polytype displays disordered 2:1 layer stacking, whereas 1M and $2M_1$ exhibit ordered stacking patterns. The 1M type features a single-layer repeat structure, while $2M_1$ has a two-layer repeat structure [19]. X-ray diffraction (XRD) is the most widely used and fundamental method in clay mineral research. It provides rapid and straightforward access to extensive, interpretable data on clay minerals, including phase identification, crystallinity analyses, investigation of the interlayer interactions, polytype identification, and quantitative or semi-quantitative analyses of clay mixtures [20]. Furthermore, it reveals their stability and transformation under various geological conditions [16,21–24].

Shortwave infrared (SWIR) instruments are widely employed for the rapid identification of minerals and compositional variations in their solid solutions [8,25]. This study utilizes SWIR spectroscopy to efficiently identify the muscovite-group minerals in altered rocks. The 2200 nm absorption feature, linked to the energy of the Al–OH bonds in muscovite-group mineral samples, serves as a crucial characteristic for mineral identification [26]. The position of the 2200 nm absorption feature is inversely correlated with Al^{VI}'s abundance in the octahedral layer, aligning with increased Si and Mg-Fe in Tschermak substitution ($Si^{IV}[Mg^{2+},Fe^{2+}]^{VI}=Al^{IV}Al^{VI}$) [27,28]. White mica minerals are classified into paragonite (a 2200 nm feature at 2180–2195 nm), muscovite (2195–2215 nm), and phengite (2215–2225 nm) [18,29]. Illite exhibits a pronounced 1910 nm absorption feature [6,30], although it cannot be reliably distinguished from muscovite in the SWIR range [31].

This study examines the relationship between the composition and spectral characteristics of muscovite in the Sinongduo low-sulfidation epithermal deposit in Tibet, with the goal of developing techniques to identify concealed mineralization. The findings suggest that this method can contribute to understanding the hydrothermal evolution of epithermal deposits, improving the exploration models, and enhancing mine planning.

2. The Geological Setting

The Sinongduo deposit is located on the eastern Gangdese Retroarc Uplift Belt (GRUB) [32] and belongs to the Southern Gangdese–Nyainqentanglha metallogenic belt (Figure 1a,b). The Sinongduo deposit underwent early skarn mineralization associated with porphyry Pb-Zn deposits during the Tethys evolution stage, followed by superimposition from the porphyry's magmatic hydrothermal Pb–Zn mineralization and finally by hydrothermal vein-type mineralization during the main Asia–India continental collisional period [33].

The regional stratigraphy is primarily composed of volcanic sedimentary deposits. The volcanic rocks can been divided into the Dianzhong Formation (E_1d), the Nianbo Formation (E_2n), and the Pana Formation (E_2p), with age dates of 64.43–61.45 Ma, 54.07 Ma, and 48.72–43.93 Ma, respectively [34]. The Dianzhong volcanic rocks host the Ag–Pb–Zn ores of the Sinongduo deposit; the dominant rock types include rhyolite porphyry, crystalline tuff, and volcanic breccia (Figure 1c).

The main Ag–Pb–Zn ore zones trend in the N-S direction, and these are hydrothermal columns and veins [3]. Silver is mainly found in the form of vein-type mineralization, with major enrichment of this kind occurring toward the top of the vein system [35].

Silicification and illitization are the most widespread hydrothermal alterations in the Sinongduo deposit, with illitization often involving silicification as part of the alteration process. The mineral assemblages identified using SWIR spectroscopy were mainly muscovite, illite, montmorillonite, and kaolinite. From the SW to the NE direction, the alteration minerals transition from muscovite and illite to a combination of muscovite, illite, and montmorillonite [36].

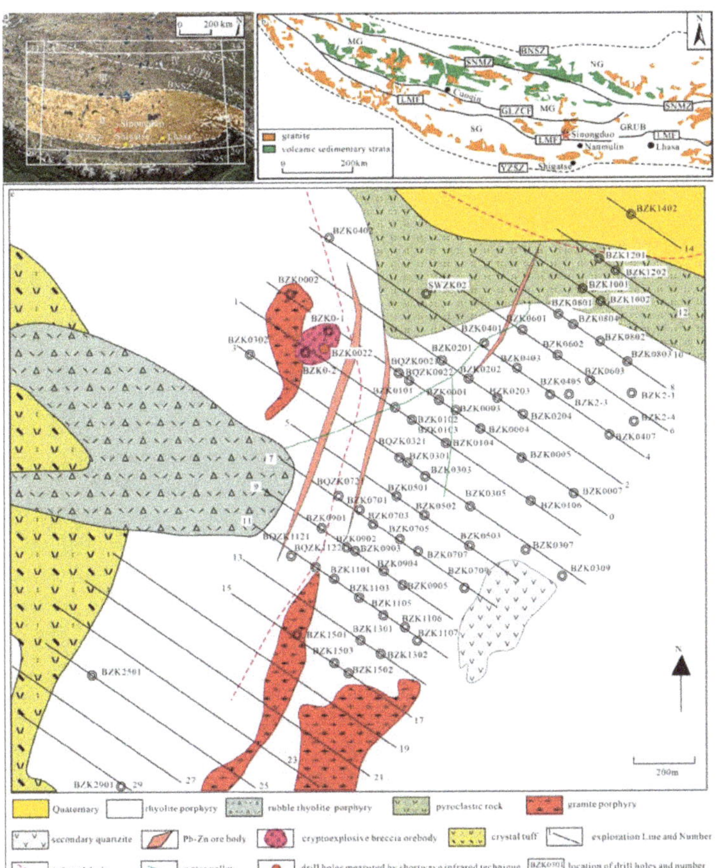

Figure 1. (**a**) Simplified map showing the location of Gangdese metallogenic belt in the Tibetan tectonic framework; (**b**) distribution of magmatic rocks in the Gangdese belt (modified after [32]); (**c**) geological map of the Sinongduo mining area (modified after [37]).

3. Sampling and Analytical Methods

The SWIR spectra were obtained from 2972 samples collected from 19 holes drilled along the SW-NE direction in the Sinongduo deposit (Table 1). The samples spanned a vertical range of 410 m and were collected at intervals of approximately 2 m. All alteration types and mineralization styles currently recognized within the Sinongduo deposit were sampled [2,35,36,38]. Selected samples were analyzed using an electron probe microanalysis (EPMA) to quantify the mineral compositions, and X-ray diffraction (XRD) was used to identify the phases (minerals) present and subsequently to analyze the crystallinity and polytype structure of the white micas. A spatial analysis was performed using ArcGIS software (ArcMap 10.6), which integrated the spectral parameters obtained from the SWIR spectroscopy and the polytype parameters derived from XRD.

Table 1. Drill hole details and summary of SWIR spectral sample statistics.

Drill Hole Number	Depth (m)	Spectral Samples	Drill Hole Number	Depth (m)	Spectral Samples
BQZK 0022	259.57	103	BZK0701	301.00	151
BZK 0-2	250.00	115	BZK0804	226.00	112
BZK0101	289.10	157	BZK0902	201.00	99
BZK0102	253.90	148	BZK1001	204.00	93
BZK0103	322.1	294	BZK1106	353.00	177
BZK0104	332	159	BZK1301	362.00	183
BZK0201	210.10	172	BZK1402	396.00	196
BZK0301	300.00	150	BZK2501	418.25	208
BK0401	301.00	142	BZK2901	401.00	185
BZK0501	205.00	128	Total	5585.02	2972

3.1. SWIR Spectral Analysis

The drill core samples were washed and air-dried for at least three days. Residual dust was removed from the dried samples prior to the analysis through the use of a brush. The absorption spectra were obtained by using an Analytical Spectral Device (ASD) portable spectrometer (with a halogen sunlight-simulating light source). The ASD's spectroscopic wavelength ranges from 350 to 2500 nm for visible to near-infrared light (the measurements were conducted at Sichuan Ruihuan Remote Sensing Environmental Technology Co., Ltd., locted in Chengdu, China, where the devices are situated). Its speed can reach 100 ms, with a wavelength accuracy of 0.5 nm for coarse rock scanning. Measurements were taken three times on different sides of a sample to obtain the average spectra. Data processing was achieved by recording the characteristic wavelengths for the Al–OH and H_2O absorption features of the white mica group, located at approximately 2200 nm and 1910 nm, respectively. Through the use of Origin software (Origon 2021) and The Spectral Geologist (TSG) software (TSG 8) to analyze the spectral features, mineral speciation assignments were made on the basis of the specific locations of the Al–OH bond absorption features and the best fits with spectra from a reference library [9].

3.2. The Basis of the Interpretation of the SWIR Spectra for the White Mica Group

Shortwave infrared analyses are based on the principles of molecular vibrational energies and the theory that molecules assume discrete and predictable energy levels [9,39]. An SWIR spectrometer with a wavelength between 1300 and 2500 nm is used in studies of white mica of the asymmetric stretching of the energy levels of the water molecules [40]. The white mica group has a similar composition and lattice structure and generates similar spectra but can be distinguished according to subtle shifts in the wavelength positions and absorption depths [41].

The SWIR studies at Sinongduo were designed to improve the delineation of alteration mineral assemblages consisting of the white mica group, which were difficult to identify visually. The key features were identified through the use of absorption minima of 1910 nm and 2200 nm as features and the best-fit white mica-group matches from a spectral library.

3.3. Bulk Geochemical Analytical Techniques

To validate the SWIR dataset, we selected seven drill cores for the EPMA and XRD measurements (the measurements were conducted at Beijing Qingchen Huanyu Petroleum Geological Technology Co., Ltd., located in Beijing, China, where the devices are situated). We specifically chose to use the EPMA-1600 (Japanese Shimadzu Co. located at Kyoto, Japan) to measure the white mica samples from the BZK0101, BZK0301, BZK0401, BZK0501, BZK1301, BZK2901, and BQZK0022 drill holes (accelerating voltage: 15 kV; beam current: 20 nA; beam spot: 5 μm). We used XRD (Japanese Rigaku Co. located at Tokyo, Japan) to conduct a quantitative analysis of the illite polytype structure and contents in the above samples (accelerating voltage: 40 kV; beam current: 100 mA; scanning range: 2.6–15° for N, T and 2.6–30° for EG). N was a natural-orientation slice, EG was an N slice composed of ethylene glycol, and T was a dehydrated slice, an N or EG slice heated to 550 °C for 2.5 h. The illite and illite–montmorillonite content were computed using the diffraction curves of the N, T and EG slices. [In our X-ray diffraction (XRD) analysis, we used the ICCD (International Centre for Diffraction Data) database for the phase identification. The specific card numbers used for the identification of the minerals were as follows: (illite: ICCD 26-0911; kaolinite: ICCD 14-0164; chlorite: ICCD 07-0075; montmorillonite: ICCD 29-1498). These card numbers were used to accurately identify the phases present in our samples, as described in the manuscript].

$$\mathrm{It} = \frac{I_{1.0nm}(EG) \times \left[\frac{h_{0.7\mathrm{nm}}(N)}{h_{0.7\mathrm{nm}}(EG)}\right]}{I_{1.0nm}(H)} \times [100 - (\mathrm{K} + \mathrm{C})] \tag{1}$$

$$\mathrm{I/S} = 100 - (\mathrm{S} + \mathrm{It} + \mathrm{K} + \mathrm{C}) \tag{2}$$

Above, I is the X-ray diffraction intensity; h is the X-ray diffraction peak; K is the kaolinite content; C is the chlorite content; It is the illite content; and I/S is the illite and montmorillonite mixture content. The calculations were performed according to the Petroleum and Natural Gas Industry Standard of the People's Republic of China (SY/T 5163-2018) [42].

The Kubler Index (KI) was determined as the full width at half maximum (FWHM) of the 10 Å (001) illite peak [43]. The KI was used to evaluate the illite crystallinity in the studied samples. For the polytype analysis, a broader scanning range of 2.6°–30° 2θ was employed, and the diffraction patterns were compared with the standard reference patterns in the ICDD database to identify the predominant polytypes.

4. Results

4.1. The SWIR Spectral Characteristics of the Muscovite-Group Minerals

According to the spectral analysis results from TSG software (TSG 8), the muscovite-group minerals display prominent absorption bands at 1400 nm, 1900 nm, and 2200 nm, with the distinction between different muscovite end members being most evident at 1900 nm. The 1900 nm absorption feature, associated with water molecules in the muscovite group, varies in depth: illite has the deepest absorption, paragonite is moderate, and muscovite is the shallowest [44]. The 2200 nm absorption feature, representing the Al-OH bonds in muscovite-group minerals, shows a minimum at 2196 nm for paragonite.

Muscovite and illite exhibit similar characteristics, with a relative maximum at 2206 nm (Figure 2a).

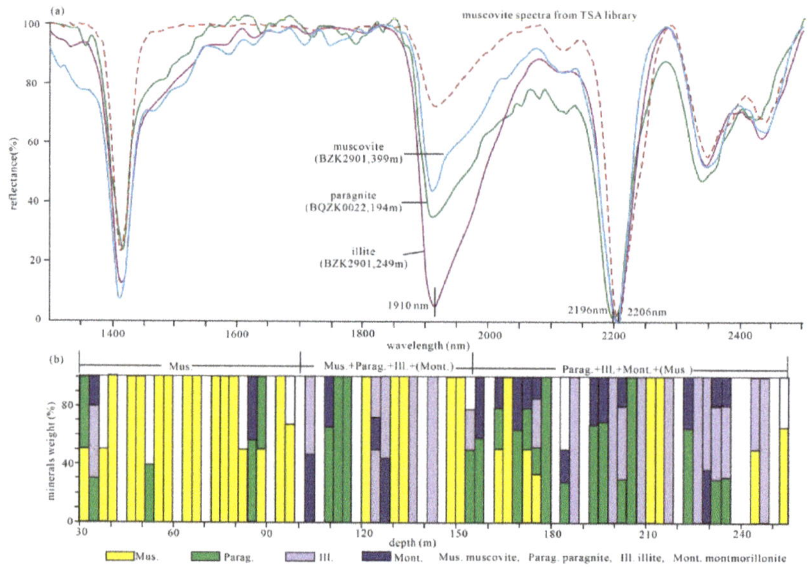

Figure 2. (**a**) illustrates that the spectra of muscovite, paragonite, and illite exhibit varying absorption depths at 1910 nm, compared with the reference library for muscovite. (**b**) presents the spatial variations in drill hole BQZK0022 as interpreted using TSG, summarizing the distribution of the mineral weights with depth and the mineral assemblage sequence: muscovite → muscovite + paragonite + illite + (montmorillonite) → paragonite + illite + montmorillonite + (muscovite).

The TSG spectral identification results indicate a substantial transformation from muscovite to paragonite as the depth increases from the top to the bottom of the drill core. The spatial distribution of the drill core shows that muscovite is mainly concentrated at the top, whereas paragonite, illite, and montmorillonite are primarily present at the bottom (Figure 2b).

The mineral assemblage progresses from the surface inward as follows: muscovite → muscovite + albite + illite + (montmorillonite) → albite + illite + montmorillonite + (muscovite). Muscovite diminishes with an increasing depth, progressively transforming into albite, illite, and montmorillonite.

4.2. The Geochemical Characteristics of Muscovite

The formula for white mica is written as $I_{0.6–1.0}$ $M_{2–3}$ $\square_{1–0}$ T_4 O_{10} A_2 [18]:

I: the interlayered cation—that is, K, Na and Ca and less commonly Cs, NH_4, Rb, and Ba;

M: the octahedral cation—that is, Fe (di- or trivalent), Mg, Al, Ti, and less commonly Mn (di- or trivalent), Zn, Cr, and V;

□: the vacancy in the octahedral sheet;

T: the tetrahedral cation—that is, Si, Al, and Fe (trivalent) and less commonly Be and B;

A: the anion—that is, OH, F, and Cl and less commonly O (oxy-micas) and S.

These calculations were made on the basis of the O = 11 anion method, assuming a total of 4 tetrahedral cations and 2 octahedral cations in the crystal structure. The cation content of the muscovite-group minerals was determined through the electron microprobe analysis (Table 2).

Table 2. This table is related to the Tschermak reaction computation, with the elemental content measured using EPMA.

Sample	SiO_2	TiO_2	Al_2O_3	FeO	MnO	MgO	CaO	Na_2O	K_2O	F	Cl	Total	Si	Al^{IV}	Al^{VI}	Ti	Fe	Mn	Mg	Ca	Na	K
BQZK0022-155	48.64	0.04	31.51	1.98	0.08	0.22	0.17	0.06	8.44	0.00	0.00	91.13	3.33	0.67	1.87	0.00	0.11	0.00	0.02	0.01	0.00	0.74
BQZK0022-98	53.07	0.06	30.32	3.82	0.15	0.39	0.07	0.04	8.79	0.00	0.00	96.69	3.45	0.55	1.76	0.00	0.21	0.01	0.04	0.00	0.00	0.73
BZK0101-120	53.62	0.06	33.27	1.69	0.24	0.51	0.13	0.07	7.81	0.26	0.00	97.67	3.39	0.61	1.86	0.00	0.09	0.01	0.05	0.01	0.01	0.63
BZK0101-280	49.84	0.07	32.33	3.00	0.43	0.36	0.10	0.05	7.63	0.00	0.01	93.82	3.31	0.69	1.85	0.00	0.17	0.02	0.04	0.01	0.00	0.65
BZK0101-80	51.04	0.09	33.23	3.32	0.16	0.32	0.11	0.06	8.09	0.00	0.01	96.43	3.31	0.69	1.85	0.00	0.18	0.01	0.03	0.01	0.00	0.67
BZK0201-209	53.19	0.10	34.00	2.47	0.17	0.24	0.02	0.03	7.49	0.13	0.00	97.83	3.36	0.64	1.89	0.00	0.13	0.01	0.02	0.00	0.01	0.60
BZK0201-88	50.26	0.08	35.43	1.18	0.10	0.08	0.05	0.06	9.46	0.00	0.00	96.71	3.24	0.76	1.93	0.00	0.06	0.01	0.01	0.00	0.01	0.78
BZK0401-120	51.59	0.05	35.83	0.79	0.07	0.22	0.14	0.05	8.16	0.04	0.00	96.95	3.28	0.72	1.96	0.00	0.04	0.00	0.02	0.01	0.01	0.66
BZK0401-180	50.07	0.03	33.30	1.66	0.11	0.40	0.16	0.02	9.43	0.01	0.01	95.19	3.29	0.71	1.87	0.00	0.09	0.01	0.04	0.01	0.00	0.79
BZK0401-277	51.83	0.14	34.15	1.43	0.30	0.44	0.13	0.09	8.65	0.11	0.01	97.27	3.31	0.69	1.88	0.01	0.08	0.02	0.04	0.01	0.00	0.70
BZK0401-97	49.42	0.12	33.01	1.53	0.09	0.33	0.01	0.06	7.77	0.02	0.00	92.36	3.31	0.69	1.91	0.01	0.09	0.00	0.03	0.00	0.00	0.66
BZK0501-187	51.98	0.12	32.47	2.09	0.15	0.53	0.13	0.05	9.02	0.00	0.00	96.53	3.36	0.64	1.83	0.01	0.11	0.01	0.05	0.01	0.01	0.74
BZK0902-142	51.54	0.01	31.72	4.01	0.04	0.13	0.15	0.07	8.63	0.00	0.00	96.30	3.37	0.63	1.81	0.00	0.22	0.00	0.01	0.01	0.01	0.72
BZK0902-38	49.75	0.07	32.64	3.24	0.08	0.07	0.07	0.04	9.39	0.00	0.01	95.36	3.29	0.71	1.84	0.00	0.18	0.00	0.01	0.00	0.00	0.79
BZK1001-152	48.46	0.06	31.83	1.24	0.38	0.39	0.08	0.08	8.50	0.05	0.00	91.06	3.31	0.69	1.88	0.00	0.07	0.02	0.04	0.01	0.01	0.74
BZK1001-75	50.27	0.14	30.30	1.92	0.20	1.05	0.03	0.08	8.44	0.03	0.00	92.43	3.39	0.61	1.80	0.01	0.11	0.01	0.11	0.00	0.01	0.73
BZK1106-153	50.42	0.03	33.94	1.96	0.10	0.26	0.15	0.02	9.26	0.03	0.01	96.17	3.28	0.72	1.88	0.00	0.11	0.01	0.03	0.01	0.01	0.77
BZK1106-205	51.69	0.07	30.74	2.18	0.12	0.92	0.14	0.05	8.68	0.03	0.00	94.62	3.41	0.59	1.79	0.00	0.12	0.01	0.09	0.01	0.01	0.73
BZK1106-332	48.95	0.07	33.22	1.60	0.09	0.42	0.15	0.04	9.69	0.03	0.00	94.24	3.26	0.74	1.87	0.00	0.09	0.00	0.04	0.01	0.01	0.82
BZK1301-182	52.27	0.09	30.01	2.84	0.06	0.81	0.13	0.03	8.25	0.07	0.02	94.57	3.44	0.56	1.78	0.00	0.16	0.00	0.08	0.01	0.01	0.69
BZK1301-262	52.71	0.10	31.76	2.78	0.15	0.77	0.22	0.06	8.98	0.00	0.03	97.55	3.38	0.62	1.79	0.01	0.15	0.01	0.07	0.01	0.00	0.74
BZK1301-62	51.40	0.02	32.81	1.98	0.07	0.21	0.01	0.03	7.68	0.12	0.00	94.33	3.37	0.63	1.90	0.00	0.11	0.00	0.02	0.00	0.01	0.64
BZK2901-162	52.31	0.06	33.07	1.65	0.13	0.77	0.11	0.08	9.14	0.10	0.01	97.40	3.35	0.65	1.84	0.00	0.09	0.01	0.07	0.01	0.01	0.75
BZK2901-322	51.13	0.07	32.32	1.78	0.10	0.97	0.12	0.03	8.63	0.13	0.00	95.28	3.34	0.66	1.83	0.00	0.10	0.01	0.09	0.01	0.01	0.72
BZK2901-82	51.09	0.10	33.24	1.49	0.14	0.72	0.10	0.07	9.05	0.04	0.00	96.04	3.31	0.69	1.86	0.00	0.08	0.01	0.07	0.01	0.01	0.75

There is a clear negative correlation between n(Si) and n(Al^{IV}) (Figure 3a) and a negative correlation between n(Al^{VI}) and n(Fe + Mg) (Figure 3b). This phenomenon is caused by Tschermak substitution ($Al^{IV} \leftrightarrow$ Si; $Al^{VI} \leftrightarrow$ Mg + Fe + Mn + Ti) within the muscovite-group minerals [45]. Tschermak substitution is considered to be governed by the concentrations of K^+, Mg^{2+}, and Fe^{2+} in hydrothermal environments, proceeding via the following reaction [45]:

$$2K_{0.7}Al_2(AlSi_3)O_{10}(OH)_2(\text{muscovitic illite}) + K^+ + 0.75Fe^{2+} + 0.75Mg^{2+} + 4.5SiO_2 + 3H_2O$$

$$\leftrightarrow 3K_{0.8}(Fe_{0.25}Mg_{0.25}Al_{1.5})(Al_{0.5}Si_{3.5})O_{10}(OH)_2(\text{phengitic illite}) + 4H^+$$

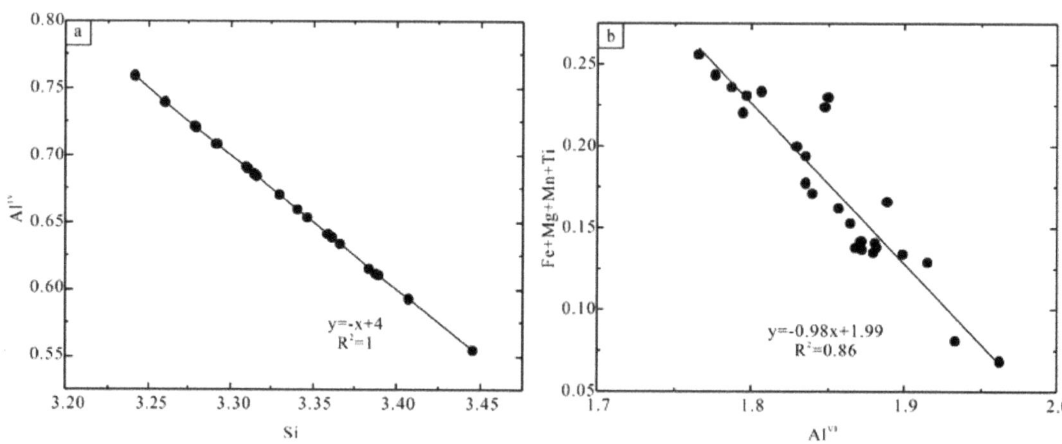

Figure 3. Tschermak substitution within muscovite-group minerals. (**a**) Substitution of Si-Al ions in the tetrahedral structure. (**b**) Substitution reactions between (Mg, Fe, Mn, Ti) and Al ions in the octahedral structure.

As a result of Tschermak substitution, the interlayer cations balance the charge, with $K^+/Na^+ + Ca^{2+} > 1$, signifying that K^+ is the dominant interlayer cation in this type of muscovite.

4.3. The XRD Characteristics of Muscovite-Group Minerals

The crystallinity, polytypism, and swelling layer value (SL) of illite were analyzed through the use of XRD tests on 13 samples collected from drill hole BZK0022 (Table 3 and Figures 4 and 5). The samples consist entirely of 100% illite. The Kübler Index (KI), inversely proportional to the crystallinity, ranges from 0.257°Δ2θ to 0.338°Δ2θ. The polytypes include 1M, 1M/$2M_1$, and $2M_1$, with 1M being the dominant form.

Table 3. This shows the illite structure and spectral characteristics.

Number	Core Depth	Rock Type	Polytype	SL (%)	KI (°2θ)	D2200/D1910	W2200 (nm)	D2200 (%)	D1910 (%)
1	BZK0022-83	Breccia-bearing tuff	1M	1.07	0.286	1.786	2198	16.4	9.18
2	BZK0022-100	Breccia-bearing tuff	1M	1.08	0.285	1.423	2196	17.5	12.3
3	BZK0022-119	Breccia-bearing tuff	$2M_1$	1.1	0.338	1.541	2193	17.1	11.1
4	BZK0022-139	Breccia-bearing tuff	1M	1.08	0.271	1.631	2198	24.3	14.9
5	BZK0022-180	Breccia-bearing tuff	$1M/2M_1$	1.05	0.29	1.985	2198	18	9.07
6	BZK0022-199	Breccia-bearing tuff	1M	1.08	0.3	2.462	2197	8.1	3.29
7	BZK0022-219	Volcanic breccia	1M	1.06	0.257	1.69	2198	19.6	11.6
8	BZK0022-239	Volcanic breccia	1M	1.05	0.261	2.019	2199	14.9	7.38
9	BZK0022-285	Volcanic breccia	$1M/2M_1$	1.08	0.285	1.706	2198	18.6	10.9
10	BZK0022-320	Volcanic breccia	$2M_1$	1.02	0.289	2.066	2196	15.1	7.31
11	BZK0022-345	Volcanic breccia	1M	1.08	0.273	1.526	2198	17.4	11.4
12	BZK0022-360	Volcanic breccia	1M	1.1	0.31	2.098	2194	21.4	10.2
13	BZK0022-379	Breccia-bearing tuff	1M	1.1	0.319	1.843	2195	18.3	9.93

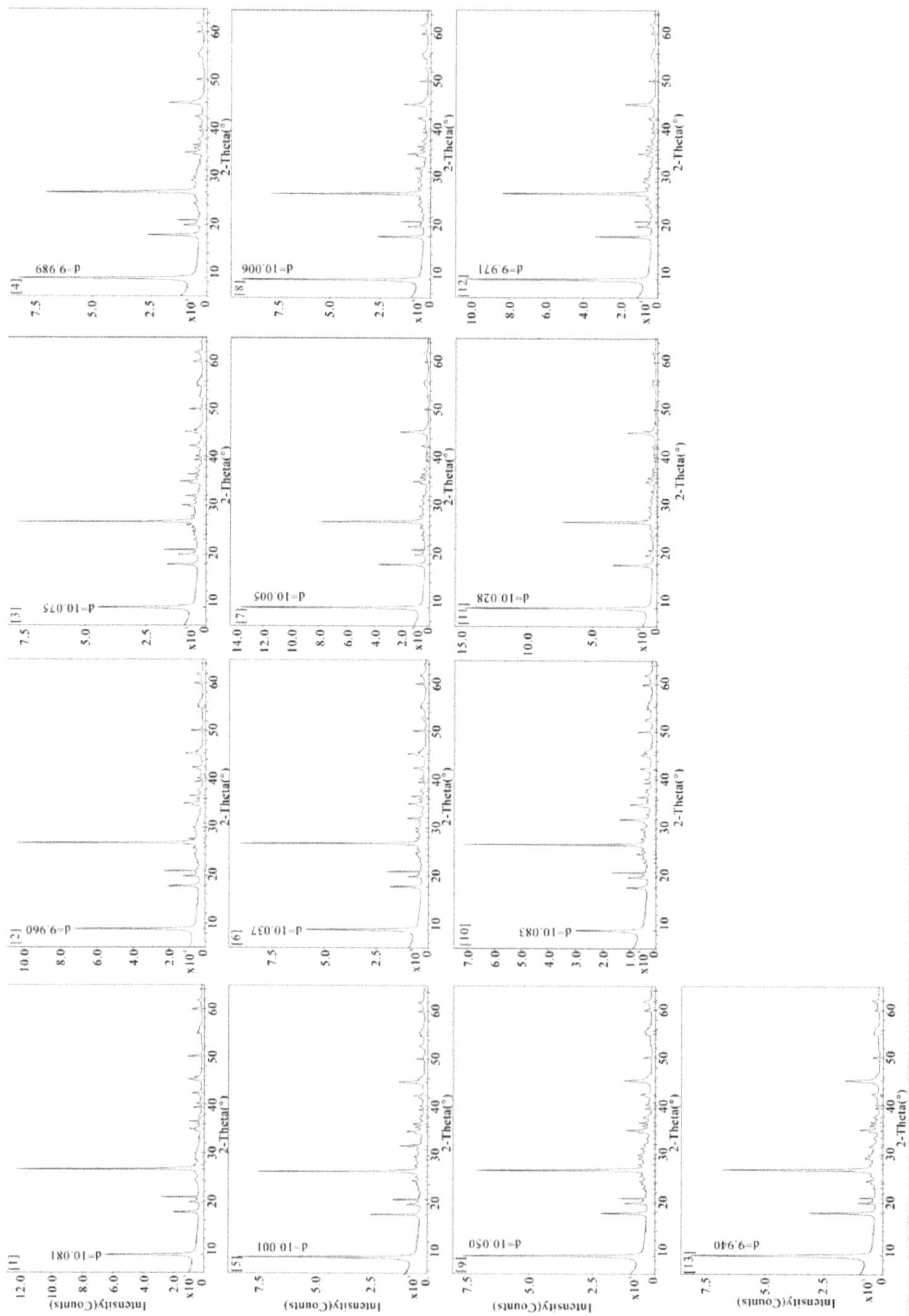

Figure 4. X-ray diffraction pattern of illite polytypism (denoting the peak position). d represents the location of illite peak.

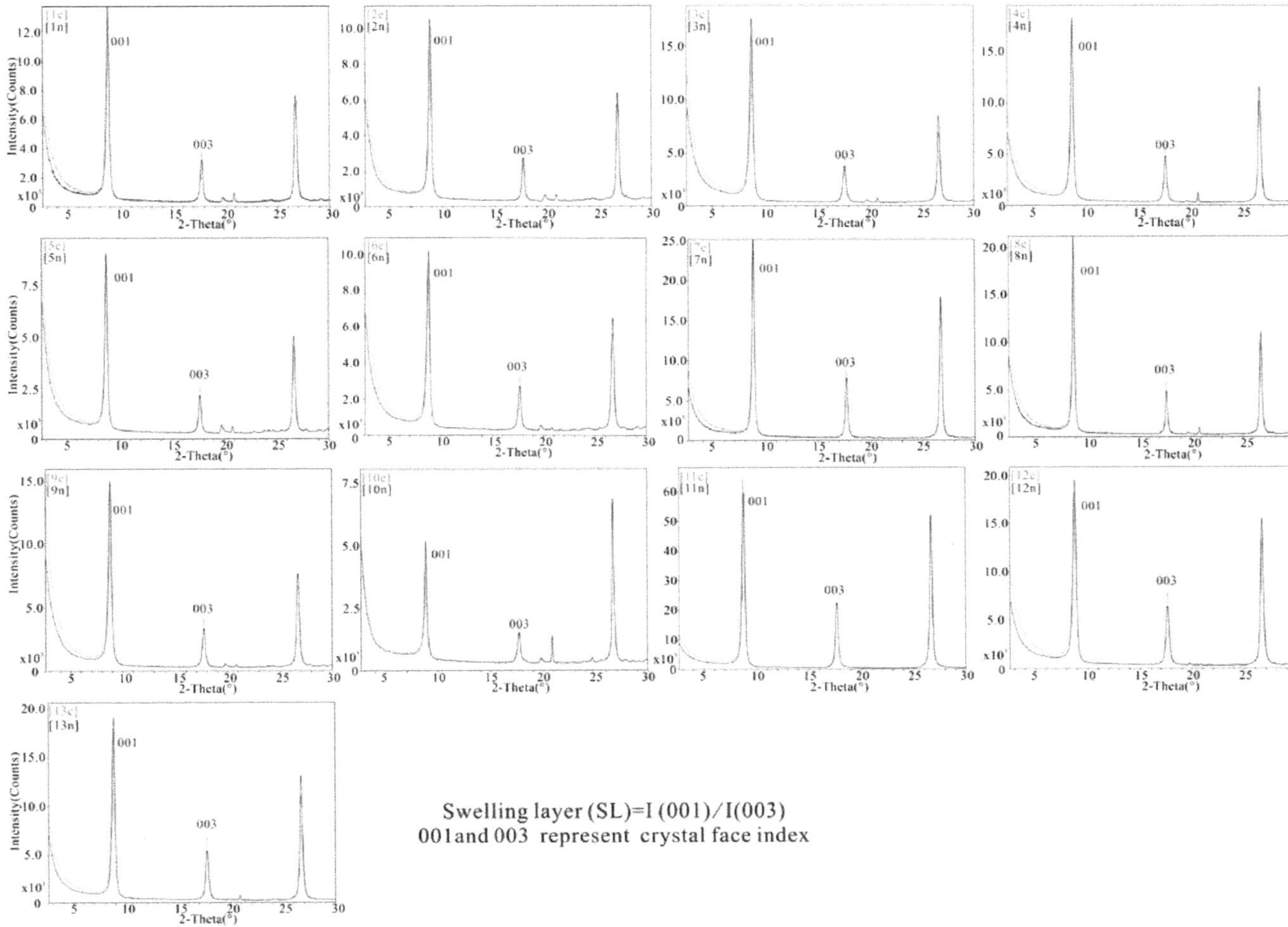

Figure 5. X-ray diffraction pattern of illite's swelling layer. e represents EG slices and n represents N slices.

5. Discussion

Various methods were utilized to identify the primary mineralogical components in the studied samples. The results indicate the dominance of white mica-group minerals, including muscovite, muscovite, and paragonite, which are key to understanding the hydrothermal alteration processes in the Sinongduo deposit. These minerals exhibit varying degrees of interlayer cation deficiency and substitution patterns, such as Tschermak substitution. This substitution involves the coupled replacement of Si^{4+} with Al^{3+} in the tetrahedral layer and the simultaneous substitution of Al^{3+} for Mg^{2+} or Fe^{2+} in the octahedral layer, reflecting the geochemical evolution of the hydrothermal system [45]. These findings provide critical insights into the mineralogical and structural evolution of the white mica-group minerals and lay the groundwork for a subsequent discussion of their geochemical behavior and cation substitution.

5.1. Classification of the Muscovite-Group Minerals

In Figure 6, the *x*-axis represents the interlayer cation content (mainly K^+), while the *y*-axis indicates the intensity of Tschermak substitution. Higher Si/Al^{IV} ratios imply Si^{4+}'s dominance in the tetrahedral layer, resulting in weaker Tschermak substitution, whereas lower ratios signify stronger substitution. Three types of interlayer-deficient muscovite were identified on the basis of the total interlayer cation content (I): Type 1: Mica with significant interlayer cation deficiency ($0.6 < I < 0.7$). At this stage, interlayer cations are greatly lacking, the Tschermak substitution is weak, and correlations include a negative relationship between I and Si/Al^{IV} and a positive one between I and sumAl, making it closer to illite. Type 2: Partially interlayer-cation-deficient mica ($0.7 < I < 0.8$). With an increasing K content, the mineral structure achieves a greater charge balance. During this stage, the Tschermak substitution intensifies, with stronger correlations between I and Si/Al^{IV} and sumAl, indicating a transitional phase from muscovite to illite, as seen in muscovite. Type 3: Mica with nearly complete interlayer cations ($0.8 < I < 0.85$). At this stage, the interlayer K is sufficient, the charge balance is optimal, and Si4+ in the tetrahedral layer is significantly replaced with Al^{3+}, yielding K^+- and Al-rich mica that resembles muscovite, including varieties such as phengite. The discussion on this type is limited due to the small dataset.

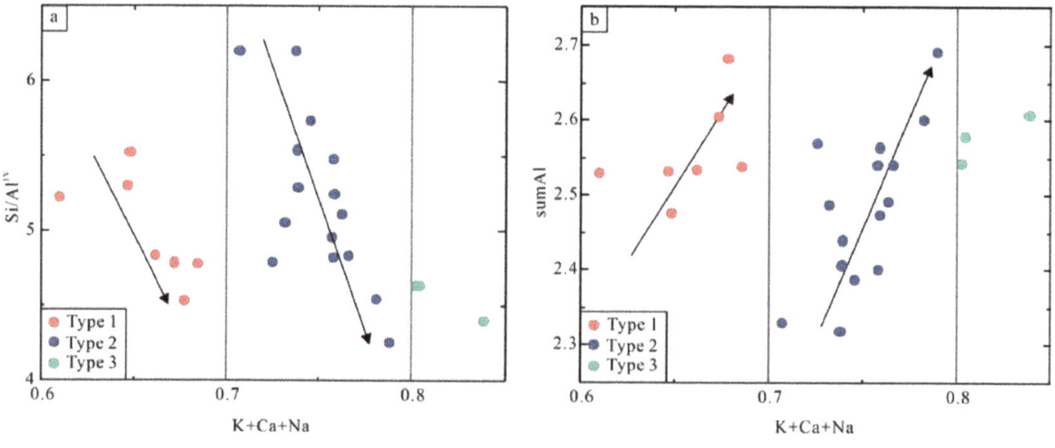

Figure 6. Classification of muscovite on the basis of the interlayer cation content. (**a**) Relationship between Si/Al^{IV} and interlayer cation content in three types of muscovite. (**b**) Relationship between sumAl and interlayer cation content in three types of muscovite.

5.2. *SWIR Characteristics of the Alteration Zones in Muscovite-Group Minerals*

The alteration in the Sinongduo hydrothermal deposit is divided into distinct zones on the basis of various physical and mineralogical properties of the muscovite-group minerals. Shortwave infrared spectroscopy (SWIR) further refines the transitional characteristics between these alteration types (Figure 7). On the basis of the wavelength shifts identified using TSG, the muscovite-group minerals are categorized into muscovite (2198–2210 nm) and paragonite (2190–2198 nm). These minerals further transform into muscovite-type illite and paragonite-type illite upon their interaction with H_2O molecules. The Ag-Pb-Zn ore bodies in the Sinongduo deposit are enveloped by an extensive, upward-propagating muscovite alteration zone, characterized by muscovite with wavelengths of 2198–2210 nm. Muscovite, as described by Sillitoe's porphyry system model [46], forms through interactions between hydrothermal fluids of a deep origin and the host rocks.

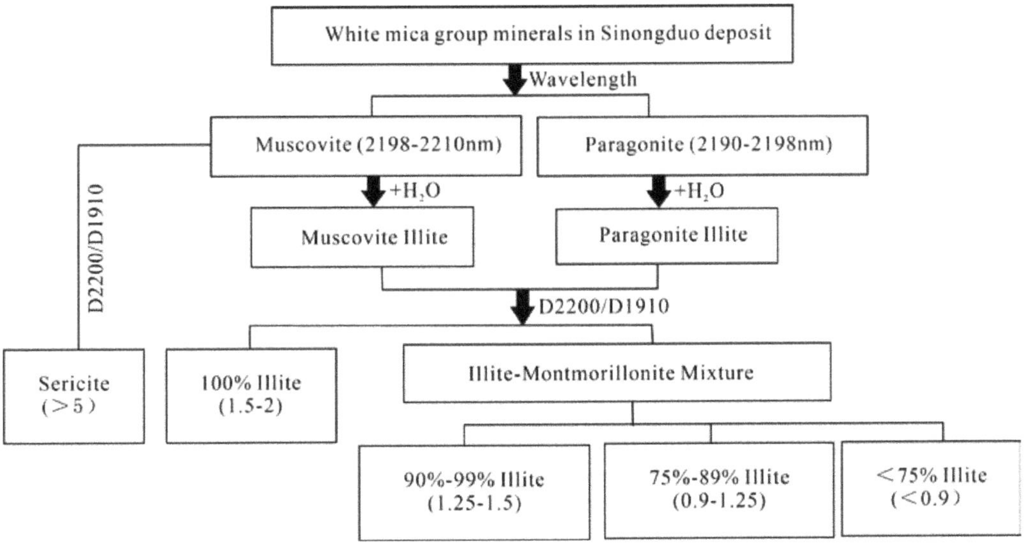

Figure 7. Tree diagram of mineral extraction on the basis of spectral characteristics.

A muscovite–paragonite–illite–montmorillonite mineralization halo formed in the lower section of the drill core (Figure 2b). This assemblage formed primarily due to the mixing of meteoric water, leading to a rapid decrease in the fluid's temperature and promoting illitization and montmorillonitization.

SWIR spectral data can be used in the field to independently identify these alteration types and confirm the alteration characteristics of the mineral distribution. From the hydrothermal center to the center of meteoric water infiltration (from bottom to top), the alteration minerals are arranged as follows: muscovite $\rightarrow$ muscovite + illite $\rightarrow$ muscovite + illite + montmorillonite $\rightarrow$ muscovite + paragonite + montmorillonite $\rightarrow$ muscovite + paragonite + illite + montmorillonite [6] (Figure 8). The illite near the hydrothermal center (W2200 $\geq$ 2198 nm) mainly forms through the transformation of muscovite, whereas the illite near the center of meteoric water infiltration (W2200 < 2198 nm) is primarily derived from paragonite.

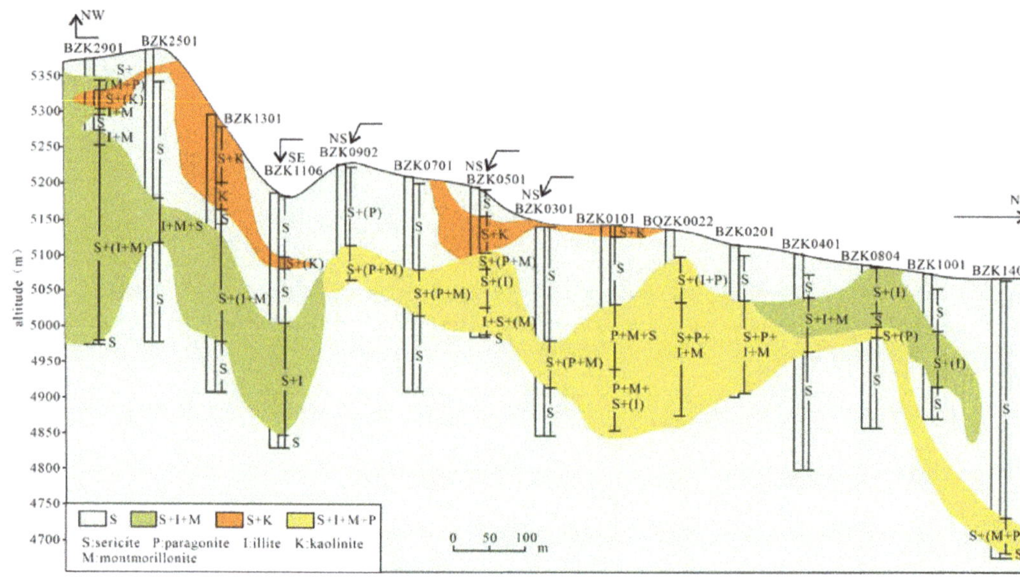

Figure 8. Alteration combination characteristics of Sinongduo deposit [6].

5.3. The Spatial Relationship Between Ag-Pb-Zn Ore Bodies and Muscovite Rock Assemblages and Their Implications for Metal Ore Formation

The shifts in the characteristic wavelength position of W2200 in the short wave spectral range indicate changes in metallogenic temperature [6]. Figure 9a illustrates that long-wavelength muscovite-group minerals (≥2200 nm) are distributed at both the top and bottom of the drill core. Vertically, the variations in W2200 suggest increased temperatures in the deep section beneath drill core BZK2901 [2], implying that deep fluids originate from the western part of the mining area. Low-sulfidation epithermal deposits are associated with volcanic activity [47]. Hence, the top-positioned long-wavelength muscovite-group minerals are linked to the volcanic processes in low-sulfidation epithermal systems, where the fluid's immiscibility induces alteration. Conversely, the muscovite-group minerals at the bottom result from alterations through the interaction of deep fluid with other minerals. The mid-section, dominated by shorter-wavelength muscovite-group minerals, is influenced by meteoric water infiltration and extends downward in the NE-SW direction [2]. This distribution highlights the potential for metal ore formation in the NE-SW direction.

Illite crystallinity can be assessed through the use of the D2200/D1910 absorption depth ratio. Although its sensitivity to changes in crystallinity decreases, the ability to quantify the mineral remains reliable when it is cross-checked with XRD results. The SWIR-IC index limits the variability in the wavelength, enhancing the precision of quantifying the illite content. The SWIR-IC increases from the top to the bottom of the drill core (Figure 9b). At the bottom, close to the heat source, muscovite exhibits strong absorption at 2200 nm and a high SWIR-IC. In the mid-section, the muscovite-group minerals combined with paragonite, illite, and montmorillonite enable the deeper penetration of water molecules into the mineral structure [2,6]. At the top, muscovite minerals combine with water to transform into illite, displaying pronounced absorption at 1910 nm and reducing the SWIR-IC (Figure 3a).

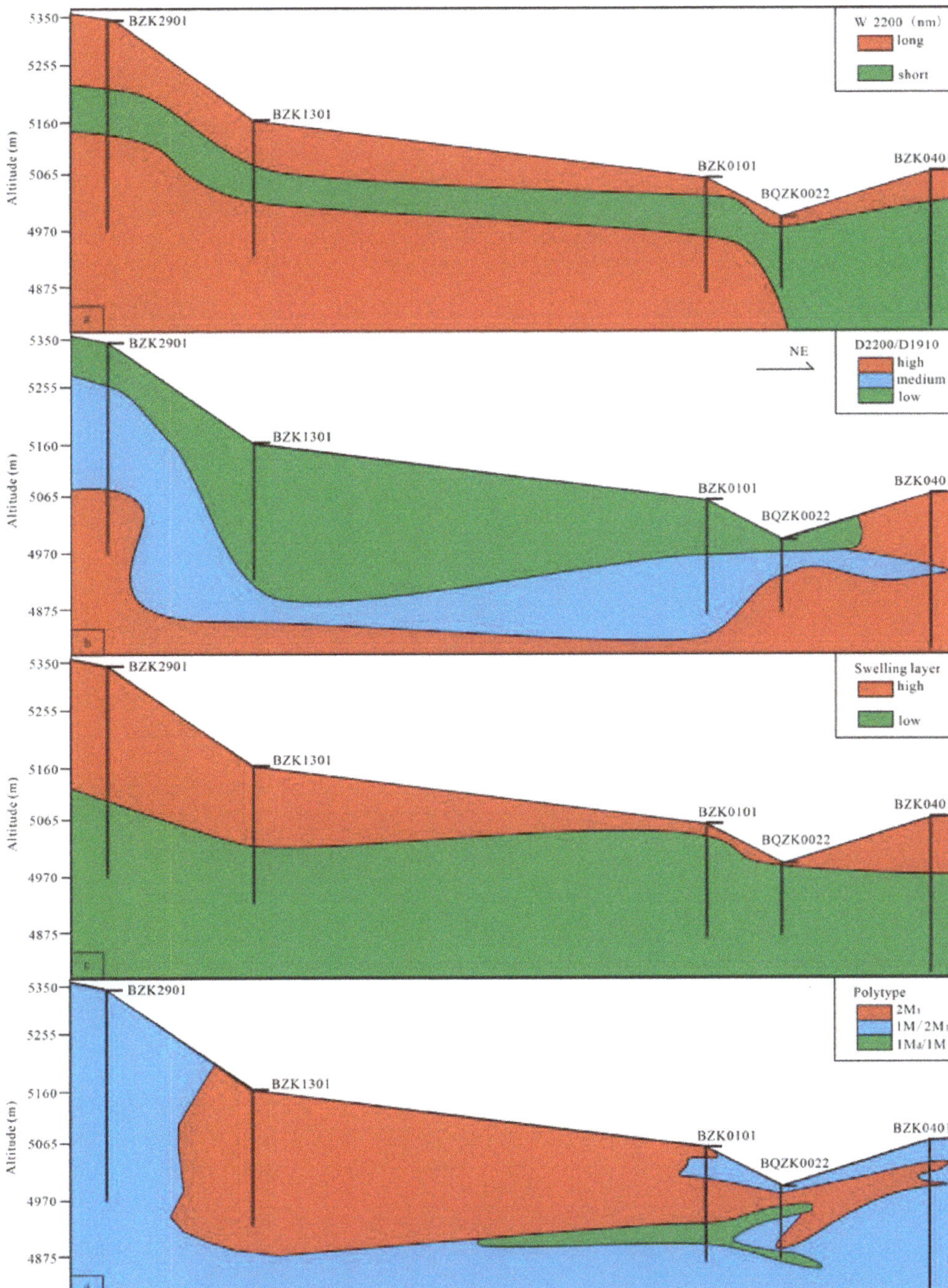

Figure 9. Map of illite characteristics on the basis of the spatial distribution of a wavelength near 2200 nm (**a**), the SWIR–IC values (**b**), the swelling layer values (**c**), and polymorphism (**d**).

The swelling layer value (SL) is a key parameter describing the interlayer swelling properties of clay minerals, where montmorillonite exhibits a higher SL than illite. Hence, the SL serves as a partial indicator of the illitization–montmorillonitization process. Figure 9c reveals that the SL values are higher at the top of the drill core than at the bottom, suggesting that illite at the top more readily transforms into montmorillonite, forming illite–montmorillonite mixed layers. At the bottom of the drill core, montmorillonite more readily transforms into illite under the influence of hydrothermal fluids of a deep origin. This pattern is a common feature, with a stronger influence of meteoric water at the top, while the bottom is predominantly influenced by deep fluid activity.

Since illite polytypes were first described, extensive research has validated the transformation sequence: $1M_d \rightarrow 1M \rightarrow 2M_1$. [19] The polytype transformation of illite involves K fixation, with the K content directly affecting the illitization rate of feldspar and muscovite in metallogenic fluids [22,23,48]. Figure 9d shows the spatial distribution of the illite polytypes: the upper region is dominated by $2M_1$, suggesting higher crystallinity and illitization rates, whereas the lower region is characterized by 1M and $2M_1$, indicating weaker illitization at greater depths.

5.4. The Shortwave Spectral Prospecting Model and Its Significance for Exploration

5.4.1. The Shortwave Spectral Prospecting Model

Through the use of the absorption characteristics of Al-OH and H_2O molecules, muscovite groups can be distinctly identified. Accordingly, a prospecting model was developed on the basis of the spectral features of muscovite-group minerals (Figure 10). The model effectively delineates the boundaries of the alteration zones. The SWIR data reveal the relationships among muscovite, illite, paragonite, and phengite, a transformation sequence also identified in the drill cores. Alterations in muscovite generally occur when granite intrudes into the sedimentary strata. The muscovite groups near the intrusive body are silicon-enriched, resulting in the formation of phengite. In distal regions from the intrusive body, the reduced temperature and pressure lead to the gradual substitution of silicates in the muscovite group with Fe and Mg ions. Cold water infiltration along the faults and fractures triggers abrupt changes in the rock fragmentation or explosion-like phenomena and pressure, causing rock explosions and forming an illite–montmorillonite (I-M) alteration zone.

The zonation of the alteration minerals is closely associated with the low-sulfidation epithermal mineralization systems under low-temperature conditions. (1) The alteration minerals are primarily composed of illite and montmorillonite, with a zoning sequence as follows—paragonite → illite → illite + montmorillonite → montmorillonite—progressing inward. (2) With sufficient water availability, paragonite transitions into illite, characterized by an enhanced 1410 nm absorption peak and a lower crystallinity. (3) Under the influence of meteoric water, sericite and paragonite interact with the water molecules to form sericite–illite or paragonite–illite. Paragonite–illite forms more readily than sericite–illite, with absorption wavelengths at 2200 nm and 2190 nm, respectively. Meteoric water, which is rainwater or surface water that has percolated through the ground, plays a crucial role in altering the minerals by introducing ions and promoting hydration, leading to the formation of illite. (4) The SWIR-IC index of illite decreases inward, reflecting its distance from the heat center or the influence of meteoric water. (5) The 1455 nm wavelength serves as a crucial spectral feature for identifying illite–montmorillonite mixed layers. (6) Illite polytypes transition inward from $1M_d \rightarrow 1M \rightarrow 2M_1$, where the 1Md type is not significantly associated with the mineralization system.

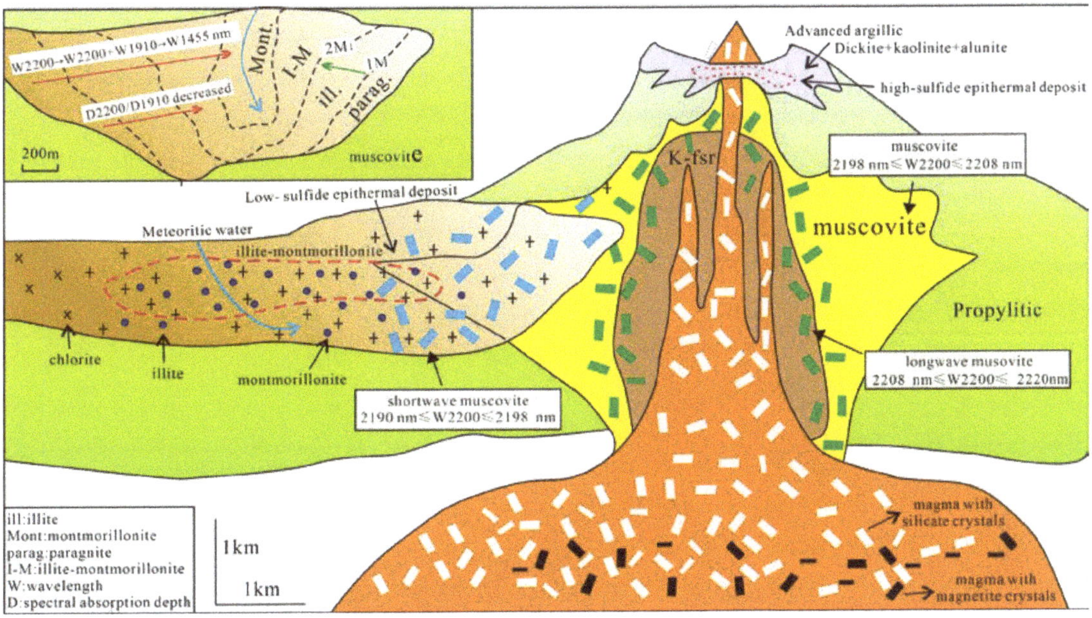

Figure 10. The SWIR spectral exploration model, illustrating the minerals identified using SWIR spectroscopy and the regulation of spectral variations.

This model establishes the correspondence between the mineralization and spectral alteration zones, offering critical guidance for exploring low-sulfidation epithermal deposits under low-temperature conditions.

5.4.2. Exploration Implications for Woruo Prospecting

The Woruo potential area, situated in the northern Sinongduo exploration zone (Figure 11a), is predominantly composed of acidic volcanic rocks (Figure 11b), with minor occurrences of chalcedony identified in the field. The SWIR analysis of the rock samples reveals that the alteration minerals are muscovite and illite. A 3D model showing the occurrence of illite samples demonstrates a decreasing trend in the SWIR-IC values in the NW-SW direction (Figure 11c). Given its position northeast of the main Sinongduo ore body, it is hypothesized that another undiscovered low-sulfidation epithermal ore body may exist near Woruo.

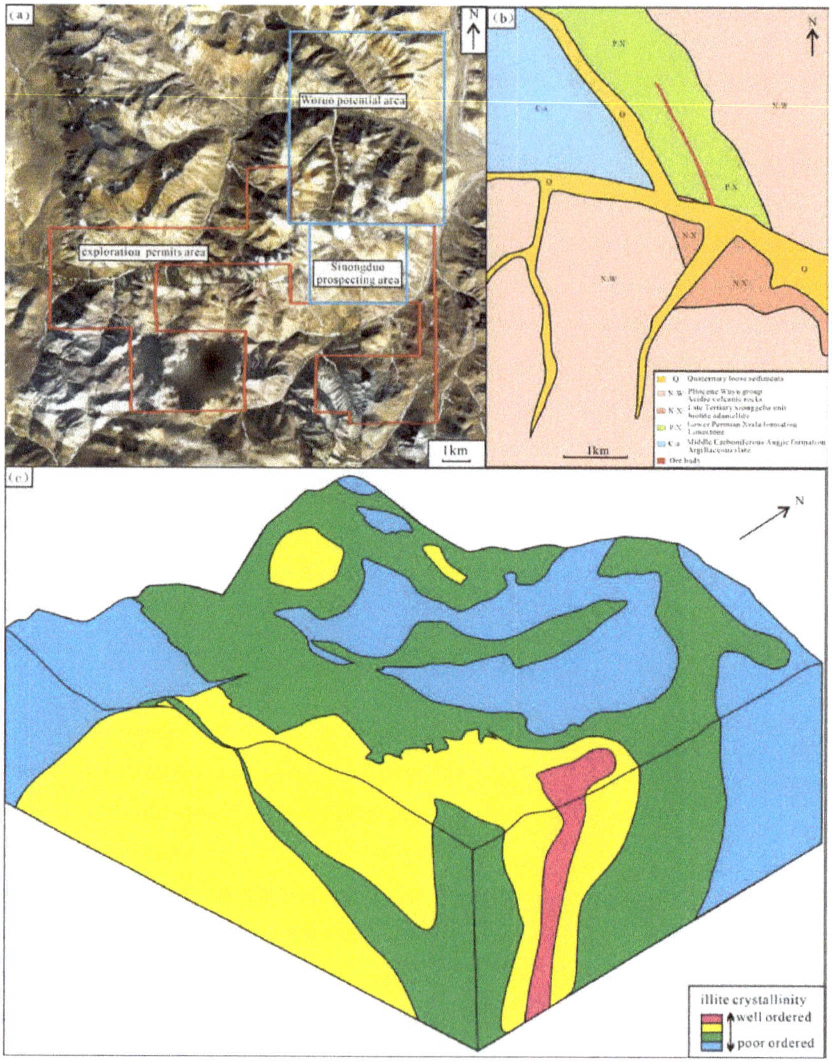

Figure 11. (**a**) High-resolution remotely sensed image of potential mineralization areas in Sinongduo and Woruo; (**b**) geological map of the Woruo region; (**c**) 3D model developed on the basis of the SWIR–IC mapping.

6. Conclusions

Muscovite-group minerals are classified into three types on the basis of their interlayer cation content: Type 1: significantly interlayer-cation-deficient mica ($0.6 < I < 0.7$); Type 2: partially interlayer-cation-deficient mica ($0.7 < I < 0.8$); Type 3: mica with nearly complete interlayer cations ($0.8 < I < 0.85$).

The SWIR analysis of the drill cores from the Sinongduo low-sulfidation deposit provides an effective means of characterizing the muscovite-group minerals and their spatial distribution. The alteration assemblage in Sinongduo transitions from the hydrothermal center outward: muscovite $\rightarrow$ muscovite + illite $\rightarrow$ muscovite + illite + montmorillonite $\rightarrow$ muscovite + paragonite + montmorillonite. The illite–montmorillonite (illite–montmorillonite in this context refers to all alteration assemblages that include both illite and montmorillonite rather than a single, distinct illite–montmorillonite assemblage) assemblage plays a crucial role in guiding exploration efforts.

The spectral model, developed from the mapping of Sinongduo and the analysis of the muscovite-group minerals, reveals spectral variations in the hydrothermal mineralization systems and displays the distribution of different mineral assemblages identified using SWIR spectroscopy, offering guidance for further exploration. Woruo, situated northeast of Sinongduo, is indicated by the spectral model as a potential site where another low-sulfidation epithermal deposit may exist.

Author Contributions: Funding acquisition: N.G. Methodology: X.L. and C.L. Project administration: S.D. Resources: N.G. Validation: W.Z. Visualization: X.L. Writing—original draft: X.L. Writing—review and editing: N.G. All authors have read and agreed to the published version of the manuscript.

Funding: This research was funded by the National Natural Science Foundation of China under the project "Spectral Characteristics of Illite in Low-Sulfidation Epithermal Deposits under the Restriction of Mineral Geochemistry," grant number 42272345, and by the National Natural Science Foundation of China (Joint Fund Project), under the project "Research on Short-Wave and Thermal Infrared Spectral Exploration Identification of Porphyry-Skarn Ore Deposits—A Case Study of the Jiama-Julong Copper Polymetallic Mining District in Tibet," grant number U2444212.

Data Availability Statement: The raw data supporting the conclusions of this article will be made available by the authors on request.

Acknowledgments: We are profoundly grateful to our advisor Na Guo, as well as our friends Siyuan Deng, Chunhao Li, and others, for their invaluable support.

Conflicts of Interest: The authors declare no conflicts of interest.

References

1. Tang, J.X.; Ding, S.; Meng, Z.; Hu, G.Y.; Gao, Y.M.; Xie, F.W.; Li, Z.; Yuan, M.; Yang, Z.Y.; Chen, G.R.; et al. The First Discovery of the Low Sulfidation Epithermal Deposit in Linzizong Volcanics, Tibet: A Case Study of the Sinongduo Ag Polymetallic Deposit. *Acta Geosci. Sin.* **2016**, *37*, 461–470.
2. Guo, N.; Guo, W.B.; Liu, D.; Huang, Y.R.; Zheng, L.; Guo, Y.N. Alteration zoning model associated with Tibetan Sinongduo epithermal deposit, the continental volcanic areas of Gangdise metallogenic belt. *Acta Petrol. Sin.* **2019**, *35*, 833–848.
3. Li, H.F.; Tang, J.X.; Ding, S.; Li, Z.; Hu, G.Y.; Xie, F.W.; Zhou, H.L.; Yang, Z.Y.; Shi, S. Fluid inclusion study and stable isotope characteristics of the Sinongduo Ag-Pb-Zn deposit, Tibet. *Acta Geosci. Sin.* **2017**, *38*, 670–686.
4. Tang, J.X.; Wang, L.Q.; Zheng, W.B.; Zhong, K.H. Ore Deposits Metallogenic Regularity and Prospecting in the Eastern Section of the Gangdese Metallogenic Belt. *Acta Geosci. Sin.* **2014**, *88*, 2545–2555. [CrossRef]
5. Hedenquist, J.W.; Lowenstern, J.B. The role of magmas in the formation of hydrothermal ore deposits. *Nature* **1994**, *370*, 519–527. [CrossRef]
6. Guo, N.; Guo, W.B.; Shi, W.X.; Huang, Y.R.; Guo, Y.N.; Lian, D.M. Characterization of Illite Clays associated with the Sinongduo low sulfidation epithermal deposit, Central Tibet using field SWIR spectrometry. *Ore Geol. Rev.* **2020**, *120*, 103228. [CrossRef]
7. Cohen, J.F. Mineralogy and Geochemistry of Hydrothermal Alteration at the Ann-Mason Copper Deposit, Nevada: Comparison of Large-Scale Ore Exploration Techniques to Mineral Chemistry. Master Thesis, Oregon State University, Corvallis, OR, USA, 2011.
8. Halley, S.; Dilles, J.H.; Tosdal, R.M. *Footprints: Hydrothermal Alteration and Geochemical Dispersion Around Porphyry Copper Deposits*; SEG Newsletter: Houston, TX, USA, 2015; pp. 1–17.
9. Harraden, C.L.; Mcnulty, B.A.; Gregory, M.J.; Lang, J.R. Shortwave infrared spectral analysis of hydrothermal alteration associated with the Pebble porphyry copper-gold-molybdenum deposit, Iliamna, Alaska. *Econ. Geol.* **2013**, *108*, 483–494. [CrossRef]
10. Wang, R.; Cudahy, T.; Laukamp, C.; Walshe, J.L.; Bath, A.; Mei, Y.; Young, C.; Roache, T.J.; Jenkins, A.; Roberts, M. White mica as a hyperspectral tool in exploration for the Sunrise Dam and Kanowna Belle gold deposits, Western Australia. *Econ. Geol.* **2017**, *112*, 1153–1176. [CrossRef]

11. Peacock, S.M.; Rushmer, T.; Thompson, A.B. Partial melting of subducting oceanic crust. *Earth Planet. Sci. Lett.* **1994**, *121*, 227–244. [CrossRef]
12. Domanik, K.J.; Holloway, J.R. The stability and composition of phengitic muscovite and associated phases from 5.5 to 11 GPa: Implications for deeply subducted sediments. *Geochim. Cosmochim. Acta* **1996**, *60*, 4133–4150. [CrossRef]
13. Blencoe, J.G.; Guidotti, C.V.; Sassi, F.P. The paragonite-muscovite solvus: II. Numerical geothermometers for natural, quasibinary paragonite-muscovite pairs. *Geochim. Cosmochim. Acta* **1994**, *58*, 2277–2288. [CrossRef]
14. Guidotti, C.V.; Sassi, F.P.; Blencoe, J.G.; Selverstone, J. The paragonite-muscovite solvus: I. PTX limits derived from the Na-K compositions of natural, quasibinary paragonite-muscovite pairs. *Geochim. Cosmochim. Acta* **1994**, *58*, 2269–2275. [CrossRef]
15. Barnes, H.L. *Geochemistry of Hydrothermal ore Deposits*; John Wiley & Sons: Hoboken, NJ, USA, 1997.
16. Srodon, J.; Eberl, D.D. Illite. *Rev. Mineral. Geochem.* **1984**, *13*, 495–544.
17. Rieder, M.; Cavazzini, G.; D'yakonov, Y.S.; Frank-Kamenetskii, V.A.; Gottardi, G.; Guggenheim, S.; Koval, P.W.; Müller, G.; Neiva, A.M.; Radoslovich, E.W. Nomenclature of the micas. *Clays Clay Miner.* **1998**, *46*, 586–595. [CrossRef]
18. Cloutier, J.; Piercey, S.J.; Huntington, J. Mineralogy, mineral chemistry and SWIR spectral reflectance of chlorite and white mica. *Minerals* **2021**, *11*, 471. [CrossRef]
19. Levinson, A.A. Studies in the mica group: Polymorphism among illites and hydrous micas. *Am. Mineral. J. Earth Planet. Mater.* **1955**, *40*, 41–49.
20. Ali, A.; Chiang, Y.W.; Santos, R.M. X-ray diffraction techniques for mineral characterization: A review for engineers of the fundamentals, applications, and research directions. *Minerals* **2022**, *12*, 205. [CrossRef]
21. Weaver, C.E. *Shale-Slate Metamorphism in Southern Appalachians*; Elsevier Science Pub Co., Inc.: Amsterdam, The Nederland, 1984.
22. Bell, T.E. Microstructure in mixed-layer illite/smectite and its relationship to the reaction of smectite to illite. *Clays Clay Miner.* **1986**, *34*, 146–154. [CrossRef]
23. Rask, J.H.; Bryndzia, L.T.; Braunsdorf, N.R.; Murray, T.E. Smectite illitization in Pliocene-age Gulf of Mexico mudrocks. *Clays Clay Miner.* **1997**, *45*, 99–109. [CrossRef]
24. Jin, Z.D.; Zhu, J.C.; Ji, J.F.; Li, F.C.; Lu, X.W. Two origins of illite at the Dexing porphyry Cu deposit, East China: Implications for ore-forming fluid constraint on illite crystallinity. *Clays Clay Miner.* **2002**, *50*, 381–387. [CrossRef]
25. Thompson, A.J.B.; Hauff, P.L.; Robitaille, A.J. *Alteration Mapping in Exploration: Application of Short-Wave Infrared (SWIR) Spectroscopy*; SEG Discovery: McLean, VA, USA, 1999; pp. 1–27. [CrossRef]
26. Clark, R.N.; King, T.V.; Klejwa, M.; Swayze, G.A.; Vergo, N. High spectral resolution reflectance spectroscopy of minerals. *J. Geophys. Res. Solid Earth* **1990**, *95*, 12653–12680. [CrossRef]
27. Duke, E.F. Near infrared spectra of muscovite, Tschermak substitution, and metamorphic reaction progress: Implications for remote sensing. *Geology* **1994**, *22*, 621–624. [CrossRef]
28. Duke, E.F.; Lewis, R.d.S. Near infrared spectra of white mica in the Belt Supergroup and implications for metamorphism. *Am. Mineral.* **2010**, *95*, 908–920. [CrossRef]
29. Vedder, W.; McDonald, R. Vibrations of the OH ions in muscovite. *J. Chem. Phys.* **1963**, *38*, 1583–1590. [CrossRef]
30. Chang, Z.S.; Hedenquist, J.W.; White, N.C.; Cooke, D.R.; Roach, M.; Deyell, C.L.; Garcia, J., Jr.; Gemmell, J.B.; McKnight, S.; Cuison, A.L. Exploration tools for linked porphyry and epithermal deposits: Example from the Mankayan intrusion-centered Cu-Au district, Luzon, Philippines. *Econ. Geol.* **2011**, *106*, 1365–1398. [CrossRef]
31. Van Ruitenbeek, F.J.; Cudahy, T.; Hale, M.; Van der Meer, F.D. Tracing fluid pathways in fossil hydrothermal systems with near-infrared spectroscopy. *Geology* **2005**, *33*, 597–600. [CrossRef]
32. Zhu, D.C.; Chung, S.L.; Mo, X.X.; Zhao, Z.D.; Niu, Y.L.; Song, B.; Yang, Y.H. The 132 Ma Comei-Bunbury large igneous province: Remnants identified in present-day southeastern Tibet and southwestern Australia. *Geology* **2009**, *37*, 583–586. [CrossRef]
33. Li, G.M.; Liu, B.; Dong, S.L.; Zhang, L. Superimposition of Cu-Fe-Pb-Zn ore concentration area in the middle of Gangdese metallogenic belt and its significance. *Miner. Depos.* **2010**, *29*, 217–218. [CrossRef]
34. Dong, G.; Mo, X.; Zhao, Z.; Wang, L.; Zhou, S. A new understanding of the stratigraphic successions of the Linzizong volcanic rocks in the Lhünzhub basin, northern Lhasa, Tibet, China. *Geol. Bull. China* **2005**, *24*, 549–557.
35. Yang, Z.Y.; Tang, J.X.; Zhang, L.J.; Zhao, X.Y.; Lang, X.H.; Zhang, C.H. Geology, geochronology, and geochemistry of Sinongduo Ag-Pb-Zn deposit in the Gangdese metallogenic belt: Implications of intermediate sulfidation mineralization in the Linzizong colcanic succession. *Earth Sci.* **2020**, *45*, 789–803. [CrossRef]
36. Guo, N.; Liu, D.; Tang, J.X.; Zheng, L.; Huang, Y.R.; Shi, W.X.; Fu, Y.; Tang, N.; Wang, C. Characteristics of alteration minerals and prospecting model revealed by shortwave infrared technique: Take Sinongduo Ag-Pb-Zn deposit as an example. *Miner. Depos.* **2018**, *37*, 556–570. [CrossRef]

37. Guo, Y.Y.; Gao, C.W.; Hao, X.L.; Tang, Y.Q.; Meng, X.M.; Zhang, Z.C.; Zhu, Y.B.; Niu, C.J.; Du, L.; Wang, X. *Summary Report on the Work of Lead-Zinc Mine in Sinongduo Mining Area, Xietongmen County, Tibet in 2015*; Zhongdi Baolian (Beijing) Land and Resources Exploration Technology Co., Ltd.: Beijing, China, 2016.
38. Ding, S. The Diagenesis and Metallogenesis of Sinongduo Epithermal Ag-Pb-Zn Deposit in Gangdese Belt, Tibet. Ph.D. Thesis, Chengdu University of Technology, Chengdu, China, 2017.
39. Burns, R.G. *Mineralogical Applications of Crystal Field Theory*; Cambridge University Press: Cambridge, UK, 1993.
40. Bishop, J.; Lane, M.; Dyar, M.; Brown, A. Reflectance and emission spectroscopy study of four groups of phyllosilicates: Smectites, kaolinite-serpentines, chlorites and micas. *Clay Miner.* **2008**, *43*, 35–54. [CrossRef]
41. Thompson, H.A.; Parks, G.A.; Brown, G.E., Jr. Dynamic interactions of dissolution, surface adsorption, and precipitation in an aging cobalt (II)-clay-water system. *Geochim. Cosmochim. Acta* **1999**, *63*, 1767–1779. [CrossRef]
42. Guo, Y.N. The Characteristics of Illite in the Sinongduo Low-Sulfide Epithermal Deposit, Tibet. Master Thesis, Chengdu University of Technology, Chengdu, China, 2020.
43. Kubler, B. Les argiles, indicateurs de métamorphisme. *Rev. Inst. Franc. Petrole.* **1964**, *19*, 1093–1112.
44. Hunt, G.R.; Ashley, R.P. Spectra of altered rocks in the visible and near infrared. *Econ. Geol.* **1979**, *74*, 1613–1629. [CrossRef]
45. Wang, L.; Percival, J.B.; Hedenquist, J.W.; Hattori, K.; Qin, K.-Z. Alteration mineralogy of the Zhengguang epithermal Au-Zn deposit, northeast China: Interpretation of shortwave infrared analyses during mineral exploration and assessment. *Econ. Geol.* **2021**, *116*, 389–406. [CrossRef]
46. Sillitoe, R.H. Porphyry copper systems. *Econ. Geol.* **2010**, *105*, 3–41. [CrossRef]
47. Corbett, G. Influence of Magmatic Arc Geothermal Systems on Porphyry–Epithermal Au–Cu–Ag Exploration Models. In Proceedings of the Terry Leach Symposium, Sydney, Australia, 17 October 2008.
48. Reynolds, R., Jr. Potassium-rubidium ratios aud polymorphism in illites and microclines from the clay size fractions of proterozoic carbonate rocks. *Geochim. Cosmochim. Acta* **1963**, *27*, 1097–1112. [CrossRef]

minerals

MDPI

Article

The Short-Wave Infrared (SWIR) Spectral Exploration Identification and Indicative Significance of the Yixingzhai Gold Deposit, Shanxi Province

Lifang Wang [1], Song Wu [1,*], Xiaodan Lai [2], Weili Yang [2], Rongliang Sun [3], Peng Liu [1], Yandong Yang [1] and Yuxin Ren [1]

[1] School of Earth Sciences and Resources, China University of Geosciences, Beijing 100083, China; wlf0020@126.com (L.W.); 3001200113@email.cugb.edu.cn (P.L.); 2101220126@email.cugb.edu.cn (Y.Y.); 2101220127@email.cugb.edu.cn (Y.R.)

[2] Zijin Mining Group Co., Ltd., Longyan 364200, China; llxd2750@163.com (X.L.); ywl5456@163.com (W.Y.)

[3] Shanxi Zijin Mining Co., Ltd., Taiyuan 034302, China; srlww11@163.com

* Correspondence: songwu@cugb.edu.cn

Abstract: The Yixingzhai gold deposit is the largest gold deposit in Shanxi Province and develops three types of mineralization: porphyry, quartz vein, and breccia. Spectral characteristic parameters of muscovite are studied by short-wave infrared (SWIR) spectral, and the exploration significance is discussed. The Al-OH wavelength of muscovite associated with porphyry mineralization gradually becomes shorter from the periphery (>2206 nm) to the center (2201–2205 nm), and the crystallinity (>2.6) gradually increases. In quartz vein mineralization, the wavelength gradually increases from the periphery (<2203 nm) to the center (2210–2211 nm), while the crystallinity does not change significantly and in a small value (<1.5). The wavelength variation range of breccia mineralization is 2198~2214 nm and is concentrated in 2201~2204 nm near the center, while the overall crystallinity is lesser than 5.5 and concentrated around 1–2.2 near the center. The wavelength and crystallinity of muscovite are mainly affected by Tschermak substitution and temperature. When the contents of Si, Fe, and Mg are low and Al^{VI} is high, the wavelength tends toward the short-wave (SW) direction, while the opposite tends toward the long-wave (LW) direction. The high crystallinity (4.1–8.4) of muscovite can be used as an indicator of porphyry gold mineralization and also provides an important indicator to explore similar types of gold deposits.

Keywords: short-wave infrared spectral; electron probe microanalysis (EPMA); exploration identification; muscovite; Yixingzhai; Shanxi

1. Introduction

In recent years, mineral spectral measurement technology has developed rapidly and has become one of the main technical methods in the field of mineral exploration, both domestically and internationally. It is widely used for extracting mineralized information from altered minerals in various deposits [1–6]. Short-wave infrared (SWIR) spectroscopy, with a wavelength range of 1300 to 2500 nm, can effectively identify minerals containing hydroxyl, amino groups, sulfates, and carbonates [7–12], as well as delineate alteration zoning [13,14] and hydrothermal mineralization centers [15–18]. Due to the portability and fast analysis speed of SWIR technology, it has been applied in ore prospecting across multiple deposits [18–22]. Porphyry and epithermal deposits are two types of magmatic-hydrothermal deposits with close spatiotemporal and genetic relationships [23]. Previous studies on porphyry-epithermal deposits have focused on geological characteristics [24], metallogenic epochs [25], and metallogenic models [26]. In recent years, some scholars have applied SWIR spectral technology to the prospecting and exploration of porphyry copper (gold-molybdenum) deposits and epithermal gold deposits, achieving good results [27–29], but research on SWIR spectral prospecting indicators for porphyry gold deposits is basically blank. For example, Harraden [29] studied the Pebble porphyry copper–gold–molybdenum deposit in Alaska, United States, and found that high-grade Cu and Au were consistent with short-wave (SW) muscovite. Tang [30] predicted in the deep of the borehole ZK2404, at an elevation of approximately 4100–4300 m, with shorter Al-OH absorption wavelengths (<2203 nm) and higher crystallinities (>5.5) in the sericite group minerals of the Tiegelongnan porphyry-epithermal copper (gold) deposits as hydrothermal centers. In addition, some studies also suggest that the closer to the center of the ore body, the longer the wavelength of muscovite minerals and the higher the crystallinity. For example, Wang [31] pointed out that the Al-OH absorption wavelengths of sericite are biased toward the long-wave (LW) direction (2208~2220 nm), which is related to the gold mineralization of the Zhengguang epithermal gold-zinc deposit. Shao [32] found a deep tectonic altered ore body near the Jiaodong Xincheng gold deposit, where illite IC (the crystallinity of muscovite group minerals in SWIR spectroscopy (SWIR-IC, also called "IC") [33,34]. The representation method is the ratio of Dep2200 to Dep1900 [35]) $\geq$ 1.2 (near 3–4.8) and sericite Pos2200 (The corresponding characteristic absorption peak of muscovite Al-OH is near 2200 nm, which is called "muscovite Al-OH absorption peak wavelength Position (Pos2200)"; the absorption depth of the corresponding absorption peak is called "muscovite Al-OH absorption peak depth Deep (Dep2200)" [30]) $\geq$ 2205 nm can be used as prospecting indicators. It can be seen that SWIR spectral technology has good indicative significance for determining the prospecting direction of deposits, but there are still differences in the infrared spectral exploration identification of different deposits, and its change mechanisms and laws need to be summarized with more examples.

The Yixingzhai gold deposit is a typical gold deposit in the Wutaishan-Hengshan gold district and the largest gold deposit in Shanxi Province [36]. The gold reserves in Yixingzhai exceeds 91 t according to [37], which has significant economic value. Many scholars have carried out various studies on the three mineralization types (porphyry, quartz vein, and breccia) in the Yixingzhai gold deposit, such as ore genesis [38,39], geochemical characteristics [36,40], metallogenic epoch [41], crystallization temperature [42], metallogenic dynamics [43], etc. However, research on exploration indicators in different mineralization types is still very limited. Therefore, the present study, based on detailed field observation and systematic petrographic work, conducted a detailed study on the alteration zoning in minerals and spectral characteristics of the Yixingzhai gold deposit using SWIR spectral technology. For the first time, SWIR spectrum exploration indicators for porphyry gold deposits were established, and as the first discovery of muscovite in porphyry gold deposits has higher crystallinity (4.1–8.4) than porphyry copper deposits (1.6–3), epithermal gold deposits (1–3), and tectonic altered gold deposits (1.1–2.7), it provides a new basis for the prospecting and exploration in the Yixingzhai as well as porphyry gold deposits.

2. Geological Background

2.1. Geological Characteristics of the Mining Area

The Yixingzhai gold deposit is located in the Wutai-Hengshan area of the central Asian orogenic belt in the North China Craton during the Paleoproterozoic era (Figure 1a). The early Precambrian basement rocks of the Hengshan terrain are termed the Hengshan

Complex, Wutai Complex, and Fuping Complex from north to south, with some geological bodies covered by the Paleoproterozoic Hutuo group, which is dominantly composed of volcanic and clastic rocks [40,44]. The lithology of the region is mainly composed of archean TTG, paleoproterozoic Hutuo group, proterozoic ultramafic rocks, and proterozoic metamorphic sedimentary rocks. The fault structures in the region are well-developed, primarily in the NW and NEE directions. The NW-trending faults are mostly formed by the vertical movement and torsion of the Yanshan crust. These faults provide structural conditions for the formation of the deposit. The NEE-trending faults are developed as Precambrian basement structures, and the low-angle bedding shear zones dominated by extrusion properties are also developed [38] (Figure 1b). Regional magmatic activities are frequent, occurring during the Wutai, Yanshan, and Himalayan periods, among which the Yanshan magmatic activity is most closely related to mineralization [36,38].

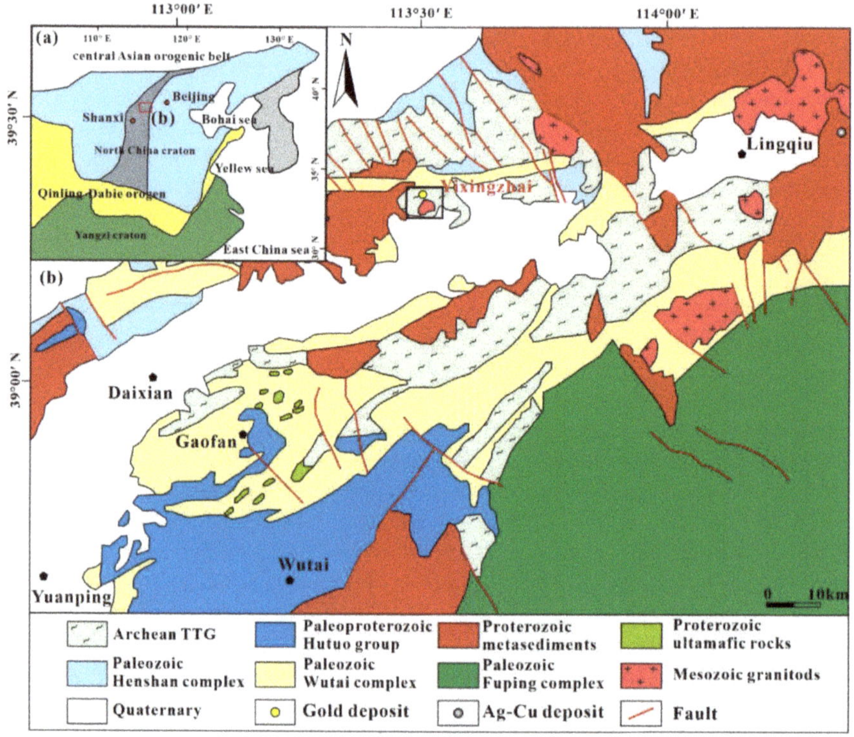

Figure 1. (**a**) North China Craton tectonic geological map (Modified after [36]), (**b**) regional geological map of the Wutai-Hengshan Terrain (Modified after [40]).

2.2. *Characteristics of the Deposit*

The Yixingzhai gold deposit is the largest gold deposit in Shanxi Province with great resource potential. It mainly develops two types of ore bodies: porphyry, quartz vein, and breccia (Figure 2). The surrounding rock of the basement of the mining area is the Hengshan complex, and the main lithologies are biotite amphibole plagioclase gneiss, amphibolite, felsic gneiss, etc. [36]. The fault system in the mining area is characterized by multi-stages, mainly developing two groups of NW-trending and NNW-trending faults. The NW-trending faults are the main rock-controlling and ore-controlling structures in the mining area. The NNW-trending faults are mainly ore-bearing structures in the mining area, which are the main places for the precipitation of ore-forming materials and control the spatial distribution of gold ore bodies [45]. The Yanshan magmatic rocks in the mining area are mainly diorite dykes, diabase dykes, acid subvolcanic rocks, the Sunzhuang diorite complex, etc. [38,46]. The Sunzhuang diorite complex is located in the south of the mining area, which is mainly composed of porphyritic granite and quartz monzodiorite [40,46]. The acid subvolcanic rocks are mainly distributed within and around the four breccia pipes of Hewan, Tietangdong, Jinjiling, and Nanmenshan, with Hewan being the largest exposed area, and the ore-bearing lithologies are mainly plagioclase quartz porphyry, quartz porphyry, etc.

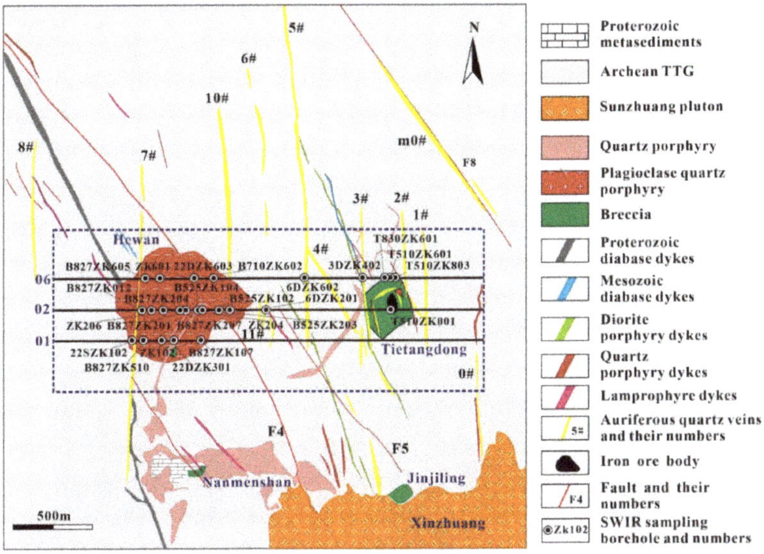

Figure 2. Geological map of Yixingzhai deposit. No. 1, No. 2, and No. 6 are geological exploration lines (Modified after [36,44]).

More than a dozen gold-bearing quartz veins of varying sizes are developed in the mining area, some of which are evenly spaced. The NW-trending fault structure controls the distribution of quartz vein gold ore bodies. According to the mineral assemblage and cross-cutting relationships, the main mineralization is divided into four stages: the quartz-sericite stage, the quartz-pyrite stage, the quartz polymetallic stage, and the calcite stage. The quartz-pyrite stage is the principal mineralization stage [47].

The Tietangdong skarn breccia pipe is the most typical breccia gold ore body in the mining area. Its composition is complex, and its economic value is high. Not only is the gold-bearing ore body found, but also an iron ore body is found. In addition, the vein rocks in the mining area are extremely developed, mainly consisting of diabase dykes and lamprophyre dykes, which are mostly irregularly interspersed throughout the mining area [36].

Previous geochronological studies have been carried out on the Yixingzhai deposit. The 40Ar/39Ar ages of muscovite in porphyry gold deposits and vein gold deposits are 139 Ma and 138 Ma, respectively [38], which are essentially consistent with the Re-Os and zircon U-Pb ages of molybdenite in quartz porphyry [36,44]. The U-Pb age of garnet in skarnized breccia is 140 Ma, and the zircon U-Pb age of quartz porphyry cutting breccia is 141 Ma [46]. The zircon ages of the Sunzhuang porphyritic granite and quartz monzodiorite are 134 Ma and 134–135 Ma, respectively [40,47]. In summary, diagenesis and mineralization ages of the Yixingzhai deposit are concentrated around 134–140 Ma.

2.3. Characteristics of Alteration and Mineralization

The main mineralization types in Yixing Village are porphyry mineralization, quartz vein mineralization, and breccia mineralization. The main lithologies in the mining area are plagioclase quartz porphyry, biotite amphibole plagioclase gneiss, breccia, etc. (Figure 2). Plagioclase quartz porphyry is mainly developed in the Hewan area, primarily composed of feldspar and quartz (Figure 3a–c). There are also small-scale breccia pipes in porphyry gold deposits, most of which are crypto-explosive breccia and amphibolite. The quartz vein gold deposits are mostly produced in the biotite amphibole plagioclase gneiss in a nearly vertical state. As the ore-bearing surrounding rock of quartz vein gold deposits, biotite amphibole plagioclase gneiss is mainly composed of quartz, biotite, and other minerals (Figure 3d,e). Breccia is mainly developed in the Tietangdong breccia pipe, and chlorite of boulders can be seen within the breccia (Figure 3f). The alteration characteristics in the mining area are pronounced, with sericitization (Figures 3a and 4a,b) generally developed, as well as silicification (Figure 3b), hematite mineralization (Figure 3c), epidotization (Figure 3d,f), chloritization (Figures 3e,f and 4a,c), carbonation (Figure 4b), etc. The metal minerals are mainly pyrite (Figure 4d), as well as a small amount of chalcopyrite (Figure 4d), sphalerite (Figure 4d), galena (Figure 4e), hematite (Figure 4f), specularite (Figure 4f), etc., among which beresitization is the most closely related to mineralization.

Figure 3. Hand specimen characteristics of Yixingzhai. (**a**) Sericitized plagioclase quartz porphyry, (**b**) silicified plagioclase quartz porphyry, (**c**) hematite-altered plagioclase quartz porphyry, (**d**) biotite amphibole plagioclase gneiss with developed chloritization, (**e**) biotite amphibole plagioclase gneiss with quartz pyrite chlorite veins, (**f**) breccia-containing chlorite, developed with epidotization; Py—pyrite; Pl—plagioclase; Qtz—quartz; Bt—biotite; Chl—chlorite; Ep—epidote.

In porphyry mineralization, muscovite is mostly altered from plagioclase, which appears fine and scaly under the microscope. The color is orange, and the particle size is about 10 μm. According to the mode of alteration, it can be divided into the complete alteration of plagioclase to sericite (Figure 5a), and the other type develops along the plagioclase cleavage (Figure 5b). In quartz vein mineralization, muscovite is associated with quartz veins or quartz + pyrite veins and is produced in the form of halo around the vein body. The particle size is mostly 50 μm to 100 μm, accompanied by calcite (Figure 5c,d). Sericite in breccia mineralization is mostly altered from biotite. The particle sizes vary, and some of the alterations are complete. The color is mainly teal and orange, and it is accompanied by chlorite (Figure 5e,f).

Figure 4. Characteristics of Yixingzhai under microscope. (**a**) Sericitization and chloritization, (**b**) muscovite alteration halo developed near quartz vein, (**c**) chloritization and epidotization, (**d**) euhedral pyrite wrapped in chalcopyrite and sphalerite, (**e**) galena surrounded by digenite; (**f**) magnetite and layered hematite enclosing fibrous specularite; Pl—plagioclase; Qtz—quartz; Chl—chlorite; Amp—amphibole; Ser—sericite; Mus—muscovite; Cal—calcite; Ep—epidote; Ccp—chalcopyrite; Sp—sphalerite; Gn—galena; Dg—digenite; Hem—hematite; Mag—magnetite; Spe—specularite. The photomicrographs (**a**–**c**) are taken in polarized transmitted light (crossed-nicols) and (**d**–**f**) in polarized reflected light.

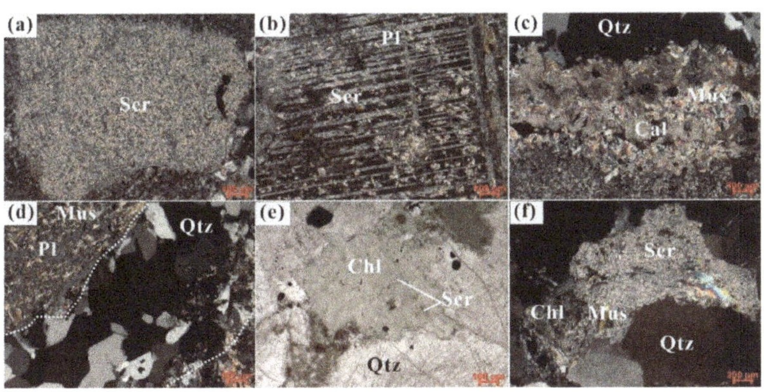

Figure 5. Microscopic characteristics of muscovite in different types of mineralization in the Yixingzhai deposit. (**a**,**b**) The alteration of plagioclase to sericite, (**c**,**d**) the sericite halo near the quartz vein, (**e**,**f**) biotite altered to sericite and chlorite; Ser—sericite; Pl—plagioclase; Qtz—quartz; Mus—muscovite; Cal—calcite; Chl—chlorite. All photomicrographs are taken in polarized transmitted light (crossed-nicols) with the exception of (**e**) in polarized reflected light.

3. Sampling and Methods

3.1. Sample Collection and Testing

In this work, the Hewan porphyry gold deposit, quartz vein gold deposit, and Tietangdong breccia gold deposit were chosen as the main research objects. According to the proximity of the ore bodies, three exploration profile lines—No. 1, No. 2, and No. 6—were selected, and five boreholes—22SZK102, B827ZK510, ZK102, 22DZK301, and B827ZK107—of Line No. 1 were selected. Line No. 2 included eleven boreholes: B827ZK102, ZK206, B827ZK204, B827ZK201, B827ZK207, B525ZK104, ZK204, B525ZK102, B525ZK203, 6DZK201, and T510ZK001, while Line No. 6 included nine boreholes: B827ZK605, ZK601,

22DZK603, B710ZK602, 6DZK602, 3DZK402, T830ZK601, T510ZK601, and T510ZK803. A total of 25 boreholes were tested using SWIR spectroscopy. The measured boreholes were recorded, photographed, and tested in detail. The test site was located in the core library of Shanxi Zijin Mining Co., Ltd., Taiyuan, China, and the total test length of boreholes was about 8700 m. After the test, the measured samples were summarized and put back into the core library in time. In this study, measurements were taken every 2 m close to the center of the ore body and every 5 m at the periphery of the ore body. Each sample was tested for three data points, and the test time was 5 to 7 s, while the time for dark minerals was appropriately extended to 10 s. Corrections were carried out every 15 to 20 min. A total of 1780 samples were tested in the study area, and a total of 5342 spectral data points were obtained.

The analysis and testing were conducted in the School of Earth Science and Resources at the China University of Geosciences (Beijing). The instrument used was the TerraSpec Halo portable mineral spectrometer, produced by ASD Company in the United States, which is designed for field geologists to quickly identify altered minerals. The detection range of the SWIR spectrum is 350 to 2500 nm, and the sampling interval is 2 nm. Before the test, the samples were washed and dried, and then the instrument was calibrated. The instrument's spectral parameter was set to 200, and the reference white was set to 400. The sample was measured when the spectral line was straight. The instrument was kept close to the sample to avoid interference caused by external light. The test points were selected to avoid, as far as possible, sulfides, quartz veins, xenoliths, and other areas [15,16].

3.2. Analytical Methods

The spectral data were processed using the software "The Spectral Geologist Version 8 (TSG 8)" developed by the Australian Federal Scientific and Industrial Research Organization (CSRIO) to automatically identify minerals and compare with the standard database, which can effectively identify minerals containing Al-OH, H_2O, Fe-OH, Mg-OH, etc. The parameters, such as the characteristic absorption peak wavelength and absorption depth of altered minerals, were extracted from the normalized reflectance spectrum (HullQuot) in the TSG software. Moscovite mainly has three characteristic groups: Al-OH, H_2O, and -OH, and the characteristic absorption peaks appear near 1400 nm, 1900 nm, and 2200 nm. The crystallinity parameters of muscovite minerals are extracted with reference to [18,48]. Finally, the final mineral species were determined by manual inspection and proofreading. In the spectrum, muscovite (sericite) and illite were not easily distinguished; thus, muscovite and illite were collectively referred to as muscovite minerals. Because the Al-OH characteristic absorption peak of the muscovite group minerals can be easily affected by kaolinite, montmorillonite, and other minerals, data affecting spectral accuracy and those with low signal-to-noise ratios in the sample were eliminated.

Electron probe microanalysis (EPMA): The test work was carried out in the electron probe laboratory of the Institute of Geology, Chinese Academy of Geological Sciences. The typical altered mineral, muscovite, in the study area was analyzed using an electron microprobe analyser (EPMA; JXA-8100, JEOL, Tokyo, Japan) with a 15 kV accelerating voltage, 20 nA probe current, and 5 μm beam diameter, and the indoor temperature was 19 °C. Before the experiment, the test samples were coated with carbon to ensure accurate testing, and the test points were selected in clean and smooth areas to avoid the influence of other impurities. SiO_2, Al_2O_3, MgO, MnO, CaO, Na_2O, K_2O, FeO, TiO_2, Cr_2O_3, and NiO were analyzed. Line and spectrometer crystal used for major element were as follows: Si (Kα, TAP), Na (Kα, TAP), K (Kα, PETJ), Al (Kα, TAP), Mg (Kα, TAP), Mn (Kα, LIFH), Fe (Kα, LIFH), Cr (Kα, LIFH), Ni (Kα, LIFH), Ti (Kα, PETJ), Ca (Kα, PETJ), F (Kα, TAP), Ba

(Kα, PETJ), and Cl (Kα, PETJ). Count times were 10 s for peak and 5 s for background per element. Natural minerals from SPI were used for standardization. ZAF corrections were carried out. The estimated precisions for major elements were ±2%.

4. Results

4.1. EPMA Results

In this study, different wavelengths of muscovite were selected for the three mineralization types, and eleven samples were selected for EPMA testing according to their relationship with the center of the ore body. The test results are shown in Table 1. According to the relationship with the center of the ore body, five samples with different wavelengths were selected in the porphyry mineralization area, with a total of 48 test points. B827ZK201-70 (2201 nm) is located in the center of the ore body, while ZK204-316 (2204 nm), B827ZK605-19 (2208 nm), B827ZK107-248 (2210 nm), and ZK601-381.56 (2214 nm) are far away from the center of the ore body. Three samples were selected for quartz vein mineralization, and a total of 20 test points were selected. The samples 6DZK602-216.1 (2211 nm) and 3DZK402-194 (2213 nm) are located in the center of the ore body, while 6DZK602-130 (2202 nm) is far away from the center of the ore body. Three samples were selected from the breccia mineralization area, and a total of 31 test points were selected. The samples T510ZK803-540 (2201 nm) and T510ZK601-19 (2208 nm) are located in the center of the ore body, while T510ZK601-231 (2204 nm) is far away.

Table 1. Average compositions (wt%) of muscovite by the electron probe microanalysis.

Types	Porphyry Mineralization		Quartz Vein Mineralization		Breccia Mineralization	
	n = 48	SEM	n = 20	SEM	n = 31	SEM
SiO_2	48.48	0.15	49.15	0.32	47.76	0.31
TiO_2	0.02	0.01	0.02	0.06	0.02	0.04
Al_2O_3	0.13	0.34	0.25	0.49	0.17	0.38
Cr_2O_3	31.08	0.00	31.42	0.00	32.70	0.00
FeO	2.40	0.15	1.41	0.22	1.46	0.14
MnO	0.05	0.00	0.06	0.00	0.06	0.00
MgO	1.44	0.14	1.69	0.18	1.87	0.20
CaO	0.05	0.01	0.04	0.02	0.05	0.02
Na_2O	11.60	0.01	11.06	0.06	11.14	0.02
K_2O	0.15	0.12	0.23	0.20	0.26	0.14
NiO	0.02	0.00	0.02	0.00	0.02	0.00
Cl	0.01	0.00	0.01	0.00	0.01	0.00
F	0.24	0.03	0.10	0.01	0.09	0.01
BaO	0.06	0.01	0.08	0.01	0.07	0.01
Total	95.65	0.05	95.45	0.07	95.59	0.07
Based on 22 oxygen						
Si	6.51	0.02	6.54	0.03	6.37	0.04
Ti	0.01	0.00	0.02	0.01	0.01	0.00
Al^{IV}	1.49	0.02	1.46	0.03	1.63	0.04
Al^{VI}	3.42	0.04	3.47	0.05	3.50	0.03
Total Al	4.91	0.05	4.93	0.07	5.14	0.06
Fe	0.27	0.02	0.16	0.03	0.16	0.02
Mn	0.00	0.00	0.00	0.00	0.00	0.00
Mg	0.29	0.03	0.34	0.04	0.37	0.04
Ca	0.00	0.00	0.00	0.00	0.00	0.00
Na	0.04	0.00	0.06	0.02	0.07	0.01
K	1.99	0.02	1.88	0.04	1.89	0.02
OH	3.90	0.01	3.96	0.01	3.97	0.00
Fe + Mg	0.56	0.04	0.49	0.05	0.54	0.05

n = number of analyzed spots. The data is the average of test points.

The muscovite content of w(SiO_2) in the porphyry mineralization is 46.48% to 50.59%, with an average of 48.48%; w(Al_2O_3) content ranges from 26.73% to 34.99%, with an average of 31.08%; w(FeO) content ranges from 0.85% to 5.57%, with an average of 2.4%; w(MgO) content ranges from 0.21% to 3.84%, with an average of 1.44%. The quartz vein mineralization of muscovite has higher w(SiO_2) (47.2% to 52.95%, with an average of 49.15%) and w(MgO) (0.24% to 2.68%, with an average of 1.69%), but lower w(FeO) (0.33% to 3.87%, with an average of 1.41%) than the porphyry mineralization. The breccia mineralization of muscovite has higher w(Al_2O_3) (29.22% to 36.02%, with an average of 32.7%) and w(MgO) (0.63% to 5.9%, with an average of 1.87%), but lower w(SiO_2) (45.21% to 50.66%, with an average of 47.76%) and w(FeO) (0.4% to 4.03%, with an average of 1.46%) than the porphyry mineralization. In general, the content of w(SiO_2) and w(Al_2O_3) in muscovite of Yixingzhai is higher, and the content of w(FeO) and w(MgO) is lower, which has the characteristics of high Si, Al, and low Fe, Mg.

4.2. Short-Wave Infrared Spectroscopy Alteration Mineral Assemblages and Alteration Zoning

4.2.1. Alteration Minerals and Alteration Assemblages

In the Yixingzhai mining area, muscovite group minerals (muscovite, phengite, paragonite, and illite), amphibole group minerals (hornblende and actinolite), chlorite group minerals (FeMg-chlorite, Mg-chlorite, and Fe-chlorite), montmorillonite minerals, kaolinite group minerals (PX-kaolinite and WX-kaolinite), carbonates (calcite and ankerite), epidote, and a small amount of saponite, biotite, and other minerals (Figure 6a) were mainly identified by SWIR. Muscovite group minerals are most widely distributed in the mining area.

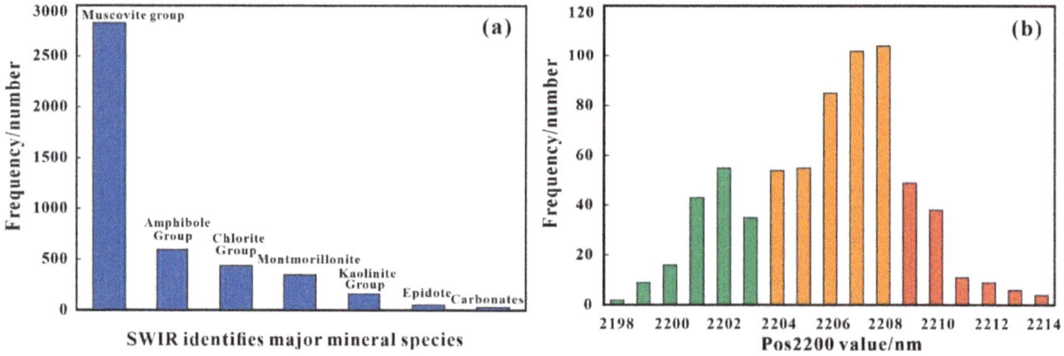

Figure 6. Histogram of the frequency distribution of different mineral species (**a**) and the Pos2200 value of muscovite (**b**) in Yixingzhai.

The value of the Pos2200 wavelength of muscovite will "drift" due to the influence of composition. According to the wavelength of Pos2200 from short to long, it can be divided into paragonite, muscovite, and phengite. When identified as paragonite and rich in Al and Na, the Pos2200 value is less than 2203 nm; when identified as phengite and poor in Al but rich in Si, the Pos2200 value is higher than 2208 nm [49]. The Pos2200 value of muscovite in Yixing ranges from 2198 to 2214 nm, with widely distributed muscovite concentrated between 2206 and 2208 nm, containing a small amount of paragonite (Pos2200 $\leq$ 2203 nm) and phengite (Pos2200 $\geq$ 2209 nm) (Figure 6b). The specific data are shown in Table S1.

Mineral assemblages of different mineralization types are different (Figure 7): In porphyry mineralization, borehole B827ZK201 is taken as an example. The main mineral assemblage consists of muscovite group minerals, accounting for about 90%, with a small amount of other clay minerals. In quartz vein mineralization and breccia mineralization, boreholes 6DZK602 and T510ZK803 are taken as examples, respectively. In these two mineralizations, muscovite group minerals are commonly combined with chlorite, epidote, amphibole, etc. The content of muscovite minerals accounts for about 30%.

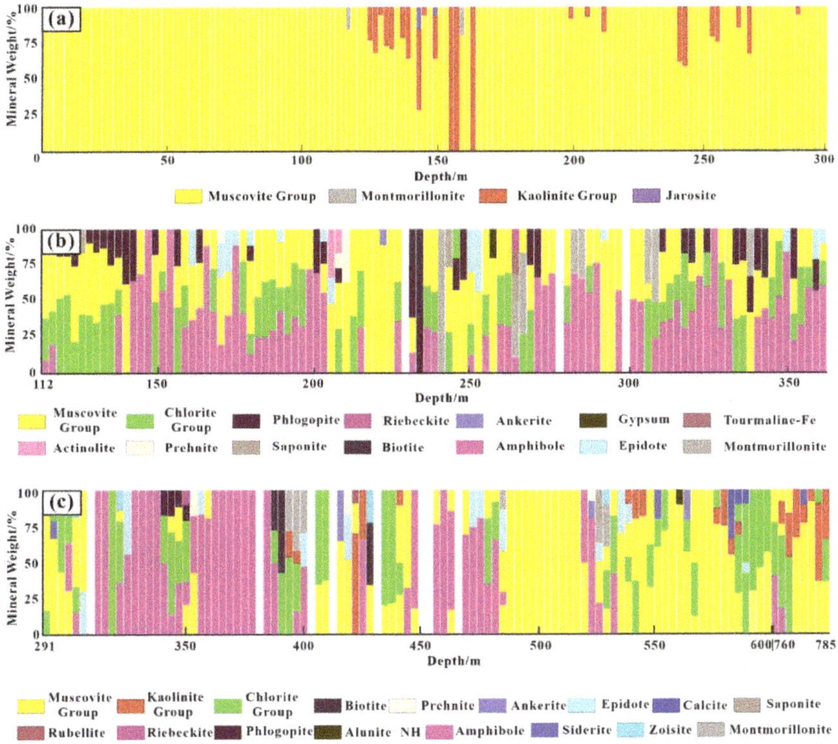

Figure 7. Typical alteration mineral assemblages for different types of mineralization in the boreholes of Yixingzhai deposit. (**a**) B827ZK201, (**b**) 6DZK602, (**c**) T510ZK803. The core is missing at 600–760 m of the borehole, and the data cannot be obtained.

4.2.2. Alteration Zoning

A detailed study of the Hewan porphyry gold ore body was carried out. From the center of the ore body to the outside, it is roughly divided into the hematite zone, kaolinization zone, strong sericitization zone, sericitization zone, and weak sericitization zone, and mineralization is closely related to sericitization (Figure 8). The hematitization and kaolinization zones are located in the center of the porphyry gold deposit. The main altered minerals are kaolinite and muscovite. The main altered minerals in the peripheral sericitization zone are muscovite and illite. The surrounding rock alterations near the quartz vein gold deposit mainly include sericitization, silicification, and chloritization. Chloritization and epidotization are generally developed in the breccia gold deposits, and skarnization occurs in the deep part near the ore body.

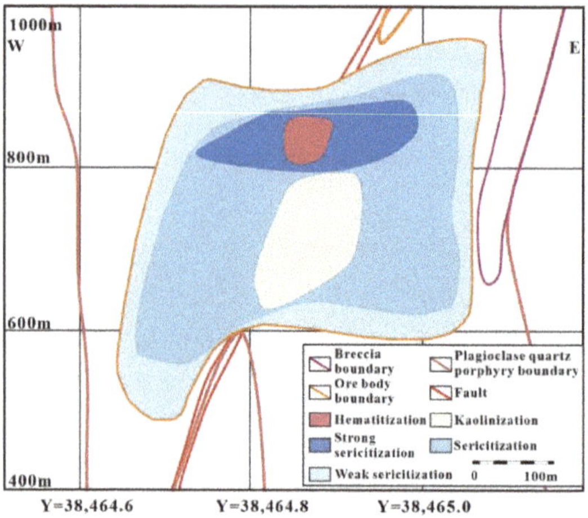

Figure 8. The alteration zoning map of the porphyry gold mineralization located on the borehole profile of the No. 2 exploration line in Yixingzhai. The figure shows the longitudinal section of No. 2 exploration line along the elevation, and the exploration line is an east-west trend.

4.3. Variation Characteristics of Muscovite Mineral Spectral Parameters

The variation characteristics of spectral parameters in porphyry mineralization are as follows: Profile Line No. 1 is far away from the ore body, and the Pos2200 value of muscovite in the porphyry is higher than 2206 nm overall, which is biased toward the LW direction (Figure 9a). The crystallinity changes significantly; the IC value in the breccia pipe is less than 2.5, while the IC value of porphyry ranges from 1.6 to 7.0 and is concentrated around 2.6 to 4.0. Near the center of the porphyry body and at an elevation of approximately 800 m, the degree of crystallinity is relatively high. The IC value of muscovite in the feldspar quartz porphyry gradually increases from deep to shallow (Figure 9b).

The variation characteristics of the spectral parameters of the porphyry body in Profile Line No. 2 are as follows: away from the ore body, the Pos2200 value of muscovite is higher than 2206 nm, which is biased toward the LW direction. The wavelength range within the ore body is 2201 to 2212 nm. The Pos2200 value of muscovite near the center of the ore body is concentrated between 2201 and 2205 nm, which is biased toward the SW direction (Figure 10a). The variation law of crystallinity is as follows: the IC value of muscovite far from the ore body in the porphyry body is low and concentrated between 1.6 and 2.5. The IC values within the ore body range from 1.6 to 8.4, but the IC value near the center of the ore body is concentrated between 2.6 and 4.0. A crystallinity anomaly area appears near the elevation of 800 m (IC > 4.1). The main lithologies of ZK206 and ZK204 are plagioclase quartz porphyry and amphibolite. The crystallinity changes significantly; the IC value of muscovite in plagioclase quartz porphyry ranges from 1.6 to 7.0 and is concentrated between 2.6 and 4.0. The IC value in plagioclase quartz porphyry gradually increases from deep to shallow. The IC value of muscovite developed in amphibolite ranges from 2.6 to 5.5 and is concentrated between 2.6 and 4.0. The IC value of muscovite developed in breccia varies greatly and is concentrated between 1.6 and 4.0 (Figure 10b).

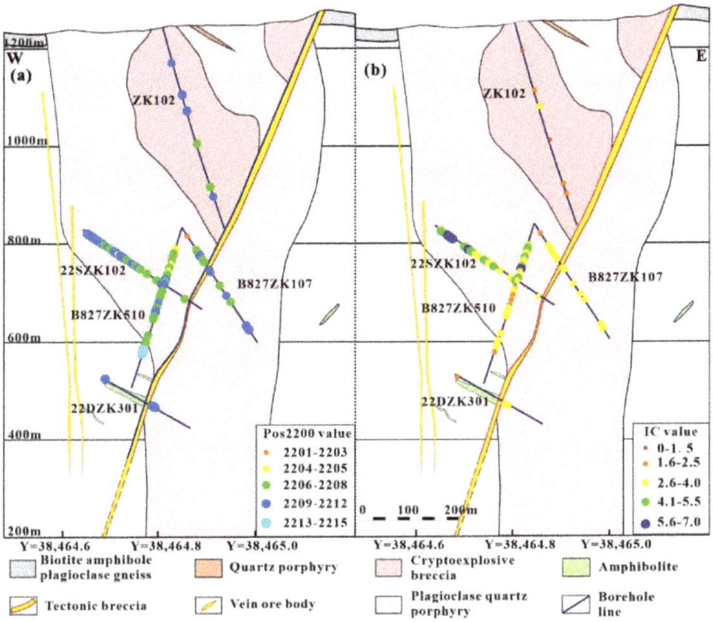

Figure 9. The distribution of muscovite Pos2200 values (**a**) and IC values (**b**) in some porphyry bodies located on the borehole profile of the No. 1 exploration line. The figure shows the longitudinal section of No. 1 exploration line along the elevation, and the exploration line is an east-west trend.

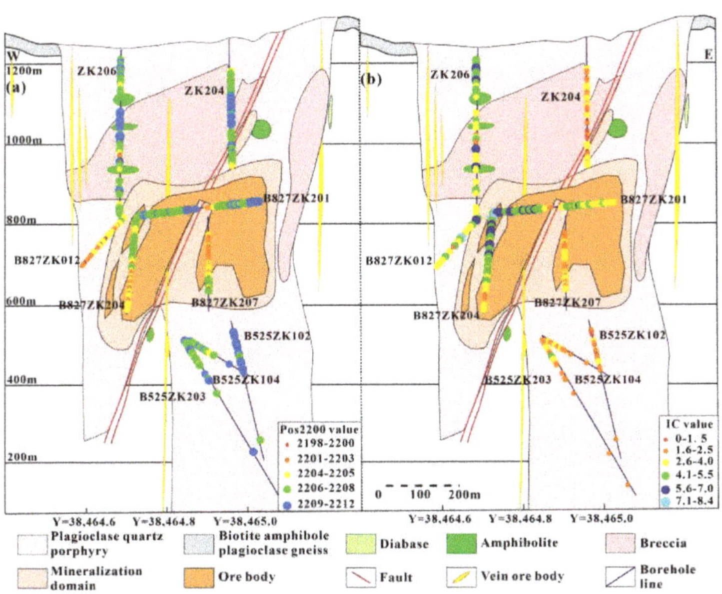

Figure 10. The distribution of muscovite Pos2200 values (**a**) and IC values (**b**) in some porphyry bodies located on the borehole profile of the No. 2 exploration line. The figure shows the longitudinal section of No. 2 exploration line along the elevation, and the exploration line is an east-west trend.

The variation characteristics of spectral parameters of the porphyry body in Profile Line No. 6 are as follows: away from the ore body, the value of muscovite Pos2200 is higher than 2206 nm, which is biased toward the LW direction. However, the wavelength near the ore body is biased toward the SW direction, similar to Profile Line No. 1 and No. 2. The borehole B710ZK602 in the ore body has a higher wavelength near the elevation of 650–700 m, which should be affected by the breccia pipe. Borehole B827ZK605, around 750–850 m, is far from the center of the ore body and has a higher wavelength (Figure 11a). The crystallinity of the porphyry body away from the ore body is concentrated between 1.6 and 2.5. The crystallinity near the ore body ranges from 1.6 to 8.4 and is concentrated between 2.6 and 4.0. The IC value of muscovite in plagioclase quartz porphyry increases gradually from deep to shallow. The IC value of muscovite in breccia, measured in boreholes ZK601 and B710ZK602, is concentrated around 1.5. The crystallinity of muscovite developed in basement gneiss in borehole 22DZK603 is low and less than 1.5 (Figure 11b).

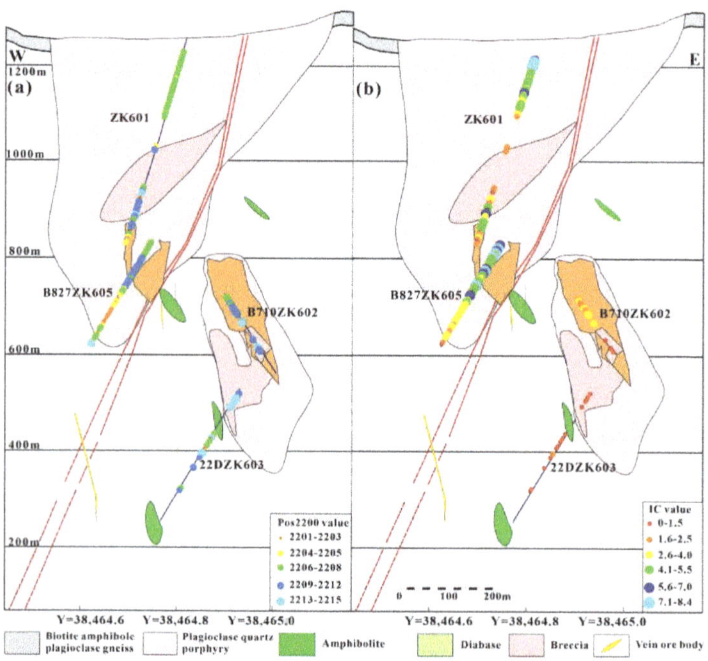

Figure 11. The distribution of muscovite Pos2200 values (**a**) and IC values (**b**) in some porphyry bodies located on the borehole profile of the No. 6 exploration line. The figure shows the longitudinal section of No. 6 exploration line along the elevation, and the exploration line is an east-west trend.

The variation characteristics of spectral parameters in quartz vein mineralization are as follows: Quartz veins of varying sizes are developed within the gneiss of the surrounding rock in Yixingzhai. The muscovite wavelength in the gneiss is generally low, typically less than 2203 nm, and tends toward the SW direction. The muscovite Pos2200 value near the quartz gold veins is higher, indicating phengite, and the wavelength is concentrated around 2211 nm (Figure 12a). The crystallinity of muscovite near the quartz gold veins is low, typically less than 1.5 (Figure 12b).

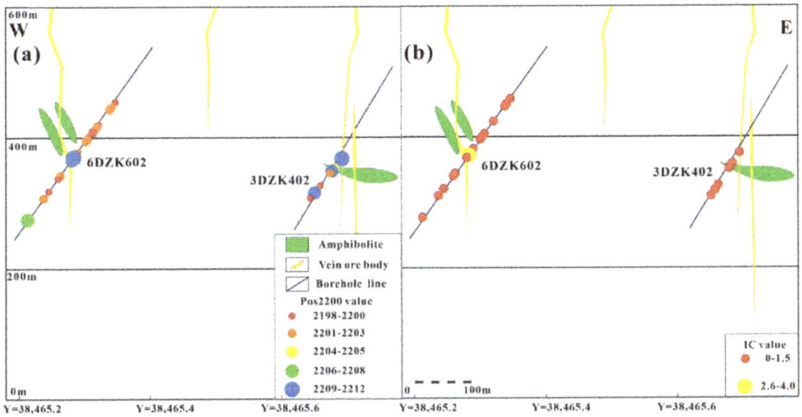

Figure 12. The distribution of muscovite Pos2200 values (**a**) and IC values (**b**) in some quartz veins located on the borehole profile of the No. 6 exploration line. The figure shows the longitudinal section of No. 6 exploration line along the elevation, and the exploration line is an east-west trend.

The variation characteristics of spectral parameters in Breccia Mineralization are as follows: The Tietangdong breccia pipe is the largest one in the Yixingzhai mining area. The SWIR spectral test of muscovite in the breccia pipe on Exploration Line No. 6 shows that the wavelength of muscovite ranges from 2198 to 2214 nm. In gneiss breccia, far away from the ore body, the Pos2200 value of muscovite ranges from 2201 to 2205 nm. The range of muscovite Pos2200 in the ore body is between 2198 and 2208 nm and concentrated in the vicinity of 2201 to 2204 nm, which is biased toward the SW direction (Figure 13a). The crystallinity of muscovite is less than 5.5 overall, and the IC value in gneiss breccia is less than 1.5. The IC value in the ore body is less than 5.5 and concentrated in the vicinity of 1 to 2.2 (Figure 13b).

The study on the variation of spectral parameters in different mineralization types shows that muscovite in porphyry mineralization gradually tends to be SW direction from the periphery to the center of the ore body, and the crystallinity near the center of the ore body gradually increases. Therefore, the Pos2200 value (2201 to 2204 nm) and crystallinity (IC > 2.6) of muscovite can be used as indicators of porphyry mineralization. Quartz vein mineralization is gradually inclined toward the LW direction from the periphery to the center of the ore body, and the crystallinity value is low. The muscovite Pos2200 value (2210 to 2211 nm) and crystallinity (IC < 1.5) can be used as indicators of quartz vein mineralization. In breccia mineralization, the ore body is close to the SW direction. The muscovite Pos2200 value (2201 to 2204 nm) and crystallinity (1 to 2.2) can be used as indicators of breccia mineralization.

The summary reveals that the Pos2200 spectral characteristic of muscovite across the three types of mineralization has similar ranges, but the average value of muscovite Pos2200 in porphyry mineralization (2206 nm) is slightly higher than in the quartz vein mineralization (2202 nm) and breccia mineralization (2203 nm). The average IC value of muscovite in porphyry mineralization (3.3) is significantly higher than in quartz vein mineralization (0.9) and breccia mineralization (1.4). It can be seen that the three mineralization types can be well distinguished by crystallinity; moreover, the maximum IC value of muscovite in porphyry can reach 8.4. The reason for the high crystallinity of muscovite needs further discussion (Figure 14).

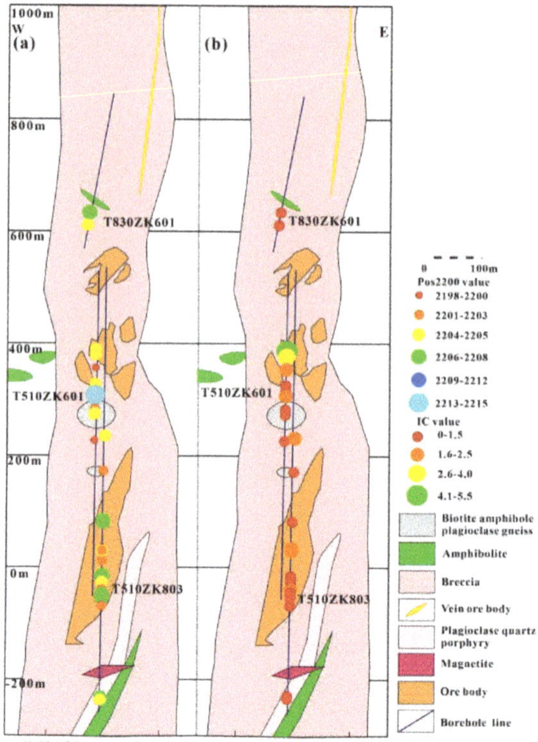

Figure 13. The distribution of muscovite Pos2200 values (**a**) and IC values (**b**) in some breccia pipe located on the borehole profile of the No. 6 exploration line. The figure shows the longitudinal section of No. 6 exploration line along the elevation, and the exploration line is an east-west trend.

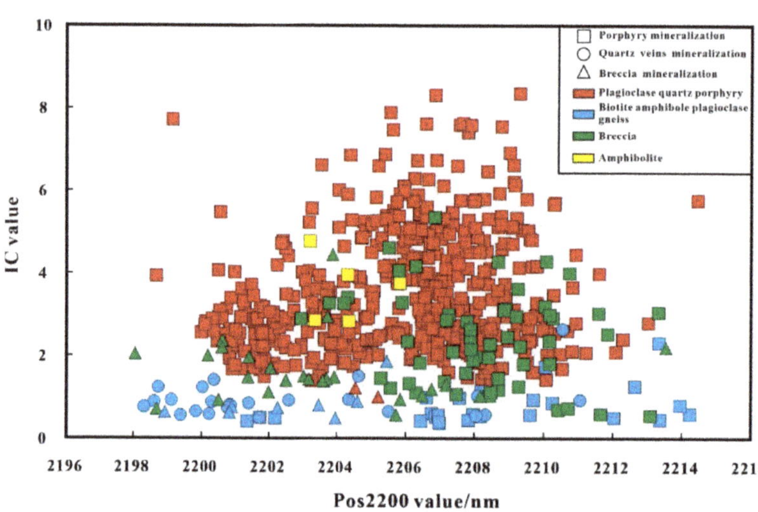

Figure 14. Scatter plot of muscovite Pos2200 values and IC values in different types of mineralization and lithologies at the Yixingzhai.

5. Discussion

5.1. The Influence Factors of Spectral Parameters Change of Muscovite

5.1.1. Components

The Al-OH characteristic absorption peak position of muscovite is affected by hydrothermal fluid temperature, pH, and other factors, mainly affected by Tschermak replacement: $^{VI}Al^{3+}$ + $^{IV}Al^{3+}$ ↔ $n^{IV}Si^{4+}$ + ^{VI} (Fe^{2+}, Mg^{2+}, Mn^{2+}) [50,51]. Due to changes in the content of Si, Al, Fe, and Mg, this results in the drift of the Al-OH characteristic absorption peak position [52,53]. The octahedral sites in muscovite are mainly occupied by Al^{VI}, Fe, and Mg, while the tetrahedral sites are mainly occupied by Al^{IV} and Si [54]. Combined with the analysis of muscovite in the three types of mineralization, it is found that Si has a negative correlation with Al^{IV} at the tetrahedral position, and there is an obvious substitution reaction (Figure 15a). At the octahedral position, Mg, Fe, and Fe + Mg are negatively correlated with Al^{VI}, and there are obvious substitution reactions (Figure 15b–d).

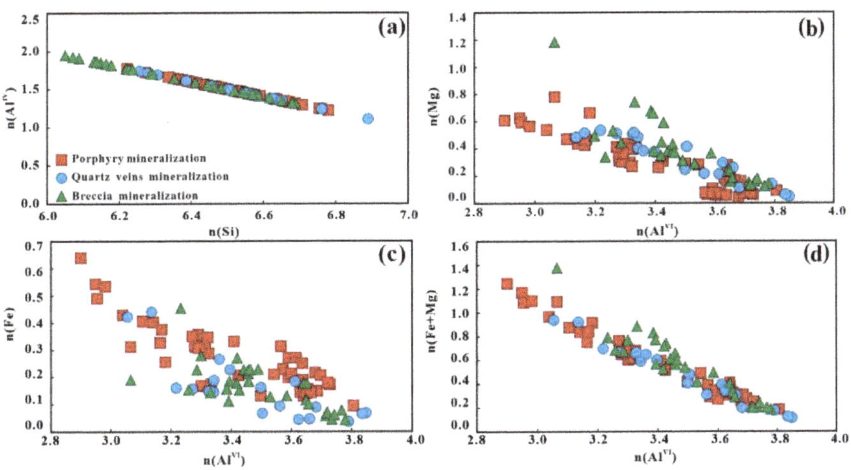

Figure 15. Correlation of n (Al^{IV}), n (Si), n (Mg), and n (Fe) in muscovite. (**a**) n (Si) vs. n (Al^{IV}) diagram, (**b**) n (Al^{VI}) vs. n (Mg) diagram, (**c**) n (Al^{VI}) vs. n (Fe) diagram, (**d**) n (Al^{VI}) vs. n (Fe + Mg) diagram.

Research has shown that the characteristic absorption peak of muscovite Al-OH is mainly related to the Al^{VI} structure in muscovite. The higher the content of Al^{VI}, the lower the Pos2200 value of muscovite and the tendency toward the SW direction. The lower the content of Al^{VI}, the higher the Pos2200 value of muscovite and the tendency toward the LW direction, which shows a negative correlation [19,35,55,56]. When the Pos2200 value of muscovite in the three mineralization types at Yixingzhai is higher, the content of Al^{VI} is lower, and the content of Si, Fe, and Mg is higher, tending toward the LW direction. When the content of Al^{VI} is higher and the content of Si, Fe, and Mg is lower, the wavelength tends toward the SW direction. The wavelength of the characteristic absorption peak of muscovite Al-OH is negatively correlated with Al^{VI} (Figure 16a) and positively correlated with Si, Fe, and Mg (Figure 16b–d). It can be seen that the wavelength of muscovite is affected by the Tschermak replacement mechanism, and the wavelength "drift" phenomenon occurs due to the different content of elemental components [35,39,57].

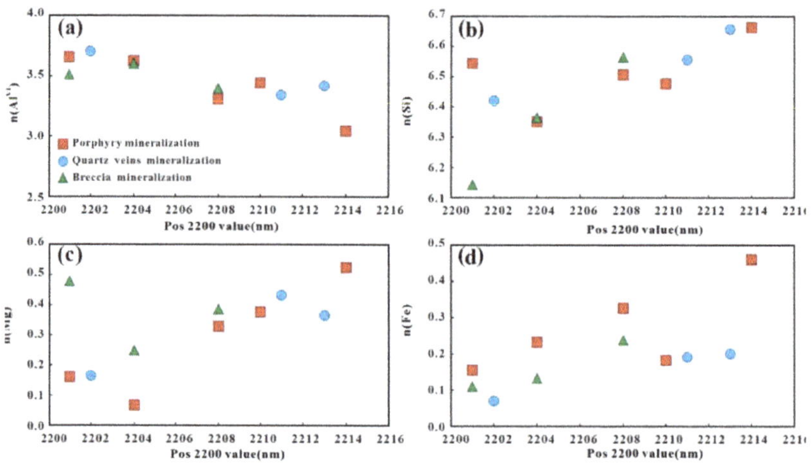

Figure 16. The relationship between muscovite chemical composition content and Pos2200 value. (**a**) Pos2200 (nm) vs. n (Al^{VI}) diagram, (**b**) Pos2200 (nm) vs. n (Si) diagram, (**c**) Pos2200 (nm) vs. n (Mg) diagram, (**d**) Pos2200 (nm) vs. n (Fe) diagram.

5.1.2. Temperature and pH

The wavelength of muscovite is affected by the Tschermak substitution reaction, and the temperature is an important factor in the substitution process. Under high-temperature conditions, muscovite has the most ideal partition model, but as the temperature decreases, the Al and K in the lattice are gradually replaced by Si and some defects [58]. According to previous studies, muscovite with a shorter wavelength is in a relatively high-temperature and high-pressure hydrothermal environment, while muscovite with a longer wavelength is in a relatively low-temperature and low-pressure hydrothermal environment [35,52,59]. Therefore, combined with the geological characteristics of Yixingzhai, previous studies have shown that the age of the porphyry gold deposit is slightly earlier than the quartz vein gold deposit [38]. The ore-forming materials and fluids related to porphyry are mainly derived from the exsolved magmatic-hydrothermal fluid in the deep magmatic chamber homologous to the Hewan porphyry [38]. Combined with spectral data, the temperature of Yixingzhai porphyry mineralization near the center of the ore body is higher, and the characteristic absorption peak wavelength of muscovite Al-OH is biased toward the SW direction. As the magmatic-hydrothermal fluid in the porphyry body moves outward along the fissure, the temperature gradually decreases, and the hydrothermal fluid fills and precipitates to form a quartz vein gold deposit. Therefore, the characteristic absorption peak wavelength of muscovite Al-OH shows the opposite trend compared with porphyry mineralization. With the outward migration of porphyry hydrothermal fluid, the temperature gradually decreases, the increase in water content leads to an increase in the 1900 nm absorption peak, and the decrease in Al content leads to a decrease in the 2200 nm characteristic absorption peak of muscovite, which results in a decrease in the IC value. Therefore, the average IC value of muscovite in porphyry mineralization (3.3) is significantly higher than that in quartz vein mineralization (0.9) and breccia mineralization (1.4). The crystallinity (IC > 2.6) of porphyry mineralization near the center of the ore body is also significantly higher than that of quartz vein mineralization (IC < 1.5) and breccia mineralization (1 to 2.2). Strong sericitization alteration occurs near the 800 m elevation of the porphyry ore body, and high-grade gold mineralization with high crystallinity (IC $\geq$ 4.1) is evenly distributed in the lower part of the hydrothermal breccia aggregate captured by the porphyry [38]. At the same time, combined with electron probe data analysis (Figure 15a), it can be seen that the crystallinity of muscovite is higher when the temperature is higher, corresponding to higher Al content and a lower Al-OH characteristic absorption peak wavelength. As the temperature decreases, crystallinity decreases, corresponding to lower Al content and a higher Al-OH characteristic absorption peak wavelength. Therefore, temperature is an important factor affecting the wavelength and crystallinity of muscovite.

In addition to the influence of temperature, the absorption wavelength of muscovite is also affected by the pH of the hydrothermal fluid. The main lithology of the Yixingzhai porphyry mineralization is plagioclase quartz porphyry, and muscovite is mainly formed by hydrolysis and alteration of feldspar. The formation process of muscovite consumes acid (H^+), and the reaction formula is $NaAlSi_3O_8$-$CaAl_2Si_2O_8$ (plagioclase) + 2HCl + KCl $\rightarrow$ K{Al_2 [$AlSi_3O_{10}$]-$(OH)_2$} (muscovite) + $2SiO_2$ (quartz) + NaCl + $CaCl_2$ [19]. As the reaction progresses, the temperature gradually decreases, so the Al-OH wavelength of muscovite away from the center of porphyry mineralization is relatively biased toward the LW direction. However, SW muscovite and LW muscovite will also convert under certain conditions, with the reaction formula being $2KAl_2$ ($AlSi_3$) O_{10} $(OH)_2$ (muscovite) + K^+ + 1.5 (Fe^{2+} and Mg^{2+}) + $4.5SiO_2$ + $3H_2O$ $\rightleftharpoons$ $3KFe_{0.5}$ ($Al_{0.5}Si_{3.5}$) O_{10} $(OH)_2$ (phengite) + $4H^+$ [50,52]. It can be concluded from the reaction formula that muscovite is formed in a relatively acidic environment, but the formation environment of phengite is relatively neutral [19]. During the migration of Yixingzhai magmatic-hydrothermal fluid from the center to the periphery, the infiltration of the fluid will be quickly neutralized by the surrounding rock, indicating that the acidic environment will be transformed into a neutral environment. Therefore, SW muscovite is mainly formed near the mineralization center, while LW phengite is mainly formed far away from the mineralization center.

5.2. Comparison with Other Deposits

The characteristics of short-wave infrared spectroscopy of muscovite group minerals have been widely used in the prospecting and exploration of porphyry deposits [15,16,18,29,55,60]. Combined with the analysis of prospecting indicators and influencing factors of spectral characteristic parameters of typical porphyry copper deposits (Table 2), the Al-OH characteristic absorption peaks of muscovite in porphyry copper deposits, such as Yulong and Pulang, are biased toward the LW direction, which can be used as a sign to indicate the mineralization center [61–63]. However, Niancun, Highland Valley, and Zhunuo porphyry copper deposits show the opposite trend [18,57,64]. The difference in Al-OH characteristic absorption peaks of muscovite is caused by the influence of chemical composition and pH in different deposit types. However, according to the analysis of the crystallinity of important spectral characteristic parameters, the maximum crystallinity of muscovite minerals in porphyry copper deposits is about 3, and the crystallinity of 1.6 to 3 is generally used as a sign of indicating the mineralization center [57,61,64]. The crystallinity of epithermal or tectonically altered rock gold deposits is generally low [65–67]. The high crystallinity of the Gangcha-Kemo gold deposit may be affected by the deep porphyry metallogenic system [27]. The crystallinity of other deposits is similar to the Yixingzhai hydrothermal veins gold deposit but lower than the porphyry gold deposit. Comprehensive analysis of the high crystallinity (4.1 to 8.4) of muscovite in the Yixingzhai porphyry gold deposit can be used as an indicator of gold mineralization. In addition, as the first discovery of muscovite in porphyry gold deposit, it has higher crystallinity than porphyry copper deposits, epithermal, and tectonically altered rock gold deposits, which provides an important basis for finding similar types of gold deposits.

Table 2. Statistical table of SWIR spectroscopy and influencing factors of muscovite group minerals in different deposits.

References	Types and Names of Deposits	Spectral Parameter Range	Exploration Indicators	The Main Factors Affecting the Spectral Parameters
[18]	Niancun porphyry copper deposit, Tibet	2192–2220 nm; 0.6–3.1	Pos2200 < 2203 nm; IC > 1.6	Temperature affects wavelength and crystallinity
[64]	Highland Valley porphyry copper deposit	-	The average of Pos2200 is 2195 nm	Chemical composition
[57]	Porphyry copper deposit in Zhunuo mining area	2191–2219 nm; 0.75–3.14	Pos2200 < 2203 nm; IC > 2.0	Tschermak substitution affects the wavelength
[61]	Yulong porphyry copper deposit	2205–2209 nm; 0–3	Pos2200: 2206–2207 nm; IC: 1.0–2.0	PH; temperature
[60]	Jiama porphyry copper deposit	-	Pos2200: 2200–2210 nm	The temperature decreases, and the aluminum content increases
[62]	Zijinshan deep porphyry copper deposit	2195–2211 nm	Pos2200: 2205–2211 nm	Tschermak substitution affects the wavelength
[63]	Pulang porphyry copper deposit, Yunnan	2198–2210 nm	Pos2200: 2205–2210 nm	Fluid properties; temperature
[27]	Gangcha-Kemo epithermal gold deposit	0–6.5 nearby	IC > 5.5	Temperature affects crystallinity
[65]	Epithermal gold deposit in Xiaotian-Mozitan Basin	2204–2220 nm; 0–3.03	IC: 1–3 nearby	Temperature affects crystallinity
[66]	Shihu altered rock gold deposit	2194–2216 nm; 0–2.4	The average of Pos2200 is 2198 nm	PH affects the wavelength, and temperature affects the crystallinity
[67]	Shaling altered rock gold deposit	2198–2220 nm; 0–2.7 nearby	Pos2200: 2207–2211 nm; IC: 1.19–2.66	Temperature and surrounding rock composition affect the wavelength; temperature affects crystallinity

6. Conclusions

(1) In the porphyry and breccia mineralization, muscovite near the mineralization center is inclined to the SW direction, while in quartz vein mineralization, it is inclined to the LW direction. The crystallinity of porphyry mineralization is significantly higher than that of quartz vein and breccia mineralization, and the crystallinity near the mineralization center gradually increases.

(2) The SWIR spectrum exploration identification of porphyry gold deposits was established for the first time. The spectral parameters of muscovite minerals in porphyry mineralization (Pos2200 2201–2204 nm and IC 1–2.2), breccia mineralization (Pos2200 2210–2211 nm and IC < 1.5), and quartz vein mineralization (Pos2200 2201–2204 nm and IC 1–2.2) can be used as indicators of mineralization centers.

(3) The wavelength and crystallinity of Yixingzhai muscovite are mainly affected by the Tschermak substitution and temperature. When the content of Si, Fe, and Mg is low and the content of Al^{VI} is high, the wavelength is biased toward the SW direction. Conversely, when the content of Al^{VI} is low and the content of Si, Fe, and Mg is high, the wavelength tends to the LW direction. Crystallinity is mainly affected by temperature; the higher the temperature, the higher the crystallinity.

(4) It is found for the first time that muscovite in porphyry gold deposits has higher crystallinity (4.1–8.4) than that of porphyry copper deposits, epithermal gold deposits, and tectonic altered gold deposits, which provides a new basis for the prospecting and exploration of similar porphyry gold deposits.
(5) SWIR spectroscopy technology can identify the altered minerals of porphyry deposits quickly and determine the exploration identification effectively, which is an effective means for future prospecting and exploration.

Supplementary Materials: The following supporting information can be downloaded at: https://www.mdpi.com/article/10.3390/min15010083/s1, Table S1: The original data of wavelength and crystallinity of muscovite from Yixingzhai. Table S2: The original data of EPMA.

Author Contributions: Conceptualization, L.W. and S.W.; methodology, L.W. and S.W.; software, L.W. and S.W.; validation, L.W. and S.W.; formal analysis, L.W.; investigation, L.W., P.L., Y.Y. and Y.R.; resources, S.W., X.L., W.Y. and R.S.; data curation, L.W. and S.W.; writing—original draft preparation, L.W. and S.W.; writing—review and editing, L.W. and S.W.; visualization, L.W.; supervision, S.W.; project administration, S.W.; funding acquisition, S.W. All authors have read and agreed to the published version of the manuscript.

Funding: This research was funded by the National Key Research and Development Program of China (2022YFC2903601).

Data Availability Statement: Data are contained within the article and Supplementary Materials.

Acknowledgments: This study owes gratitude to the three reviewers for their valuable comments and the rock core warehouse management staff of Shanxi Zijin Mining Co., Ltd., for their assistance in transporting the rock cores and spectral testing. Here, we have gratitude for everyone who has helped with this article.

Conflicts of Interest: Xiaodan Lai and Weili Yang are employees of Zijin Mining Group Co., Ltd., Fujian. The paper reflects the views of the scientists and not the company. Rongliang Sun is an employee of Shanxi Zijin Mining Co., Ltd. The paper reflects the views of the scientists and not the company.

References

1. Laukamp, C.; Legras, M.; Montenegro, V.; Windle, S.; Mcfarlane, A. Grandite-based resource characterization of the skarn-hosted Cu-Zn-Mo deposit of Antamina, Peru. *Miner. Depos.* **2022**, *57*, 107–128. [CrossRef]
2. Neal, L.C.; Wilkinson, J.J.; Mason, P.J.; Chang, Z. Spectral characteristics of propylitic alteration minerals as a vectoring tool for porphyry copper deposits. *J. Geochem. Explor.* **2018**, *184*, 179–198. [CrossRef]
3. Mathieu, M.; Roy, R.; Launeau, P.; Cathelineau, M.; Quirt, D. Alteration mapping on drill cores using a HySpex SWIR-320m hyperspectral camera: Application to the exploration of an unconformity-related uranium deposit (Saskatchewan, Canada). *J. Geochem. Explor.* **2017**, *172*, 71–88. [CrossRef]
4. Carrino, T.A.; Crósta, A.P.; Toledo, C.L.B.; Silva, A.M. Unveiling the hydrothermal mineralogy of the Chapi Chiara gold prospect, Peru, through reflectance spectroscopy, geochemical and petrographic data. *Ore Geol. Rev.* **2015**, *64*, 299–315. [CrossRef]
5. Hao, J.Y.; Duan, L.A.; Zhang, Y.; Zhao, H.T.; Shao, Y.J.; Guo, Y.C.; Wang, X.; Song, S.L. Machine learning on white mica short-wave infrared (SWIR) spectral data in the Tengjia Au deposit, Jiaodong peninsula (Eastern China): A prospecting indicator for lode gold deposits. *Ore Geol. Rev.* **2024**, *173*, 106230. [CrossRef]
6. Ducart, D.F.; Crosta, A.P.; Souza Filho, C.R.; Coniglio, J. Alteration mineralogy at the Cerro La Mina epithermal prospect, Patagonia, Argentina; field mapping, short-wave infrared spectroscopy, and ASTER images. *Econ. Geol. Bull. Soc. Econ. Geol.* **2006**, *101*, 981–996. [CrossRef]
7. Thompson, A.J.B.; Hauff, P.L.; Robitaille, A.J. *Alteration Mapping in Exploration: Application of Short-Wave Infrared (SWIR) Spectroscopy*; SEG Discovery: Littleton, CO, USA, 1999; pp. 1–27.
8. Swayze, G.A.; Clark, R.N. Infrared-Spectra And Crystal-Chemistry of Scapolites—Implications For Martian Mineralogy. *J. Geophys. Res. Solid Earth* **1990**, *95*, 14481–14495. [CrossRef]
9. Hunt, G.R. Spectral signatures of particulate minerals in the visible and near infrared. *Geophysics* **1977**, *42*, 501–513. [CrossRef]

10. Hunt, G.R.; Ashley, R.P. Spectra of altered rocks in the visible and near infrared. *Econ. Geol. Bull. Soc. Econ. Geol.* **1979**, *74*, 1613–1629. [CrossRef]
11. Hunt, G.R.; Wynn, J.C. Visible and near-infrared spectra of rocks from chromium-rich areas. *Geophysics* **1979**, *44*, 820. [CrossRef]
12. Gaffey, S.J. Spectral Reflectance of Carbonate Minerals in the Visible and Near-Infrared (0.35–2.55 Microns)—Calcite, Aragonite, and Dolomite. *Am. Miner.* **1986**, *71*, 151–162.
13. Leng, C.B.; Wang, D.Z.; Yu, H.J.; Tian, F.; Zhang, X.C. Mapping hydrothermal alteration zones with short wavelength infrared (SWIR) spectra and magnetic susceptibility at the Pulang porphyry Cu-Au deposit, Yunnan, SW China. *Miner. Depos.* **2024**, *59*, 699–716. [CrossRef]
14. Mahdieh, H.Z.; Tangestani, M.H.; Francisco, V.R.; Yusta, I. Spectral characteristics of minerals in alteration zones associated with porphyry copper deposits in the middle part of Kerman copper belt, SE Iran. *Ore Geol. Rev.* **2014**, *62*, 191–198.
15. Ren, H.; Zheng, Y.Y.; Wu, S.; Zhang, X.; Ye, J.W.; Chen, X.D. Short-wavelength infrared characteristics and in-dications of exploration of the Demingding copper-molybdenum deposit in Tibet. *J. Earth Sci.* **2020**, *45*, 930–944.
16. Tian, F.; Leng, C.B.; Zhang, X.C.; Tian, Z.D.; Zhang, W.; Guo, J.H. Application of short-wave infrared spectroscopy in Gangjiang porphyry Cu-Mo deposit in Nimu ore field, Tibet. *J. Earth Sci.* **2019**, *44*, 2143–2154.
17. Wang, R.; Cudahy, T.; Laukamp, C.; Walshe, J.L.; Bath, A.; Mei, Y.; Young, C.; Roache, T.J.; Jenkins, A.; Roberts, M.; et al. White Mica as a Hyperspectral Tool in Exploration for the Sunrise Dam and Kanowna Belle Gold Deposits, Western Australia. *Econ. Geol.* **2017**, *112*, 1153–1176. [CrossRef]
18. Yang, Z.M.; Hou, Z.Q.; Yang, Z.S.; Qu, H.C.; Li, Z.Q.; Liu, Y.F. Application of short wavelength infrared (SWIR) technique in exploration of poorly eroded porphyry Cu district: A case study of Niancun ore district Tibet. *Miner. Depos.* **2012**, *31*, 699–717.
19. Zhang, Z.X.; Dai, J.J.; Wang, X.G.; Hu, Z.H.; Wan, X.; Peng, B.; Fu, M.H.; Zhao, L.X. Shortwave infrared characteristics of muscovite from giant Shimensi tungsten deposit in northern Jiangxi Province and implication to exploration. *Miner. Depos.* **2023**, *42*, 116–127.
20. Camilo, U.; Maher, K. White Mica Geochemistry of the Copper Cliff Porphyry Cu Deposit: Insights from a Vectoring Tool Applied to Exploration. *Econ. Geol.* **2018**, *113*, 1269–1295.
21. Thimsen, E.; Sadtler, B.; Berezin, M.Y. Shortwave-infrared (SWIR) emitters for biological imaging: A review of challenges and opportunities. *Nanophotonics* **2017**, *6*, 1043–1054. [CrossRef]
22. Laakso, K.; Peter, J.M.; Rivard, B.; White, H.P. Short-wave infrared spectral and geochemical characteristics of hydrothermal alteration at the Archean Izok Lake Zn-Cu-Pb-Ag volcanogenic massive sulfide deposit, Nunavut, Canada; application in exploration target vectoring. *Econ. Geol. Bull. Soc. Econ. Geol.* **2016**, *111*, 1223–1239. [CrossRef]
23. Shan, S.Q.; Xie, G.Q.; Liu, W.Y.; Zheng, J.H.; Xing, B. Hydrothermal alteration and sulfide zoning characteristics of the Čukaru peki super-large porphyry-epithermal copper-gold deposit in western Tethys, serbia and its implication for exploration. *Geotecton. Metallog.* **2023**, *47*, 1085–1109.
24. Yan, Y.F.; Tan, J.; Li, Y.H.; Ruan, S.K. Geological characteristics of epithermal gold deposits and status quo of researches in China. *Resour. Environ. Eng.* **2007**, *21*, 7–11.
25. Zhu, X.F.; Chen, Y.J.; Wang, P.; Zhang, C.; Cai, Y.L.; Deng, K.; Xu, Q.W.; Li, K.Y. Zircon U-Pb age, geochemistry and Hf isotopes of the causative porphyry from the Bilihe porphyry gold deposit, Inner Mongolia. *Front. Earth Sci.* **2018**, *25*, 119–134.
26. Wang, Q.; Tang, J.X.; Chen, Y.C.; Hou, J.F.; Li, Y.B. The metallogenic model and prospecting direction for the Duolong super large copper (gold) district, Tibet. *Acta Petrol. Sin.* **2019**, *35*, 879–896. [CrossRef] [PubMed]
27. Wang, S.H.; Shen, J.F.; Peng, Z.D.; Xu, L.W.; Niu, G.; Liu, H.M.; Li, J.C.; Du, B.S.; Liu, J.J. Altered Rock characteristics and gold prospecting of Gangcha-Kemo area in Gansu province. *Geoscience* **2021**, *35*, 579–588.
28. Xu, C.; Chen, H.Y.; White, N.; Qi, J.P.; Zhang, L.J.; Zhang, S.; Duan, G. Alteration and mineralization of Xinan Cu-Mo ore deposit in Zijinshan orefield, Fujian Province, and application of short wavelength infra-red technology (SWIR) to exploration. *Min. Depos.* **2017**, *36*, 1013–1038.
29. Harraden, C.L.; Mcnulty, B.A.; Gregory, M.J.; Lang, J.R. Shortwave infrared spectral analysis of hydrothermal alteration associated with the Pebble porphyry copper-gold-molybdenum deposit, Iliamna, Alaska. *Econ. Geol. Bull. Soc. Econ. Geol.* **2013**, *108*, 483–494. [CrossRef]
30. Tang, N.; Lin, B.; Li, Y.B.; Wang, Y.Y.; Li, J.J. Application of short wavelength infrared spectroscopy in porphyry-epithermal system: A case study of Tiegelongnan super-large copper (gold) deposit, Tibet. *Acta Geol. Sin.* **2021**, *95*, 2613–2627.
31. Wang, L.; Percival, J.B.; Hedenquist, J.W.; Hattori, K.; Qin, K. Alteration Mineralogy of the Zhengguang Epithermal Au-Zn Deposit, Northeast China: Interpretation of Shortwave Infrared Analyses During Mineral Exploration and Assessment. *Econ. Geol.* **2021**, *116*, 389–406. [CrossRef]

32. Shao, X.W.; Peng, Y.M.; Wang, G.W.; Zhao, X.Y.; Tang, J.Y.; Huang, L.L.; Liu, X.N.; Zhao, X.D. Application of SWIR, XRF and thermoelectricity analysis of pyrite in deep prospecting in the Xincheng gold orefield, Jiaodong Peninsula. *Front. Earth Sci.* **2021**, *28*, 236–251.
33. Doublier, M.P.; Roache, T.; Potel, S. Short-wavelength infrared spectroscopy; a new petrological tool in low-grade to very low-grade pelites. *Geology* **2015**, *38*, 1031–1034. [CrossRef]
34. Yang, K.; Browne, P.R.L.; Huntington, J.F.; Walshe, J.L. Characterising the hydrothermal alteration of the Broadlands-Ohaaki geothermal system, New Zealand, using short-wave infrared spectroscopy. *J. Volcanol. Geotherm. Res.* **2001**, *106*, 53–65. [CrossRef]
35. Jones, S.; Herrmann, W.; Gemmell, J.B. Short Wavelength Infrared Spectral Characteristics of the HW Horizon: Implications for Exploration in the Myra Falls Volcanic-Hosted Massive Sulfide Camp, Vancouver Island, British Columbia, Canada. *Econ. Geol.* **2005**, *2*, 273–294. [CrossRef]
36. Zhang, L.Z. Element geochemistry and age of the altered-porphyry gold mineralization in Yixingzhai gold de-posit, Fanshi city, Shanxi province. *Geol. Miner. Resour. South China* **2018**, *34*, 134–141.
37. Bai, X.M. Geological characteristics, geochronology, and prospecting direction of the Tietangdong cryptoexplosive breccia pipe in the Yixingzhai Gold Deposit, Shanxi Province. *Gold* **2022**, *43*, 9–18.
38. Liu, X.G. Metallogenic characteristies and genesis analysis of Yixingzhai gold deposit in the Fanshi County of Shanxi, China. *Miner. Resour. Geol.* **2022**, *36*, 299–309.
39. Zhang, J.Q.; Li, S.R.; Santosh, M.; Luo, J.Y.; Li, C.L.; Song, J.Y.; Lu, J.; Liang, X. The genesis and gold mineralization of the crypto-explosive breccia pipe in the Yixingzhai gold region, central North China Craton. *Geol. J.* **2020**, *55*, 5664–5680. [CrossRef]
40. Li, S.R.; Santosh, M.; Zhang, H.F.; Luo, J.Y.; Zhang, J.Q.; Li, C.L.; Song, J.Y.; Zhang, X.B. Metallogeny in response to lithospheric thinning and craton destruction: Geochemistry and U–Pb zircon chronology of the Yixingzhai gold deposit, central North China Craton. *Ore Geol. Rev.* **2014**, *56*, 457–471. [CrossRef]
41. Zhang, L.Z.; Gao, W.S.; Deng, X.D. Geology and Geochronology of Magmatic–Hydrothermal Breccia Pipes in the Yixingzhai Gold Deposit: Implications for Ore Genesis and Regional Exploration. *Minerals* **2024**, *14*, 496. [CrossRef]
42. Sun, R.L. Mineralization temperature calculated from tetrahedrite thermodynamic: Implication for vein-type gold mineralization in the Yixingzhai gold deposit, Shanxi. *South China Geol.* **2023**, *39*, 581–587.
43. Ye, R.; Tu, G.C.; Zhao, L.S.; Shen, Y.L.; Luo, L. Modes of occurrence of Au-mineral particles in Yixingzhai gold deposit and their metallogenic dynamic significance. *Miner. Depos.* **2002**, *21*, 278–282.
44. Zhang, J.Q.; Li, S.R.; Santosh, M.; Niu, S.D.; Li, Q.; Lu, J. The magmatic–hydrothermal mineralization systems of the Yixingzhai and Xinzhuang gold deposits in the central North China Craton. *Ore Geol. Rev.* **2017**, *88*, 416–435. [CrossRef]
45. Li, C.L.; Li, S.R.; Luo, J.Y.; Zhang, J.Q.; Song, J.Y. Characteristies of gold-bearing minerals in the Yixingzhai gold deposit, Fanshi County, Shanxi Province. *Geol. China* **2011**, *38*, 119–128.
46. Zhang, L.Z.; Chen, L.; Wang, G.P.; Deng, X.D.; Li, J.W. Garnet U-Pb dating constraints on the timing of breccia pipes formation and genesis of gold mineralization in Yixingzhai gold deposit, Shanxi province. *J. Earth Sci.* **2020**, *45*, 108–117.
47. Zhang, J.Q.; Li, S.R.; Santosh, M.; Li, Q.; Niu, S.D.; Li, Z.D.; Zhang, X.G.; Jia, L.B. Timing and origin of Mesozoic magmatism and metallogeny in the Wutai-Hengshan region; implications for destruction of the North China Craton. *J. Asian Earth Sci.* **2015**, *113*, 677–694. [CrossRef]
48. Ross, P.S.; Bourke, A.; Schnitzler, N.; Conly, A. Exploration Vectors from near Infrared Spectrometry near the McLeod Volcanogenic Massive Sulfide Deposit, Matagami District, Québec. *Econ. Geol.* **2019**, *114*, 613–638. [CrossRef]
49. Guo, N.; Huang, Y.R.; Zheng, L.; Tang, N.; Fu, Y.; Wang, C. Alteration zoning and prospecting model of epithermal deposit revealed by shortwave infrared technique: A case study of Tiegelongnan and Sinongduo Deposits. *Acta Geosci. Sin.* **2017**, *38*, 767–778.
50. Halley, S.; Dilles, J.H.; Tosdal, R.M. *Footprints: Hydrothermal Alteration and Geochemical Dispersion Around Porphyry Copper Deposits*; SEG Discovery: Littleton, CO, USA, 2015; pp. 1–17.
51. Ernst, W.G. Significance of phengitic micas from low-grade schists. *Am. Miner.* **1963**, *48*, 1357–1373.
52. Xu, Q.J.; Ye, F.W.; Zhang, Z.X.; Zhang, C.; Li, X.C.; Li, R.W.; Tian, C. Short-wavelength infrared spectroscopy and EPMA of white mica from Baiyanghe uranium district in Xinjiang. *World Nucl. Geosci.* **2024**, *41*, 47–62.
53. Hunt, G.R. Near-infrared (1.3–2.4 mm) spectra of alteration minerals—Potential for use in remote sensing. *Geophysics* **1979**, *44*, 1974–1986. [CrossRef]
54. Tappert, M.C.; Rivard, B.; Giles, D.; Tappert, R.; Mauger, A. The mineral chemistry, near-infrared, and mid-infrared reflectance spectroscopy of phengite from the Olympic Dam IOCG deposit, South Australia. *Ore Geol. Rev.* **2013**, *53*, 26–38. [CrossRef]
55. Herrmann, W.; Blake, M.; Doyle, M.; Huston, D.; Kamprad, J.; Merry, N.; Pontual, S. Short Wavelength Infrared (SWIR) Spectral Analysis of Hydrothermal Alteration Zones Associated with Base Metal Sulfide Deposits at Rosebery and Western Tharsis, Tasmania, and Highway-Reward, Queensland. *Econ. Geol.* **2001**, *96*, 939–955. [CrossRef]

56. Duke, E.F. Near infrared spectra of muscovite, Tschermak substitution, and metamorphic reaction progress; implications for remote sensing. *Geology* **1994**, *22*, 621–624. [CrossRef]
57. Zheng, S.L.; Wu, S.; Zheng, Y.Y.; Chen, L.; Wei, C.; Huang, P.C.; Jing, X.K.; Yu, Z.Y.; He, C.T.; Ren, H. Identifying potential porphyry copper mineralization at the Zhu'nuo ore-cluster district in western Gangdese, southern Tibet: Insights from shortwave infrared (SWIR) spectrometry and geochemical anomalies. *Ore Geol. Rev.* **2022**, *151*, 105202. [CrossRef]
58. He, W.; Lin, B.; Song, Y.X.; Fang, X.; Wang, Q.; Yang, H.H.; Li, F.Q. Mineralogical characteristics and implications of muscovite from the Tiegelongnan super-large copper (gold) deposit in Tibet. *Acta Geol. Sin.* **2023**, *97*, 1938–1955.
59. Zhang, S.T.; Chu, G.B.; Cheng, J.M.; Zhang, Y.; Tian, J.; Li, J.P.; Sun, S.Q.; Wei, K.T. Short wavelength infrared (SWIR) spectroscopy of phyllosilicate minerals from the Tonglushan Cu-Au-Fe deposit, Eastern China: New exploration indicators for concealed skarn orebodies. *Ore Geol. Rev.* **2020**, *122*, 103516. [CrossRef]
60. Guo, N.; Guo, K.; Zhang, T.T.; Liu, T.H.; Hu, B.; Wang, C.W. Hydrothermal alteration distribution model of the Jiama (Gyama) copper-polymetallic deposit based on shortwave technique. *Acta Geosci. Sin.* **2012**, *33*, 641–653.
61. Tian, C.H.; Yang, L.Q.; He, W.Y.; Zhang, S.Y.; Liu, S.T.; Wu, C. Characteristics of short wave infrared spectroscopy of sericite group minerals and their implications for exploration in the Yulong porphyry copper deposit, Tibet. *Sediment. Geol. Tethyan Geol.* **2022**, *42*, 40–49.
62. Qiu, J.H.; Liu, W.Y.; Chen, J.W.; Lai, X.D.; Zhong, X.H.; Li, J.Y.; Long, H. Alteration mineralogy, characteristics and shortwave infrared spectroscopy of white mica in the Zijinshan ore field, Fujian Province: Implications for porphyry Cu prospecting. *Ore Geol. Rev.* **2024**, *168*, 106065. [CrossRef]
63. Guo, D.X.; Liu, X.; Zhang, H.L.; Zhang, Z.G. The infrared spectroscopy characteristics of alteration and mineralization in the porphyry copper deposit in pulang, Yunnan province. *Rock Miner. Anal.* **2021**, *40*, 698–709.
64. Lypaczewski, P.; Rivard, B.; Lesage, G.; Byrne, K.; D'Angelo, M.; Lee, R.G. Characterization of Mineralogy in the Highland Valley Porphyry Cu District Using Hyperspectral Imaging, and Potential Applications. *Minerals* **2020**, *10*, 473. [CrossRef]
65. Li, X.C.; Zhang, H.D.; Hua, B.; Zhang, Z.; Lu, L. A study on metallogenic indication of epithermal deposit based on short wave infrared spectroscopy. *Geol. Anhui* **2023**, *33*, 133–138.
66. Yao, Y.Z.; Zhu, Y.C.; Liu, J.C.; Li, W.Q. Footprints of ore fluid pathway and implications to mineral exploration in the Shihu Gold Deposit, North China: Evidence from short wave infrared spectroscopy of illitic alteration rocks. *J. Geochem. Explor.* **2021**, *229*, 106833. [CrossRef]
67. An, B.G. Study on the Characteristics of Short-Wave Infrared Spectrum in the Alteration Zone of Shaling Gold Deposit, Shandong. Master's Thesis, Northeastern University, Shenyang, China, 2020.

minerals

MDPI

Article

Geology, Petrology and Geochronology of the Late Cretaceous Klaza Epithermal Deposit: A Window into the Petrogenesis of an Emerging Porphyry Belt in the Dawson Range, Yukon, Canada

Well-Shen Lee [1,*], Daniel J. Kontak [1], Patrick J. Sack [2,*], James L. Crowley [3] and Robert A. Creaser [4]

1 Harquail School of Earth Sciences, Sudbury, ON P3E 2E3, Canada
2 Yukon Geological Survey, 91807 Alaska Hwy, Whitehorse, YT Y1A 0R3, Canada
3 Department of Geosciences, Boise State University, 1295 University Drive, Boise, ID 83706, USA
4 Department of Earth and Atmospheric Sciences, University of Alberta, Edmonton, AB T6G 2E3, Canada
* Correspondence: wellshenlee@gmail.com (W.-S.L.); patrick.sack@yukon.ca (P.J.S.)

Abstract: Geologic understanding of the richly mineralized Dawson Range gold belt (DRGB) in the central Yukon, Canada is hindered by: (1) limited outcrop exposure due to thick soil cover; and (2) low resolution age-constraints despite a long history of porphyry Cu–Au–Mo deposit (PCD) exploration. Here, the well-preserved Klaza Au–Ag–Pb–Zn porphyry–epithermal deposit is used as a type-example of Late Cretaceous magmatic–hydrothermal mineralization to address the complex metallogeny of the DRGB. U–Pb zircon dating defines four magmatic pulses of Late Triassic to Late Cretaceous ages with the latter consisting of the Casino (80–72 Ma) and Prospector Mt. (72–65 Ma) suites. The Casino suite has five phases of intermediate-to-felsic calc-alkaline composition, correspond with older (77 Ma) porphyry mineralization, and displays evidence of magma mingling. The intermediate-to-mafic, slightly alkalic Prospector Mt. suite shows evidence of mingling with the youngest Casino suite phases, correlates with younger (71 Ma), intermediate-sulfidation epithermal and porphyry-type mineralization, and shoshonitic basalts of the Carmacks Group. Zircon trace element data suggest a common melt source for these suites; however, the younger suite records features (e.g., high La/Yb) that indicate a higher pressure melt source. The results from this study highlight the Prospector Mt. suite as a historically overlooked causative magma event linked to Au-rich PCDs in the DRGB and extends the temporal window of PCD prospectivity in this area. The transition from mid-Cretaceous Whitehorse suite magmas to Late Cretaceous Casino-Prospector Mt. suite magmas is proposed to reflect a transition from subduction to localized extension, which is becoming more recognized as a common characteristic of productive porphyry belts globally.

Keywords: telescoped porphyry; Dawson Range; metallogeny; petrogenesis; paragenesis; Casino; Prospector Mt.; Klaza; Yukon; post-subduction; slab rollback

1. Introduction

The North American Cordillera represents a long-lived orogen, active since the Neoproterozoic, which involved many episodes of accretion, rifting and varied magmatism with a variety of spatially- and temporally-related hydrothermal mineralization types [1–3]. Consequently, the area is well known for a wide spectrum of ore-deposit types that include porphyry, epithermal, both volcanogenic- and sedimentary-rock-hosted massive sulfide deposits, MVT and orogenic gold deposits [1,2]. Numerous studies have documented the nature and origin of these ore deposit types at various scales which collectively have contributed to furthering the understanding of metallogeny of this and analogous settings [4–12]. As expected, however, details of processes relevant at more refined scales for different districts require more concentrated and focused studies. This is the case for the Late Cretaceous evolution of the Yukon segment of the Canadian Cordillera [1–3].

The Late Cretaceous Klaza Au–Ag–Pb–Zn–(Cu) epithermal deposit has a current indicated resource of 4.457 Mt containing 686,000 oz Au, 14,071,000 oz Ag, 73,268,000 lbs Pb and 92,107,000 lbs Zn at grades of 4.8 g/t Au and 98 g/t Ag, 0.7% Pb, and 0.9% Zn [13] and a positive Preliminary Economic Assessment indicates its likely future mining potential [14]. Although porphyry-type targets, and thus potentially more significant mineral resources, have long been speculated to be present on the Klaza property [14–18], compelling evidence for such mineralization has not been presented prior to this study, which is based on the work of Lee (2021). This paper uses an extensive database provided by recent (2011–2020) exploration to propose a revised tectonic and petrogenetic model for the Klaza system. The present study is based on new field constraints derived from logging 12,000 m of core, related regional surface mapping, and follow-up work. The latter work comprised petrographic studies, whole rock lithogeochemistry, geochronology (U–Pb zircon, Re–Os molybdenite, Ar–Ar muscovite) and in situ (LA–ICP–MS) zircon dating, high-precision zircon dating (CA–TIMS), and geochemistry. In addition, an extensive company dataset of metal assays allowed for the examination of metal zonation at the deposit scale. This information is collectively used to improve the metallogenic model in an emerging porphyry Cu–Au district in Yukon.

2. Regional Geology and Mineralizing Events

The DRGB stretches from the Yukon–Alaska border to within 50 km of Carmacks in Yukon, Canada ([1]; Figure 1A). However, the known porphyry Cu–Au deposits (PCDs), prospects, and occurrences are at present restricted to the mid-to-southern Dawson Range area (Figure 1B). This preferential distribution of porphyry and epithermal systems could be a function of varying exhumation extents (and thus preservation potential; see discussion section) or accessibility for exploration activities in the DRGB.

Basement rocks underlying most magmatic–hydrothermal systems in this area comprise Devonian and older metamorphosed volcaniclastic and sedimentary rocks of the parautochthonous Yukon–Tanana terrane (YTT; Figure 1B). The YTT re-accreted to the Laurentian margin between the Late Permian and Early Triassic during west-dipping subduction on its eastern flank [2], now represented by the Slide Mt. Terrane and the Whitewater Fault [19]. However, when subduction polarity switched in the Late Triassic, it resulted in the east-dipping subduction of the Kula plate on the western flank of the YTT [2]. Consequently, intermediate calc-alkalic Minto (ca. 204–195 Ma) and Long Lake (ca. 188–183 Ma; [20]) suite plutonic bodies of granodiorite to monzonite composition intrude the contact between the YTT and the Stikine terrane. These rocks host rafts of the migmatized Late Triassic porphyry Cu–Au–Ag systems (Minto, Carmacks Copper) in the Minto Copper Belt [21]. This Late Triassic to Early Jurassic magmatic pulse coincides with the well-known and significant porphyry Cu–Au–Mo endowment further south in British Columbia [22].

The westward migration at ca. 120 Ma of the aforementioned volcanic arc gave rise to intermediate to felsic, metaluminous, calc-alkaline Whitehorse suite (ca. 120–98 Ma) granodioritic bodies that intrude the contact between the Minto-Long Lake suites and the YTT (Figure 1B). The Mt. Nansen Group (ca. 115 Ma) andesites, which form a coeval extrusive counterpart to the Whitehorse suite rocks, are documented to overlie and are in-turn intruded by younger phases of the Whitehorse magmatic pulse [23]. Mineralization correlated with the extensive Whitehorse suite magmatism in the DRGB includes the Pattinson, Idaho Creek, and Antoniuk prospects [1].

East-dipping subduction continued in the Late Cretaceous and resulted in the emplacement of the intermediate calc-alkaline Casino suite rock (ca. 80–72 Ma; [1]) with the same magmatic locus as the Whitehorse suite rocks (Figure 1B). This Late Cretaceous intrusive suite is significantly less voluminous than the earlier plutonic suites and importantly was emplaced as high-level porphyry dikes, plugs, and breccias; they are inferred to be the causative magmas for several porphyry–epithermal systems in the southern DRGB [1]. The more significant centres of mineralization related to this magmatism include (see Figure 1B): (1) Casino with the resource of 1057 Mt at 0.2% Cu, 0.23 g/t Au and 0.022% Mo [1]; (2) nucleus with an indicated resource of 31 Mt at 0.65 g.t Au, 0.07% Cu, and 0.70% Ag [24]; and (3) revenue with an indicated resource of 11.4 Mt at 0.38 g/t Au, 0.12% Cu, 2.4 g/t Ag, 0.016% Mo, and 0.008% W [24]. Many PCDs in the DRGB host significant Cu–Au–Mo resources and occur with genetically associated epithermal extensions [25,26].

The youngest magmatism of 72–67 Ma [1], termed the Prospector Mt. suite, consists of intermediate, subalkaline intrusive rocks that are coeval with the Mg-rich shoshonitic Carmacks Group (ca. 70 Ma) basalts [27]. Some Ag-rich polymetallic veins with intermediate sulfidation (IS) epithermal characteristics are associated with these intrusions (e.g., Prospector Mt., Sixtymile River: [1,28]). Several tectonic models were proposed to address this magmatic pulse: (1) plume-related magmatism [29]; (2) lithospheric delamination [30]; and (3) slab break-off [31].

Metallogenic epochs in the DRGB are defined by varying timeframes in the literature (as discussed above). In this contribution, we use new age constraints to modify the original groupings from Allan et al. (2013) and redefine three metallogenic epochs correlating with magmatism and mineralization in the DRGB as: (1) mid-Cretaceous Whitehorse (120–90 Ma); (2) early-Late Cretaceous Casino (80–72 Ma); and (3) late-Late Cretaceous Prospector Mt. (72–65 Ma).

2.1. The Mount Nansen Gold Corridor (MNGC)

The Mount Nansen Gold Corridor (MNGC; Figure 1C) comprises approximately thirty mineral occurrences with the overall geological setting interpreted as an NW-trending horst structure [16]. Most of the mineral occurrences are hosted in steeply dipping NW-trending fault structures within Whitehorse suite granodiorite host rocks, Casino suite porphyritic dikes, Minto suite foliated granitoids, or YTT metasedimentary rocks; all of which form the basement to the area. NE-trending sinistral faults, which appear to post-date and offset NW-trending structures, are proposed to likely be syn-kinematic with the NW-trending dextral faults [1]. These NE-trending sinistral faults were likely active in the latest Late Cretaceous [1], with the normal component conducive to forming epithermal-type veins. This hypothesis is supported by recent U–Pb calcite data [32], which present Prospector Mt. age (ca. 72 Ma) calcite veins hosted in hydrothermal veins and faults from the Freegold Mt. district (FMGD; Figure 1B).

The MNGC comprises two clusters of gold-rich polymetallic veins: (1) the Klaza cluster (NW; Figure 1C); and (2) the past-producing Brown-McDade cluster (SE; Figure 1C). Both clusters form the intermediate sulfidation epithermal extensions of a central porphyry complex [16]. The central porphyry complex located south of Klaza comprises the Kelly, Cyprus, and Rusk drilled prospects (Figure 1C). Selby et al. [33,34] proposed a 72–67 Ma age of formation for the MNGC systems (Cyprus/Rusk) based on the dating of molybdenite

(Re–Os) and hydrothermal K-feldspar (Ar–Ar). Mortensen et al. [17] proposed a post-77 Ma age of formation for the Klaza system and a ca. 115 Ma age for the Brown-McDade system, both based on co-spatial relationships with porphyritic dikes and their corresponding in situ U–Pb zircon ages.

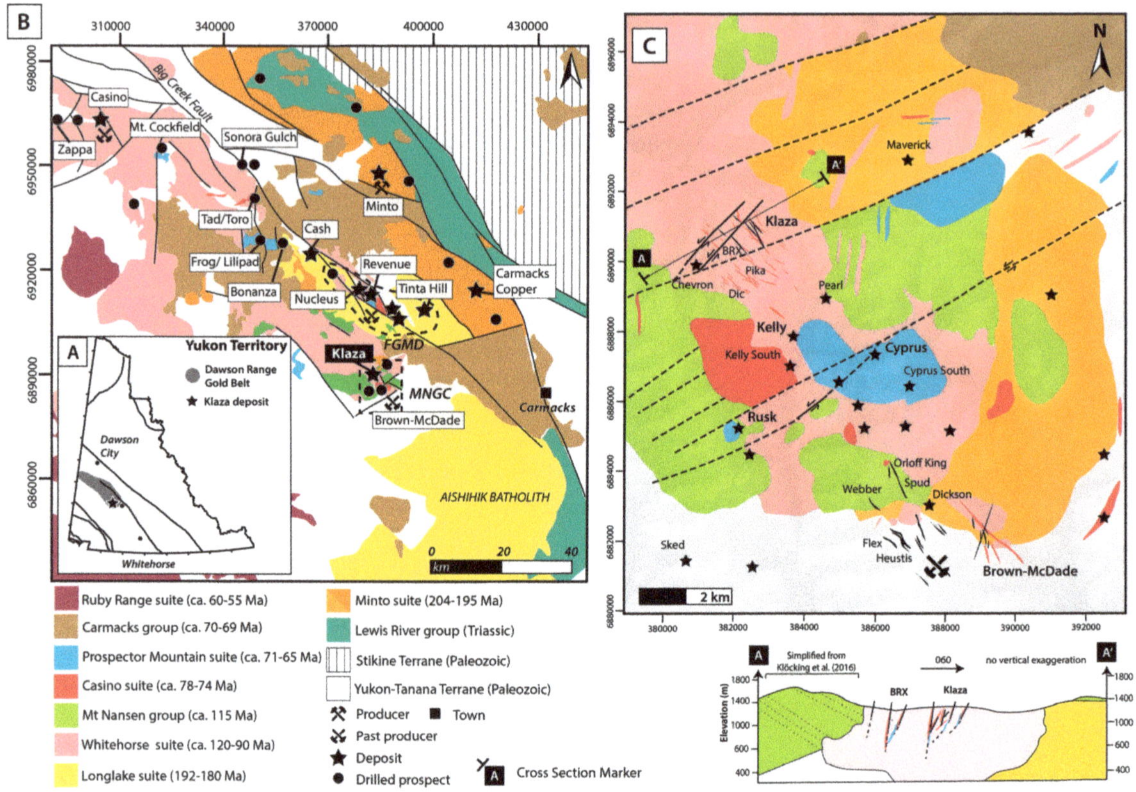

Figure 1. Regional- and local-scale geological maps of the study area. (**A**) Inset map showing the location of the Klaza deposit and the Dawson Range Gold Belt relative to cities in the Yukon. (**B**) Simplified geologic map of the southern Dawson Range redrawn and modified from Yukon Geological Survey [35]. The locations of the Mt. Nansen Gold Corridor (MNGC) and Freegold Mt. District (FGMD) are indicated. All mineral occurrences displayed in this map are either porphyry or epithermal systems. (**C**) Geologic map of the Mt. Nansen Gold Corridor modified from Sack et al. [36] and Lee et al. [37]. The locations of the Klaza deposit and past-producing Brown-McDade mine are indicated. A cross-section of (**A**,**A′**) matching the colour scheme of this map is provided. UTM Zone 08, Datum: NAD 83.

2.2. Magmatic–Hydrothermal Paragenesis and Alteration

A detailed description of the mineralogical, textural, and geochemical characteristics of the intrusive suites best seen in the Klaza deposit area is presented by Lee et al. [25] whereas its detailed hydrothermal history was summarized by Lee, W.-S., et al. [37]. This information is summarized in chronological order to form a framework for the analytical results presented herein. A shorthand nomenclature (i1 to i5) is used to simplify referencing of intrusive phases, as documented in the drill core, in paragenetic order, whereas stages (1 to 4) are used to define the hydrothermal events. Furthermore, a shorthand prefix of "p" and "v" (e.g., p-Pro and v-Pro) is used to indicate porphyry-related and epithermal vein-related alteration assemblages, respectively.

Four intrusive magmatic suites are present in the MNGC: (1) early Triassic Minto (Figure 2A); (2) mid-Cretaceous Whitehorse (120–90 Ma; Figure 2C,D); (3) Late Cretaceous Casino (80–72 Ma; Figure 2E–I); and (4) Late Cretaceous Prospector Mt. (72–65 Ma; Figure 2J). Geochemical fingerprinting of these suites using lithogeochemistry indicates they are all characterized by intermediate-to-evolved, calc-alkaline compositions and have a common lower-crust melt source [25].

The Minto suite intrusive rocks are medium-to-coarse-grained porphyritic foliated hornblende-granodiorite with K-feldspar phenocrysts. This suite is observed in outcrop NNE of the Klaza deposit where they unconformably underly Mt. Nansen Group volcanic rocks and are intruded by Whitehorse suite plutonic rocks [36]. Mt. Nansen Group andesitic rocks (Figure 2B) post-date the Minto suite, flank the Klaza deposit area (Figure 1C), and are intruded by the Whitehorse intrusive suite (i1; oldest intrusive suite documented in drill core) which is multiphase and consists of biotite granodiorite (Figure 2C) to biotite tonalite (Figure 2D). This unit is common at depth in the Klaza deposit and in the Kelly zone. The i1 phase is cut by subsequent magmatic units (i.e., dikes) and forms a primary host for hydrothermal (epithermal and porphyry) mineralization. The Casino suite consists of at least five phases (i2 to i4; Figure 2E–I) spanning hornblende diorite to biotite granodiorite compositions with porphyritic to equigranular textures, whereas currently only one phase is identified for the Prospector Mt. suite (described below as i5; Figure 2J). The Carmacks Group basalts (Figure 2K) do not occur in the deposit area but overlie the previously described lithologies in exposures ~ 10 km east of the Klaza deposit [36]. A summary of crosscutting relationships between the various intrusive phases and vein types is provided in Figure 2L.

The Casino suite phases (i2 to i3c), which are confined to the Kelly zone, display evidence of magma mingling as evidenced by gradational contacts (Figure 3A), cross-cutting relationships (Figure 3B–D), embayed quartz (Figure 3E), and sieve-textured plagioclase (Figure 3F,G). These phases also host porphyry-type mineralization (see below). In contrast, the i4 phase (Figure 2I) is a plagioclase porphyritic granodiorite that occurs as 10 cm to 3 m-wide dikes hosted in NW-trending structures throughout the deposit area. Mineral modes for each intrusive phase are provided in Table 1.

Table 1. Mineral modes for major intrusive phases based on field and petrographic observations of least altered samples.

Intrusive Phase	% Quartz	% Plagioclase	% K-Feldspar	% Biotite	% Hornblende	% Sericite *
i1	25–40	30–45	0–20	10–15	1–8	5–15
i2	30	40	?	0–1	0–5	20–30
i3a	0–5	30–40	0	0–5	5–10	5–10
i3b	0–5	45–50	0	20–25	30	5–10
i3c	25–30	30–40	?	7–10	2–5	30–50
i4	15–20	25–30	?	5–10	0–1	30–50
i5	0–5	?	?	5–10	0–5	50–65

Sericite * = secondary, alteration-related mineral phase consisting of muscovite and illite inferred to be replacing K-feldspar and lesser plagioclase. ? = no primary K-feldspar preserved, hence unable to estimate modal abundance.

Porphyry-type mineralization and alteration (Stage 1): Four vein types localized to the Kelly zone are grouped together and these define the stage 1 event (Figure 2L). Intrusive phases observed at Klaza (i1 to i5) are described in their chronological order of formation based on cross cutting relationships; representative examples are shown in Figure 3. As the veins have features typical of high-temperature magmatic–hydrothermal events and generally conform to those described in typical calc-alkaline porphyry deposit settings, the nomenclature of Seedorff et al. [38] and Sillitoe [39] is herein adopted and used to describe them.

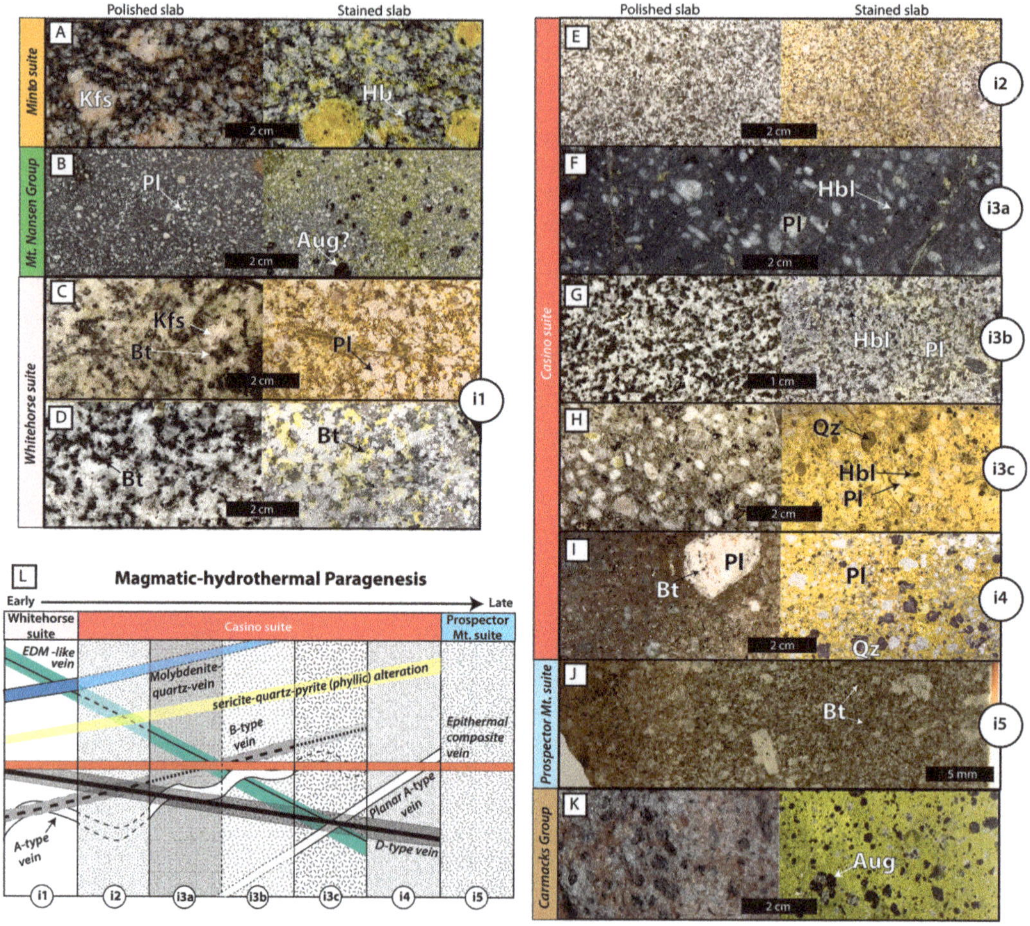

Figure 2. Polished (**left**) and stained (**right**; sodium cobaltinitrite) samples of important igneous rocks at Klaza. Note the unit name is provided on the left side of images. (**A**) Minto suite granodiorite with subhedral K-feldspar phenocrysts (surface sample). (**B**) Mt. Nansen Group andesite from adjacent to deposit area (surface sample). (**C**) Whitehorse suite (WS) hornblende-biotite granodiorite having a light propylitic overprint (Eastern BRX zone: KL-17-374; 75 m). (**D**) WS biotite-hornblende tonalite (KL-15-286; 93 m). (**E**) Casino suite (CS) monzogranite to granite (Kelly zone: KL-16-314; 117 m). (**F**) CS plagioclase(-biotite) phyric diorite (i3a phase). (**G**) CS hornblende diorite (Kelly zone: KL-16-314; 331 m). (**H**) CS plagioclase-quartz-biotite phyric granodiorite (Kelly zone: KL-16-314; 358 m). (**I**) Plagioclase-quartz(-biotite) phyric granodiorite (Kelly zone: KL-16-314; 453 m). (**J**) Prospector Mt. suite plagioclase phyric biotite diorite (Kl-14-193; 248 m). (**K**) Carmacks Group basalt (surface sample). (**L**) Timeline with crosscutting relationships among various intrusive phases and high- to low-temperature vein types at the Klaza deposit and Kelly prospect. Mineral abbreviations in accordance with Whitney and Evans [40].

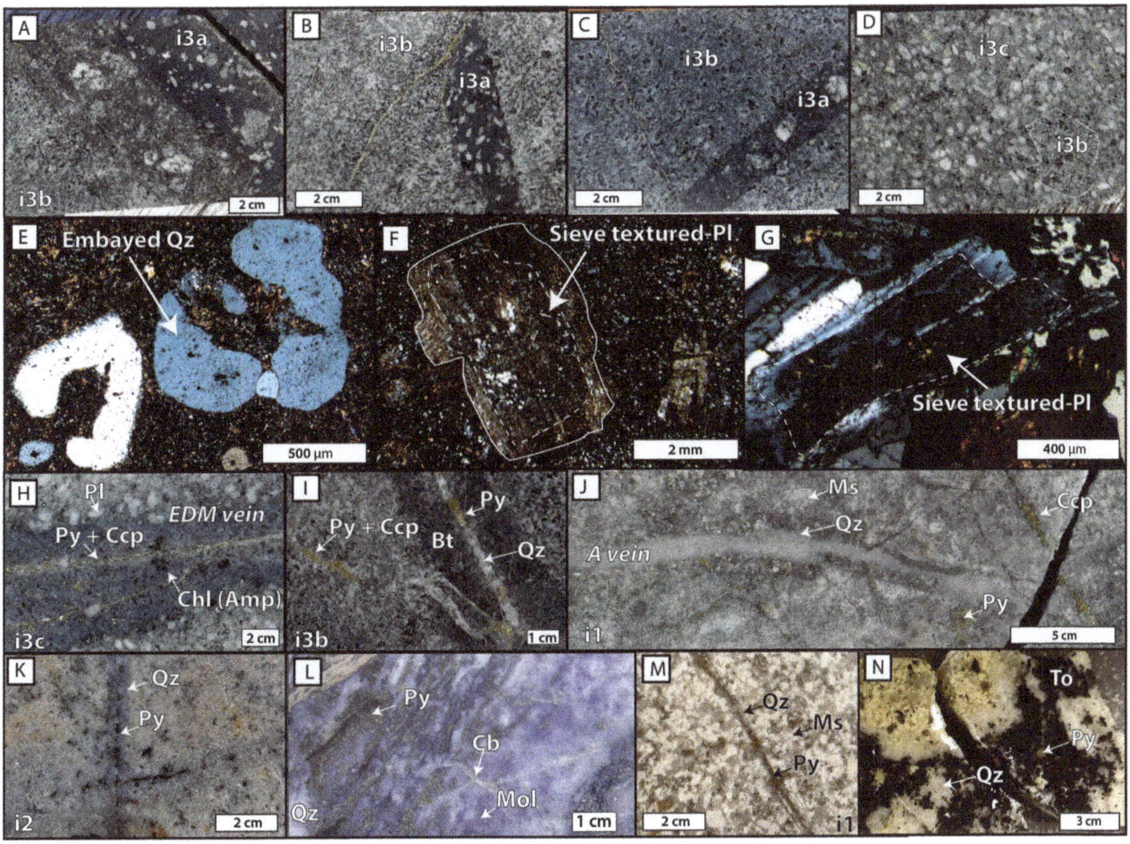

Figure 3. Images of drill core and polished thin sections (pts) in crossed nicols of various units in the Klaza deposit setting. (**A**) Gradational contact between i3a and i3b phases (Kelly zone: KL-16-314; 358.2 m depth). (**B**) Clast of i3a phase within i3b (Kelly zone: KL-16-314; 370.2 m depth). (**C**) Dike of i3a cutting i3b phase (Kelly zone: KL-16-314; 352.65 m depth). (**D**) Clast of i3b phase in the i3c phase (Kelly zone: KL-16-314; 358.8 m depth). (**E**) Pts image of embayed quartz grains in i4 phase dike. Note that the matrix is mostly sericite and biotite (surface sample). (**F**) Pts image of a relict sieve-textured plagioclase grain in a sericite-altered groundmass in i4 phase dike (Kelly zone: KL-16-314; 446.5 m). (**G**) Pts image of sieve-textured plagioclase grain intergrown with biotite and hornblende in i3b phase (Kelly zone: KL-16-314; 332.2m). (**H**) EDM-type vein overprinted by propylitic alteration with later fractures lined by pyrite and chalcopyrite (Kelly zone: KL-16-314; 442.3 m depth). (**I**) EDM-type vein cored by an A-type vein (Kelly zone: KL-16-314; 358 m depth). (**J**) A-type vein overprinted by phyllic alteration (Kelly zone: KL-16-314; 379.19 m depth). (**K**) B-type vein with pyrite overprinted by phyllic alteration (Kelly zone: KL-16-314; 114 m depth). (**L**) Quartz–molybdenite vein cut by epithermal-type pyrite and carbonate veins (Kelly zone: KL-16-314; 431 m depth). (**M**) D-type vein boarded by phyllic alteration (Central Klaza zone: KL-15-240; 146 m depth). (**N**) Tourmaline–quartz–muscovite–pyrite vein with associated phyllic alteration (Central Klaza zone: KL-12-133; 327.75 m depth). Mineral abbreviations in accordance with Whitney and Evans [40].

The earliest veins are considered to equate to the early dark micaceous-(EDM) type as they are characterized by their dark biotite–muscovite nature (Figure 3H,I). They are 4 to 10 cm in width, are replaced by chlorite during subsequent propylitic overprinting (Figure 3H), and are commonly lined by quartz-pyrite ± chalcopyrite due to later reopening coincident with formation of later A- (Figure 3J) and D- (Figure 3M) type veins [41]. A less common variant of these veins is an earlier biotite (EB) vein type which has a K-feldspar selvage [37].

Sinuous A-type veins (Figure 3I,J) are uncommon relative to EDM veins. They are comprised of cloudy, granular quartz, are ~1 cm wide with pyrite–muscovite selvages, and rarely have chalcopyrite. A weak Fe-carbonate overprint on these veins is attributed to a later epithermal event (see below). B-type veins (Figure 3K) comprise centre-line pyrite–quartz stockwork veins and these are seen throughout the Kelly zone, but are most commonly hosted in the i2 phase of this zone. Potassic alteration is observed deep in the Central Klaza zone as a rare, magnetite–biotite ± K-feldspar alteration (see below). Although the A- and B-type veins are typically associated with potassic alteration in porphyry settings [38,39], these veins are commonly overprinted by phyllic (p-Phy) and propylitic (p-Pro) alteration in the Kelly zone (Figure 3J,K).

Quartz molybdenite veins (Figure 3L) are 3 to 10 cm wide, are hosted in the i1 phase, and are commonly cut by pyrite–arsenopyrite–carbonate veinlets associated with the later epithermal stages; consequently, these veins have a combination of phyllic (p-Phy and v-Phy) alteration haloes. The D-type veins (quartz–muscovite–pyrite; Figure 3M) are narrow (i.e., hairline-width) but have well-developed phyllic (p-Phy) alteration haloes of cm-scale. These veins have several variations, including a quartz–muscovite–pyrite–tourmaline ± rhodochrosite, which can form as breccia zones in the Central Klaza zone (Figure 3N).

Propylitic (p-Pro) alteration contains an assemblage of chlorite–hematite ± epidote ± pyrite. Chlorite replacing magmatic hornblende and biotite is the dominant alteration mineral, whereas trace hematite occurs in magmatic K-feldspar which can lend it a red to orange appearance. The assemblage is extensive on the deposit scale with a minor overprint even seen on the least altered samples. This assemblage occurs syn-to-post D-type vein formation, and, therefore, overprints stage 1 veins and p-Phy alteration.

Prospector Mt. suite: A Prospector Mt.-age dike (ca. 72 Ma) was first reported by Mortensen et al. [17] amongst other Casino-age dikes; importantly this age has since been reproduced in this study (see below). This previously unrecognized intrusive phase (i5) is highly altered and only pseudomorphs of primary feldspar minerals are preserved amongst remnant primary biotite, secondary illite and calcite (Figure 2J). The i5 phase, which has a slightly alkalic nature [18], is seen as biotite–plagioclase-bearing dioritic dike rocks which occur co-spatially with the i4 phase; however, this phase is less voluminous than the i4 phase. It differs from the i4 phase in its lack of resorbed quartz phenocrysts and displays an overall lower silica content; however, it also displays features of magma mingling with the i4 phase (Figure 4A). The i5 phase is cut (Figure 4B,C) and brecciated (Figure 4D) by Stage 2b (see below) epithermal mineralization, therefore, it forms an upper time constraint to the mineralization.

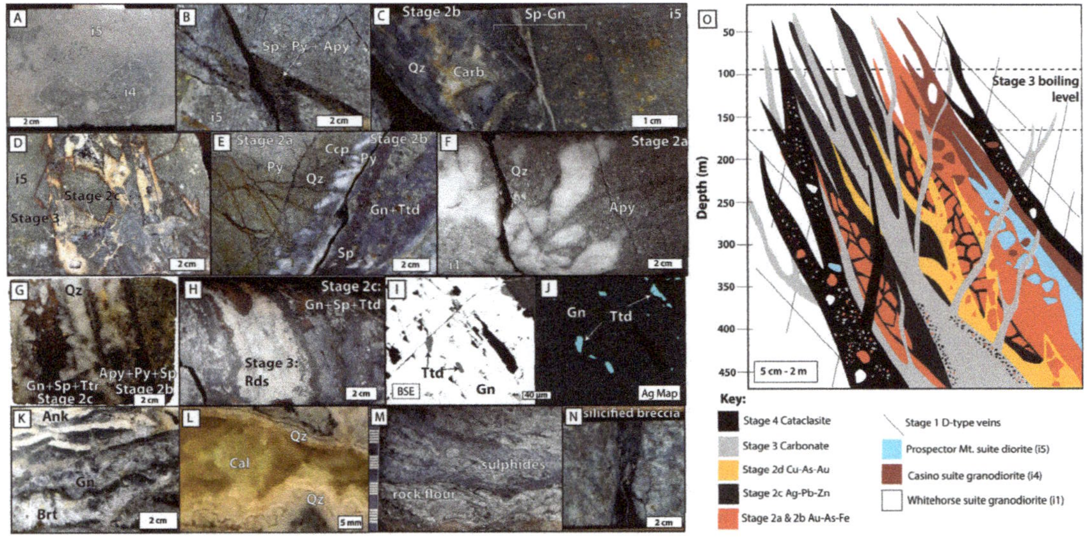

Figure 4. Rock slabs and backscatter electron (BSE) images of rock type and veins at Klaza along with a schematic diagram summarizing the evolution of the setting. (**A**) Anhedral i4 dike clast within i5 dike (Kelly zone: KL-16-309; 395.66 m depth). (**B**) Stage 2b vein cutting i5 dike (Central Klaza zone: KL-11-12; 195.8 m depth). (**C**) Stage 2b and stage 3 vein at the contact between an i5 dike and i1 granodiorite (Central Klaza zone: KL-11-12; 176.5 m depth). (**D**) Prospector Mt. suite dike (i5) brecciated and cemented by Stage 2c (Western Klaza zone: KL-14-178; 96.5 m depth). (**E**) Stage 2a massive pyrite cut by Stage 2b cloudy quartz–pyrite–sphalerite–arsenopyrite vein and Stage 2c galena–sphalerite–tetrahedrite (Central BRX zone: KL-14-428; 57 m depth). (**F**) Stage 2a massive arsenopyrite and prismatic Stage 2a quartz (Western BRX zone: KL-17-398; 123 m depth). (**G**) Stage 2b banded arsenopyrite–pyrite–sphalerite–quartz vein with a centre-fill Stage 2c galena–sphalerite–tetrahedrite–quartz vein (Central Klaza zone: KL-14-193; 271.86 m depth). (**H**) Stage 3 rhodochrosite vein with sphalerite cutting Stage 2c galena–sphalerite–tetrahedrite–quartz (Western BRX zone: KL-17-398; 123.5 m depth). (**I**) BSE image of Stage 2c galena with tetrahedrite inclusions (Central Klaza zone: KL-11-12; 221.19 m). (**J**) SEM X-ray map of (**I**) showing the Ag-rich nature of tetrahedrite. (**K**) Bladed barite hosted in quartz-carbonate alongside banded ankerite veins from stage 3 (Central Klaza zone: KL-11-18; 181.90 m). (**L**) Colloform- and moss-textured quartz-carbonate vein from stage 3 (Central Klaza zone: KL-12-133; 148.65 m depth). (**M**) Fault gouge containing milled vein and wallrock material (Western BRX zone: KL-17-401; 82 m depth). (**N**) Silicified breccia in i1 granodiorite (Central Klaza zone: KL-12-133; 352.20 m depth). (**O**) A schematic diagram of the Klaza composite vein substages and their relationship to magmatic and high-T vein phases. Mineral abbreviations in accordance with Whitney and Evans [40].

Epithermal-type mineralization and alteration (Stage 2): The stage 2 veins constitute the main-ore stage at Klaza and comprises three sub-stages (Figure 4) with the mineralogy zoned laterally (on the deposit scale) with respect to the inferred heat source, thus from hotter (SE) to colder (NW). Detailed petrographic observations and supporting micro-analytical data for these hydrothermal stages are beyond the scope of this study and are instead discussed elsewhere [42].

Stage 2a veins are massive pyrite ± arsenopyrite veins that are cut by later white, cloudy (due to abundant fluid inclusions) quartz ± chalcopyrite ± enargite veins (Figure 4E,F). The abundance of arsenopyrite increases towards the NW, away from the heat source (Figure 4F), whereas the abundance of both chalcopyrite and enargite increases toward the heat source (SE). Stage 2b veins are banded arsenopyrite–sphalerite–pyrite-bearing smoky quartz veins, their textures indicating repeated dilation and fill of these antitaxial-like veins (Figure 4E,G). Petrographic and LA-ICP-MS analysis indicate that Au occurs as electrum inclusions and also as a refractory phase (i.e., invisible Au) in arsenopyrite and pyrite from both Stage 2a and Stage 2b [42]. Stage 2c veins comprise quartz–galena–sphalerite–sulfosalt ± chalcopyrite fill (Figure 4G,H). These veins cut Stage 2a veins and locally form jigsaw breccias. Quartz occurs as euhedral crystals occluding space following semi-massive infill by galena and sphalerite. Intergrown sulfosalts include several Ag-bearing phases such as tetrahedrite, pyrargyrite, and argentite (now acanthite). Galena is also argentiferous and contains intergrowths of boulangerite, argentite, and Ag–tetrahedrite (Figure 4I,J). The sulfosalts are zoned laterally and vertically, thus Ag- and Pb-rich phases occur in the Western Klaza zone (NW), whereas the Ag-content of tetrahedrite decreases with depth. In addition, chalcopyrite abundance increases with depth.

Epithermal carbonate veins (Stage 3): Sulfides hosted in carbonate veins are minor contributors to the overall Ag–Pb–Zn budget of the ore system. This vein stage cuts stage 1 and stage 2 assemblages and occurs as breccia (Figure 4D) or narrow veins (Figures 3L and 4C,H) and vein swarms. The carbonate type varies and includes ankerite, dolomite, calcite, and rhodochrosite with dolomite being the most abundant phase. Clasts of Pb–Zn sulfide phases are commonly observed in this stage (Figure 4H); however, primary galena and sphalerite are observed to be intergrown with bladed barite phases (Figure 4K), colloform, and moss-textured carbonate (Figure 4L). The bladed barite and colloform ± moss-textured carbonate phases, features commonly attributed to fluid boiling [43], are observed at shallow depths (75–125 m) and generally post-date the deposition of the precious- and base metals seen in Stage 2.

Cataclasis (Stage 4): Stage 4 is syn- to post-stage 3 and occurs in the form of extensive shearing and cataclasis along the composite-vein-hosting fault structure. Vein-hosting faults display evidence of post-mineral movement, such as clay-altered fault gouges on the margin of mineralized quartz veins (Figure 4M). This cataclasis is seen as polymictic breccias with sub-rounded clasts of earlier vein material of all stages, the various intrusive phases (i1, i4, and i5), silicified breccias (Figure 4N), and rock flour.

The epithermal event forms wide, complex, composite veins due to their strong structural controls. The noted textural relationships between substages and vein stages described above are summarized in Figure 4O. Alteration associated with the epithermal stages is described below.

A transect depicting the typical changes in alteration mineralogy from wall rock to mineralized veins that relates to the epithermal type mineralization is shown in Figure 5A–C and summarized graphically in Figure 5D. Stage 2 veins are associated with a proximal, phyllic (v-Phy; muscovite–illite–quartz–pyrite ± sphalerite ± galena) alteration halo. This assemblage occurs in very narrow (cm-scale) intervals beside composite epithermal veins (Figure 5A,C). This halo is bleached white because of extensive muscovite alteration resulting from acid-alteration near the fluid-infiltrating structure. Fine-grained sphalerite and galena are observed in addition to pyrite and hairline D-type quartz–pyrite–muscovite veins.

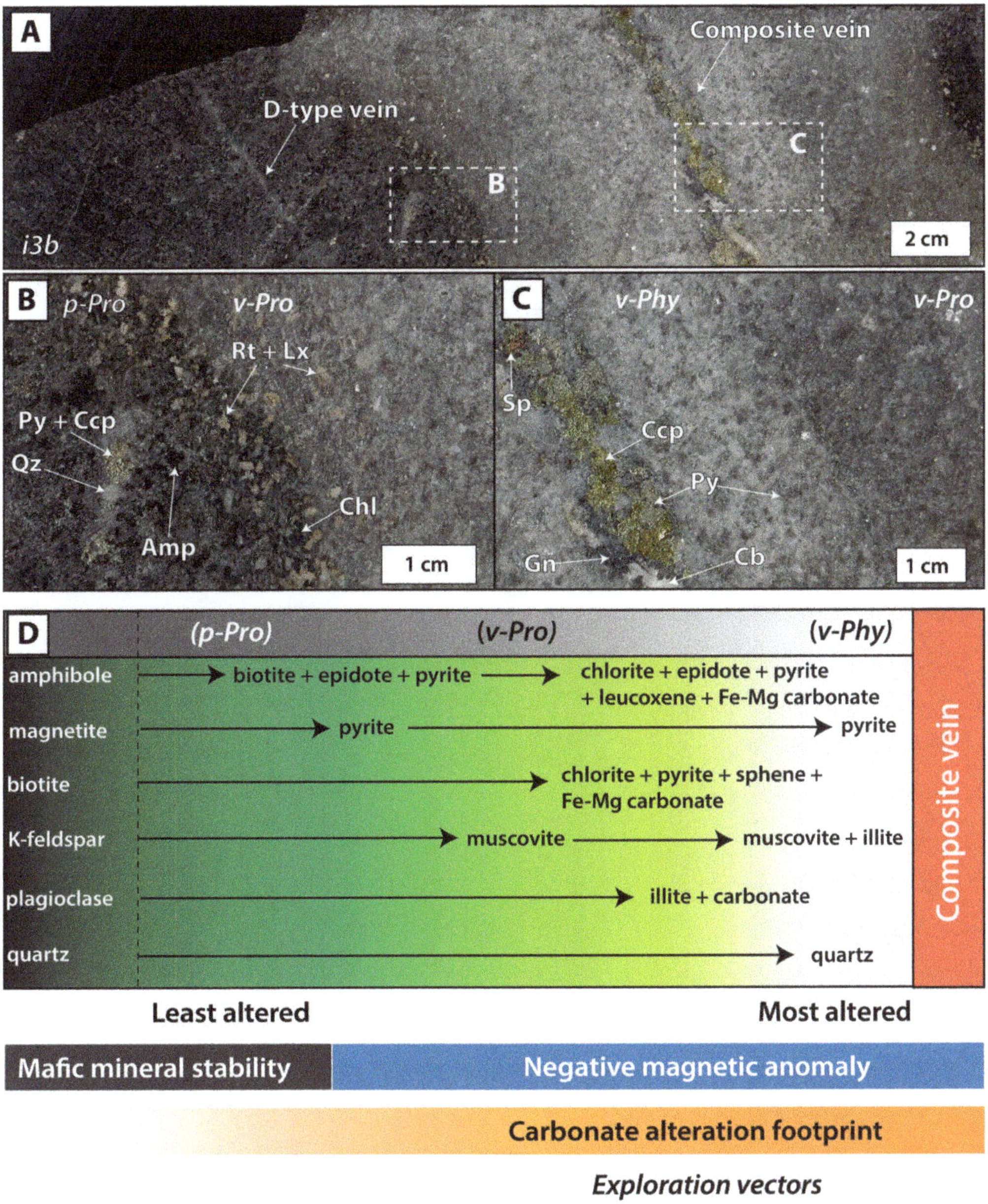

Figure 5. Summary of veins and alteration at the Klaza deposit setting. (**A**) Representative slab of unit i3b (i.e., a hornblende diorite) showing example of alteration types present in the vicinity of composite

epithermal veins. (**B**) Close up view of porphyry-related propylitic (p-Pro) and epithermal vein-related propylitic (v-Pro) alteration. The boundary between these alteration types is defined by the alteration of biotite and amphibole to rutile and leucoxene. (**C**) Close up view of epithermal vein-related phyllic (v-Phy) and v-Pro alteration. (**D**) Schematic diagram depicting the transitional mineral stabilities of primary magmatic phases among the different alteration types. Mineral abbreviations in accordance with Whitney and Evans [40].

The epithermal veins are also associated with a distal, propylitic (v-Pro) alteration halo. This assemblage mainly consists of chlorite–epidote–Fe–Mg–carbonate–rutile and illite–muscovite intergrowths. It is characterised by a 1–2 cm wide chlorite–leucoxene (rutile) band on its outer margin and the limit of mafic minerals (hornblende–biotite–magnetite). Importantly, the lack of magnetite in this alteration allows related vein and alteration zones to be modeled using magnetic susceptibility measurements. Its width (4 m to several m) varies according to degree of brecciation of the host, which is a function of structural preparation. The Fe–Mg–carbonate is more extensive and is observed up to 8 m from the sulfide veins. This footprint is best seen in oxidized drill-core (Figure 6) due to carbonate oxidation (bright orange; Figure 6C,E) or using carbonate staining protocols [44].

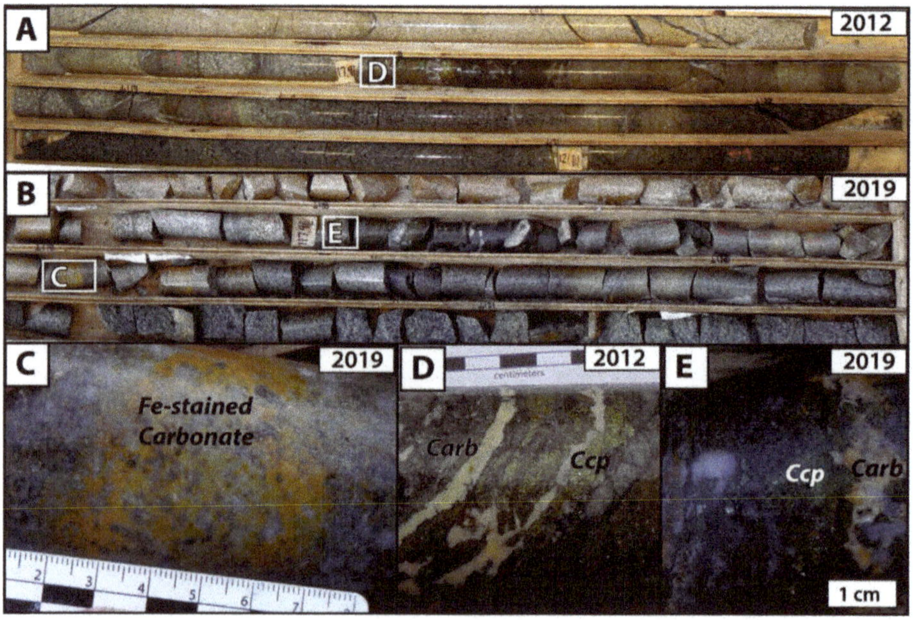

Figure 6. Comparison of drillhole KL-12-114 (110 to 121.5 m depth) from the Klaza setting as seen in 2012 (**A**) versus in 2019 (**B**) to highlight the orange staining from the oxidation of the Fe-carbonate alteration. (**C**) Close-up view of the Fe-stained carbonate. (**D**,**E**) Comparison of carbonate colour in the same piece of core from 2012 (**D**) and 2019 (**E**).

3. Methodology

A detailed magmatic–hydrothermal paragenesis is used to provide geologic context for all analytical techniques discussed below. The methodologies are elaborated in detail in Supplementary Text S1 (Ar–Ar muscovite and Re–Os molybdenite geochronology), Supplementary Text S2 (CA-TIMS zircon geochronology), and Supplementary Text S3 (and LA-ICP-MS U–Pb zircon geochronology and trace element analysis).

3.1. Field Work and Paragenesis

Sampling and field work totalling four months were conducted at Klaza in the 2017 and 2019 field seasons. The paragenesis of magmatic and hydrothermal events was established through a combination of drill-core logging, which provided crosscutting relationships, detailed petrography (transmitted and reflected light complemented with scanning electron microscopy and energy dispersive spectroscopy (SEM-EDS)), whole-rock lithogeochemistry, and geochronology (U–Pb zircon (CA-TIMS and LA ICP-MS), Re–Os molybdenite, and $^{40}Ar/^{39}Ar$ muscovite). A list of samples and their locations is provided in Supplementary Table S7.

3.2. Geochronology

$^{40}Ar/^{39}Ar$ in muscovite: Three representative samples of phyllic- (muscovite–pyrite–quartz) altered rocks from the Klaza deposit were collected and made into thin sections to verify the presence of muscovite. Subsequently, samples were irradiated at the Oregon State TRIGA Reactor, Corvallis, Oregon, USA, and analysed by the furnace step-wise heating method at the Nevada Isotope Geochronology Laboratory at the University of Nevada Las Vegas (UNLV), USA, following the procedures outlined in Zamora-Vega et al. [45].

Re–Os in molybdenite: Two samples of molybdenite-bearing hydrothermal veins were collected from drillcore at the Kelly zone (Klaza) and from the Flex zone (Brown-McDade) in 2020 were submitted to the Re–Os Crustal Geochronology Laboratory at the University of Alberta with the employed methods for Re–Os analysis described in detail by Selby and Creaser [46].

U–Pb in zircon (zircon separation): Twelve samples of feldspar porphyritic intrusive dikes from the Klaza deposit were collected and sent to Overburden Drilling Management Laboratories in Ottawa, Canada, where they underwent mineral separation using electric-pulse disaggregation. Zircon grains were extracted through the use of a shaking table and micropanning the table concentrate followed by hand-picking of the pan concentrate at Laurentian University.

U–Pb zircon (CA-TIMS): Four samples were sent to the Isotope Geology Laboratory at Boise State University, USA, in 2018 for CL imaging and LA-ICP-MS analysis. Subsets of zircons from each sample were then selected from the LA-ICP-MS-analysed zircons and dated by chemical abrasion isotope dilution thermal ionization mass spectrometry (CA-TIMS; [47]).

U–Pb zircon (LA-ICP-MS): Ten samples were sent to the Isotope Geology Laboratory at the University of Alberta, Canada, in 2018for CL imaging and LA-ICP-MS analysis following the methods of Simonetti et al. [48]. Each analytical run includes periodic analysis of a primary reference material interspersed through the analytical session, to which the raw data for unknown samples were normalized to correct for instrumental bias and within-run drift. Secondary reference materials were also analysed, and the results are in agreement with standard values (Supplementary Table S5).

3.3. Zircon Trace Element Analysis

Zircon trace element analyses were conducted at the Mineral Exploration Research Centre Isotope Geochemistry Lab (MERC-IGL), at Laurentian University, Sudbury, Canada. The zircons analysed were those used for dating (see above) and the analysis overlapped pre-existing laser ablation spots (40 μm). Reference materials were analysed multiple times at the beginning and end of each session, and once every ten unknowns throughout the session. Three seconds at the beginning and two second at end of the ablation period were excluded from the spectra in order to minimize potential fractionation effects. Iolite v3.6 [49,50] was used for data reduction.

4. Results

4.1. $^{40}Ar/^{39}Ar$ Muscovite Geochronology

Three samples of Whitehorse suite granodiorite with secondary muscovite were analysed using the $^{40}Ar/^{39}Ar$ stepwise heating method to determine the timing of phyllic (p-Phy) alteration. The full results are provided in Supplementary Table S1, whereas age spectra are summarized in Figure 7. The plateau ages given follow the criteria of Fleck et al. [51].

Sample KZ-1 is a stage 1 muscovite–pyrite–quartz D-vein (ddh KL-12-133, 441 m depth) from the Central Klaza zone. It yielded a plateau age of 79.13 ± 0.30 Ma (Figure 7A; MSWD = 0.24) based on 70.9% of the gas released. The low-temperature gas fraction reflects a slight thermal overprint at <76 Ma.

Sample KZWS17-03 is a muscovite–pyrite–quartz alteration halo (Figure 3M) associated with a stage 1 D-vein from the Central Klaza zone (ddh KL-15-240, 146 m depth). It yielded a plateau age of 77.74 ± 0.79 Ma (Figure 7B; MSWD = 1) based on 63.1% of the gas released.

Sample KZWS17-56 is a stage 1 (early) muscovite–tourmaline–quartz–pyrite vein/breccia from the Central Klaza zone (ddh KL-12-133, 93 m depth). It yielded a plateau age of 77.66 ± 0.71 Ma (Figure 7C; MSWD = 0.4) based on 78.6% of the gas released.

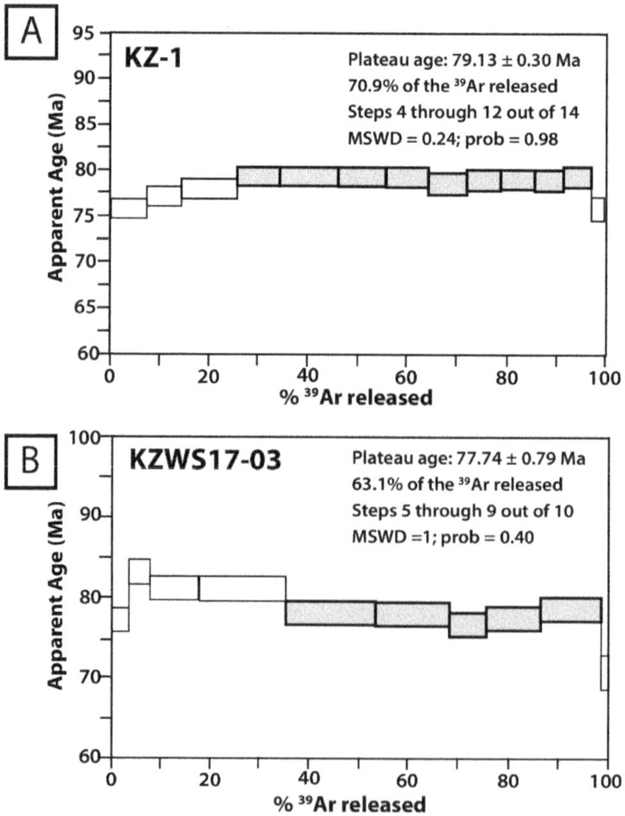

Figure 7. *Cont.*

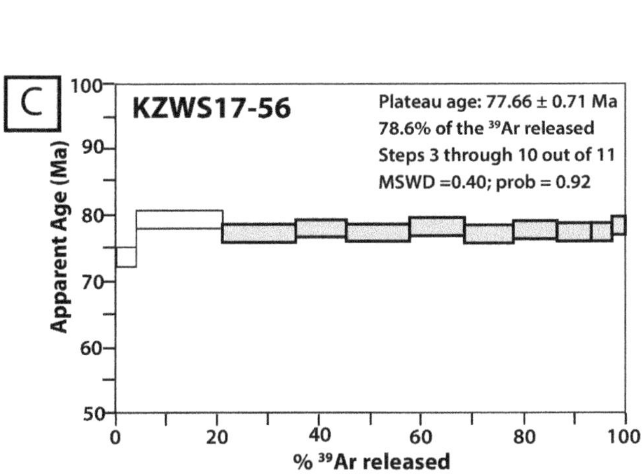

Figure 7. Summary of Ar–Ar age spectra for muscovite obtained from step-wise heating of pure mineral separates. Note that the grey denotes the steps used in the plateau age calculations. (**A**) KZ-1 (KL-12-133; 442 m). (**B**) KZWS17-03 (KL-15-240; 146 m). (**C**) KZWS17-56 (KL-12-133; 93 m depth).

4.2. Re–Os Molybdenite Geochronology

The results for the two samples analysed are tabulated in Table 2. A sample of white quartz vein with pyrite and molybdenite mineralization was taken from the Flex zone (DDH260B-160) of the Brown-McDade cluster (Figure 1C). This sample yielded an age of 108.5 ± 0.5 Ma. Replicate analyses showed similar, reproducible Re–Os dates with no significant difference.

A sample from the Kelly zone (KL-481, 387.1m; Figure 1C) contained blebby-to-disseminated molybdenite in a sinuous (A-type?) quartz vein, which is cut by a planar quartz vein. This sample yielded a Re–Os age of 76.3 ± 0.4 Ma.

Table 2. Compilation of Re–Os molybdenite geochronology results.

Sample ID	**Re (ppm)**	**±2σ**	**^{187}Re (ppm)**	**±2σ**	**^{187}Os (ppb)**	**±2σ**	**Model Age (Ma)**	**±2σ (Ma)**
DDH260B-160 FLEX	1085	3	681.8	1.9	1233.4	0.3	108.5	0.5
DDH260B-160 FLEX RPT [1]	1139	3	715.6	2.0	1302.1	0.1	109.1	0.5
DDH260B-160 FLEX-MAG [2]	35.08	0.10	22.05	0.06	40.19	0.08	109.3	0.5
KL481-387.1m	6.452	0.018	4.056	0.011	5.161	0.014	76.3	0.4

[1] Replicate analysis. [2] Analysis of magnetic mineral separate with 5% molybdenite.

4.3. U–Pb Geochronology

Representative samples of the intrusive phases were processed for CA-TIMS (four samples) and LA-ICP-MS (five samples) U–Pb zircon geochronology. The results are tabulated in Supplementary Tables S2 and S5, respectively, whereas images of the dated samples and relevant plots of the age data are presented in Figures 8 and 9. We use the term "date" for analytical results, whereas "age" refers to the interpretation.

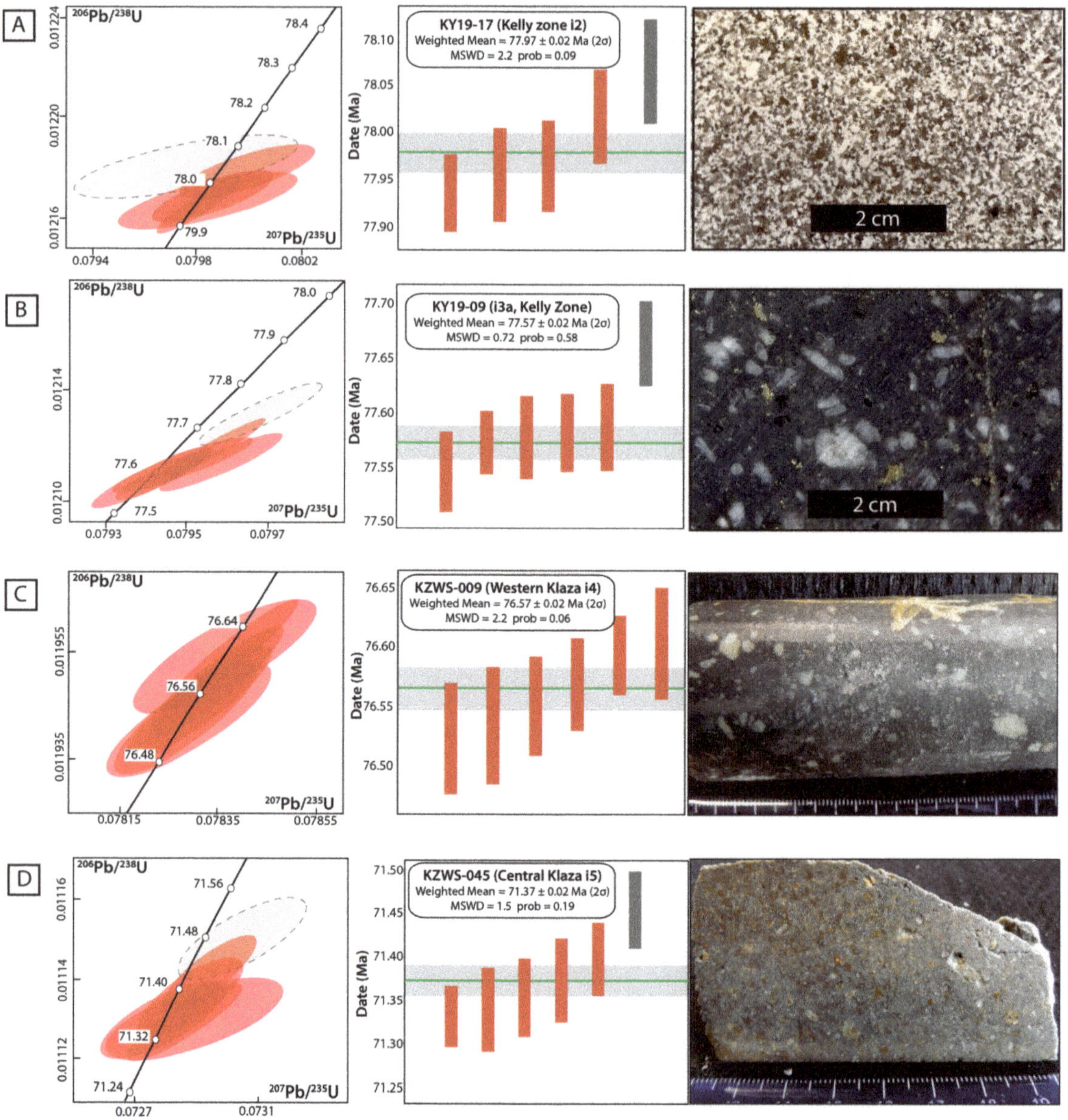

Figure 8. Summary of results for CA-TIMS U–Pb zircon geochronology for intrusive phases i2 (**A**), i3a (**B**), i4 (**C**), and i5 (**D**). The diagrams on the left are Concordia plots. Middle diagrams are weighted mean plots, and images on the right show pieces of representative material from the drillhole sample interval that was submitted for dating. Red ellipses in the Concordia plots and red bars in weighted mean plots denote the data for the youngest crystalizing zircons in the sample (autocrysts). The grey zone in the weighted mean plots represents the 2σ uncertainty limits whereas the green line represents the weighted mean. Images are also shown of the dated phases.

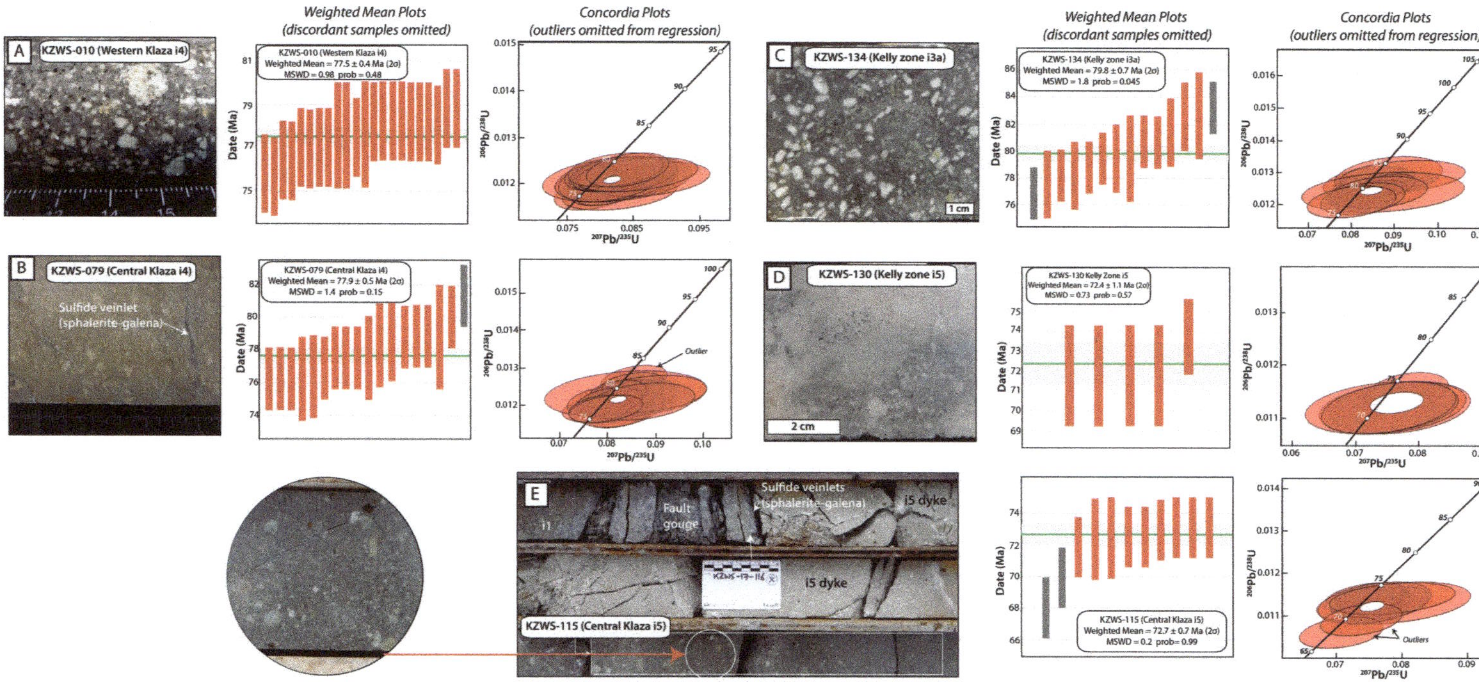

Figure 9. Summary of results from in situ LA-ICP-MS U–Pb dating of zircon for various intrusive rocks (**A–E**) in the Klaza setting shown in both weighted mean plots and Concordia diagrams. Red bars in the weighted mean plots denote data used for calculating the weighted mean age constraints. The grey zone in the weighted mean plots represents the 2σ uncertainty limits whereas the green line represents the weighted mean. Images are also shown of the dated phases.

CA-TIMS analysis:

Twenty-five zircon grains from KY-19-17 (i2 phase of Casino suite; Figure 8A) from the Kelly zone yielded LA-ICP-MS dates of 100.5 ± 6.1 to 67.2 ± 5.7 Ma. Five grains were analysed by CA-TIMS, with the four youngest dates yielding a weighted mean of 77.97 ± 0.02 Ma (Mean Square of Weighted Deviation (MSWD) = 2.2, probability of fit = 0.09) which is the interpreted igneous crystallization age. The single grain with a date of 78.05 ± 0.07 is interpreted as containing an inherited component.

Sixty-two zircon grains from KY-19-09 (i3a phase of Casino suite; Figure 8B) from the Kelly zone yielded LA-ICP-MS dates of 79.4 ± 2.4 to 67.4 ± 4.4 Ma. Six grains were analysed by CA-TIMS, with the five youngest dates yielding a weighted mean of 77.57 ± 0.02 Ma (MSWD = 0.72, probability of fit = 0.58) which is interpreted as the igneous crystallization age. The other grain with a date of 77.73 ± 0.06 Ma is interpreted as containing an inherited component.

Sample KZWS-009 is a plagioclase–quartz–hornblende–biotite–phyric granodiorite dike (i4 phase of Casino suite; Figure 8C) from the Western Klaza zone. Twenty-one grains were analysed by LA-ICP-MS and yielded dates of 83.3 ± 3.2 to 73.0 ± 2.5 Ma whereas 15 grains yielded equivalent dates with a weighted mean of 76.8 ± 1.4 Ma (MSWD = 1.4, probability of fit = 0.14). CA-TIMS on six grains yielded equivalent dates with a weighted mean of 76.57 ± 0.02 Ma (MSWD = 2.2, probability of fit = 0.06) which is interpreted as the igneous crystallization age.

Sample KZWS-045 is a plagioclase–biotite–phyric diorite dike (i5 phase of the Prospector Mt. suite; Figure 8D) sampled from the Central Klaza zone. Twenty-six grains analysed by LA-ICP-MS yielded dates of 81.3 ± 5.0 to 67.0 ± 2.0 Ma, whereas nineteen grains yielded equivalent dates with a weighted mean of 72.2 ± 1.3 Ma (MSWD = 1.6, probability of fit = 0.05). CA-TIMS was performed on six grains, the five youngest of which yielded equivalent dates with a weighted mean of 71.37 ± 0.02 Ma (MSWD = 1.5, probability of fit = 0.19) which is interpreted as the igneous crystallization age. The other date of 71.48 ± 0.05 is from a grain that is interpreted as having an inherited component.

LA-ICP-MS analysis: For these samples, the results are all summarized in Figure 9 where data are presented in both weighted mean and Concordia plots. In all samples, the former age (where applicable) based on the youngest zircon(s) dates is interpreted to represent the best estimate of the time (i.e., age) of crystallization for the host samples.

KZWS-010 is an i4 phase plagioclase quartz–porphyritic biotite–phyric dike from the Western Klaza zone. Twenty-two of the thirty analyses yielded concordant dates between 75 ± 2 and 79 ± 2 Ma (Figure 9A). A weighted mean date of 77.5 ± 0.4 Ma (MSWD = 0.98) is yielded for all concordant dates in the sample.

KZWS-079 is an i4 phase plagioclase–quartz porphyritic dike from the Central Klaza zone. Of the 25 zircons analysed, 18 were concordant with dates between 76 ± 2 and 81 ± 2 Ma (Figure 9B). The oldest date is interpreted to be from an inherited zircon (xenocryst) and treated as an outlier. The weighted mean date of 77.9 ± 0.5 Ma (MSWD = 1.4) is yielded for all concordant dates in the sample.

KZWS-134 is a porphyritic plagioclase phyric dike (i3a phase) from the Kelly zone. The 14 (out of 25 analysed zircons) concordant dates range between 77 ± 2 and 84 ± 2 Ma (Figure 9C). Two concordant dates are flagged as outliers and are interpreted to be: (a) antecrystic (older date) and a potentially younger autocryst at 77± 2 Ma, which overlaps in age with CA-TIMs dates from sample KY-19-09; or (b) a result of Pb loss. Although the weighted mean date (79.8 ± 0.7 Ma; MSWD = 1.8) for all concordant zircons (excluding the two outliers) is a plausible crystallization age of the sample, the crystallization age for the i3a phase is likely to be slightly younger at ~77.5 Ma as resolved by CA-TIMS dating.

KZWS-130 is an i5 plagioclase phyric diorite dike from the Kelly zone (Figure 9D). The analysis of 13 zircons yielded three clusters of dates: (1) three between 90 ± 2 and 115 ± 2 Ma which equate to the Whitehorse plutonic suite; (2) five between 74 ± 2 and 78 ± 2 Ma which equate to Casino suite magmatism; and (3) five with similar dates of 71 ± 2 and 72 ± 2 Ma for which only the youngest dates are concordant. The crystallization age of this sample is interpreted to be the weighted mean age calculated from the five concordant dates at 72.4 ± 1.1 Ma (MSWD = 0.7).

Sample KZWS-115 is from an i5 phase diorite dike which is cut by 1 cm-wide sulfide veins (sphalerite–galena) and is in contact with a 10 cm-wide fault gouge that contains clasts of sulfides and vein material. The dike contains argillic alteration up to 2 m from the fault gouge (Figure 9E). The 27 analysed zircons yielded a range of dates: (1) five discordant zircons with mid-Cretaceous dates between 90 ± 2 and 115 ± 2 Ma which equate to the Whitehorse plutonic suite; (2) three to four discordant zircons between 82 ± 2 and 88 ± 2 Ma that do not equate to known pulses of magmatism in the region and which may, therefore, reflect Pb loss; and (3) the 11 remaining zircons range from 75 ± 2 to 68 ± 2 Ma with the majority of these concordant and yielding a date of ca. 72 Ma with the youngest at 68 ± 2 Ma. Among these 11 concordant zircons, the two with the youngest dates are flagged as outliers and could be the youngest autocrysts in this intrusive phase but are not included in the weighted mean age of 72.7 ± 0.7 Ma (MSWD = 0.2). As the dike is cut by sulfide veins, it provides a maximum age for epithermal mineralization.

4.4. Zircon Trace Element Geochemistry

Zircon grains used for in situ dating were also analysed for trace elements which included REE. As with dating, the points selected for analysis were based on CL imaging to selected areas of magmatic growth and to avoid xenocrystic areas or where post-crystallization alteration may have occurred [52,53]. Based on the results of U–Pb dating, samples are grouped according to their ages and hence intrusive suites rather than simply by samples or host due to the issue of inheritance. A summary of igneous zircon provenance is provided in Figure 10, whereby the inheritance of both xenocrystic and antecrystic zircons is illustrated.

Screening of data: The zircon analyses (N = 128; see Supplementary Table S6, Sheet A) were filtered to remove anomalous data based on the criteria of Lu et al. [54], which includes: (1) apatite contamination (La > 2.5 ppm; N = 29); and (2) titanium contamination from titanite (Ti > 50 ppm; N = 12). Of the remaining zircon data (N = 87), 19 are Prospector Mt. age, 51 are Casino age, seven are Whitehorse suite age (112–110 Ma), and eight are a younger pulse of Whitehorse suite rock (100 to 88 Ma). It is noted here that the dates for individual zircon U–Pb analyses have uncertainties of ± 2–3 Ma, therefore, the distinction between Casino- and Prospector Mt.-age zircons may be limited in extent.

At current detection limits, the chondrite-normalized REE profiles of zircons from each intrusive suite appear relatively similar (i.e., distinct Ce anomalies, enriched HREEs, and slight Sm anomalies in Late Cretaceous-aged zircons; Figure 11). However, trace element plots (Figures 12 and 13) show distinct differences in zircon types that can be grouped by their age. Thus, in the following discussion, three groups of zircon ages are frequently referenced (Figure 12F): (1) older Casino age (CAZ-1; 84–77 Ma); (2) younger Casino age (CAZ-2; 76–74 Ma); and (3) Prospector Mt. age (PMZ; 72–68 Ma).

Fractionation trends: The progressive fractionation in each suite is tracked via the change in trace elements in zircon using several proxies, these being Hf, Y, U, Th and REE. Thus, progressive differentiation is reflected by an increase in Hf [55–57] or decreasing Th/U ratio [56–59]. In addition, the crystallization of amphibole removes Y from the melt and thereby the Hf/Y ratio increases [59–61]. In this regard, all three zircon groups show chemical signatures indicative of amphibole crystallization (Figure 12A), as supported by the presence of amphibole in both Casino and Prospector Mt. suite rocks.

The CAZ-1 is less evolved relative to the CAZ-2 based on Hf/Y and Yb/Gd_n versus Th/U plots (Figure 12A,B), which is further supported by the diorite (i3a) versus granodioritic (i3c and i4) compositions of these suites, respectively; the PMZ is intermediate compared to these suites. The Yb/Gd_n ratios of zircons from all three groups are low (<8; Figure 12C) and, as noted by Richards et al. [61], such values suggest a role of garnet fractionation in the melt history (i.e., high-pressure), but the PMZ is clearly more enriched in Gd_n for a given Yb_n value (Figure 12D). CAZ-1, CAZ-2, and PMZ display similar ranges of Sm/Ce_n (Figure 12C); however, CAZ-2 displays elevated Yb/Gd ratios relative to PMZ for a given Sm/Ce_n.

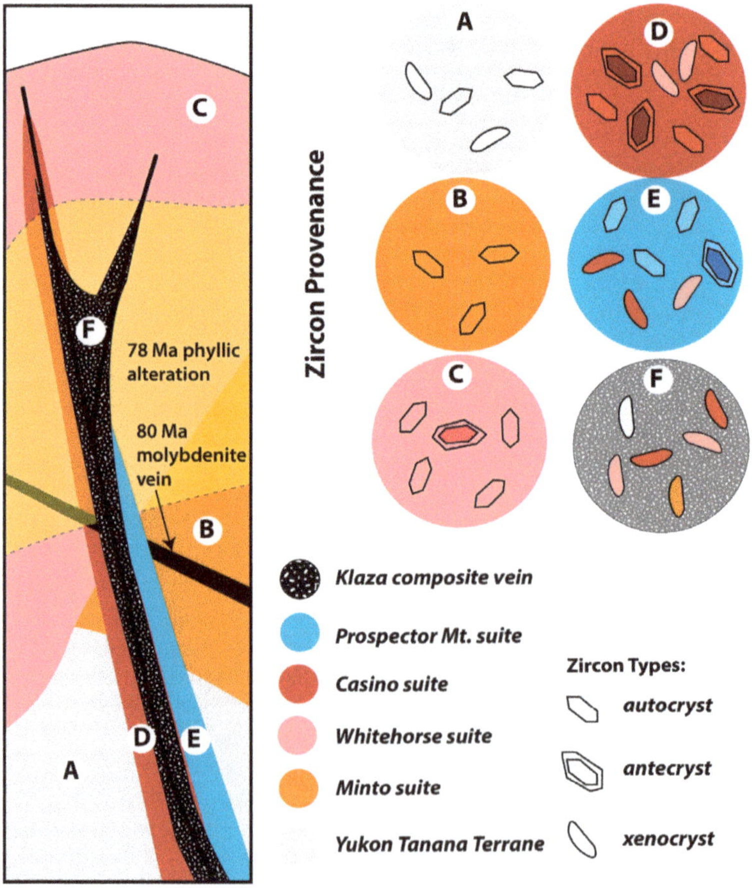

Figure 10. Schematic diagram depicting the provenance of igneous zircon in the different magmatic phases of the Klaza setting. This diagram illustrates the complexities of the magmatic environment of the Klaza system, where the Whitehorse suite, Casino suite, and Prospector Mountain suite share a common reservoir, resulting in the inheritance issues described above. Care must be taken by the geochronologist to ensure that only autocrysts are used in constraining the crystallization age of the intrusive phase.

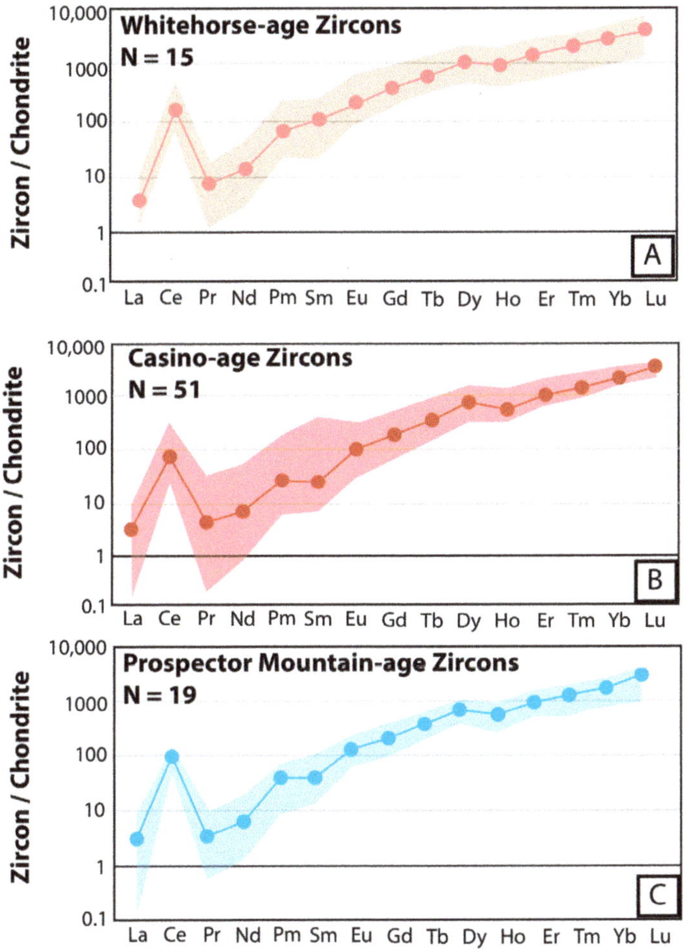

Figure 11. Plots of chondrite-normalized REE values for each group of zircons showing the range in grey and averages marked by dots. The diagrams are arranged in the order of Whitehorse suite (**A**), Casino suite (**B**) and Prospector Mountain suite (**C**). Chondrite data from Sun and McDonough [62].

Nb/Ta ratio: The Nb/Ta ratio is commonly used to assess the extent of fractionation of magmas whereby lower Nb/Ta ratios are correlated with more evolved rocks due to the partitioning of Nb and Ta commensurate with rutile and amphibole crystallization [63,64]. The results show a clear separation of the PMZ data which have higher (i.e., less evolved) Nb/Ta ratios (3 to 6) relative to CAZ-1 and CAZ-2 (Figure 12E). The data also show a diversity in magmatic evolution among the Whitehorse suite zircons.

Titanium-in-zircon thermometer: Ti-in-zircon thermometry [65] yielded temperature estimates that range from 900 ± 60 to 600 ± 40 °C (Figure 13A and Supplementary Table S6, Sheet B). As expected, there is a negative trend between temperature and degree of fractionation with the fields for CAZ-1 and CAZ-2 overlapping (Figure 13A) and there is no relationship with zircon age (Figure 13B).

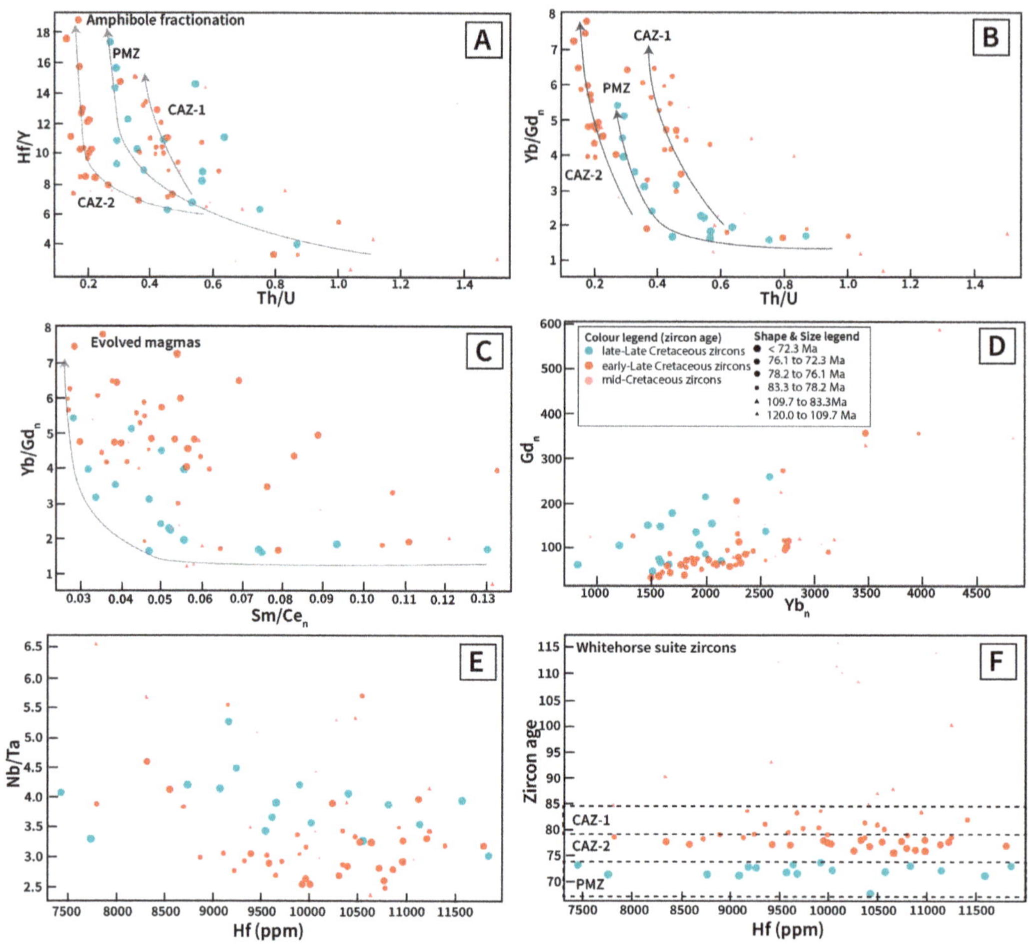

Figure 12. Summary of the trace element chemistry of zircons from different phases at the Klaza setting. (**A**) Binary plot comparing Hf/Y with Th/U. (**B**) Binary plot comparing Yb/Gd with Th/U. (**C**) Binary plot comparing Yb/Gd with Sm/Ce. (**D**) Binary plot comparing Gd with Yb. (**E**) Binary plot comparing Nb/Ta with Th/U. (**F**) Binary plot comparing calculated zircon U–Pb ages with Th/U.

Ce and Eu content: The Ce and Eu contents of zircons are used as proxies for estimating fO_2 conditions of magmas [54,66,67], with various methods used to calculate this parameter. For example, Zhong et al. (2019) discuss the disadvantages of using Ce_n/Ce^* based on $Ce^* = \sqrt{La_n \times Pr_n}$) versus $Ce^* = Nd_n{}^2/Sm_n$) (cf., Loader et al., 2017). Here, the approach of Zhong et al. (2019) was evaluated and the results (Supplementary Table S6, Sheet C and Supplementary Text S4) show the $Ce_n/Ce^*{}_C$ values are significantly higher compared with the values using the Loader et al. (2017) approach. The results based on Zhong et al. [68] are considered to be the best estimate of Ce anomalies since they address problems of overestimation in Ce_n/Ce^* using the method of Loader et al. [69] and, in addition, it avoids complications due to detection limits on La and Pr measurements in zircon. The Eu_n/Eu^* values were calculated with the standard approach of $Eu^* = \sqrt{Sm_n \times Gd_n}$.

The results for PMZ and CAZ-1 show a range in Eu_n/Eu^* values between 0.4 and 1.1, whereas CAZ-2 data cluster between 0.45 and 0.6 and there is no apparent linear relationship with temperature (Figure 13C). However, the large spikes in Eu_n/Eu^* values between 725 and 775 °C in some PMZ and CAZ-1 zircons are noted and thus indicate that the process (or processes) acted to change this value in some of the melts.

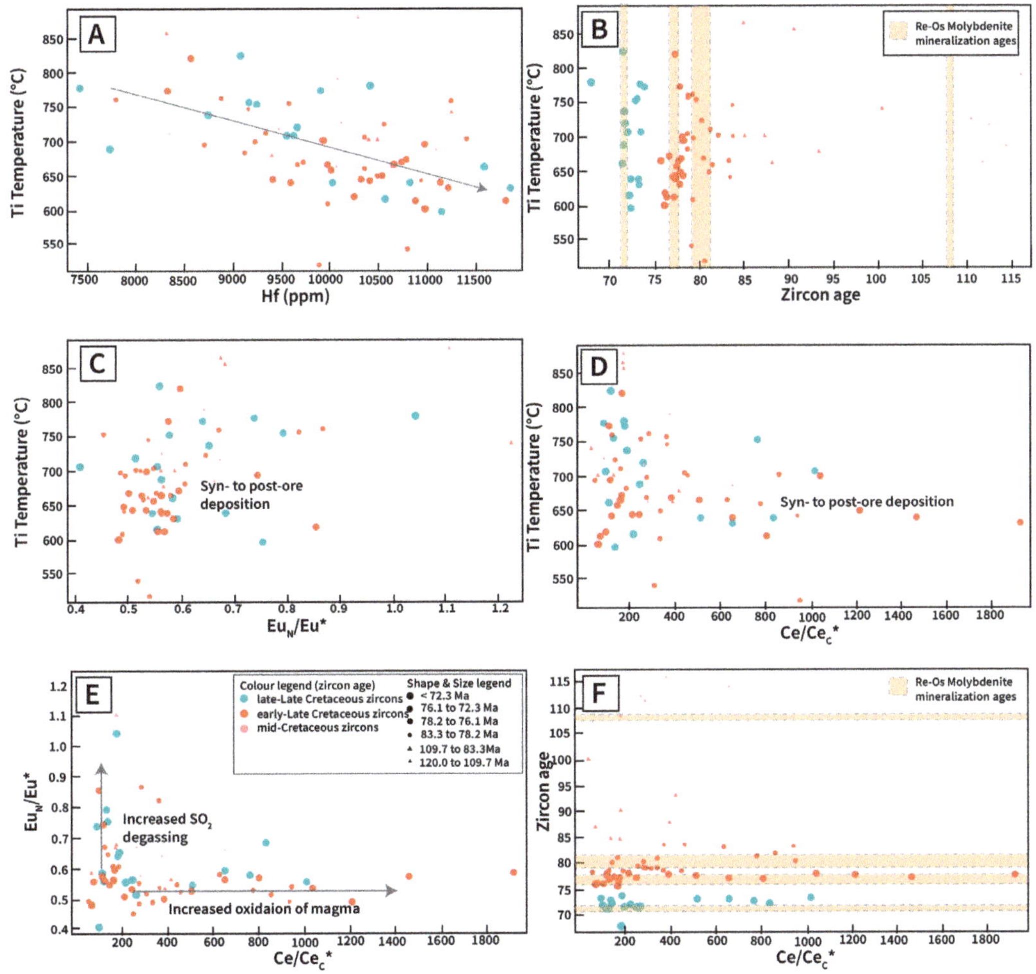

Figure 13. Summary of age, inferred temperatures of formation, and geochemical data for zircons from the different phases in the Klaza study area. (**A–D**) Binary plots comparing the calculated temperature based on Ti in zircon to Hf (**A**), calculated age (**B**), Eu/Eu* (**C**), and Ce/Ce*C (**D**). (**E**) Binary plot comparing Eu/Eu* with Ce/Ce*C. (**F**) Binary plot comparing calculated U–Pb ages with Ce/Ce*C.

The calculated Ce_n/Ce^*_C values of the zircons are also evaluated in regard to temperature (Figure 13D) and Eu_n/Eu^* values (Figure 13E). Two trends are noted in the data: (1) increasing Ce/Ce^*_C at generally lower temperatures, which may plausibly reflect the role of volatiles; and (2) increasing Ce/Ce^*_C at lower Eu_N/Eu^* values. In addition, the change in the calculated Ce/Ce^*_C parameter as a function of age (Figure 13F) indicates that the largest range, and highest values, are for the younger Casino and Prospector Mt. suites with values for the older Whitehorse suite samples much lower. The range

Ce_n/Ce^*_C values, therefore, reflect the relevant process (or processes) that appear to be time-dependent in regard to the nature of magmatism.

5. Discussion

Magmatism and hydrothermal mineralization at Klaza straddle the boundary between two of the most productive magmatic suites (by Cu–Au–Ag resource; Figure 14) in the DRGB. In addition, the Klaza-Brown-McDade system and the Freegold Mt. District (FGMD; Figure 1B) show characteristics of superimposed porphyry systems, thus harbouring increased potential for enriched Cu–Au–Mo ore zones [25].

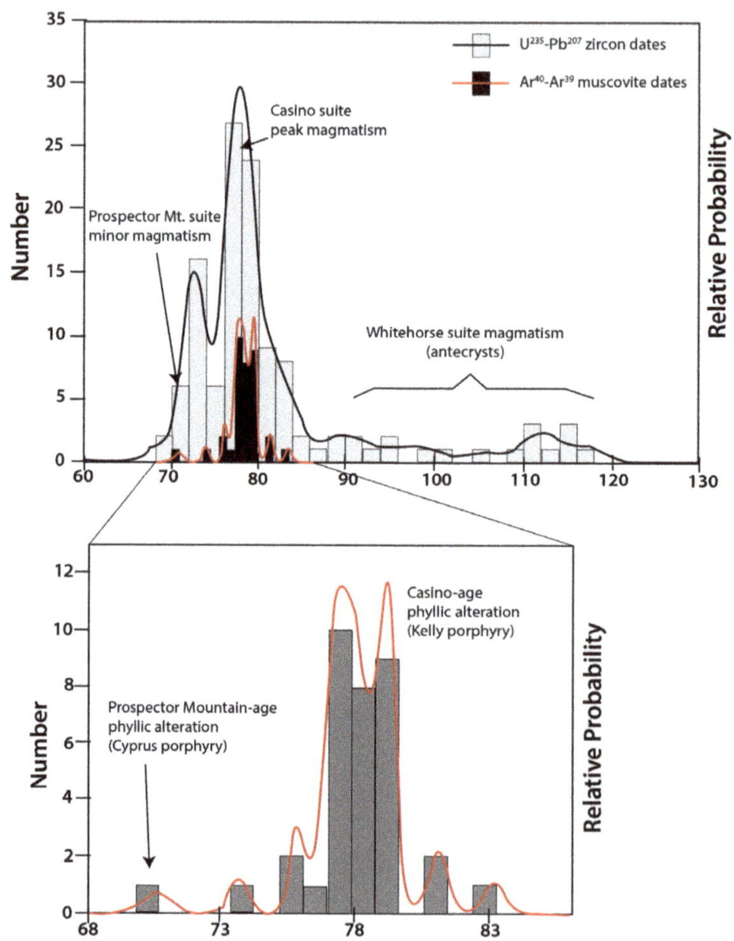

Figure 14. Histogram and relative probability plots for the various geochronological data (U–Pb zircon and Ar–Ar muscovite) from this study.

Our observational (i.e., field and petrography), geochemical-, and geochronological-based paragenetic reconstructions are now used to discuss the petrogenesis and evolution of the Klaza Igneous Complex (KIC) and the genesis of the Klaza superimposed porphyry–epithermal system. The KIC spans a highly protracted history of periodic magmatic activity (Late Triassic to Late Cretaceous) for a trans-crustal magmatic system. Magma mingling and a change in tectonic regime over time are suggested to have been responsible for the overall chemical evolution of these magmas, the implications of which are discussed below.

5.1. Three Superimposed Magmatic–Hydrothermal Systems

Hart and Langdon [16] and Mortensen et al. [17] discussed the similarities in hydrothermal mineralogy, NW-trending structures (Figure 1C), and deposit styles between the Klaza and Brown-McDade systems, and suggested linkages between the two systems, despite cross-cutting relationships between hydrothermal mineralization and dikes of mid-Cretaceous and Late Cretaceous ages differing between the two systems [17].

Geochronologic results (Re–Os in molybdenite ages) from this study (Table 2) provide definitive evidence to demonstrate that the Flex hydrothermal system (part of Brown-McDade) is of mid-Cretaceous age. Combined with Re–Os molybdenite ages from Lee et al. [25], the data support the superposition of three magmatic–hydrothermal events (Figure 15): (1) mid-Cretaceous Brown-McDade (Whitehorse suite); (2) Late Cretaceous Kelly (Casino suite); and (3) Late Cretaceous Cyprus-Klaza (Prospector Mt. suite).

Evidence of said superposition can be observed in the field where mid-Cretaceous-age dikes [17] spatially coincide with ca. 76–78 Ma phyllic alteration related to the alteration footprint of the Kelly porphyry system (this study). The stockwork quartz–chalcopyrite–molybdenite veins of the Kelly system are in turn overprinted by phyllic alteration inferred to be from the Klaza–Cyprus epithermal–porphyry system. The Klaza veins cut Prospector Mountain suite-aged dikes (ca. 71 Ma; this study) whereas the Cyprus porphyry yields a molybdenite age of ca. 71 Ma [25].

5.2. Petrogenesis of the Klaza Igneous Complex

The rocks from the Klaza deposit display effects of alteration overprint even in the least altered samples. Potassium and Na were demonstrated to be highly mobile and susceptible to the effects of alteration through grant isocons in Lee [18]. Because of this, K and Na in TAS (total alkali silica) diagrams were unsuitable to be used as the only discriminator of rock compositions. Niobium is a known indicator of tectonic setting due to its behaviour as an incompatible element, with significant enrichment in alkalic melts derived from plume-related magmas. Lithogeochemical data presented by Lee et al. [25] show a progressive increase in Nb content within the intrusive rocks from mid-Cretaceous to Late Cretaceous. This progressive increase in Nb, which is used as a proxy for alkalinity, was attributed to changes in the tectonic environment with the magma source remaining the same. The protracted interval of hydrothermal productivity from mid-Cretaceous to Late Cretaceous demonstrated in this study (Figure 15) supports the above interpretation.

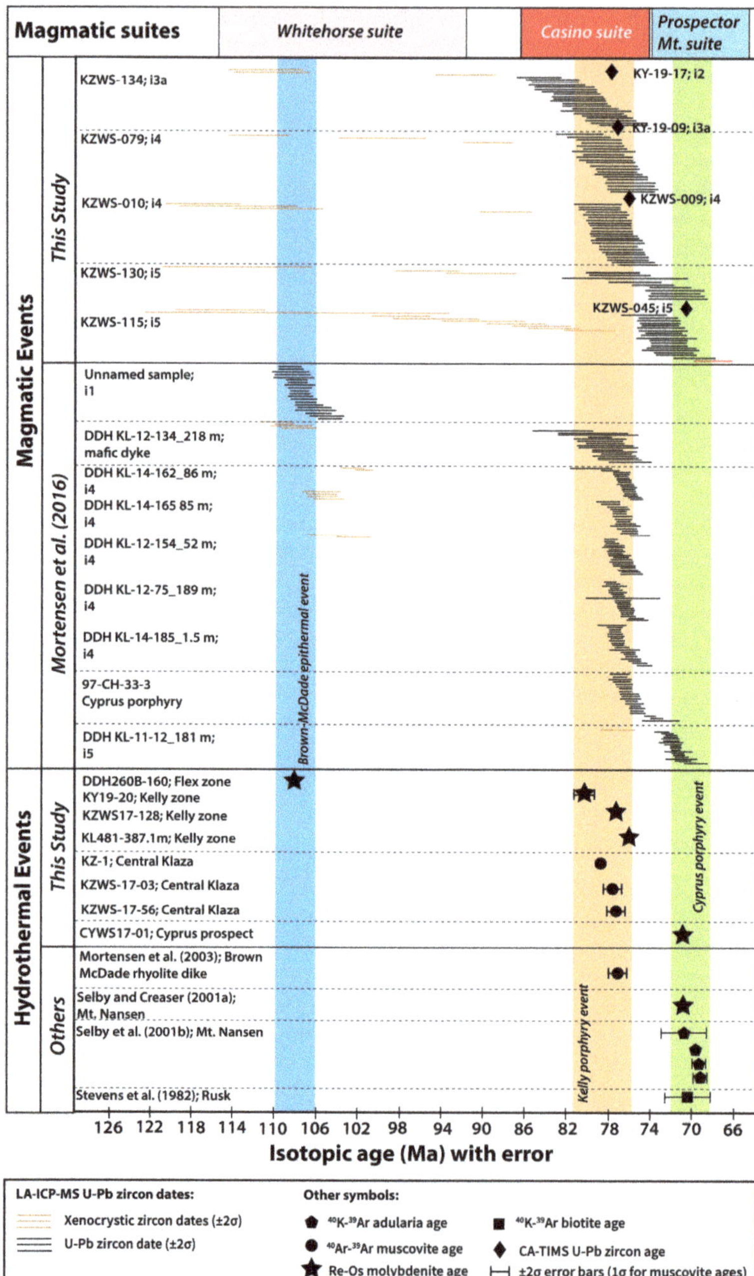

Figure 15. Compilation of geochronological data relevant to the Klaza study area based on this study and data from previous workers. Cited geochronological data are from Mortensen et al. [17,70], Selby and Creaser [33], Selby et al. [34], and Stevens et al. [71].

The Late Cretaceous magmas display several key features indicative of productive porphyry Cu systems: (1) a highly protracted magmatic history; (2) evidence for magma mingling; and (3) dynamic changes in magmatic composition with time. *Protracted magmatic history*: Geochronological constraints of Late Cretaceous magmatism are reflected in an apparent ~10 myr U–Pb zircon record from 80 to 70 Ma. A histogram of the relevant age data depicts a bimodal spread (Figure 14), with peaks corresponding to times of Casino (77–79 Ma) and Prospector Mt. (~72 Ma) age magmatic pulses. Productive porphyry deposits globally display evidence of protracted magmatic–hydrothermal activity [38,39,61,72–74], as seen in the regional setting of the study area. The unusually prolonged period of magmatic activity in mineralized porphyry settings, and more specifically at Klaza, provides opportunities for magma mingling (see below), compositional evolution (i.e., fractionation), and at multiple porphyry-related mineralizing events (Re–Os molybdenite; Figure 15 and [25]).

Magma chamber dynamics: Lee et al. [25] proposed a tectonic model whereby the KIC formed as a result of multiple emplacement events in a trans-crustal magmatic system [75] during the Late Cretaceous. This model addresses the similar whole rock geochemical footprint (REE and trace elements) of the Casino and Prospector Mt. suites in the KIC, with a suggested common deep crustal (MASH; melting–assimilation–storage–homogenization) source [73,75]. Magma mingling is documented as gradational contacts between Casino suite phases (i3a and i3b) in drill core samples (Figure 3A), whereas magma mingling is documented in drill core between the youngest phases of the Casino suite (ca. 73 Ma) and older phases (ca. 72 Ma) of Prospector Mt. suite magmas (Figure 4A).

Zircon trace element data for the Casino and Prospector Mt. suites show strong similarities, with the main differences lying in the MREE and HREE content of the magmas (Figure 11B,C). Since CAZ-1 and PMZ are temporally distinct, the similarities in trace element content and fractionation behaviour may reflect magma mingling and/or partial melting from a common deep reservoir. The trace element data (Hf/Y versus Th/U) also indicate the Casino and Prospector Mt. suites display characteristics of amphibole fractionation (Figure 12A); however, the low Yb/Gd values in CAZ-1, CAZ-2, and PMZ suggest a role of garnet fractionation in the melt history (Figure 12B) due to a deep source (see [73] for discussion). The observed variations in the geochemical compositions of the zircons in each suite (Figure 12) suggest an evolution related to the time and depth of partial melting.

The Ti-in zircon data or temperature and fluctuations in the Eu and Ce anomalies (Figure 13) are attributed to variations of temperature and oxidation states in the magma related to the recharging of the magmatic reservoir and fluid discharge events. Thus, the recorded variation in zircon temperature is interpreted to reflect the periodic influx of juvenile magma with commensurate higher temperatures [76,77], whereas the decrease in temperature likely records fractionation and/or possible loss of heat during magma degassing related to ore deposition events. Regarding the latter, the noted changes in Eu_n/Eu^* values mostly occur in PMZ and CAZ-1 zircons between 725 and 775 °C, as was noted in zircons elsewhere and attributed to SO_2 degassing of the magma [78] or reduced water content related to ore deposition [79]. Furthermore, that all zircons in the sample suites have Eu_n/Eu^* values above 0.4 suggests that SO_2 degassing and fluid content may be the dominant process impacting Eu distribution in the Klaza intrusive rocks; this has strong implications for ore deposition as volatile-rich magmas are documented around the world to be associated with productive ore systems ([73] and examples cited within). We further note that although Green and Pearson [80] proposed that strong positive Eu anomalies ($Eu_n/Eu^* > 1$) can be caused by plagioclase accumulation in the rock or fractionation of hornblende, Lee, R.G., et al. [79] proposed that higher Eu_n/Eu^* values may be the result of increasing volatile content in the melt or magma mingling. Evidence for the latter hypothesis is observable in the drill core where intrusive phases are in gradational contact (Figure 3A).

Although increased Ce anomalies in zircon are generally correlated to increased oxygen fugacity in the magma [66], the increase in Ce_n/Ce_C^* values of the Casino and Prospector Mt. suite rocks cannot be attributed to oxygen fugacity changes alone, since they correspond to lower Eu_n/Eu^* values (Figure 13E). Lee, R.G., et al. [79] propose zircon with these geochemical characteristics either formed from volatile-rich, crystal-poor magmas [81], or formed in a magma chamber that experienced high-temperature melt influx from a source that had yet to crystalize zircon and titanite. Such an event would enrich these zircons in Ce (and its Ce_n/Ce_C^*) and other trace elements. In regard to the present study, both hypotheses are likely since these suites are biotite–amphibole-bearing and display textural evidence of a dynamic magma chamber. Importantly, this implies the observed spikes in Ce/Ce_C^* values that correlate with porphyry mineralization events (Figure 13F) may be coincident with magma influx/recharge events.

Magmatic evolution: Lithogeochemical data for the various intrusive suites presented in Lee et al. [25] display a large compositional range for the Whitehorse, Casino and Prospector Mt. suites. The Whitehorse and Casino suite rocks generally have intermediate-to-felsic, calc-alkaline compositions (andesite to granodiorite), whereas the Prospector Mt. suite rocks record intermediate to slightly mafic and alkalic compositions (andesite to diorite). This apparent trend of increasing alkalinity and mafic composition in the evolution of these magmas is posited to reflect a change from subduction-derived magmas in the mid-Cretaceous to possibly extensional, non-subduction-related magmatism in the Late Cretaceous. This change in magmatism is recorded in other porphyry settings globally (e.g., Cripple Creek, Colorado, [82]; Northern Luzon, Philippines, [83]; Northern Chile, [84]; Western Tethyan Belt, [85]).

Based on whole-rock geochemical data, the Whitehorse and Minto suites display plagioclase fractionation-controlled negative Eu anomaly, whereas the Casino and Prospector Mt. suites (also plagioclase-bearing) have Eu_N/Eu^* values near unity. This feature is commonly interpreted, in particular for the causative magmas in mineralized porphyry settings, to be an artifact of highly oxidized magmas such that Eu is present as Eu^{3+} which precludes the development of a negative Eu anomaly [61,86]. Furthermore, both of the Late Cretaceous suites plot in the field of adakite-like rocks (ALR; La/Yb versus Yb plot), supporting zircon trace element results that suggest garnet was present in the melt source and whereby both the Casino and Prospector Mt. suites were likely sourced at greater depths [87] than the Whitehorse and Minto suite rocks.

Magma mingling and compositional changes are also suspected to contribute to the variety of zircon ages in each sample. The terminology for zircon provenance, as defined in Olierook et al. [88], is applied below. A significant amount of inherited zircon (i.e., xenocrystic) was identified in this study (Figure 15), as was also noted by Mortensen et al. [17]. As described by [89], the provenance of minerals in a trans-crustal magmatic system can be challenging to decipher since crystals can include phenocrysts that grew in the latter stage of magma evolution (autocrysts), antecrysts from earlier stages involving crystal cycling and related pulses, and/or xenocrysts from previous, unrelated magma (or the wallrock/basement). In addition, all the aforementioned types can be remobilized multiple times.

Zircons displaying mid-Cretaceous ages (Figure 15) are interpreted as xenocrysts sourced from the Whitehorse suite granodiorite which hosts these dikes whereas zircons with Late Triassic and Permian ages are interpreted as originating from the Minto suite and the metamorphosed basement, respectively. The distinction between antecrystic and autocrystic zircons is challenging, as it is in any study [88–90], particularly when fluctuations in the composition of the Late Cretaceous magmas can enhance or reduce zircon saturation and thus its consequent growth or dissolution. Zircon morphology (Supplementary Figures) displays a combination of bright and dark anhedral cores that are rimmed by dark oscillatory zones. As such, the youngest concordant zircons in each sample are interpreted as autocrystic. The remaining zircons in each sample are termed antecrystic and reflect crystallization in the same magma chamber but at an earlier time in its evolution.

The change in magma composition (towards more mafic, alkalic compositions) with time could explain the larger proportion of antecrystic zircons relative to the youngest concordant autocrystic zircons (Figure 10). Additionally, this could also explain the apparent historic sampling bias of Casino-age zircons by previous workers (Figure 15).

5.3. Assembly of the Klaza Igneous Complex and Superimposed Porphyry System

Herein we combine the tectonic and genetic model with the results from Lee et al. [25] with new data from this study to reconstruct the formation of the KIC and MNGC. We propose a subduction-initiated arc setting from 204 to 80 Ma (Figure 16A) with a steepening of the subducting slab (Kula plate) through westward rollback in the Late Cretaceous (80 to 70 Ma), following on the earlier work of Nelson et al. [2] at a more regional scale for the Northern Canadian Cordillera. The emplacement of the Minto and Whitehorse suite rocks is discussed here in addition to the KIC to provide tectonic context for the discussion.

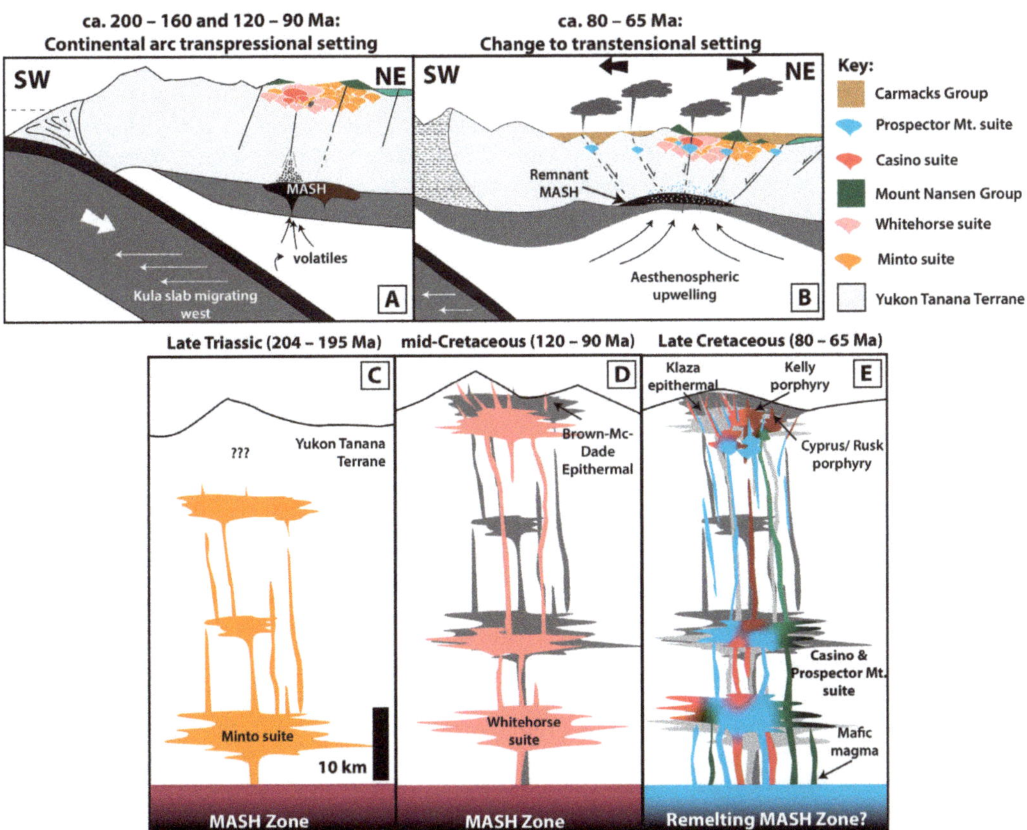

Figure 16. (**A**) Schematic tectonic model from the Late Triassic to the mid-Cretaceous (90 Ma), based on models proposed by Monger and Gibson [91]. (**B**) Schematic tectonic model for the Late Cretaceous

(80–65 Ma) supported by the model proposed by Gabrielse et al. [92]. (**C–E**) Schematic petrogenetic evolution of the district from the Late Triassic to the latest Cretaceous. Note the inset key in the upper right for various magmatic events.

Late Triassic to mid-Cretaceous (90 Ma): Arc-related subduction initiates partial melting of the hydrated mantle to form a MASH zone at the base of the crust. The Minto suite rocks were emplaced into the Yukon Tanana Terrane between 204 and 195 Ma at ~25 km depth (hornblende thermobarometry; [93]; Figure 16C). The Whitehorse suite is emplaced utilizing the same structures as the Minto suite to form a composite batholith in the MNGC (Figure 16D). The Brown-McDade epithermal system formed at shallow levels in the MNGC during this mid-Cretaceous magmatism at ca. 108 Ma.

A change in tectonic setting in the Late Cretaceous (80 Ma) led to the emplacement of the less voluminous Casino suite rocks (i.e., dikes and small bodies). Note that both the Minto and Whitehorse suites contain compositional variations and are themselves multiphase batholiths [1,20]. However, the complexities of these intrusive bodies are not discussed here or reflected in Figure 16 since insufficient zircon data were collected in the study area. The reader is referred to Sack et al. [20] for a regional compilation of whole rock and zircon data for Triassic–Jurassic intrusions in Yukon.

Late Cretaceous (80 to 72 Ma): The Casino suite magmas are bimodal in composition (felsic and mafic phases) and are inferred to have mingled at depth; they were emplaced as small pulses over a period of 8 myr (Figure 16E). As these suites are geochemically similar to the Whitehorse and Minto suites, it is posited that they were sourced from recycled MASH material. A dynamic magma chamber with frequent mafic influx events contributed to changing magma compositions. Mafic, amphibole- and plagioclase-rich phases (i3a, i3b) and felsic (i2) to intermediate phases (i3c) are emplaced in the Kelly zone, whereas intermediate to felsic dikes were emplaced at Klaza (see Figure 1C). Two pulses of molybdenite-bearing porphyry (Cu–Au–Mo) mineralization occur at 80 and 77 Ma, respectively, correlating to the emplacement of the Kelly porphyry.

The hydrothermal fluids associated with the Kelly porphyry formed a large, ca. 79 Ma phyllic footprint which overprints some of the currently intersected porphyry mineralization (Figure 3J,K). This footprint is reflected in the histogram of Ar–Ar muscovite dates in Figure 14, with a peak ranging from 77 to 80 Ma. This phyllic halo is interpreted to be an overprinting event coinciding with the collapsing thermal cell of the porphyry system.

Late Cretaceous (72 to 65 Ma): Current resolutions in geochronology data depict a continuous transition from Casino to Prospector Mt. suite magmatism (Figure 17), with the change in magma composition occurring gradually between 80 and 65 Ma ([25]; this study). This compositional change is hypothesized to reflect the gradual change from a subduction setting to a local extensional setting, as documented in other metallogenic settings and related to asthenospheric upwelling and re-melting of the pre-existing MASH zone [61,83,85,94]. The Prospector Mt. suite magma, which has a chemistry (e.g., depleted HREE signature) that reflects melt generation in a garnet-stable zone (i.e., deep), comprises a more mafic, slightly alkaline composition that mingled with the remnant, younger Casino suite magmas to generate a mixed magmatic signature (Figure 16E). The resulting hybrid Prospector Mt. suite magmas were emplaced at shallower, epizonal depths as narrow diorite dikes between 72 and 65 Ma.

Regionally, the Late Cretaceous (~80 Ma to 65 Ma) is described to be a time of tectonic change, with: (1) documented early signs of movement of the Tintina fault [92]; and (2) a transition point from a dominantly compressional setting with crustal thickening and mountain building activity towards a dominantly strike–slip setting [91].

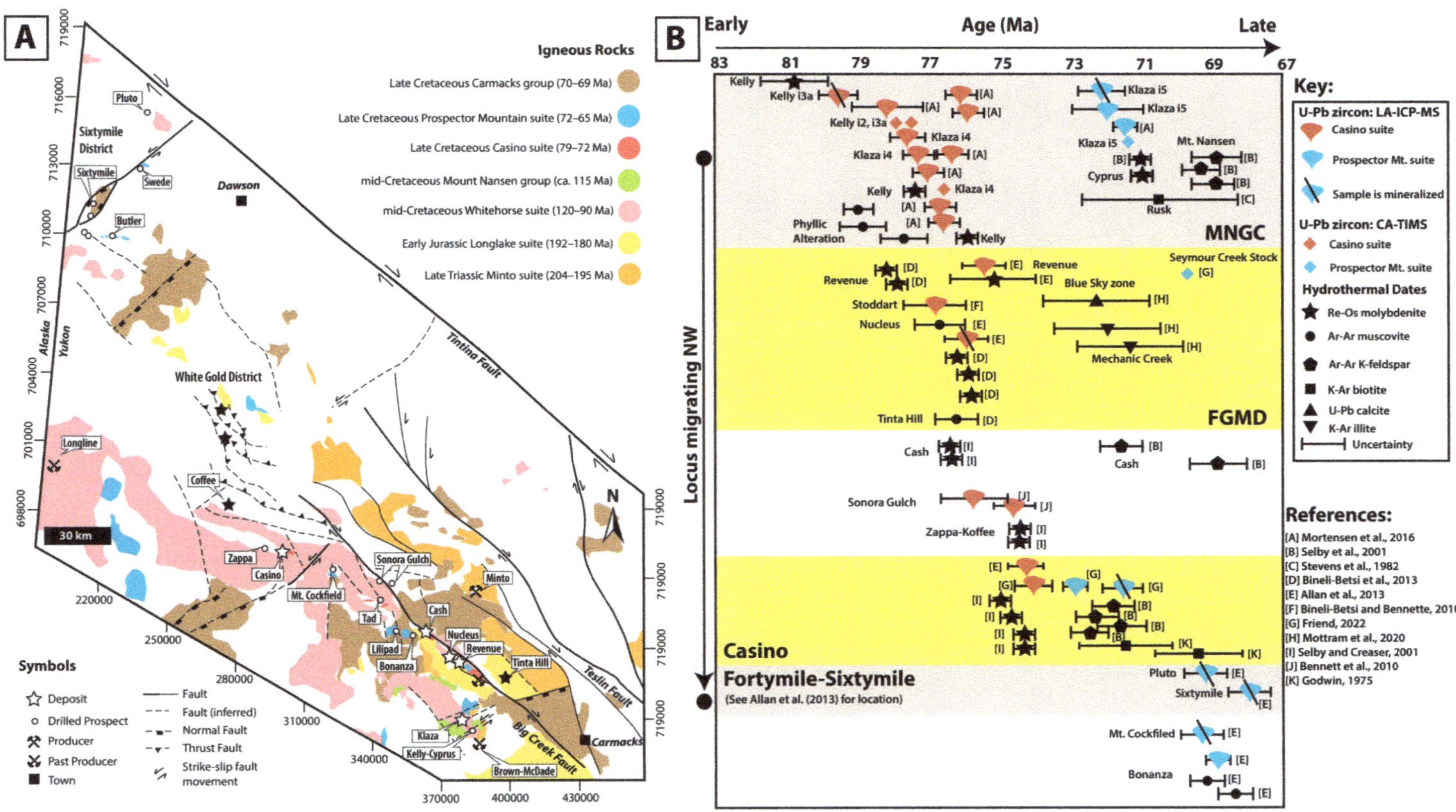

Figure 17. (**A**) Simplified geological map of the Dawson Range including the Sixtymile area. Only igneous rocks comprising ages discussed in this contribution are displayed. (**B**) Compilation of geochronologic data for other Late Cretaceous porphyry–epithermal occurrences in the Dawson Range Gold Belt. Regional geochronology data sourced from Allan et al. [1], Mortensen et al. [17], Mottram et al. [32], Selby and Creaser [33], Selby et al. [34], Stevens et al. [71], Bineli-Betsi et al. [95], Bineli-Betsi and Bennette [96], Friend [97], Bennett et al. [98], and Godwin [99].

Locally, this Late Cretaceous magmatic event is accompanied by porphyry mineralization as established from ca. 71 Ma Re–Os molybdenite ages [25,33] at Rusk and Cyprus, respectively. Although the footprint of Prospector Mt. related alteration is less significant in comparison to the Casino-age footprint, it cannot be dismissed since thermal overprints of this age are suggested in the low-temperature part of the Ar–Ar age-spectra (Figure 14) and Re–Os molybdenite ages (Figure 15). The Prospector Mt.-age alteration is also recorded in Ar–Ar K-feldspar [34] and K–Ar biotite [71] ages (Figure 15).

Lee et al. [25] proposed the above tectono-magmatic setting is conducive to forming alkalic-type epithermal Au and porphyry Cu–Au systems. Thus, the Cyprus and Rusk porphyry systems were emplaced within the MNGC with the possibility of overlapping with pre-existing Kelly porphyry-related mineralization at depth. The Klaza and Brown-McDade intermediate sulfidation epithermal veins were emplaced distally to the Cyprus–Rusk porphyry complex at NW and SE locations, respectively (Figure 1C). Based on the information presented here, we now know that the Brown-McDade system was emplaced in the mid-Cretaceous, whereas the Klaza system was emplaced as the Late Cretaceous, distal equivalent to the Cyprus–Rusk porphyry complex, with the magmatic and hydrothermal components of the system occupying the same NW trending structures that hosted mid-Cretaceous dikes and mineralization.

6. Exploration Implications

(1) The high-precision ages assigned to each intrusive phase (Figure 8) match the paragenetic order prescribed to each phase (i2 to i5), demonstrating that a consistent, paragenetically relevant logging scheme can be built for the Klaza deposit, which will likely improve grade-control models in the future.

(2) Complexities in the petrogenesis of the Casino and Prospector Mt. suite rocks have resulted in unequal distributions of zircon populations and produced a geochronologic record biased towards Casino (80–72 Ma) suite ages (Figure 14).

(3) Hydrothermal ages documenting Prospector Mt.-age alteration (see Figure 14) are present in the older (pre-2010) literature; however, many of these dates were interpreted as outlier values or minerals that were thermally reset due to emplacement of the Carmacks Group basalts [1,17].

(4) Evidence of Prospector Mt.-age porphyry mineralization (Re–Os molybdenite; [25,33]) from the Cyprus prospect (MNGC) provides strong support for the validity of the hydrothermal dates yielded from biotite, hydrothermal K-feldspar, and muscovite from other localities in the DRGB (Figure 17B).

(5) Current regional maps do not document the presence of the Carmacks Group basalts near localities where Prospector Mt.-age alteration is recorded (Figure 1B). Thermochronological studies [100,101] also demonstrate that it is highly unlikely that the Carmacks Group basalts underwent significant erosion since their emplacement at 70 Ma.

(6) Geochronology data from other porphyry occurrences in the DRGB (Figure 17B) suggests: (i) regional magmatism in the DRGB occurs nearly continuously between 80 and 67 Ma; (ii) the boundary between Casino and Prospector Mt. suite rocks cannot be easily distinguished on the basis of geochronology alone; (iii) the locus of Casino suite magmatism and hydrothermal mineralization appears to migrate along the Big Creek Fault from the MNGC in the SE towards the Casino deposit in the NW; and (iv) the transition into Prospector Mt. suite magmatism occurs at the Casino deposit, and then proceeds further NW into the Sixtymile district and Fortymile district (Pluto), prior to being emplaced as isolated plugs in the MNGC, the FGMD, at Mt. Cockfield,

and at Bonanza. Mineralization and magmatism correlated with the Prospector Mt. suite is also documented in eastern Alaska [1,102].

(7) The new geochronological results presented in this study and Lee et al. [25] enable the temporal window for PCD prospectivity in the DRGB to be extended by 7 myr, encompassing Prospector Mt. suite rocks, currently identified to be sourced from environments conducive for forming PCDs.

(8) Geochronological data combined with whole rock and zircon geochemical results also indicate strong similarities between the Casino and Prospector Mt. suites, with hybridized signatures varied by temporal factors. The results from this study suggest that these suites may be one-and-the-same, with the youngest a slightly evolved counterpart of the same magma source.

(9) The geochronological data from the DRGB, when viewed together from a regional perspective (Figure 17), depicts a large (~200 km-long), protracted magmatic–hydrothermal event stretching from the town of Carmacks (Yukon) into eastern Alaska. This long-lived magmatic–hydrothermal system produced oxidized, hydrous magmas, and is strongly correlated with regional, orogen-parallel, transcrustal strike–slip structures (Big Creek Fault), hence containing significant tectonomagmatic building blocks [75] required to form porphyry Cu–Au districts.

(10) The scale of this Late Cretaceous porphyry activity in the DRGB (~200 km-long) is comparable to the metallogenic belt of northern Chile. When overlain, the DRGB would encompass the Escondida district [103], Chuquicamata-El Abra [104], and Quebrada Blanca (note also that Late Cretaceous magmatism and mineralization also extends ~600 km further south into British Columbia and northeast into Alaska). Therefore, it is important for regional explorers to adopt a belt-scale approach to exploration in the DRGB as a metallogenic corridor.

(11) Findings from this study can be further applied to other porphyry-productive metallogenic districts globally (e.g., Western Tethyan Belt, Turkey, Northern Luzon, Philippines, Northern Chile) in the search for superimposed hypogene ore shells.

(12) Should sampling biases similar to that discussed for the DRGB be identified, previously overlooked intrusive suites can be revaluated for prospectivity, thus potentially increasing the number of prospective mineral targets in established districts.

7. Conclusions

This study area experienced two changes in regard to broad-scale geotectonics in the mid-Cretaceous and Late Cretaceous which translated into varied arc-type magmatism and related mineralization. The nature and subsequent evolution of these magmas combined with their high-level crustal setting generated favourable fertile causative intermediate to felsic, oxidized, hydrous magmas which culminated in the formation of porphyry–epithermal mineralization at three times: (1) high-to-intermediate sulphidation epithermal deposit formation at Brown-McDade at ca. 108 Ma; (2) typical calc-alkaline arc-type Casino suite magmas at 80–72 Ma that gave rise to forming porphyry Cu–Au–Mo deposits; and (3) a subsequent back-arc extensional environment at ca. 72 Ma that formed an environment conducive to forming intermediate-sulfidation Au–Ag–Pb–Zn epithermal systems related to the Prospector Mt. magmas.

The Casino and Prospector Mt. suite magmas show evidence of magma mingling and mafic magma influx corresponding with mineralization events (Re–Os molybdenite). Both magmas show hybridized signatures indicative of a common source. The overall structural controls in the DRGB increase the likelihood of the superposition of the two aforementioned metallogenic events to form mineral deposits with enriched ore shells. The same structural controls benefited the Klaza system by allowing the focusing of hydrothermal fluids through long-lived crustal-scale structures to form wide, Au-rich vein intersections.

Findings from this study can be used to inform regional explorers in the DRGB and in other porphyry belts globally. Petrogenic and geochronologic constraints on causative and non-prospective intrusive suites in these districts should be revaluated so that opportunities for new discoveries are not missed due to human bias.

Supplementary Materials: The following supporting information can be downloaded at: https://www.mdpi.com/article/10.3390/min15010038/s1, Figures: Zircon Morphology; Table S1: Ar-Ar Muscovite Results; Table S2: CA-TIMS U-Pb results; Table S3: Boise University LA-ICP-MS data; Table S4: Zircon Trace Element Chemistry machine configuration; Table S5: University of Alberta LA-ICP-MS data; Table S6: Zircon Trace Element Data; Table S7: Sample locations; Table S8: Regional Geochronology Compilation; Text S1: Ar-Ar and Re-Os Geochronology; Text S2: Zircon Analytical Methods; Text S3: Zircon Analytical Methods Part 2; Text S4: Ce Comparison Plots.

Author Contributions: Conceptualization, W.-S.L.; methodology, W.-S.L., J.L.C. and R.A.C.; validation, W.-S.L., J.L.C. and R.A.C.; formal analysis, W.-S.L., J.L.C. and R.A.C.; investigation, W.-S.L.; resources, D.J.K. and P.J.S.; data curation, W.-S.L., J.L.C. and R.A.C.; writing—original draft preparation, W.-S.L.; writing—review and editing, W.-S.L., D.J.K. and P.J.S.; visualization, W.-S.L.; supervision, D.J.K. and P.J.S.; project administration, D.J.K.; funding acquisition, W.-S.L. and P.J.S. All authors have read and agreed to the published version of the manuscript.

Funding: The authors are grateful for in-kind financial and logistical support from the Yukon Geological Survey (YGS) to conduct fieldwork at the Klaza property in 2019. We also thank Rockhaven Resources Ltd. (RK) for in-kind financial and logistical support to visit and sample at the Klaza property in 2017 and 2019. We acknowledge financial support from the Targeted Geoscience Initiative (TGI 5) for the early phase of this study.

Data Availability Statement: Data are contained within the article and Supplementary Materials.

Acknowledgments: The work presented in this paper is a result of the lead author's PhD thesis at Laurentian University. Jeff Marsh is thanked for his contributions in zircon trace element chemical analysis at Laurentian University. Andy DuFrane is thanked for his contributions in LA-ICP-MS U–Pb zircon analysis at the University of Alberta. Four anonymous reviewers are thanked for their time and feedback, which greatly improved this manuscript. This is Yukon Geological Survey contribution #070.

Conflicts of Interest: W.-S.L. is currently employed by Galore Creek Mining Corporation (GCMC). The paper reflects the views of the authors and not GCMC. We declare that the research was conducted in the absence of any commercial or financial relationships with GCMC that could be construed as a potential conflict of interest.

References

1. Allan, M.M.; Mortensen, J.K.; Hart, C.J.; Bailey, L.A.; Sánchez, M.G.; Ciolkiewicz, W.; McKenzie, G.G.; Creaser, R.A. Magmatic and metallogenic framework of west-central Yukon and eastern Alaska. In *Tectonics, Metallogeny, and Discovery: The North American Cordillera and Similar Accretionary Settings*; Society of Economic Geologists: Littleton, CO, USA, 2013; Volume 17, pp. 111–168.
2. Nelson, J.L.; Colpron, M.; Israel, S. Tectonics and Metallogeny of the Northern Cordillera of British Columbia, Yukon and Alaska. In *Tectonics, Metallogeny, and Discovery: The North American Cordillera and Similar Accretionary Settings*; Society of Economic Geologists: Littleton, CO, USA, 2013; Volume 17, pp. 53–110.
3. Colpron, M.; Crowley, J.L.; Gehrels, G.; Long, D.G.F.; Murphy, D.C.; Beranek, L.; Bickerton, L. Birth of the northern Cordilleran orogen, as recorded by detrital zircons in Jurassic synorogenic strata and regional exhumation in Yukon. *Lithosphere* **2015**, *7*, 541–562. [CrossRef]
4. Corriveau, L. Iron oxide copper-gold deposits: A Canadian perspective. In *Mineral Deposits of Canada: A Synthesis of Major Deposit-types, District Metallogeny, the Evolution of Geological Provinces, and Exploration Methods*; Geological Association of Canada, Mineral Deposits Division, Special Publication: St. Johns, NL, Canada, 2007; Volume 5, pp. 307–328.

5. Dubé, B.; Gosselin, P.; Mercier-Langevin, P.; Hannington, M.; Galley, A. Gold-rich volcanogenic massive sulphide deposits. In *Mineral Deposits of Canada: A Synthesis of Major Deposit-Types, District Metallogeny, the Evolution of Geological Provinces, and Exploration Methods*; Geological Association of Canada, Mineral Deposits Division, Special Publication: St. Johns, NL, Canada, 2007; Volume 5, pp. 75–94.
6. Galley, A.G.; Hannington, M.; Jonasson, I. Volcanogenic massive sulfide deposits. In *Mineral Deposits of Canada: A Synthesis of Major Deposit-Types, District Metallogeny, the Evolution of Geological Provinces, and Exploration Methods*; Geological Association of Canada, Mineral Deposits Division, Special Publication: St. Johns, NL, Canada, 2007; Volume 5, pp. 141–161.
7. Goodfellow, W.D.; Lydon, J.W. Sedimentary-exhalative (SEDEX) deposits. In *Mineral Deposits of Canada: A Synthesis of Major Deposit-Types, District Metallogeny, the Evolution of Geological Provinces, and Exploration Methods*; Geological Association of Canada, Mineral Deposits Division, Special Publication: St. Johns, NL, Canada, 2007; Volume 5, pp. 163–183.
8. Hart, C.J.R. Reduced intrusion-related gold systems. In *Mineral Deposits of Canada: A Synthesis of Major Deposit-Types, District Metallogeny, the Evolution of Geological Provinces, and Exploration Methods*; Geological Association of Canada, Mineral Deposits Division, Special Publication: St. Johns, NL, Canada, 2007; Volume 5, pp. 95–112.
9. Paradis, S.; Hannigan, P.; Dewing, K. Mississippi Valley-type lead-zinc deposits. In *Mineral Deposits of Canada: A Synthesis of Major Deposit-Types, District Metallogeny, the Evolution of Geological Provinces, and Exploration Methods*; Geological Association of Canada, Mineral Deposits Division, Special Publication: St. Johns, NL, Canada, 2007; Volume 5, pp. 185–203.
10. Peter, J.M.; Layton-Matthews, D.; Piercey, S.; Bradshaw, G.; Paradis, S.; Boulton, A. Volcanic-hosted massive sulphide deposits of the Finlayson Lake District, Yukon. In *Mineral Deposits of Canada: A Synthesis of Major Deposit-Types, District Metallogeny, the Evolution of Geological Provinces, and Exploration Methods*; Geological Association of Canada, Mineral Deposits Division, Special Publication: St. Johns, NL, Canada, 2007; Volume 5, pp. 471–508.
11. Sinclair, W.D. Porphyry Deposits. In *Mineral Deposits of Canada: A Synthesis of Major Deposit-Types, District Metallogeny, the Evolution of Geological Provinces, and Exploration Methods*; Geological Association of Canada, Mineral Deposits Division, Special Publication: St. Johns, NL, Canada, 2007; Volume 5, pp. 223–243.
12. Taylor, B.E. Epithermal gold deposits. In *Mineral Deposits of Canada: A Synthesis of Major Deposit-Types, District Metallogeny, the Evolution of Geological Provinces, and Exploration Methods*; Geological Association of Canada, Mineral Deposits Division, Special Publication: St. Johns, NL, Canada, 2007; Volume 5, pp. 113–139.
13. Ross, A.A.; Kirchner, I.; Martin, C.J.; Dumala, M.R.; Methven, G. *Technical Report Describing Updated Diamond Drilling, Metallurgical Testing and Mineral Resources on the Klaza Property, Yukon, Canada*; NI-43-101 Technical Report; Rockhaven Resources Ltd.: Vancouver, BC, Canada, 2018.
14. Ross, A.A.; Kirchner, I.; Martin, C.J.; Dumala, M.R.; Methven, G.; Molavi, M.; Warren, D.; Borntraeger, B. *Technical Report Describing Updated Diamond Drilling, Metallurgical Testing and Mineral Resources on the Klaza Property, Yukon, Canada*; NI-43-101 Technical Report; Rockhaven Resources Ltd.: Vancouver, BC, Canada, 2020; 263p.
15. Lamb, J. The Geology and Mineralogy of the Brown McDade Mine. Master's Thesis, University of British Columbia, Vancouver, BC, Canada, 1947; 65p.
16. Hart, C.J.R.; Langdon, D.M. Geology and mineral deposits of the Mount Nansen camp, Yukon. In *Yukon Exploration and Geology 1997*; Government of Yukon: Whitehorse, YT, Canada, 1997; pp. 129–138.
17. Mortensen, J.K.; Hart, C.J.R.; Tarswell, J.; Allan, M.M. U-Pb zircon age and Pb isotopic constraints on the age and origin of porphyry and epithermal vein mineralization in the eastern Dawson Range, Yukon. In *Yukon Exploration Geology 2015*; Yukon Geological Survey: Whitehorse, YT, Canada, 2016; pp. 165–185.
18. Lee, W.-S. Metallogeny and Characterization of late Cretaceous Superimposed Porphyry Cu-Au-Mo and Epithermal Au-Ag Systems in the Dawson Range, Yukon, Canada: Case Study on the Klaza Deposit. Ph.D. Thesis, Laurentian University, Sudbury, ON, Canada, 2021; 272p. Available online: https://zone.biblio.laurentian.ca/handle/10219/3667 (accessed on 1 December 2022).
19. Klepacki, D.W.; Wheeler, J.O. *Stratigraphic and Structural Relations of the Milford, Kaslo and Slocan Groups, Goat Range, Lardeau and Nelson Map Areas, British Columbia*; Geological Survey of Canada, Paper, 85-1A; Geological Survey of Canada: Ottawa, ON, Cananda, 1985; pp. 277–286.
20. Sack, P.J.; Colpron, M.; Crowley, J.L.; Ryan, J.J.; Allan, M.M.; Beranek, L.P.; Joyce, N.L. Atlas of Late Triassic to Juras-sic plutons in the Intermontane terranes of Yukon. In *Yukon Geological Survey Open File 2020-1*; Yukon Geological Survey: Whitehorse, YT, Canada, 2020; 365p.
21. Kovacs, N.; Allan, M.M.; Crowley, J.L.; Colpron, M.; Hart, C.J.R.; Zagorevski, A.; Creaser, R.A. Carmacks Copper Cu-Au-Ag Deposit: Mineralization and Post ore Migmatization of a Stikine Arc Porphyry Copper System in Yukon, Canada. *Econ. Geol.* **2020**, *115*, 1413–1442. [CrossRef]
22. Logan, J.M.; Mihalynuk, M.G. Tectonic controls on early Mesozoic paired alkaline porphyry deposit belts (Cu-Au + Ag-Pt-Pd-Mo) within the Canadian Cordillera. *Econ. Geol.* **2014**, *109*, 827–858. [CrossRef]

23. Klöcking, M.; Mills, L.; Mortensen, J.; Roots, C. Geology of mid-Cretaceous volcanic rocks at Mount Nansen, central Yukon, and their relationship to the Dawson Range batholith. In *Yukon Geological Survey Open File 2016-25*; Yukon Geological Survey: Whitehorse, YT, Canada, 2016; 37p.
24. Sim, R.C.; Davis, B. *Freegold Mt. Project*; NI-43-101 Technical Report; Premium Nickel Resources Ltd.: Toronto, ON, Canada, 2020; 193p.
25. Lee, W.-S.; Kontak, D.J.; Richards, J.P.; Barresi, T.; Creaser, R.A. Superimposed Porphyry systems in the Dawson Range, Yukon. In *Tectonomagmatic Influences on Metallogeny and Hydrothermal Ore Deposits: A Tribute to Jeremy P. Richards*; Society of Economic Geologists: Littleton, CO, USA, 2021; Volume 1, pp. 21–48.
26. Mottram, C.M.; Kellett, D.A.; Barresi, T.; Chapman, G.G.; Halle, J. Tracking the porphyry-epithermal mineralization transition using U-Pb carbonate dating. *Geology* **2024**, *52*, 723–728. [CrossRef]
27. Smuk, K. Metallogeny of Epithermal Gold and Base Metal Veins of the Southern Dawson Range, Yukon. Master's Thesis, McGill University, Montreal, QC, Canada, 1998; p. 155.
28. Glasmacher, U.; Freidrich, G. Volcanic-hosted epithermal gold-sulfide mineralization and associated enrichment processes, Sixtymile River area, Yukon Territory, Canada. In *Yukon Geology*; Government of Yukon: Whitehorse, YT, Canada, 1992; Volume 3, pp. 271–291.
29. Johnston, S.T.; Wynne, P.J.; Francis, D.; Hart, C.J.R.; Enkin, R.J.; Engebretson, D.C. Yellowstone in Yukon: The Late Cretaceous Carmacks Group. *Geology* **1996**, *24*, 997–1000. [CrossRef]
30. Mortensen, J.K.; Hart, C.J.R. Late and postaccretionary Mesozoic magmatism and metallogeny in the northern Cordillera, Yukon and east-central Alaska. In *Geological Society of America Abstracts with Program*; The Geological Society of America: Boulder, CO, USA, 2010; Volume 42, p. 676.
31. Ciolkiewicz, W.; Mortensen, J.; Ryan, J.; Hart, C. Space-time composition patterns of the Late Cretaceous magmatism in west-central Yukon and east-central Alaska: Insights into the tectonic evolution of the northern Cordillera. In Proceedings of the Cordilleran Tectonics Workshop, Victoria, BC, Canada, 24–26 February 2012; Schedule and Abstracts. p. 15.
32. Mottram, C.M.; Kellett, D.A.; Barresi, T.; Zwingmann, H.; Friend, M.; Todd, A.; Percival, J.B. Syncing fault rock clocks: Direct comparison of U-Pb carbonate and K-Ar illite fault dating methods. *Geology* **2020**, *48*, 1179–1183. [CrossRef]
33. Selby, D.; Creaser, R.A. Late and mid-Cretaceous mineralization in the Northern Canadian Cordillera: Constraints from Re-Os molybdenite dates. *Econ. Geol.* **2001**, *96*, 1461–1467. [CrossRef]
34. Selby, D.; Nesbitt, B.E.; Creaser, R.A.; Reynolds, P.H.; Muehlenbachs, K. Evidence for a nonmagmatic component in potassic hydrothermal fluids of porphyry Cu-Au-Mo systems, Yukon, Canada. *Geochim. Cosmochim. Acta* **2001**, *65*, 571–587. [CrossRef]
35. Yukon Geological Survey. Yukon Digital Bedrock Geology. Available online: http://www.geology.gov.yk.ca/update_yukon_bedrock_geology_map.html (accessed on 20 August 2019).
36. Sack, P.J.; Eriks, N.; van Loon, S. Revised geological map of Mount Nansen area (NTS 115/I3 and part of 115I/2). In *Yukon Geological Survey, Open File*; 2 Sheets: Sheet 1 Scale 1:50 000 and Sheet 2 Scale 1:20 000; Yukon Geological Survey: Whitehorse, YT, Canada, 2022.
37. Lee, W.-S.; Kontak, D.J.; Richards, J.P.; Sack, P. Updated geology and porphyry copper potential of the Klaza deposit, Mount Nansen district (Yukon MINFILE 115I 067). In *Yukon Exploration and Geology 2019*; Yukon Geological Survey: Whitehorse, YT, Canada, 2020; pp. 75–97.
38. Seedorff, E.; Dilles, J.H.; Proffett, J.M.; Einaudi, M.T.; Zurcher, L.; Stavast, W.J.; Johnson, D.A.; Barton, M.D. Porphyry Deposits: Characteristics and Origin of Hypogene Features. In *Economic Geology 100th Anniversary Volume*; Society of Economic Geologists: Littleton, CO, USA, 2005; pp. 251–298.
39. Sillitoe, R.H. Porphyry Copper Systems. *Econ. Geol.* **2010**, *105*, 3–41. [CrossRef]
40. Whitney, D.L.; Evans, B.W. Abbreviations for names of rock-forming minerals. *Am. Mineral.* **2010**, *95*, 185–187. [CrossRef]
41. Monecke, T.; Monecke, J.; Reynolds, T.J. The influence of CO_2 on the solubility of quartz in single-phase hydrothermal fluids: Implications for the formation of stockwork veins in porphyry copper deposits. *Econ. Geol.* **2019**, *114*, 1195–1206. [CrossRef]
42. Lee, W.-S.; Kontak, D.J.; Petts, D.C.; Jackson, S.E. Mineral chemistry-driven protocol to unravel the complex paragenesis of the Klaza (Yukon, Canada) superimposed porphyry epithermal system. *Ore Geol. Rev.* **2022**, *143*, 27. [CrossRef]
43. Moncada, D.; Mutchler, S.; Neito, A.; Reynolds, T.J.; Rimstidt, J.D.; Bodnar, R.J. Mineral textures and fluid inclusion petrography of the epithermal Ag–Au deposits at Guanajuato, Mexico: Application to exploration. *J. Geochem. Explor.* **2012**, *114*, 20–35. [CrossRef]
44. Hitzman, M.W. Routine staining of drill core to determine carbonate mineralogy and distinguish carbonate alteration textures. *Miner. Depos.* **1999**, *39*, 794–798. [CrossRef]
45. Zamora-Vega, O.; Richards, J.P.; Spell, T.; Dufrane, S.A.; Williamson, J. Multiple mineralization events in the Zacatecas Ag-Pb-Zn-Cu-Au district, and their relationship to the tectonomagmatic evolution of the Mesa Central, Mexico. *Ore Ge-Ology Rev.* **2018**, *102*, 519–561. [CrossRef]

46. Selby, D.; Creaser, R.A. Macroscale NTIMS and microscale LA-MC-ICP-MS Re-Os isotopic analysis of molybdenite: Testing spatial restrictions for reliable Re-Os age determinations, and implications for the decoupling of Re and Os within molybdenite. *Geochim. Cosmochim. Acta* **2004**, *68*, 3897–3908. [CrossRef]
47. Mattinson, J.M. Zircon U-Pb chemical abrasion ("CA-TIMS") method: Combined annealing and multi-step partial dis-solution analysis for improved precision and accuracy of zircon ages. *Chem. Geol.* **2005**, *220*, 47–66. [CrossRef]
48. Simonetti, A.; Heaman, L.M.; Chacko, T.; Banerjee, N.R. In situ petrographic thin section U-Pb dating of zircon, monazite, and titanite using laser ablation-MC-ICP-MS. *Int. J. Mass Spectrom.* **2006**, *253*, 87–97. [CrossRef]
49. Patton, C.; Woodhead, J.D.; Hellstrom, J.C.; Hergt, J.M.; Greig, A.; Mass, R. Improved laser ablation U-Pb zircon geochronology through robust downhole fractionation correction. *Geochem. Geophys. Geosyst.* **2010**, *11*, 36. [CrossRef]
50. Patton, C.; Hellstrom, J.; Paul, B.; Woodhead, J.; Hergt, J. Iolite: Freeware for the visualisation and processing of mass spectrometric data. *J. Anal. At. Spectrom.* **2011**, *26*, 2508–2518. [CrossRef]
51. Fleck, R.J.; Sutter, J.F.; Elliot, D.H. Interpretation of discordant 40Ar/39Ar age spectra of mesozoic tholeiites from antarctica. *Geochim. Cosmochim. Acta* **1977**, *41*, 15–32. [CrossRef]
52. Bellousova, E.A.; Griffin, W.L.; O'Reilly, S.Y. Zircon Crystal Morphology, Trace Element Signatures and Hf Isotope Composition as a Tool for Petrogenetic Modelling: Examples From Eastern Australian Granitoids. *J. Petrol.* **2006**, *47*, 329–353. [CrossRef]
53. Corfu, F.; Hanchar, J.H.; Hoskin, P.W.O.; Kinny, P. Atlas of Zircon Textures. *Rev. Mineral. Geochem.* **2003**, *53*, 469–500. [CrossRef]
54. Lu, Y.; Loucks, R.R.; Fiorentini, M. Zircon compositions as a pathfinder for porphyry Cu ± Mo ± Au deposits. In *Tectonics and Metallogeny of the Tethyan Orogenic Belt*; Economic Geology Special Publication 19; Society of Economic Geologists: Littleton, CO, USA, 2016; pp. 329–347.
55. Watson, E.B.; Wark, D.A.; Thomas, J.B. Crystallization thermometers for zircon and rutile. *Contrib. Mineral. Petrol.* **2006**, *151*, 413–433. [CrossRef]
56. Wooden, J.L.; Mazdab, F.K.; Barth, A.P. Temperatures (Ti) and compositional characteristics of zircon: Early observations using high mass resolution on the USGS Stanford SHRIMP-RG. *Geochim. Cosmochim. Acta* **2006**, *70*, 707. [CrossRef]
57. Clairborne, L.L.; Miller, C.F.; Wooden, J.L. Trace element composition of igneous zircon: A thermal and compositional record of the accumulation and evolution of a large silicic batholith, Spirit Mt., Nevada. *Contrib. Mineral. Petrol.* **2010**, *160*, 511–531. [CrossRef]
58. Clairborne, L.L.; Miller, C.F.; Walker, B.A. Tracking magmatic processes through Zr/Hf ratios in rocks and Hf and Ti zoning in zircons: An example from the Spirit Mt. batholith, Nevada batholith, Nevada. *Mineral Mag.* **2006**, *70*, 517–543.
59. Gagnevin, D.; Daly, J.S.; Kronz, A. Zircon texture and chemical composition as a guide to magmatic processes and mixing in a granitic environment and coeval volcanic system. *Contrib. Mineral. Petrol.* **2010**, *159*, 579–596. [CrossRef]
60. Moore, C.M.; Carmichael, I.S.E. The hydrous phase equilibria (to 3 kbar) of an andesite and basaltic andesite from west-ern Mexico: Constraints on water content and conditions of phenocryst growth. *Contrib. Mineral. Petrol.* **1998**, *130*, 304–319. [CrossRef]
61. Richards, J.P.; Spell, T.; Rameh, E.; Razique, A.; Feltcher, T. High Sr/Y Magmas Reflect Arc Maturity, High Mag-matic Water Content, and Porphyry Cu ± Mo ± Au Potential: Examples from the Tethyan Arcs of Central and Eastern Iran and Western Pakistan. *Econ. Geol.* **2012**, *107*, 295–332. [CrossRef]
62. Sun, W.; McDonough, W.F. Chemical and isotopic systematics of oceanic basalts: Implications for mantle composition and processes. *Geol. Soc. Lond. Spec. Publ.* **1989**, *42*, 313–345. [CrossRef]
63. Tang, M.; Lee, C.A.; Chen, K.; Erdman, M.; Costin, G.; Jiang, H. Nb/Ta systematics in arc magma differentiation and the role of arclogites in continent formation. *Nat. Commun.* **2019**, *10*, 235. [CrossRef] [PubMed]
64. Li, L.; Xiong, X.L.; Liu, X.C. Nb/Ta Fractionation by Amphibole in Hydrous Basaltic Systems: Implications for Arc Magma Evolution and Continental Crust Formation. *J. Petrol.* **2017**, *58*, 3–28. [CrossRef]
65. Ferry, J.M.; Watson, E.B. New thermodynamic models and revised calibrations for the Ti-in-zircon and Zr-in-rutile thermometers. *Contrib Mineral Petrol.* **2007**, *154*, 429–437. [CrossRef]
66. Ballard, J.R.; Palin, J.M.; Campbell, I.H. Relative oxidation states of magmas inferred from Ce"IV"/Ce"III" in zircon: Ap-plication to porphyry copper deposits of northern Chile. *Contrib. Mineral. Petrol.* **2002**, *144*, 347–364. [CrossRef]
67. Shen, P.; Hattori, K.; Pan, H. Oxidation condition and metal fertility of granitic magmas: Zircon trace-element data from porphyry Cu deposits in the central Asian orogenic belt. *Econ. Geol.* **2015**, *110*, 1861–1878. [CrossRef]
68. Zhong, S.; Seltmann, R.; Qu, H.; Song, Y. Characterization of the zircon Ce anomaly for estimation of oxidation state of magmas: A revised Ce/Ce* method. *Mineral. Petrol.* **2019**, *113*, 755–763. [CrossRef]
69. Loader, M.A.; Wilkinson, J.J.; Armstrong, R.N. The effect of titanite crystallisation on Eu and Ce anomalies in zircon and its implications for the assessment of porphyry Cu deposit fertility. *Earth Planet. Sci. Lett.* **2017**, *472*, 107–119. [CrossRef]

70. Mortensen, J.K.; Appel, V.L.; Hart, C.J.R. Geological and U-Pb age constraints on base and precious metal vein systems in the Mount Nansen area, eastern Dawson Range, Yukon. In *Yukon Exploration and Geology 2002*; Exploration and Geological Services Division, Yukon Region, Indian and Northern Affairs Canada: Whitehorse, YT, Canada, 2003; pp. 165–174.
71. Stevens, R.D.; Delabio, R.N.; Lachance, G.R. *Age Determinations and Geological Studies, K-Ar Isotopic Ages*; Geological Survey of Canada, Paper 82-2; Geological Survey of Canada: Ottawa, ON, Cananda, 1982; 52p.
72. Lund, K.; Aleinikoff, J.N.; Kunk, M.J.; Unruh, D.M.; Zeihen, G.D.; Hodges, W.C.; Du Bray, E.A.; O'Neill, J.M. SHRIMP U-Pb and 40Ar/39Ar Age Constraints for Relating Plutonism and Mineralization in the Boulder Batholith Region, Montana. *Econ. Geol.* **2002**, *97*, 241–267. [CrossRef]
73. Richards, J.P. High Sr/Y Arc Magmas and Porphyry Cu ± Mo ± Au Deposits: Just Add Water. *Econ. Geol.* **2011**, *106*, 1075–1081. [CrossRef]
74. Correa, K.J.; Rabbia, O.M.; Hernandez, L.B.; Selby, D.; Astengio, M. The Timing of Magmatism and Ore Formation in the El Abra Porphyry Copper Deposit, Northern Chile: Implications for Long-Lived Multiple-Event Magmatic-Hydrothermal Porphyry Systems. *Econ. Geol.* **2016**, *111*, 1–28. [CrossRef]
75. Richards, J.P. Tectono-Magmatic Precursors for Porphyry Cu-(Mo-Au) Deposit Formation. *Econ. Geol.* **2003**, *98*, 1515–1533. [CrossRef]
76. Harrison, M.T.; Watson, B.E.; Aikman, A.B. Temperature spectra of zircon crystallization in plutonic rocks. *Geology* **2007**, *35*, 635–638. [CrossRef]
77. Tapster, S.; Condon, D.J.; Naden, J.; Noble, S.R.; Petterson, M.G.; Roberts, N.M.W.; Saunders, A.D.; Smith, D.J. Rap-id thermal rejuvenation of high-crystallinity magma linked to porphyry copper deposit formation; evidence from the Koloula Porphyry Prospect, Solomon Islands. *Earth Planet. Sci. Lett.* **2016**, *442*, 206–217. [CrossRef]
78. Dilles, J.H.; Kent, A.J.R.; Wooden, J.L. Zircon compositional evidence for sulfur degassing from ore-forming arc magmas. *Econ. Geol.* **2015**, *110*, 241–251. [CrossRef]
79. Lee, R.G.; Byrne, K.; D'Angelo, M.; Hart, C.J.R.; Hollings, P.; Gleeson, S.A.; Alfaro, M. Using zircon trace element composition to assess porphyry copper potential of the Guichon Creek batholith and Highland Valley Copper deposit, south-central British Columbia. *Miner. Depos.* **2020**, *56*, 215–238. [CrossRef]
80. Green, T.H.; Pearson, N.J. Experimental determination of REE partition coefficients between amphibole and basaltic to andesitic liquids at high pressure. *Geochim. Cosmochim. Acta* **1985**, *49*, 1465–1468. [CrossRef]
81. Erdmann, S.; Wodicka, N.; Jackson, S.E.; Corrigan, D. Zircon textures and composition: Refractory recorders of magmatic volatile evolution? *Contrib. Mineral. Petrol.* **2013**, *165*, 45–71. [CrossRef]
82. Kelley, K.D.; Ludington, S. Cripple Creek and other alkaline-related gold deposits in the southern Rocky Mountains, USA: Influence of regional tectonics. *Miner. Depos.* **2002**, *37*, 38–60. [CrossRef]
83. Hollings, P.; Wolfe, R.; Cooke, D.R.; Waters, P.J. Geochemistry of Tertiary Igneous Rocks of Northern Luzon, Philippines: Evidence for a Back-Arc Setting for Alkalic Porphyry Copper-Gold Deposits and a Case for Slab Roll-Back? *Econ. Geol.* **2011**, *106*, 1257–1277. [CrossRef]
84. Richards, J.P.; Lopez, G.P.; Zhu, J.-J.; Creaser, R.A.; Locock, A.J.; Mumin, A.H. Contrasting Tectonic Settings and Sulfur Contents of Magmas Associated with Cretaceous Porphyry Cu ± Mo ± Au and Intrusion-Related Iron Oxide Cu-Au De-posits in Northern Chile. *Econ. Geol.* **2016**, *112*, 295–318. [CrossRef]
85. Rabayrol, F.; Hart, C.J.R.; Thorkelson, D.J. Temporal, spatial and geochemical evolution of late Cenozoic post-subduction magmatism in central and eastern Anatolia, Turkey. *Lithos* **2019**, *336*, 67–96. [CrossRef]
86. Frey, F.A.; Green, D.H.; Roy, S.D. Integrated Models of Basalt Petrogenesis: A Study of Quartz Tholeiites to Olivine Melilitites from South Eastern Australia Utilizing Geochemical and Experimental Petrological Data. *J. Petrol.* **1978**, *19*, 463–513. [CrossRef]
87. Richards, J.P.; Kerrich, R. Special Paper: Adakite-Like Rocks: Their Diverse Origins and Questionable Role in Metallogenesis. *Econ. Geol.* **2007**, *102*, 537–576. [CrossRef]
88. Olierook, H.K.H.; Kirkland, C.L.; Szilas, K.; Hollis, J.A.; Gardiner, N.J.; Steenfelt, A.; Jiang, Q.; Yakymchuk, C.; Evans, N.J.; McDonald, B.J. Differentiating between inherited and autocrystic zircon in granitoids. *J. Petrol.* **2020**, *61*, egaa081. [CrossRef]
89. Gilmer, A.K.; Sparks, R.S.J.; Blundy, J.D.; Rust, A.C.; Hauff, F.; Hoernle, K.; Spencer, C.J.; Tapster, S. Petrogenesis and Assembly of the Don Manuel Igneous Complex, Miocene–Pliocene Porphyry Copper Belt, Central Chile. *J. Petrol.* **2018**, *59*, 1067–1108. [CrossRef]
90. Bickerton, L.; Kontak, D.J.; Murphy, J.B.; Kellett, D.A.; Samson, I.A.; Marsh, J.H.; Dunning, G.; Stern, R. The age and origin of the South Mountain Batholith (Nova Scotia, Canada) as constrained by zircon U–Pb geochronology, geochemistry, and O–Hf isotopes. *Can. J. Earth Sci.* **2022**, *59*, 418–454. [CrossRef]
91. Monger, J.W.H.; Gibson, H.D. Mesozoic-Cenozoic deformation in the Canadian Cordillera: The record of a "Conti-nental Bulldozer"? *Tectonophysics* **2019**, *757*, 153–169. [CrossRef]

92. Gabrielse, H.; Murphy, D.C.; Mortensen, J.K. Cretaceous and Cenozoic dextral orogen-parallel displacements, magmatism, and paleogeography, north-central Canadian Cordillera. In *Paleogeography of the North American Cordillera: Evidence For and Against Large-Scale Displacements*; Special Paper 46; Geological Association of Canada: St. John's, NL, Canada, 2006; pp. 255–276.
93. Topham, M.J.; Allan, M.M.; Mortensen, J.K.; Hart, C.J.R.; Colpron, M.; Sack, P.J. Crustal depth of emplacement of the Early Jurassic Aishihik and Tatchun batholiths, west central Yukon. In *Yukon Exploration and Geology 2015*; Yukon Geological Survey: Whitehorse, YT, Canada, 2016; pp. 233–251.
94. Richards, J.P. Postsubduction porphyry Cu-Au and epithermal Au deposits: Products of remelting of subduction-modified lithosphere. *Geology* **2009**, *37*, 247–250. [CrossRef]
95. Bineli Betsi, T.; Lentz, D.; Chiaradia, M.; Kyser, K.; Creaser, R.A. Genesis of the Au-Bi-Cu-As, Cu-Mo ± W, and base-metal Au-Ag mineralization at the Mt. Freegold (Yukon, Canada): Constraints from Ar-Ar and Re-Os geochronology and Pb and stable isotope compositions. *Miner. Depos.* **2013**, *48*, 991–1017. [CrossRef]
96. Bineli Betsi, T.; Bennett, V. New U-Pb age constraints at Freegold Mt.: Evidence for multiple phases of polymetallic mid- to Late Cretaceous mineralization. In *Yukon Exploration and Geology 2009*; Government of Yukon: Whitehorse, YT, Canada, 2010; pp. 57–84.
97. Friend, M.A. Tectonomagmatic Framework for Late Cretaceous Post-Subduction Metallogeny in the Dawson Range, West-Central Yukon. Master's Thesis, University of British Columbia, Vancouver, BC, Canada, 2022.
98. Bennett, V.; Schulze, C.; Ouellette, D.; Pollries, B. Deconstructing complex Au-Ag Cu mineralization, Sonora Gulch project, Dawson Range: A Late Cretaceous evolution to the epithermal environment. In *Yukon Exploration and Geology 2009*; Government of Yukon: Whitehorse, YT, Canada, 2010; pp. 23–45.
99. Godwin, C.I. Geology of the Casino Porphyry Copper-Molybdenum Deposit, Dawson Range, Y.T. Ph.D. Thesis, University of British Columbia, Vancouver, BC, Canada, 1975; 274p.
100. Bineli-Betsi, T.; Lentz, D.; Mcinnes, B.; Evans, N. Emplacement ages and exhumation rates for intrusion-hosted Cu-Mo-Sb-Au mineral systems at Freegold Mt. (Yukon, Canada): Assessment from U-Pb, Ar-Ar, and (U-Th)/He geochronometers. *Can. J. Earth Sci.* **2012**, *49*, 653–670. [CrossRef]
101. Moher, M. (U-Th)/He Thermochronology of the Aishihik Batholith, Central Yukon: Evidence for Stable Crust in the Cretaceous. Master's Thesis, University of Ottawa, Ottawa, ON, Canada, 2018; 83p.
102. Kreiner, D.C.; Holm-Denoma, C.; Pianowski, L.; Flood, Z.; Stevenson, D.; Graham, G.E.; Vazquez, J.; Creaser, R.A. Geochronology and Mapping Constraints on the Time-Space Evolution of the Igneous and Hydrothermal Systems in the Taurus Cu-Mo District, Eastern Alaska. *Econ. Geol.* **2023**, *118*, 745–778. [CrossRef]
103. Richards, J.P.; Boyce, A.J.; Pringle, M.S. Geologic Evolution of the Escondida Area, Northern Chile: A Model for Spatial and Temporal Localization of Porphyry Cu Mineralization. *Econ. Geol.* **2001**, *96*, 271–305. [CrossRef]
104. Campbell, I.H.; Ballard, J.R.; Palin, J.M.; Allen, C.; Faunes, A. U-Pb Zircon Geochronology of Granitic Rocks from the Chuquicamata-El Abra Porphyry Copper Belt of Northern Chile: Excimer Laser Ablation ICP-MS Analysis. *Econ. Geol.* **2006**, *101*, 1327–1344. [CrossRef]

minerals

MDPI

Article

An Archean Porphyry-Type Deposit: Cu-Au Mineralization Associated with the Chibougamau Tonalite–Diorite Pluton, Abitibi Greenstone Belt, Canada

Alexandre Crépon [1], Lucie Mathieu [1,2,*], Daniel J. Kontak [3,*], Jeffrey Marsh [3] and Michael A. Hamilton [4]

1 Centre d'Etude sur les Ressources Minérales, Université du Québec à Chicoutimi (UQAC), Chicoutimi, QC G7H 2B1, Canada; alexandrecrepon.geology@gmail.com
2 Consultation Geox Inc., Val-d'Or, QC J9P 0C1, Canada
3 Harquail School of Earth Sciences, Laurentian University, Sudbury, ON P3E 2C6, Canada; jmarsh@laurentian.ca
4 Jack Satterly Geochronology Laboratory, Department of Earth Sciences, University of Toronto, Toronto, ON P3E 6H5, Canada; mahamilton@es.utoronto.ca
* Correspondence: lmathieu.geology@protonmail.com (L.M.); dkontak@laurentian.ca (D.J.K.)

Abstract: The Neoarchean diorite- and tonalite-dominated Chibougamau pluton (Canada) is ideal for case studies dedicated to the petrogenesis and timing of emplacement of fertile magmatic systems and associated Cu-Au porphyry systems. Using whole-rock analyses, geochronology, and zircon chemistry, it is determined that an early magmatic phase (pre-2714 Ma) is derived from a dioritic magma with a moderate fO_2 (ΔFMQ 0 to +1), which is optimal for transporting Au and Cu, and that diorite is a potentially fertile magma. Field descriptions indicate that the main mineralizing style consists of sulfide-filled hairline fractures and quartz–carbonate veins. This is likely the consequence of fluid circulation facilitated by a well-developed diaclase network formed following the intrusion of magma at about 4–7 km depth in a competent hosting material. The petrographic features of fluid inclusions (FIs), considered with their microthermometric data and evaporate mound chemistry, suggest the exsolution of early CO_2-rich fluids followed by the unmixing of later aqueous saline fluids characterized by a magmatic signature (i.e., Na-, Ca-, Fe-, Mn-, Ba-, and Cl-F). The type of magmatism and its oxidation state, age relationships, the nature of mineralization, and fluid chemistry together support a model whereby metalliferous fluids are derived from an intermediate hydrous magma. This therefore enforces a porphyry-type metallogenic model for this Archean setting.

Keywords: porphyry mineralization; Chibougamau pluton; Archean; greenstone belt; geochronology; zircon chemistry; fluid inclusions

1. Introduction

Archean greenstone belts are assemblages of supracrustal rocks consisting of basaltic lava flows and other volcanic rocks, as well as sedimentary rocks. The intrusive systems, with regard to volume, are dominated by tonalite–trondhjemite–granodiorite (TTG) suites [1,2], whereas the lesser volume tonalite–trondhjemite–diorite (TTD) suites are metallogenically more important as they are related to porphyry systems such as the Chester intrusive complex [3,4] and the Chibougamau pluton [5], both of which are located in the Abitibi greenstone belt (AGB) in the southern part of the Superior craton, Canada.

In greenstone belts such as the AGB, historically mined commodities have come from volcanogenic massive sulfides (VMSs) and orogenic gold systems, which have dominated and continue to dominate exploration and mining efforts [6]. In contrast, the porphyry-type systems are viewed as being of lesser importance due to their rarity, although they have been shown to be of economic value, e.g., the Côté Gold deposit [3,4]. It is apparent, therefore, that much remains to be done to better elucidate mineralizing processes related to Archean magmatic–hydrothermal systems. This contribution is a case study focused on the Chibougamau pluton, located in the northeastern corner of the AGB, and is intended to address the latter issue by providing insight into several relevant and important aspects: (1) the nature of the magmatic processes that apparently produced fertile causative plutons; (2) the features of the Cu-Au mineralized zones; and (3) the physicochemical characteristics of the mineralizing hydrothermal fluids.

The Archean Chibougamau pluton is a polyphase intrusion made of tonalite and diorite intrusive phases. Recent investigations have determined that these intrusive phases are sourced from distinct reservoirs, i.e., crustal (basalt) and metasomatized mantle sources [5]. However, the dynamics of magma emplacement, i.e., continuous magmatism over a long period of time versus discrete magmatic pulses, remain unknown and is addressed here using in situ U-Pb zircon geochronology. It remains unclear whether one or more of the intrusive phases is able to carry Au and Cu through the crust, and this is addressed using the oxygen fugacity (fO_2) parameter derived from zircon chemistry. A tentative calculation of the relative volume of magma involved during the building of the pluton is also performed.

The mineralization genetically related to the Chibougamau pluton has long been recognized as porphyry-type [7]. Exploration and mapping efforts by the private sector and by the Geological Survey of Québec (Ministère de l'Énergie et des Resources Naturelles—MERN) have provided a wealth of details on the physical characteristics of the numerous mineralized settings. This information has been compiled and is herein evaluated to provide insights into mineralizing processes. In addition, a pilot study performed on one such mineralized setting (Cedar Bay) utilized fluid inclusions (FIs) to assess the nature of the ore fluids and, via microthermometry and evaporate-mound analysis (i.e., salt mounds obtained by overheating FIs), provided the quantification of fluid chemistry.

Overall, therefore, the FI study is intended as a proof of concept, the conclusions of which can potentially be extended to the rest of the mineralized areas related to the Chibougamau pluton. Additionally, the results documenting the nature of the mineralizing fluid in this Archean setting are contrasted with fluids that typify different ore systems [8,9].

The integration of the three aforementioned aspects of this study provides the basis for assessing the metallogenic model that best accounts for Au-Cu mineralization. Thus, the data collected are used to comment on the dynamics of the magmatic systems, on the fertility of the main intrusive phases, on the mineralizing style, and on the physicochemical characteristics of the mineralizing fluid. It is concluded that, whereas the study area is atypical in regard to metallogeny (see below), the Cu-Au mineralization related to the Chibougamau pluton displays many characteristics that typify post-Archean porphyry systems [10–12], accommodating our observations best. The objective of this study is to gain a better understanding of an Archean porphyry-type mineralization. In the first part of this contribution, the physical characteristics of mineralized areas are reviewed to highlight the uniqueness of Archean porphyry systems. The petrogenesis of the causative magma, its capacity to carry Au and Cu through the crust, and the dynamics of magma emplacement are then addressed using whole-rock analyses, zircon chemistry, and U-Pb zircon geochronology. Finally, the physicochemical characteristics of the mineralizing fluid are addressed via a pilot study of FIs.

2. Geological Setting

2.1. Geology of the Chibougamau Area

The AGB of the southern part of the Superior craton, Canada (Figure 1), along with other greenstone belts of this age records a magmatic–tectonic evolution that can be summarized into two main periods: synvolcanic and syntectonic [13,14]. The synvolcanic period was characterized by sustained magmatic activity that produced large volumes of dominantly mafic and tholeiitic volcanic rocks (lava plains and central volcanoes), intermediate to felsic volcanic rocks with calc-alkaline affinities, localized tonalite–trondhjemite–granodiorite (TTG) intrusive suites, and several layered tholeiitic intrusions [15–17]. In contrast, the syntectonic period was dominated by compressional deformation following the development of linear (E-W-striking) sedimentary basins and the injection of small volumes of K-rich magmas (e.g., monzodiorite, monzonite, syenite, sanukitoid) into the upper crust [15,18,19].

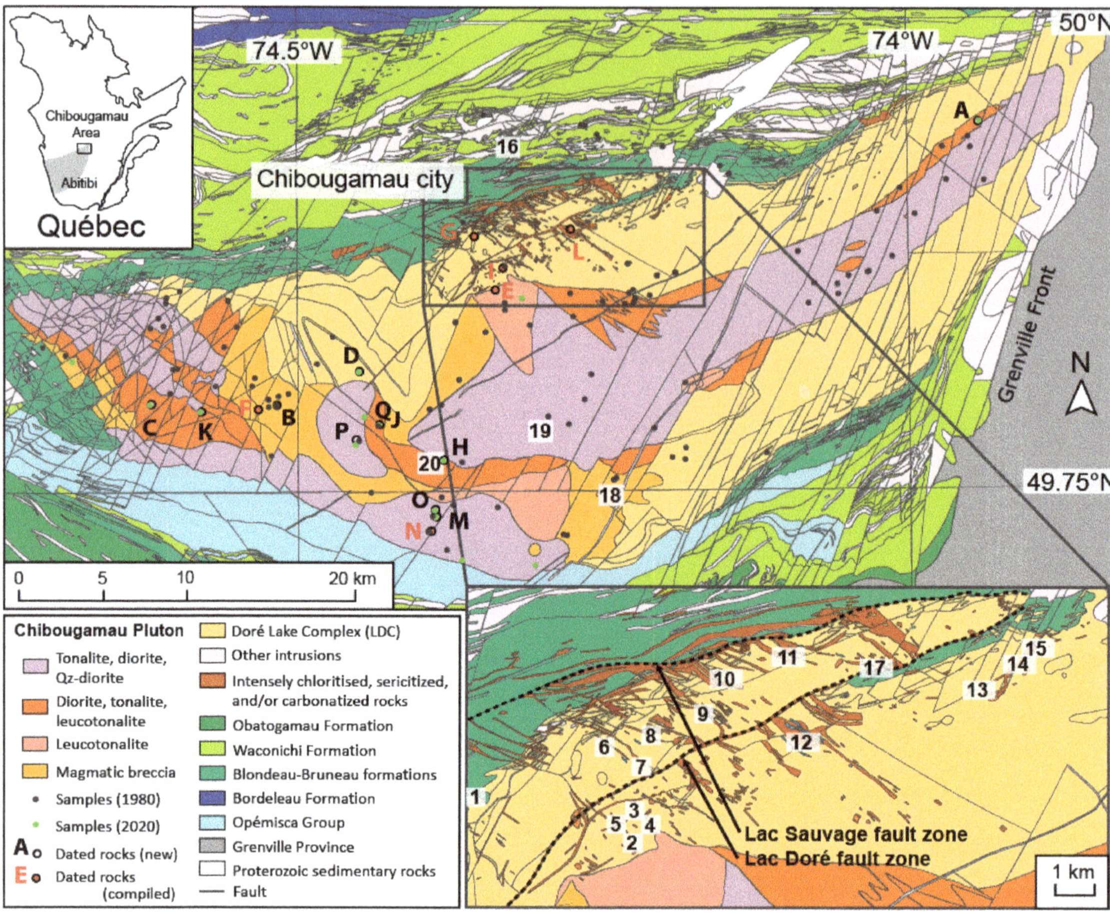

Figure 1. Geological map of the Chibougamau pluton and hosting material modified from the Ministère de l'Energie et des Ressources Naturelles (MERN), Québec (projection UTM NAD83 zone 18N). The inset is a zoom on Central Camp, which is the historical mining camp of the Chibougamau area. The letters refer to ages listed in Table 1. The numbers locate deposits and showings listed in Table 2.

The Chibougamau pluton, found in the northeastern corner of the AGB, is in the eastern part of the Matagami–Chibougamau belt that makes up the northern part of the AGB (Figure 1). In this belt, the synvolcanic period extends from >2790 to 2710 Ma [20] and is dominated by the Roy Group, which comprises two volcanic cycles. Cycle 1 is dominated by tholeiitic lava flows of the Obatogamau Formation [21] and by the intermediate to felsic volcanic rocks of the ca. 2730 to 2727 Ma Waconichi Formation [22–24]. Coeval magmatic activity formed the 2728.3 +1.2/−1.1 Ma Lac Doré Complex (LDC) layered intrusion [18,24,25]. Cycle 2 consists of the volcanic, volcanoclastic, and sedimentary rocks of the 2724.4 ± 1.2 Ma Bruneau, <2721 Ma Blondeau, and undated Bordeleau formations [20,26,27] (Figure 1). Coeval magmatism produced the three ultramafic to mafic sills of the Cummings Complex, with the uppermost sill dated at 2716.7 +1.0/−0.4 Ma [24], and the early intrusive phases of the Chibougamau pluton (see below).

The Chibougamau pluton may equate to the sub-volcanic complex of the central volcano of volcanic cycle 2 [28]. This central volcano developed over an evolved oceanic crust, possibly as N-S shortening initiated [14]. The Chibougamau pluton forms the core of the Chibougamau anticline and was likely intruded along a E-W trending structure within the rocks of the LDC [5]. This polyphase pluton is dominated by (1) amphibole-rich diorite referred to as meladiorite; (2) quartz–diorite; (3) biotite-tonalite; and (4) biotite-bearing leucotonalite that may, in part, correspond to trondhjemite. The main primary minerals of these intrusive phases are plagioclase, quartz, amphibole, and biotite, with additional rare K-feldspar present [5,29,30]. Compiled age data indicate that the pluton is composite with ca. 2718–2714 Ma phases coeval with volcanic cycle 2, as well as later ca. 2705–2695 Ma phases emplaced during the syntectonic period (Table 1).

The upper part of the Roy Group is overlain by the ca. 2704–2692 Ma Opémisca Group [22,31], which consists of two E-W-striking sedimentary basins located north and south of the Chibougamau pluton. The southern basin has discordant contact with this pluton, thus indicating it was exhumed prior to basin formation [27]. Deformation may have begun toward the end of the synvolcanic period [14], but is most pronounced during the syntectonic phase in the AGB [15]. The main compressional stage induced N-S shortening prior to the stabilization of the craton (cratonisation) and was related to a series of deformation events collectively referred to as the Kenoran orogeny [32], which affected the craton from north to south between 2.72 and 2.68 Ga [33]. In the Chibougamau area, the D_1 event produced open folds along the NNW-SSE axis, whereas the main shortening event of D_2, along with its waning stage D_3, induced km-scale open folding (E-W-striking fold axis) and the tightening of pre-existing structures (e.g., Chibougamau anticline). The D_2 event also produced penetrative foliation, E-W to NW-SE dextral reverse shear zones, and major faults, and was coeval with the peak of greenschist facies metamorphism [15,34]. The D_3 event induced reverse movement along NE-SW shear zones (Figure 1). The D_4 event formed NNE-SSW to N-S, striking dextral and sinistral faults, and was related to the ca. 1.1 Ga Grenville orogeny [15,35].

Table 1. Compiled and new ages for the Chibougamau pluton. Sample locations are available in Table S2C (new ages) and Table S2D (compiled ages).

Figure 1	Rock	Age	Reference
A	ACR-29-1 (tonalite)	2728.6 ± 1.5 Ma	This study
B	ACR-33-1 (diorite)	2725.2 ± 1.5 Ma	This study
C	ACR-32-1 (tonalite)	2722.5 ± 1.9 Ma	This study
D	ACR-21-1 (diorite dyke)	2722 ± 37 Ma	This study
E	Tonalite intrusive phase	2718 ± 2 Ma	[36]
F	Diorite intrusive phase	2716 +3/−2 Ma	[37,38]
G	Tonalite intrusive phase	2715 ± 1 Ma	[37]
H	ACR-6-1 (diorite)	2714.8 ± 0.93 Ma	This study
I	Tonalite porphyritic dyke	2714 +3/−2 Ma	[37,38]
J	ACR-11-5 (diorite)	2713.4 ± 2.5 Ma	This study
K	ACR-16-2 (diorite)	2710.4 ± 7.7 Ma	This study
L	Quartz–diorite dyke	2705.1 +1.7/−1.2 Ma	[37,38]
M	ACR-24-3 (tonalite)	2702.2 ± 1.2 Ma	This study
N	Tonalite porphyritic dyke	2701.7 ± 2.9 Ma	[39]
O	ACR-19-3 (tonalite)	2699.5 ± 1.3 Ma	This study
P	ACR-31-1/20ACR-0031A (tonalite)	2696.9 ± 1 Ma	This study
Q	ACR-11-2/20ACR-0011C (monzodiorite)	2695.4 ± 1 Ma	This study

2.2. *Chibougamau Cu-Au Mineralization*

The Chibougamau pluton is associated with several past producing Cu-Au deposits and associated prospects that are mostly located north of the pluton in Central Camp (Figure 2). This historical mining center developed from the 1930s to the late 1990s and exploited high-grade Cu-Au deposits, which are again being evaluated for production. As summarized by the Doré Copper Mining Corp, Toronto, Canada (https://www.dorecopper.com, accessed on 1 November 2024), representative historical underground operations are Cedar Bay (3.86 Mt @ 1.63 wt.% Cu, 3.3 g/t Au), Copper Rand (16.5 Mt @ 1.8 wt.% Cu, 2.8 g/t Au), and Joe Mann (4.5 Mt @ 0.25 wt.% Cu, 8.26 g/t Au), whereas current resource delineation indicates a high potential for further mining, e.g., Corner Bay, where indicated (2.68 Mt @ 2.66 wt.% Cu, 0.26 g/t Au) and inferred (5.86 Mt @ 3.53 wt.% Cu, 0.27 g/t Au) resources are delineated.

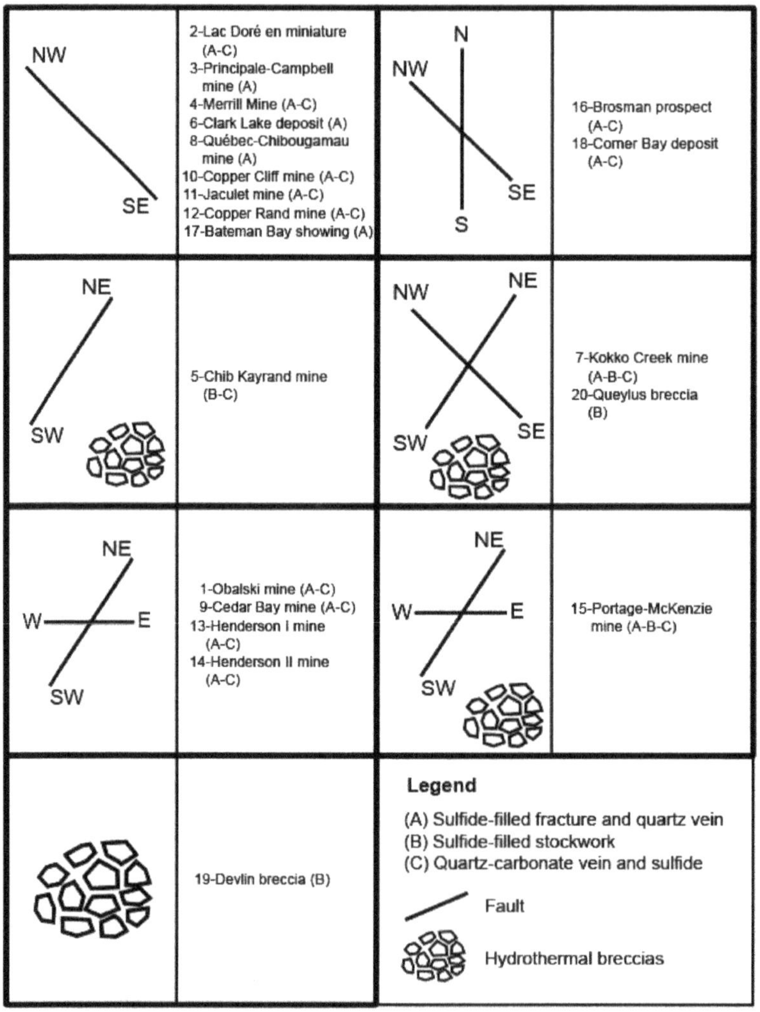

Figure 2. Summary of the structural features and distribution of various Cu-Au mineralizing styles A, B, and C (see text for discussion) associated with the Chibougamau pluton.

South of Central Camp, two of the main prospects, Devlin and Queylus, are located in intrusive phases of the Chibougamau pluton, whereas the third (Corner Bay) is hosted by the LDC (Table 2). The Brosman prospect, whose genetic relationship with the Chibougamau pluton is unclear, is located north of Central Camp in rocks of the Blondeau Formation and of the Cummings Complex (Figure 1). A magmatic origin for the mineralizing fluids in Central Camp has been inferred from earlier fluid inclusion studies, noting the elevated salinity (qualitative as salinity was not quantified) of mixed $CaCl_2$—NaCl fluids [40]. The chemistry of pyrite, as determined from in situ LA-ICP-MS analysis, displays a polymetallic signature (e.g., Bi and Te enrichment), suggestive of a magmatic–hydrothermal origin for the fluids [41–43].

The Cu-Au mineralization related to the Chibougamau pluton displays a range of features, as has been widely reported in previous studies [20,44–74]. The latter has been compiled (Supplementary Material S1) and is summarized below.

The mineralization both intersects and is intersected by aphanitic to feldspar porphyritic dykes (tonalite and diorite) that are genetically related to the Chibougamau pluton, as determined by field relationships and whole-rock chemistry [54,75,76]. Additionally, the ages of ca. 2714–2715 Ma observed for two of the dykes relate them to an intrusive phase of the pluton (Table 1). Thus, an important observation is that these dykes are considered to be inter-mineral dykes and provide evidence for the overlap of magmatic and hydrothermal activity [11].

In most mineralized zones, with the exception of the hydrothermal breccia of the Devlin prospect, the distribution of sulfide and alteration minerals is controlled by fractures and shear zones with various orientations. In Central Camp, fractures are generally NW-oriented (Figure 2). Other fractures and shear zones, in and south of this area, are NE-oriented and sets are oriented NW-SE and N-S, NW-SE and NE-SW, and E-W and NE-SW (Figure 2). Hydrothermal breccia is well developed at Devlin and is also observed in Central Camp for 4 (out of 17) compiled mineralized zones.

Alteration halos associated with mineralization comprise intense proximal types restricted to a few meters around the sulfide-rich zone, as well as distal types. The main proximal alterations are chloritization, sericitization, and quartz (veining) $\pm$ carbonatization. Potassic metasomatism is best documented for the Queylus breccia, where a microcline $\pm$ biotite assemblage is noted to occur. Mineralized quartz veins are frequently reported (see below) and occur with carbonate, magnetite, chlorite, white and dark micas, and microcline (Supplementary Material S1). Iron metasomatism also generated a large amount of hydrothermal magnetite in some mineralized zones (e.g., Obalski mine). Distal alterations are of the propylitic and phyllic types, and their intensity gradually increases with increasing proximity to orebodies. Additionally, hematization (red staining) is common and of irregular distribution. In the Chibougamau pluton, propylitic alteration is the most frequently reported type, as manifested by the replacement of Fe-Mg-bearing magmatic minerals by chlorite, epidote, and carbonates, accompanied by veinlets of chlorite, epidote, and quartz $\pm$ carbonates.

The main sulfide phases in mineralized intersections are pyrite, chalcopyrite, and pyrrhotite. These can be associated with magnetite $\pm$ molybdenite. There are three types of mineralizing styles: (1) style A is dominated by 'smoked' (darken) quartz, and sulfide-filled fractures and/or shear zones; (2) style B corresponds to sulfide-filled stockworks; and (3) style C consists of quartz (white) and carbonate veins associated with disseminated sulfides and rare visible gold (Figure 2).

Style A sulfide filled-fractures are NW-oriented planar zones with thicknesses of 0.5 to 2 m (Copper Rand mine). Locally, these thickness values can be up to 50 m (Portage mine). These zones are filled with pyrite, chalcopyrite $\pm$ pyrrhotite, and minor amounts of magnetite, molybdenite, sphalerite, and galena. Disseminated pyrite is also seen around the mineralized fractures and the main proximal alteration is chloritization.

Style B sulfide-filled stockworks overprint the magmatic breccia unit of the Chibougamau pluton (Figure 1) and are associated with hydrothermal breccia (Devlin, Queylus) or with NW- and NE-oriented fractures (Figure 2). The mineralized stockworks consist of pyrite and chalcopyrite, with lesser amounts of pyrrhotite, magnetite, and molybdenite, as well as minor quantities of sphalerite and galena. Disseminated chalcopyrite and pyrite ± magnetite are also reported. The main alterations are chloritization, sericitization, and K-feldspar alteration (Queylus).

Table 2. Main physical attributes of Cu-Au deposits and prospects associated with the Chibougamau pluton. Additional details for these mineralized areas are available in Supplementary Material S1.

Names	Mineralization	Reference
(1) Obalski mine	quartz–carbonates Cu(-Au)-mineralized veins and magnetite zones	[62]
(2) *Lac Doré en miniature* (outcrop)	Cu-Au showing	[53,71,72]
(3) Principale-Campbell mine	Cu(-Au) mineralization hosted by quartz–sulfide-filled fractures	[47,77]
(4) Merril mine	shear zone-hosted Cu(-Au)-mineralized quartz veins and hydrothermal breccia	[47,53,54,57,72]
(5) Chib-Kayrand mine	shear zone-hosted quartz–sulfide-filled fractures, Cu(-Au)-mineralized quartz–carbonates veins and hydrothermal breccia	[47,71]
(6) Clark Lake deposit	Cu(-Mo)-mineralized quartz veins and hydrothermal breccia	[52–54,72]
(7) Kokko Creek mine	quartz–sulfide-filled fractures and mineralized Cu(-Au) quartz–carbonate veins and hydrothermal breccia	[69,70]
(8) Québec-Chibougamau mine	Cu(-Au) mineralization in shear zone	[78]
(9) Cedar Bay mine	Cu(-Au) mineralization in shear zone and quartz-carbonate veins	[59–61]
(10) Copper Cliff mine	quartz–sulfide-filled fractures and mineralized Cu(-Au) quartz–carbonates veins	[58]
(11) Jaculet mine	Cu(-Au)-mineralized quartz–carbonates veins	[63]
(12) Copper Rand mine	Cu-Au mineralization hosted in quartz–sulfide-filled fractures and Cu(-Au)-mineralized carbonates–quartz–magnetite veins	[48,49]
(13) Henderson I mine (14) Henderson II mine	shear zone-hosted quartz- and sulfide-filled fractures and Cu(-Au)-mineralized quartz–carbonates veins	[64–68,79]
(15) Portage-McKenzie mine	shear zone-hosted Cu-Au mineralized quartz veins, hydrothermal breccias and Cu(-Au)-mineralized quartz–carbonates	[50]
(16) Brosman prospect	shear zone-hosted sulfide-filled quartz fractures and Cu(-Au)-mineralized quartz–carbonates veins	[56]
(17) Bateman Bay showing	Cu(-Au) mineralization in shear zone	[63]
(18) Corner Bay deposit	Cu(-Au)-mineralized quartz veins and Cu-Mo mineralized quartz–carbonates veins	[51,52]
(19) Devlin breccia	Cu-mineralized (disseminated sulfide) hydrothermal breccia	[73,74]
(20) Queylus breccia	Cu-bearing hydrothermal breccia	[55]

Style C veins cut type A and B mineralization. These veins, hosted by shear zones and extensional fractures, contain calcite and siderite, as well as tourmaline and sulfides (pyrite, chalcopyrite, pyrrhotite, molybdenite ± sphalerite, galena, and magnetite). Some veins display tourmaline-rich zones, which alternate with quartz–dominated areas (e.g., Portage-McKenzie mine), suggesting multiple episodes of dilation and mineral precipitation [80]. Proximal alteration (few cm up to 30 m from veins) corresponds to sericitization and chloritization. Style C veins and associated gold (free gold and pyrite-associated gold) is attributed to the remobilization of earlier magmatic–hydrothermal mineralization by the regional D_2 deformation event, i.e., similar to orogenic style of Au mineralization in the AGB in general [40,54,64,68,71,81].

3. Methodology

3.1. Sampling and Whole-Rock Analyses

The mapping of the Chibougamau pluton performed by the MERN during the 1980s [29,30], as compiled and georeferenced recently [5], was integrated into this study. This dataset comprises outcrop descriptions (i.e., mineralogy of the main unit, presence of dykes, and shape of the contacts between and among intrusive units) and geolocations, as well as geochemical analyses performed in the 1980s [5]. In 2019, powders (n = 113) of samples collected by D. Racicot (see above) were re-analyzed by the MERN as part of this study. This was performed by ALS Chemex Labs Ltd. (Vancouver, BC, Canada) with major and minor elements analyzed by inductively coupled plasma atomic emission spectroscopy (ICP-AES), following acid digestion, with detection limits of 0.01 wt%. Trace elements were also analyzed by ICP-AES following a four-acid digestion of powders with a detection limits from 0.01 to 5 ppm (Table S2A).

Additional fieldwork was performed in 2020 to describe and sample additional outcrops (Supplementary Material S2). Thick sections (50 μm thick) of eleven lithologies used for dating were produced for petrographic observations performed using an Olympus BX53M petrographic microscope equipped with a motorized stage and an SC50 camera. Also, 21 samples (~2 kg) were collected and analyzed by ALS Chemex Labs Ltd. for major, minor, and trace elements, using the same procedure and with the same detection limits as noted (Table S2B).

3.2. Zircon Geochronology

3.2.1. MC-ICP-MS Geochronology

Zircon concentrates were produced for nine samples (Table S2C) by the Overburden Drilling Management Ltd. (ODM) laboratory (Ottawa, ON, Canada). Rocks were crushed using the electric pulse disaggregation method and the heavy minerals were separated using a shaking table and heavy liquids with densities of 3.0 and 3.3 g/cm^3. Pan concentrates containing the zircon fraction were poured onto a tap, mounted onto epoxy resin, and fine-polished at UQAC, Chicoutimi, Quebec. In both samples, zircon grains had distinct sizes, with the largest (heaviest) fraction separated gravitationally from the smaller zircon grains during the preparation of epoxy mounts; these larger (~70–200 μm) and smaller (<100 μm) grains were mounted separately. Mounts were imaged in natural light using an Olympus BX53M microscope at UQAC. In addition, backscattered electron (BSE) and cathodoluminescence (CL) images were obtained using a Tescan Vega 3 SEM at the MERC-Isotope Geochemistry Laboratory (MERC-IGL) of Laurentian University (Sudbury, ON, Canada) (Supplementary material S3). Images were used to avoid altered and fractured grains, in addition to apatite and other types of inclusions, during subsequent laser ablation analysis.

Zircon U-Th-Pb isotopes and trace element analyses were conducted on ~300 zircons per sample at MERC-IGL, Laurentian University. Laser ablation sampling was performed using a Photon Machines Analyte G2 ArF excimer laser, with a 193 nm wavelength, a <5 ns pulse width, and a HelEx II cell. The ablation duration was 30 s, with laser fluence of 2 J/cm^2 and a 7 Hz repetition rate, leaving estimated ablation pit depths of <15 μm. Overall, 60 s of background were measured at the beginning and end of the analytical session, with 30 s of background measured between each ablation. Helium carrier gas flows through the ablation cell were 0.525 L/min (cup) and 0.1 L/min (cell), with 1.18 L/min Ar and 8 mL/min N_2 makeup gas added downstream of the cell. The ablated aerosol was split downstream of the sample cell, so that U-Th-Pb and trace elements measurements could be conducted simultaneously.

The U-Th-Pb isotopes measurements were conducted using a Thermo Scientific Neptune Plus multicollector (MC) ICP-MS, equipped with Jet interface and nine Faraday cups. All MC-ICP-MS analyses were conducted in low resolution static mode to ensure maximum sensitivity and stability. All cups were coupled with 10^{11} Ω amplifiers, with the exception of L2, which was coupled with the 10^{12} Ω amplifier for better precision on the low-intensity ^{207}Pb signal. Ion beam drift is typically <0.05 amu, with the flat top peak width typically standing at ~0.2 amu, ensuring stable on-peak measurements throughout long-duration analytical sessions. Cool gas and Ar auxiliary gas flows were set at 16 L/min and 0.75 L/min, respectively, and RF power was set to 1200 W. Trace element measurements were conducted using a Thermo Scientific iCap-TQ ICP-MS in a single quad mode to ensure maximum sensitivity for the low to intermediate mass range. The masses' measured and associated dwell times are shown in Supplementary Material S4. Cool gas, auxiliary gas, and RF power were set at 14 L/min, 0.8 L/min, and 1550 W, respectively.

The raw U-Th-Pb and trace element data were synchronized in Iolite. This was synchronized along with the laser log file, prior to processing, so that the same time slices (selections) for each analysis were used in the calculations. Baseline subtraction, instrumental drift, and downhole fractionation corrections were performed with the U-Pb Geochronology reduction scheme (DRS) implemented within Iolite v4 [82,83], with U-Th-Pb isotope ratios normalized to the zircon reference material OGC (OG1; 3465 ± 1 Ma; [84]). The OGC reference material was analyzed three times at the beginning and end of the session, and once every ten unknowns throughout the session. Typically, three seconds at the beginning and one second at end of the ablation period were excluded from the selections to minimize potential fractionation effects, leaving ~26 s of signal for integration. Within-run variance in the measured ratios for OGC (i.e., the additional percent error required to achieve MSWD = 1) was propagated into the 2-standard-error (2SE) uncertainty for all unknowns.

Verification reference materials DD91-1 (2682 $\pm$ 1 Ma; [85]), OGC and Maniitsoq (3008.7 $\pm$ 0.66 Ma; [86]) were analyzed repeatedly throughout the session to ensure the accuracy of the U-Pb ratios. Measured dates were identical to the published TIMS ages, and low uncertainty and MSWD were obtained. Trace element data were processed using the Internal Standard DRS method within the Iolite v4 software and were normalized to the synthetic glass NIST610. An assumed stoichiometric concentration of 15.284 wt.% Si was used for all zircon to account for differing ablation characteristics between the glass and zircon. The synthetic glass NIST612 and zircon 91500 (1065 Ma; [87,88]) were used to verify the accuracy of trace elements analyses. Analytical results and operating parameters are provided in Supplementary material S3.

3.2.2. CA-ID-TIMS (Chemical Abrasion Isotope Dilution Mass Spectrometry) Geochronology

Samples 20ACR-0011C and 20ACR-0031A (2–3 kg each) were processed at the Jack Satterly Geochronology Laboratory (JSGL) at the University of Toronto. The samples underwent conventional jaw crushing, followed by grinding using a Bico^{TM} disk mill. The initial separation of heavy minerals was carried out by passing the ground sample and derived heavy concentrates were subsequently derived multiple times over a shaking water table. Further processing employed density separations with methylene iodide and involved standard paramagnetic separations using a Frantz^{TM} isodynamic separator. Final sample selection was achieved via handpicking in alcohol under a binocular microscope, and we chose the freshest and least cracked core- and inclusion-free grains of zircon for isotope dilution thermal ionization mass spectrometry. Detailed zircon selection, chemical abrasion pretreatment techniques (modified after Mattinson [89]), U and Pb chemical separation, and full analytical methods were performed following the protocol described in Hamilton et al. [90].

The weights of mineral fractions chosen for ID-TIMS analyses were estimated from scaled digital photomicrographs using the density of zircon. Zircon fractions were given brief ultrasonic treatment and washed before dissolution. A mixed ^{205}Pb-^{235}U isotopic spike was added to the dissolution capsules during sample loading. Zircon grains were dissolved using concentrated HF in Teflon bombs at 205 °C for 3–4 days; they were then dried and re-dissolved in 3N HCl overnight [91]. Uranium and lead were isolated in 50 μL anion exchange columns using HCl elutions, dried down, and then loaded onto outgassed rhenium filaments with silica gel [92]. Lead and UO_2 were analyzed on a VG354 mass spectrometer using a Daly collector in pulse counting mode. The mass discrimination correction for this detector was constant, at 0.07%/AMU. The dead time of the Daly system was 16 ns for Pb during the analytical period, monitored using the SRM982 Pb standard.

Mass spectrometer (ID-TIMS) data were reduced using in-house software (UtilAge program) coded by D. Davis (JSGL, University of Toronto). All common Pb was assigned to procedural blank (see footnote to Supplementary Material Table S5A). The plotting of Concordia curves and averaging of age results were carried out using the Isoplot 3.71 add-in for MS Excel [93]. The uncertainties in the U-decay constants, though shown on the Concordia diagram, were not factored into the calculated weighted average age (analytical data are provided in Supplementary material S5).

3.3. *Fluid Inclusions (FI)*

The fluid inclusions (FIs) were studied using nine samples from high-grade quartz–chalcopyrite–pyrite $\pm$ carbonate veins of the Cedar Bay mine (Figure 1). Several of the samples were from assays intervals (0.5 m length) from drillhole CRD-20-4C, which ran between 4.1 and 4.8 wt% Cu, with one at 14 wt% Cu, and between 0.1 and 3.6 g/t Au. Petrographic study performed to characterize this mineralization and associated alteration, in addition to identifying FIs and fluid inclusion assemblages (FIAs) [94,95], was complete using both transmitted and reflected light on thick sections (150 μm) with an Olympus BX-51 microscope equipped with a Q-Imaging digital capture system in the Fluid Inclusion Lab at Laurentian University. Of the nine samples, five containing the best FIs and FIAs were used for microthermometry, which was performed using a Linkham THMSG600 heating–freezing stage with an automated controller unit. The instrument was calibrated using synthetic FI standards for the triple point of CO_2 (−56.6 °C) and H_2O (0 °C), and for the critical point of H_2O (374 °C). Repeated runs of standards indicate a reproducibility of ± 0.1 °C for low-temperature and ± 2 °C for high-temperature runs, using heating rates of 1 °C/min and 3–5 °C/min, respectively, near phase changes. The following temperatures were recorded: (1) freezing point FI (Tf); (2) first melting (Tfm); (3) melting of hydrohalite (TmHH); (4) melting of ice (Tm_{ice}); (5) melting (Tm_{CO2}) and homogenization (Th_{CO2}) of CO_2; (6) melting of halite (TmH) or other solids; and (6) final homogenization (Th) that was always the homogenization of the liquid (Th → L) for L_{H2O}-V_{H2O} or, in some cases, TmH where halite was present (see Table S6A in Supplementary material S6). Salinities were calculated using Tm_{ice} and TmH [95]. Bulk composition, in terms of H_2O:NaCl:$CaCl_2$, was calculated where TmHH was available [96].

The fluid chemistry was further evaluated using the evaporate mound analysis (EMA) method, whereby salt mounds (i.e., precipitates) are obtained by overheating FIs [97]. Prior to heating, samples were soaked several times in acetone and dried to preclude surface contamination. The chips were then heated to 465 °C using a Linkham stage at a rate of 50 °C/min, and then rapidly cooled to room temperature. This procedure produced abundant isolated mounds for the two studied samples. Quartz chips were subsequently mounted on double-sided carbon tape adhered to a petrographic glass slide, carbon-coated, and then imaged and analyzed using the same SEM-EDS noted above. The EDS analysis used a 20 kV accelerating voltage, a 1.2 nA beam current, a working distance ranging from 15 to 18 mm, and an acquisition time of 5 s. Imaging (BSE and CL) used a 15 kV accelerating voltage and a 1.0 nA beam current. The EMA was performed using both point and raster modes [97] and data were processed using the Burker software. The chemical data were normalized to 100 wt% without SiO_2 to correct for interference from the quartz substrate.

4. Results

4.1. Petrography of Igneous Rocks

The 2020 samples are dominated by plagioclase (±sericitized), amphibole, biotite, and quartz with minor amounts of amphiboles, epidote, apatite, titanomagnetite, titanite, and zircon. Similar mineral assemblages characterize the 1980 sample suite [5,98]. A more detailed petrographic study was performed on the eleven samples used for geochronology (Figure 3). These eleven samples are dominated (35–50 vol%; Table S2C) by euhedral and extensively hydrolyzed plagioclase that is most altered in mafic samples (Figure 3a). Hornblende is an important phase (10–20 vol%) in intermediate rocks (Figure 3b), dominates (55 vol%) mafic rocks (Figure 3a), and is a minor phase (5 vol%) in felsic samples (Figure 3c,d). Hornblende is generally replaced by actinolite and chlorite, especially in the hydrated mafic samples (Figure 3a), and is with biotite in intermediate to felsic rocks (Figure 3b). Quartz, an interstitial phase most abundant in felsic rocks (to 45 vol%), is associated with orthoclase (5 vol%; Table S2C). Epidote occurs as a secondary mineral, although coarse epidote inclusions in amphibole grains (sample ACR-19-3) may have a magmatic origin. The main oxide is titanomagnetite (2–5 vol%), which occurs as magnetite with ilmenite oxy-exsolutions and as isolated magnetite and ilmenite grains. Ilmenite is generally rimmed by titanite or is entirely replaced by titanite in rocks with <1–2 vol% oxides prior to regional greenschist facies metamorphism (D_2 event).

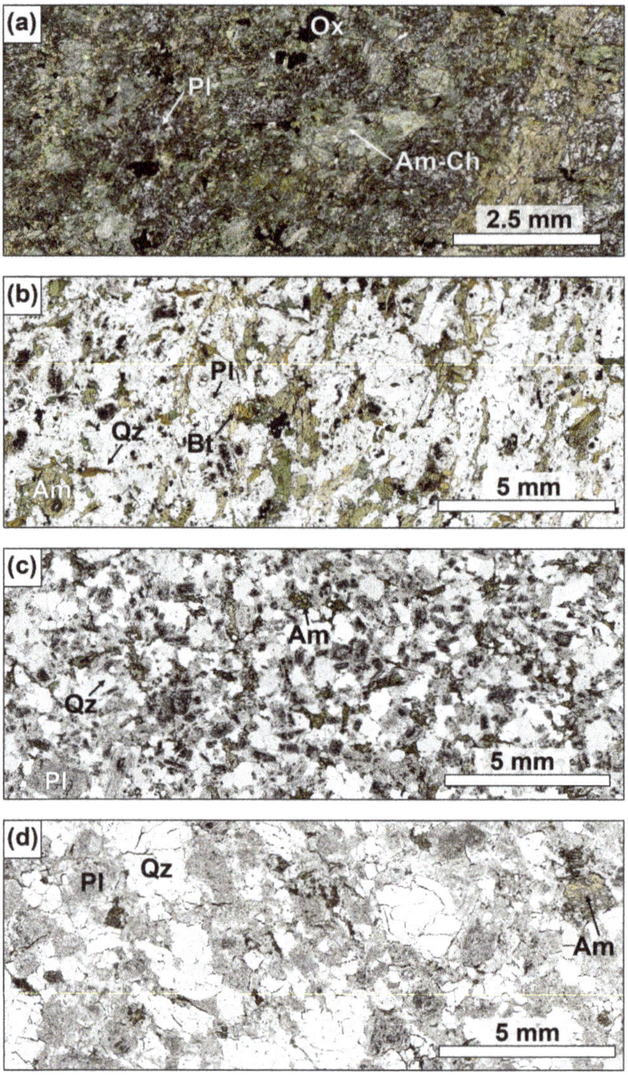

Figure 3. Diorite samples ACR-16-2 (**a**) and ACR-6-1 (**b**), and tonalite samples ACR-32-1 (**c**) and ACR-24-3 (**d**) observed in plane-polarized light. Abbreviations: amphibole (Am); chlorite (Ch); biotite (Bt); oxide (Ox); plagioclase (Pl); quartz (Qz).

4.2. Petrography of Mineralized Samples

Mineralized samples from the main mineralization style (style A) contain sulfide and darkened quartz veins in fractures or shear zones. The samples consist of quartz–carbonate veins intersected by stringers of sulfides (pyrite-chalcopyrite) and by veins of phyllosilicates (chlorite and greenish biotite), indicating that quartz veining occurred early and possibly prior to the main Cu mineralizing event.

In the veins, quartz has a granular texture and displays variable degrees of strain, i.e., subgrains and undulose extinction. Remnant quartz grain domains of subhedral to euhedral shapes are preserved, particularly where quartz is intergrown with carbonate (Figure S6A), with such areas inundated with FIs (see below). Cathodoluminescence imaging of two mineralized samples did not show relict primary zoning, as is common in hydrothermal quartz, but did reveal (1) light- and dark-colored domains associated with a mosaic-like texture (Figure S6A) and (2) bright CL zones, defining late brittle–ductile micro-shear zones overprinting earlier dark CL quartz (Figure S6A). The domains in quartz–carbonate veins that were selected for microthermometry displayed the least modified post-formation textures, limiting the risk of post-entrapment modification of FIs.

4.3. Geochemistry

Samples from the 1980s field campaign, re-analyzed in 2019, as well as recently acquired (2020) samples display similar petrological and chemical characteristics. All samples are thus integrated into the single dataset presented here. The rocks are classified as meladiorite, diorite, tonalite, and leucotonalite on the basis of the quartz/amphibole ratio and following the classification proposed by D. Racicot [5]. Dykes are undifferentiated magma injections which, for the most part, emanate from the Chibougamau pluton. The pluton also contains enclaves of gabbro (xenoliths from the LDC), granophyre (LDC xenoliths), hornblendite (plagioclase-poor meladiorite), and pegmatite (Table S2A,B). The limited scatter of data away from an apparent differentiation trend on the total alkali silicate (TAS) diagram (Figure 4a) confirms the results of earlier work indicating that the pluton is dominated by fresh to weakly altered rocks [5].

The intrusive phases have a large range of SiO_2 (~45 to 80 wt%) and correspond to mafic to intermediate (referred to as the Diorite phase hereafter) and felsic (Tonalite) rocks (Figure 4a, Table 3). Both phases are differentiated according to the distribution of samples on the TAS and Na-K-Ca diagrams (Figure 4a,b), are Al-rich (>15–16 wt% Al_2O_3), and equate to metaluminous (Diorite phase) and peraluminous (Tonalite phase) rocks (Figure 4c). The few samples enriched in K possibly reflect localized hydrothermal alteration (i.e., sericitization), whereas other samples follow the trondhjemite differentiation trend (Figure 4b). The dykes belong to both intrusive phases, according to their silica content, and have similar chemistries (Table 3). The main intrusive phases and the dykes are subalkaline (see TAS diagram in Figure 4a), with a calc-alkaline affinity (Figure 4b). The few samples with a tholeiitic to transitional affinities likely correspond to xenoliths from the LDC that were misclassified as diorite in the field (Figure 4d). Most rocks fall in the field defined by TTG suites, with the Diorite phase displaying systematically lower $(La/Yb)_N$ compared to the Tonalite phase (Figure 4e).

Table 3. Main chemical characteristics of the Diorite and Tonalite intrusive phases, as defined by field observations. The whole dataset is available in Table S2A,B.

	Diorite Phase				Tonalite Phase			
	Pluton		Dyke		Pluton		Dyke	
	Average	Std *	Average	Std	Average	Std	Average	Std
SiO_2 (wt.%)	55.07	7.31	54.19	5.73	70.36	3.51	69.55	2.14
Al_2O_3	16.66	1.92	16.38	2.01	15.36	1.70	15.84	0.79
Fe_2O_{3T}	8.02	3.13	8.01	2.05	2.47	1.14	2.49	1.09
MgO	5.34	3.34	5.53	2.61	0.90	0.58	0.83	0.41
MnO	0.12	0.05	0.12	0.04	0.03	0.02	0.03	0.04
CaO	8.14	2.18	8.55	2.59	2.72	1.25	2.65	1.33
Na_2O	3.25	0.96	2.40	1.36	4.48	1.05	4.51	1.45
K_2O	0.86	0.51	0.48	0.73	1.73	1.14	1.92	1.94
TiO_2	0.72	0.41	0.75	0.25	0.23	0.14	0.21	0.10
P_2O_5	0.23	0.20	0.23	0.14	0.09	0.07	0.07	0.03
Cr_2O_3	0.03	0.04	0.03	0.03	0.01	0.00	0.01	0.00
SrO	0.06	0.03	0.03	0.00	0.06	0.05	0.00	0.00
BaO	0.03	0.03	0.01	0.01	0.06	0.03	0.00	0.00
LOI	1.78	1.04	3.56	2.64	1.43	0.80	1.86	1.34
Total	100.23	0.65	100.19	0.59	99.82	0.85	99.96	0.48
Zr (ppm)	95.78	51.65	89.83	39.38	94.54	52.93	79.58	23.64

* Std—standard deviation.

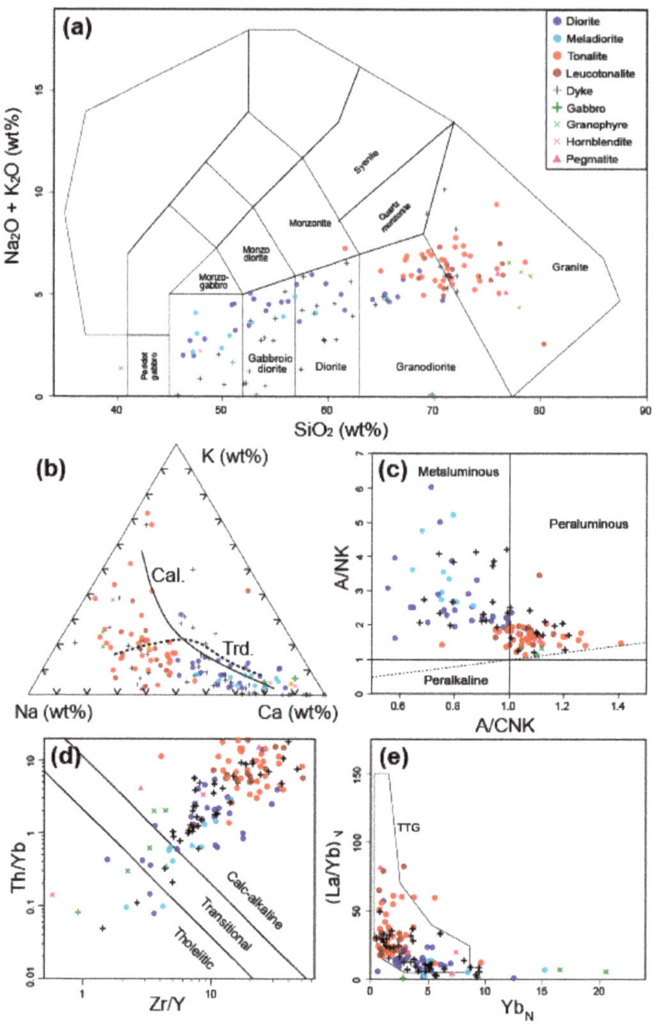

Figure 4. Whole-rock chemistry of the main intrusive phases, xenoliths, and dykes of the Chibougamau pluton displayed in (**a**) the total alkali silica (TAS) diagram [99]; (**b**) the Na-K-Ca ternary diagram, showing the calc-alkaline (Cal.) and trondhjemite (Trd.) trends [100]; (**c**) the Al/(Na + K) versus Al/(Ca + Na + K) diagram (oxides molar proportions) of [101]; (**d**) the Zr/Y versus Th/Yb binary diagram [102]; and (**e**) the Yb_N versus $(La/Yb)_N$ discrimination diagram for TTG suites [103].

Harker diagrams show continuous trends between the Diorite and Tonalite phases (Figure 5a–i), with MgO, CaO, TiO_2, P_2O_5, and FeO_T decreasing and Na_2O increasing with increasing SiO_2 content (Figure 5b–h). In the Diorite phase, Al_2O_3 does not correlate with SiO_2 and Al_2O_3 is negatively correlated with silica in Tonalite (Figure 5a). Potassium has low abundance in Diorite and displays variable but generally higher values in Tonalite (Figure 5e). For Zr, positive and negative correlations with silica content are observed for the Diorite and Tonalite phases, respectively (Figure 5i).

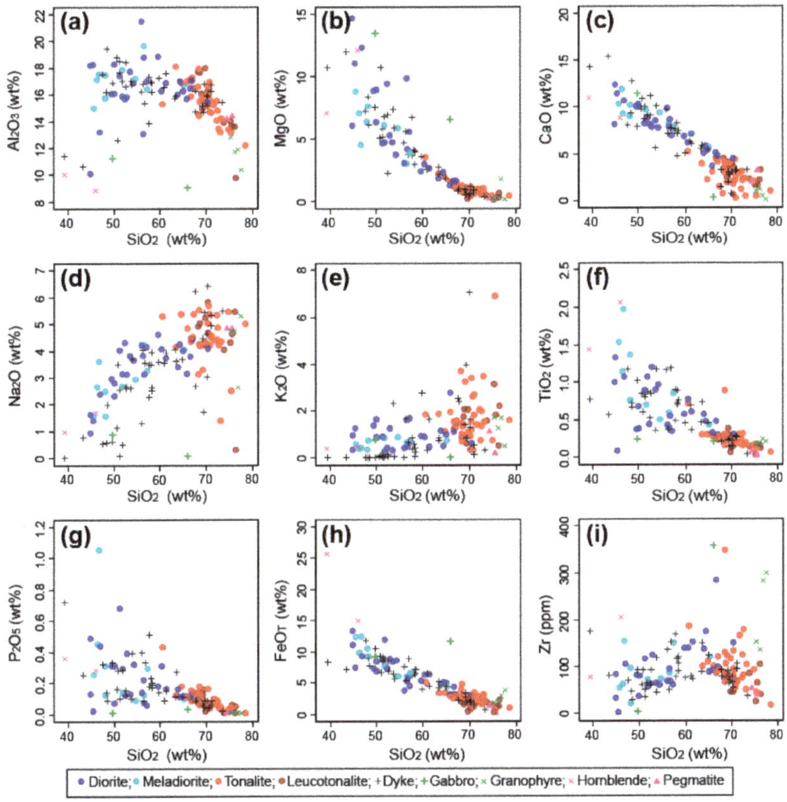

Figure 5. Whole-rock chemistry of the main intrusive phases, xenoliths, and dykes of the Chibougamau pluton displayed in Harker-type diagrams for Al_2O_3 (**a**), MgO (**b**), CaO (**c**), Na_2O (**d**), K_2O (**e**), TiO_2 (**f**), P_2O_5 (**g**), FeO_T (**h**), and Zr (**i**).

The intrusive phases, dykes included, generally display "hockey-stick" profiles in chondrite-normalized REE diagrams (Figure 6a–c). The xenoliths from the LDC (gabbro and granophyre) display flat to weakly fractionated REE profiles, as expected for tholeiitic intrusive rocks (not shown on diagram and available as Supplementary Material S2). In contrast, the pluton samples are characterized by moderate LREE enrichment, HREE depletion, and an absence of any Eu anomaly, except for 5 of the 123 samples that display positive Eu anomalies (Figure 6a–c). The Diorite phase is more HREE-enriched than the Tonalite, whereas both display similar LREE contents (Figure 6a,b). Consequently, the Diorite phase is less fractionated than the Tonalite phase, with $(La/Yb)_N < 24$ for Diorite and $(La/Yb)_N$ between 20 and 80 for Tonalite (Figure 6d).

A previous study has shown that field names can be unreliable and that rocks can be reclassified based on the $(La/Yb)_N$ [1]. Reclassifying the rocks (dykes included) based on their $(La/Yb)_N$ confirms that the main difference among the rock types is their HREE content. Most of the samples selected for geochronological investigation (see below) are moderately fractionated, with samples ACR-19-3, ACR-24-3, and 20ACR-0011C being the most fractionated (Figure 6e,f). On an extended multi-element diagram, the rock groups display similar profiles, with those with $(La/Yb)_N > 24$ generally having more pronounced negative Nb, Ta, and Ti anomalies (Figure 6g).

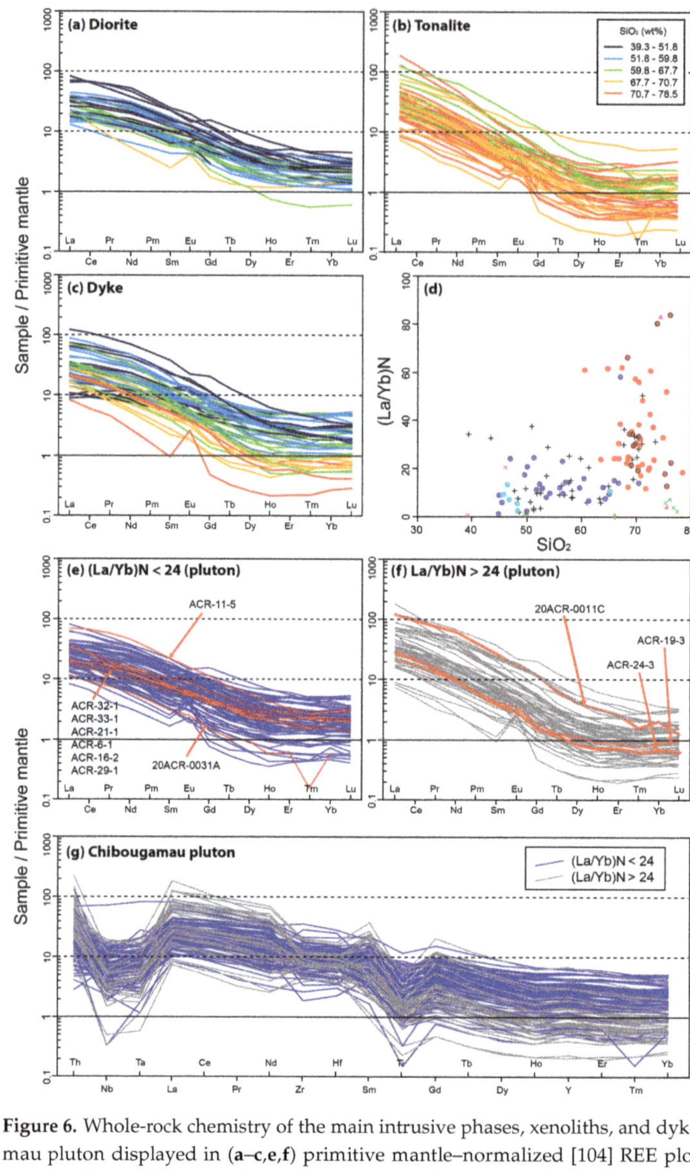

Figure 6. Whole-rock chemistry of the main intrusive phases, xenoliths, and dykes of the Chibougamau pluton displayed in (**a**–**c**,**e**,**f**) primitive mantle–normalized [104] REE plots, (**d**) a $(La/Yb)_N$ versus SiO_2 (wt.%) binary diagram, and a (**g**) spidergram modified from Pearce [105]. For (**d**), the legend is the same as for Figure 5. The following rock types are displayed: (**a**) 32 samples from the Diorite phase identified based on field observations (diorite, meladiorite, hornblendite) and using the Zr/Y versus Th/Yb binary diagram (Figure 4c); (**b**) 47 samples attributed to the Tonalite phase on the basis of field observations (tonalite and leucotonalite); and (**c**) 44 dykes identified using field observations and the Zr/Y versus Th/Yb binary diagram (Figure 4c). Reclassified samples are also displayed: (**e**) REE diagram for samples with $(La/Yb)_N$ <24 (n = 70); (**f**) REE diagram for samples with $(La/Yb)_N$ >24 (n = 53); and (**g**) spidergram for all samples (n = 123).

4.4. Zircon U-Pb Geochronology

4.4.1. MC-ICP-MS Geochronology

Zircon grains were recovered from nine samples. For four samples (ACR-6-1, ACR-19-3, ACR-24-1, ACR-32-1), the hundreds of small (50–100 μm, up to 200 μm) grains mounted in epoxy are euhedral and, in CL, display delicate oscillatory zoning with few transgressive zones and core regions that can be distinguished in a few grains (Supplementary Material S3). The other samples (ACR-11-5, ACR-33-1, ACR-16-2, ACR-21-1) returned a limited amount of small (50–100 μm) euhedral to anhedral grains that, in CL, mostly display thickened to weakly expressed oscillatory zoning. Sample ACR-29-1 returned larger zircon grains (100–500 μm) with a euhedral habit and oscillatory zoning in CL that is mostly irregular and locally thickened (Supplementary Material S3).

As noted earlier, zircons were separated gravitationally into large (~70–200 μm and above) and small (<100 μm) size fractions for analysis. However, as both size fractions returned similar chemistries and ages (Figures 7–9), the results were integrated into a same dataset to obtain dates. For all samples, a filter criterion for pristine grains (i.e., low La content) was applied and the results are summarized as follows.

For sample ACR-6-1, one zircon population is identified (Figure 7a). Filtered analyses (La < 0.1 ppm) are mostly concordant and return an intercept age of 2714.8 ± 0.93 Ma (MSWD = 1.7) (Figure 7b).

For sample ACR-11-5, La-poor grains define one zircon population (Figure 7c), with the filtered analyses (La < 1 ppm) mostly concordant and giving an intercept age of 2713.4 ± 2.5 Ma (MSWD = 2.2) (Figure 7d).

For sample ACR-19-3, La-poor grains define one zircon population (Figure 7e), with the filtered analyses (La < 0.1 ppm) concordant and giving an intercept age of 2699.5 ± 1.3 Ma (MSWD = 1.6) (Figure 7f).

For sample ACR-33-1, La-poor grains define one main zircon population, with three older grains defining an unsatisfactory statistical population (Figure 8a). For the former, the filtered analyses (La < 1 ppm) are mostly concordant and give an intercept age of 2725.2 ± 1.5 Ma (MSWD = 1.4) after removing five grains (red in inset of Figure 8b) that fall outside the main statistical population (Figure 8b).

For sample ACR-24-3, zircon grains define one zircon population (Figure 8c) with the filtered analyses (La < 0.02 ppm) being concordant and giving an intercept age of 2702.2 ± 1.2 Ma (MSWD = 1.5) (Figure 8d).

For sample ACR-32-1, La-poor grains define one main zircon population (Figure 8e). Filtered zircon analyses (La < 0.1 ppm) are concordant and return an intercept age of 2722.5 ± 1.9 Ma (MSWD = 1.6) (Figure 8f).

For sample ACR-29-1, La-poor grains define one zircon population (Figure 7e). Filtered zircon analyses (La < 0.1 ppm) are mostly concordant and an intercept age of 2728.6 ± 1.5 Ma (MSWD = 0.67) is obtained after removing four grains (see inset of Figure 9a) that fall outside of the main statistical population (Figure 9b).

For sample ACR-16-2, only a few grains were recovered. Filtered zircon analyses (La < 1.6 ppm, median is 0.12 ppm La) are mostly concordant and return an intercept age of 2710.4 ± 7.7 Ma (MSWD = 0.38) (Figure 9c).

For sample ACR-21-1, an intercept age of 2722 ± 37 Ma (MSWD = 0.87) was calculated using five grains that contain 0.4–0.6 ppm La (Figure 9d). These ages extend the duration of magmatism in the Chibougamau pluton (Table 1; Figure 10) (see Section 5).

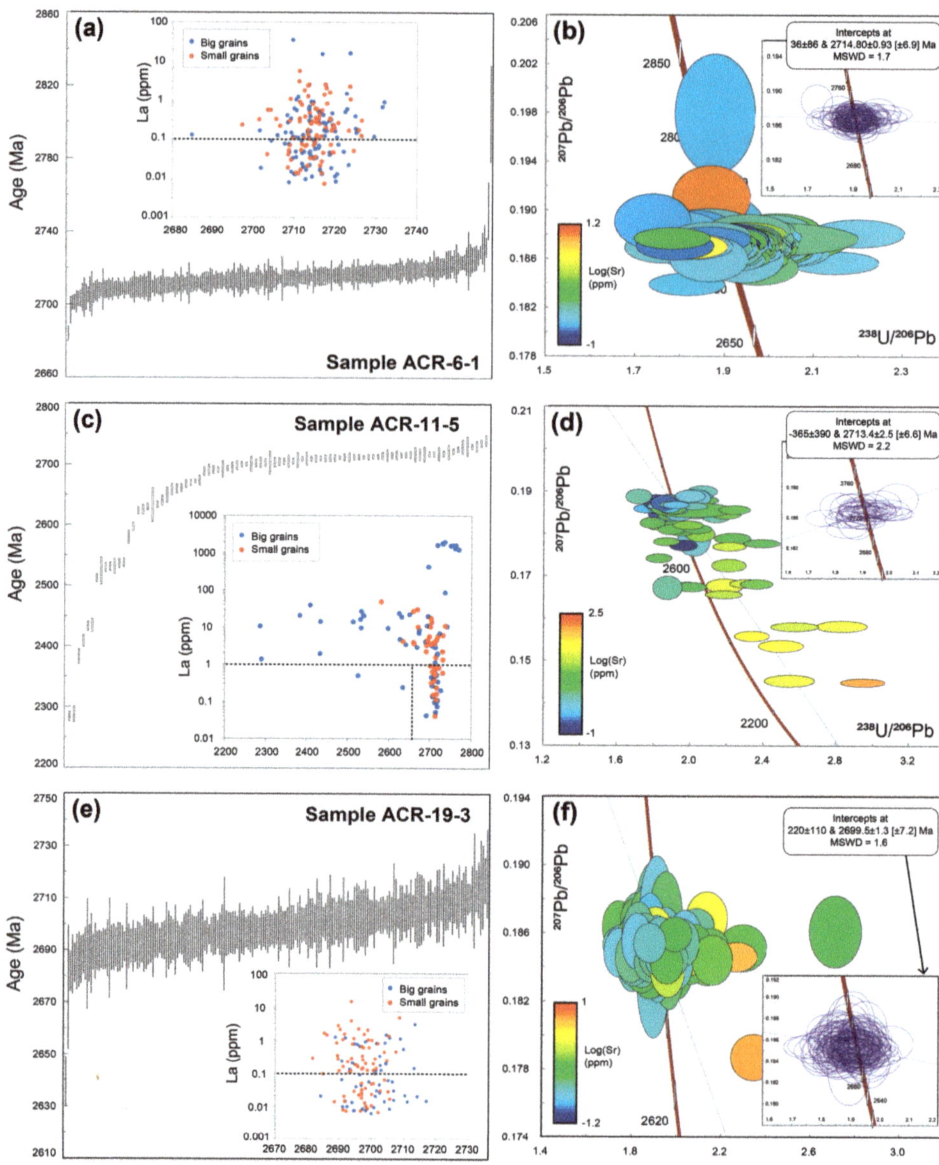

Figure 7. Zircon U-Pb age spectra (**left**) and Tera–Wasserburg (TW) Concordia diagrams (**right**) for samples (**a**) ACR-6-1, (**b**) ACR-11-5, and (**c**) ACR-19-3. Age spectra inset shows relationship between individual 207Pb/206Pb dates and La concentrations. The threshold value for data filtering is defined by departure from vertical array. Ellipses in TW Concordia diagrams are colored for Sr concentration (ppm). The inset shows data conforming to filter criteria for each sample, from which sample crystallization ages are calculated.

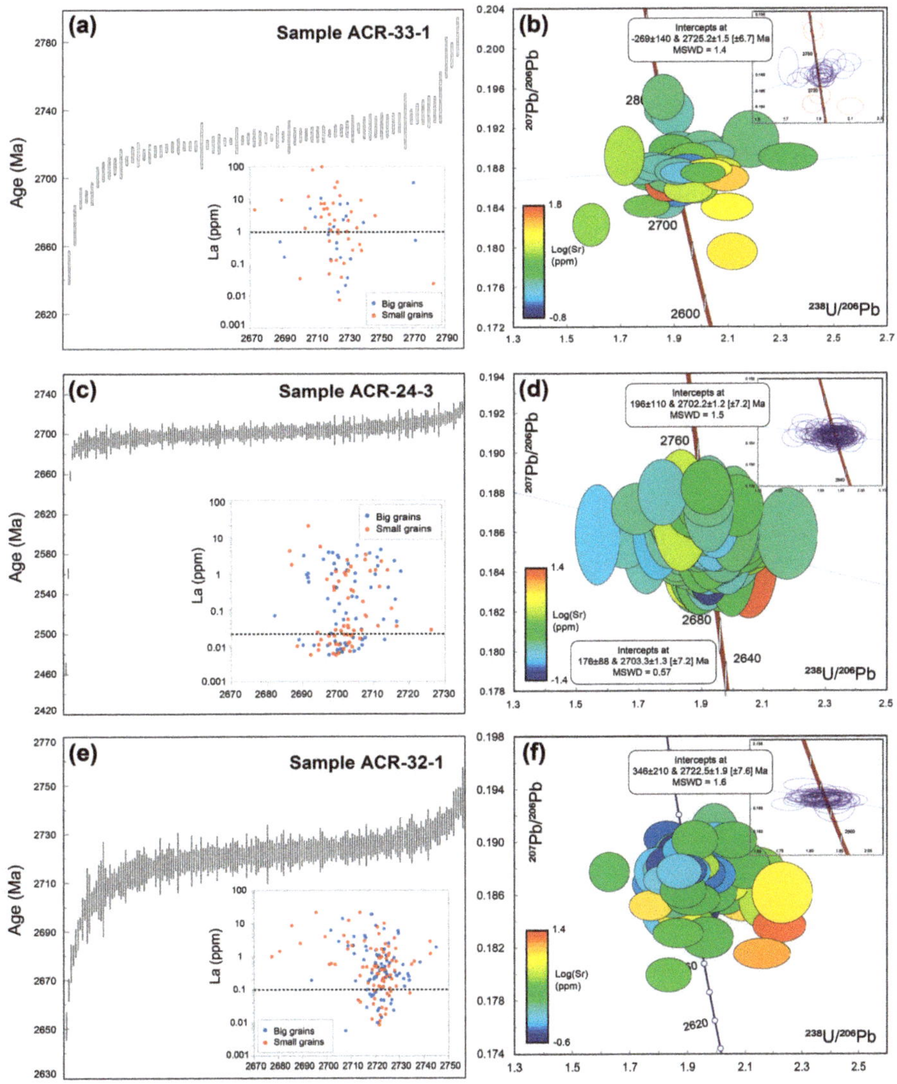

Figure 8. Summary of the LA-ICP-MS zircon U-Pb data in cumulative plots of $^{207}Pb/^{206}Pb$ ages (**left**) and Tera–Wasserburg (TW) diagrams (**right**) for samples (**a**) ACR-33-1, (**b**) ACR-24-3, and (**c**) ACR-32-1. Age spectra insets show relationship between individual $^{207}Pb/^{206}Pb$ dates and La concentrations, with the threshold value for data filtering defined by a departure from the vertical array. Ellipses in TW diagrams are colored for Sr concentration (ppm), with insets showing data conforming to filter criteria for each sample, from which zircon crystallization ages are calculated.

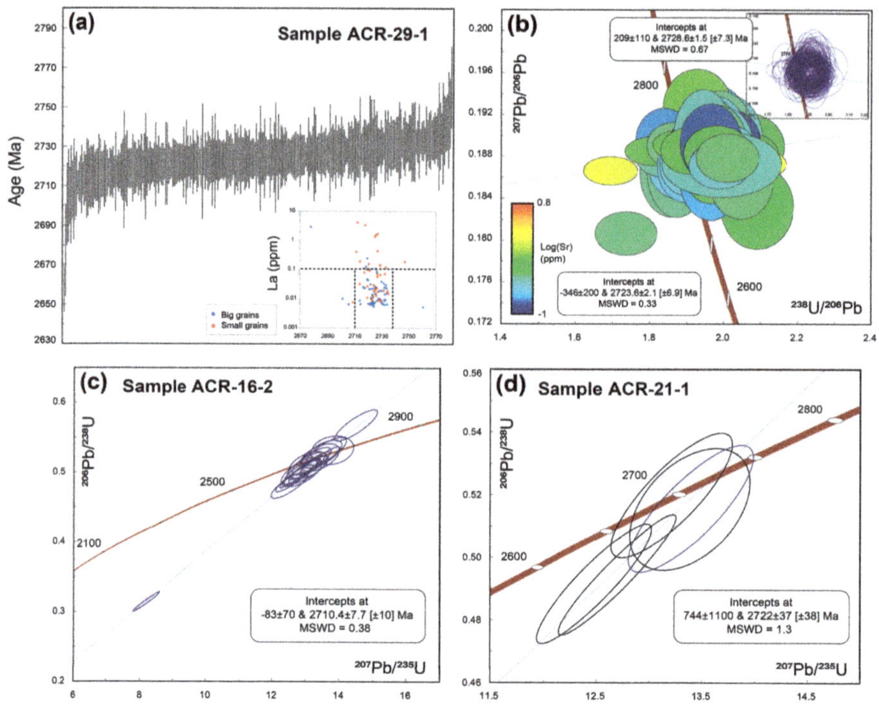

Figure 9. (**a**,**b**) Summary of the LA-ICP-MS zircon U-Pb data in a cumulative plot of ^{207}Pb/^{206}Pb ages (**left**) and Tera–Wasserburg (TW) diagram (**right**) for sample ACR-29-1. The age spectra inset in (**a**) shows relationship between individual ^{207}Pb/^{206}Pb dates and La concentrations, with the threshold value for data filtering defined by a departure from the vertical array. Ellipses in TW diagram in (**b**) are colored for Sr concentration (ppm), with the inset showing data conforming to filter criteria from which the zircon crystallization age was calculated. (**c**,**d**) Conventional Concordia diagrams for the few zircon grains recovered from samples (**c**) ACR-16-2 and (**d**) ACR-21-1.

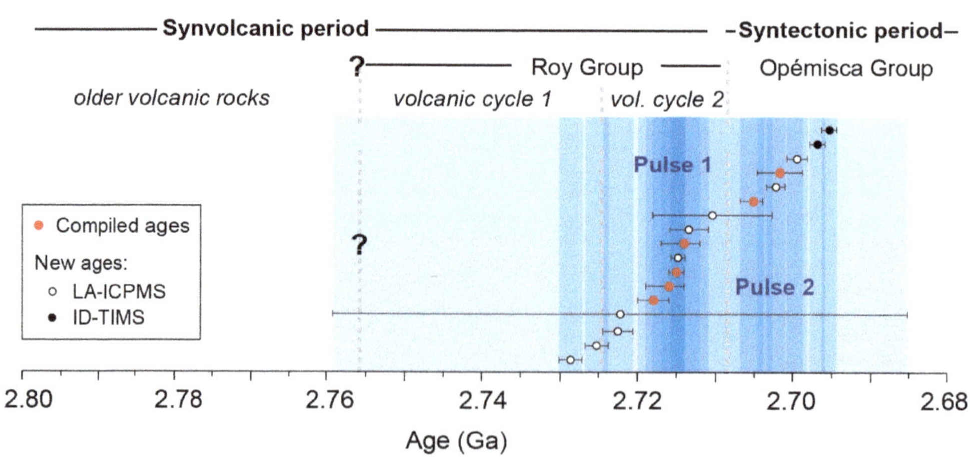

Figure 10. Compiled and new age data for the Chibougamau pluton (Table 1). The blue rectangles display individual ages and associated errors. The '?' sign indicates that the onset of volcanic cycle 1 is poorly constrained.

4.4.2. CA-ID-TIMS Geochronology

Sample 20ACR-0011C comes from an amphibole-bearing quartz–monzodiorite phase observed on an outcrop that displays intrusive contacts among the diorite, tonalite, and trondhjemite phases. Sample 20ACR-0011C consists of 63 vol% elongated euhedral plagioclase, 5 vol% interstitial quartz, 12 vol% sub-euhedral K-feldspar, and 20 vol% amphibole ±biotite–magnetite–apatite. Sample 20ACR-0031A is an altered (epidotized) tonalite phase located near outcrops of tonalite, diorite, and quartz–diorite intrusive phases. Sample 20ACR-0031A consists of 65 vol% euhedral hydrolyzed plagioclase, 27 vol% interstitial quartz, 2 vol% K-feldspar, 2–5 vol% of chloritized biotite, and 1 vol% epidote.

The zircon population recovered from sample 20ACR-0011C is dominated by clear and colorless to pale pink sub- to euhedral grains and broken grain fragments ranging in size from approximately 150 to over 500 microns in terms of maximum dimensions. Aspect ratios are commonly 2–3:1 with square cross sections, and rarely reach up to length/breadth ratios of 6:1 (Supplementary Material Figure S5B). Three chemically abraded single-grain fractions were selected for TIMS analysis (Figure S5B), and the results are presented in Supplementary Material Table S5A and Figure 11. Data for two fractions are perfectly concordant, while the data for a third are 0.7% discordant, with all three analyses having a narrow range of $^{207}Pb/^{206}Pb$ ages between 2693.7 and 2695.5 Ma. The weighted average $^{207}Pb/^{206}Pb$ age calculated for the two most strictly concordant fractions (Z1, Z3) is 2695.4 ± 1.0 Ma (MSWD = 0.03; probability of fit = 87%; Figure 11), which is interpreted to represent the primary crystallization age for quartz–monzodiorite.

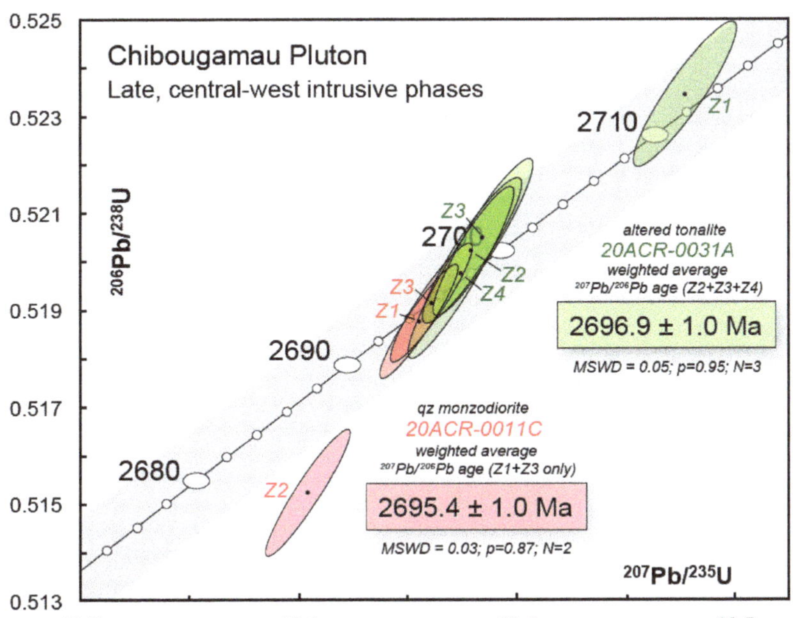

Figure 11. Combined Concordia diagram showing CA-ID-TIMS dating results (U-Pb, zircon) for two samples (20ACR-0011C and 20ACR-0031A) of late intrusive phases of the Chibougamau pluton.

The zircon population of the 20ACR-0031A sample comprises nearly gem-quality, water-clear, mostly stubby, sharply faceted euhedral prismatic grains ranging up to approximately 200 microns in maximum dimension (Supplementary Material Figure S5B). Rare crystal forms are elongated up to 5:1. Four fractions of representative, chemically abraded single grains were selected and analyzed from this sample (Figure S5B, Table S5A). Isotopic analyses for three of the fractions (Z2, Z3, Z4) were completely overlapping with themselves and Concordia and together yielded a weighted average $^{207}Pb/^{206}Pb$ age of 2696.9 ± 1.0 Ma (MSWD = 0.05; probability of fit = 95%; Figure 11), which we considered a robust age estimate for the crystallization of this tonalite. A fourth fraction (Z1) was also concordant but had an older $^{207}Pb/^{206}Pb$ age of 2710.7 Ma, and this likely reflected a mixed age due to the presence of a small, slightly older inherited component which was not optically visible.

These two new statistically overlapping ID-TIMS age determinations for intrusive units in the central western part of the composite Chibougamau pluton currently represent the youngest ages reported to date from this extensive intrusive suite. The new dates thereby indicate that magmatism extended for at least ~20 My (from ca. 2718 to 2695 Ma) within the area occupied by the Chibougamau pluton (Figure 10) (see Section 5).

4.5. Zircon Chemistry

The fO_2 parameter is calculated using the chemistry of the nine zircon populations, LA-ICP-MS ages and the method of Loucks [106]. This empirical method requires the measurement of the U, Ce, and Ti contents of zircon to evaluate the fO_2 parameter. The crystallization age of zircon, as reported above, is used to estimate the eU parameter, which is the age-corrected initial U content of zircon [106]. Overall, the samples exhibit moderate oxidation (~0 to 2 ΔFMQ), expressed in terms of deviation from the fayalite–magnetite–quartz (FMQ) buffer (Figure 12a). We observe that the fO_2 parameter increases with increasing La content for values above 1 ppm, indicating that grains containing >1 ppm La are altered and return unreliable fO_2 values (Figure 12a). Calculating statistics for grains containing <0.1 ppm La, a value selected because it is more conservative than 1 ppm La, returns statistically meaningful results for samples ACR-6-1, ACR-19-3, ACR-24-1, ACR-32-1, and ACR-29-1, which contain many (>>10) pristine zircon grains (Figure 12b). These samples returned the following fO_2 values, presented as median ± standard deviation values: ΔFMQ +0.23 ± 0.54 (ACR-29-1), ΔFMQ +0.80 ±0.75 (ACR-32-1), ΔFMQ +0.89 ± 0.54 (ACR-6-1), ΔFMQ +1.33 ± 0.43 (ACR-24-3), and ΔFMQ +1.21 ± 0.38 (ACR-19-3). The other samples (ACR-11-5, ACR-33-1, ACR-16-2, ACR-21-1) had <10 pristine zircon grains and returned fO_2 values that are relatively consistent with the values reported above, but that were considered less statistically meaningful (Figure 12b). We observed that older zircon grains returned fO_2 values close to the FMQ buffer and that younger zircon grains returned fO_2 values above FMQ +1.

4.6. Fluid Inclusions (FI)

In this study, the fluid inclusion assemblage (FIA) approach is used whereby discrete fluid trapping events are assessed with respect to the timing of FI entrapment as primary (P), secondary (S), and pseudosecondary (PS) types [94,95]. Additionally, the term indeterminate (I) is used where FIs do not conform to the latter types (e.g., isolated or 3D arrays of FIs) [95,107] or lack the textural evidence (i.e., crystal growth faces) necessary for classification.

As mentioned above, two samples were imaged using an SEM (CL images). In both samples, the presence of bright CL spots is attributed to the presence of open FIs lined with grinding powder and indicate that quartz contains (1) FIs aligned along linear features (fractures) and commonly surrounded by dark CL zones, which are S-type FIAs (Figure S6A), and (2) randomly distributed FIs in dark CL domains, which are considered I-type FIAs (Figure S6A).

In addition to FIAs, several types of FI (1a, 1b, 2, 3, 4, and 5) are identified (Figure 13, Table 4). We note that, whereas there may be a large overall range in microthermometric data for an FI type, the ranges are small for a given FIA, as observed elsewhere [108]. Additionally, for many of the cooled (to <100 °C) FI populations, metastability was an issue (i.e., the FI did not freeze), which was likely related to high salinities and Ca enrichment [109]. This was especially the case for FI types 1a and 1b (see below). The size of FIs is highly variable for all types, except for type 5 (see below), and ranges from <3–5 μm to 20–30 μm.

The shapes of the FIs vary greatly due to post-entrapment modification or decrepitation, as seen elsewhere [95].

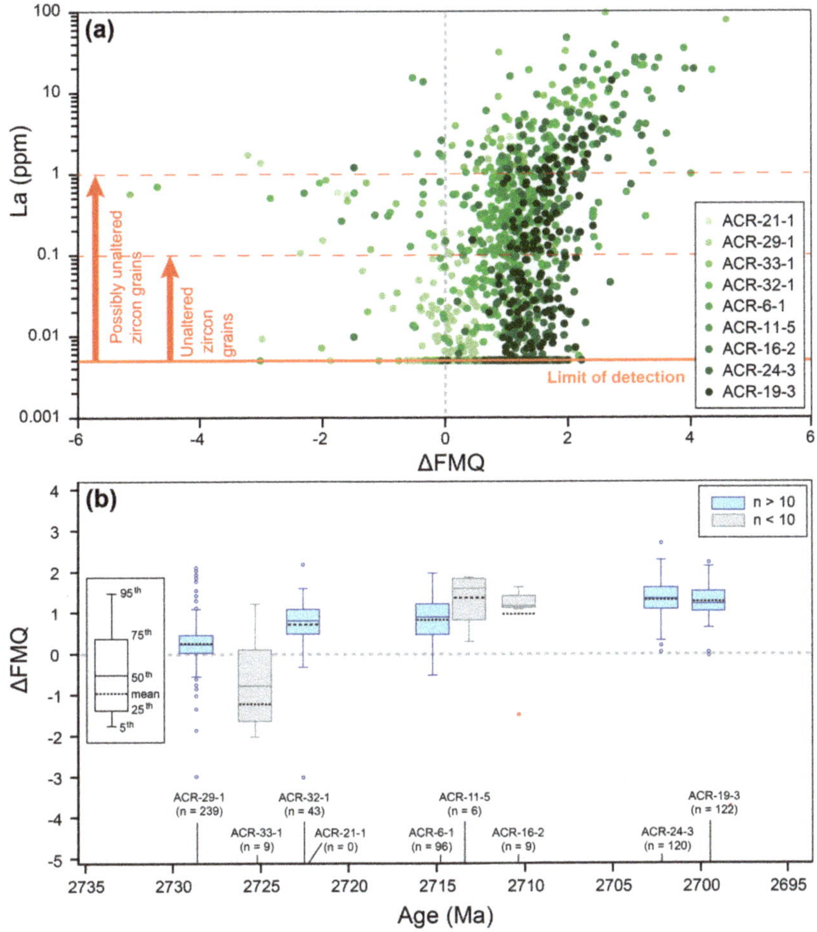

Figure 12. Oxidation state of the studied magmatic system, as estimated using zircon chemistry and the method of Loucks [106]. The oxidation state, expressed in terms of deviation from the fayalite–magnetite–quartz (FMQ) buffer, is displayed using the (**a**) La versus ΔFMQ binary diagram showing the entire dataset and (**b**) ΔFMQ versus age diagram, where only those zircon grains containing <0.1 ppm La are displayed using box plots.

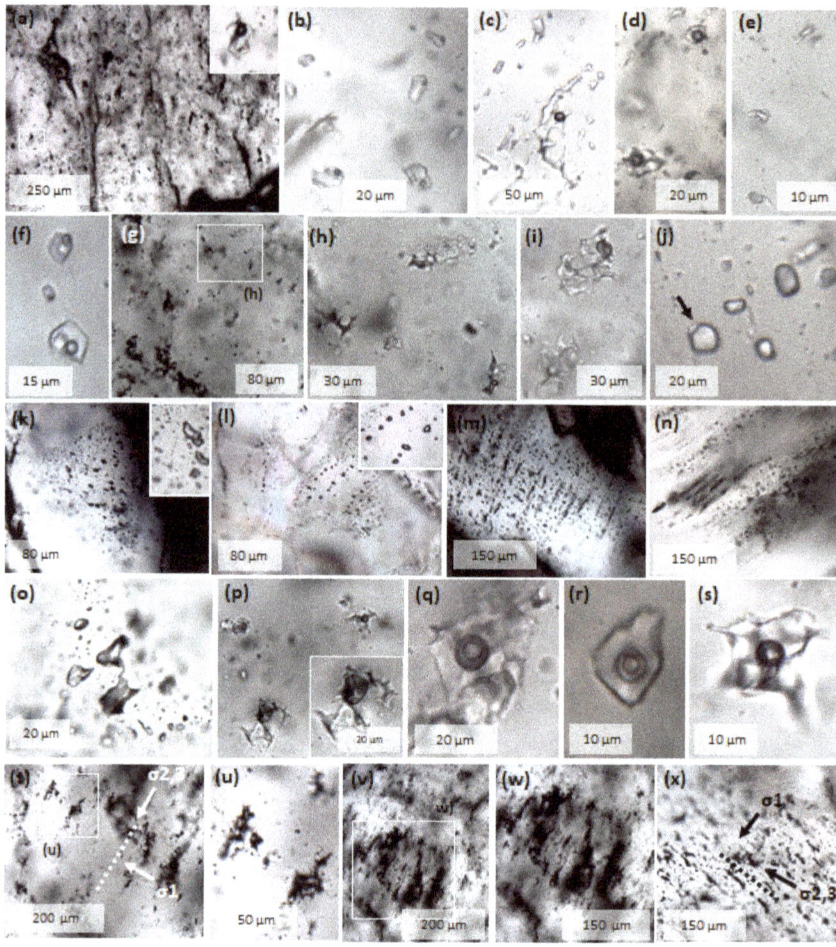

Figure 13. Images of fluid inclusions (FIs) in quartz taken in plane-polarized light. The following abbreviations are used: vapor (V), solid (S), and liquid (L); (**a**) quartz inundated with randomly distributed FIs (I type fluid inclusion assemblage—FIA). Some FIs display faint flattening. The inset shows a large type 1 FI (I type); (**b**) a type 1a FI (L_{H2O}-V_{H2O}) with a low degree of fill (F < 5–10); (**c**) large (i.e., >50–100 µm) and necked type 1a FIs (L_{H2O}-V_{H2O}) with low degrees of fill; (**d**,**e**) an isolated type 1b FI (L_{H2O}-V_{H2O}) with a high degree of fill (F > 30–40); and (**f**) a type 2 FI (L_{H2O}-V_{H2O}-Halite) with a variably developed decrepitate texture. The white box is enlarged in (**g**); (**g**,**h**) FIs with irregular shapes and with abundant S and low F values; (**j**) isolated type 3 FIs (V_{H2O}). The picture is taken at −10 °C. The arrow points to a white phase (ice) at the top of one inclusion; (**k**) a type 3 V-rich FI aligned along a planar structure and in the core of a quartz grain; (**l**) a V-rich FI decorating quartz crystal face (P type FIA); (**m**) a type 3 V-rich FI with FIs flattened and aligned; (**n**) a type 3 V-rich FI decorating planar features; (**o**) a type 3 V-type FI at the interface of sutured quartz grains, with the necking of large FIs; (**p**) a decrepitated type 4 FI (L_{H2O}-L_{CO2}) with central L_{CO2} and an enlarged FI in the inset; (**q**,**r**) a three-phase type 4 FI (L_{H2O}-L_{CO2}-V_{CO2}) of irregular and equant shapes; (**s**) a type 4 FI (L_{H2O}-L_{CO2}-V_{CO2}) with halite; (**t**,**u**) decrepitated and flattened FIs with preferential alignment and the inferred orientation of strain axes (σ1 and σ2,3); (**v**,**w**) examples of decrepitated and flattened FIs; and (**x**) V-rich FIAs showing alignment with the inferred orientation of strain axes. The symbols in (**t**) and (**x**) are used to denote the strain ellipse and orientation of the maximum (σ1), intermediate (σ2), and minimum (σ1) stresses, i.e., the orientation of the flattening plane containing σ2 and σ3.

Table 4. Summary of the microthermometric properties of fluid inclusions (FI) in this study.

	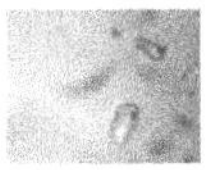	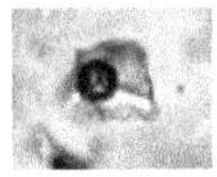	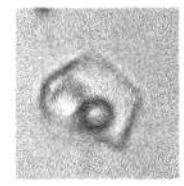	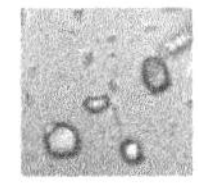	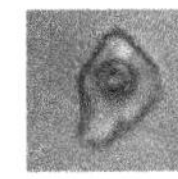	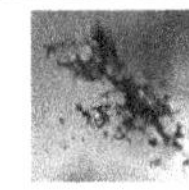
FI Type	1a	1b	2	3	4	5
FIA	I, S, PS	I, S, PS	I, S, PS	P, I, PS, S	I	S
Phases at 25 °C	L-V	L-V	L-V-H ± Solids	V	L_{H2O}-L_{CO2} ± H	x
Chemistry	L_{H2O}-V_{H2O}-NaCl-$CaCl_2$	L_{H2O}-V_{H2O}-NaCl-$CaCl_2$	L_{H2O}-V_{H2O}-NaCl-$CaCl_2$	V_{H2O}	H_2O-CO_2-NaCl	H_2O-CO_2-NaCl (?)
Occurrence	Common	Rare to moderate	Common	Common	Rare to common	Common
$\%V_{H2O}$ or L_{CO2}	<10–15	>15	5–10 to 40	>90	15–20	x
Tf	−80 to −90 °C	−60 to −90 °C	x	X	x	
Tfm	−50 to −60 °C	−50 to −40 °C	x	X	x	x
Tmice	−6 to −45 °C, but most <−22 °C	−19 to −35 °C	x	−6 °C (n = 1 FIA)	−7.7 to−15 °C	x
$TmCO_2$					−58 to −60	−59 °C (n = 1)
TmHH	−25 to −50 °C	−25 to −40 °C	x	X	X	x
$ThCO_2$					17 to 30 °C for one FIA at 12 °C	
Salinity (wt.% eq. NaCl)	21 to 30	24 to 29	29 to 50	9.2	11.3 to 18.6; >30	x
Th(L)	85 to 170 °C	205 to 400 °C; and >325 to 380 °C for some FI	115 to 400 °C	X	325 °C (n = 1 FIA)	x
Th(H)	X	X	130 °C (n = 1 FIA), most 275->400 °C	X	x	x

Type 1 FIs are the most common types and can locally inundate quartz (Figure 13a). These are two-phase aqueous (L_{H2O}-V_{H2O}) FIs and occur as I, S, and PS types of FIAs. Variable L:V ratios are used to define type 1a FIs (<10–15% V; Figure 13b,c) and type 1b FIs (>15% V; Figure 13d,e). A large range of Th is obtained for type 1a (85 to 170 °C) and type 1b FIs (205 to 400 °C for most) (Table 4). There is no apparent trend between salinity and Th values (Figure S6B). Both types have high salinities (21 to 30 and 24 to 29 wt.% eq. NaCl for types 1a and 1b FIs, respectively) and are variably Ca-rich, as determined from Tm_{ice} and TmHH relationships (Figure S6B).

Type 2 FIs are aqueous with solid phases (*L_{H2O}-V_{H2O}-Halite ± Solids*), with V ≤ from 5–10 to 40%. The main solid is halite, with small sub-rounded unidentified solids also rarely observed (Figure 13f–i). The variable phase proportions commonly noted likely occur due to necking. These FIs are common and occur as I, S, and PS types of FIA. The Th values vary widely from 115 to 400 °C, with salinities of 29 to 50 wt.% eq. NaCl, as determined from ThH. Some multi-solid FIs with irregular shapes, including decrepitated textures, did not show detectable phase changes after heating to 450 °C.

Type 3 FIs (*V_{H2O}-rich*) with >90% V (Figure 13j–o) are abundant and variably occur (1) as isolated clusters in the core of quartz grains (Figure 13j,k); (2) as decorating grain surfaces (Figure 13l); (3) as defining subparallel planes (Figure 13m); (4) along ductile micro-planar features (Figure 13n) and grain boundaries (Figure 13o); or (5) on planes transcending quartz grain boundaries. They are considered as P, I, PS, and S in origin. As no phase changes were noted during cooling (to −196 °C), these FIs are considered V_{H2O} versus mixed COHN types. In one FIA, a narrow rim in a FI nucleated solids on cooling (Figure 13i) that melted at −6 °C, equating to a salinity of 9.2 wt.% eq. NaCl.

Type 4 aqueous–carbonic FIs (L_{H2O}-L_{CO2}-V_{CO2} ± *Halite*) vary from rare to common. They are considered as mostly I type FIAs based on their random distribution. They can display either two (L_{H2O}-L_{C2O}) or three (L_{H2O}-L_{CO2}-V_{CO2}) phases (Figure 13p–r). Halite may also occur in this aqueous–carbonic FI (Figure 13s). The volume of CO_2 is similar in FIs with and without halite, pointing to similar X_{CO2} despite differing salinities. These FIs are small with equant to irregular shapes (Figure 13r), or much larger with highly irregular shapes, having scalloped outlines (Figure 13p,q); these latter FIs can be abundant (Figure 13t,u). The irregular shape of the FI is attributed to post-entrapment modification (i.e., decrepitation, see below). Commonly, the smaller FIs populate areas around the latter and are plausibly neonates that define decrepitate clusters [110]. The larger and irregularly shaped FIs can contain 20 to 80% of the carbonic phase. The Tm_{CO2} values are between −58 to −60 °C and are thus lower than those obtained for pure CO_2 (−56.6 °C). This indicates the presence of another dissolved gas, likely CH_4 (X_{CH4} < 0.05). For Th_{CO2} (always to L_{CO2}), a range of 17 to 30 °C is obtained for one FIA at 12 °C. Based on the latter and on the estimated carbonic phase present, the X_{CO2} is generally 0.05 to 0.15. The salinities for the FI lacking halite, as determined from Tm_{ice} and given that Tm_{clath} was not determined, are between 11.3 and 18.6 wt.% eq. NaCl. No Th(H) measurements were obtained on the FI containing halite but, based on their sizes, salinity was estimated to be ca. 30 wt.% NaCl (Table 4).

Type 1, 2, and 4 FIs commonly have irregular shapes. Where the original type of fill cannot be identified due to extreme modification, these decrepitated FIs are referred to as type 5. Their textures record post-entrapment re-equilibration to new densities, i.e., isochoric path [95], with changes in their internal pressure inducing either implosion or explosion [95]. As noted above, these large FIs can be surrounded by haloes of smaller FIs, referred to as neonates, and this is common for type 4 FIs. Type 5 FIs are abundant in all samples (Figure 13a,g,t), the texture of these FIs varies from subtle or incipient (Figure 13h,i,p) to very well developed (Figure 13t,u), and they can define subparallel arrays (Figure 13v,w).

As noted above, type 3 FIs form V-rich FIAs that commonly define planes (Figure 13m,x). The shapes and distributions of these inclusions can be used to make inferences about the strain ellipsoid [111] and about the possible flattening plane ($\sigma2$–$\sigma3$) (Figure 13t,x).

The chemistry of FIs was quantified using evaporate mounds (EMs) in quartz chips taken from the two samples also used for microthermometry. These EMs occurred either randomly (I type FIA) or as elongated trails (S-type FIA) (Figure 14a–c). Chemical data were acquired using both point (n = 122) and raster (n = 40) modes on the SEM instrument and related software. The former is used for smaller EMs (<5 μm) or to identify solid phases; the latter provides more representative compositions of the EM that undergoes fluid fractionation upon mound formation [97]. Representative images of EM and equivalent X-ray maps (Na, Ca, S, F) are presented (Figure 14d,e) and chemical values are summarized in binary (Figure 15a–d) and ternary (Figure 15e–g) plots, as well as box plots (Figure S6C). The dominant cations are Na and Ca, which are negatively correlated (Figure 15a,b,e). The EMs contain a lesser amount (<5 wt.%) of Fe and Mn based on data normalized to 100 wt.%. Barium is detected in 12% (n = 20) of the analyzed EMs and K is much less abundant than Na and Ca (Figure 15e) and is generally less abundant than Fe and Mn (Figure 15f). The dominant anion in the EM is Cl, but significant amounts of both S and F are also observed (Figure 15c,d,g). Calcium is positively correlated with S and F (Figure 15c,d), especially at higher concentrations, suggesting the occurrence of Ca-bearing fluoride and sulfate phases.

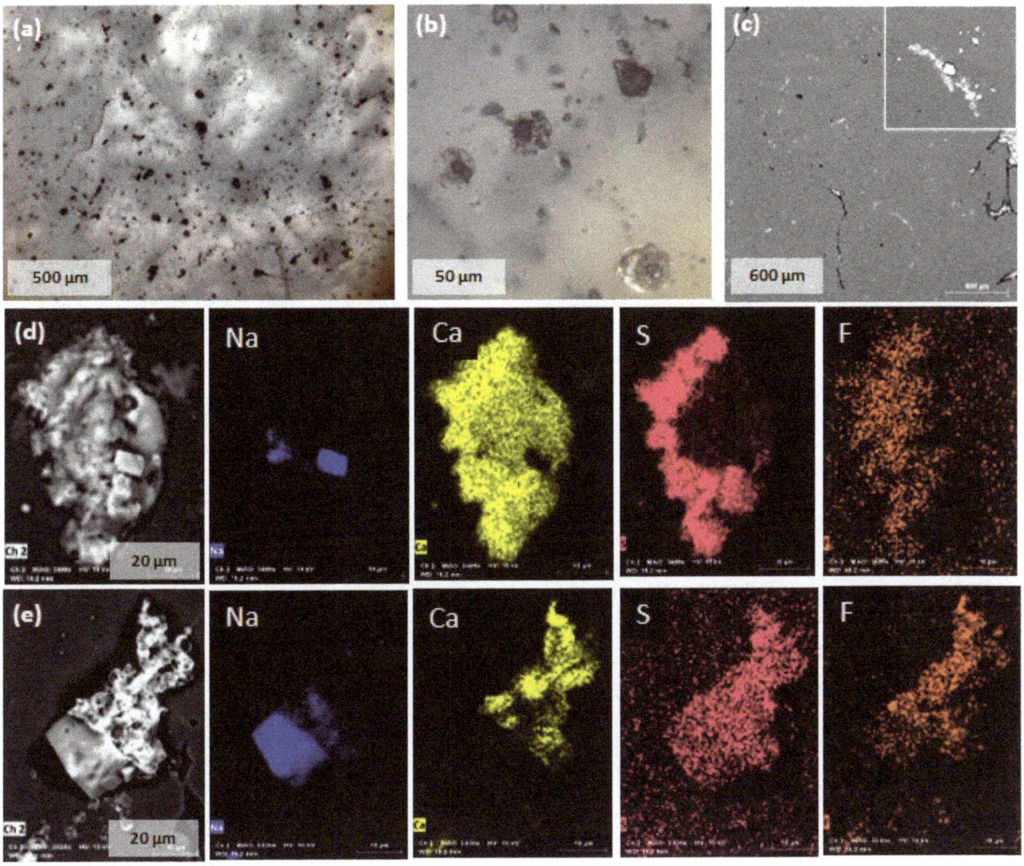

Figure 14. Images of evaporate mounds (EMs) seen in reflected light (RL) and backscattered electron (BSE) images; (**a**) a low-magnification RL image illustrating the abundance and the random distribution of EM; (**b**) a close-up image, in RL, showing the large size typical of the EMs analyzed; (**c**) BSE image showing EMs as either isolated sub-rounded types or elongate trails of decrepitated secondary FI. The inset shows an example of the latter type of EM with cubes of halite; (**d**) BSE image of a single EM with a bulk composition (normalized to 100 wt.%) of 4 Na, 43.8 Ca, 0 K, 36.8 Cl, 11.8 S, 0.8 F, 2.2 Fe, and 0.8 Mn, and X-ray maps shown in false colors; (**e**) a BSE image of an EM with a bulk composition (normalized to 100 wt.%) of 31 Na, 13.6 Ca, 0.3 K, 43.3 Cl, 1.4 S, 3.2 F, 2.6 Fe, 1.6 Mn, and 2.8 Ba, and X-ray maps shown in false colors.

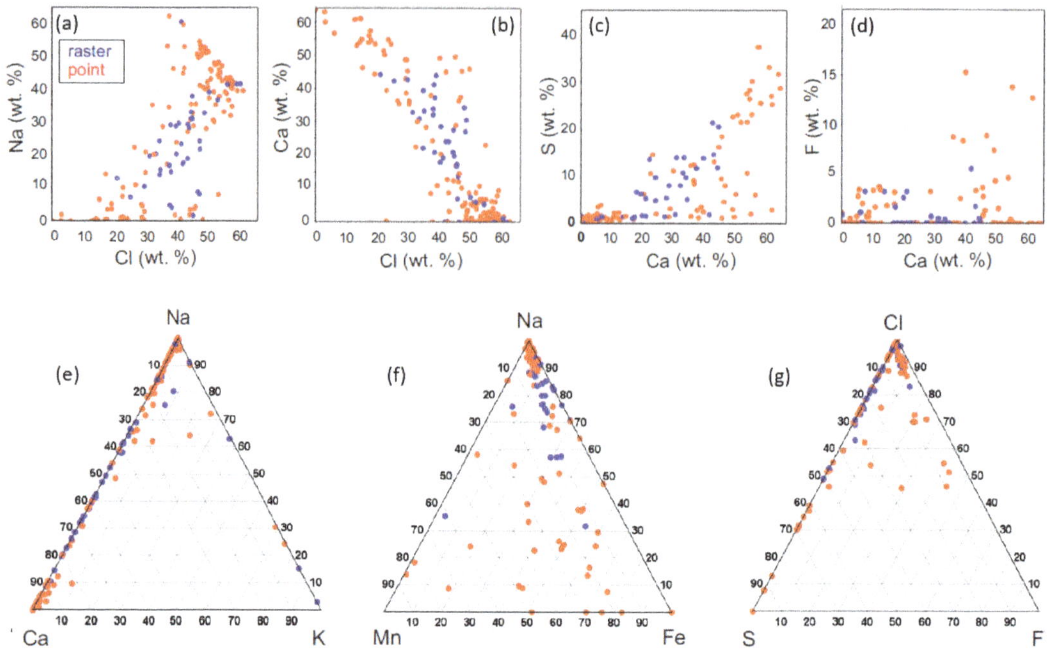

Figure 15. Chemical data for evaporate mound (EM) analysis in binary and ternary plots. Note that the data, separated into raster and point analyses (see text), are normalized to 100 wt.% after correcting for interference from the quartz substrate (see text for explanation). Note how raster data are grouped compared to the point data.

5. Discussion

5.1. Mineralizing Styles and Metallogenic Model

In Central Camp, most deposits and showings can be classified as mineralized styles A and C, with a subordinate amount (3 out of 15 mineralized areas) classified as style B (i.e., breccia-dominated) (Figure 2). Style A is dominant and is characterized by sulfide-filled fractures and shear zones associated with quartz–carbonate (darken quartz) veins, disseminated sulfides (pyrite, chalcopyrite, and pyrrhotite), and hydrothermal magnetite. This mineralization is contemporaneous to the magmatic event that produced the Chibougamau pluton. This was determined using the ages and chemistry of the inter-mineral dykes emanating from the pluton [54,75,76]. These dykes constitute an important part of the argument for a porphyry style of mineralization related to the Chibougamau pluton [72]. Style A is related to the porphyry stage. Style C is related to a late-stage remobilization of an unquantified, but likely minor, part of the mineralization (Au especially) into quartz–carbonate veins (white quartz) during the syntectonic period [81,112].

Sulfide-filled fractures are generally NW-oriented in Central Camp (Figure 2), although a subset of mineralized fractures is NE-oriented (e.g., Chib-Kayrand and Kokko Creek deposits), especially in the eastern part of Central Camp (Henderson I and II, and Portage-McKenzie mines), where NE- and E-oriented structures dominate. In the rest of the AGB, the E-W direction is noted to be the dominant structural trend and corresponds to the orientation of faults and foliation planes (e.g., Lac Sauvage fault; Figure 1) developed during regional shortening events (mostly D_2–D_3). Because the E-W direction is rare in Central Camp, the sulfide-filled structures had to develop prior to regional shortening. The NE direction also corresponds to the orientation of the Lac Doré fault zone that extends through the central part of Central Camp (Figure 1). This fault zone has an apparent dextral displacement of 1 km, with an unquantified vertical component, and it is part of a series of NE-oriented brittle to brittle–ductile faults that displace E-W foliation planes and other structures developed in the Chibougamau area during the D_2 event [18]. These NE faults predate the D_4 event (Grenville orogeny) and may correspond to structures reactivated on several occasions [18]. It cannot be excluded that these NE-structures predate regional shortening (D_2 event). The NW direction is uncommon in the Chibougamau area [18].

The dominant NW-oriented and subordinate NE-oriented sulfide-filled structures of Central Camp are thus unlikely to be associated with a regional event and instead reflect a local process. These structures are interpreted as a network of fractures (diaclases) developed in the rocks hosting the Chibougamau pluton (i.e., rocks of the LDC) prior to the porphyry stage. Diaclases may have developed within the competent rocks of the LDC, which are mostly anorthosites in Central Camp, in response to doming induced by the inflation of the Chibougamau pluton. Diaclases may have efficiently drained fluids in Central Camp, preventing the build-up of fluid overpressure and limiting the development of hydrothermal breccia.

Outside of Central Camp, mineralized zones display a range of orientations attributed to the orientation of local diaclase networks. Also, breccias are abundant in two of the four mineralized areas and this points to inefficient drainage of fluids in the southern part of the pluton, which may explain the scattered distribution of mineralized systems in the south. Also, style A mineralization is observed outside of Central Camp, suggesting variably efficient fluid drainage. Style C is ubiquitous, suggesting that the remobilization of earlier ores during the syntectonic period impacted the bulk of the mineralization associated with the Chibougamau pluton.

The hydrothermal alteration associated with the Cu-Au mineralization documented here comprises phyllic and propylitic types, which are commonly described in porphyry systems [10,11,113]. Also typical of porphyry systems and noted by the aforementioned authors is K-metasomatism (i.e., biotite and rare K-feldspar), which is a proximal alteration that affected a limited volume of rock in the Chibougamau area. Iron metasomatism (hydrothermal magnetite) and sulfidation (mostly pyrite and chalcopyrite) are restricted to fracture infilling. The limited extent of high-temperature hydrothermal alteration (K and Fe metasomatism) could be a consequence of the well-developed network of diaclases that drained fluids efficiently and thus limited the degree of prolonged interaction with host rocks. We conclude that the rheology (competent anorthosite units of the LDC) and structure (orthogonal diaclases) of the country rocks that host the volatile- and metal-bearing magma intrusions that released mineralizing magmatic fluids is paramount in explaining the unique physical characteristics (i.e., sulfide-filled fractures) of the Cu-Au porphyry style of mineralization associated with the Chibougamau pluton.

5.2. Intrusive Phases

The Chibougamau pluton is made of two coeval intrusive phases: (1) a mafic to intermediate phase (referred to as the Diorite phase here) consisting of hornblende-gabbro and diorite rocks that are, respectively, designated as meladiorite and diorite in the field; and (2) a tonalite to $\pm$ trondhjemite phase designated as tonalite or leucotonalite in the field and referred to as the Tonalite phase here [5,98]. Both intrusive phases are differentiated, as reflected by the continuous range of SiO_2 values obtained from whole-rock analyses (Figure 4a). Limited data scattering (K excepted) shown in the TAS and Harker diagrams (Figures 4a and 5) suggests that the sampled rocks were not significantly modified by hydrothermal alteration. It is thus concluded that the chemical composition of these samples can provide insights into magmatic processes.

In the field, the Diorite phase, identified as amphibole-rich and quartz-poor, contains 39.3 to 67.0 wt.% SiO_2. Because rock identification in the field can be challenging, samples were reclassified using the $(La/Yb)_N$ criteria and following the same rationale as previous studies of AGB magmatism [1]. Using the $(La/Yb)_N < 24$ criteria (Figure 6d,e), 70 of 123 samples were reclassified and shown to be characterized by weakly fractionated REE profiles (i.e., moderate LREE enrichment and HREE depletion), with 44.9 to 78.5 wt.% SiO_2.

These samples are metaluminous to slightly peraluminous with a calc-alkaline affinity (Figure 4c,e), are extensively differentiated (Figure 16a), and have moderately negative Ta, Nb, and Ti anomalies (Figure 6g). Despite some geochemical scattering attributed to variable Na and K contents in the most felsic phases and minor enrichment in K as differentiation progresses, most data show a K-poor calc-alkaline trend (Figure 16b). This suite may correspond to a magma series that differentiated from a parental magma that gave rise to most of the intrusive rocks of the Chibougamau pluton.

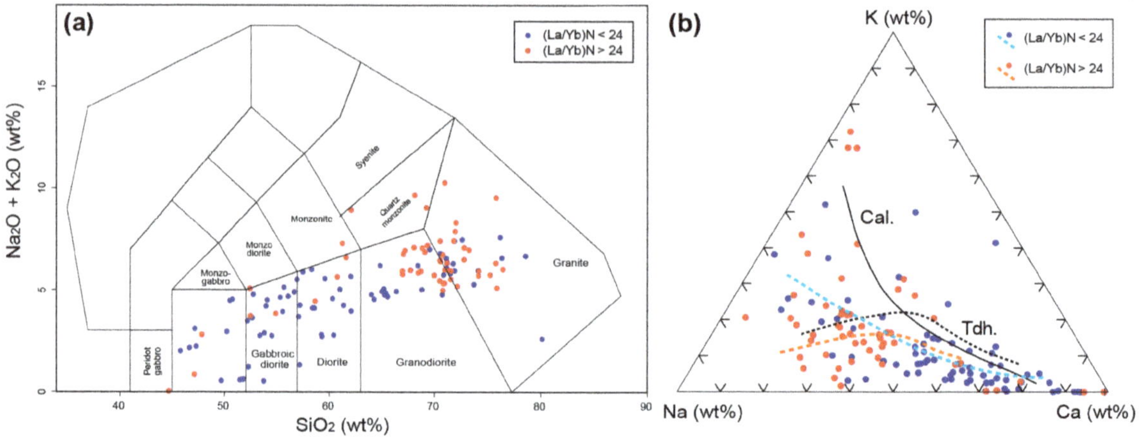

Figure 16. Samples of the Diorite and Tonalite phases (dykes and main intrusion, see Figure 6) classified using the $(La/Yb)_N$ ratio (n = 123) and displayed in (**a**) the total alkali silica (TAS) diagram [99] and (**b**) the Na-K-Ca ternary diagram. Note that, in this plot, the Tonalite phase is inferred to follow a trondhjemite (Tdh.) trend, while the Diorite phase follows a calc-alkaline (Cal.) trend as defined by [100].

The Tonalite phase, identified in the field as both amphibole- and biotite-poor, contains 63.4 to 76.2 wt.% SiO_2 and generally displays extensive HREE depletion. Using the $(La/Yb)_N > 24$ criteria, 53 out of 123 samples are reclassified as belonging to the Tonalite phase, with most samples containing >65 wt.% SiO_2 and 11 samples containing 39.4 to 61 wt.% SiO_2. Samples in this phase are peraluminous and calc-alkaline (Figure 4c,e), and display pronounced negative Ta, Nb, and Ti anomalies (Figure 6g). Some samples are enriched in K, which could be a consequence of hydrothermal alteration, but most have a low K/Na ratio and define a trondhjemite differentiation trend (Figure 16b). The felsic samples of the Tonalite phase thus display the characteristics of a moderately differentiated TTG melt. Samples with <63 wt.% SiO_2 do not fall within the chemical and petrological definition for TTG intrusive suites [2], but instead these intermediate to mafic rocks could correspond to magmas from the Diorite phase having anomalous REE profiles due to, for example, the accumulation of a LREE-enriched epidote, as documented in the diorite-dominated intrusion of the Regnault project [114].

Following on from the above and discarding samples with <63 wt.% SiO_2 reduces the Tonalite phase to 43 samples and increases the Diorite phase dataset to 80 samples. Given that D. Racicot, who collected most of the samples discussed here, performed a well-distributed sampling within the pluton [5], sample distribution is considered a good proxy for volumes of the intrusive phases. From this, it is concluded that the Chibougamau pluton is made of ~62% Diorite phase and ~38% Tonalite phase. Furthermore, we concur with a previous study [5] in stating that the Tonalite phase is a typical TTG melt derived from the deep partial melting of hydrated basaltic material [2], whereas the Diorite phase has a large mantle component, and this has important metallogenetic implications commented on below.

5.3. Geochronology

Six U-Pb TIMS ages were available for the Chibougamau pluton from previous studies, with this study providing an additional eleven samples with zircon U-Pb ages based on LA-ICP-MS (n = 9) and CA-ID-TIMS (n = 2) (Table 1). Most zircon grains recovered from these eleven samples are small (50–100 μm, up to 250 μm) except for the up to 500 μm long grains recovered from sample ACR-29-1. The latter is an orthogneiss located in the eastern part of the Chibougamau pluton and next to the Grenville front, whereas the other samples are of igneous origin and were collected from the western and central parts of the pluton (Figure 1). Sample ACR-29-1 displays amphibolite-grade metamorphism related to the Neoproterozoic Grenville orogeny, but this metamorphic event is unlikely to have entirely recrystallized (and increase the size of) the zircon grains given their Archean ages. The eastern part of the pluton may correspond to an older (Table 1), deeper, and more slowly crystallized part of the intrusive system compared to that exposed to the west, which was later uplifted during the Grenville orogeny.

The textures of the newly dated zircon grains were used to classify the samples into two groups: (1) Group 1 zircon is generally pristine with a large number of grains recovered (samples ACR-6-1, ACR-19-3, ACR-24-1, ACR-32-1, ACR-29-1, 20ACR-0011C, and 20ACR-0031A); (2) Group 2 zircon is extensively altered with only a few grains recovered. Consequently, the age and redox conditions calculated using Group 1 zircon are considered more reliable than parameters extracted from Group 2 zircon.

Based on a compilation of the new and earlier age dates, a relatively continuous magmatic event, from ca. 2.73 to 2.69 Ga, is identified, with two periods of intense magma production referred to as pulses (Figure 10). Magmatic pulse 1 was initiated at ~2728 Ma but mostly spanned from ~2720 Ma to ~2714 Ma, whereas pulse 2 extended from ~2707 Ma to ~2695 Ma (Figure 10). The time interval between these two main intrusive stages corresponds to the transition between the synvolcanic and syntectonic period within the AGB. It also reflected temporal changes for the Chibougamau pluton, possibly accompanied by early regional shortening [14], which likely impacted the nature of the plumbing system feeding magma to the pluton.

Seven of the new zircon ages of >2710 Ma are obtained for rocks with $(La/Yb)_N$ ratio < 24 and 45 to 67 wt.% SiO_2. These rocks belong to the Diorite phase identified previously. Among the five compiled samples with >2714 Ma ages, three are identified as tonalite in the field and one is identified as diorite (Table 1). Additionally, one whole-rock analysis from the Sigeom dataset [115] is matched to one of the dated sites, which is characterized by its $(La/Yb)_N$ of 37 and 68.9 wt.% SiO_2 (Table S2D). However, given the large number of intrusive contacts documented in the Chibougamau pluton [98], it is unclear whether the latter is from a dated phase. Thus, based on historical and new sampling, magmatic pulse 1 (~2720–2714 Ma) and pre-2710 Ma magmatism is considered to consist exclusively of (or at least be dominated by) the Diorite phase. This phase is characterized as weakly to moderately oxidized, with the oldest phase having a redox value of ΔFMQ +0.23 $\pm$ 0.54 (sample ACR-29-1) and the main intrusive phase (pulse 1) having fO_2 of ΔFMQ +0.80 $\pm$ 0.75 (ACR-32-1) to +0.89 $\pm$ 0.54 (ACR-6-1) (Figure 13b).

Two LA-IPC-MS and one of the new ID-TIMS returned ages of <2710 Ma, which are from samples with $(La/Yb)_N$ ratios of 24, 42, and 21, and silica contents of 65.9, 72.5, and 70.3 wt.% SiO_2, respectively, and these equate to the Tonalite phase identified previously. Again, samples for the earlier compiled ages were compared to the Sigeom dataset with the following results. The sample dated 2705.1 +1.7/−1.2 Ma [116] is located near two quartz–diorite dykes with $(La/Yb)_N$ ratio of 9 and 14, and containing 43.4 and 57.8 wt.% SiO_2, respectively. In contrast, the sample dated 2701.7 $\pm$ 2.9 Ma [39] is located near tonalite and trondhjemite rocks with $(La/Yb)_N$ ratio of 28 and 36 and containing 70.4 and 68.9 wt.% SiO_2, respectively (Table S2D). It is thus concluded that pulse 2 (~2707–2697 Ma) is dominated by (or consists entirely of) the Tonalite phase. This phase is more oxidized than the Diorite phase, with fO_2 values of ΔFMQ +1.33 $\pm$ 0.43 (ACR-24-3) and +1.21 $\pm$ 0.38 (ACR-19-3).

The youngest age (2695.4 ± 1 Ma, ID-TIMS) is obtained for sample 20ACR-0011C, a quartz–monzodiorite rock with a $(La/Yb)_N$ ratio of 61 and a silica content of 60.5 wt.% SiO_2. This sample may belong to a magmatic event related to other late and K-bearing intrusions in the Chibougamau area, e.g., the 2700.2 ± 1.6 Ma quartz–monzodiorite and granodiorite of the Muscocho pluton [117] and the 2707.6 ± 1.4 Ma tonalite and syenite of the Lac Line stock [118]. The relationship between this late magmatism and the Chibougamau pluton remains unclear and will not be discussed further here.

Increasing oxidation state during the syntectonic period has been documented in the Chibougamau area and in the rest of the Superior craton by recent studies [119,120]. This is interpreted to reflect the increasing involvement of metasomatized mantle and contamination by supracrustal components (e.g., sedimentary rocks) in the source regions of the magmas produced during the syntectonic period. This increased oxidation state has been documented for the alkaline series and other K-enriched magmas, and it may also characterize TTG-like magmas. Indeed, TTG-like melts produced during the syntectonic period are enriched in K [19] and are more oxidized (this study) than the more voluminous TTG magmas [2] emplaced during the synvolcanic period.

5.4. Implications for the Metallogenic Model

The Cu-Au porphyry style of mineralization in the Chibougamau pluton is related to the early magmatic event that culminated between ~2720 and 2714 Ma (pulse 1), with the possible exception of the Queylus and Devlin breccias, which may instead be associated with later magmatism that culminated at ~2707 to 2697 Ma (pulse 2) [5,72]. Most of the Cu-Au mineralization developed during the synvolcanic stage and is thus genetically related to the Diorite phase (mostly diorite and meladiorite in the field). This intrusive phase displays several characteristics favorable to porphyry-style mineralization, including a mantle-sourced, moderately oxidized mafic parental magma that subsequently differentiated extensively and a significant volume, with the Diorite phase representing about 62% of the Chibougamau pluton while being emplaced before the Tonalite phase.

While data are lacking, it is posited that Archean mantle sources contained more Cu and Au than basaltic reservoirs (source of TTG suites) and thus transferred more metals to the causative magma. The subsequent extensive differentiation favors enrichment of Cu and Au into moderately oxidized residual melts [121,122] that can later exsolve Cu-, Au-, and S-enriched mineralizing fluids [123]. Importantly, the fO_2 values obtained from >2714 Ma old zircon grains are also close to the optimal condition required for Au transportation in mafic and S-bearing magmas [124]. We conclude therefore that the Diorite phase of the Chibougamau pluton formed an optimum causative intrusive phase in the context of porphyry-type deposits [11] and that it is the most likely source of mineralizing fluids. Diorite-dominated plutons are much less abundant than TTG intrusive suites in the Abitibi greenstone belt [1] and attention should focus on these diorite-dominated intrusions and on tonalite–trondhjemite–diorite (TTD) suites when exploring for Cu-Au deposits in the Superior craton.

5.5. Fluid Inclusions (FIs)

The studied FIs come from style A mineralization, with the quartz-hosting FIs formed before the precipitation of chalcopyrite and pyrite. This is consistent with observations made from the nearby Corner Bay mineralization [43] and conforms to conclusions drawn from other porphyry systems, where in at least some cases chalcopyrite tends to form in areas where quartz has dissolved in response to pressure variation, with such veins referred to as type C [6]. In this section, FIs will be used to discriminate between hydrothermal fluids derived from the Chibougamau pluton (porphyry stage, synvolcanic period) and metamorphic fluids that likely circulated during the syntectonic period, as has been performed in other contexts [9,125].

5.5.1. Fluid Types, Timing of Entrapment, and Implications of Textures

Petrographic research, including CL imaging, points to FIs preserved in the least deformed parts of quartz veins, i.e., quartz unmodified by post-formation stress-induced recrystallization. The FIs are therefore considered representative of pre-deformation fluid history. The main FIAs observed were classified as S, PS, and I types, with the P type being rare mostly because the necessary criteria (FIs decorating primary growth zones) could not be verified. However, an abundance of FIs in I-type FIAs may correspond to P-type FIAs [95]. In addition, whereas S- and PS-type FIAs may post-date the crystallization of quartz, they can still be synchronous with quartz and sulfide precipitation occurring elsewhere in the system.

The FIs are classified as type 1 (1a and 1b) and types 2 to 5 based on shape and phase present (Table 4). Each FI type has similar microthermometric properties, even if they occur as distinct populations (FIAs), except for types 1 and 2 which overlap. This indicates that different fluid types were trapped, which is typical in porphyry settings given their longevity and protracted fluid histories from early high temperatures to later lower temperatures. These characteristics produce a mixture of fluid types, varying from high to low salinities [9,10,123]. Although several FI types are commonly observed in the same area (i.e., field of view), there is no unequivocal evidence of co-entrapment (e.g., occurrence on same plane or in the same growth zone), which contrasts with what is commonly noted for porphyry-type systems [123]. Thus, although it is difficult to interpret in regard to the relative timing of entrapment, the presence of type 2 FIs (L-V-Halite) and type 3 FIs (V-rich; noted to decorate faces of quartz crystals and thus P type) is consistent with unmixing of a parental fluid to give two complementary fluids. This may account in part for the distribution of FIs in the salinity–temperature space (Figure S6B). Also, mixing of types 2 and 3 may account for type 1b FIs, which have salinity values between those of types 2 and 3 FIs. In contrast, type 1a FIs represents the lowest-temperature fluid and, given their saline- and Ca-rich nature, reflect the trapping of external circulating fluids that equilibrated with the Ca-rich wall rocks. Lastly, the presence of type 4 FIs (H_2O-CO_2) as a distinct population, commonly I-type FIAs, suggests the entrapment of a homogenous or one-phase fluid. Type 4 FIs may represent the earliest fluid type, as has been reported in other porphyry settings (see below).

Type 5 FIs are characterized by their decrepitate texture. Such FIs are noted in orogenic gold systems [110,126–129] and are attributed to pressure cycling ($\Delta P = P_{lithostatic} \leftrightarrow P_{hydrostatic}$) based on analogous textures reproduced experimentally [130,131]. However, similar textures have recently been reported in porphyry settings [132,133]. Decrepitated FIs tend to display preferential alignment, and they define subparallel trails that can be used to infer the differential stress and the orientation of the principal stress axis [111,134]. The alignment of types 3 and 5 FIs occurs in quartz grains that display little evidence of strain (i.e., no subgrains or undulose extinctions) (Figure 13l,s,u). Thus, we interpret the decrepitated FIs to reflect entrapment during periods of dominantly $P_{lithostatic}$ conditions and post-entrapment modification as P decreased towards $P_{hydrostatic}$. The noted alignment of decrepitated FIs reflects the orientation of the ambient deviatoric stress in the system as the FIs were modified. A similar interpretation of decrepitated FIs in young porphyry systems has been formulated [135,136].

5.5.2. Chemistry of FI

The chemistry of the mineralizing fluid is deduced from microthermometry and EM analysis. Results indicate that types 1a, 1b, and 2 FIs represent Ca-rich fluids with a continuum from low- to high salinity (8 to 50 wt.% eq. NaCl), whereas type 3 FIs are low-density and CO_2-rich inclusions plausibly related to unmixing (see above) and type 4 FIs are H_2O-CO_2 inclusions with $X_{CO2} < 0.1$–0.05. The occurrence of CO_2-bearing fluids is noted in other porphyry-type settings [137–140]. In orogenic gold systems, metamorphic fluids are H_2O-CO_2 fluids with $X_{CO2} = 0.1$–0.2 and may comprise a minor, and likely post-ore, saline H_2O type of FI [9].

The EM data confirm that the solute is dominated by Ca and Na, with lesser but significant amounts of Fe, Mn, and Ba (Figure 15). The main anion is Cl with lesser but significant amounts of S and F. These chemical characteristics are consistent with a magmatic–hydrothermal system according to compiled FI chemical data [9], while fluids involved in orogenic gold systems are Na-rich and lack F according to another compilation [127]. Fluorine is a signature of magmatic–hydrothermal systems according to EM data [97,141–144]. The fluids documented here are comparable to those identified in the ca. 2740 Ma Côté Gold Au(-Cu) porphyry-type deposit hosted in the Chester intrusive complex of the AGB, which are variably saline (7 to 35 wt.% eq. NaCl) and correspond to H_2O-CO_2-NaCl-$CaCl_2$ fluids containing Na, Ca, K, Fe, and Mn, with significant F and S contents [145]. Both the Chibougamau pluton (and its anorthosite-dominated host rocks) and the Chester complex are dominated by Ca-rich rocks (e.g., diorite), and thus the Ca enrichment observed in FIs is likely a consequence of fluid–rock exchanges (see arrow in Figure S6B). We conclude that the chemistry of the FIs studied here is consistent with a magmatic–hydrothermal setting (porphyry) and that the circulation of metamorphic fluids during the syntectonic period did not significantly impact the FIs contained in the studied quartz vein.

5.5.3. Constraining P-T of Fluid Entrapment

The nature of the FIs and related thermometric data provide the basis for making some inferences in regard to the P and T of fluid entrapment. Type 4 FIs (H_2O-CO_2), which represents the earliest fluid trapped, have a consistent bulk density. These FIs were trapped as a homogenous fluid in the one phase region of the P-T space rather than on or below its solvus. Trapping occurred at a minimum P of ca. 600–700 bars (i.e., on the solvus), as determined using a Th value of 325 °C (obtained for one FIA) and by calculating the solvus for a fluid with $X_{CO2} = 0.10$ and a salinity of 6 wt.%, following on from previous work [146]. However, the presence of biotite-bearing altered rocks and of hypogene chalcopyrite mineralization [10] point to a temperature above 325 °C and to high pressure, and this values can be inferred from isochoric projections of said fluid to about 4 km at 400 °C and 7 km at 500 °C [146]. In other porphyry-type settings, the entrapment of H_2O-CO_2 fluids is considered to happen during the earliest stages and to occur at depths like these obtained here [139,140].

The subsequent entrapment of types 1b and 2 FI, which may overlap with sulfide mineralization, indicates minimum temperature conditions of 400 °C based on the maximum Th values. The pressure is constrained, assuming that type 1b and 2 FIs coexist with type 4 FIs and were trapped on the solvus; this indicates pressures of about 300–400 bars [147]. This pressure is much lower than the pressure of 600–700 bars obtained for type 4 FI. This is attributed to changing P-T conditions during the evolution of the porphyry mineralizing system [123,133,148], with (1) a transition for $P_{lithostatic}$ to $P_{hydrostatic}$ synchronous with the mineralizing event and (2) the cooling of the fluid that induced saturation and the subsequent precipitation of Cu sulfides.

6. Conclusions

This contribution is a preliminary early attempt at quantifying magmatic and mineralizing processes for an Archean porphyry-type mineralized setting. The geochemical and geochronological results strongly suggest that the Chibougamau pluton is dominated by a diorite magma (Diorite phase) with moderate fO_2 (~ΔFMQ 0 to +1) that was generated towards the end of the synvolcanic period, with magmatism culminating at ca. 2720–2714 Ma. Porphyry mineralization related to this voluminous intrusive phase had an optimal fO_2 for transporting Au and Cu. In contrast, the subsequent magmatism that culminated at ca. 2707–2697 Ma had a smaller volume, was a TTG-like magma (Tonalite phase) with a higher fO_2 (ΔFMQ + 1 to +2), and was only related to minor (or to no) mineralization.

The main mineralizing style, which consists of sulfide-filled fractures and dark quartz and carbonate veins (style A), derives from the development of a well-drained hydrothermal system circulating in a pre-existing diaclase network developed in the competent rocks of the LDC, the main hosting material of the Chibougamau pluton. The unusual aspect of the mineralized bodies, with sulfide-filled hairline fractures (style A) dominating over hydrothermal breccia (style B), is thus a consequence of the structural context and should not be used to distinguish Archean from post-Archean mineralizing processes in porphyry systems. Also, a minor part of the mineralization, present as visible gold in white quartz veins, corresponds to mineralization styles A and B, remobilized by metamorphic fluids (style C) during regional deformation and metamorphism that affected the Superior craton prior to its stabilization at the end of the Archean.

The nature and chemistry of the FIs point to a mineralizing fluid with a clear magmatic signature, as indicated by chemical data, i.e., Ca-, Na-, and Cl-rich fluid with a lesser but significant amount of Fe, Mn, and Ba, as well as S and F. The abundance of Ca in the fluid is likely a consequence of fluid–rock interactions and reflects the Ca-enriched nature of the hosting material (LDC), which should not be considered when comparing the studied Archean mineralization to post-Archean systems. A variety of FI types are reported, with high- and low-salinity FIs and early CO_2-rich inclusions (type 4). The latter is a characteristic that the Chibougamau area shares with some modern porphyry systems [139,140]. Evidence for fluid unmixing has also been obtained, with co-occurring type 2 FI (L-V-Halite) and type 3 FI (V-rich). We do note that for the limited number of samples considered, metamorphic fluids typically related to orogenic gold systems do not dominate the samples. This study shows, however, the potential for FIs derived from Archean quartz to provide reliable insights into ancient mineralizing processes, and such investigations should be extended to Central Camp and to other Archean porphyries.

Supplementary Materials: The following supporting information can be downloaded at: https://www.mdpi.com/article/10.3390/min14121293/s1, Table S1: Physical characteristics of the Cu-Au mineralization spatially associated with the Chibougamau pluton; Table S2: whole-rock analyses (compiled and new samples) and modal proportions; Table S3: CL images of zircon grains targeted for LA-ICP-MS analysis; Table S4: LA-ICP-MS zircon chemistry (trace element and geochronological results); Table S5: CA-ID-TIMS U-Pb zircon isotopic data and grain images for selected intrusive rocks from the Chibougamau Pluton; Table S6: texture of quartz and fluid inclusions (FI), and diagrams showing the results of FI analyses.

Author Contributions: Conceptualization, L.M. and D.J.K.; methodology, A.C., L.M., D.J.K., J.M. and M.A.H.; software, A.C., L.M., D.J.K., J.M. and M.A.H.; validation, A.C., L.M., D.J.K., J.M. and M.A.H.; formal analysis, A.C., L.M., D.J.K. and M.A.H.; investigation, A.C., L.M. and D.J.K.; resources, L.M. and D.J.K.; data curation, A.C., L.M., D.J.K. and M.A.H.; writing—original draft preparation, A.C., L.M., and D.J.K.; writing—review and editing, L.M., D.J.K. and M.A.H.; supervision, L.M. and D.J.K.; funding acquisition, L.M. and D.J.K. All authors have read and agreed to the published version of the manuscript.

Funding: This study was undertaken as part of the Metal Earth project (Laurentian University) investigation of the Chibougamau area. This research was funded by the Canada First Research Excellence Funds and federal/provincial/industry partners (http://merc.laurentian.ca/research/metal-earth/; accessed on 1 January 2022). This project was also funded by the NSERC (Natural Sciences and Engineering Research Council) Discovery Grant to L. Mathieu (Reference number RGPIN-2018-06325).

Data Availability Statement: Datasets are available as Supplementary Materials to this contribution.

Acknowledgments: Warm thanks are addressed to three anonymous reviewers who greatly contributed to improve this contribution. The authors are indebted to the many collaborators of the Metal Earth and NSERC projects. Many thanks to François Leclerc, MERN, for making the re-analyses of historical rocks samples from the Chibougamau pluton possible. This is Metal Earth contribution number MERC-ME-2024-31.

Conflicts of Interest: Lucie Mathieu is president of Consultation Geox Inc., Val-d'Or, Québec, Canada. The paper reflects the views of the author, as formulated when she was a university rofessor, and not the company.

References

1. Mathieu, L.; Crépon, A.; Kontak, D.J. Tonalite-Dominated Magmatism in the Abitibi Subprovince, Canada, and Significance for Cu-Au Magmatic-Hydrothermal Systems. *Minerals* **2020**, *10*, 242. [CrossRef]
2. Moyen, J.-F.F.; Martin, H. Forty Years of TTG Research. *Lithos* **2012**, *148*, 312–336. [CrossRef]
3. Katz, L.R.; Kontak, D.J.; Dubé, B.; McNicoll, V.; Creaser, R.; Petrus, J.A. An Archean Porphyry-Type Gold Deposit: The Côté Gold Au (-Cu) Deposit, Swayze Greenstone Belt, Superior Province, Ontario, Canada. *Econ. Geol.* **2020**, *116*, 47–89. [CrossRef]
4. Katz, L.R.; Kontak, D.J.; Dubé, B.; McNicoll, V. The Geology, Petrology, and Geochronology of the Archean Côté Gold Large-Tonnage, Low-Grade Intrusion-Related Au (±Cu) Deposit, Swayze Greenstone Belt, Ontario, Canada. *Can. J. Earth Sci.* **2017**, *54*, 173–202. [CrossRef]
5. Mathieu, L.; Racicot, D. Petrogenetic Study of the Multiphase Chibougamau Pluton: Archaean Magmas Associated with Cu-Au Magmato-Hydrothermal Systems. *Minerals* **2019**, *9*, 174. [CrossRef]
6. Monecke, T.; Mercier-Langevin, P.; Dubé, B.; Frieman, B.M. Geology of the Abitibi Greenstone Belt. *Rev. Econ. Geol.* **2017**, *19*, 7–49.
7. Pilote, P.; Cimon, J.; Dion, C.; Kirkham, R.; Robert, F.; Sinclair, W.D.; Daigneault, R. *Les Gisements de Type Cu-Au Porphyrique de La Région de Chibougamau*; MRN report DP 93-02; Ministère des Ressources Naturelles: Québec, QC, Canada, 1993.
8. Steele-MacInnis, M.; Lecumberri-Sanchez, P.; Marshall, D.; Kontak, D.J. Contributions of Fluid Inclusions to Genetic Models for Mineral Deposits. In *Fluid and Melt Inclusions: Applications to Geologic Processes*; Mineralogical Association of Canada: Québec, QC, Canada, 2020.
9. Bodnar, R.J.; Lecumberri-Sanchez, P.; Moncada, D.; Steele-MacInnis, M. Fluid Inclusions in Hydrothermal Ore Deposits. In *Treatise on Geochemistry*, 2nd ed.; Elsevier: Oxford, UK, 2014; pp. 119–142.
10. Seedorff, E.; Dilles, J.H.; Proffett, J.M.; Einaudi, M.T.; Zurcher, L.; Stavast, W.J.A.; Johnson, D.A.; Barton, M.D. Porphyry Deposits: Characteristics and Origin of Hypogene Features. *Econ. Geol.* **2005**, *100*, 251–298.
11. Sillitoe, R.H. Porphyry Copper Systems. *Econ. Geol.* **2010**, *105*, 3–41. [CrossRef]
12. Sinclair, W.D. Porphyry Deposits. In *Mineral Deposits of Canada: A Synthesis of Major Deposit-Types, District Metallogeny, the Evolution of Geological Provinces, and Exploration Methods*; Geological Association of Canada, Mineral Deposits Division: St. John's, NL, Canada, 2007; Volume 5, pp. 223–243.
13. Laurent, O.; Martin, H.; Moyen, J.F.; Doucelance, R. The Diversity and Evolution of Late-Archean Granitoids: Evidence for the Onset of "Modern-Style" Plate Tectonics between 3.0 and 2.5 Ga. *Lithos* **2014**, *205*, 208–235. [CrossRef]
14. Mathieu, L.; Snyder, D.B.; Bedeaux, P.; Cheraghi, S.; Lafrance, B.; Thurston, P.; Sherlock, R. Deep into the Chibougamau Area, Abitibi Greenstone Belt: Structure of a Neoarchean Crust Revealed by Seismic Reflection Profiling. *Tectonics* **2020**, *39*, e2020TC006223. [CrossRef]
15. Daigneault, R.; St-Julien, P.; Allard, G.O. Tectonic Evolution of the Northeast Portion of the Archean Abitibi Greenstone Belt, Chibougamau Area, Quebec. *Can. J. Earth Sci.* **1990**, *27*, 1714–1736. [CrossRef]
16. Thurston, P.C.; Ayer, J.A.; Goutier, J.; Hamilton, M.A. Depositional Gaps in Abitibi Greenstone Belt Stratigraphy: A Key to Exploration for Syngenetic Mineralization. *Econ. Geol.* **2008**, *103*, 1097–1134. [CrossRef]
17. Beakhouse, G.P. *The Abitibi Subprovince Plutonic Record: Tectonic and Metallogenic Implications*; Open File report 6268; Ontario Geological Survey: Sudbury, ON, Canada, 2011.
18. Daigneault, R.; Allard, G.O. *Le Complexe Du Lac Doré et Son Environnement Géologique (Région de Chibougamau-Sous-Province de l'Abitibi)*; MRN report MM-89-03; Ministère des Ressources Naturelles: Québec, QC, Canada, 1990; ISBN 2551123313.
19. Moyen, J.-F. Archean Granitoids: Classification, Petrology, Geochemistry and Origin. *Geol. Soc. Lond. Spec. Publ.* **2019**, *489*, 15–49. [CrossRef]
20. Leclerc, F.; Roy, P.; Pilote, P.; Bédard, J.H.; Harris, L.B.; McNicoll, V.J.; van Breemen, O.; David, J.; Goulet, N. *Géologie de La Région de Chibougamau*; MERN report RG 2015-03; Ministère de l'Énergie et des Ressources Naturelles: Québec, QC, Canada, 2017.
21. Boucher, A.; Mathieu, L.; Hamilton, M.M.A.; Bedeaux, P.; Daigneault, R.R. Petrogenesis and Economic Potential of the Obatogamau Formation, Chibougamau Area, Abitibi Greenstone Belt. *Can. J. Earth Sci.* **2021**, *58*, 519–541. [CrossRef]
22. Leclerc, F.; Bédard, J.H.; Harris, L.B.; McNicoll, V.J.; Goulet, N.; Roy, P.; Houle, P. Tholeiitic to Calc-Alkaline Cyclic Volcanism in the Roy Group, Chibougamau Area, Abitibi Greenstone Belt—Revised Stratigraphy and Implications for VHMS Exploration. *Can. J. Earth Sci.* **2011**, *48*, 661–694. [CrossRef]
23. Legault, M.I. *Environnement Métallogénique Du Couloir de Fancamp Avec Emphase Sur Les Gisements Aurifères de Chevrier, Région de Chibougamau, Québec*; Université du Québec à Chicoutimi: Chicoutimi, QC, Canada, 2003.
24. Mortensen, J.K. U–Pb Geochronology of the Eastern Abitibi Subprovince. Part 1: Chibougamau–Matagami–Joutel Region. *Can. J. Earth Sci.* **1993**, *30*, 11–28. [CrossRef]
25. Allard, G.O. *Doré Lake Complex and Its Importance to Chibougamau Geology and Metallogeny*; MRN report DPV-386; Ministère des Ressources Naturelles: Québec, QC, Canada, 1976.
26. Davis, D.; Simard, M.; Hammouche, H.; Bandyayera, D.; Goutier, J.; Pilote, P.; Leclerc, F.; Dion, C. *Datations U-Pb Effectuées Dans Les Provinces Du Supérieur et de Churchill En 2011–2012*; MERN report RP-2014-05; Ministère de l'Énergie et des Ressources Naturelles: Québec, QC, Canada, 2014.
27. Leclerc, F.; Harris, L.B.; Bédard, J.H.; Van Breemen, O.; Goulet, N. Structural and Stratigraphic Controls on Magmatic, Volcanogenic, and Shear Zone-Hosted Mineralization in the Chapais-Chibougamau Mining Camp, Northeastern Abitibi, Canada (1,2). *Econ. Geol.* **2012**, *107*, 963–969. [CrossRef]

28. Dimroth, E.; Rochelau, M.; Mueller, W.; Archer, P.; Brisson, H.; Fortin, G.; Jutras, M.; Lefebvre, C.; Piché, M.; Pilote, P.; et al. Paleogeographic and Paleotectonic Response to Magmatic Processes: A Case History from the Archean Sequence in the Chibougamau Area, Quebec. *Geol. Rundsch.* **1985**, *74*, 11–32. [CrossRef]
29. Racicot, D. *Géochimie et Métallogénie de La Partie Orientale Du Pluton de Chibougamau*; Ministère Des Ressources Naturelles et de La Faune, Québec: Québec, QC, Canada, 1980.
30. Racicot, D. *Géochimie et Métallogénie Des Parties Occidentale et Centrale Du Pluton de Chibougamau*; Ministère Des Ressources Naturelles et de La Faune, Québec: Québec, QC, Canada, 1981.
31. David, J.; Davis, D.W.; Goutier, J.; Legault, M.; Roy, P. *Datations U-Pb Effectuées Dans La Sous-Province de l'Abitibi En 2005–2006*; RP 2007-01; Ministère Des Ressources Naturelles et de La Faune, Québec: Québec, QC, Canada, 2007.
32. Stockwell, C.H. *Revised Precambrian Time Scale for the Canadian Shield*; Geological Survey of Canada: Ottawa, ON, Canada, 1972; pp. 52–72.
33. Percival, J.A.; Skulski, T.; Sanborn-Barrie, M.; Stott, G.M.; Leclair, A.D.; Corkery, M.T.; Boily, M. Geology and Tectonic Evolution of the Superior Province, Canada. In *Tectonic Styles in Canada: The Lithoprobe Perspective*; Geological Association of Canada: St. Johns, NL, Canada, 2012; Volume 49, pp. 321–378.
34. Dimroth, E.; Mueller, W.; Daigneault, R.; Brisson, H.; Poitras, A.; Rocheleau, M. Diapirism during Regional Compression: The Structural Pattern in the Chibougamau Region of the Archean Abitibi Belt, Quebec. *Geol. Rundsch.* **1986**, *75*, 715–736. [CrossRef]
35. Baker, D.J. The Metamorphism and Structural History of the Grenville Front near Chibougamau. Ph.D. Thesis, University of Georigia, Athens, GA, USA, 1980. Unpublished.
36. Krogh, T.E. Improved Accuracy of U-Pb Zircon Ages by the Creation of More Concordant Systems Using an Air Abrasion Technique. *Geochim. Cosmochim. Acta* **1982**, *46*, 637–649. [CrossRef]
37. Pilote, P.; Dion, C.; Joanisse, A.; David, J.; Machado, N.; Kirkham, R.V.; Robert, F. Géochronologie Des Minéralisations d'affiliation Magmatique de l'Abitibi, Secteurs Chibougamau et de Troilus-Frotet: Implications Géotectoniques. In *Programme et Résumés, Séminaire D'information Sur la Recherche Géologique*; MRN report, DV-97-03; Ministère des Ressources Naturelles: Québec, QC, Canada, 1997; p. 47.
38. Joanisse, A. Datation de La Carbonatite Du Lac Shortt. Bachelor's Thesis, Université du Québec à Montréal, Montréal, QC, Canada, 1994. Unpublished.
39. McNicoll, V.; Dubé, B.; Goutier, J.; Mercier-Langevin, P.; Dion, C.; Monecke, T.; Ross, P.-S.; Thurston, P.; Pilote, P.; Bédard, J.; et al. *Nouvelles Datations U-Pb Dans Le Cadre Du Projet ICG-3 Abitibi/Plan Cuivre: Incidences Pour l'Interprétation Géologique et l'Exploration Des Métaux Usuels*; MRN report DV-2008-06; Ministère des Ressources Naturelles: Québec, QC, Canada, 2008.
40. Guha, J.; Leroy, J.; Guha, D. Significance of Fluid Phases Associated with Shear Zone Cu-Au Mineralization in the Doré Lake Complex, Chibougamau, Quebec. *Bull. Mineral.* **1979**, *102*, 569–576.
41. Mathieu, L. Detecting Magmatic-Derived Fluids Using Pyrite Chemistry: Example of the Chibougamau Area, Abitibi Subprovince, Québec. *Ore Geol. Rev.* **2019**, *114*, 103–127. [CrossRef]
42. Robert, F. Syenite-Associated Disseminated Gold Deposits in the Abitibi Greenstone Belt, Canada. *Miner. Depos.* **2001**, *36*, 503–516. [CrossRef]
43. Mathieu, L.; Savard, D.; Kulynych-Rinta, A. Chemical Mapping of Trace Elements in Pyrite Provides Insight into Mineralizing Processes: The Example of the Neoarchean Cu–Au Porphyry System of the Chibougamau Area, Canada. *Geol. Ore Depos.* **2024**, *66*, 454–470. [CrossRef]
44. Pilote, P.; Kirkham, R.V.; Robert, F.; Daigneault, R.; Magnan, M. *Développement d'un District à Minéralisations de Type Cu-Au (Mo) Porphyriques Dans La Région de Chibougamau et Implications Métallogéniques*; DV94-04; Ministère Des Ressources Naturelles et de La Faune, Québec: Québec, QC, Canada, 1995.
45. Guha, J.; Chown, E.H.; Archambault, G.; Barnes, S.-J.; Brisson, H.; Daigneault, R.; Dion, C.; Dubé, B.; Mueller, W.U.; Pilote, P. Metallogeny in Relation to Magmatic and Structural Evolution of an Archean Greenstone Belt: Chibougamau Mining District. In *Gold and Base-Metal Mineralization in the Abitibi Subprovince, Canada, with Emphasis on the Québec Segment*; Hoe, S.E., Robert, F., Groves, D.I., Eds.; The University of Western Australia: Perth, Australia, 1990; pp. 121–166.
46. Larouche, C.P. *Technical Review & Evaluation of the Exploration Potential of the "LAC CHIBOUGAMAU" Mining Properties in McKenzie, Roy, Obalski and Lemoine Townships, Abitibi Mining District Province of Québec NTS 32G-16*; Ministère de l'Énergie et des Ressources Naturelles: Québec, QC, Canada, 2012; pp. 1–175.
47. Pilote, P.; Robert, F.; Sinclair, W.D.; Kirkham, R.V.; Daigneault, R. Minéralisations de Type Porphyriques et Filoniennes Dans Le Secteur Du Lac Doré-Les Secteurs Du Lac Clark et de l'Île Merrill. In *Géologie et Évolution Métallogénie de la Région de Chibougamau: Des Gîtes de Type de Type Cu-Au-Mo Porphyriques Aux Gisements Filoniens Mésothermaux Aurifères*; Ministère des Ressources Naturelles et de la Faune, Québec: Québec, QC, Canada, 1996; pp. 69–92.
48. Magnan, M.; Blais, A.; Daigneault, R.; Pilote, P.; Robert, F. La Mine Copper Rand. In *Géologie et Évolution Métallogénie de la Région de Chibougamau: Des Gîtes de Type de Type Cu-Au-Mo Porphyriques Aux Gisements Filoniens Mésothermaux Aurifères*; Ministère des Ressources Naturelles et de la Faune, Québec: Québec, QC, Canada, 1996; pp. 93–102.
49. Magnan, M.; Pilote, P.; Daigneault, R. *Minéralisations et Altérations à La Mine Copper Rand, Chibougamau*; MRN report ET-98-01; Ministère des Ressources Naturelles: Québec, QC, Canada, 1999.

50. Tessier, A.C.; Hodgson, C.J.; Blais, A.; Larouche, V.; Houle, P.; Lulin, J.-M. Le Gisement Cupro-Aurifère de La Mine Portage. In *Géologie et Évolution Métallogénie de la Région de Chibougamau: Des Gîtes de Type de Type Cu-Au-Mo Porphyriques Aux Gisements Filoniens Mésothermaux Aurifères*; Ministère des Ressources Naturelles et de la Faune, Québec: Québec, QC, Canada, 1996; pp. 103–114.
51. Bellavance, Y.; Lachance, J.-P.; Trudeau, Y.; Tessier, A.C. Le Gisement Cuprifère de Corner Bay. In *Géologie et Évolution Métallogénie de la Région de Chibougamau: Des Gîtes de Type de Type Cu-Au-Mo Porphyriques Aux Gisements Filoniens Mésothermaux Aurifères*; Ministère des Ressources Naturelles et de la Faune, Québec: Québec, QC, Canada, 1996; pp. 121–130.
52. Pilote, P.; Allard, G.O.; Bellavance, Y.; Blais, A.; Boily, B.; Brisson, H.; Chown, E.H.; Daigneault, R.; Corta, H.d.; Dion, C.; et al. *Géologie et Métallogénie Du District Minier de Chapais-Chibougamau: Nouvelle Vision Du Potentiel de Découverte*; Ministère Des Ressources Naturelles et de La Faune, Québec: Québec, QC, Canada, 1998.
53. Pilote, P.; Dion, C.; Morin, R. *Géologie et Évolution Métallogénique de La Région de Chibougamau: Des Gîtes de Type Cu-Au-Mo Porphyriques Aux Gisements Filoniens Mésothermaux Aurifères*; MRN report MB-96-14; Ministère des Ressources Naturelles: Québec, QC, Canada, 1996.
54. Sinclair, W.D.; Pilote, P.; Kirkham, R.V.; Robert, F.; Daigneault, R. A Preliminary Report of Porphyry Cu-Mo-Au and Shear Zone-Hosted Cu-Au Deposits in the Chibougamau Area, Québec. In *Current Research 1994-C.*; Geological Survey of Canada: Ottawa, ON, Canada, 1994; pp. 303–309.
55. Pilote, P. Les Brèches Du Canton Queylus. In *Géologie et Évolution Métallogénie de la Région de Chibougamau: Des gîtes de Type de Type Cu-Au-Mo Porphyriques Aux Gisements Filoniens Mésothermaux Aurifères*; Ministère des Ressources Naturelles et de la Faune, Québec: Québec, QC, Canada, 1996; pp. 136–139.
56. Côté-Mantha, O.; Daigneault, R.; Gaboury, D.; Chartrand, F.; Pilote, P. Geology, Alteration, and Origin of Archean Au-Ag-Cu Mineralization Associated with the Synvolcanic Chibougamau Pluton: The Brosman Prospect, Abitibi Greenstone Belt, Canada. *Econ. Geol.* **2012**, *107*, 909–934. [CrossRef]
57. Pilote, P.; Dion, C.; Cimon, J.; Sinclair, W.D.; Daigneault, R.; Magnan, M. *Les Minéralisations de Type Cu-Mo-Au Porphyriques et Les Gisements Filoniens Du Lac Doré, Chibougamau: Processus Hydrothermaux Distincts Ou Évolution d'un Même Évènement Minéralisateur*; DV94-09; Ministère Des Ressources Naturelles et de La Faune, Québec: Québec, QC, Canada, 1994.
58. Christmann, P. *Etude Métallogénique de La Mine Copper Cliff (Cu-Au), Chibougamau, Comté d'Abitibi-Est, Province de Québec, Canada*; Ministère de l'Énergie et des Ressources Naturelles: Québec, QC, Canada, 1979; p. 396.
59. Miller, R.J.M. Wall-Rock Alteration at the Cedar Bay Mine, Chibougamau District, Quebec. *Econ. Geol.* **1961**, *56*, 321–330. [CrossRef]
60. Howe, R.W. *Report on Cedar Bay Property, Lac Aux Dorés, Chibougamau District*; MERN report GM-00547; Ministère de l'Énergie et des Ressources Naturelles: Québec, QC, Canada, 1949.
61. Hinse, R. *Report on the Cedar Bay Mineralization*; MERN report GM-09899; Ministère de l'Énergie et des Ressources Naturelles: Québec, QC, Canada, 1957.
62. Larouche, P. *Geological and Structural Mapping Carried out on the OBALSKI MINE Property (CM-507) 32G-16, Prepared by Tomagold Corp*; MERN report GM-69909; Ministère de l'Énergie et des Ressources Naturelles: Québec, QC, Canada, 2017.
63. Rioux, L. *Propriété Bateman Bay Indice Jaculet/C3, Canton McKenzie (32G16), Rapport Sur Les Travaux de Cartographie D'échantillonage En Rainures*; GM70510; Ministère de l'Énergie et des Ressources Naturelles: Québec, QC, Canada, 2018.
64. Guha, J.; Archambault, G.; Leroy, J. A Correlation between the Evolution of Mineralizing Fluids and the Geomechanical Development of a Shear Zone as Illustrated by the Henderson 2 Mine, Quebec. *Econ. Geol.* **1983**, *78*, 1605–1618. [CrossRef]
65. Vollo, N.B. *Preliminary Report on the Geology of the Henderson Copper Deposit, Chibougamau, Québec*; GM 09900-A.; Ministère de l'Énergie et des Ressources Naturelles: Québec, QC, Canada, 1960.
66. Zuckerkandel, W. *Henderson 1-S/3 Mine*; GM 44630; Ministère de l'Énergie et des Ressources Naturelles: Québec, QC, Canada, 1987.
67. Zuckerkandel, W. *Henderson Mine, Joint Exploration Project 1986–1988*; GM48404; Ministère de l'Énergie et des Ressources Naturelles: Québec, QC, Canada, 1989.
68. Guha, J.; Koo, J. Role of Fluid State Mobilization during Metamorphism of the Henderson Ore Bodies, Chibougamau, Quebec, Canada. *Can. J. Earth Sci.* **1975**, *12*, 1516–1523. [CrossRef]
69. Malouf, P.M. *Rapport Sur Kokko Creek Mining Corporation Limited*; GM 01146-A.; Ministère de l'Énergie et des Ressources Naturelles: Québec, QC, Canada, 1951.
70. Leblanc, M.; Larouche, C.P. *Project: Kokko Creek; Results on the Prospecting and Channel Sampling Programs Completed During the Fall of 2013*; GM68395; Ministère de l'Énergie et des Ressources Naturelles: Québec, QC, Canada, 2014.
71. Robert, F. Timing Relationships between Cu-Au Mineralization, Dykes, and Shear Zones in the Chibougamau Camp, Northeastern Abitibi Subprovince, Quebec. In *Current Research 1994-C.*; Geological Survey of Canada: Ottawa, ON, Canada, 1994; pp. 287–294.
72. Pilote, P. Le Camp Minier de Chibougamau et Le Parautochtone Grenvillien: Métallogénie, Métamorphisme et Aspects Structuraux. In Proceedings of the Livret-Guide D'excursion B1; Joined anual meeting, Geological Association of Canada—Mineralogical Association of Canada (GAC-MAC), Montréal, QC, Canada, 15 May 2006; p. 138.
73. Pilote, P. Le Gisement Cuprifère de Devlin. In *Géologie et Évolution Métallogénie de la Région de Chibougamau: Des Gîtes de Type de Type Cu-Au-Mo Porphyriques Aux Gisements Filoniens Mésothermaux Aurifères*; Ministère des Ressources Naturelles et de la Faune, Québec: Québec, QC, Canada, 1996; pp. 131–136.

74. Lange-Brard, F. *Etude Des Minéralisations, Alterations et Phases Fluides Associées Au Gite Devlin (Chibougamau, Quebec, Canada)*; Université du Québec à Chicoutimi: Québec, QC, Canada, 1984; p. 133.
75. Maillet, J. Pétrographie et Géochimie Des Dykes Du Camp Minier de Chibougamau. Master's Thesis, Université du Québec à Chicoutimi, Chicoutimi, QC, Canada, 1978. Unpublished.
76. Robert, F. Vein Fields in Gold Districts: The Example of Val d'Or, Southeastern Abitibi Subprovince, Québec. In *Current Research 1994-C.*; Geological Society of London: London, UK, 1994; pp. 295–302.
77. Jeffrey, W.G. The Geology of the Campbell Chibougamau Mine. Ph.D. Thesis, McGill University, Montreal, QC, Canada, 1959.
78. Duquette, G. *Fiche de Gîte, Mine Québec-Chibougamau*; GM 25128; Ministère de l'Énergie et des Ressources Naturelles: Québec, QC, Canada, 1969.
79. Malouf, S.E. *Preliminary Report on the Geology of the Henderson Copper Deposit*; MRN Report GM 09900-B.; Ministère Des Ressources Naturelles: Québec, QC, Canada, 1960.
80. Sibson, R.H. Conditions for Fault-Valve Behaviour. *Geol. Soc. Lond. Spec. Publ.* **1990**, *54*, 15–28. [CrossRef]
81. Tessier, A.; Hodgson, C.J.; Lulin, J.M.; Blais, A. *Les Contrôles Des Minéralisations Cu-Au Pré-Tectoniques et Au (Cu) Tardi-Tectoniques à La Mine Portage, Chibougamau, Québec*; MRN report DV 94-09; Ministère des Ressources Naturelles: Québec, QC, Canada, 1994.
82. Paton, C.; Hellstrom, J.; Paul, B.; Woodhead, J.; Hergt, J. Iolite: Freeware for the Visualisation and Processing of Mass Spectrometric Data. *J. Anal. Spectrom.* **2011**, *26*, 2508–2518. [CrossRef]
83. Petrus, J.A.; Kamber, B.S. VizualAge: A Novel Approach to Laser Ablation ICP-MS U-Pb Geochronology Data Reduction. *Geostand. Geoanalytical Res.* **2012**, *36*, 247–270. [CrossRef]
84. Stern, R.A.; Bodorkos, S.; Kamo, S.L.; Hickman, A.H.; Corfu, F. Measurement of SIMS Instrumental Mass Fractionation of Pb Isotopes during Zircon Dating. *Geostand. Geoanalytical Res.* **2009**, *33*, 145–168. [CrossRef]
85. Davis, D.W. U–Pb Geochronology of Archean Metasedimentary Rocks in the Pontiac and Abitibi Subprovinces, Quebec, Constraints on Timing, Provenance and Regional Tectonics. *Precambrian. Res.* **2002**, *115*, 97–117. [CrossRef]
86. Marsh, A.D.; Parker, W.G.; Stockli, D.F.; Martz, J.W. Regional Correlation of the Sonsela Member (Upper Triassic Chinle Formation) and Detrital U-Pb Zircon Data from the Sonsela Sandstone Bed near the Sonsela Buttes, Northeastern Arizona, USA, Support the Presence of a Distributive Fluvial System. *Geosphere* **2019**, *15*, 1128–1139. [CrossRef]
87. Wiedenbeck, M.; Hanchar, J.M.; Peck, W.H.; Sylvester, P.; Valley, J.; Whitehouse, M.; Kronz, A.; Morishita, Y.; Nasdala, L.; Fiebig, J. Further Characterisation of the 91500 Zircon Crystal. *Geostand. Geoanalytical Res.* **2004**, *28*, 9–39. [CrossRef]
88. Wiedenbeck, M.; Alle, P.; Corfu, F.Y.; Griffin, W.L.; Meier, M.; Oberli, F.V.; von Quadt, A.; Roddick, J.C.; Spiegel, W. Three Natural Zircon Standards for U Th Pb, Lu Hf, Trace Element and REE Analyses. *Geostand. Newsl.* **1995**, *19*, 1–23. [CrossRef]
89. Mattinson, J.M. Zircon U-Pb Chemical Abrasion ("CA-TIMS") Method: Combined Annealing and Multi-Step Partial Dissolution Analysis for Improved Precision and Accuracy of Zircon Ages. *Chem. Geol.* **2005**, *220*, 47–66. [CrossRef]
90. Hamilton, M.A.; Álvaro, J.J.; Barr, S.M.; Jensen, S.; Johnson, S.C.; Palacios, T.; van Rooyen, D.; White, C.E. U–Pb Zircon Ages from Tuffaceous Beds in the Terreneuvian to Cambrian Series 2 Sections of Avalonian Southern New Brunswick, Canada: New Constraints on Chronostratigraphic Correlations and the Cambrian Time Scale. *Geol. Soc. Lond. Spec. Publ.* **2024**, *542*, SP542–2022. [CrossRef]
91. Krogh, T.E. A Low-Contamination Method for Hydrothermal Decomposition of Zircon and Extraction of U and Pb for Isotopic Age Determinations. *Geochim. Cosmochim. Acta* **1973**, *37*, 485–494. [CrossRef]
92. Gerstenberger, H.; Haase, G. A Highly Effective Emitter Substance for Mass Spectrometric Pb Isotope Ratio Determinations. *Chem. Geol.* **1997**, *136*, 309–312. [CrossRef]
93. Ludwig, K.R. *User's Manual for Isoplot 3.71 a Geochronological Toolkit for Excel*; Berkeley Geochronological Center Special Publication: Berkeley, CA, USA, 2009; Volume 4, p. 72.
94. Goldstein, R.H.; Reynolds, T.J. *Systematics of Fluid Inclusions in Diagenetic Minerals*; SEPM Society for Sedimentary Geology: Tulsa, OK, USA, 1994.
95. Bodnar, R.J. Introduction to Aqueous-Electrolyte Fluid Inclusions. In *Fluid inclusions: Analysis and interpretation*; Mineralogical Association of Canada: Québec, QC, Canada, 2003; Volume 32, pp. 81–100.
96. Steele-Macinnis, M.; Esposito, R.; Bodnar, R.J. Thermodynamic Model for the Effect of Post-Entrapment Crystallization on the H2O-CO2 Systematics of Vapor-Saturated, Silicate Melt Inclusions. *J. Petrol.* **2011**, *52*, 2461–2482. [CrossRef]
97. Kontak, D.J. Analysis of Evaporate Mounds as a Complement to Fluid-Inclusion Thermometric Data: Case Studies from Granitic Environments in Nova Scotia and Peru. *Can. Mineral.* **2004**, *42*, 1315–1329. [CrossRef]
98. Racicot, D.; Chown, E.H.; Hanel, T. Plutons of the Chibougamau-Desmaraisville Belt: A Preliminary Survey. In *Chibougamau, Stratigraphy and Mineralization*; Guha, J., Chown, E.H., Eds.; Canadian Institute of Mining and Metallurgy: Westmount, QC, Canada, 1984; Volume 34, pp. 178–197.
99. Middlemost, E.A.K. Naming Materials in the Magma/Igneous Rock System. *Earth-Sci. Rev.* **1994**, *37*, 215–224. [CrossRef]
100. Barker, F.; Arth, J.G. Generation of Trondhjemitic-Tonalitic Liquids and Archean Bimodal Trondhjemite-Basalt Suites. *Geology* **1976**, *4*, 596–600. [CrossRef]
101. Shand, S.J. *Eruptive Rocks: Their Genesis, Composition, Classification, and Their Relation to Ore Deposits with a Chaper on Meteorites*; Hafner Pub Co.: Frederick, MD, USA, 1943.
102. Ross, P.S.; Bédard, J.H. Magmatic Affinity of Modern and Ancient Subalkaline Volcanic Rocks Determined from Trace-Element Discriminant Diagrams. *Can. J. Earth Sci.* **2009**, *46*, 823–839. [CrossRef]

103. Martin, H. Effect of Steeper Archean Geothermal Gradient on Geochemistry of Subduction-Zone Magmas. *Geology* **1986**, *14*, 753–756. [CrossRef]
104. Hofmann, A.W. Chemical Differentiation of the Earth: The Relationship between Mantle, Continental Crust, and Oceanic Crust. *Earth Planet. Sci. Lett.* **1988**, *90*, 297–314. [CrossRef]
105. Pearce, J.A. Geochemical Fingerprinting of Oceanic Basalts with Applications to Ophiolite Classification and the Search for Archean Oceanic Crust. *Lithos* **2008**, *100*, 14–48. [CrossRef]
106. Loucks, R.R.; Fiorentini, M.L.; Henriquez, G.J. New Magmatic Oxybarometer Using Trace Elements in Zircon. *J. Petrol.* **2020**. [CrossRef]
107. Roedder, E. Fluid Inclusions. *Rev. Mineral.* **1984**, *12*, 644.
108. Fall, A.; Bodnar, R.J. How Precisely Can the Temperature of a Fluid Event Be Constrained Using Fluid Inclusions? *Econ. Geol.* **2018**, *113*, 1817–1843. [CrossRef]
109. Wilkinson, J.J.; Baker, M.J.; Cooke, D.R.; Wilkinson, C.C. Exploration Targeting in Porphyry Cu Systems Using Propylitic Mineral Chemistry: A Case Study of the El Teniente Deposit, Chile. *Econ. Geol.* **2020**, *115*, 771–791. [CrossRef]
110. Tuba, G.; Kontak, D.J.; Choquette, B.G.; Pfister, J.; Hastie, E.C.G.; van Hees, E.H.P. Fluid Diversity in the Gold-Endowed Archean Orogenic Systems of the Abitibi Greenstone Belt (Canada) I: Constraining the PTX of Prolonged Hydrothermal Systems. *Ore Geol. Rev.* **2021**, *135*, 104221. [CrossRef]
111. Tarantola, A.; Diamond, L.W.; Stunitz, H. Modification of Fluid Inclusions in Quartz by Deviatoric Stress I: Experimentally Induced Changes in Inclusion Shapes and Microstructures. *Contrib. Mineral. Petrol.* **2010**, *160*, 825–843. [CrossRef]
112. Tessier, A.C.; Hodgson, C.J.; Lulin, J.M. Formation of a Late-Tectonic, Mesothermal-Type Gold-Quartz Vein Deposit by Remobilization of Metals from a Magmatic-Hydrothermal Copper-Gold Deposit at the Portage Copper-Gold Mine, Chibougamau, Québec, Canada. In Proceedings of the Precambrian '95, International Conference on Tectonics and Metallogeny of Early/Mid Proterozoic Orogenic Belts, Montreal, QC, Canada, 28 August–1 September 1995; p. 131.
113. Lowell, J.D.; Guilbert, J.M. Lateral and Vertical Alteration-Mineralization Zoning in Porphyry Ore Deposits. *Econ. Geol.* **1970**, *65*, 373–408. [CrossRef]
114. Mathieu, L.; MacDonald, F. Petrography and Geochemistry of the Intrusive Rocks at the Diorite-Hosted Regnault Au Mineralization. *Minerals* **2022**, *12*, 128. [CrossRef]
115. SIGÉOM Système d'information Géominière Du Québec. Ministère de l'Énergie et des Ressources naturelles, Québec, Canada. 2020. Available online: https://sigeom.mines.gouv.qc.ca/signet/classes/I1102_indexAccueil?l=f (accessed on 1 June 2020).
116. Joanisse, A. Géochronologie Des Minéralisations Filoniennes à Cu-Au Retrouvées Dans Le Complexe Du Lac Doré. Master's Thesis, Université du Québec à Montréal, Montréal, QC, Canada, 1998.
117. Mathieu, L.; Madon, B.; Hamilton, M.A. Physico-Chemical Parameters of Neoarchean Syntectonic Magmatism: The Example of the Muscocho Pluton, Abitibi Subprovince. *Ore Geol. Rev.* **2020**, *125*, 103670. [CrossRef]
118. Côté-Mantha, O. Architecture et Origine Du Système de Minéralisation Polymétallique Du Secteur Lac Line, Région de Chibougamau, Québec. Ph.D. Thesis, Université du Québec à Chicoutimi, Chicoutimi, QC, Canada, 2009. Unpublished.
119. Madon, B.; Mathieu, L.; Marsh, J.H. Oxygen Fugacity and Volatile Content of Syntectonic Magmatism in the Neoarchean Abitibi Greenstone Belt, Superior Province, Canada. *Minerals* **2020**, *10*, 966. [CrossRef]
120. Mole, D.R.; Thurston, P.C.; Marsh, J.H.; Stern, R.A.; Ayer, J.A.; Martin, L.A.J.; Lu, Y.J. The Formation of Neoarchean Continental Crust in the South-East Superior Craton by Two Distinct Geodynamic Processes. *Precambrian Res.* **2021**, *356*, 106104. [CrossRef]
121. Richards, J.P. Magmatic to Hydrothermal Metal Fluxes in Convergent and Collided Margins. *Ore Geol. Rev.* **2011**, *40*, 1–26. [CrossRef]
122. Richards, J.P. The Oxidation State, and Sulfur and Cu Contents of Arc Magmas: Implications for Metallogeny. *Lithos* **2015**, *233*, 27–45. [CrossRef]
123. Heinrich, C.A. The Chain of Processes Forming Porphyry Copper Deposits—An Invited Paper. *Econ. Geol.* **2024**, *119*, 741–769. [CrossRef]
124. Botcharnikov, R.E.; Linnen, R.L.; Wilke, M.; Holtz, F.; Jugo, P.J.; Berndt, J. High Gold Concentrations in Sulphide-Bearing Magma under Oxidizing Conditions. *Nat. Geosci.* **2011**, *4*, 112–115. [CrossRef]
125. Steele-MacInnis, M.; Manning, C.E. Hydrothermal Properties of Geologic Fluids. *Elements* **2020**, *16*, 375–380. [CrossRef]
126. Kontak, D.J.; Gourcerol, B.; Petrus, J.; Thurston, P.C. Insights into the Gold Metallogeny of the Meguma Terrane of Nova Scotia, Canada, from LA-ICP-MS Arsenopyrite Geochemistry. *Atl. Geol.* **2018**, *54*.
127. Kontak, D.J.; Tuba, G. How Can Fluid Inclusion Studies Better Constrain Orogenic Gold Deposit Models: Case Studies from the Superior Province and Meguma Terrane, Canada [Ext. Abs.]. In Proceedings of the Society for Geology Applied to Mineral Deposits Annual Meeting, QuebecM, QC, Canada, 5 February 2017.
128. Zoheir, B.; El-Wahed, M.A.; Pour, A.B.; Abdelnasser, A. Orogenic Gold in Transpression and Transtension Zones: Field and Remote Sensing Studies of the Barramiya-Mueilha Sector, Egypt. *Remote. Sens.* **2019**, *11*, 2122. [CrossRef]
129. Taylor, R.D.; Monecke, T.; Reynolds, T.J.; Monecke, J. Paragenesis of an Orogenic Gold Deposit: New Insights on Mineralizing Processes at the Grass Valley District, California. *Econ. Geol.* **2021**, *116*, 323–356. [CrossRef]
130. Sterner, S.M.; Bodnar, R.J. Synthetic Fluid Inclusions VII. Re Equilibration of Fluid Inclusions in Quartz during Laboratory Simulated Metamorphic Burial and Uplift. *J. Metamorph. Geol.* **1989**, *7*, 243–260. [CrossRef]

131. Vityk, M.O.; Bodnar, R.J. Textural Evolution of Synthetic Fluid Inclusions in Quartz during Reequilibration, with Applications to Tectonic Reconstruction. *Contrib. Mineral. Petrol.* **1995**, *121*, 309–323. [CrossRef]
132. Tsuruoka, S.; Monecke, T.; Reynolds, T.J. Evolution of the Magmatic-Hydrothermal System at the Santa Rita Porphyry Cu Deposit, New Mexico, USA: Importance of Intermediate-Density Fluids in Ore Formation. *Econ. Geol.* **2021**, *116*, 1267–1284. [CrossRef]
133. Sun, Y.; Li, B.; Sun, F.; Ding, Q.; Wang, B.; Li, Y.; Wang, K. Mineralization Events in the Xiaokele Porphyry Cu (-Mo) Deposit, NE China: Evidence from Zircon U-Pb and K Feldspar Ar-Ar Geochronology and Petrochemistry. *Resour. Geol.* **2020**, *70*, 254–272. [CrossRef]
134. Diamond, L.W.; Tarantola, A. Interpretation of Fluid Inclusions in Quartz Deformed by Weak Ductile Shearing: Reconstruction of Differential Stress Magnitudes and Pre-Deformation Fluid Properties. *Earth Planet. Sci. Lett.* **2015**, *417*, 107–119. [CrossRef]
135. Sun, M.; Monecke, T.; Reynolds, T.J.; Yang, Z. Understanding the Evolution of Magmatic-Hydrothermal Systems Based on Microtextural Relationships, Fluid Inclusion Petrography, and Quartz Solubility Constraints: Insights into the Formation of the Yulong Cu-Mo Porphyry Deposit, Eastern Tibetan Plateau, China. *Miner. Depos.* **2021**, *56*, 823–842.
136. Chang, J.; Li, J.-W.; Audétat, A. Formation and Evolution of Multistage Magmatic-Hydrothermal Fluids at the Yulong Porphyry Cu-Mo Deposit, Eastern Tibet: Insights from LA-ICP-MS Analysis of Fluid Inclusions. *Geochim. Cosmochim. Acta* **2018**, *232*, 181–205. [CrossRef]
137. Rusk, B.G.; Reed, M.H.; Dilles, J.H. Fluid Inclusion Evidence for Magmatic-Hydrothermal Fluid Evolution in the Porphyry Copper-Molybdenum Deposit at Butte, Montana. *Econ. Geol.* **2008**, *103*, 307–334. [CrossRef]
138. Audétat, A.; Li, W. The Genesis of Climax-Type Porphyry Mo Deposits: Insights from Fluid Inclusions and Melt Inclusions. *Ore Geol. Rev.* **2017**, *88*, 436–460. [CrossRef]
139. Mao, W.; Rusk, B.; Yang, F.; Zhang, M. Physical and Chemical Evolution of the Dabaoshan Porphyry Mo Deposit, South China: Insights from Fluid Inclusions, Cathodoluminescence, and Trace Elements in Quartz. *Econ. Geol.* **2017**, *112*, 889–918. [CrossRef]
140. Yang, K.; Bodnar, R.J. Orthomagmatic Origin for the Ilkwang Cu–W Breccia-Pipe Deposit, Southeastern Kyongsang Basin, South Korea. *J. Asian Earth Sci.* **2004**, *24*, 259–270. [CrossRef]
141. Kontak, D.J.; Kyser, K. A Fluid Inclusion and Isotopic Study of an Intrusion-Related Gold Deposit (IRGD) Setting in the 380 Ma South Mountain Batholith, Nova Scotia, Canada: Evidence for Multiple Fluid Reservoirs. *Miner. Depos.* **2011**, *46*, 337–363. [CrossRef]
142. Tweedale, F.; Hanley, J.; Kontak, D.; Rogers, N. Petrographic Observations and Evaporate-Mound Analysis of Quartz-Hosted Fluid Inclusions: Applications to Assess Metal Fertility in Granites. *Atl. Geol.* **2014**, 50.
143. Alimohammadi, M.; Alirezaei, S.; Kontak, D.J. Application of ASTER Data for Exploration of Porphyry Copper Deposits: A Case Study of Daraloo-Sarmeshk Area, Southern Part of the Kerman Copper Belt, Iran. *Ore Geol. Rev.* **2015**, *70*, 290–304. [CrossRef]
144. Kerr, M.J.; Hanley, J.J.; Kontak, D.J.; Morrison, G.G.; Petrus, J.; Fayek, M.; Zajacz, Z. Evidence of Upgrading of Gold Tenor in an Orogenic Quartz-Carbonate Vein System by Late Magmatic-Hydrothermal Fluids at the Madrid Deposit, Hope Bay Greenstone Belt, Nunavut, Canada. *Geochim. Cosmochim. Acta* **2018**, *241*, 180–218. [CrossRef]
145. Gourcerol, B.; Kontak, D.J.; Thurston, P.C.; Petrus, J.A. Application of LA-ICP-MS Sulfide Analysis and Methodology for Deciphering Elemental Paragenesis and Associations in Addition To Multi-Stage Processes in Metamorphic Gold Settings. *Can. Mineral.* **2018**, *56*, 39–64. [CrossRef]
146. Schmidt, C.; Bodnar, R.J. Synthetic Fluid Inclusions: XVI. PVTX Properties in the System H2O-NaCl-CO2 at Elevated Temperatures, Pressures, and Salinities. *Geochim. Cosmochim. Acta* **2000**, *64*, 3853–3869. [CrossRef]
147. Driesner, T.; Heinrich, C.A. The System H2O–NaCl. Part I: Correlation Formulae for Phase Relations in Temperature–Pressure–Composition Space from 0 to 1000 C, 0 to 5000 Bar, and 0 to 1 XNaCl. *Geochim. Cosmochim. Acta* **2007**, *71*, 4880–4901. [CrossRef]
148. Monecke, T.; Monecke, J.; Reynolds, T.J.; Tsuruoka, S.; Bennett, M.M.; Skewes, W.B.; Palin, R.M. Quartz Solubility in the H2O-NaCl System: A Framework for Understanding Vein Formation in Porphyry Copper Deposits. *Econ. Geol.* **2018**, *113*, 1007–1046. [CrossRef]

minerals

MDPI

Article

The Paleoproterozoic Raimunda Porphyry-Type Gold Deposit, Juruena Mineral Province, Amazonian Craton (Brazil): Constraints Based on Petrological, Fluid Inclusion and Stable Isotope Data

Adriana Araujo Castro Lopes [1,2,*] and Márcia Abrahão Moura [1]

[1] Universidade de Brasília (UnB), Instituto de Geociências, Campus Universitário Darcy Ribeiro, Brasília 70910-900, DF, Brazil; mamoura@unb.br
[2] Universidade Federal do Oeste do Pará (UFOPA), Campus Universitário de Juruti, Juruti 68170-000, PA, Brazil
* Correspondence: adriana.castro@ufopa.edu.br

Abstract: The Juruena Mineral Province is an emerging world-class mineral province in the southern Amazonian Craton, due to numerous Au-Cu and base metal deposits, such as the Raimunda deposit, related to Novo Mundo 2.03–1.98 Ga I-type calc-alkaline granites. Its hydrothermal alteration zones comprise Na-metasomatism, microclinization, propylitic and sericitic alteration, silicification, a sulfide stage, and late carbonate alteration. The disseminated mineralization, associated with the sulfide stage, the main mineralization stage, is represented by gold inclusions and fracture-filling grains in pyrite, chalcopyrite, and Cu-Bi sulfides. Chlorite geothermometer and fluid inclusion data indicate temperature conditions of 325–380 °C for the mineralization. The coexistence of high-temperature aqueous and aqueous-carbonic fluid inclusions, based on a microthermometric study of fluid inclusions, reveals a mixing of medium-saline hot fluids with cooler and low-saline fluid. The $\delta^{18}O_{fluid}$ (3.11–7.86‰) and $\delta^{34}S_{py}$ data (−1.4–0.1‰) are coherent to the magmatic origin of the mineralizing fluid. Gold was initially transported as chlorine complexes in a hot, high-salinity, acidic, and oxidized fluid from the magma chamber, and later as H_2S^- complexes. The chemical-physical instability during fluid ascent is interpreted as a triggering factor for ore precipitation. The results offer valuable insights into the genesis of porphyry–Au deposits and their implications for prospecting in the Amazonian Craton.

Keywords: Paleoproterozoic; Amazonian Craton; fluid inclusions; stable isotope; porphyry–Au deposit

1. Introduction

The Amazonian Craton (Brazil) has undergone a series of calc-alkaline granitogenic events over its geological history, some of which led to gold and other metal mineralization [1]. The Tapajós and Juruena Mineral provinces are the major metallogenic provinces in Brazil, located in the southern Amazonian Craton. The Juruena Province, especially, has been an important gold producer in the country. Its main mineralization is predominantly spatially and temporally related to Paleoproterozoic calc-alkaline granites formed in volcanic-arc settings [2–5].

In the Juruena Mineral Province, mainly in the surroundings of the cities Nova Santa Helena, Peixoto de Azevedo, Matupá, Guarantã do Norte, and Novo Mundo, primary gold mineralization associated mainly with granite systems and, subordinately, with volcanic sequences occur either disseminated or hosted into veins and stockwork-veins [5,6].

The emplacement of these bodies into continental magmatic arcs has been essential for forming disseminated, vein-type, and Au ± Cu ± base metals stockwork-type magmatic-hydrothermal systems. The systematic and integrative study of these orebodies points out to a magmatic-hydrothermal model comparable to that of porphyry gold deposits [7], low-, intermediate-, and high-sulfidation epithermal systems [8], and structurally controlled [9] placed at variable crustal levels.

The Novo Mundo granite complex, which hosts the Raimunda deposit, the main object of this study, is in the homonymous municipality in the mid-northern region of the state of Mato Grosso, east of the Juruena Mineral Province. This granite body hosts several mineral deposits, including the Raimunda deposit, located in its northwestern region.

The percolation of hydrothermal fluids promotes anomalous and systematic mineralogical and geochemical changes in the host rocks, the extent and magnitude of which depend on their alteration state, porosity, permeability, and degree of chemical reactivity of the host rocks. The fluid-rock interaction generates alteration footprints that may assist in determining the exploratory guides in mature terrains [10].

Despite recent advances in knowledge about Proterozoic mineralizing events in the Amazonian Craton, prospecting complex Paleoproterozoic gold-copper deposits in Juruena Mineral province is a challenge, especially due to the few studies that have investigated the origin of mineralizing fluids using stable isotopic data to constrain the genesis of mineral deposits in the southern region of the Amazonian Craton.

This paper integrates new petrological (i.e., drill core samples, whole-rock assay results), ore mineral chemistry (e.g., chlorite, muscovite, sulfides and gold), fluid inclusion, and stable isotope data (O and S) from a preserved Paleoproterozoic deposit, located in the southern Amazonian Craton, and proposes a metallogenetic model for the Raimunda deposit and other analogous deposits in Brazil and worldwide.

2. Geological Settings

The Amazonian Craton has been a tectonically stable area since the Grenvillian Orogeny (1.2 Ga). It has been understood for decades as the result of the successive amalgamation of accretionary arcs from the Archean to the Paleo- and Mesoproterozoic [11,12]. However, newer alternative concepts and models have been proposed based on new isotopic and geophysical data (Figure 1).

Based on the progressively younger igneous bodies striking NE, at 1.88 Ga, the central region of the Amazonian Craton underwent a period of flat-slab subduction, which caused the migration of arc-related units towards the Central Amazonian block [13]. The sequence related to the Paleoproterozoic Tapajós Province arc was developed on an Archean basement and the well-defined E-W-striking lineament, which extends from the Tapajós Province to the Carajás Province and represents the ancient cratonic basement [14]. The lack of fold belts and/or metamorphic terranes, as well as the orogenic and post-orogenic tectonics, led Juliani et al. [15] to suggest weakly extensional to slightly compressive settings for the region.

Magnetotelluric profiles carried out in the eastern region of the province found a high-resistivity anomaly approximately 20 km deep, which is ascribed to the Peixoto de Azevedo Domain, surrounded by conductive anomalies related to the Statherian magmatism and mafic enclaves [16]. Furthermore, this domain presents a controversial tectonic evolution linked to long-lived shear deformation comparable to that of the Tapajós Province [17].

In the southern segment of the Amazonian Craton, there are two extensive Paleoproterozoic domains: an older one, named Peixoto de Azevedo (2.05 to 1.98 Ga), and a younger one, named Juruena (1.85–1.75 Ga). The Peixoto de Azevedo Domain comprises the Cuiú-Cuiú Magmatic Arc and post-magmatic arc rocks, mainly composed of Paleo- to Mesoproterozoic volcano-plutonic sequences formed in arc settings [6,18,19].

The eastern segment of the Juruena Province is mainly composed of oxidized, medium- to high-K calc-alkaline, volcanic-plutonic, and meta- to peraluminous rocks. These rocks mostly belong to the magnetite series (I-type granites) and are volcanic, subvolcanic, and alkaline granites [20].

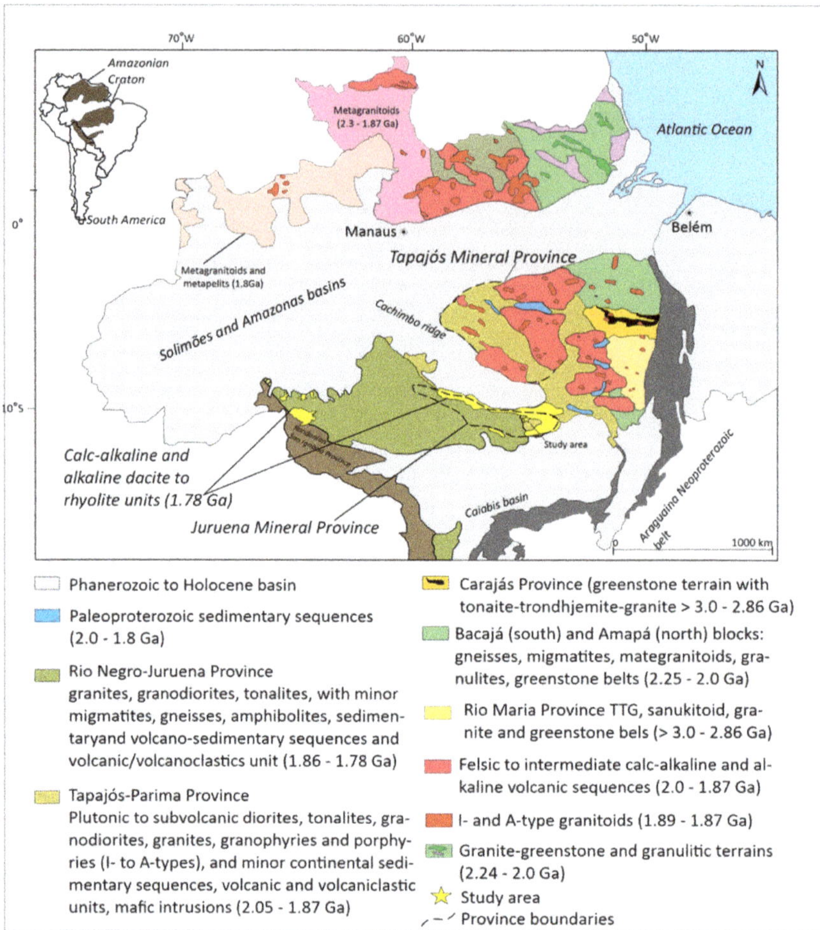

Figure 1. Geological and tectonic environment of the Juruena Mineral Province in the Amazonian Craton, based on [5,21]. Geochronological provinces based on [11].

2.1. Geology of the Deposit

The exploration area of Novo Mundo Paleoproterozoic Complex exhibits surface coverage that contains dozens of mineral deposits that host Paleoproterozoic Au mineralization [22], Au-Cu-Mo [9], and Cu [23] along a NW-SE-striking lineament.

The Novo Mundo complex intrudes Cuiú-Cuiú Archean rocks (granitic- to tonalitic gneisses and migmatites) and shows U–Pb data between 2.03 and 1.98 Ga [24]. It has an approximately 60-km diameter and is composed mainly of granites, along W-NW, which is compatible with the preferred orientation of the major regional lineaments mapped from the oldest terranes in that region [3], although there is no evidence of structural controls [19].

The Raimunda deposit is associated with monzonite, biotite granodiorite, monzogranite and syenogranite facies. The last two are the main host rocks of the deposit. The granites are cut by dikes of basaltic–andesite and andesite compositions.

The overview of the deposit is shown in Figure 2A,B, with a detailed view of the open-pit constructed by artisanal miners in the late 1990s and early 2000s for ore extraction. Figure 2C–E show rare blocks of outcrops classified as syenogranite, monzogranite and mafic bodies, respectively, all found around the deposit. Contact between the granitic facies is rarely observed. In drill holes, subangular to rounded microgranular enclaves are locally registered, especially monzonite enclaves into syenogranite.

Figure 2. Raimunda deposit outcrops, Novo Mundo, Juruena Mineral Province, Brazil. (**A**,**B**) Photograph of old ore extraction pit area, highlighting the location of some of the drill holes; (**C**) Syenogranite block showing potassic feldspar phenocrysts; (**D**) Monzogranite block found around the area; (**E**) Mafic rocks blocks.

Although the study area is in a region with local NW-SE-striking ductile-brittle shear zones (Novo Mundo Shear Zone–NMSZ) [25], which host several gold-bearing mineral deposits in the eastern region of the Juruena Mineral Province [6] and display a high density of microfractures, the rocks of the Raimunda deposit have not undergone significant deformation. The strongly fractured and crushed appearance of the minerals results from the action of the hydrothermal system over those rocks. Centimeter-to-millimeter-thick veins and veinlets are mostly non-oriented and contain alteration minerals (e.g., quartz, chlorite, muscovite, epidote, calcite, pyrite and rutile).

2.1.1. Monzonite Facies

The monzonite is dark gray, medium-grained, inequigranular, and exhibits a phaneritic texture (Figure 3A). It contains andesine crystals (An_{40-30} tending to An_{25}) (20–25%), orthoclase (18–20%), quartz (18–20%), biotite (3–5%), and hornblende (2–5%). Accessory minerals are magnetite (2%), titanite (1%), apatite 1%), and zircon (<0.1%), as well as hydrothermal minerals, such as epidote (1–2%), chlorite, muscovite, and carbonate (~5%) (Figure 4A).

Figure 3. Host rocks from the Raimunda deposit. (**A**) Monzonite, showing microclinization and veinlets composed of Cal, Chl, Qz and Ser with no preferred direction pattern; (**B**) Weakly microclinized monzogranite, cut by veinlets without preferential orientation of Qz + Chl + Ms + opaques; (**C**) Microclinized syenogranite, cut by venules of Cb, Qz, and Ms; (**D**) Microclinized granodiorite, cut by venules of Qz + Chl + Ms. Abbreviations: Cal: calcite; Chl: chlorite; Ms: muscovite; Qz: quartz.

Figure 4. Hydrothermal alteration features in drill core samples from Raimunda gold deposit. (**A**) Monzonite micrographic texture, with portions of agglomerated and recrystallized quartz, along

with Pl and Kfs, in addition to Ms, Chl, and opaque minerals; (**B**) Megacrystal of Qz strongly fractured in monzogranite rock, occurring in clusters of hydrothermal minerals containing Chl, Ser, and opaque minerals in zones of strong silicification; (**C**) Monzogranite alteration cluster containing Ms, Chl, Rt, and mafic minerals; (**D**) Monzogranite showing pervasive silicification, along with either monomineralic with only quartz or containing other phases; massive sulfide levels are observed by cutting the sample; (**E**) Granodiorite showing recrystallized Qz, Pl, and Kfs together with mafic minerals such as Amp, Bt, Chl, Ep, and mass of Ms; (**F**) Cluster of recrystallized minerals in alteration mass containing Pl, Qz, Fks, in addition to Ms, Chl in syenogranite sample; (**G**) Mass of hydrothermal minerals in syenogranite containing Ms, Qz, Rt lamellae, and opaque minerals. Abbreviations: Amp: amphibole; Bt: biotite; Chl: chlorite; Ep: epidote; Kfs: potassium feldspar; Ms: muscovite; Pl: plagioclase; Py: pyrite; Qz: quartz; Rt: rutile; Ser: sericite/fine muscovite.

2.1.2. Monzogranite Facies

The monzogranite is the more representative facies of the deposit. It has a greenish-gray color and is composed of quartz (25–30%), albite (15–20%), orthoclase (20–25%), and biotite (3–5%); accessory minerals (magnetite < 1%), and hydrothermal minerals (muscovite, chlorite, carbonate, epidote, and rutile, 12–15%) (Figure 3B). Microtextures show widespread veins and veinlets through the rock in association with potassic alteration (Figure 4B–D). Locally, open-space filling textures ("comb texture") are also observed, especially proximal to or in sulfide-bearing strongly hydrothermalized zones. Quartz megacrysts (0.5–3 mm) are restricted to strongly silica-altered zones (Figure 4B).

2.1.3. Syenogranite Facies

The syenogranite is reddish to pinkish and, in places, displays a porphyritic texture with quartz megacrystals/phenocrysts (Figure 3C). It is composed of quartz (30–35%), orthoclase (20–25%), albite (15–20%), and biotite (1–2%). Accessory magnetite (<1%), apatite (<0.1%), and zircon (<0.1%) feature as inclusions in quartz, plagioclase, and potassium feldspar. Muscovite, carbonate, and epidote commonly replace feldspar and plagioclase (12–15%) (Figure 4F,G). These samples also contain quartz megacrystals (0.5–3.5 mm) in strongly silicified zones, which are similar to those of monzogranites.

2.1.4. Granodiorite Facies

The granodiorite is greyish, medium-grained, composed of quartz (25–30%), andesine (An_{40-30} tending to An_{25}) (25–30%), orthoclase (15–20%), biotite (2–5%), and hornblende (1–3%) (Figure 3D). The accessory minerals are magnetite (1–3%), titanite (1%), apatite (<0.1%), and zircon (<0.1%); and the secondary minerals are muscovite, carbonate, epidote, and rutile (10–12%) (Figure 4E).

3. Materials and Methods

3.1. Fieldwork, Sampling, and Petrography

Three fieldwork campaigns were carried out for mapping and sampling. In total, 80 samples were collected for petrographic analysis to identify the main gangue and ore mineral associations of hydrothermal alteration and ore-bearing zones, as well as their patterns and styles [26,27]. This enables a better understanding of their evolution throughout time in the system.

The polished thin sections and polished sections were made at the thin-section laboratory of the Geosciences Institute of the University of Brasília (LABLAM-UnB). The doubly polished thin sections were made at the thin-section laboratory of the Petrografia-BR company (Contagem–MG). The petrographic descriptions were carried out at the Microscopy Laboratory of the UnB Geosciences Institute using an Olympus BX-50 petrographic microscope (Olympus Corporation, Tokyo, Japan), as well as at the UnB Fluid Inclusions Laboratory (LIF-UnB) using an Olympus BX-50 petrographic microscope.

3.2. Electron Microprobe

The mineral chemistry analyses were determined using a JEOL® JXA-8230 Electron Probe Micro-analyzer (EPMA) (JEOL Ltd., Tokyo, Japan), at the Institute of Geosciences, University of Brasília. For silicates, the chemical analyses were obtained using an acceleration voltage of 15 kV, a beam current of 10 nA, and a nominal beam diameter of 1 µm in the spot mode. The counting time for each element during the analysis was 10 s on the peak and 5 s on the background. Interference correction was implemented for Ti and V. Matrix effects were corrected using an in-house ZAF program. The natural and synthetic calibration standards used were albite (Na), forsterite (Mg), orthoclase (Al, Si, K), topaz (F), vanadinite (Cl, V), apatite (P), andradite (Ca, Fe), pyrophanite (Ti, Mn), barite (Ba), Ni oxide (Ni), and vanadinite (V).

For the sulfides and gold, an accelerating voltage of 20 kV, a beam current of 20 nA, and a counting time of 10 s, except for selenium (25 s) were used; the counting time for backgrounds was always half of the used-on peaks values. The X-ray Kα-lines were used for sulfur, selenium, copper, iron, nickel, and cobalt; Lα for arsenic, silver, antimony, palladium, cadmium, tellurium, and molybdenum; and Mα for lead, mercury, gold, and platinum; Mα and Mβ were used for bismuth. The natural minerals applied as standards in both electron probes were pyrite (Fe, S), chalcopyrite (Cu), galena (Pb), and stibnite (Sb), and the native elements gold, silver, platinum, palladium, bismuth, copper, and antimony. The synthetic standards were gallium arsenide (As), zinc selenide (Zn, Se), and cinnabar (Hg). The following detection limits were obtained (in wt.%): 0.018 for gold, 0.012 for silver, 0.008 for palladium, 0.016 for platinum, 0.032 for bismuth, 0.025 for lead, 0.016 for copper, 0.013 for iron, 0.017 for mercury, 0.025 for arsenic, 0.009 for antimony, 0.017 for selenium, 0.005 for sulfur, 0.012 for cadmium, 0.013 for nickel and cobalt, 0.017 for tellurium, and 0.030 for molybdenum.

3.3. Chlorite Geothermometry

The studies of formation temperature and pressure conditions at the Raimunda deposit were complemented using a chlorite-independent geothermometer. The structural formulas of chlorite and muscovite were calculated based on 28 and 11 oxygen atoms, respectively. The temperatures are based on the cationic proportion of tetrahedral aluminum (Al^{IV}) in chlorites, which allows the crystallization temperature to be calculated using the following equation: T (°C) = 321.98 × ($Al^{IV}/2$) − 61.92 [28,29]. In total, 18 analyses of chlorite were carried out on carbon-coated polished thin sections at the LME-UnB using a JEOL® JXA-8230 EPMA (JEOL Ltd., Tokyo, Japan). The WDS operated under the conditions of an acceleration voltage of 15 Kv, a beam current of 10 nA, and a count time of 10 s.

3.4. Fluid Inclusions

The study of fluid inclusions was performed on doubly polished thin sections from six different mineralized samples. Petrography of fluid inclusions was carried out at room temperature (±25 °C) under a Zeiss Axioplan 2 petrographic microscope (Zeiss, Oberkochen, Germany) according to the criteria of [30–32].

Microthermometric measurements were obtained with a LINKAM THMSG600 heating-freezing system (Linkam Scientific, Salfords, Surrey, UK) mounted on an Olympus BX-50 petrographic microscope (Olympus Corporation, Tokyo, Japan), at the UnB Fluid Inclusions Laboratory (LIF-UnB). Rock fragments were observed at variable temperatures between −195 °C and 600 °C under a 10× to 100× objective lens following the procedures of [31].

The calibration was carried out using synthetic H_2O–NaCl two-phase and H_2O–CO_2 three-phase fluid inclusions at speed rates of 2 to 15 °C/min, with an estimated accuracy of ±0.5 °C for freezing (±25 at −150 °C), and ±2 °C for heating (above 350 °C).

3.5. Stable Isotopes

The study of stable isotopes was performed on minerals of representative mineralized samples from the Raimunda deposit. Sample selection and sample preparation took place at the sample preparation laboratory at the University of Brasília (LPA-UnB). The pure mineral concentrates were handpicked under a binocular microscope at LME-UnB.

Analyzes were performed at the Queen's Facility for Isotope Research (QFIR) laboratory at Queen's University, Ontario, Canada. Results are reported as delta (δ) values per mil (‰) relative to V-SMOW (Vienna Standard Mean Ocean Water) for oxygen, and V-CDT (Canyon Diablo Troilite) for sulfur.

Oxygen isotope data were obtained for nine samples, between 20 and 28 mg each, using the Thermo Finnigan Delta V Advantage and Finnigan Delta XP mass spectrometers (Thermo Fisher Scientific, Waltham, MA, USA). The analytical protocol applies the laser fluorination method to release oxygen from minerals through the reaction with BrF_5 by using a CO_2 laser as a heat source. It is then either converted into CO_2 through the reaction with graphite at 750 °C, or it is collected as O_2 in a molecular sieve and analyzed as O_2. The analytical uncertainties for $\delta^{18}O$ were lower than 0.1 and 0.16‰.

For sulfur, six samples of pyrite were prepared, each weighing between 80 and 95 mg. For this analysis, the SO_2 produced through combustion of about 0.5 mg of sample is loaded into tin capsules reacting at 1080 °C. The gases were transported through current and reduced by contact with high-purity copper wires. The SO_2 was eventually measured using gas chromatography. Analyzes were performed using a Flash Elemental Analyzer (EA) coupled with a ThermoFisher MAT 253 Isotope Ratio Mass Spectrometer (IRMS) (Thermo Fisher Scientific, Waltham, MA, USA). Analytical uncertainties are estimated at 0.2‰.

4. Results

4.1. Hydrothermal Alteration and Mineralization

The Au (±Cu) Raimunda deposit sits in the NW of the Novo Mundo Complex, being hosted by variably hydrothermally altered granite facies (i.e., monzonite, biotite granodiorite, monzogranite, and syenogranite).

The types of hydrothermal alteration identified at the Raimunda deposit, and their respective mineral assemblages, are summarized in Table 1. The hydrothermal system and the associated mineralization are mainly observed through pervasive zones, and, subordinately, along veins and veinlets and the more affected rocks are the monzo- to syenogranite facies.

Based on overprinting relationships, mineralogy, and alteration assemblages, seven alteration zones are recognized: (1) Early sodic metasomatism (albite); (2) Microclinization (orthoclase/microcline-quartz1-hematite); (3) Early or intermediate propylitic alteration (epidote-chorite1-carbonate-rutile-titanite-pyrite1); (4) Intermediate to proximal sericitic alteration (muscovite-chlorite2-quartz2-pyrite2-gold1); (5) Silicification (quartz3); (6) Main mineralization phase: sulfide stage (gold2-pyrite3-chalcopyrite-bismuthinite and other Bi- and Cu sulfides); and (7) Late-stage carbonate alteration and chloritization (carbonate-chlorite3-quartz4).

The alterations occur throughout the ore-bearing host granite at the deposit. At the field scale, and even when analyzing the different drill cores, it is not possible to macroscopically delineate hydrothermal alteration zones.

The early Na-metasomatism alteration modified the essential mineralogy of the granites. The oligoclase is completely or partially replaced by albite and the magmatic biotite is locally preserved.

Subsequently, microclinization represents another outward alteration halo surrounding the previous one, and it is commonly observed at the periphery of the alteration profile. Microclinization commonly occurs within all granite facies; although more intensely in the syenogranite, through the alteration of K-feldspar and the formation of microcline rims. The orthoclase is microclinized and the feldspar and biotite crystals are partially to completely altered to fine muscovite. Hydrothermal quartz grains (quartz1) are generally milky.

Table 1. Hydrothermal alteration stages and paragenetic evolution table of the Raimunda Au (±Cu) deposit, Juruena Mineral Province, Brazil.

Event	Pre-Mineralization			Mineralization Stages			Post-Mineralization
Mineral \ Alteration Stages	Na-Metassomatism	Microclinization	Propylitic Alteration	Sericitic Alteration (Ms-Qz2-Chl2-Py2-Au1)	Silicification (Qz3)	Sulfide Stage (Py3-Au2-Ccp-Cc-Bin)	Carbonate Alteration (Cal-Chl3-Qz4)
Quartz 1		Abundant	Minor				
Quartz 2				Minor			
Quartz 3					Abundant		
Quartz 4							Abundant
K- Feldspar		Abundant					
Albite	Abundant						
Magnetite		Minor	Minor	Traces			
Titanite			Minor				
Rutile			Minor	Minor			
Chlorite		Traces	Minor	Minor			Minor
Muscovite		Traces	Traces	Abundant		Minor	
Calcite		Traces	Abundant				Abundant
Epidote			Minor				
Pyrite 1		Minor					
Pyrite 2				Abundant			
Pyrite 3						Abundant	
Chalcopyrite				Minor		Abundant	
Chalcocite						Minor	
Larosite						Minor	
Bismuthinite						Minor	
Hodrushite/krupkaite/wittchenite						Minor	
Gold				Minor		Abundant	

Traces ---------- Minor ——— Abundant ▬▬▬

Temporal evolution ⟶

The propylitic alteration can be noticed in all granitic facies of the Raimunda deposit but is more perceptive in non-mineralized samples, including monzonite and granodiorite facies. It is pervasive, affecting especially feldspar crystals, which are partially or completely replaced by epidote-chlorite1-carbonate. It is common to occur in veins and/or veinlets with epidote-chlorite1-carbonate crossing the samples and rutile, titanite, and fine-grained pyrite1 in agglomerates mass.

Sericitic alteration is marked by its pale green color throughout all granite facies, where they occur pervasively or selectively altered and become more intense towards the monzogranite facies. The alteration mineral assemblage is mainly composed of muscovite, chlorite2 (Chl2), pyrite2 (Py2), quartz2 (Qz2), and chalcopyrite. The main feature is the texture obliterated by the replacement of feldspars with fine muscovite, along with disseminated fine-grained pyrite and quartz, locally. In this case, pyrite (Py2) occurs disseminated either into quartz veins and veinlets or associated with quartz2 and muscovite. This phase was identified as the first stage of mineralization (i.e., muscovite, pyrite2, quartz2, and gold1). Euhedral to subhedral pyrite is classified as pyrite2, which is poorly fractured and displays a low degree of fracture filling, occurring disseminated in host rocks and, locally, into veins or veinlets that cut through the rock (Figure 5A). Gold (Au1) occurs either as inclusions or as fracture-fills in pyrite2.

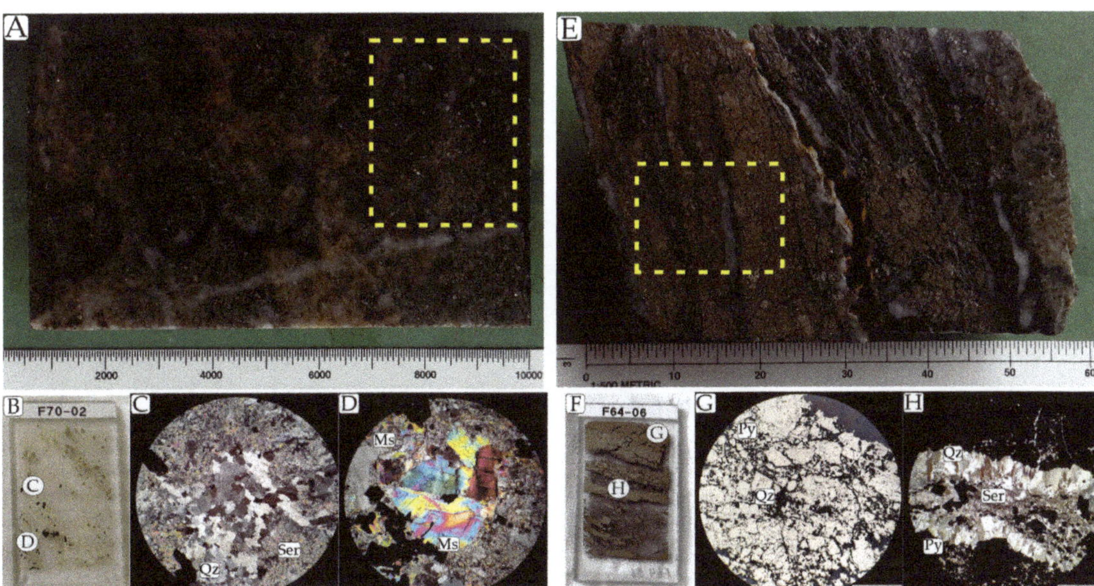

Figure 5. Hydrothermal alteration features and mineralized samples from the Raimunda Au (±Cu) deposit. (**A**) Drill hole sample with pervasive silicification, generalized sulfidation, chloritization, and sericitic alteration; (**B**) Polished thin of the yellow delimited field in (**A**); (**C**) photomicrograph showing strong silicification with recrystallized Qz surrounded by intense sericitic alteration; (**D**) Ms lamellae with recrystallized Qz, opaques, and sericitic alteration; (**E**) Drill hole sample highlighting pressure relief features, vuggy silica in quartz veins/veinlets, and levels of massive sulfide interspersed with monomineralic Qz veinlets; (**F**) Polished thin of the yellow delimited field in (**E**); (**G**) Massive sulfide levels showing Py crystals strongly fractured; (**H**) Sulfide veinlets, with Qz and Ms. Abbreviations: Ms: muscovite; Py: pyrite; Qz: quartz; Ser: sericite/fine muscovite.

Silicification occurs pervasively or as veins and veinlets (Figure 5C). The veins/veinlets consist of microcrystalline quartz (massive and mosaic texture—Qz3), associated with sulfides and other minerals, such as muscovite and carbonate.

The sulfide stage is both disseminated and forming massive sulfide zones that crosscut the rocks. At the deposit, that corresponds to the main mineralizing stage with a high content of associated sulfides (i.e., pyrite3, chalcopyrite, chalcocite, bismuthinite, hodrushite, krupkaite, wittchenite) and gold2. The pyrite3 occurs in massive sulfide levels, strongly fractured, and with a high degree of fracture filling (Figure 4E,G). Gold (Au2) occurs as fracture-fills in pyrite3 and as seldom inclusions in quartz3. The samples highlight pressure relief features, vuggy silica in quartz veins/veinlets, and levels of massive sulfide (Figure 5E,F).

Late hydrothermal alteration is represented by carbonate alteration. It is less expressive than the last two stages and restricted to the final stages of hydrothermal alteration. In general, potassium feldspar and albite are affected by carbonate alteration. Carbonate-quartz4-chlorite3 veins and veinlets crosscut the rock samples (Figure 5E).

4.2. Ore and Hydrothermal Mineral Chemistry

Tables S1–S7 display EPMA analyses of pyrite, chalcopyrite, gold, copper-rich minerals, bismuth-rich minerals, chlorite and muscovite which can be found in the Supplementary Material. The selection considered the variable styles of mineralization (veins/veinlets and disseminated) for each mineral as well as their ore mineral paragenesis.

4.2.1. Sulfides and Gold

The analysis of pyrite focused on types 2 (Py2) and 3 (Py3), which are associated with the mineralization stages at the deposit (Table S1, Figure 6). Py3 (Figure 6A) contains slightly higher contents of Au (Py3: 0–0.09%; Py2: 0.02–0.07%), Te (Py3: 0–0.24%; Py2: 0–0.12%), and Mo (Py3: 0.01–0.17%; Py2: 0–0.08%). Furthermore, there are higher contents of Pb (Py3: 0.1–0.34%; Py2: 0.19–0.23%), as well as higher contents of Cu (Py3: 0–0.82%; Py2: 0–0.04%).

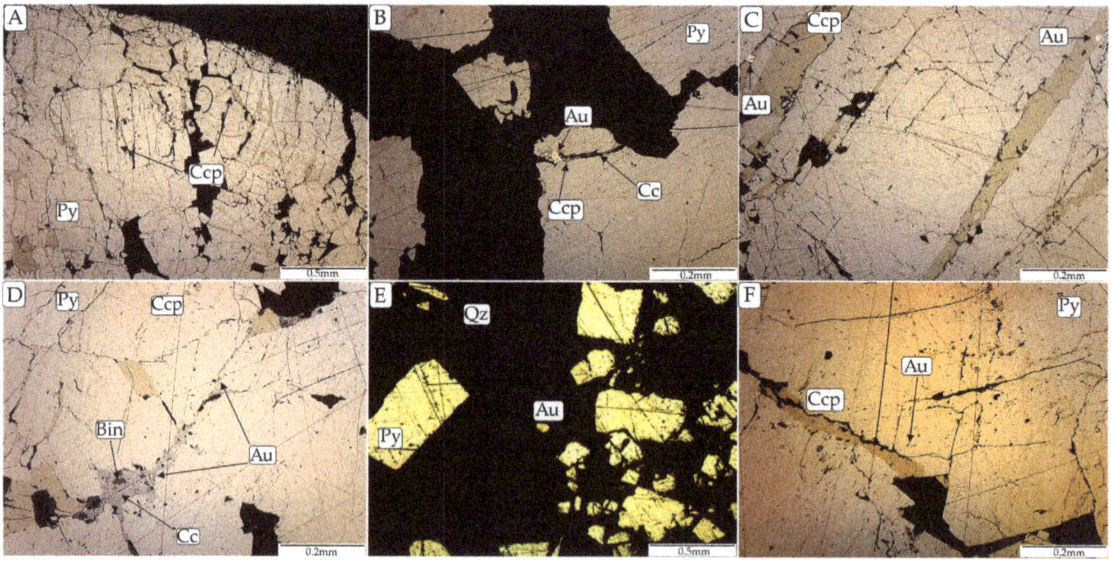

Figure 6. Ore mineral association in the Raimunda Au (±Cu) deposit, Novo Mundo, Juruena Mineral Province, Brazil. The images were obtained in reflected light and with parallel polarizers. (**A**) Massive sulfide texture, with anhedral and strongly fractured Py filled with Ccp, Cc, Bin, Au, and other copper sulfides; (**B–D**) Strongly fractured pyrite filled by Au-Ccp-Cc-Bin and Cu sulfides association; (**E**) Au Included in Qz-Py association; (**F**) Au included in Py crystals strongly fractured and filled with other sulfides. Abbreviations: Au: gold; Bin: bismuthinite; Ccp: chalcopyrite; Cc: chalcocite; Qz: quartz; Py: pyrite.

The significant amount of these elements in Py3 compared to Py2 reveals the genetic differences between those types based on chemical evidence, which had been previously corroborated by textural features. This points out variable timings and fluctuating element availability throughout ore precipitation.

The chalcopyrite occurs in fracture-fills and inclusions in Py2 and, more intensely, in Py3 (Figure 6A,C,D,F). Chemically, there are no significant differences between them, with average Au contents of 0–0.06%, Ag contents of 0–0.16%, Te contents of 0–0.23%, Mo contents of 0–0.12%, Pb contents of 0.04–0.30% and Bi contents of 0–0.23%.

Chalcocite (Cu_2S) occurs together with chalcopyrite and other copper sulfides, such as larosite ($(Cu, Ag)_{21}(Pb, Bi)_2 S_{13}$), filling fractures in pyrite3. Bismuthinite is the most abundant bismuth mineral and, likewise, occurs as fracture-fills. Other Cu- and Bi-rich sulfides identified as part of the mineral paragenesis at the deposit are hodrushite ($Cu_8Bi_{12}S_{22}$), krupkaite ($PbCuBi_3S_6$), and wittchenite (Cu_3BiS_3) (Figure 6D).

All sulfides are in equilibrium with Py3 making up the main mineralization stage at the Raimunda deposit. Gold occurs either as inclusions (Au1) or fracture-fills (Au2) (Figure 6B–F). In both cases, gold inclusions and fracture-fills in pyrite showed no significant compositional variation between them (Table S3, Supplementary Material). Fracture-filling gold grains (Au2) display average Au values of 57.03–66.77%, and Ag values of 28.45–28.74%, while gold inclusions yield Au contents of 65.91–66.80% and Ag contents of 27.90–28.70%. Therefore, their chemistry reveals relative constant silver values and variable gold values, which are related to Cu, Te, and Mo input.

4.2.2. Chlorite Chemistry and Geothermometry

The chemical composition of chlorite contributes to the study of hydrothermal alteration processes to delineate the former physical–chemical conditions (T, P, fO_2, pH, and host rock composition) of the geological environment at the time of its formation, which makes it an important indicator of the rock formation temperature [28,33].

Table S6 (Supplementary Material) presents 18 microprobe analyses of chlorite (i) after biotite (Chl1), (ii) in equilibrium with muscovite and epidote (sericitic alteration) (chl2), and (iii) in association with the late-stage carbonatization (Chl3). The compositional variation in chlorite might be related to the partial replacement of igneous biotite by chlorite, which had been depicted from textural features and is used for a classification scheme.

Magnesium-rich chlorite yields Fe/(Fe + Mg) ratios varying between 0.33 and 0.56, and Si/Al ratio varying between 1.02 and 1.39. Si atoms range between 5.24 and 5.98 a.p.f.u., and Al^{IV} ranges from 2.02 to 2.76 a.p.f.u. Iron-rich chlorite exhibits Fe/(Fe + Mg) ratios of 0.67–0.84 and less varying Si/Al ratios of 1.03–1.09. Si atoms range between 5.31 and 5.51 a.p.f.u., and Al^{IV} values fall within the range of 2.49–2.70 a.p.f.u. Based on the classification of [34], these compositions plot at the transition between clinochlore-chamosite (Figure 7A,B).

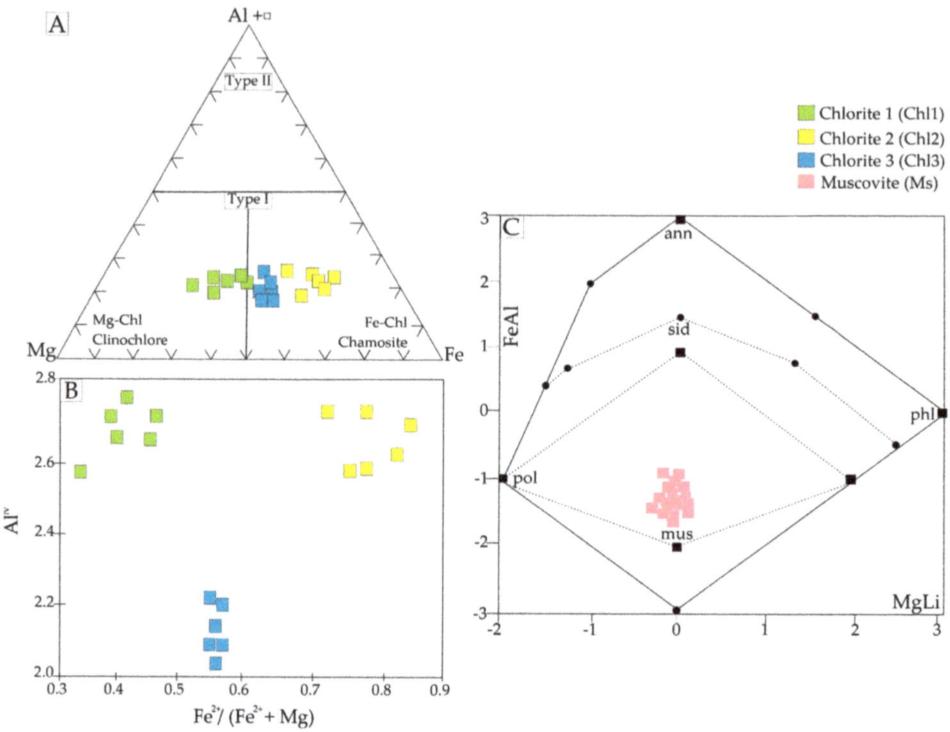

Figure 7. Classification diagrams of chlorite and muscovite. (**A**) Ternary diagram Mg, Al + □ (□ corresponds to the octahedral vacancy) and Fe, allocating the Fe-rich (chamosite) and Mg-rich (clinochlore) types; (**B**) Diagram showing the covariance between the Fe/(Fe + Mg) ratio and Al^{IV} for the chlorites of the deposit; (**C**) FeAl vs. MgLi classification plot of muscovite by [35], placing the three mica species in solid solution. The samples fall into the muscovite field. Abbreviations: Ann: annite; Chl: chlorite; Ms: muscovite; Phl: phlogopite; Pol: polylithionite; Sid: siderophyllite.

Empirical geothermometers were proposed by several authors [28,29,34,36,37]. Each one applies correction criteria based on the characteristics of the given chlorite, as follows: (i) tetrahedral Al (Al^{IV}) contents [28,29] and (ii) corrected Al^{IV} contents based on the relationship between Al^{IV} and Fe/(Fe + Mg) ratios [36–39]. Empirical geothermometers used in this study are summarized in Table S6 (Supplementary Material). The peak temperature was obtained through the equation of Cathelineau [29]. The values calculated according to Cathelineau and Nieva [28] are similar to those calculated according to Cathelineau [29]. The lowest temperatures were obtained using the equation of Zang and Fyfe [37]. Based on the features and compositions of the different types of chlorites here identified, the formation temperatures estimated through Cathelineau [29] were selected.

The chlorite geothermometry data are coherent with the petrographic data: the hydrothermal alteration had its onset with propylitic alteration, marked by the replacement of biotite by chlorite (Chl1) along with the formation of other Fe-rich minerals (e.g., rutile) at temperatures between 355 and 382.4 °C. As alteration progressed, Chl2, which is associated with sericitic alteration, began to nucleate, at temperatures between 338.9 and 372.8 °C, under the conditions where the first mineralization stage took place. During late hydrothermal alteration, Chl3 was predominant in association with late quartz and carbonate veinlets, at temperatures ranging between 263.3 and 297.87 °C.

4.2.3. Muscovite

Electron microprobe data and the chemical formula of muscovite from the Raimunda deposit are available in Table S7 (Supplementary Material). The muscovite samples chosen for analysis are those associated with the first (1) mineralization stage (sericitic alteration) due to their greater occurrence. The 16 spot analyses reveal a relatively homogeneous composition with small variations in Al_2O_3 (29.8–31.4%), MgO (0.6–1.18%), FeO (1.54–5%), and Na_2O (0.13–0.4%) [35] (Figure 7C).

4.3. Fluid Inclusions Data

4.3.1. Petrography and Classification

The fluid inclusion study was carried out on quartz grains from the mineralization stage (Qz2 and Qz3) and late alteration stage (Qz4), using doubly polished thin sections from the Raimunda deposit. Primary and secondary fluid inclusions were described according to the fluid inclusion assemblage (FIA) criteria proposed by [40], which is based on the classification criteria of [31].

Based on the distribution, number of phases present at room temperature, patterns of occurrence, phase transitions throughout heating and freezing runs, and Raman spectrometry data, nine subtypes of fluid inclusions were recognized. The main characteristics are summarized in Table 2.

Type 1 (t1) are two-phase (liquid + vapor) aqueous liquid-rich primary (t1-a) and secondary (t1-b) inclusions. Type t1-a inclusions show clear evidence of primary fluid, and occur in Qz3, which is associated with sulfide stage, as clusters or isolated inclusions with a liquid fill ration (F) of 0.70–0.8. They are usually colorless, with a light-colored bubble surrounded by a dark rim. They are rounded, oval, and, more rarely, irregularly shaped, varying between 8–20 μm (Figure 8A,C). Type t1-b inclusions encompass secondary inclusions in Qz2 associated with sericitic alteration. They mostly occur along trails and subordinately as clusters. They are colorless and display an F of 0.7–0.8. The light-colored bubble moves at room temperature in some samples. They range between 3 and 12 μm and are rounded, sub-rounded, elongated, and, more rarely, rectangular.

Type 2 (t2) are primary, single-phase carbonic vapor-rich, two-phase (L_{H2O} + V_{CO2}) and three-phase (L_{H2O} + L_{CO2} + V_{CO2}) aqueous-carbonic fluid inclusions that occur in Qz3. Single-phase inclusions are gray, with an F of 100% CO_2, varying between 5 and 15 μm. Two-phase and three-phase inclusions display bubbles with a dark rim, varying between 5 and 20 μm. They are sub-rounded, elongated, and, more locally, irregularly-shaped. Their F is 30–50 vol% (Figure 8A–C). They occur as single inclusions or in clusters being oval to irregularly shaped. Three-phase inclusions usually display CO_2 bubbles only during reheating, before the homogenization temperature for CO_2 is reached. They are considered part of the same fluid inclusion assemblage (FIA).

Type 3 encompasses three-phase primary (t3-a) and three-phase secondary (t3-b) inclusions (L_{H2O} + V_{H2O} + S_1) containing daughter minerals and multiphase-solid phases (L_{H2O} + V_{H2O} + S_1 + S_2). The t3-a inclusions correspond to the primary in Qz4, associated with late-carbonatization. It varies between 10 and 20 μm, and is elongated, and sub-rounded; colorless to greenish. They contain cubic solid phases (daughter mineral) that are colorless, isotropic, and interpreted as halite based on petrographic and microthermometric data as well as in the lack of a Raman signal. Overall, the F for both subtypes is 15–20 vol% (vapor), 15–20 vol% (solid), and 60–70 vol% (liquid).

The t3-b fluid inclusion is secondary in Qz3 and occurs as trails and/or aligns associated with sulfide stage. They vary between 8 and 22 μm and are square-shaped, elongated, sub-rounded, and greenish. The cubic solid phases (daughter mineral) are colorless and, locally, also occur as a trapped solid (opaque mineral) in the fluid inclusion (Figure 8D). The F for both subtypes is 15–20 vol% (vapor), 15–20 vol% (solid), and 60–70 vol% (liquid) (Figure 8D,E).

Table 2. Petrographic and microthermometric data of fluid inclusions from the Raimunda deposit, Juruena Mineral Province, Brazil. Abbreviations: L—liquid; V—vapor; S—daughter mineral (saturation solid); Ts—trapped solid. *Te*: eutectic temperature; Tm_{hh}—hydrohalite melting temperature; Tm_{ice}—ice melting temperature; Tm_{CO2}—CO_2 melting temperature; Tm_{cla}—clathrate melting temperature; Th_{CO2}—CO_2 homogenization temperature; Th_t—total homogenization temperature.

Fluid Inclusion Type	Host Rock	Hydrothermal Alteration Stage	Mineral Association	Petrographic Characteristics	Microthermometric Data	Model System
Type 1-a (L + V); *n* = 28	Monzogranite; syenogranite	Sulfide stage	Py3 + Ccp + Qz3 + Au2	Primary; random; colorless; ellipsoidal, oval to irregularly shaped; 8–20 µm; F: 0.70–0.8	*Te* = −24 to −21 °C; Tm_{ice} = −6.4 to −0.8 °C; Th_t = 233.5 to 471.2 °C; salinity = 1.396–9.72 wt.% NaCl eq.; density; 0.409–0.895 g/cm^3	H_2O-NaCl
Type 1-b (L + V); *n* = 25	Monzogranite; syenogranite	Sericitic alteration	Ms + Qz2 + Au1	Secondary; planar grouping and trails; oblate; 3–12 µm; F: 0.7–0.8	Tm_{ice} = −4.2 to −0.5 °C; Th_t = 200.5 to 297 °C; salinity = 0.879–6.737 wt.% NaCl eq.; density = 0.715–0.918 g/cm^3	H_2O-NaCl
Type 2 (V; L + V; L + L + V); *n* = 40	Monzogranite	Sulfide stage	Py3 + Ccp + Qz3 + Au2	Primary; random distribution or in clusters; grayish; dark bubble; elongated and oval; 5–20 µm; F: 30–50	Tm_{CO2} = −57.3 to −56°C; Tm_{ice} = −3.8 to −2.9 °C; Tm_{cla} = 6.2–8.5 °C; Th_{CO2} = 19–30.1 °C; Th_t = 210 to 364.7 °C; salinity = 3.00–5.94 wt.% NaCl eq.; density = 0.87–0.88 g/cm^3	H_2O-NaCl-CO_2
Type 3-a (L + V; L + V + S); *n* = 3	Monzogranite	Carbonate alteration	Cal + Chl3 + Qz4	Primary; random distribution; square-shaped, elongated, and sub-rounded; colorless or greenish; 8–22 µm; F: 60–70; with one cubic crystal (halite daughter mineral)	Tm_{hh} = −10.7 to −3.8 °C; Th_t = 290.5 to 367.5 °C; salinity = 38.88–50.25 wt.% NaCl + $CaCl_2$ eq.; density = 0.78–0.82 g/cm^3	H_2O-NaCl-$CaCl_2$
Type 3-b (L + V; L + V + S + Ts); *n* = 19	Monzogranite	Sericitic alteration	Py3 + Ccp + Qz3 + Au2	Secondary; planar grouping and trails; oblate square-shaped, elongated, and sub-rounded; 3–12 µm; F: 60–80; with colorless cubic crystals (halite daughter mineral) and trapped solid locally	*Te*: −55 to −51.2 °C; Tm_{ice} = −43.2 to −37.8 °C; Tm_{hh} = −10.6 to −1.1 °C; Th_t = 100.6 to 251.5 °C; salinity = 33.02–38.21 wt.% NaCl + $CaCl_2$ eq.; density = 0.87–0.97 g/cm^3	H_2O-NaCl-$CaCl_2$
Type 4-a (L + V); *n* = 19	Monzogranite; syenogranite	Silicification; Sulfide stage	Py3 + Ccp + Qz3 + Au2	Primary; clusters and isolated; colorless; 8–20 µm; F: 0.8–0.85	Tm_{ice} = −5.4 to −1.0 °C; Th_t = 199.2 to 288 °C; salinity = 1.57–8.41 wt.% NaCl eq.; density = 0.74–0.92 g/cm^3	H_2O-NaCl
Type 4-b (L + V); *n* = 41	Monzogranite; syenogranite	Sericitic alteration	Ms + Qz2 + Au1	Secondary; trails/aligned along micro-fractures in the quartz; 8–20 µm; F: 0.8–0.85	Tm_{ice} = −4.7 to −0.1 °C; Th_t = 101 to 195 °C; salinity = 0.53–6.59 wt.% NaCl eq.; density = 0.87–0.97 g/cm^3	H_2O-NaCl
Type 5 (L + V); *n* = 8	Monzogranite; syenogranite	Sericitic alteration	Ms + Qz2 + Au1	Secondary; irregular to sub-rounded shaped; colorless; 15 and 50 µm; F: 0.75–0.8	Tm_{ice} = −4.2 to −0.8 °C; Th_t = 98 to 190 °C; salinity = 1.39–6.74 wt.% NaCl eq.; density = 0.92–0.97 g/cm^3	H_2O-NaCl
Type 6 (L; V)	Monzogranite; syenogranite	All assemblages		Colorless to greenish; rounded/oval to irregular; 4 and 20 µm; it occurs in all samples studied		H_2O-NaCl

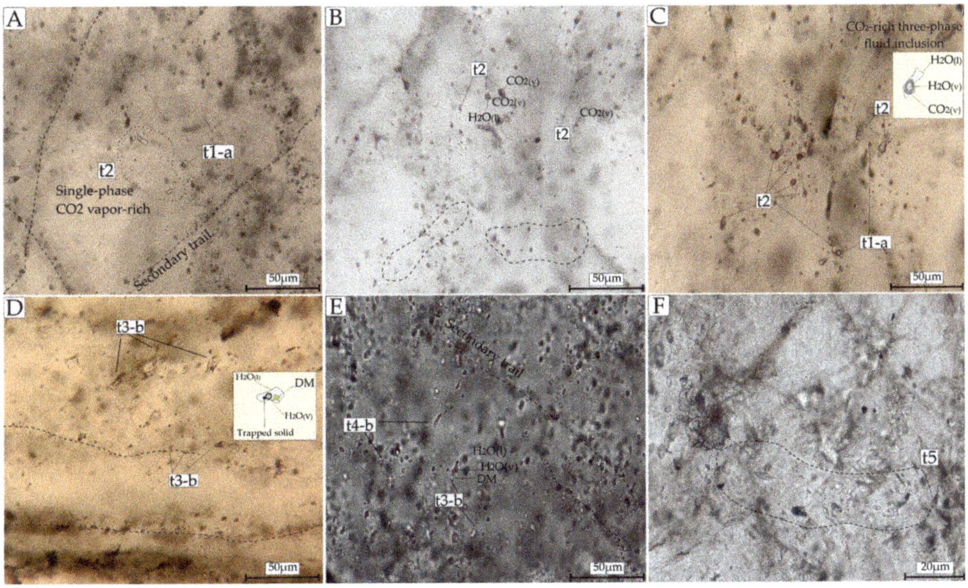

Figure 8. Fluid inclusions mapped in samples from the Raimunda deposit, Juruena Mineral Province, Brazil. (**A**) t1-a inclusions, primary liquid-rich aqueous and t2, primary, single-phase carbonic vapor-rich inclusions, occurring in clusters or isolated; (**B**) t2 inclusions, single-phase carbonic and two-phase aqueous-carbonic; (**C**) t1-a fluid inclusions occurring together with t2 fluid inclusions, two-phase and three-phase aqueous-carbonic; (**D**) tracks of secondary inclusions of type t3-b, three-phase (liquid, vapor and halite crystal– DM daughter mineral); (**E**) 4-b fluid inclusions, two-phase aqueous, and t3-b, three-phase aqueous inclusions; (**F**) t5 occurring in parallel tracks/lines and inclusions.

Type 4 (t4) are liquid-rich two-phase aqueous inclusions. Primary fluid inclusions (t4-a) in Qz3 are associated with sulfide stage. Secondary inclusions (t4-b) predominate in all samples studied and are associated with sericitic alteration (Qz2). In both types, the liquid phase is dominant, with an F of 0.8–0.85 and with a vapor bubble content of 15–20 vol%. They are colorless, 8 to 20 μm, and display rounded, sub-rounded, elongated, and irregular shapes. They are generally found as trails/aligned along micro-fractures in the crystal (t3-b), clusters, and isolated (t3-a) (Figure 8E).

Type 5 (t5) are secondary, liquid-rich two-phase aqueous (L + V) inclusions (Figure 8F). The degree of fill (F) is 0.75–0.8. They are irregular to sub-rounded, vary between 15 and 50 μm, and occur only in Qz2 which is associated with sericitic alteration.

Type 6 (t6) are single-phase liquid-rich and/or vapor-rich inclusions. They are colorless to greenish, rounded/oval to irregular, and locally occur along trails varying between 4 and 20 μm. Vapor-rich inclusions are dark, irregular, and vary between 2 and 15 μm. They occur in all samples studied.

4.3.2. Microthermometric Data

Table 2 summarizes the fluid inclusions microthermometric data of the Raimunda deposit. Microthermometry was carried out on all types of inclusions in quartz grains previously selected according to petrographic criteria. A total amount of 182 valid measurements were obtained (Table S8, Supplementary Material). The salinity and density of types 1, 4, and 5 aqueous fluid inclusions were calculated based on the equation of Bodnar [40], using the final ice melting temperature (Tm_{ice}) to estimate salinity in the H_2O–NaCl system. For the fluid inclusions interpreted to be of the system H_2O–NaCl–$CaCl_2$, salinities were calculated using the Steele-Macinnis et al. equation [41]. The salinities in H_2O–NaCl–CO_2

inclusions (type 2) were estimated using the final CO_2-clathrate melting temperatures [42], and the densities were estimated using the equations of Parry 1986 [43].

Type t1-a shows the temperature of the first visible melting (*Te*) ranging between −24.5 and −21.0 °C and encompasses inclusions with the final ice melting temperatures (Tm_{ice}) between −6.4 and −0.8 °C (Figure 9A). The total homogenization temperature (Th_t), into the liquid phase, took place between 233.5 and 471.2 °C (Figure 9C), with average values of 325–380 °C. The salinities data are between 1.40 and 9.72 wt.% NaCl eq. and the density values between 0.41 and 0.89 g/cm^3. In type t1-b, Tm_{ice} took place between −6.4 and −0.5 °C (Figure 9B), with an average from −4.2 to −0.5 °C and the Th_t varies from 200.5 to 297 °C (Figure 9D). The salinity varies from 0.88 to 6.74 wt.% NaCl eq. and densities of 0.71–0.92 g/cm^3.

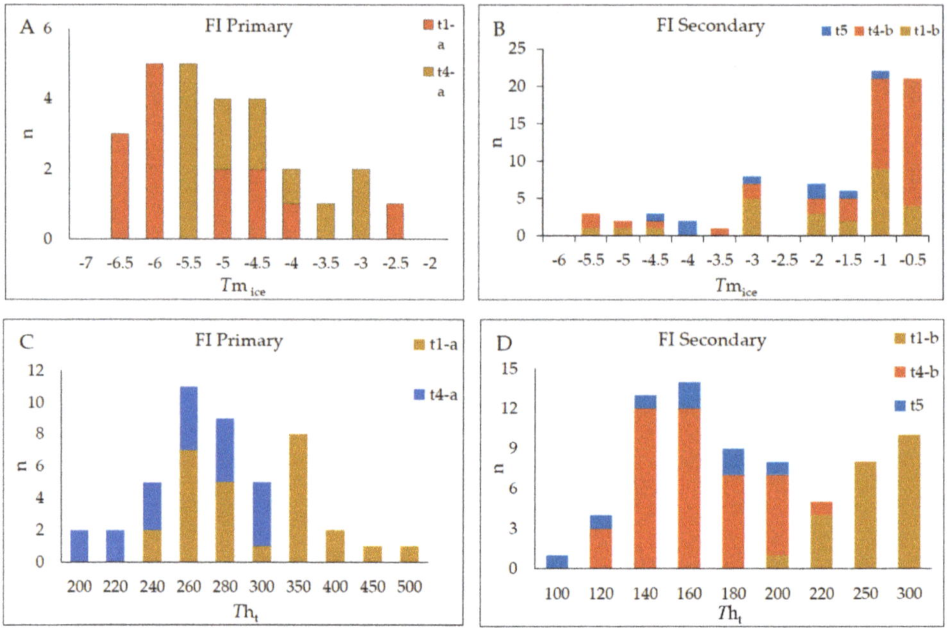

Figure 9. Microthermometric histograms of H_2O–NaCl aqueous fluid inclusions from the Raimunda deposit, Novo Mundo, Juruena Mineral Province. (**A**,**B**) Variation of Tm_{ice} vs. frequency of primary (t1-a and t4-a) and secondary (t1-b, t4-b and t5) inclusions; (**C**,**D**) Th_t of primary (t1-a and t4-a) and secondary (t1-b, t4-b and t5) inclusions.

Type 2 has average CO_2 melting temperature (Tm_{CO2}) values ranging between −56.8 and −56.6 °C for two-phase and three-phase aqueous-carbonic inclusions (Figure 10A). For carbonic inclusions, Tm_{CO2} is −57.3 to −57 °C. These values vary to either more positive or more negative values compared to the pure H_2O-CO_2 system (−56.6 °C). This points out the presence of other volatile compounds apart from CO_2 (e.g., CH_4 and N_2). The clathrate melting temperatures (Tm_{Cla}) are between 6.2 and 8.5 °C and the CO_2 homogenization temperatures (Th_{CO2}) vary between 26.8 and 30.1 °C for aqueous-carbonic inclusions, and between 19 and 26 °C for single-phase carbonic inclusions. Th_t varies from 210 to 390 °C, but the main range for three-phase (L_{H2O} + L_{CO2} + V_{CO2}) inclusions take place between 280 and 340 °C; and, for two-phase (L_{H2O} + V_{CO2}) inclusions, the main temperature range is 300–364.7 °C. Salinities are between 3.00 and 5.94 wt.% NaCl eq. and the densities range from 0.87 to 0.88 g/cm^3.

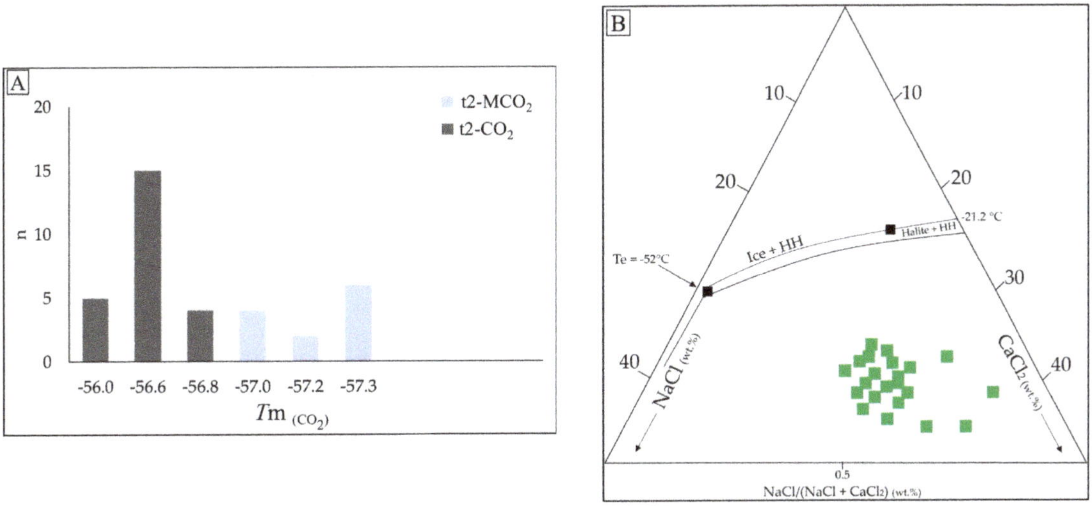

Figure 10. (**A**) Fluid inclusions microthermometric Tm_{CO2} histogram vs. frequency; (**B**) Ternary diagram showing the composition of t3 fluid inclusions (green square boxes) in terms of mass percent equivalent in NaCl and $CaCl_2$ [41].

Type t3-a primary inclusions showed Tm_{hh} values ranging between −10.7 and −3.8 °C. These inclusions homogenized into the liquid and vapor phase between 290.5 and 367.5 °C (Th_t). The salinities are between 38.88 and 50.25 wt.% NaCl + $CaCl_2$ eq. with densities varying from 0.78 to 0.82 g/cm^3.

Type t3-b inclusions show the first observed melting temperature range between −55 to −51.2 °C, interpreted as a eutectic temperature (*Te*) and suggesting the composition of the H_2O-NaCl-$CaCl_2$ system [41]. The final melting of ice is between −43.2 and −37.8 °C, and the hydrohalite dissolution (Tm_{hh}) occurs between −10.7 and −1.1 °C, with a main interval between −9 to −5 °C. During heating, the vapor bubble homogenizes first (132 to 202 °C), and, as temperature increases, daughter minerals homogenize (Th_t) in the range of 192–278.6 °C, with an average between 210 and 250 °C. The salinities show range values of 33.02–38.21 wt.% NaCl + $CaCl_2$ eq. and density values range from 0.87 to 0.97 g/cm^3. The H_2O–NaCl–$CaCl_2$ ternary diagram plots the composition for type t3–b inclusions at the Raimunda deposit based on Steele-MacInnis et al. [41] (Figure 10B).

Type 4-a primary inclusions display Tm_{ice} values between −5.4 and −1.0 °C (Figure 9A) and Th_t took place in the range between 200 and 288 °C (Figure 9C). The corresponding salinity of 1.56–8.41 wt.% NaCl eq. and the density was estimated between 0.74 and 0.92 g/cm^3. Type t4-b secondary inclusions yielded Tm_{ice} values in the range of −4.7 and −0.1 °C (Figure 9B). This type homogenized into the liquid phase between 101 and 195 °C (Figure 9D). The salinity is between 0.53 and 6.59 wt.% NaCl eq. with densities between 0.74 and 0.92 g/cm^3.

In type 5, the Tm_{ice} values fall between −4.2 and −0.8 °C and total homogenization temperatures (Th_t) into the liquid phase range between 98 and 190 °C. The salinity is between 1.39–6.74 wt.% NaCl eq. and densities between 0.92 and 0.97 g/cm^3 (Figure 9B,D). This type homogenized both into the liquid and vapor phase and the latter showed some inclusions crepitating while other inclusions did not return to their original state before heating.

4.4. Stable Isotopes Data

Stable isotope data are summarized in Table 3. The oxygen data were obtained for Qz2, which is associated with sericitic alteration (Ms-Qz2-Py2-Au1), and for Qz3, which is related to the sulfide stage (Py3-Au2-Ccp-Cc-Bin and Cu sulfides); sulfur data were obtained for Py2 (sericitic alteration) and Py3 (sulfide stage), both associated with the mineralization stages at the Raimunda deposit (Table 1).

Table 3. Isotopic composition of hydrothermal minerals and fluids from the Raimunda deposits, Novo Mundo granite, Juruena Mineral Province, Brazil.

Sample	Mineral	Hydrothermal Alteration Stage	$\delta^{34}S_{Py}$ (‰)	$\delta^{18}O_{Qz}$ (‰)	$\delta^{18}O_{H2O}$ (‰) [44]	Estimated Isotopic Equilibrium Temperature (°C) from Fluid Inclusion Data
F75-05	Qz2	Sericitic alteration		9	3.11–3.99	340–370
F70-02	Qz2	Sericitic alteration		9.6	3.72–4.59	340–370
F73-02	Qz2	Sericitic alteration		10.3	4.72–5.29	340–370
F70-03	Py2	Sericitic alteration	−0.5			340–370
F70-03	Qz2	Sericitic alteration		11.5	5.62–6.50	340–370
F73-03	Py3	Sulfide stage	0.1			325–380
F73-03	Qz3	Sulfide stage		12.6	6.22–7.86	325–380
F48-A15	Py3	Sulfide stage	−1			325–380
F48-A15	Qz3	Sulfide stage		12.2	5.82–7.46	325–380
F70-01	Py3	Sulfide stage	−0.1	11.1	4.73–6.36	325–380
F78-05	Py3	Sulfide stage	−0.7			325–380
F78-05	Qz3	Sulfide stage		11.6	5.23–6.85	325–380
F73-04	Py3	Sulfide stage	−1.4			325–380
F73-04	Qz3	Sulfide stage		10.7	4.32–5.95	325–380

The $\delta^{18}O$ values of quartz vary between 9.0 and 12.6‰. The isotopic values of the fluid ($\delta^{18}O_{H2O}$) were calculated using the total homogenization temperature range (Th_t) of primary t1 fluid inclusions (325–380 °C) as the isotopic equilibrium temperature for the main mineralized phase, sulfide stage, and the chlorite geothermometric temperature (340–370 °C) for the sericitic alteration stage.

These fluid inclusions are associated with the main mineralization stage at the Raimunda deposit. For the quartz-water system, the oxygen isotopic fractionation factor and the isotopic balance equation were used [44]; for oxygen, the following formula was applied to determine the temperature range between 250 and 500 °C: $1000\ \ln\alpha_{Qz\text{-}Water} = 3.34 \times (10^6T^{-2}) - 3.31$.

$\delta^{34}S_{py}$ values for sulfides slightly varied between −1.4 and +0.1‰. Pyrite associated with sericitic alteration (Py2) yielded a $\delta^{34}S$ value of −0.5‰; pyrite from massive sulfide zones (Py3) provided $\delta^{34}S$ values from −1.4 to +0.1‰. The isotopic fractionation factor between pyrite–H_2S was retrieved from [45].

5. Discussion

5.1. Tectonic Settings of Host Rocks, Hydrothermal Zonation and Mineralization

The Au (±Cu) Raimunda hydrothermal deposit is hosted in the monzogranite (2029 ± 4 Ma) and syenogranite (1987 ± 7.4 Ma to 1964 ± 1 Ma) facies, which are part of the Cuiú-Cuiú granite complex (2.03 and 1.98 Ga), eastern region of the Juruena Mineral Province [24]. These rocks are related to a plutonic-volcanic event in volcanic arc settings [46].

The zones more intensely affected by hydrothermal alteration are indicative of ore precipitation at intermediate to shallower levels. This is corroborated by pressure-release structures and vuggy quartz as well as breccias and brittle-like structures typical of shallower depths (Figure 5E).

Gold-rich mineral deposits in Novo Mundo were studied by [19,22]. These authors found disseminated gold hosted in intensely hydrothermally-altered syenogranite and monzogranite crosscut by andesitic to basaltic dikes. Those authors observed that mineralization was associated with intense phyllic alteration overprinting strong pervasive potassic alteration [22]. These other deposits share strong similarities with the Raimunda deposit, which might be indicative of similar mineralization events in the region. Other neighboring mineral deposits are hosted in 1.99–1.97 Ga granites [22].

In Novo Mundo, the mineralogical composition, the magmatic and hydrothermal textural features, the mineral parageneses, the cutting and overprinting relationships among host rocks and alteration and mineralization zones, microthermometric data, as well as chlorite and muscovite chemistry, allow us to propose the following paragenetic sequences for the Raimunda deposit presented in Table 1.

The evolution of the mineralizing hydrothermal system goes through the development of sericitic alteration (muscovite + pyrite2 + quartz2 + gold1), a lower-temperature stage compared to the previous ones (Na-metasomatism, microclinization, and propylitic alteration). Sericitic alteration represents the first mineralization stage, with a temperature estimate between 340 and 370 °C based on chlorite chemistry (Table S6, Supplementary Material).

This stage is followed by the main ore precipitation stage, with average temperature estimates between 325 and 380 °C, based on the homogenization temperature of type 1-a fluid inclusions. The following post-mineralization stage is marked by cooling and de- compression of hydrothermal fluids [47] and is represented by chloritization–carbonate alteration, with average temperatures of 262–280 °C, based on chlorite geothermometer (Chl3).

5.2. Origin, Hydrothermal System, and Evolution of Fluid Composition

Isotopic and fluid inclusion data indicate an evolution from hot, medium-salinity fluid (>450 °C and $\leq$10 wt.% NaCl eq.) to cooler, medium- to low-salinity fluids (<300 °C and $\leq$5 wt.% NaCl eq.).

The petrographic characteristics and fluid inclusion microthermometric data reveal the existence of two fluids: a low to medium salinity H_2O–NaCl–CO_2 aqueous-carbonic fluid (3.0–5.94 wt.% NaCl eq.) and a low-temperature high-salinity H_2O–NaCl–$CaCl_2$ aqueous late fluid (33.02–50.25 wt.% NaCl + $CaCl_2$ eq.), and that make up the fluid system of the Au ($\pm$Cu) Raimunda deposit.

The CO_2 content in aqueous-carbonic and carbonic fluid inclusions might be related to hydrolysis during hydrothermal alteration of the host rock (propylitic alteration), which consumed water from the fluid, enriching it in CO_2. The high homogenization temperatures observed at these inclusions (up to 472 °C) corroborate this interpretation.

Aqueous fluids are divided into two chemical systems, based on the relationship *Te* versus salinity: (1) H_2O-NaCl fluid in Qz2 and Qz3, with *Te* between −24.5 °C and −21.4 °C, and salinity between 0.87 and 9.72 wt.% NaCl eq.; (2) H_2O-NaCl-$CaCl_2$ fluid in Qz3 and Qz4, with the first melting temperature observed, interpreted as a *Te*, between −55 and −51.2 °C, and salinity between 33.02–50.25 wt.% NaCl + $CaCl_2$ eq. (Table S8, Supplementary Material).

The saturated fluid displays homogenization temperatures Th_t (290.5–367.5 °C, for t3-a; 200–250 °C, for t3-b main interval) with complete dissolution of the daughter mineral, thereby indicating homogeneous entrapment. And, despite the occurrence of two primary fluids (H_2O–CO_2 and H_2O–NaCl) deriving from a single fluid and the variable homogenization into the vapor and liquid phases (t3 and t5), theses inclusion types do not belong to the same assembly (FIA), and the homogenization occurs in a medium to a large range of temperature, which is compatible with a homogeneous entrapment.

The occurrence of brines in aqueous and aqueous-carbonic primary fluid inclusions can yield the fluid mixing hypotheses, leading to brine dilution and an increase in salinity for aqueous-carbonic fluid, therefore, explaining the presence of the late-carbonate alteration at the deposit. Fluid mixing may have taken place at the time of the formation of Qz3, with ongoing brine activity through the formation of Qz4 at a lower temperature, when the aqueous-carbonic fluid began to reduce its activity and precipitate calcite, which is corroborated by late hydrothermal alteration (Cal + Chl3).

Thus, the evolution of the medium-salinity and very hot system (>470 °C) is linked to the stages of hydrothermal alteration prior to mineralization (sodic metasomatism, microclinization, and propylitic alteration); and the medium to low salinity (<10 wt.% NaCl eq.) cool system or progressively cooler (<380 °C) is associated with the main to late alteration stages sericitic alteration and contemporaneous sulfide stage associated with silica veins and carbonate alteration). Furthermore, the transition from microclinization into late carbonate alteration can suggest a progressive increase in pH values [48–50], which likely lasted throughout ore precipitation.

The calculated isotopic composition of the hydrothermal fluid responsible for both mineralization stages shows $\delta^{18}O_{H2O}$ values for stage 2 in the range between +4.32 and +7.86‰ (sulfide stage), while those calculated for stage 1 vary between +3.11 and +6.49‰ (sericitic alteration) (Table 3), which is coherent to isotopic compositions of fluids associated with felsic magmas [51] and are compatible with the magmatic compositional isotopic range expected for world-class porphyry-type magmatic-hydrothermal deposits, with a homogeneous isotopic fluid ($\delta^{18}O_{H2O}$, 6–10‰) [52–55]. The results are similar to paragenetic and fluid inclusion data, and point to mixing between magmatic and meteoric fluids, that have values within the $\delta^{18}O$ range of −10 to −20‰ according to Craig [56] but remained predominantly with characteristics of magmatic fluid [57].

Figure 11 compares $\delta^{18}O_{H2O}$ values, based on temperatures provided by chlorite chemistry, which is associated with sericitic alteration and the first mineralization stage, and $\delta^{18}O_{H2O}$ values based on homogenization temperatures from fluid inclusions associated with the main mineralization stage at the Raimunda deposit. The two groups show an isotopic range similar to those obtained for magmatic-hydrothermal deposits with mantle-derived fluid.

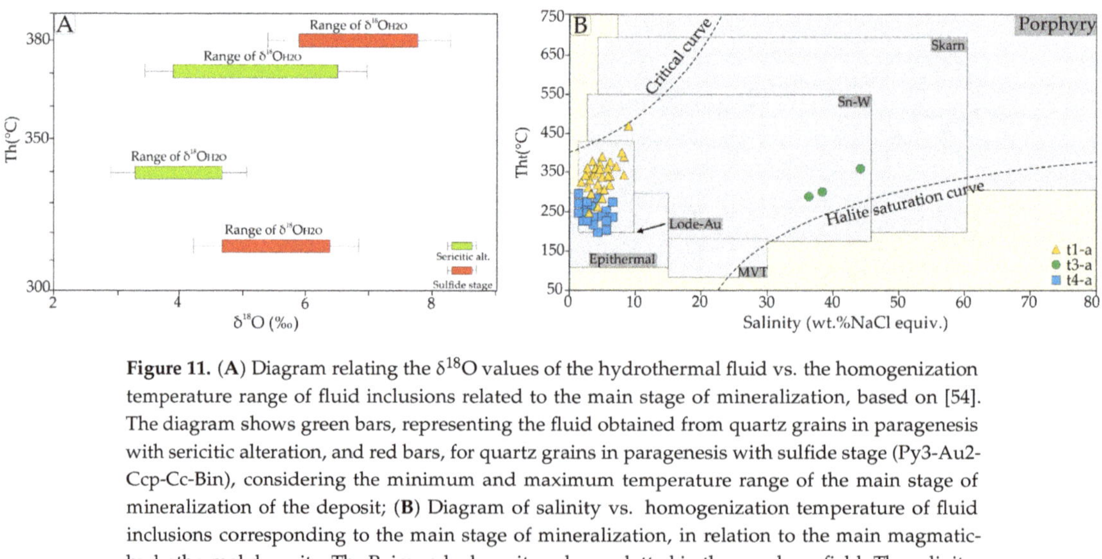

Figure 11. (**A**) Diagram relating the $\delta^{18}O$ values of the hydrothermal fluid vs. the homogenization temperature range of fluid inclusions related to the main stage of mineralization, based on [54]. The diagram shows green bars, representing the fluid obtained from quartz grains in paragenesis with sericitic alteration, and red bars, for quartz grains in paragenesis with sulfide stage (Py3-Au2-Ccp-Cc-Bin), considering the minimum and maximum temperature range of the main stage of mineralization of the deposit; (**B**) Diagram of salinity vs. homogenization temperature of fluid inclusions corresponding to the main stage of mineralization, in relation to the main magmatic-hydrothermal deposits. The Raimunda deposit analyses plotted in the porphyry field. The salinity-temperature curves and fields are based on [58].

The coexistence of t1-a and t2 primary inclusions at the deposit, with similar temperatures, can indicate simultaneous fluid entrapment. Phase changes among the different types of inclusions may suggest that fluid immiscibility preceded their entrapment; homogenization temperatures decrease from type t1-a to t4-a on the salinity vs. homogenization temperatures diagram (Figure 11B), which is likely due to progressive cooling and/or mixing with relatively lower-temperature fluids, such as meteoric water [58].

$\delta^{34}S$ data directly reflects its source given that under proper redox and pH conditions for precipitation, much or all of the sulfur available in the system may have been converted into H_2S at the first mineralization stage [59]. In turn, once the second mineralization stage is composed of sulfides only, the isotopic composition of sulfur in pyrite matches that of the sulfur source [59].

Figure 12 correlates $\delta^{34}S$ values of the Raimunda deposit (−1.4 to +0.1‰) with the isotopic ranges of several deposits in the Tapajós and Juruena Provinces, as well as with world-class deposits and deposits and geological environments available in the literature and are compatible with those of other world-class magmatic-hydrothermal deposits [60–62].

Reference values of $\delta^{34}S$ from [63], combined with petrological and mineral textural evidence at the Raimunda deposit, are indicative of mineralization of magmatic origin, which may be related to the fluids that hydrothermally altered the Novo Mundo granite. This is corroborated by the similarity between $\delta^{34}S$ values and hydrothermal patterns at the Raimunda deposit and those reported for other world-class magmatic-hydrothermal deposits and/or granite-related mineral deposits [64–68].

The occurrence of primary gold in the Juruena Mineral Province demonstrates a close spatial relationship with (i) oxidized I-type calc-alkaline granites, (ii) volcanic rocks, and (iii) volcano-sedimentary sequences linked to continental magmatic arcs [4]. The common occurrence of native Au as inclusions in pyrite and chalcopyrite is indicative of early Au oversaturation in the fluid which we relate to sulfide precipitation and phase separation, destabilizing the $AuHS^0$ or $Au(HS)^{2-}$ complex, leading to the accumulation of Au particles [67,69].

At the Raimunda deposit, the ore is composed of native gold and has a mantle-like polymetallic signature (Au ± Ag ± Cu ± Bi ± Te ± Pb) associated with pyrite, chalcopyrite, chalcocite, larosite, bismuthinite and other bismuth sulfides emplaced at a zone of intense sulfidation.

This copper-rich sulfide ore assemblage suggests that the Raimunda deposit formed under fO_2 conditions above the PPM buffer, pyrrhotite + pyrite + magnetite, which is close to $mSO_2/mH_2S = 1$ [69,70]. It is compatible with moderate to high fO_2 conditions where fluids oscillate from low to high temperatures; the latter being represented by moderate-salinity aqueous-carbonic fluids (up to ≤370 °C and up to ≤10 wt% NaCl eq.). This scenario suggests that ore precipitation is linked and/or similar to porphyry magmatic-hydrothermal systems [71].

Hence, our hypothesis is that during the final stage of granite crystallization (~2.03–1.98 Ga), high-salinity and higher-temperature single-phase fluids, with an intermediate density, were exsolved from a magma at high depth. These fluids went through immiscibility and separated into H_2O-NaCl and CO_2-rich fluids. During their ascent, fluids decompressed and cooled down [56], while ions (e.g., Au^+, Ag^+, Te^{2-} and S^{2-}) precipitated along favorable structures, under proper P–T conditions, giving rise to the first mineralization stage.

Other ore-forming elements (i.e., Au, Cu^+, Ag^+, Fe^{2+}, S^{2-} and Te^{2+}) were exsolved from the magmatic fluid and precipitated along microfractures and stockwork sulfide veins.

This last stage gave rise to the main stage of mineralization at the deposit, which is coherent with the ore mineral paragenesis.

The ore precipitation was triggered by the cooling of magmatic fluid expelled from magmatic chambers at shallower depths [67]. This model outlines that the input of meteoric water would only affect peripheral and/or post-mineralization hydrothermal alteration, such as propylitic alteration [57,67], without playing a significant role in the main mineralization stage, and the homogeneous $\delta^{18}O_{fluid}$ values (3.11–7.86‰) corroborates this scenario. The relative depth of gold precipitation and, consequently, deposit formation can be helpful to preserve the Raimunda deposit until these days (Figure 13).

All these features are compatible with those of porphyry magmatic-hydrothermal deposits, commonly found in Phanerozoic volcano-plutonic settings, e.g., [72]. Although they lack some classic features of Phanerozoic analogous surrounding the Raimunda deposit, which may be due to variable erosion levels, it is likely associated with mineralizing events reported earlier for the Tapajós-Parima Paleoproterozoic Province [8].

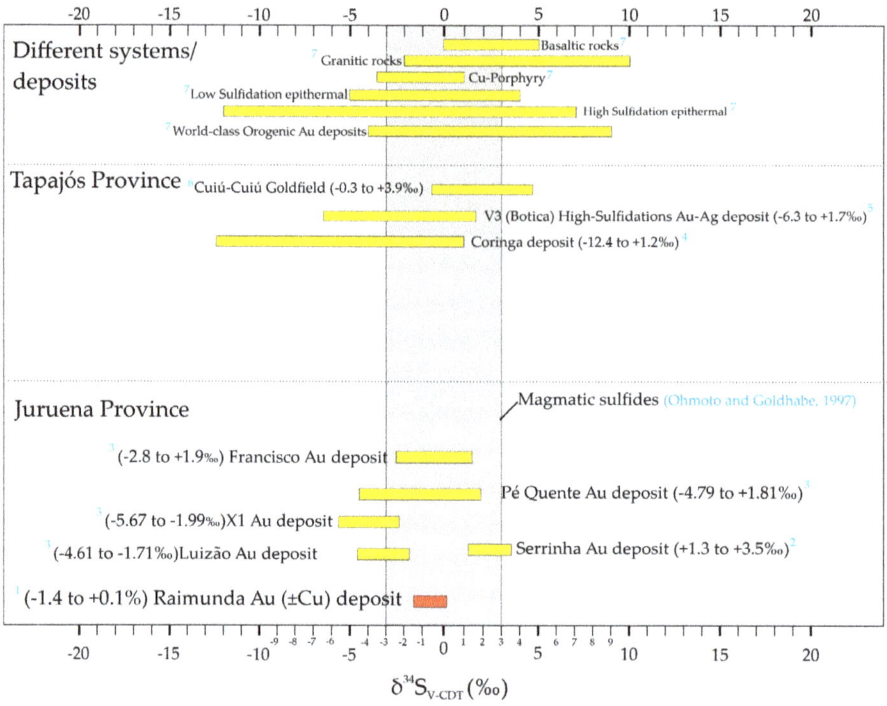

Figure 12. Sulfur isotope values ($\delta^{34}S_{CDT}$) of different geological mineral deposits, isotopic data from the Tapajós and Juruena Province, Amazonian Craton. Numbers: 1—this study; 2—[2]; 3—[19]; 4—[8]; 5—[68]; 6—[71]; 7—Mineral deposits from Shanks [51]. The magmatic sulfides range (−3 to +3‰) corresponds to [62].

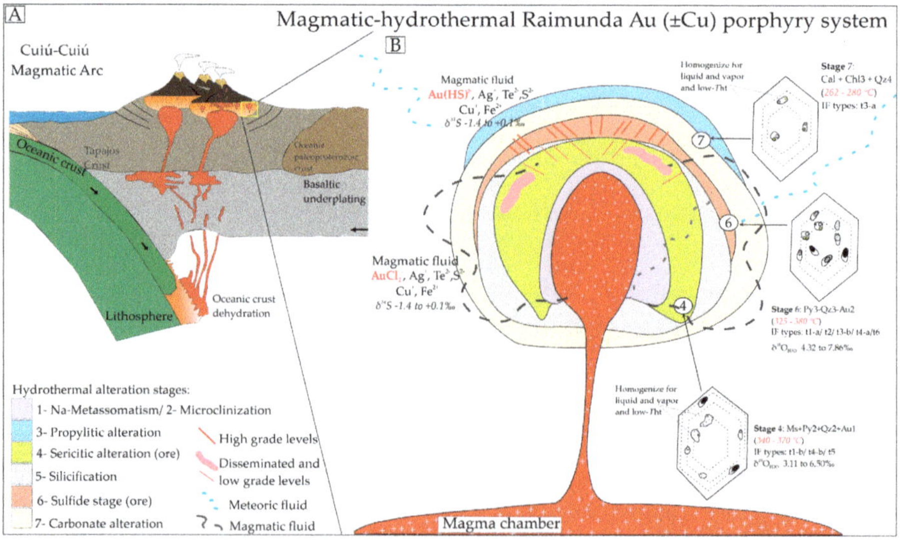

Figure 13. Graphic mineralization model showing the magmatic and hydrothermal periods of the Raimunda Au (±Cu) deposit, Juruena Mineral Province, Amazonian Craton. (**A**) The schematic

representation of granite generation to Novo Mundo granites [24] during the evolution of the Paleoproterozoic Cuiú-Cuiú magmatic arc (2.03–1.96 Ga); (**B**) The hydrothermal and fluid evolution, indicating the main fluid stages, temperatures and isotope composition of Raimunda deposit. The zonation is only illustrative, as there is an overlap of stages in the Raimunda system.

6. Conclusions

The current study allows us to draw the following conclusions about the Au (±Cu) Raimunda deposit:

The gold mineralization is hosted in arc I-type calc-alkaline, peraluminous to metaluminous granitic rocks, subdivided into four granite facies (monzonite, monzogranite, biotite granodiorite, and syenogranite), with U–Pb zircon ages of 2.03–1.98 Ga, named Novo Mundo Complex, eastern region of the Juruena Mineral Province, Amazonian Craton.

The hydrothermal system resembles that of mineral deposits previously studied in the Novo Mundo Complex and in the eastern region of the province, which reveals an important regional association.

Hydrothermal alteration data and mineral and fluid chemistry data allow us to establish the following relationships: (a) the initial stages of alteration indicate temperatures > 470 °C; (b) progressive evolution of the hydrothermal system led to a temperature decrease that culminated into the first mineralization stage associated with sericitic alteration (340 and 370 °C); (c) it was followed by the main stage of ore precipitation, sulfide stage (Py3-Ccp-Cc-Bin-Au2), with average temperatures of 325–380 °C, as demonstrated by the homogenization temperature of type 1-a primary fluid inclusions; (d) the post-mineralization stage, marked by carbonate alteration and chloritization (Chl3), yielded average temperatures from 262 to 280 °C for the system.

Au (±Cu) mineralization occurs as Ag-rich (25–28%, electrum) gold inclusions and fracture-fills in pyrite, and the main mineralization stage mineralization at the deposit is composed of the following sulfide assemblage: pyrite, chalcopyrite, chalcocite, larosite, bismuthinite, hodrushite, krupkaite, and wittchenite.

The fluid inclusion study identified nine different subtypes of inclusion, varying from H_2O–NaCl, H_2O–NaCl–CO_2, and H_2O–NaCl–$CaCl_2$ model systems. Based on the relationship between *T*e versus salinity, it was possible to identify two chemical systems in the Raimunda deposit: (1) H_2O–NaCl fluid in Qz2 and Qz3, with *T*e between −24.5 and −21 °C and salinity between 0.87 and 9.72 wt.% NaCl eq.; and (2) H_2O–NaCl–$CaCl_2$ fluid in Qz3 and Qz4, with *T*e between −55.3 and −51.2 °C, and salinity of 33.02–50.25 wt.% NaCl + $CaCl_2$ eq.

The Raimunda deposit evolved from essentially high to medium temperature (≤472 °C) and medium salinity (≤10 wt.% NaCl eq.) magmatic fluids, mixed with fluids meteoric ($Th_{t:}$ 98–190 °C; lower salinity: 1.39–6.74 wt.% NaCl eq.).

Ore precipitated under considerably high fS_2 and fO_2 conditions, and relatively medium to high temperatures (e.g., homogenization temperature of t1-a primary fluid inclusion suggests ~325 to 380 °C).

The isotopic composition of the mineralizing fluid points out to magmatic fluid, with $\delta^{18}O_{H2O}$ values of 3.11–7.86‰, indicating magmatic origin with a meteoric fluid contribution. The main stage of ore deposition was triggered by the instability of magmatic fluids with $\delta^{34}S_{py}$ values between −1.4 to +0.1‰, similar to several regional and world-class magmatic-hydrothermal deposits.

The overall features of the Raimunda deposit are genetically compatible with those of magmatic-hydrothermal systems related to Au-mineralized porphyry formed from oxidized granitic magmas. These findings have implications for mineral exploration of gold and associated metals in the Juruena Mineral Province and Tapajós Province, Amazonian Craton.

Supplementary Materials: The following supporting information can be downloaded at: https://www.mdpi.com/article/10.3390/min14121185/s1, Table S1: Electron microprobe analyses results for pyrite crystals to Raimunda Au (±Cu) deposit, Novo Mundo, Juruena Mineral Province, Brazil, Table S2: Electron microprobe analyses results for chalcopyrite to Raimunda Au (±Cu) deposit, Novo Mundo, Juruena Mineral Province, Brazil, Table S3: Electron microprobe analyses results for gold to Raimunda Au (±Cu) deposit, Novo Mundo, Juruena Mineral Province, Brazil, Table S4: Electron microprobe analysis of chalcocite crystals and other copper-rich minerals in samples from Raimunda Au (±Cu) deposit, Novo Mundo, Juruena Mineral Province, Brazil, Table S5: Electron microprobe analysis of bismutinita and other copper-rich minerals in samples from Raimunda Au (±Cu) deposit, Novo Mundo, Juruena Mineral Province, Brazil, Table S6: Electron microprobe analyses results for chlorite from the Raimunda Au (±Cu) deposit, Novo Mundo, Juruena Mineral Province, Brazil. Abrreviations: CAT: Cathelineau (1988) [29]; CAT&N: Cathelineau & Nieva (1985) [28]; Z&F: Zang & Fyfe (1995) [37], Table S7: Electron microprobe analyses results for muscovite to Raimunda Au (±Cu) deposit, Novo Mundo, Juruena Mineral Province, Brazil, Table S8: Summary of microthermometric data of fluid inclusions from the Raimunda deposit, Novo Mundo, Juruena Mineral Province.

Author Contributions: Conceptualization, A.A.C.L. and M.A.M.; methodology, A.A.C.L.; software, A.A.C.L.; validation, A.A.C.L. and M.A.M.; formal analysis, A.A.C.L.; investigation, A.A.C.L. and M.A.M.; resources, A.A.C.L.; data curation, M.A.M.; writing—original draft preparation, A.A.C.L.; writing—review and editing, M.A.M.; supervision, M.A.M.; project administration, M.A.M.; funding acquisition, M.A.M. All authors have read and agreed to the published version of the manuscript.

Funding: This research was funded by the Coordenação de Aperfeiçoamento de Pessoal de Nível Superior—Brazil (CAPES)—Grant numbers 001 and by the Conselho Nacional de Desenvolvimento Científico e Tecnológico, Brazil (CNPq) (Grant numbers 426769/2016-3), provided funding for research.

Data Availability Statement: Data is contained within the article and Supplementary Materials.

Acknowledgments: The authors are grateful to mining companies Nexa Resource and Bemisa Holding S.A. for their logistical support and donation of materials to carry out this study; CAPES—Coordenação de Aperfeiçoamento de Pessoal de Nível Superior (Brazil) for the financial resources for assistance of laboratory analysis; CNPq—Conselho Nacional de Desenvolvimento Científico e Tecnológico, Brazil, for the financial resources for assistance of laboratory and field analysis.

Conflicts of Interest: The authors declare no conflicts of interest.

References

1. Dardenne, M.A.; Schobbenhaus, C. *Metalogênese do Brasil*, 1st ed.; Universidade de Brasília: Brasília, Brazil, 2001; p. 394.
2. Moura, M.A.; Botelho, N.F.; Olivo, G.R.; Kyser, T.K. Granite-related Paleoproterozoic, Serrinha gold deposit, Southern Amazonia, Brazil: Hydrothermal alteration, fluid inclusion and stable isotope constraints on genesis and evolution. *Econ. Geol.* **2006**, *101*, 585–605. [CrossRef]
3. Paes de Barros, A.J. Granitos da Região de Peixoto de Azevedo—Novo Mundo e Mineralizações Auríferas Relacionadas—Província Aurífera Alta Floresta (MT). Ph.D. Thesis, Instituto de Geociências, Universidade Estadual de Campinas, Campinas, Brazil, 2007.
4. Juliani, C.; Vasquez, M.L.; Klein, E.L.; Villas, R.N.; Echeverri-Misas, C.M.; Santiago, E.S.B.; Monteiro, L.V.S.; Carneiro, C.C.; Fernandes, C.M.D.; Usero, G. Metalogenia da Província Tapajós. In *Metalogênese das Províncias Tectônicas Brasileiras*, 1st ed.; Silva, M.G., Jost, H., Kuyumajian, R.M., Eds.; CPRM-Serviço Geológico do Brasil: Brasília, Brazil, 2014; pp. 51–90.
5. Juliani, C.; Assis, R.R.; Monteiro, L.V.S.; Fernandes, C.M.D.; Martins, J.E.Z.S.; Costa, J.R.C. Gold in Paleoproterozoic (2.1 to 1.77 Ga) Continental Magmatic Arcs at the Tapajós and Juruena Mineral Provinces (Amazonian Craton. Brazil): A New Frontier for the Exploration of Epithermal-Porphyry and Related Deposits. *Minerals* **2021**, *11*, 714. [CrossRef]
6. Mesquita, M.J.; Gomes, M.E.B.; Moreira, I.C.; Paes, R.A.S.; Hevelyn, E.S.M.; Matos, J.H.; Ruggiero, A.; Primo, G.; Ducart, D.F.; Poggi, L.; et al. Paleoproterozoic gold deposits at Alta Floresta Mineral Province, Brazil: Two overprinted mineralizing events? In *Recent Advances in Understanding Gold Deposits: From Orogeny to Alluvium*; Torvela, T., Lambert-Smith, J.S., Chapman, R.J., Eds.; Special Publications; Geological Society of London: London, UK, 2022; Volume 516, pp. 109–154. [CrossRef]
7. Jamasmie, C. Anglo American 'Encouraged' by Drilling Results from Copper Asset in North Brazil. Available online: https://www.mining.com/anglo-american-encouraged-drilling-results-copper-asset-north-brazil/ (accessed on 18 November 2024).
8. Guimarães, S.B.; Klein, E.L.; Harris, C.; Costa, I.S.L. Metallogenesis of the orosirian epithermal coringa gold-silver (Cu-Pb-Zn) deposit, southeastern Tapajós mineral Province, Amazonian craton, Brazil. *Ore Geol. Rev.* **2021**, *128*, 103908. [CrossRef]

9. Poggi, L.; Ducart, D.F.; Mesquita, M.J.; Moreira, I.C.; Gomes, M.E.B.; Filho, C.R.S. New Insights into the Evolution and Footprints of the Paraíba Au-Cu-Mo Deposit, Alta Floresta Mineral Province (Brazil), through Integration of Spectral and Conventional Methods. *Minerals* **2022**, *12*, 1327. [CrossRef]
10. Gaillard, N.; Williams-Jones, A.E.; Clark, J.R.; Salvi, S.; Perrouty, S.; Linnen, R.L.; Olivo, G.R. The use of lithogeochemistry in delineating hydrothermal fluid pathways and vectoring towards gold mineralization in the Malartic district, Québec. *Ore Geol. Rev.* **2020**, *120*, 103351. [CrossRef]
11. Tassinari, C.C.G.; Macambira, M.J.B. A evolução tectônica do Cráton Amazônico. In *Geologia do Continente Sul-Americano: Evolução da obra de Fernando Flávio Marques de Almeida*; Mantesso Neto, V., Ed.; Beca: São Paulo, Brazil, 2004; pp. 471–485.
12. Santos, J.O.S.; Hartmann, L.A.; Gaudette, H.E.; Groves, D.I.; McNaughto, N.; Fletcher, I.R. A new understanding of the provinces of the Amazonian Craton based on integration of sector mapping and U-Pb and Sm-Nd geochronology. *Gondw. Res.* **2000**, *3*, 453–488. [CrossRef]
13. Fernandes, C.M.D.; Juliani, C.; Monteiro, L.V.S.; Lagler, B.; Echeverri Misas, C.M. High-K calc-alkaline to A-type fissure-controlled volcano-plutonism of the São Félix do Xingu region, Amazonian craton, Brazil: Exclusively crustal sources or only mixed Nd model ages? *J. S. Am. Earth Sci.* **2011**, *32*, 351–368. [CrossRef]
14. Carneiro, C.D.C.; Juliani, C.; Carreiro-Araújo, S.A.; Monteiro, L.V.S.; Crósta, A.P.; Fernandes, C.M.D. New crustal framework in the Amazon Craton based on geophysical data: Evidences of deep east–west trending suture zones. *IEEE Geosci. Remote Sens. Lett.* **2019**, *16*, 20–24. [CrossRef]
15. Juliani, C.; Fernandes, C.M.D.; Monteiro, L.V.S. Características da subducção, paleoclima e eventos erosivos paleoproterozoicos (2.1–1.88 Ga) e seus efeitos na estruturação da parte sul do Cráton Amazônico. In *Simpósio de Geologia da Amazônia*, 14th ed.; Gorayeb, P.S.S., Teixeira, S.G., Eds.; Sociedade Brasileira de Geologia–Núcleo Norte: Marabá, Brazil, 2015; p. 81. (In Portuguese)
16. Fernandes, C.M.; Fontes, S.L.; La Terra, E.F.; Miquelutti, L.G.; Maurya, V.P. Electrical crustal structure of Alta Floresta Gold Province eastern sector, SW Amazon Craton, Brazil. *J. S. Am. Earth Sci.* **2018**, *86*, 244–258. [CrossRef]
17. Cassini, L.V.; Moyen, J.; Juliani, C. Orosirian magmatism in the Tapajós Mineral Province (Amazonian Craton): The missing link to understand the onset of Paleoproterozoic tectonics. *Lithos* **2020**, *356–357*, 105350. [CrossRef]
18. Alves, C.L.; Rizzotto, G.J.; Rios, F.S.; Barros, M.A.S. The Orosirian Cuiú-Cuiú magmatic arc in Peixoto de Azevedo domain, Southern of Amazonian Craton. *J. South Am. Earth Sci.* **2020**, *102*, 102648. [CrossRef]
19. Alves, C.L.; Rizzotto, G.J.; Rios, F.S.; Gonçalves, G.F. *Áreas de Relevante Interesse Mineral-Projeto Evolução Crustal e Metalogenia da Província Mineral Juruena-Teles Pires*; CPRM: Goiânia, Brazil, 2019; ISBN 978-85-7499-547-2.
20. Assis, R.R. Depósitos Auríferos Associados ao Magmatismo Félsico da Província de Alta Floresta (MT), Cráton Amazônico: Litogeoquímica, Idade das Mineralizações e Fonte dos Fluidos. Ph.D. Thesis, Instituto de Geociências, Universidade Estadual de Campinas, Campinas, Brazil, 2015.
21. Lagler, B.; Juliani, C.; Fernandes, C.M.D.; da Cruz, R.S.; Vieira, D.A.S. Paleoproterozoic volcanic caldera in the Amazonian Craton, northern Brazil: Stratigraphy, lithofacies characterization, and lithogeochemical constraints. *J. S. Am. Earth Sci.* **2019**, *95*, 102252. [CrossRef]
22. Assis, R.R.; Xavier, R.P.; Creaser, R.A. Linking the Timing of Disseminated Granite-Hosted Gold-Rich Deposits to Paleoproterozoic Felsic Magmatism at Alta Floresta Gold Province, Amazon Craton, Brazil: Insights from Pyrite and Molybdenite Re-Os Geochronology. *Econ. Geol.* **2017**, *112*, 1937–1957. [CrossRef]
23. Teixeira, N.; Borda, M.; Ganade, C.; Gaspar, J.C.; Chelle-Michou, C.; Porto, F. *Jaca Cu-Au (Mo) Pórfiro: Geocronologia e Isotopia do Sistema Magmático Mineralizante*; Sumário Executivo; CPRM: Goiânia, Brazil, 2023; 7p.
24. Testa, L.G.C. Geologia, Petrologia e Geocronologia dos Granitos da Região de Novo Mundo (MT) e Implicações para a Evolução Magmática e Regional. Master's Thesis, University of Brasília, Brasília, Brazil, 2018.
25. Miguel, E., Jr. Mineralizações Auríferas do Lineamento Peru-Trairão Província Aurífera de Alta Floresta—MT, Controle Estrutural e idade U–Pb das Rochas Hospedeiras. Master's Thesis, Instituto de Geociências, Universidade Estadual de Campinas, Campinas, Brazil, 2011.
26. Pirajno, F. *Hydrothermal Processes and Mineral Systems*, 1st ed.; Springer: Berlin, Germany, 2009; p. 1273.
27. Hedenquist, J.W.; Arribas, A. 2022. Exploration Implications of Multiple Formation Environments of Advanced Argillic Minerals. *Econ. Geol.* **2022**, *117*, 609–643. [CrossRef]
28. Cathelineau, M.; Nieva, D. A chlorite solid solution geothermometer. The Los Azufres geothermal system (Mexico). *Contrib. Miner. Petrol.* **1985**, *91*, 235–244. [CrossRef]
29. Cathelineau, M. Cation site occupancy in chlorites and illites as a fuction of temperature. *Clay Miner.* **1988**, *23*, 471–485. [CrossRef]
30. Roedder, E. Fluid inclusions. In *Reviews in Mineralogy*, 1st ed.; Mineralogical Society of America: Blacksburg, VA, USA, 1984; Volume 12.
31. Shepherd, T.J.; Rankin, A.H.; Alderton, D.H.M. *A Practical Guide to Fluid Inclusions Studies*; Blackie and Sons: Glasgow, UK; London, UK, 1985; p. 253.
32. Van den Kerkhof, A.M.; Hein, U.F. Fluid inclusion petrography. *Lithos* **2001**, *55*, 27–47. [CrossRef]
33. Chinchilla, D.; Arroyo, X.; Merinero, R.; Piña, R.; Nieto, F.; Ortega, L.; Lunas, R. Chlorite geothermometry applied to massive and oscillatory-zoned radiated Mn-rich chlorites in the Patricia Zn-Pb-Ag epithermal deposit (NE, Chile). *Appl. Clay Sci.* **2016**, *134*, 210–220. [CrossRef]
34. Hey, M.H. A new review of chlorites. *Miner. Mag.* **1954**, *30*, 277–292. [CrossRef]

35. Tischendorf, G.; Forster, H.-J.; Gottesmann, B.; Rieder, M. True and brittle micas: Composition and solid-solution series. *Miner. Mag.* **2007**, *71*, 285–320. [CrossRef]
36. Kranidiotis, P.; MacLean, W.H. Systematics of chlorite alteration at the Phelps Dodge massive sulfide deposit, Matagami, Québec. *Econ. Geol.* **1987**, *82*, 1898–1911. [CrossRef]
37. Zang, W.; Fyfe, W.S. Chloritization of the hydrothermally altered bedrock at the Igarapé Bahia gold deposit, Carajás, Brazil. *Miner. Depos.* **1995**, *30*, 30–38. [CrossRef]
38. Jowett, E.C. Fitting iron and magnesium into the hydrothermal chlorite geothermometer. In Proceedings of the GAC/MAC/SEG Joint Annual Meeting, Program with Abstracts, Toronto, Canada, 27–29 May 1991; Volume 16, p. A62.
39. Xie, X.; Byerly, G.R.; Ferrell, R.E. IIb trioctahedral chlorite from the Barberton greenstone belt: Crystal structure and rock composition constraints with implications for geothermometry. *Contrib. Mineral. Petrol.* **1997**, *126*, 275–291. [CrossRef]
40. Goldstein, R.H.; Reynolds, T.J. *Systematics of Fluid Inclusions in Diagenetic Minerals*; SEPM (Society for Sedimentary Geology) Short Course: Tulsa, OK, USA, 1994; p. 213.
41. Steele-MacInnis, M.; Bodnar, R.J.; Naden, J. Numerical model to determine the composition of H_2O–NaCl–$CaCl_2$ fluid inclusions based on microthermometric and microanalytical data. *Geochim. Cosmochim. Acta* **2022**, *75*, 21–40. [CrossRef]
42. Bodnar, R.J. Revised equation and table for determining the freezing point depression of H_2O-NaCl solutions. *Geochim. Cosmochim. Acta* **1993**, *57*, 683–684. [CrossRef]
43. Parry, W.T. Estimation of XCO_2, P and fluid inclusion volume from fluid inclusion temperature measurements in the system NaCL-CO_2-H_2O. *Econ. Geol.* **1986**, *81*, 1009–1013. [CrossRef]
44. Matsuhira, Y.; Goldsmith, J.R.; Clayton, R.N. Oxygen isotopic fractionation in the system quartz-albite-anorthite-water. *Geochim. Cosmochim. Acta* **1979**, *43*, 1131–1140. [CrossRef]
45. Ohmoto, H.; Rye, R.O. Isotopes of sulfur and carbon. In *Geochemistry of Hydrothermal Ore Deposit*, 3rd ed.; Barnes, H.L., Ed.; Wiley: Washington, DC, USA, 1979; pp. 509–561.
46. Guimarães, S.B.; Klein, E.L. Geochemical and isotopic constraints on the host rocks of the magmatic-hydrothermal Coringa gold-silver (Cu-Pb-Zn) deposit of the Tapajós mineral province, Amazonia craton, Brazil. *J. S. Am. Earth Sci.* **2020**, *103*, 102726. [CrossRef]
47. Monecke, T.; Monecke, J.; Reynolds, T.J.; Tsuruoka, S.; Bennett, M.M.; Skewes, W.B.; Palin, R.M. Quartz solubility in the H_2O-NaCl system: A framework for understanding vein formation in porphyry copper deposits. *Econ. Geol.* **2018**, *113*, 1007–1046. [CrossRef]
48. Baker, T. Emplacement depth and carbon dioxide-rich fluid inclusions in intrusion-related gold deposits. *Ec. Geol.* **2002**, *97*, 1111–1117. [CrossRef]
49. Chiaradia, M. Gold endowments of porphyry deposits controlled by precipitation efficiency. *Nat. Commun.* **2020**, *11*, 248. [CrossRef]
50. Reed, M.H. Hydrothermal alteration and its relationship to ore fluid composition. In *Geochemistry of Hydrothermal Ore Deposits*; Barnes, H.L., Ed.; John Wiley: New York, NY, USA, 1997; pp. 303–365.
51. Shanks III, W.C. Stable isotope geochemistry of mineral deposits. In *Treatise on Geochemistry*; Elsevier: Amsterdam, The Netherlands, 2014; pp. 59–85. [CrossRef]
52. Li, L.; Sun, F.; Bai, Y.; Wang, C.; Qian, Y.; Sun, Y.; Tan, S. Identification of hydrothermal alteration and mineralization in the Sancha magmatic Cu-Ni-Au sulfide deposit, NW China: Implications for timing and genesis of mineralization. *Ore Geol. Rev.* **2022**, *143*, 104770. [CrossRef]
53. Sheppard, S.M.F. Characterization and isotopic variations in natural waters. In *Stable Isotopes in High-Temperature Geological Processes*; Valley, J.W., Taylor, H.P., Jr., O'Neil, J.R., Eds.; Walter de Gruyter GmbH & Co KG: Berlin, Germany, 1986; Volume 16, pp. 165–183.
54. Zeng, G.; Xiang, W.; Zhang, Z.; Wang, J.; Jiang, J.; Wu, F. Fluid inclusions, H-O-S-Pb isotopes, and Rb-Sr geochronology of the Augaro gold deposit, West Eritrea, NE Africa: Implications for ore genesis. *J. S. Am. Earth Sci.* **2023**, *206*, 105029. [CrossRef]
55. Zhang, Z.; Wang, Y.; Liu, j.; Lin, S.; Zhang, F.; Zhang, W.; Bo, Z.; Li, H. Sulfur isotope, fluid inclusions, and quartz solubility model decipher multistage hydrothermal fluids evolution and ore precipitation mechanism at the Nantai porphyry Mo deposit, Qinling, Central China. *Chem. Geol.* **2023**, *617*, 121273. [CrossRef]
56. Craig, H. Isotopic variations in meteoric waters. *Science* **1961**, *133*, 1702–1703. [CrossRef] [PubMed]
57. Fekete, S.; Weis, P.; Driesner, T.; Bouvier, A.S.; Baumgartner, L.; Heinrich, C.A. Contrasting hydrological processes of meteoric water incursion during magmatic-hydrothermal ore deposition: Na oxygen isotope study by ion microprobe. *Earth Planet. Sci. Lett.* **2016**, *451*, 263–271. [CrossRef]
58. Wilkinson, J.J. Fluid inclusions in hydrothermal ore deposits. *Lithos* **2001**, *55*, 229–272. [CrossRef]
59. Cooke, D.R.; Deyell, C.L.; Waters, P.J.; Gonzales, R.I.; Zaw, K. Evidence for magmatic-hydrothermal fluids and ore-forming processes in epithermal and porphyry deposits of the Baguio district, Philippines. *Econ. Geol.* **2011**, *106*, 1399–1424. [CrossRef]
60. Xiong, S.F.; Gong, Y.J.; Jiang, S.Y.; Zhang, X.J.; Li, Q.; Zeng, G.P. Ore genesis of the Wusihe carbonate-hosted Zn-Pb deposit in the Dadu River Valley district, Yangtze Block, SW China: Evidence from ore geology, S-Pb isotopes, and sphalerite Rb-Sr dating. *Miner. Deposita* **2018**, *53*, 967–979. [CrossRef]
61. Du, J.; Audétat, A. Early sulfide saturation is not detrimental to porphyry Cu-Au formation. *Geology* **2020**, *48*, 519–524. [CrossRef]
62. Ohmoto, H.; Goldhabe, M. Sulfur and carbon isotopes. In *Geochemistry of Hydrothermal Ore Deposits*; Barnes, H., Ed.; Willey: New York, NY, USA, 1997; pp. 517–612.

63. Sinclair, W. Porphyry deposits. In *Mineral Deposits of Canada—A Synthesis of Major Deposit Types, District Metallogeny, the Evolution of Geological Provinces, and Exploration Methods*; Goodfellow, W.D., Ed.; Special Publication; Geological Association of Canada, Mineral Deposits Division: St. Johns, NL, Canada, 2007; Volume 5, pp. 223–243, ISBN 978-1-897095-24-9.
64. Wilson, A.J.; Cooke, D.R.; Harper, B.J.; Deyell, C.L. Sulfur isotopic zonation in the Cadia district, southeastern Australia: Exploration significance and implications for the genesis of alkalic porphyry gold–copper deposits. *Miner. Depos.* **2007**, *42*, 465–487. [CrossRef]
65. Richards, J.P. Postsubduction porphyry cu-au and epithermal Au deposits: Products of remelting of subduction-modified lithosphere. *Geology* **2009**, *37*, 247–250. [CrossRef]
66. Lambert-Smith, J.S.; Allibone, A.; Treloar, P.J.; Lawrence, D.M.; Boyce, A.J.; Fanning, M. Stable C.O.; Isotope, S. Record of Magmatic-Hydrothermal Interactions Between the Faleme Fe Skarn and the Loulo Au Systems in Western Mali. *Econ. Geol.* **2020**, *115*, 1537–1558. [CrossRef]
67. Sillitoe, R.H. Porphyry copper systems. *Econ. Geol.* **2010**, *105*, 3–41. [CrossRef]
68. Juliani, C.; Rye, R.O.; Nunes, C.M.; Snee, L.W.; Silva, R.H.C.; Monteiro, L.V.; Neto, A.A. Paleoproterozoic high-sulfidation mineralization in the Tapajós gold province, Amazonian Craton, Brazil: Geology, mineralogy, alunite argon age, and stable-isotope constraints. *Chem. Geol.* **2005**, *215*, 95–125. [CrossRef]
69. Hoss, A.; Haase, K.M.; Keith, M.; Klemb, R.; Melfos, V.; Gerlach, L.; Pelloth, F.; Falkenberg, J.J.; Voudouris, P.; Strauss, H.; et al. Magmatic and hydrothermal evolution of the Skouries Au-Cu porphyry deposit, northern Greece. *Ore Geol. Rev.* **2024**, *173*, 1062332024. [CrossRef]
70. Audétat, A.; Simon, A.C. Magmatic controls on porphyry copper genesis. *Econ. Geol.* **2012**, *16*, 553–572. [CrossRef]
71. Assunção, R.F.S.; Klein, E.L. The Moreira Gomes deposit of the Cuiú-Cuiú goldfield: Fluid inclusions and stable isotope constraints and implications for the genesis of granite-hosted gold mineralization in the Tapajós Gold Province, Brazil. *J. S. Am. Earth Sci.* **2014**, *49*, 85–105. [CrossRef]
72. Groves, D.I.; Bierlein, F.P. Geodynamic setting of mineral deposit systems. *J. Geol. Society* **2007**, *164*, 19–30. [CrossRef]

minerals

MDPI

Article

Evaluation of Granite Fertility Utilizing Porphyry Indicator Minerals (Zircon, Apatite, and Titanite) and Geochemical Data: A Case Study from an Emerging Metallogenic Province in the Taimyr Peninsula, Siberian High Arctic

Stepan V. Berzin [1], Dmitry L. Konopelko [1,*], Sergei V. Petrov [1], Vasiliy F. Proskurnin [2], Evgeny I. Berzon [2], Mikhail Yu. Kurapov [1], Tamara A. Golovina [3], Natalya Ya. Chernenko [4], Vasiliy S. Chervyakovskiy [5], Roman S. Palamarchuk [6] and Elena M. Andreeva [1]

1 Institute of Earth Sciences, St. Petersburg State University, 7/9 University Emb., St. Petersburg 199034, Russia; sbersin@yandex.ru (S.V.B.)
2 A.P. Karpinsky Russian Geological Research Institute, 74 Sredniy Prospect, St. Petersburg 199106, Russia
3 JSC «Polymetal», 2, Narodnogo Opolcheniya Prospect, St. Petersburg 198216, Russia
4 LLC "Polar Forwarding Company", 101 Alekseeva St., Krasnoyarsk 660052, Russia
5 A.N. Zavaritskiy Institute of Geology and Geochemistry, UB RAS, 15 Akademika Vontsovskogo St., Yekaterinburg 620110, Russia
6 South Urals Research Center of Mineralogy and Geoecology, UB RAS, The Ilmeny State Reserve, Miass 456317, Russia
* Correspondence: konopelko@inbox.ru or d.konopelko@spbu.ru

Abstract: The Taimyr Peninsula in the Russian High Arctic comprises a late Paleozoic-early Mesozoic collisional belt where several porphyry-type mineralization occurrences were identified during the last decade, making this area a potential exploration target for Cu-Mo deposits. In order to further evaluate the metallogenic potential of the poorly outcropped northeastern part of Taimyr, samples from seven granitoid intrusions were investigated in this study aimed to evaluate the granite fertility based on petrography, geochemistry, and composition of porphyry indicator minerals (zircon, apatite, and titanite). The studied intrusions represent small to moderate-sized bodies (40–800 km^2) composed of biotite ($\pm$amphibole) quartz monzonites, granodiorites, granites, and biotite leucogranites that formed in the course of late Paleozoic-early Mesozoic tectono-magmatic events at the Siberian margins. The late Carboniferous Tessemsky massif represents suprasubduction granitoid series, while the Pekinskiy, Shirokinskiy, Dorozhinskiy, Kristifensenskiy, and Yuzhno-Lodochnikovskiy massifs are correlated with the early Triassic Siberian Traps LIP. The rocks of intrusions comprise a relatively uniform geochemically, predominantly magnesian, slightly peraluminous, calc-alkaline high-K amphibole-bearing I-type granitoid series with adakitic affinity, where Triassic plume-related granitoids inherit geochemical signatures of Carboniferous supra-subduction granitoids, and all rock types are marked by enrichment in LILE and negative Ta, Nb, and Ti anomalies. It is suggested that the adakitic geochemical characteristics of the Taimyr granites are a result of derivation from a relatively homogeneous mafic lower crustal source that formed at the stage of Carboniferous continental subduction and continued to produce granitic melts in the course of the early Mesozoic magmatic evolution. Whole rock geochemistry and composition of porphyry mineral indicators (zircon, apatite, and titanite) indicate that the Taimyr granites crystallized from oxidized water-saturated magmas at moderate temperatures, with the majority of samples showing characteristics typical for porphyry-fertile granites worldwide (fO_2 = ΔFMQ +1 to +3 with zircon Eu/Eu* > 0.4 and apatite SO_3 > 0.2 wt.%). Data from Dorozhinskiy, Kristifensenskiy, Pekinskiy, and Tessemskiy intrusions fully match geochemical criteria for porphyry-fertile granitoids, and these massifs are considered the most prospective for Cu-Mo mineralization. Granites from Shirokinskiy and Yuzhno-Lodochnikovskiy intrusions only partially match compositional constraints for fertile melts and can be considered as second-tier exploration targets. Finally, available data for the Simsovsky massif preclude its classification as a porphyry-fertile body. These conclusions are in line with previously developed exploration criteria for the northeastern Taimyr, showing that geochemical indicators of granite-fertility can be used on a regional scale in parallel with other exploration methods.

Keywords: Taimyr; granite; fertility; porphyry mineralization; porphyry indicator minerals; zircon; apatite; titanite; Russian High Arctic

1. Introduction

Growing demand for copper is driven by industrial development and projected copper shortages in the future. Currently, about 40% of the world's copper is mined from the porphyry copper deposits, which represent an important source of this metal and primary exploration targets [1,2]. Therefore, mining companies all over the world allocate significant funds for the recognition of new areas potentially prospective for porphyry copper deposits. The Hercynian collisional belt of the Taimyr Peninsula in the Russian Arctic is a region where several occurrences of porphyry-type mineralization have been discovered recently, making this remote area a new, promising exploration target [3]. Porphyry-type mineralization recognized in the Taimyr Peninsula associates with the late Paleozoic-early Mesozoic granitoids formed in supra-subduction and post-collisional tectonic settings, which are favorable for porphyry deposits [4,5]. Mo, Cu, Au, and base metal mineralization was identified in granitoid intrusions comprising the "Taimyr Porphyry Belt" stretching from west to east for more than 1000 km [3–5] (Figure 1). As a result of recent exploration projects, including geophysical and geochemical mapping and limited pilot drilling, several mineralized districts within the Taimyr Peninsula with significant estimated forecast resources of Mo, Cu, and Au have been recognized [6]. However, further on-land prospecting was hampered by the high costs of exploration in the poorly outcropped Arctic areas, and a special project was initiated aimed at discriminating between barren and fertile granitoid intrusions based on the geochemical studies and porphyry indicator minerals. This approach represents an actively developing branch of research utilizing the composition of igneous minerals (zircon, apatite, titanite, magnetite, etc.) to discriminate between porphyry and non-porphyry environments (the system "fertility") in order to assess the presence of a possible porphyry system within particular granitoid intrusions [7]. It was shown that the chemical composition of accessory minerals in granites can be used as an indicator of magma oxidation state, content of fluids, and degree of magma fractionation [7–13]. These characteristics are applied to distinguish fertile granitoid systems where the high oxygen fugacity of magma led to a greater activity of hydrothermal fluids, providing a favorable environment to form a porphyry deposit [7]. In this paper, we present whole-rock geochemistry and zircon, apatite, and titanite mineral chemistry data for seven granitoid intrusions from the northeastern part of the Taimyr Peninsula. Based on these results, we assess the crystallization temperature, oxygen fugacity, and fluid content in the magma of the studied intrusions, allowing for the recognition of targets potentially prospective for porphyry copper mineralization. The new data also characterize the geochemical affinity and sources of supra-subduction, post-collisional, and plume-related granitoid series of the Taimyr Peninsula and facilitate better understanding of the tectonic and metallogenic evolution of this region during the late Paleozoic–early Mesozoic times.

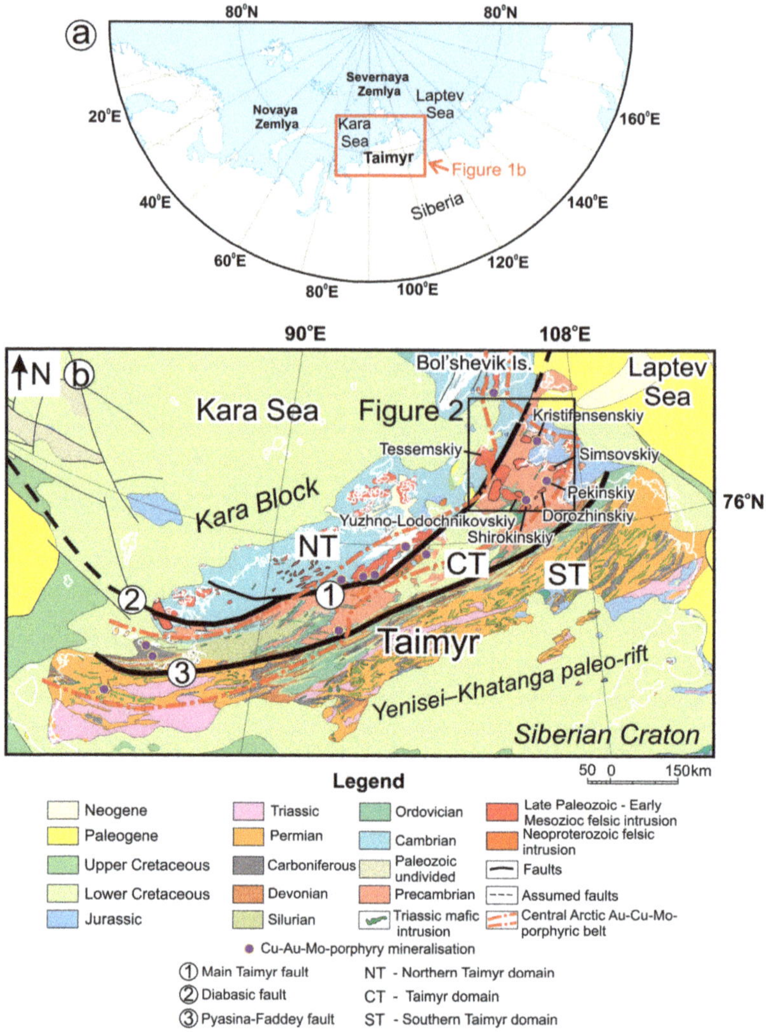

Figure 1. (**a**) Map of the Arctic region showing location of the Taimyr Peninsula. (**b**) Simplified geological map of the Taimyr Peninsula and Severnaya Zemlya Archipelago with location of the studied intrusions; modified after [14,15]. Contour of the Taimyr porphyry belt and porphyry mineralization after [4].

2. Geology of the Taimyr Peninsula

The Taimyr Peninsula and Severnaya Zemlya Archipelago comprise a ca. 440,000 km^2 land mass located on the northern edge of Siberia between the Laptev and Kara marginal seas of the Arctic Ocean. The Yenisei–Khatanga paleo-rift that formed in the course of the Mesozoic Cimmerian orogeny separates these vast pre-Mesozoic terranes from the Siberian Craton located to the south (Figure 1).

The pre-Mesozoic basement of the Taimyr–Severnaya Zemlya fold belt is subdivided into three major tectonic units or domains, namely the Southern, Central, and Northern Taimyr Domains (Figure 1). The Main Taimyr regional fault and its western continuation called the Diabasic fault separate the Northern Taimyr Domain from the Central Domain, while the Pyasina–Faddey fault separates the Central and Southern Taimyr domains. The regional faults formed in the course of southward thrusting of terranes onto the northern margin of the Siberian Craton and represent complex tectonic lineaments generally dipping to the northeast (Figure 1) [16–18].

The Northern Taimyr Domain and Severnaya Zemlya Archipelago comprise uplifted parts of the Kara microcontinent (Kara Block), where its pre-Mesozoic basement is outcropping and available for study. Within the Northern Taimyr Domain, the basement rocks are predominantly represented by the early Paleozoic sediments that formed in a passive margin geodynamic setting. The sediments are intensely deformed and metamorphosed under greenschist-to-amphibolite-facies conditions [17,19]. They were originally mapped as Proterozoic in age [16,20,21]. However, recent detrital zircon studies have shown that the depositional ages of these metasediments varied from Cambrian to Ordovician [22–26]. Deformed early Paleozoic basement rocks of the Northern Taimyr Domain are crosscut by voluminous Carboniferous to Triassic granitoid intrusions that occupy approximately 20% of its territory at the present-day erosion surface [14,19,27–37].

The basement of the Central Taimyr Domain consists of Meso- and Neoproterozoic sediments, ophiolitic blocks, and igneous rocks that were intensely deformed and metamorphosed under greenschist- to amphibolite-facies conditions. These rocks are overlain by Ediacaran to Silurian sediments intercalating with felsic metavolcanics [17,19,38,39]. The Precambrian basement of the Central Taimyr Domain was initially interpreted as a collage of diverse terranes accreted to the continental margin of Siberia during the Neoproterozoic [16,20,40–46]. An alternative point of view suggests that the Central Domain was accreted to the margin of Siberia in the Mesoproterozoic, and in the Neoproterozoic it developed as an active continental margin [47,48]. In the latest Neoproterozoic, this terrane was transformed into a passive margin, where sedimentation in a shelf-type environment continued until the late Paleozoic. The late Paleozoic-early Mesozoic magmatic complexes of the Central Taimyr Domain are represented by the early Permian–Triassic granites and early Triassic doleritic dykes and sills, associated with synchronous magmatism of the Siberian Traps Large Igneous Province (LIP), which is interpreted to form in response to the formation of a Siberian mantle plume at the Permo-Triassic boundary [14,17,37,39,49,50]. Jurassic to lower Cretaceous deposits have limited distribution and fill local depressions in both the Northern and Central Taimyr domains [19].

The Southern Taimyr Domain is made up of sedimentary piles that formed at the Siberian passive margin since the early Paleozoic. The Ordovician to Triassic shelf-type sedimentary strata, exposed within the Southern Domain, are intruded by the latest Permian–early Triassic mafic dykes and sills associated with the Siberian Traps LIP and variably deformed in the course of subsequent Mesozoic events. The degree of deformation decreases to the south: the northern part of the Southern Taimyr Domain is characterized by steep thrusts and folds of predominantly southern vergency, while its southern part is less deformed and covered by Mesozoic sediments of the Yenisei–Khatanga paleo-rift [17,21,51,52].

The general structure of the Taimyr–Severnaya Zemlya fold-and-thrust belt formed as a result of late Carboniferous–early Permian collision between the Kara Block and Siberia [14,16,33,35,42,53]. After accretion of the Kara Block to the northern margin of the Siberian Craton, the compressional regime associated with convergence between the two continents terminated in the early Permian and was followed by a regional extension that affected the northern Siberian margins at the Permian–Triassic boundary [54–58]. In Taimyr, this extension was manifested by the opening of the Yenisei–Khatanga paleo-rift that was filled with early Mesozoic sediments, which were variably deformed in the late Jurassic–early Cretaceous when convergence triggered by the regional Cimmerian orogenic events resumed (Figure 1).

3. Late Paleozoic–Early Mesozoic Granitoid Series and Associated Porphyry-Type Mineralization

Several authors [14,20,28,35,36,41,59] have shown that supra-subduction magmatic series developed in the Northern Taimyr Domain and the Severnaya Zemlya Archipelago have ages varying between ca. 345 and 285 Ma (early Carboniferous to early Permian). After collision, the Taimyr Peninsula was affected by a post-collisional magmatic pulse that triggered the emplacement of granitoid intrusions over the whole area across terrane boundaries. After the termination of post-collisional magmatic activity at ca. 265 Ma, the Taimyr experienced extension, and voluminous igneous complexes related to the Permo-Triassic Siberian Traps LIP were emplaced between 251 and 228 Ma. Consequently, the late Paleozoic–early Mesozoic granitoid series of the Taimyr Peninsula evolved in various geodynamic environments during an extended time span from Carboniferous to Triassic. However, in spite of their diverse geodynamic origin, most of the Taimyr granitoids display geochemical and metallogenic features characteristic of magmatic series of continental arcs [59]. Accordingly, the Cu-Mo mineralization occurrences have been identified in association with granitoids of all three stages of the late Paleozoic–early Mesozoic magmatism in the Taimyr Peninsula [3–5]. Based on association with porphyritic granitoids and character of alteration in the host rocks, these occurrences were defined by Proskurnin et al. (2021) [4] as copper porphyry-type mineralization according to classification schemes of Dill (2010) [60] and Robb (2020) [61]. It was also shown that this area did not experience significant erosion during the late Mesozoic and Cenozoic times, which is favorable for the preservation of porphyry deposits [4].

Carboniferous supra-subduction porphyritic granites have formed at the southern active continental margin of the Kara Block and host several occurrences of Au and Mo mineralization located in the islands of Severnaya Zemlya Archipelago [4,5]. In particular, porphyry-type Mo mineralization was found in association with Carboniferous granites at Cape Palets and Cape Taimyr in the southern part of the Bolshevik Island [4]. Similar Carboniferous granites, which are widely distributed in the Northern Taimyr Domain, are represented in this study by the Tessemskiy massif (Figure 2).

Early Permian post-collisional granitoid intrusions developed in the Northern and Central Taimyr domains are also considered as possible exploration targets for porphyry-type mineralization based on the recent discovery of the Shumyaschyeye porphyry-type Mo-Au occurrence associated with porphyritic granite of the Kamenisty massif in the Northern Taimyr Domain. This occurrence is represented by molybdenite, pyrite, chalcopyrite, and arsenolite mineralization in quartz-sericite veins in the endocontact zone of the intrusion [4].

Another granitoid complex productive for porphyry-type deposits comprises the late Permian–Triassic plume-related granites and syenites associated with the Siberian Traps LIP developed in all Taimyr domains across terrane boundaries [4,59]. These intrusions host several Mo–Au–Cu-porphyry and polymetallic mineralization occurrences, including three of them located within this study area: Porphyrovoe, Kristifensen, and Pekinskiy (Figure 2). The Porphyrovoe Mo occurrence is associated with a small satellite intrusion of the Yuzhno-Lodochnikovskiy massif composed of porphyritic granite. In this occurrence, molybdenite, chalcopyrite, pyrite, pyrrhotite, and magnetite are disseminated in brecciated and altered rhyolite-porphyry and upper Proterozoic clastic sediments in the contact zone of the porphyritic granite intrusion (Figure 3) [4]. The Kristifensen Mo–Cu–Au porphyry-type occurrence is also hosted in the contact zone of the homonymous granite massif. In the Pekinskiy massif, chalcopyrite, pyrite, and molybdenite mineralization was described in small stocks of altered porphyritic granite of the second intrusive phase [5] and in brecciated altered granitoids in the southwestern part of intrusion [4]. Photographs illustrating exposure and structure of characteristic mineralized granitoids from known porphyry-type occurrences in the Taimyr Peninsula are presented in Figure 3.

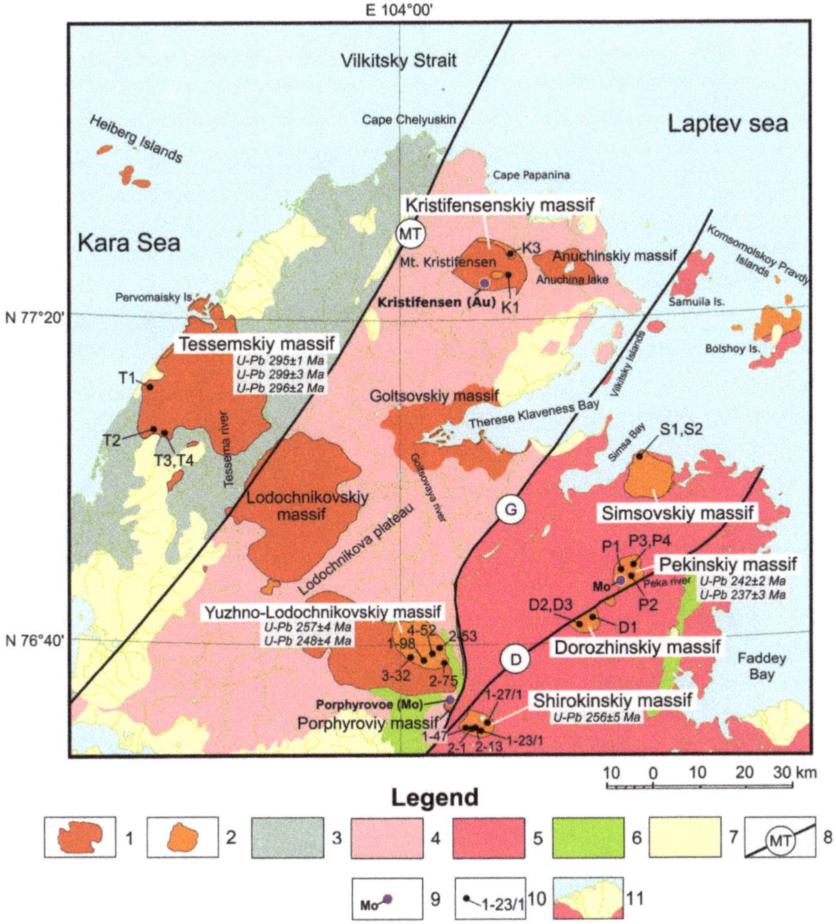

Figure 2. Simplified geological map of northeast of the Taimyr Peninsula after [19,62] with clarifications from the authors, with location of studied late Paleozoic–early Mesozoic granite massifs and with sample location. Legend: 1, 2—massifs of late Paleozoic–early Mesozoic granitoids: (1) moderately alkaline granodiorite-granite and (2) granite-leucogranite; 3—North Kara domain with Ediacaran–Cambrian (?) terrigenous flyschoid formations; 4, 5, 6—Central Taimyr domain with accretionary Neoproterozoic (1), Meso-Neoproterozoic (2) volcano-sedimentary rocks, including ophiolites and paleo-island arc complexes, and Ediacaran–Middle Paleozoic (3) platform carbonate-terrigenous carbonaceous sediments; 7—Jurassic-Cenozoic covered sediments; 8—main faults: Main Taimyr fault (MT), Goltsovsky fault (G), Dorozhninsky fault (D); 9—Cu–Au–Mo-porphyry mineralization; 10—sample location; 11—coastline and rivers. U-Pb zircon ages from [4,5,14,17,35,39,59].

Several Mo, W, Cu, Zn, As, and Ag geochemical anomalies in loose sediments were identified within and around granitoid intrusions in this study area and elsewhere in the Taimyr Peninsula that, in combination with known bedrock occurrences, allowed to estimate the total forecast resources of the Taimyr Porphyry Belt, comprising 2,490,000 tonnes of Mo, 585,000 tonnes of Cu, and 250 tonnes of Au [4].

Figure 3. Outcrop and sample photographs of mineralized granitoids from Shirokinskiy massif (**a**–**f**) and Porphyrovoe Mo occurrence (**g**) in the northeastern part of the Taimyr Peninsula: (**a**) outcrop of mineralized granite where rusty colors indicate presence of oxidized ore minerals; (**b**) sharp-angled fragments of mineralized porphyry granitoid; (**c**) altered granite with chalcopyrite mineralized veins; (**d**) typical porphyritic granite; (**e**) altered granite with pyrite-chalcopyrite mineralization; (**f**) hydrothermal quartz-pyrite-chalcopyrite vein-type mineralization in altered granite; (**g**) quartz vein with sulfide mineralization in altered granite containing 0.72 wt.% Cu and 0.2 wt.% Pb. Abbreviations: Ccp—chalcopyrite, Py—pyrite.

4. Description of the Studied Intrusions and Sampling

In order to further evaluate the prospectivity of the northeastern Taimyr area shown in Figure 2, granite samples from the Kristifensenskiy, Pekinskiy, Dorozhinskiy, Tessemskiy, Simsovskiy, Shirokinskiy, and Yuzhno-Lodochnikovskiy massifs were investigated in this study utilizing methodology based on porphyry indicator minerals and geochemical data.

The Tessemskiy massif with an age of 299–295 Ma [35,59] represents a suprasubduction granitoid complex located in the Northern Taimyr Domain on the shore of the Kara Sea (Figure 2). It has an area of about 650 km^2, irregular shape, and sharp sub-vertical contacts. The Tessemskiy massif consists of biotite-amphibole granodiorite and is crosscut by a small linear body of Triassic muscovite leucogranite exposed at the shore of the Kara Sea (Figure 2). The massif intrudes Cambrian metasedimentary rocks and is cut by the early

Triassic dolerite dikes. Four granite samples from the Tessemskiy massif were collected for this study.

The Shirokinskiy massif (256 Ma) [4] represents one of the earliest intrusions formed in response to a Siberian mantle plume in the Central Taimyr Domain. The massif intrudes upper Silurian–middle Devonian sediments [4]. It comprises a small stock (ca. 40 km^2) of granodiorite and porphyritic granite. Five samples were collected for this work.

The adjacent Yuzhno-Lodochnikovskiy intrusion is located in the Central Taimyr Domain and has a similar age of ca. 250 Ma, corresponding to the Permian-Triassic boundary [14,39]. It is an average-sized intrusion (ca. 800 km^2) composed of amphibole-biotite and biotite granodiorites, quartz monzonites, and granites. Five samples were chosen for investigation.

The Pekinskiy massif, dated at ca. 240 Ma [5,59], is located in the Central Taimyr Domain (Figure 2) and represents a typical granitoid intrusion associated with a Siberian mantle plume. It is a relatively small granitoid body occupying an area of ca. 70 km^2. The intrusion is composed of granitoid rocks emplaced in three phases: the first phase comprises coarse-grained and porphyritic granites and granosyenites, the second—porphyritic biotite granite, and the third—differently oriented dikes and veins of aplite, pegmatoid granite, and less commonly pegmatite. Four samples of porphyritic biotite granites and porphyritic quartz monzodiorites were collected from the Pekinskiy massif for analysis.

The average-sized Kristifensenskiy massif (ca. 160 km^2), emplaced in the Central Taimyr Domain, was not geochronologically investigated and is tentatively attributed to Triassic plume-related granites [17]. The intrusion is composed of porphyritic biotite granites, from which two samples were chosen for analysis.

Finally, the least studied Simsovskiy massif represents a ca. 130 km^2 stock composed of porphyritic biotite granites, from which only 2 granite samples were available for this work.

Because petrochemical classification of the rocks was one of the aims of this study, sampling of altered varieties was avoided, where possible, and most samples represented fresh porphyritic or even-grained granitoids of main intrusive phases.

5. Analytical Procedures

Major and trace elements in granitoid samples were analyzed using ICP-OES Agilent 5100 and ICP-AES Optima 8300 equipment at the SGS Vostok Limited, Chita, Russia, and using an ICP-AES iCAP PRO, ICP-MS iCAP-Qc, and Sulfur/Carbon Analyzer SC-144 at the Gipronickel Institute, St. Petersburg, Russia. Interlaboratory comparison of the data for Ca, Ti, Mn, K, Rb, Sr, Ce, and Cu revealed a good agreement of the analytical results from the two labs.

Zircon, apatite, and titanite grains were recovered from rock samples by conventional mineral separation procedures and handpicked under a binocular microscope. Mineral grains together with reference materials were mounted into 25 mm epoxy resin discs and polished to expose the interiors of the grains.

Internal structures of the mineral grains mounted into resin discs were imaged using a Tescan MIRA LMS SEM (TESCAN GROUP, Brno-Kohoutovice, Czech Republic) equipped with backscattered electron (BSE) detectors and INCA Energy 450 X-Max 80 (Oxford Instruments, Abingdon, Oxfordshire, UK) microanalysis system at the Common Use Center "Geoanalyst" Zavaritsky Institute of Geology and Geochemistry of the Urals Branch of the Russian Academy of Sciences (IGG), Yekaterinburg, Russia.

The composition of zircon, apatite, and titanite was studied utilizing a Cameca SX100 electron microprobe (EPMA) equipped with backscattered electron (BSE) and cathodoluminescence (CL) detectors and five wavelength dispersive (WDS) spectrometers at the IGG, Yekaterinburg, Russia. Operating conditions of the microprobe were: 15 kV accelerating voltage, beam current of 10 to 20 nA, 2 μm beam diameter, natural and synthetic minerals, and glasses were used as calibration standards.

In-situ zircon, apatite, and titanite analyses were performed utilizing a NexION 300S quadrupole ICP-MS spectrometer coupled to a NWR 213 laser ablation attachment synchronized with the mass spectrometer (LA-ICP-MS) at the IGG, Yekaterinburg, Russia. All equipment used is housed in a cleanroom of ISO 7 class. Helium (grade A, TU 0271–135–31323949–2005) [63–65] was used to transport the sample material to the mass spectrometer after ablation; argon (GOST 10157-2016) [63–65] was used to transfer the sample to the torch of the mass spectrometer. In determining the microelement composition of zircon, apatite, and titanite, synthetic glasses NIST SRM 612 and 610 (Standard Reference Material of the National Institute of Standards and Technology, Gaithersburg, MD, USA) were used as an external standard [63–65]. The analyzed elements included Li, Be, Na, Mg, Al, Si, Ca, Sc, Ti, V, Cr, Fe, Mn, Co, Ni, Cu, Zn, Ga, Ge, As, Rb, Sr, Y, Zr, Nb, Mo, Ag, Cd, In, Sn, Sb, Cs, Ba, REE, Hf, Ta, W, Tl, Pb, Bi, Th, and U. The mass spectrometer drift was corrected using NIST SRM 610 glass measured by the "bracketing technique" through 10 craters; a NIST SRM 612 glass was used as a secondary reference material. The results were processed using the GLITTER V. 4.4.4 software and the CaO content of apatites and SiO_2 content of zircon, titanite, and synthetic glasses as an internal standard. The following operation parameters of the equipment were used: (1) power of the high-frequency generator of the mass spectrometer ~1500 W; number of scanning cycles 1; number of replicas 500; flow rates of plasma-forming, auxiliary, and sample-supplying argon at a connecting tube length of the mass spectrometer with an LA attachment of 1.5 m 18.00, ~1.2, and 0.94 dm^3/min, respectively; ground delay time 10 ms; (2) laser radiation energy density of the LA attachment 10.5–11.5 J/cm^2; pulse repetition frequency 10 Hz; laser spot diameter 25 μm; helium transport flow rate 400 cm^3/min; laser heating time 20 s; its operation time 50 s; trajectory of the laser beam along the apatite grain is a "point" ablation, trajectory along the glasses is a line (speed 70 μm/s, length ~700 μm, number of passes 4).

Analytical spots for LA-analysis were chosen at a distance from mineral inclusions identified at the BSE and CL images, as shown in Supplementary Figure S1. In order to exclude signals from mineral inclusions located under the surface of the grains, additional filtering for outliers was carried out for each element signal when processing LA-ICP-MS results. Those parts of the signals in which a significant change in the signals of non-formula elements was recorded were excluded from the calculated range. An example of the selection of a signal range for calculations of the element concentrations is shown in Supplementary Figure S1.

6. Results

Whole rock samples and compositions of zircon, apatite, and titanite were investigated in most of the studied intrusions. A few exceptions include samples from the Shirokinskiy massif, which did not produce titanite separates; the Yuzhno-Lodochnikovskiy massif, from which only accessory apatite was available for analysis; and the Simsovsky massif, where only zircon separates were available for this study.

6.1. Whole Rock Geochemistry

Major and trace elements were analyzed in 25 granitoid samples from the Kristifensenskiy, Pekinskiy, Dorozhinskiy, Tessemskiy, Shirokinskiy, and Yuzhno-Lodochnikovskiy massifs located in the Northeastern Taimyr (Figure 2). Coordinates of sampling points, petrographic description and modal composition of the rocks, and analytical results are given in Supplementary Table S1. Although the studied intrusions formed in different geotectonic settings, it was shown that they comprise a relatively uniform geochemical series [59], and, therefore, their geochemical characteristics are described below collectively.

Analyzed samples of the Taimyr granitoids comprise fresh, unaltered rock types characterized by relatively low L.O.I. values generally below 1 wt.% (Supplementary Table S1). The rocks have SiO_2 concentrations in the range of 62.2 to 75 wt.% and generally plot in the fields of granite, granodiorite, and quartz monzonite in a TAS classification diagram (Figure 4a). The majority of samples are slightly peraluminous and have ASI ($Al_2O_3/(Na_2O + K_2O)_{mol.}$) values between 1.0 and 1.1 (Figure 4a). The rocks plot in the fields of calc-alcalic, alcali-calcic, and alkalic series on the ($Na_2O + K_2O$–CaO) vs. SiO_2 diagram and in the fields of high-K and shoshonitic series on a diagram K_2O vs. SiO_2 (Figure 4c,d). Most samples consistently plot in the field of calc-alkaline series on a $FeO_{tot}/(FeO_{tot} + MgO)$ vs. SiO_2 diagram, with the exception of several leucogranite samples from the Dorozhinskiy, Pekinskiy, and Tessemskiy massifs representing highly fractionated vein varieties that are strongly enriched in Fe relative to Mg and plot in the field of tholeiitic series (Figure 4e).

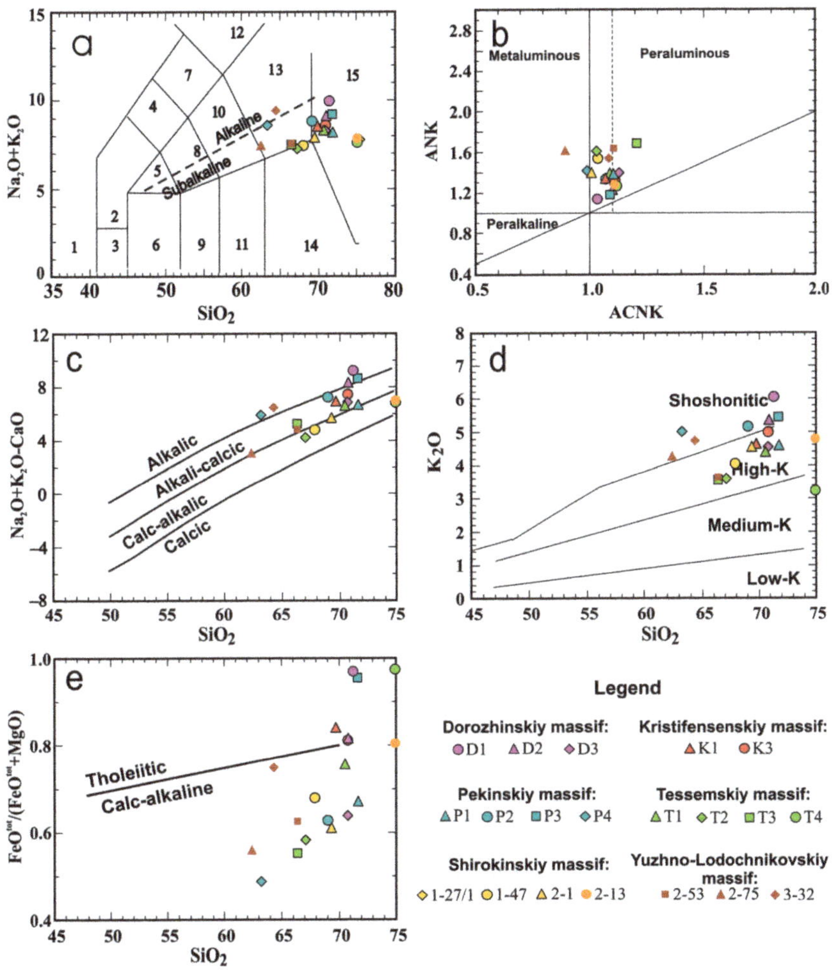

Figure 4. Granites of the Taimyr Peninsula on classification and discrimination diagrams: (**a**) $Na_2O + K_2O$ vs. SiO_2, wt.% (TAS) diagram, fields after [66,67]: 1—foidolite, 2—foid gabbro, 3—peridotitic gabbro, 4—foid monzodiorite, 5—monzogabbro, 6—gabbro, 7—foid monzosyenite, 8—monzodiorite, 9—gabbroic diorite, 10—monzonite, 11—diorite, 12—foid syenite, 13—syenite and quartz monzonite, 14—granodiorite, 15—granite; (**b**) ANK vs. ACNK diagram, ANK = $Al_2O_3/(Na_2O + K_2O)_{mol.}$, ACNK = $Al_2O_3/(Na_2O + K_2O + CaO)_{mol.}$, fields after [68]; (**c**) ($Na_2O + K_2O$–CaO) vs. SiO_2, wt.% diagram, fields after [69]; (**d**) K_2O vs. SiO_2, wt.% diagram, fields after [70]; (**e**) $FeO_{tot}/(FeO_{tot} + MgO)$ vs. SiO_2, wt.% diagram, fields after [69].

REE spectra of granitoids from the Dorozhinskiy, Kristifensenskiy, Pekinskiy, and Tessemskiy massifs are characterized by moderate enrichment in LREE with $(La/Yb)_N$ ratios varying from 4 to 18 and relatively small negative Eu anomalies (Figure 5a,c). Granites from the Shirokinskiy and Yuzhno-Lodochnikovskiy massifs have more fractionated REE patterns with $(La/Yb)_N$ ratios from 14 to 72 and less pronounced negative Eu anomalies (Figure 5e). Multicationic diagrams, presented in Figure 5b,d,f, show enrichment of the studied granites in LIL elements and pronounced negative Ta-Nb-Ti anomalies typical for subduction-related granitoids [71]. Correspondingly, on discrimination diagrams Rb vs. Y + Nb and Y vs. Nb, shown in Figure 6, samples of the Taimyr granites predominantly plot in the field of volcanic arc granite and marginally in the field of syn-collisional granite.

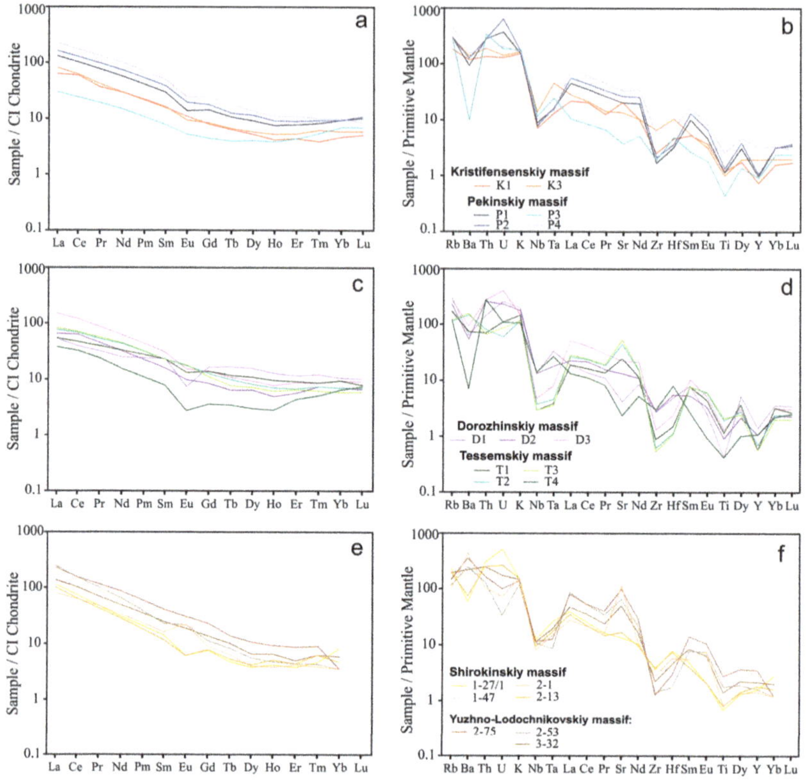

Figure 5. Chondrite-normalized REE patterns (**a**,**c**,**e**) and primitive mantle-normalized trace element diagrams (**b**,**d**,**f**) of granites from the Taimyr Peninsula: (**a**,**b**) Kristifensenskiy and Pekinskiy massifs; (**c**,**d**) Dorozhinskiy and Tessemskiy massifs; (**e**,**f**) Shirokinskiy and Yuzhno-Lodochnikovskiy massifs. Normalizing values from [72].

Berzin et al. (2024) and Kurapov et al. (2024) [59,73] have noted that granitoid rocks of the Northeastern Taimyr are geochemically similar to high-potassium continental adakites. In line with their observations, all granitoids investigated in this work are characterized by high Sr/Y ratios and relatively low Y concentrations, and plot in the field of adakite on a discrimination diagram Sr/Y vs. Y, shown in Figure 7a. On a discrimination diagram, La/Yb vs. Yb samples from the Shirokinskiy and Yuzhno-Lodochnikovskiy massifs consistently plot in the field of adakite, while samples from the Pekinskiy, Kristifensenskiy, Dorozhinskiy, and Tessemskiy massifs have lower La/Yb ratios and only marginally plot in the adakite field, with the majority of samples falling into the field of normal arc lavas (Figure 7b).

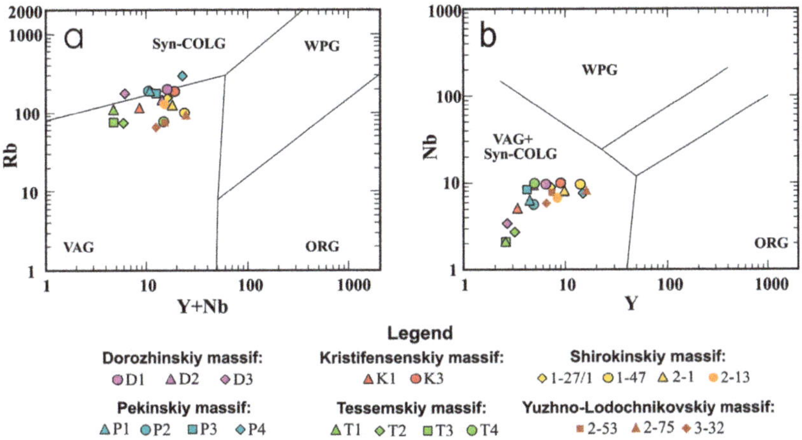

Figure 6. Granites of the Taimyr Peninsula on discrimination diagrams of [74]: (**a**) Rb vs. Y + Nb; (**b**) Y vs. Nb. Abbreviations: VAG—volcanic arc granites; WPG—within-plate granites; Syn-COLG—syn-collisional granites.

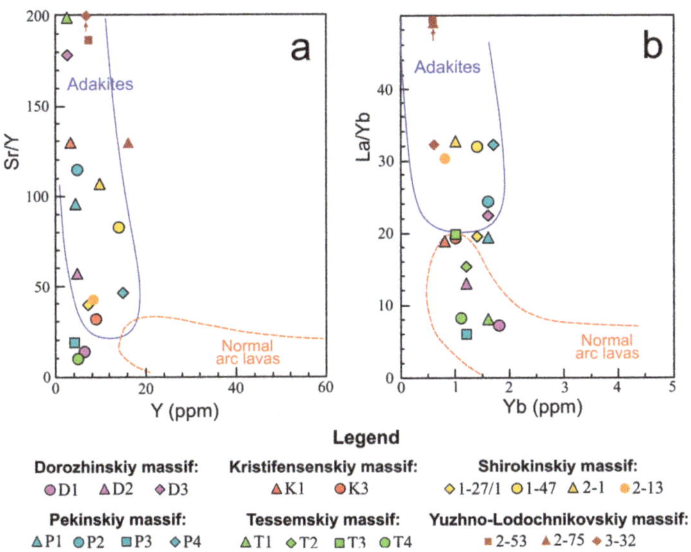

Figure 7. Granites of the Taimyr Peninsula on discrimination diagrams that are used to distinguish adakite from normal arc lavas: (**a**) Sr/Y vs. Y and (**b**) La/Yb vs. Yb. After [75], field boundaries after [76].

6.2. Porphyry Indicator Minerals

6.2.1. Zircon

Analyzed concentrations of trace elements in zircon grains from Dorozhinskiy, Kristifensenskiy, Pekinskiy, Simsovskiy, Shirokinskiy, and Tessemskiy massifs (65 analyzed grains from 14 rock samples) are presented in Supplementary Table S1. Zircon grains recovered from the studied samples show relatively little variation and comprise a population of magmatic zircon with oscillatory zoning.

Unaltered zircon grains with oscillatory zoning characteristic for magmatic zircon without visible mineral inclusions [77] were chosen for in-situ LA-ICP-MS analysis. Some of the elevated concentrations of LREE may result from the sampling of apatite (<0.5%), titanite, or monazite inclusions present within zircon grains in the course of LA analysis [78]. Therefore, analyses of zircon with La concentration above 5 ppm and $(Sm/La)_N$ ratios below 5, falling in the field of hydrothermal zircon on discrimination diagram $(Sm/La)_N$ vs. La [77], shown in Figure 8, were excluded from calculations of crystallization conditions.

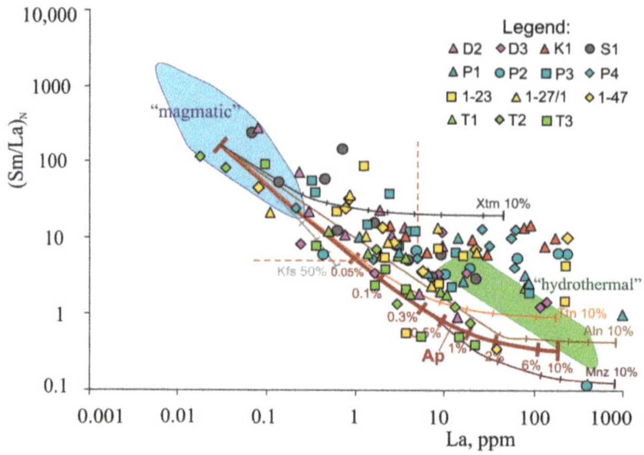

Figure 8. Composition of zircon from the Taimyr granites on $(Sm/La)_N$ vs. La discrimination diagram for magmatic and hydrothermal zircon after [77] showing mixing curves between zircon and mineral inclusions after [78]. Abbreviations: Ap—apatite, Ttn—titanite, Xtm—xenotime, Aln—allanite, Mnz—monazite, and Kfs—K-feldspar.

Additional criteria to distinguish and exclude contaminated analyses of zircon were: (1) concentration of Fe above 5000 ppm reflecting contamination by magnetite; (2) concentration of Mg above 5000 ppm and Al above 5000 ppm reflecting contamination by silicate minerals; and (3) concentration of Ti above 200 ppm reflecting contamination by rutile [79,80].

The concentration of Ta in all analyzed zircon grains is below 0.2 ppm (Supplementary Table S1), which indicates that the Eu/Eu* ratio in zircon was not affected by the early crystallization of titanite [81,82]. This conclusion is in line with frequently observed zircon inclusions in titanite, showing that zircon crystallized prior to titanite from the melt. It means that the trace element composition of the analyzed zircon grains was not affected by co-crystallization of titanite and can be used to estimate the redox state and water saturation of the melts.

Zircon grains from the studied intrusions demonstrate similar HREE-enriched REE patterns with variably pronounced positive Ce anomalies and relatively small negative Eu anomalies (Figure 9), and their $(Eu/Eu^*)/Y^*10^4$ ratios vary from 1 to 11, (Ce/Nd)/Y from 0.001 to 0.1, and Dy/Nd from 2 to 100 (Figure 10), which is characteristic for porphyry-fertile granitoids [83,84].

The temperature of zircon crystallization was calculated utilizing Ti-in-zircon thermometer [85]. Estimated crystallization temperatures for most samples from the Dorozhinskiy, Kristifensenskiy, Shirokinskiy, Pekinskiy, Tessemskiy, and Simsovskiy massifs generally fall in the range of 650–780 °C (Supplementary Table S1) and correspond to the temperature range below 750 °C characteristic for porphyry copper mineralization [84]. Slightly higher temperatures in the range of 730–800 °C were found for samples P3 and P4 from the Pekinskiy intrusion.

The oxidation state of the melt was estimated using an oxybarometer of Loucks et al. (2020) [86] based on concentrations of Ce, U, and Ti in zircon. Calculated average ΔFMQ values for samples from the Dorozhinskiy, Pekinskiy, and Shirokinskiy massifs, given in Supplementary Table S1, vary from 1.0 to 2.1 (±0.6) and indicate an oxidized environment characteristic for porphyry-fertile granitoids [84]. Samples from the

Kristifensenskiy, Tessemskiy, and Simsovskiy massifs yielded slightly lower ΔFMQ values in the range of 0.5–1.0 (±0.6), which are typical for barren intrusions [84].

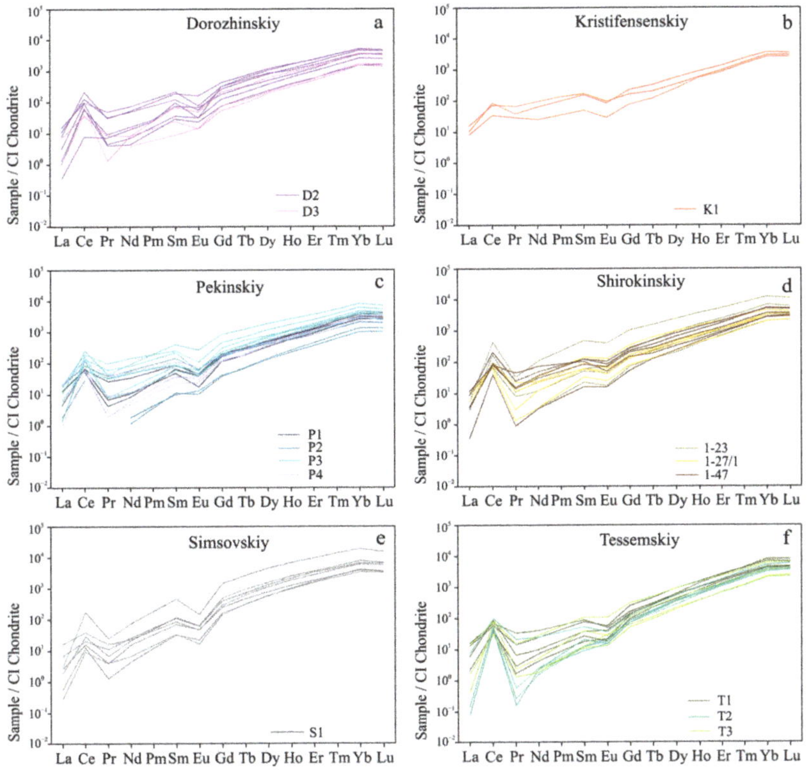

Figure 9. Chondrite-normalized REE patterns in zircon from the Taimyr granites: (**a**) Dorozhinskiy massif; (**b**) Kristifensenskiy massif; (**c**) Pekinskiy massif; (**d**) Shirokinskiy massif; (**e**) Simsovskiy massif; (**f**) Tessemskiy massif. Normalizing values from [72].

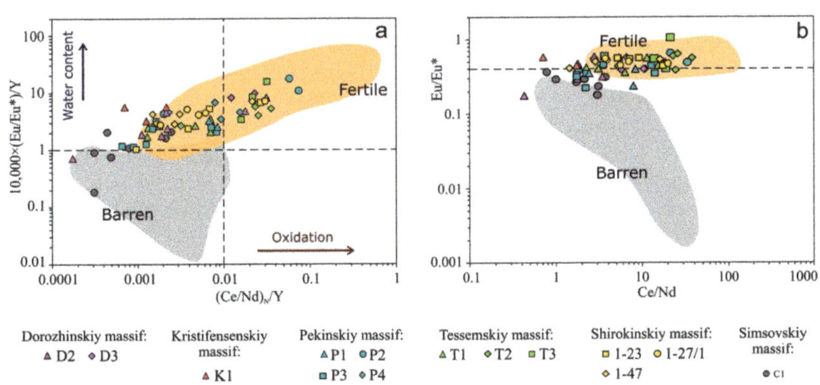

Figure 10. Composition of zircon from the Taimyr granites on discrimination diagrams with fields for barren and fertile magmatic systems: (**a**) 10,000 × Eu/Eu*/Y vs. $(Ce/Nd)_N/Y$; (**b**) Eu/Eu* vs. Ce/Nd. After [83,84].

6.2.2. Apatite

The chemical compositions of accessory apatite from the Dorozhinskiy, Kristifensenskiy, Pekinskiy, Shirokinskiy, Tessemskiy, and Yuzhno-Lodochnikovskiy massifs (135 analyzed grains from 15 rock samples) are given in Supplementary Table S1 and in Figure 11.

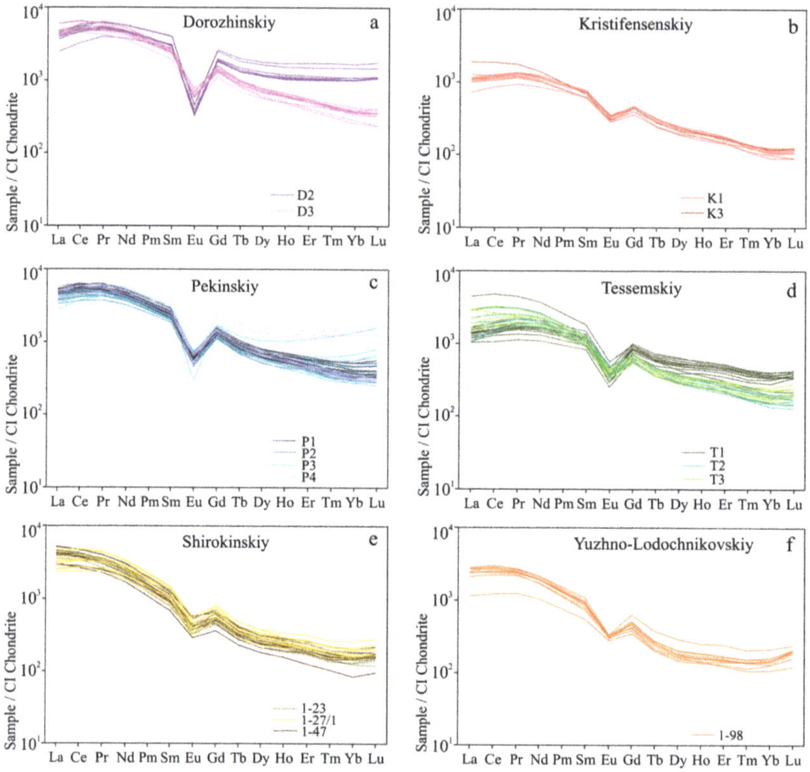

Figure 11. Chondrite-normalized REE patterns in apatite from the Taimyr granites: (**a**) Kristifensenskiy massif; (**b**) Pekinskiy massif; (**c**) Dorozhinskiy massif; (**d**) Tessemskiy massif; (**e**) Shirokinskiy massif; (**f**) Yuzhno-Lodochnikovskiy massif. Normalizing values from [72].

All analyzed apatite grains contain elevated fluorine concentrations in the range 2.61–4.15 wt.% characteristic for fluorapatite. Concentrations of Cl are generally in the range <0.03–0.07 wt.% except sample 1–47 from the Shirokinskiy massif with higher Cl contents of 0.35–0.59 wt.%, and concentrations of SO_3 are also relatively high in the range <0.09–0.31 wt.% typical for Cu-mineralized intrusions (Figure 12a) [87].

Grains of accessory apatite from the studied intrusions have LREE-enriched REE patterns with variably pronounced negative Eu anomalies (Figure 11) and elevated (Eu/Eu*) ratios similar to adakitic porphyry-fertile granitoids of Tibet (Figure 12b) [83]. Elevated Sr contents in apatite (up to 2000) ppm result in high Sr/Y values (up to 7.4), which is also characteristic for apatites from Cu-Mo porphyry systems [88,89].

The Mn-in-apatite oxybarometer by Miles et al. (2014) [90] was used to estimate the redox conditions of apatite crystallization. Analyzed apatite grains from all samples have Mn contents in the range 260–1200 ppm, yielding oxygen fugacity ($\log fO_2$) values between −10.4 and −12.1 (±2.8) (Supplementary Table S1). Only one out of four samples from the Pekinskiy massif (sample P3) has elevated Mn content (850–2700 ppm), resulting in slightly lower calculated $\log fO_2$ values between −13 and −16, corresponding to less oxidized conditions.

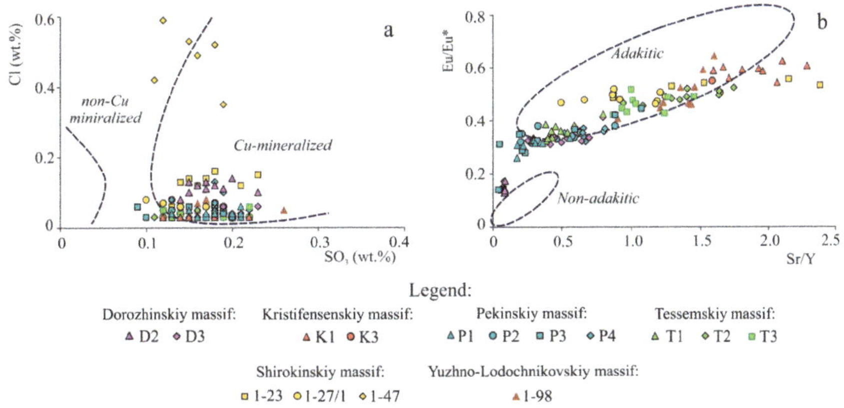

Figure 12. Composition of apatite from the Taimyr granites on discrimination diagrams: (**a**) Cl vs. SO_3; (**b**) Eu/Eu* vs. Sr/Y after [87–89].

6.2.3. Titanite

The chemical compositions of accessory titanite were analyzed in 8 samples from the Dorozhinskiy, Kristifensenskiy, Pekinskiy, and Tessemskiy massifs (in total 37 analytical spots, Supplementary Table S1). Because zircon inclusions in titanite are common, analyses of titanite with Zr concentrations above 2000 ppm were considered contaminated by zircon and excluded from calculations of crystallization conditions [91,92].

Most of the analyzed titanite grains have low $(Al/Fe)_{mol.}$ ratios (0.8 to 1.6), low Nb/Ta ratios (6.2 to 12), and high contents of REE (10,300 to 24,200 ppm) and Y (2000 to 9000 ppm), as well as high Th/U (0.6 to 6.4) (Figure 13). REE spectra of titanite are characterized by moderate enrichment in LREE (LREE/HREE ratios between 2 and 11) with pronounced negative Eu and relatively small negative Ce anomalies (Figure 14). All these features are typical for titanite of magmatic origin [93–95]. In the whole dataset (Figure 13, Supplementary Table S1), only 3 titanite grains from sample T3, collected from the Tessemskiy massif, have relatively high Nb/Ta ratios (29 to 49), low Th/U ratios (1.0 to 1.5), and relatively low contents of REE (8200 to 11,000 ppm) and Y (200 to 1000 ppm) that may indicate hydrothermal or late magmatic origin of these grains [95].

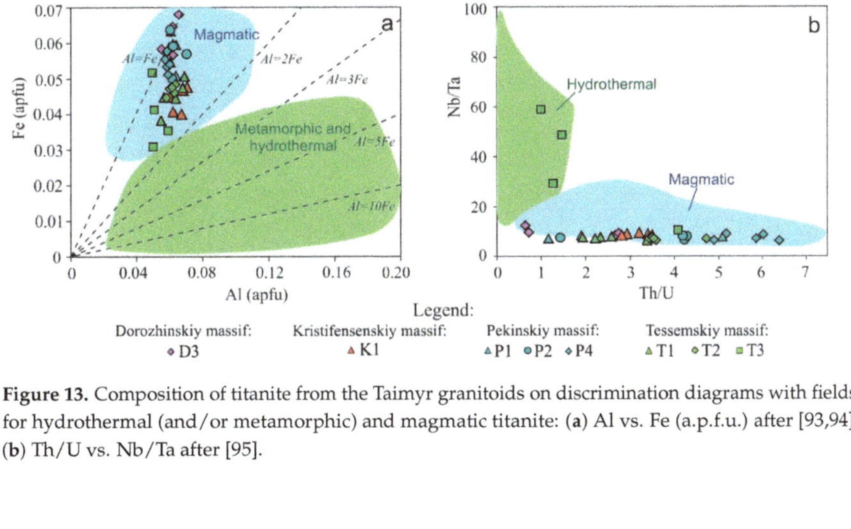

Figure 13. Composition of titanite from the Taimyr granitoids on discrimination diagrams with fields for hydrothermal (and/or metamorphic) and magmatic titanite: (**a**) Al vs. Fe (a.p.f.u.) after [93,94]; (**b**) Th/U vs. Nb/Ta after [95].

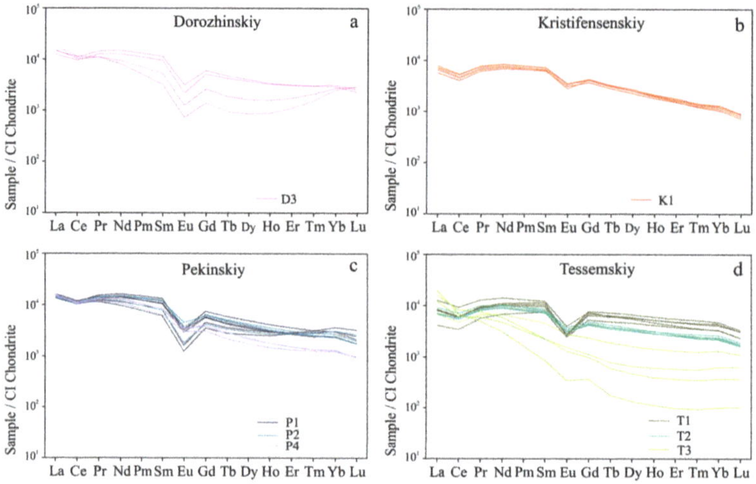

Figure 14. Chondrite-normalized REE patterns in titanite from the Taimyr granites: (**a**) Dorozhinskiy massif; (**b**) Kristifensenskiy massif; (**c**) Pekinskiy massif; (**d**) Tessemskiy massif. Normalizing values from [72].

The crystallization pressure of titanite was estimated utilizing an empirical barometer of [96], based on titanite Al_2O_3 content that yielded a pressure of ca. 220 MPa for all analyzed granite samples (Supplementary Table S1). Assuming that crystallization pressure was equal to 0.22 GPa, the temperature of crystallization was estimated using the Zr-in-titanite geothermometer of Haydenet et al. (2008) [97]. Calculated crystallization temperatures for all studied samples are in the range of 642 to 716 °C (±20 °C) that closely matches the temperatures obtained by Ti-in-zircon geothermometry (Supplementary Table S1) and confirm a magmatic origin of the titanite.

All analyzed titanites, except 3 hydrothermal grains from sample T3, have relatively high contents of REE + Y (13,000 to 31,000 ppm) (Figure 14), Ga (59 to 165 ppm), and Ta (180 to 1100 ppm), Nb (1300 to 7800 ppm), as well as high Nb/Ta < 11.3 and Zr/Hf < 17.5 ratios (Figure 15, Supplementary Table S1) that are typical for porphyry-fertile granitoid intrusions according to [98]. As seen in Figure 15, most analytical data consistently plot in the fields of fertile granitoids on the discrimination diagram Ta vs. Nb, while on the diagram REE + Y/1000 vs. LREE/HREE, the data are significantly scattered, with only titanites from the Pekinskiy massif plotting in the field of fertile intrusions.

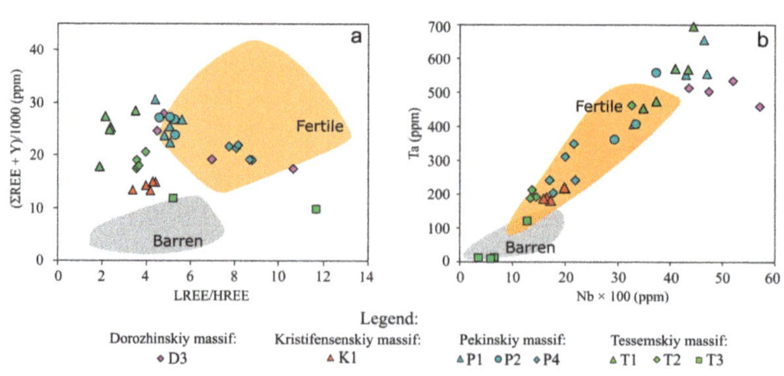

Figure 15. Composition of titanite from the Taimyr granites on discrimination diagrams with fields for barren and fertile magmatic systems: (**a**) (ΣREE + Y)/1000 ppm vs. LREE/HREE; (**b**) Ta (ppm) vs. Nb × 100 ppm after [98].

7. Discussion

7.1. Geochemical Constraints on the Petrogenesis and Fertility of the Taimyr Granites

The studied granitoid complexes of the northeastern Taimyr Peninsula include small to medium-sized intrusions (40–800 km^2) composed of biotite (±amphibole) quartz monzonites, granodiorites and granites comprising the main intrusive phase, and minor late biotite leucogranites. The presence of porphyritic varieties points to a hypabyssal character of the rocks. The late Carboniferous Tessemsky granite formed in a suprasubduction geodynamic setting, while the other studied intrusions are supposedly associated with the formation of the Siberian Traps LIP at the Permo-Triassic boundary [59].

The intrusions comprise predominantly magnesian (Figure 5d), slightly peraluminous (Figure 5c), high-K calc-alkaline (Figure 5b), and amphibole-bearing granitoid rocks, marked by diagnostic LILE (e.g., Cs, Th, Ba, K, Pb) enrichment and negative anomalies in HFSEs (e.g., Nb, Ta, Ti) (Figure 6), which are typical features of subduction-related granites elsewhere (Figures 5 and 6) [71,99,100]. In general, the northeastern Taimyr granitoids can be classified as evolved high-K I-type series, which is favorable for porphyry-type mineralization and hosts a number of world-class porphyry copper-gold deposits, such as Bingham in the USA, Grasberg in Indonesia, Northparkes in Australia, Oyu Tolgoi in Mongolia, and Peschanka in Russia [1,101].

Although the studied intrusions emplaced in different geotectonic settings within a time span of ca. 50 Ma, they comprise a relatively geochemically uniform complex. This is in line with the conclusions of Kurapov et al. (2024) [59], who used a larger geochemical dataset to show that the late Paleozoic-early Mesozoic granitoids of Taimyr formed a continuous magmatic series where Triassic plume-related granitoids inherited geochemical signatures of Carboniferous supra-subduction magmatic rocks, indicating a similar magma source for these intrusions.

Another characteristic feature of the Taimyr granites noted by Berzin et al. (2024) and Kurapov et al. (2024) [59,73] is their geochemical similarity to adakites. The term "adakite" was proposed by M.S. Drummond and M.J. Defant (1990) [75] for intermediate-acid volcanic rocks from Adak Island in the Aleutian arc, in which major and trace element composition implied an origin by melting of subducted oceanic basalts. Compositionally, adakites are rich in silica (SiO_2 > 56 wt.%) and sodium (K_2O/Na_2O < 0.5) with high Sr (≥400 ppm) and low Y (≤18 ppm) and Yb (≤1.9 ppm) concentrations and high Sr/Y (>40) and La/Yb (>20) ratios [75,76,102]. In the last decades, potassium-rich magmatic rocks with adakitic trace element characteristics have been repeatedly reported from post-collisional geodynamic environments, such as in Tibet and Himalaya after 50 Ma, and referred to as high potassium continental adakites [102–107]. Most of the Taimyr granite samples investigated in this study have K_2O/Na_2O ratios in the range of 0.3–6.9 (Supplementary Table S1) in combination with high Sr/Y (>40) and La/Yb (>20) ratios (Figure 7), and plot in the fields of High-K and shoshonitic series in the K_2O vs. SiO_2 diagram (Figure 4), indicating their similarity to high potassium continental adakites characteristic for post-collisional settings.

Adakites rich in Na_2O are usually interpreted to be generated by partial melting of MORB-type oceanic crust at eclogite-facies conditions [108], while potassium-rich continental adakites are generally attributed to partial melting of thickened and/or hydrated mafic lower crust [109]. Experimental studies have shown that trace element geochemical features of continental and arc-related adakites (fractionated REE patterns with depletion in MREE, depletion in Y relative to Sr, and inconspicuous Eu anomaly (Eu/Eu* = 0.8~0.9)) can be explained by crystallization of amphibole as an early fractionation phase at the expense of plagioclase in hydrous arc magma under relatively high-pressure conditions (>0.7 GPa) within the field of garnet stability [110–112] or by partial melting of a mafic source containing garnet as a residual mineral at middle to lower crustal depths [113,114]. The latter model fits well with the observed characteristics of the Taimyr granites, where fractionated REE patterns point to origination from a garnet-bearing protolith, while depletion in MREE, evident from slightly concave-up REE patterns, and a lack of significant negative Eu anomalies (Figure 5) indicate derivation from a mafic garnet ±amphibole-rich and plagioclase-poor source. The origin of the Taimyr granites from the lower continental crust is further supported by their Sr-Nd isotopic signatures, with εNd varying from −4 to +3.5 and $^{87}Sr/^{86}Sr_i$. ratios generally in the range 0.7040–0.7090 [59]. These variably mixed isotopic signatures are characteristic for lower crustal materials elsewhere [99] and were reported for similar continental adakites derived from the thickened lower crust of Tibet.

Thus, it can be tentatively suggested that adakitic geochemical characteristics of the Taimyr granites are a result of derivation from a relatively homogeneous source represented by a thickened mafic lower crust that formed at the stage of Carboniferous continental subduction [14,17,20,31,33,35,36,41,53] and continued to produce granitic melts during the early Permian post-collisional event and later in response to the development of a Siberian mantle plume at the Permo-Triassic boundary. It was shown that oxidized, water-saturated adakitic melts produced from thickened lower crust in similar geotectonic settings are highly productive with respect to Mo-Cu porphyry systems [115,116].

7.2. *Constraints from the Porphyry Indicator Minerals*

Porphyry Cu-Mo-Au deposits are commonly related to small stocks of acid to intermediate porphyritic rocks crystallized under oxidized and water-saturated conditions at moderate temperatures [2,111,115]. The composition of porphyry indicator minerals, such as zircon ($ZrSiO_4$), apatite ($Ca_5(PO_4)_3F$), and titanite ($CaTi(SiO_4)O$), is widely used to constrain the oxidation state of magma, content of volatiles, and degree of magma fractionation in order to distinguish porphyry-fertile granitoid intrusions [7,8,11,12,117–121].

7.2.1. Redox Conditions of the Melt

The redox conditions of andesitic-rhyolitic magmas are an important fertility indicator because high oxygen fugacity can prevent the early crystallization of sulfides, which ensures sufficient accumulation of ore-forming elements in residual melts for later mineralization [111,122].

Based on mineralogical composition, Proskurnin et al. (2021) [4] classified the granites of the Taimyr Porphyry Belt as magnetite-series, which is indicative of derivation from oxidized magmas. Further assessment of the oxidation state can be conducted utilizing the composition of porphyry indicator minerals, such as zircon. Experimental studies have shown that Ce and Eu anomalies in REE spectra of accessory zircon mainly depend on the oxygen fugacity of the melt, with stronger positive Ce anomalies and moderately negative Eu anomalies characteristic for zircon crystallized from oxidized magmas [123–125]. In particular, distinctive Eu anomalies, expressed as Eu/Eu* ratios in zircon > 0.4, characterize the fertile magma batches associated with world-class porphyry Cu–Mo±Au systems in Central Asia, Chile, Peru, and North America [118,119,122,126]. However, Ce/Ce* is often difficult to estimate accurately due to low concentrations of La and Pr in zircon, e.g., [80,118]. As an alternative, Loader et al. (2017,2022) [81,82] proposed to use chondrite-normalized Ce/Nd ratio as an estimate of the Ce anomaly in zircon. Following this approach, based on a global dataset, Lu et al. (2016) [83] have shown that zircons from fertile intrusions have higher 1000 × (Eu/Eu*)/Y (>1) and (Ce/Nd)/Y (>0.01), and lower Dy/Yb (<0.3) ratios compared to barren plutons as a result of early garnet and hornblende crystallization that depleted concentration of Y in the residual melt. Most of the analyzed zircon grains from the Taimyr granites plot in the field of fertile magmatic suites in a diagram 1000 × (Eu/Eu*)/Y vs. (Ce/Nd)/Y, with the exception of data for the Simsovskiy massif, which entirely plots in the field of barren granites (Figure 10a). However, in a diagram Eu/Eu* vs. Ce/Nd, only data from the Tessemskiy, Shirokinskiy, and Kristifensenskiy intrusions have Eu/Eu* > 0.4 and consistently plot in the field of fertile suites, while half of the analyses from the Pekinskiy and Dorozhinskiy massifs, as well as all analytical points from the Simsovskiy massif, show Eu/Eu* < 0.4 and plot slightly below the field of fertile granites (Figure 10b).

A new tool for quantitative estimation of magmatic oxidation state was presented by Loucks et al. (2020) [86], who developed a novel oxybarometer based on Ce, U, and Ti concentrations in zircon. Applying this method, the ΔFMQ values have been calculated for all analyzed zircon grains from the Taimyr granites (Supplementary Table S1). Zircon grains from the Dorozhinskiy, Pekinskiy, and Shirokinskiy massifs yielded ΔFMQ values in the range from +1.0 to +2.1 indicative of derivation from oxidized melts and consistent with the range of ΔFMQ +1 to +3 established for fertile porphyry systems worldwide [84] and references therein. Zircons from the Kristifensenskiy and Tessemskiy massifs yielded varying and generally lower ΔFMQ values in the range of +0.5 to +1.1, and zircons from the Simsovskiy massif yielded ΔFMQ values below +0.5 typical for reduced, barren magmatic suites. In addition to the zircon oxybarometer, the Mn-in-apatite oxybarometer by Miles et al. (2014) [90] was used to estimate redox conditions of apatite crystallization in the studied granites. Mn contents in the analyzed apatite grains vary in the range from 260 to 1200 ppm and yield oxygen fugacity ($logfO_2$) values between −10.4 and −12.1 (±2.8) (Supplementary Table S1). One exception is represented by apatites recovered from sample P3 from the Pekinskiy massif that are relatively enriched in Mn (850–2700 ppm) and yield slightly lower $logfO_2$ values between −13 and −16, corresponding to less oxidized conditions. Because the Mn content of apatite can depend on the melt composition, the Mn-in-apatite oxybarometer was used as an additional estimate of the redox conditions in combination with the zircon oxybarometer [127–129].

7.2.2. Estimates of Crystallization Temperature

The temperature of zircon crystallization was calculated utilizing the Ti-in-zircon thermometer of Watson et al. (2006) [85]. Zircons from the Dorozhinskiy, Kristifensenskiy, Shirokinskiy, Pekinskiy, Tessemskiy, and Simsovskiy massifs record relatively low temperatures between 650 °C and 780 °C and generally correspond to the temperature range below 750 °C characteristic for porphyry-fertile systems [84] and references therein. Zircons from samples P3 and P4 from the Pekinskiy intrusion yielded slightly higher crystallization temperatures in the range 730–800 °C. Additionally, the temperature was estimated with the Zr-in-titanite geothermometer of Haydenet et al. (2008) [97], which yielded titanite crystallization temperatures in the range of 640 to 720 °C that closely match the temperatures obtained by Ti-in-zircon geothermometry (Supplementary Table S1) and is consistent with crystallization of both minerals from relatively cool water-saturated magmas favorable for porphyry formation [84].

7.2.3. Implications for the Granite-Fertility from Apatite Composition

Apatite is a good tracer of fluids in a magmatic system because it incorporates volatiles directly into its crystal lattice [130,131]. All analyzed apatite crystals from the Taimyr granites are relatively rich in F (generally in the range 2.2–4.2 wt.%) and poor in Cl (mostly $\leq$ 0.15 wt.%, up to 0.5 wt.% in sample 1–47), corresponding to the fluorapatite group [131,132]. Sulfur is another important volatile species that can be incorporated in apatite. In silicic oxidized melts, the sulfur valence changes from S^{2+} to S^{6+}, and sulfate ions replace P and/or Ca in the apatite lattice [132,133]. It was shown that apatite, crystallized at reduced conditions, tends to have lower sulfur contents regardless of the amount of sulfur dissolved in the parental melt [134–136]. In accordance with this, apatite from porphyry-fertile granites has elevated SO_3 contents commonly exceeding 0.15–0.2 wt.% [87,137]. The analyzed SO_3 concentrations in apatite recovered from the Taimyr granites generally fall in the range of 0.09–0.60 wt.%, which is consistent with apatite compositions from porphyry-fertile granitoids elsewhere and from Cu-mineralized high-potassium adakites of eastern Tibet in particular (Figure 12a).

Although trace element concentrations in accessory apatite generally reflect the composition of the associated melt, primary geochemical signatures of the melt can be better preserved in this mineral compared to the whole rock samples that are often affected by secondary alteration [11]. In particular, similar to Mn, which is used in apatite oxybarometer, other redox-sensitive trace elements in apatite, such as Eu, can also serve as a proxy to the redox state of parental magma [90]. Because the ionic radius of Eu^{3+} is similar to that of Ca^{2+} [132], Eu^{3+} can be easier incorporated in the apatite lattice compared to Eu^{2+}

and other REE^{2+}, and, therefore, apatite crystallizing from oxidized magma that predominantly contains Eu^{3+} would concentrate more Eu and show less pronounced negative Eu anomalies expressed as elevated Eu/Eu* values [110,138]. As seen in a diagram Eu/Eu* vs. Sr/Y (Figure 12b), apatite from the studied Taimyr granites has elevated Eu/Eu* ratios in the range of 0.3–0.6, indicating crystallization under oxidized conditions, and high Sr/Y ratios typical for adakitic rocks. These results provide further evidence for the similarity of the Taimyr granites with Cu-mineralized continental adakites of Tibet and are in line with conclusions derived from the whole rock and zircon geochemical data.

7.2.4. Additional Constraints on the Granite-Fertility from Composition of Titanite

Because titanite is susceptible to post-magmatic alteration processes [93,94], it was important to distinguish and filter out hydrothermally altered titanite grains from the dataset used in this study. When zircon-contaminated analyses with Zr concentrations above 2000 ppm were filtered out, the remaining data were plotted in Al vs. Fe (a.p.f.u.) and Th/U vs. Nb/Ta diagrams that are used to discriminate between magmatic and hydrothermal titanite (Figure 13). The diagrams show that all analyzed titanite grains plot in the fields of magmatic origin with exception of sample T3 from the Tessemsky massif, which has high Nb/Ta ratios (Figure 13b) and "disturbed" REE patterns, i.e., significantly depleted in HREE compared to other samples from this intrusion (Figure 14). Therefore, titanite from sample T3 was classified as hydrothermal, and corresponding analytical data were excluded from further consideration.

Similar to apatite, abundances of redox-sensitive trace elements in titanite can be used to estimate the redox conditions of its crystallization [98,139]. It was shown that, compared to Ce^{4+} and Eu^{2+}, Ce^{3+} and Eu^{3+} can easier substitute Ca in titanite because their ionic radii are similar to those of Ca^{2+}. In the same way, Fe^{3+} and Ga^{3+} would preferably occupy the Ti sites of titanite, compared to Fe^{2+} and Ga^{2+} [92,98,140]. Consequently, the increase in oxygen fugacity, which increases concentrations of Ce^{4+}, Eu^{3+}, Fe^{3+}, and Ga^{3+} in the melt, facilitates incorporation of more Fe, Eu, and Ga in the titanite crystal lattice and suppresses incorporation of Ce. Taking this into account, the obtained analyses of titanite from the Taimyr granites were evaluated in terms of their Al/Fe, Eu/Eu*, and Ce/Ce* ratios and Ga enrichment and compared with data reported for porphyry-fertile granitoids of Tibet (Figure 15). Concentrations of Ga in the analyzed titanite vary from 70 to 160 ppm, and their Al/Fe ratios are in the range between 0.8 and 1.7 (Supplementary Table S1), which is consistent with the composition of titanite from Cu-mineralized oxidized granitoids of Tibet [98]. REE patterns of the Taimyr titanite are characterized by noticeable negative Ce anomalies (Figure 14), and their Eu/Eu* ratios fall in the range of 0.3–0.9 (Supplementary Table S1), closely corresponding to Eu/Eu* ratios of titanites from porphyry-fertile granitoids of Tibet [98]. Titanite from the mineralized Tibetan intrusions is also shown to have much higher Ta, Nb, and ΣREE + Y contents and higher LREE/HREE ratios compared to barren granitoids from the same area [98,139]. The discrimination diagrams (ΣREE + Y)/1000 vs. LREE/HREE and Ta vs. Nb, presented in Figure 15, demonstrate that titanite from the Taimyr granites generally plots in the fields of fertile compositions, indicating their similarity with Cu-mineralized adakitic granitoids from the India-Asia collision zone.

8. Conclusions

Granitoid intrusions of the northeastern Taimyr Peninsula comprise small to moderate-sized bodies composed of biotite (±amphibole) quartz monzonites, granodiorites, granites, and leucogranites. The intrusions formed in the course of late Paleozoic-early Mesozoic tectono-magmatic evolution of the Siberian margins: the late Carboniferous Tessemsky massif represents suprasubduction granitoid series, while the Dorozhinskiy, Kristifensenskiy, Pekinskiy, Shirokinskiy, and Yuzhno-Lodochnikovskiy massifs are correlated with the early Triassic magmatism of the Siberian Traps Large Igneous Province.

The studied granitoids comprise a geochemically relatively uniform series where Triassic plume-related granitoids inherited geochemical signatures of Carboniferous supra-subduction magmatic rocks. The series includes predominantly magnesian, slightly peraluminous high-K calc-alkaline amphibole-bearing I-type granitoids, marked by enrichment in LILE and negative anomalies of Nb, Ta, and Ti typical of subduction-related rocks. The rocks have high Sr/Y and La/Yb ratios characteristic for high-potassium continental adakites of collisional zones. It is suggested that adakitic geochemical characteristics of the Taimyr granites are derived from a relatively homogeneous source represented by a thickened mafic lower crust that formed at the stage of the Carboniferous continental subduction and continued to produce granitic melts in the course of the early Mesozoic tectono-thermal events.

Whole rock geochemical data and compositions of porphyry mineral indicators (zircon, apatite, and titanite) show that granitoids of the northeastern Taimyr crystallized from oxidized and water-saturated magmas at moderate temperatures and generally fit the exploration criteria established for porphyry-fertile granitoids worldwide (ΔFMQ +1 to +3 with zircon Eu/Eu* > 0.4 and apatite SO_3 > 0.2 wt.%). The obtained results also help to distinguish the most prospective targets among the seven studied intrusions. Data from the Dorozhinskiy, Kristifensenskiy, Pekinskiy, and Tessemskiy intrusions fully match geochemical criteria for porphyry-fertile granitoids, and these massifs are considered the most prospective for Cu-Mo mineralization. Granites from the Shirokinskiy and Yuzhno-Lodochnikovskiy intrusions only partially match compositional constraints for fertile melts and can be considered as second-tier exploration targets. Finally, although mineral chemistry data for the Simsovsky massif were obtained only for zircon, estimated crystallization conditions for this intrusion preclude its classification as a porphyry-fertile body. These conclusions are in line with previously developed exploration criteria for the northeastern Taimyr, showing that geochemical indicators of granite-fertility can be used on a regional scale in parallel with other exploration methods.

Supplementary Materials: The following is available online at https://www.mdpi.com/article/10.3390/min14111065/s1. Supplementary Table S1: Sample description and analytical data, Supplementary Figure S1: Representative BSE and CL-images of the studied zircon grains showing positioning of LA analytical spots at a distance from visible mineral inclusions.

Author Contributions: Conceptualization, S.V.B., D.L.K. and S.V.P.; formal analysis, D.L.K., S.V.B. and S.V.P.; funding acquisition, D.L.K., S.V.B., S.V.P. and M.Y.K.; investigation, S.V.B., V.F.P., E.I.B., S.V.P., M.Y.K. and R.S.P.; methodology, S.V.B., D.L.K. and S.V.P.; resources, S.V.P., T.A.G., N.Y.C., V.S.C. and E.M.A.; supervision, D.L.K. and S.V.B.; visualization, S.V.B. and D.L.K.; writing—original draft, D.L.K. and S.V.B.; writing—review and editing, S.V.B., D.L.K. and S.V.P. All authors have read and agreed to the published version of the manuscript.

Funding: This work was supported by the Russian Science Foundation, Project # 23-27-00283 to D.L.K., S.V.B., S.V.P. and M.Y.K.

Data Availability Statement: Data are contained within the article and Supplementary Materials.

Acknowledgments: We are grateful to many colleagues for their help and cooperation during field work in Taimyr and at the stage of laboratory studies. Olga Plotinskaya provided important and constructive comments on the early version of the text. We are thankful for the detailed and valuable comments of three anonymous reviewers that helped to improve the original version of the manuscript.

Conflicts of Interest: Natalya Ya. Chernenko is employee of LLC "Polar Forwarding Company. The paper reflects the views of the scientists and not the company.

References

1. Sillitoe, R.H. Porphyry Copper Systems. *Econ. Geol.* **2010**, *105*, 3–41. [CrossRef]
2. Sillitoe, R.H. Geological Criteria for Porphyry Copper Exploration. *Acta Geol. Sin.* **2014**, *88* (Suppl. 2), 597–598. [CrossRef]
3. Proskurnin, V.F.; Petrov, O.V.; Romanov, A.P.; Kurbatov, I.I.; Gavrish, A.V. Gold-bearing copper-molybdenum-porphyry belt of the Taimyr and Severnaya Zemlya. In *Scientific and Methodological Basis for Forecasting, Prospecting, Evaluation of Diamond Deposits, Precious and Non-Ferrous Metals, Proceedings of the IX International Scientific and Practical Conferences, Dnipro, Ukraine, 2–3 December 2019*; FSBI "TSNIGRI": Moscow, Russia, 2019; p. 133.

4. Proskurnin, V.F.; Petrov, O.V.; Romanov, A.P.; Kurbatov, I.I.; Gavrish, A.V.; Proskurnina, M.A. Central Arctic gold-bearing copper-molybdenum-porphyric belt. *Reg. Geol. I Metallog.* **2021**, *85*, 31–49. [CrossRef]
5. Berzon, E.I.; Proskurnin, V.F.; Chernenko, N.Y.; Maltseva, C.V. Prospects for identifying porphyry copper deposits on the Chelyuskin Peninsula. In *Proizvodstvennye i Metodicheskie Aspekty Geologorazvedochnyh Rabot. Geologicheskiy Vestnik;* Trushin, S.I., Oseckij, A.I., Solopov, Y.A., Eds.; JSC "Polymetal UK": St. Petersburg, Russia, 2021; Volume 9, pp. 3–16.
6. Proskurnin, V.F.; Proskurnina, M.A.; Kurapov, M.Y.; Gavrish, A.V.; Brusnitsyn, A.I.; Shatov, V.V. Verkhnetareysky gold-copper-porphyry potential ore center of Central Taimyr. In *Scientific and Methodological Basis for Forecasting, Searching, and Evaluation of Diamond, Precious and Non-Ferrous Metal Deposits, Proceedings of the XIII International Scientific and Practical Conference, Voronezh, Russia, 18–19 April 2024;* FSBI "TSNIGRI": Moscow, Russia, 2024; pp. 316–320.
7. Wilkinson, J.J.; Cooke, D.R.; Baker, M.J.; Chang, Z.; Wilkinson, C.C.; Chen, H.; Fox, N.; Hollings, P.; White, N.C.; Gemmell, J.B.; et al. Porphyry indicator minerals and their mineral chemistry as vectoring and fertility tools. *Geol. Surv. Can. Open File* **2017**, 8345. [CrossRef]
8. Ballard, J.R.; Palin, J.M.; Campbell, I.H. Relative oxidation states of magmas inferred from Ce(IV)/Ce(III) in zircon: Application to porphyry copper deposits of northern Chile. *Contrib. Miner. Petrol.* **2002**, *144*, 347–364. [CrossRef]
9. Belousova, E.A.; Griffin, W.L.; O'Reilly, S.Y.; Fisher, N.I. Apatite as an indicator mineral for mineral exploration: Trace-element compositions and their relationship to host rock type. *J. Geochem. Explor.* **2002**, *76*, 45–69. [CrossRef]
10. Ward, R. Isotope Geochemistry of Zircons from the Guichon Batholith, Highland Valley Copper Deposit: Southern British Columbia: Relationship Between Ce^{4+}/Ce^{3+} in Zircon, Oxidation State in Magmas and Ore Genesis. Bachelor's Thesis, Carleton University, Ottawa, ON, Canada, 2008.
11. Mao, M.; Rukhlov, A.S.; Rowins, S.M.; Spence, J.; Coogan, L.A. Apatite trace element compositions: A robust newtool for mineral exploration. *Econ. Geol.* **2016**, *111*, 1187–1222. [CrossRef]
12. Cooke, D.R.; Agnew, P.; Hollings, P.; Baker, M.; Chang, Z.; Wilkinson, J.J.; White, N.C.; Zhang, L.; Thompson, J.; Gemmell, J.B.; et al. Porphyry Indicator Minerals (PIMS) and Porphyry Vectoring and Fertility Tools (PVFTS)–Indicators of Mineralization Styles and Recorders of Hypogene Geochemical Dispersion Halos. In Proceedings of the Exploration 17: Sixth Decennial International Conference on Mineral Exploration, Toronto, ON, Canada, 22–25 October 2017; pp. 457–470.
13. Pan, L.-C.; Hu, R.-Z.; Bi, X.-W.; Li, C.; Wang, X.-S.; Zhu, J.-J. Titanite major and trace element compositions as petrogenetic and metallogenic indicators of Mo ore deposits: Examples from four granite plutons in the southern Yidun arc, SW China. *Amer. Mineral.* **2018**, *103*, 1417–1434. [CrossRef]
14. Vernikovsky, V.A.; Vernikovskaya, A.E.; Proskurnin, V.F.; Matushkin, N.Y.; Proskurnina, M.A.; Kadilnikov, P.; Larionov, A.N.; Travin, A.V. Late Paleozoic–Early Mesozoic Granite Magmatism on the Arctic Margin of the Siberian Craton during the Kara-Siberia Oblique Collision and Plume Events. *Minerals* **2020**, *10*, 571. [CrossRef]
15. Ershova, V.B.; Anfinson, O.; Prokopiev, A.V.; Khudoley, A.K.; Stockli, D.F.; Faleide, J.I.; Gaina, C.; Malyshev, N.A. Detrital Zircon (U-Th)/He Ages from Paleozoic Strata of the Severnaya Zemlya Archipelago: Deciphering Multiple Episodes of Paleozoic Tectonic Evolution within the Russian High Arctic. *J. Geodyn.* **2018**, *119*, 210–220. [CrossRef]
16. Zonenshain, L.P.; Kuz'min, M.I.; Natapov, L.M. *Plate Tectonics of the USSR Territory;* Nedra: Moscow, Russia, 1990.
17. Vernikovsky, V.A. The Geodynamic Evolution of the Taimyr Folded Area. *Geol. Pac. Ocean* **1996**, *12*, 691–704.
18. Drachev, S.S.; Malyshev, N.A.; Nikishin, A.M. Tectonic history and petroleum geology of the Russian Arctic Shelves: An overview. *Geol. Soc. Lond. Petrol. Geol. Conf. Ser.* **2010**, *7*, 591–619. [CrossRef]
19. Kachurina, N.V.; Makariev, A.A.; Makarieva, E.M.; Gavrish, A.V.; Orlov, V.V.; Sakharov, A.A.; Dymov, V.A.; Kozlov, S.A. *State Geological Map of the Russian Federation. Scale 1:1 000 000 (Third Generation). Sheet T-45-48 Cape Chelyuskin. Explanatory Letter;* Makariev, A.A., Makarieva, E.M., Eds.; Kartfabrika VSEGEI: St. Petersburg, Russia, 2013.
20. Pogrebitsky, Y.E. *Paleotectonic Analysis of the Taimyr Folded System;* Nedra: Leningrad, Russia, 1971; 248p.
21. Bezzubtsev, V.V.; Zalyaleev, R.M.; Sakovich, A.B. *Geological Map of Mountainous Taimyr. Scale 1:500 000. Explanatory Note;* Krasnoyarskgeologiya: Krasnoyarsk, Russia, 1986; 177p.
22. Ershova, V.B.; Prokopiev, A.V.; Khudoley, A.K.; Shneider, G.V.; Andersen, T.; Kallerund, K.; Makariev, A.A.; Kolchanov, D.A. U-Pb Dating of Detrital Zircons from the Lower Paleozoic Deposits of the North Kara Basin. *Dokl. Earth Sci.* **2015**, *464*, 444–447. [CrossRef]
23. Ershova, V.B.; Prokopiev, A.V.; Khudoley, A.K.; Sobolev, N.N.; Petrov, E.O. Detrital zircon ages and provenance of the Upper Paleozoic successions of Kotel'ny Island (New Siberian Islands archipelago). *Lithosphere* **2015**, *7*, 40–45. [CrossRef]
24. Ershova, V.B.; Prokopiev, A.V.; Nikishin, V.A.; Khudoley, A.K.; Malyshev, N.A.; Nikishin, A.M. New Data on Upper Carboniferous-Lower Permian Deposits of Bol'shevik Island, Severnaya Zemlya Archipelago. *Polar Res.* **2015**, *34*, 24558. [CrossRef]
25. Ershova, V.B.; Prokopiev, A.V.; Khudoley, A.K.; Proskurnin, V.F.; Andersen, T.; Kullerud, K.; Stepunina, M.A.; Kolchanov, D.A. New U–Pb Isotopic Data for Detrital Zircons from Metasedimentary Sequences of Northwestern Taimyr. *Dokl. Earth Sci.* **2017**, *474*, 613–616. [CrossRef]
26. Ershova, V.B.; Prokopiev, A.V.; Khudoley, A.K.; Andersen, T.; Kullerud, K.; Kolchanov, D.A. U–Pb Age and Hf Isotope Geochemistry of Detrital Zircons from Cambrian Sandstones of the Severnaya Zemlya Archipelago and Northern Taimyr (Russian High Arctic). *Minerals* **2020**, *10*, 36. [CrossRef]

27. Proskurnin, V.F. New volcano-plutonic association of Severnaya Zemlya and features of its metal content. In *Subsoil of Taimyr. Collection of Scientific Papers*; Samoilov, A.G., Malich, N.S., Eds.; Committee of the Russian Federation on Geology and Subsoil Use "TAIMYRGEOLKOM": Norilsk, Russia, 1995; pp. 93–100.
28. Vernikovsky, V.A.; Neimark, L.A.; Ponomarchuk, V.A.; Vernikovskaya, A.E.; Kireev, A.D.; Kuz'min, D.S. Geochemistry and Age of Collision: Granitoides and Metamorphites of the Kara Microcontinent (Northern Taimyr). *Russ. Geol. Geophys.* **1995**, *36*, 46–60.
29. Gor, Y.G.; Egorov, V.N.; Kabankov, V.I.; Kuzmin, V.G.; Lopatin, B.G.; Markovsky, V.A.; Musatov, E.E.; Proskurnin, V.F.; Sobolevskaya, R.F.; Tuganova, E.V.; et al. *State Geological Map of the Russian Federation. Scale 1:1 000 000 (New Series). Sheet T-45-47 (October Revolution Is.) and T-48-50 (Bol'shevik Is.). Explanatory Letter*; Malich, N.S., Ed.; Publishing House of the St. Petersburg Cartographic Factory VSEGEI: St. Petersburg, Russia, 2003; 213p.
30. Lorenz, H.; Gee, D.G.; Whitehouse, M.J. New Geochronological Data on Palaeozoic Igneous Activity and Deformation in the Severnaya Zemlya Archipelago, Russia, and Implications for the Development of the Eurasian Arctic Margin. *Geol. Mag.* **2007**, *144*, 105–125. [CrossRef]
31. Vernikovsky, V.A.; Sal'nikova, E.; Kotov, A.B.; Ponomarchuk, V.A.; Kovach, V.P.; Travin, A.V.; Yakovleva, S.Z.; Bereznaya, N.G. Age of Post-Collisional Granitoids of Northern Taimyr: U-Pb, Sm-Nd, Rb-Sr, and Ar-Ar Data. *Dokl. Akad. Nauk. Akad. Nauk* **1998**, *363*, 375–378.
32. Kuzmin, V.G.; Avdyunichev, V.V.; Gavrish, A.V.; Lazurkin, D.V.; Proskurnin, V.F.; Smirnov, A.N.; Ushakov, V.I.; Fokin, V.I.; Shulga, Y.D. *Severnaya Zemlya. Geological Structure and Minerageny*; Gramberg, I.S., Ushakov, V.I., Eds.; Publishing House VNIIOkeangeologiya: St. Petersburg, Russia, 2000; 187p.
33. Kurapov, M.Y.; Ershova, V.B.; Makariev, A.A.; Makarieva, E.V.; Khudoley, A.K.; Luchitskaya, M.V.; Prokopiev, A.V. Carboniferous Granitoid Magmatism of Northern Taimyr: Results of Isotopic-Geochemical Study and Geodynamic Interpretation. *Geotectonics* **2018**, *52*, 225–239. [CrossRef]
34. Proskurnina, M.A.; Proskurnin, V.F.; Remizov, D.N.; Larionov, A.N. Ring Intrusions of the Bespamyatnaya Areal: Manifestations of Shoshonite-Latite Magmatism in Northern Taimyr. *Reg. Geol. Metallog.* **2019**, *76*, 5–22.
35. Kurapov, M.Y.; Ershova, V.B.; Khudoley, A.; Luchitskaya, M.V.; Makariev, A.A.; Makarieva, E.M.; Vishnevskaya, I.A. Late Palaeozoic Magmatism of Northern Taimyr: New Insights into the Tectonic Evolution of the Russian High Arctic. *Int. Geol. Rev.* **2020**, *63*, 1990–2012. [CrossRef]
36. Kurapov, M.Y.; Ershova, V.B.; Khudoley, A.K.; Makariev, A.A.; Makarieva, E.M. The First Evidence of Late Ordovician Magmatism of the October Revolution Island (Severnaya Zemlya Archipelago, Russian High Arctic): Geochronology, Geochemistry and Geodynamic Settings. *Nor. J. Geol.* **2020**, *100*, 1–15. [CrossRef]
37. Kurapov, M.Y.; Ershova, V.B.; Khudoley, A.K.; Luchitskaya, M.V.; Stockli, D.F.; Makariev, A.A.; Makarieva, E.M.; Vishnevskaya, I.A. Latest Permian–Triassic Magmatism of the Taimyr Peninsula: New Evidence for a Connection to the Siberian Traps Large Igneous Province. *Geosphere* **2021**, *17*, 2062–2077. [CrossRef]
38. Sobolevskaya, R.F. New data on the stratigraphy of Cambrian sediments of Central Taimyr. In *Trudy NIIGA*; NIIGA: Leningrad, Russia, 1959; Volume 105, pp. 44–49.
39. Khudoley, A.K.; Verzhbitsky, V.E.; Zastrozhnov, D.A.; O'Sullivan, P.; Ershova, V.B.; Proskurnin, V.F.; Tuchkova, M.I.; Rogov, M.A.; Kyser, T.K.; Malyshev, S.V.; et al. Late Paleozoic—Mesozoic Tectonic Evolution of the Eastern Taimyr-Severnaya Zemlya Fold and Thrust Belt and Adjoining Yenisey-Khatanga Depression. *J. Geodyn.* **2018**, *119*, 221–241. [CrossRef]
40. Makhlaev, L.V.; Korobova, N.I. *Genetic Granitoid Series of the Precambrian of Taimyr*; Krasnoyarsk Book Publishing House: Krasnoyarsk, Russia, 1972; 130p.
41. Proskurnin, V.F. Magmatic formations of the Taimyr-Severozemelskaya folded system, their ore content and geodynamic features of formation. In *Ore Content of Magmatic Formations of Siberia*; SNIIGiMS: Novosibirsk, Russia, 1991; pp. 33–39.
42. Ufland, A.K.; Natapov, L.M.; Lopatin, V.M. About the tectonic nature of the Taimyr. *Geotectonics* **1991**, *6*, 76–93.
43. Metelkin, D.V.; Vernikovsky, V.A.; Kazansky, A.Y.; Bogolepova, O.K.; Gubanov, A.P. Paleozoic History of the Kara Microcontinent and Its Relation to Siberia and Baltica: Paleomagnetism, Paleogeography and Tectonics. *Tectonophysics* **2005**, *398*, 225–243. [CrossRef]
44. Vernikovsky, V.A.; Vernikovskaya, A.E.; Pease, V.L.; Gee, D.G. *Neoproterozoic Orogeny Along the Margins of Siberia*; Geological Society: London, UK, 2005. [CrossRef]
45. Metelkin, D.V.; Vernikovsky, V.A.; Matushkin, N.Y. Arctida between Rodinia and Pangea. *Precambrian Res.* **2015**, *259*, 114–129. [CrossRef]
46. Matushkin, N.Y.; Metelkin, D.V.; Vernikovsky, V.A.; Proskurnin, V.F.; Abashev, V.V.; Larionov, A.N.; Petrushkov, B.S. New Evidence of the Independent Precambrian Tectonic Historyof the Faddey Terrane of the Central Taimyr Accretionary Belt. *Dokl. Earth Sci.* **2023**, *511*, 636–644. [CrossRef]
47. Proskurnin, V.F.; Vernikovsky, V.A.; Metelkin, D.V.; Petrushkov, B.S.; Vernikovskaya, A.E.; Gavrish, A.V.; Bagaeva, A.A.; Matushkin, N.Y.; Vinogradova, N.P.; Larionov, A.N. Rhyolite–granite association in the Central Taimyr zone: Evidence of accretionary-collisional events in the Neoproterozoic. *Russ. Geol. Geophys.* **2014**, *55*, 18–32. [CrossRef]
48. Priyatkina, N.; Collins, W.J.; Khudoley, A.K.; Zastrozhnov, D.A.; Ershova, V.B.; Chamberlain, K.; Shatsillo, A.; Proskurnin, V.F. The Proterozoic Evolution of Northern Siberian Craton Margin: A Comparison of U–Pb–Hf Signatures from Sedimentary Units of the Taimyr Orogenic Belt and the Siberian Platform. *Int. Geol. Rev.* **2017**, *59*, 1632–1656. [CrossRef]

49. Vernikovsky, V.A.; Pease, V.L.; Vernikovskaya, A.E.; Romanov, A.P.; Gee, D.G.; Travin, A.V. First Report of Early Triassic A-Type Granite and Syenite Intrusions from Taimyr: Product of the Northern Eurasian Superplume? *Lithos* **2003**, *66*, 23–36. [CrossRef]
50. Khudoley, A.K.; Chamberlain, K.R.; Schmitt, A.K.; Harrison, T.M.; Prokopyev, A.V.; Sears, D.W.; Veselovsky, R.V.; Proskurnin, V.F. New data on U-Pb dating of mafic intrusions of Taimyr, northern and southeastern Siberia based on baddelite and their implications for the tectonics and stratigraphy of the region. In *Isotope Systems and Time of Geological Processes, Proceedings of the IV Russian Conference on Isotope Geochronology, St. Petersburg, Russia, 2–4 June 2009*; IP Katalkina: St. Petersburg, Russia, 2009; Volume II, pp. 243–245.
51. Ravitch, M.G.; Chaika, L.A. Small intrusions of the Byranga ridge (*Taimyr peninsula*). In *Trudy NIIGA*; NIIGA: Leningrad, Russia, 1959; Volume 88, 147p.
52. Augland, L.E.; Andresen, A.; Corfu, F.; Daviknes, H.K. Late Ordovician to Silurian Ensialic Magmatism in Liverpool Land, East Greenland: New Evidence Extending the Northeastern Branch of the Continental Laurentian Magmatic Arc. *Geol. Mag.* **2012**, *149*, 561–577. [CrossRef]
53. Pease, V.L.; Kuzmichev, A.B.; Danukalova, M.K. The New Siberian Islands and Evidence for the Continuation of the Uralides, Arctic Russia. *J. Geol. Soc.* **2014**, *172*, 1–4. [CrossRef]
54. Khain, V.E.; Sokolov, B.A.; Kleshchev, K.A.; Shein, V.S. Tectonic and Geodynamic Setting of Oil and Gas Basins of the Soviet Union. *Am. Assoc. Pet. Geol. Bull.* **1991**, *75*, 313–325.
55. Saunders, A.D.; England, R.W.; Reichow, M.K.; White, R.V. A Mantle Plume Origin for the Siberian Traps: Uplift and Extension in the West Siberian Basin, Russia. *Lithos* **2005**, *79*, 407–424. [CrossRef]
56. Dobretsov, N.L.; Polyansky, O.P.; Reverdatto, V.V.; Babichev, A.V. Dynamics of the Arctic and Adjacent Petroleum Basins: A Record of Plume and Rifting Activity. *Russ. Geol. Geophys.* **2013**, *54*, 888–902. [CrossRef]
57. Afanasenkov, A.P.; Nikishin, A.M.; Unger, A.V.; Bordunov, S.I.; Lugovaya, O.V.; Chikishev, A.A.; Yakovishina, E.V. The Tectonics and Stages of the Geological History of the Yenisei–Khatanga Basin and the Conjugate Taimyr Orogen. *Geotectonics* **2016**, *50*, 161–178. [CrossRef]
58. Krivolutskaya, N.A.; Latyshev, A.V.; Dolgal, A.S.; Gongalsky, B.I.; Makarieva, E.M.; Makariev, A.A.; Svirskaya, N.M.; Bychkova, Y.V.; Yakushev, A.I.; Asavin, A.M. Unique PGE-Cu-Ni Noril'sk Deposits, Siberian Trap Province: Magmatic and Tectonic Factors in Their Origin. *Minerals* **2019**, *9*, 66. [CrossRef]
59. Kurapov, M.Y.; Konopelko, D.L.; Biske, Y.S.; Proskurnin, V.F.; Petrov, S.V.; Proskurnina, M.A.; Berzon, Y.Y.; Ershova, V.B.; Berzin, S.V.; Stepanov, S.Y. Geodynamic settings of Late Paleozoic–Early Mesozoic granitoid magmatism at the Arctic continental margins: Insights from new geochronological and geochemical data from the Taimyr peninsula. *Minerals* **2024**, *14*, 423. [CrossRef]
60. Dill, H.J. The "chessboard" classification scheme of mineral deposits: Mineralogy and geology from aluminum to zirconium. *Earth-Sci. Rev.* **2010**, *100*, 1–420. [CrossRef]
61. Robb, L. *Introduction to Ore-Forming Processes*, 2nd ed.; Wiley-Blackwell: Hoboken, NJ, USA, 2020; 496p.
62. Markovsky, V.A.; Kabankov, V.Y.; Sobolevskaya, R.F.; Proskurnin, V.F.; Shneider, G.V.; Lazareva, L.N.; Gavrilov, A.V. *State Geological Map of the Russian Federation. Scale 1:200 000. Taimyr Series. Sheets T-47-XXVIII, XXIX, XXX (Marga River), T-48-XIX, XX, XXI (Chelyuskin Peninsula), T-48-XXII, XXIII, XXIV (Samuila Island), T-48-XXV, XXVI, XXII (Lodochnikov Plateau), T-48-XXVIII, XXIX, XXX (Simsa Bay). Explanatory Note*; Makariev, A.A., Makarieva, E.M., Eds.; Publishing House of the St. Petersburg Cartographic Factory VSEGEI: St. Petersburg, Russia, 2000.
63. Zaitceva, M.V.; Pupyshev, A.A.; Shchapova, J.V.; Votyakov, S.L. U-Pb dating of zircons using NexION 300S quadrupole mass spectrometer with inductively coupled plasma and NWR 213 attachment for laser ablation. *Analitika i Kontrol'* **2016**, *20*, 294–306. [CrossRef]
64. Chervyakovskaya, M.V.; Chervyakovskiy, V.S.; Votyakov, S.L. LA-ICP-MS analysis of trace elements in silicate minerals on ICP-MS NEXION 300s mass spectrometer with NWR 213 attachment for laser ablation: Methodological aspects. *Geodyn. Tectonophys.* **2022**, *13*, 0605. [CrossRef]
65. Chervyakovskaya, M.V.; Chervyakovskiy, V.S.; Pupyshev, A.A.; Votyakov, S.L. Determination of the Trace Element Composition and U–Pb Dating of Apatite by Inductively Coupled Plasma Laser Ablation Mass Spectrometry on a NexION 300S spectrometer with an NWR 213 Attachment. *J. Anal. Chem.* **2024**, *79*, 1252–1268. [CrossRef]
66. Middlemost, E.A.K. Naming Materials in the Magma/Igneous Rock System. *Earth Sci. Rev.* **1994**, *37*, 215–224. [CrossRef]
67. Sharpenok, L.N.; Kostin, A.E.; Kukharenko, E.A. TAS-diagram sum of alkalis–silica for chemical classification and diagnostic of plutonic rocks. *Regionalnaya Geol. I Metallogenia* **2013**, *56*, 40–50.
68. Maniar, P.D.; Piccoli, P.M. Tectonic discrimination of granitoids. *Geol. Soc. Amer.* **1989**, *101*, 635–643. [CrossRef]
69. Frost, B.R.; Frost, C.D. A geochemical classification for feldspathic igneous rocks. *J. Petrol.* **2008**, *49*, 1955–1969. [CrossRef]
70. Le Maitre, R.W.; Bateman, P.; Dudek, A.; Keller, J.; Lameyre, J.; Le Bas, M.J.; Sabine, P.A.; Schmid, R.; Sorensen, H.; Streckeisen, A.; et al. *A Classification of Igneous Rocks and Glossary of Terms*; Blackwell: Oxford, UK, 1989.
71. Pearce, J.A. A User's Guide to Basalt Discrimination Diagrams. In *Trace Element Geochemistry of Volcanic Rocks: Applications for Massive Sulphide Exploration*; Wyman, D.A., Ed.; Short Course Notes; Geological Association of Canada: St. John's, NL, Canada, 1996; Volume 12, pp. 79–113.
72. Sun, S.S.; McDonough, W.F. Chemical and Isotopic Systematics of Oceanic Basalts: Implications for Mantle Composition and Processes. *Geol. Soc. Lond. Spec. Publ.* **1989**, *42*, 313–345. [CrossRef]

73. Berzin, S.V.; Konopelko, D.L.; Petrov, S.V.; Kurapov, M.Y.; Golovina, T.A.; Chernenko, N.Y.; Chervyakovskiy, V.S. Utilizing Compositions of Zircon and Apatite for Prospecting of Cu-Mo-Au-Porphyry Mineralization in the Pekinsky and Tessemsky Granitoid Massifs of the Taimyr-Severozemelskaya Folded Area. *Lithosphere* **2024**, *24*, 547–565. [CrossRef]
74. Pearce, J.A.; Harris, N.B.W.; Tindle, A.G. Trace Element Discrimination Diagrams for the Tectonic Interpretation of Granitic Rocks. *J. Petrol.* **1984**, *25*, 956–983. [CrossRef]
75. Drummond, M.S.; Defant, M.J. A Model for Trondhjemite-Tonalite-Dacite Genesis and Crustal Growth via Slab Melting: Archean to Modern Comparisons. *J. Geophys. Res.* **1990**, *95*, 21503–22152. [CrossRef]
76. Richards, J.P.; Kerrich, R. Special Paper: Adakite-like Rocks: Their Diverse Origins and Questionable Role in Metallogenesis. *Econ. Geol.* **2007**, *102*, 537–576. [CrossRef]
77. Hoskin, P.W.O. Trace-element composition of hydrothermal zircon and the alteration of hadean zircon from the Jack Hills, Australia. *Geochim. Cosmochim. Acta* **2005**, *69*, 637–648. [CrossRef]
78. Zhong, S.; Feng, C.; Seltmann, R.; Li, D.; Qu, H. Can magmatic zircon be distinguished from hydrothermal zircon by trace element composition? The effect of mineral inclusions on zircon trace element composition. *Lithos* **2018**, *314–315*, 646–657. [CrossRef]
79. Belousova, E.A.; Griffin, W.L.; O'Reilly, S.Y.; Fisher, N.I. Igneous zircon: Trace element composition as an indicator of source rock type. *Contrib. Mineral. Petrol.* **2002**, *143*, 602–622. [CrossRef]
80. Belousova, E.A.; Griffin, W.L.; O'Reilly, S.Y. Zircon crystal morphology, trace element signatures and Hf isotope composition as a tool for petrogenetic modelling: Examples from eastern Australian granitoids. *J. Petrol.* **2006**, *47*, 329–353. [CrossRef]
81. Loader, M.A.; Wilkinson, J.J.; Armstrong, R.N. The effect of titanite crystallisation on Eu and Ce anomalies in zircon and its implications for the assessment of porphyry Cu deposit fertility. *Earth Planet. Sci. Let.* **2017**, *472*, 107–119. [CrossRef]
82. Loader, M.A.; Nathwani, C.L.; Wilkinson, J.J.; Armstrong, R.N. Controls on the magnitude of Ce anomalies in zircon. *Geochim. Cosmochim. Acta* **2022**, *328*, 242–257. [CrossRef]
83. Lu, Y.-J.; Loucks, R.R.; Fiorentini, M.; McCuaig, T.C.; Evans, N.J.; Yang, Z.-M.; Hou, Z.-Q.; Kirkland, C.L.; Parra-Avila, L.A.; Kobussen, A. Zircon compositions as a pathfinder for porphyry Cu $\pm$ Mo $\pm$ Au Deposits. *Soc. Econ. Geol. Spec. Pub.* **2016**, *19*, 329–347.
84. Pizarro, H.; Campos, E.; Bouzari, F.; Rousse, S.; Bissig, T.; Gregoire, M.; Riquelme, R. Porphyry indicator zircons (PIZs): Application to exploration of porphyry copper deposits. *Ore Geol. Rev.* **2020**, *126*, 1–18. [CrossRef]
85. Watson, E.B.; Wark, D.A.; Thomas, J.B. Crystallization thermometers for zircon and rutile. *Contrib. Mineral. Petrol.* **2006**, *151*, 413–433. [CrossRef]
86. Loucks, R.R.; Fiorentini, M.L.; Henríquez, G.J. New magmatic oxybarometer using trace elements in zircon. *J. Petrol.* **2020**, *61*, egaa034. [CrossRef]
87. Pan, L.-C.; Hu, R.-Z.; Bi, X.-W.; Wang, Y.; Yan, J. Evaluating magmatic fertility of Paleo-Tethyan granitoids in eastern Tibet using apatite chemical composition and Nd isotope. *Ore Geol. Rev.* **2020**, *127*, 103757. [CrossRef]
88. Cao, M.; Li, G.; Qin, K.; Seitmuratova, E.Y.; Liu, Y. Major and trace element characteristics of apatites in granitoids from C. Kazakhstan: Implications for petrogenesis and mineralization. *Res. Geol.* **2012**, *62*, 63–83. [CrossRef]
89. Zafar, T.; Rehman, H.U.; Mahar, M.A.; Alam, M.; Oyebamiji, A.; Rehman, S.U.; Leng, C.B. A critical review on petrogenetic, metallogenic and geodynamic implications of granitic rocks exposed in north and east China: New insights from apatite geochemistry. *J. Geodynamics* **2020**, *136*, 101723. [CrossRef]
90. Miles, A.J.; Graham, C.M.; Hawkesworth, C.J.; Gillespie, M.R.; Hinton, R.W.; Bromiley, G.D. Apatite: A new redox proxy for silicic magmas. *Geochim. Cosmochim. Acta* **2014**, *132*, 101–119. [CrossRef]
91. Higgins, J.B.; Ribbe, P.H. Crystal-chemistry and space groups of natural and synthetic titanites. *Amer. Mineral.* **1976**, *61*, 878–888.
92. Tiepolo, M.; Oberti, R.; Vannucci, R. Trace-element incorporation in titanite: Constraints from experimentally determined solid/liquid partition coefficients. *Chem. Geol.* **2002**, *191*, 105–119. [CrossRef]
93. Aleinikoff, J.N.; Wintsch, R.P.; Fanning, C.M.; Dorais, M.J. U–Pb geochronology of zircon and polygenetic titanite from the Glastonbury Complex, Connecticut, USA: An integrated SEM, EMPA, TIMS, and SHRIMP study. *Chem. Geol.* **2002**, *188*, 125–147. [CrossRef]
94. Cao, M.; Qin, K.; Li, G.; Evans, N.J.; Jin, L. In situ LA-(MC)-ICP-MS trace element and Nd isotopic compositions and genesis of polygenetic titanite from the Baogutu reduced porphyry Cu deposit, Western Junggar, NW China. *Ore Geol. Rev.* **2015**, *65*, 940–954. [CrossRef]
95. Fu, Y.; Sun, X.M.; Zhou, H.Y.; Lin, H.; Yang, T.J. In-situ LA-ICP–MS U–Pb geochronology and trace elements analysis of polygenetic titanite from the giant Beiya gold–polymetallic deposit in Yunnan Province, Southwest China. *Ore Geol. Rev.* **2016**, *77*, 43–56. [CrossRef]
96. Erdmann, S.; Wang, R.; Huang, F.; Scaillet, B.; Zhao, K.; Liu, H.; Chen, Y.; Faure, M. Titanite: A potential solidus barometer for granitic magma systems. *Comptes Rendus Geosci.* **2019**, *351*, 551–561. [CrossRef]
97. Hayden, L.A.; Watson, E.B.; Wark, D.A. A thermobarometer for sphene (titanite). *Contrib. Mineral. Petrol.* **2008**, *155*, 529–540. [CrossRef]
98. Xu, L.; Bi, X.; Hu, R.; Tang, Y.; Wang, X.; Xu, Y. LA-ICP-MS mineral chemistry of titanite and the geological implications for exploration of porphyry Cu deposits in the Jinshajiang–Red River alkaline igneous belt, SW China. *Mineral. Petrol.* **2015**, *109*, 181–200. [CrossRef]
99. Konopelko, D.; Eklund, O. Timing and Geochemistry of Potassic Magmatism in the Eastern Part of the Svecofennian Domain, NW Ladoga Lake Region, Russian Karelia. *Precambrian Res.* **2003**, *120*, 37–53. [CrossRef]

100. Konopelko, D.; Biske, Y.S.; Kullerud, K.; Ganiev, I.; Seltmann, R.; Brownscombe, W.; Mirkamalov, R.; Wang, B.; Safonova, I.; Kotler, P.; et al. Early Carboniferous metamorphism of the Neoproterozoic South Tien Shan-Karakum basement: New geochronological results from Baisun and Kyzylkum, Uzbekistan. *J. Asian Earth Sci.* **2019**, *177*, 275–286. [CrossRef]
101. Müller, D.; Groves, D.I. *Potassic Igneous Rocks and Associated Gold-Copper Mineralization*, 5th ed.; Springer International Publishing: Cham, Switzerland, 2019; 398p. [CrossRef]
102. Konopelko, D.; Seltmann, R.; Dolgopolova, A.; Safonova, I.; Glorie, S.; De Grave, J.; Sun, M. Adakite-like granitoids of Songkultau: A relic of juvenile Cambrian arc in Kyrgyz Tien Shan. *Geosci. Front.* **2021**, *12*, 141–160. [CrossRef]
103. Chung, S.-L.; Liu, D.; Ji, J.; Chu, M.-F.; Lee, H.-Y.; Wen, D.-J.; Lo, C.-H.; Lee, T.-Y.; Qian, Q.; Zhang, Q. Adakites from continental collision zones: Melting of thickened lower crust beneath southern Tibet. *Geology* **2003**, *31*, 1021–1024. [CrossRef]
104. Hou, Z.Q.; Gao, Y.F.; Qu, X.M.; Rui, Z.Y.; Mo, X.X. Origin of Adakitic Intrusives Generated during Mid-Miocene East-West Extension in Southern Tibet. *Earth Planet. Sci. Lett.* **2004**, *220*, 139–155. [CrossRef]
105. Wang, Q.; McDermott, F.; Xu, J.F.; Bellon, H.; Zhu, Y.T. Cenozoic K-Rich Adakitic Volcanic Rocks in the Hohxil Area, Northern Tibet: Lower-Crustal Melting in an Intracontinental Setting. *Geology* **2005**, *33*, 465. [CrossRef]
106. Li, Y.; Li, X.; Wang, C.; Wei, Y.; Chen, X.; He, J.; Xu, M.; Hou, Y. Miocene adakitic intrusions in the Zhongba terrane: Implications for the origin and geochemical variations of post-collisional adakitic rocks in southern Tibet. *Gondwana Res.* **2017**, *41*, 65–76. [CrossRef]
107. Lin, C.; Zhang, J.J.; Wang, X.X.; Huang, T.L.; Zhang, B.; Fan, Y.S. Himalayan Miocene adakitic rocks, a case study of the Mayum pluton: Insights into geodynamic processes within the subducted Indian continental lithosphere and Himalayan mid-Miocene tectonic regime transition. *Geol. Soc. Amer. Bul.* **2021**, *133*, 591–611. [CrossRef]
108. Rapp, R.P.; Watson, E.B. Dehydration Melting of Metabasalt at 8-32 Kbar: Implications for Continental Growth and Crust-Mantle Recycling. *J. Petrol.* **1995**, *36*, 891–931. [CrossRef]
109. Qian, Q.; Hermann, J. Partial Melting of Lower Crust at 10-15 Kbar: Constraints on Adakite and TTG Formation. *Contrib. Mineral. Petrol.* **2013**, *165*, 1195–1224. [CrossRef]
110. Müntener, O.; Kelemen, P.B.; Grove, T.L. The role of H_2O during crystallization of primitive arc magmas under uppermost mantle conditions and genesis of igneous pyroxenites: An experimental study. *Contrib. Mineral. Petrol.* **2001**, *141*, 643–658. [CrossRef]
111. Richards, J.P. The oxidation state, and sulfur and Cu contents of arc magmas: Implications for metallogeny. *Lithos* **2015**, *233*, 27–45. [CrossRef]
112. Nathwani, C.L.; Loader, M.A.; Wilkinson, J.J.; Buret, Y.; Sievwright, R.H.; Hollings, P. Multi-stage arc magma evolution recorded by apatite in volcanic rocks. *Geology* **2020**, *48*, 323–327. [CrossRef]
113. Moyen, J.F. High Sr/Y and La/Yb Ratios: The Meaning of the "Adakitic Signature". *Lithos* **2009**, *112*, 556–574. [CrossRef]
114. Moyen, J.F. The Composite Archaean Grey Gneisses: Petrological Significance, and Evidence for a Non-Unique Tectonic Setting for Archaean Crustal Growth. *Lithos* **2011**, *123*, 21–36. [CrossRef]
115. Seedorff, E.; Dilles, J.H.; Proffett, J.M., Jr.; Einaudi, M.T.; Zurcher, L.; Stavast, W.J.A.; Johnson, D.A.; Barton, M.D. Porphyry deposits: Characteristics and origin of hypogene features. *Econ. Geol.* **2005**, *100*, 251–298. [CrossRef]
116. Sun, W.; Huang, R.; Li, H.; Hu, Y.; Zhang, C.; Sun, S.; Zhang, L.; Ding, X.; Li, C.; Zartman, R.E.; et al. Porphyry deposits and oxidized magmas. *Ore Geol. Rev.* **2015**, *65*, 97–131. [CrossRef]
117. Imai, A. Variation of Cl and SO_3 contents of microphenocrystic apatite in intermediate to silicic igneous rocks of Cenozoic Japanese island arcs: Implications for porphyry Cumetallogenesis in theWestern Pacific Island arcs. *Res. Geol.* **2004**, *54*, 357–372. [CrossRef]
118. Shen, P.; Hattori, K.; Pan, H.; Jackson, S.; Seitmuratova, E. Oxidation condition and metal fertility of granitic magmas: Zircon trace element data from porphyry Cu deposits in the Central Asian Orogenic Belt. *Econ. Geol.* **2015**, *110*, 1861–1878. [CrossRef]
119. Lee, R.G.; Dilles, J.H.; Tosdal, R.M.; Wooden, J.L.; Mazdab, F.K. Magmatic evolution of granodiorite intrusions at the El Salvador Porphyry copper deposit, Chile, based on trace element composition and U/Pb age of zircons. *Econ. Geol.* **2017**, *112*, 245–273. [CrossRef]
120. Sun, S.J.; Yang, X.Y.; Wang, G.J.; Sun, W.D.; Zhang, H.; Li, C.Y.; Ding, X. In situ elemental and Sr-O isotopic studies on apatite from the Xu-Huai intrusion at the southern margin of the North China Craton: Implications for petrogenesis and metallogeny. *Chem. Geol.* **2019**, *510*, 200–214. [CrossRef]
121. Cassini, V.; Moyen, J.-F.; Cellier, G.; de Freitas, B.; Juliani, C.; Laurent, O. Towards the fertility trend: Unraveling the economic potential of igneous suites through whole-rock and zircon geochemistry (example from the Tapajós mineral Province, Northern Brazil). *Ore Geol. Rev.* **2022**, *142*, 104643. [CrossRef]
122. Dilles, J.H.; Kent, A.J.R.; Wooden, J.L.; Tosdal, R.M.; Koleszar, A.; Lee, R.G.; Farmer, L.P. Zircon compositional ev-idence for sulfur-degassing from ore-forming arc magmas. *Econ. Geol.* **2015**, *110*, 241–251. [CrossRef]
123. Burnham, A.D.; Berry, A.J. An experimental study of trace element partitioning between zircon and melt as a function of oxygen fugacity. *Geochim. Et Cosmochim. Acta* **2012**, *95*, 196–212. [CrossRef]
124. Burnham, A.; Berry, A. The effect of oxygen fugacity, melt composition, temperature and pressure on the oxidation state of cerium in silicate melts. *Chem. Geol.* **2014**, *366*, 52–60. [CrossRef]
125. Trail, D.; Watson, E.B.; Tailby, N.D. Ce and Eu anomalies in zircon as proxies for the oxidation state of magmas. *Geochim. Et Cosmochim. Acta* **2012**, *97*, 70–87. [CrossRef]

126. Chelle-Michou, C.; Chiaradia, M.; Ovtcharova, M.; Ulianov, A.; Wotzlaw, J.-F. Zircon petrochronology reveals the temporal link between porphyry systems and the magmatic evolution of their hidden plutonic roots (the Eocene Coroccohuayco deposit, Peru). *Lithos* **2014**, *198–199*, 129–140. [CrossRef]
127. Stokes, T.N.; Bromiley, G.D.; Potts, N.J.; Saunders, K.E.; Miles, A.J.; EIMF. The effect of melt composition and oxygen fugacity on manganese partitioning between apatite and silicate melt. *Chem. Geol.* **2019**, *506*, 162–174. [CrossRef]
128. Bromiley, G.D. Do concentrations of Mn, Eu and Ce in apatite reliably record oxygen fugacity in magmas? *Lithos* **2021**, *384–385*, 105900. [CrossRef]
129. Wang, H.; Cai, K.; Sun, M.; Xia, X.-P.; Lai, C.-K.; Li, P.; Wan, B.; Zhang, Z. Apatite as a magma redox indicator and its application in metallogenic research. *Lithos* **2022**, *422–423*, 106749. [CrossRef]
130. Harlov, D.E. Apatite: A fingerprint for metasomatic processes. *Elements* **2015**, *11*, 171–176. [CrossRef]
131. Webster, J.D.; Piccoli, P.M. Magmatic apatite: A powerful, yet deceptive. *Mineral. Elements* **2015**, *11*, 177–182. [CrossRef]
132. Piccoli, P.M.; Candela, P.A. Apatite in igneous systems. *Rev. Mineral. Geochem.* **2002**, *48*, 255–292. [CrossRef]
133. Parat, F.; Holtz, F.; Klügel, A. S-rich apatite-hosted glass inclusions in xenoliths from La Palma: Constraints on the volatile partitioning in evolved alkaline magmas. *Contrib. Miner. Petrol.* **2011**, *162*, 463–478. [CrossRef]
134. Jugo, P.J.; Wilke, M.; Botcharnikov, R.E. Sulfur K-edge XANES analysis of natural and synthetic basaltic glasses: Implications for S speciation and S content as function of oxygen fugacity. *Geochim. Cosmochim. Acta* **2010**, *74*, 5926–5938. [CrossRef]
135. Konecke, B.A.; Fiege, A.; Simon, A.C.; Linsler, S.; Holtz, F. An experimental calibration of a sulfur-in-apatite oxybarometer for mafic systems. *Geochim. Et Cosmochim. Acta* **2019**, *265*, 242–258. [CrossRef]
136. Tang, J.; Li, A.P.; Xu, W.L. Geochronology and geochemistry of late Carboniferous-Middle Jurassic magmatism in the Helong area, NE China: Implications for the tectonic transition from the Paleo-Asian oceanic to circum-Pacific regime. *Geol. J.* **2020**, *55*, 1808–1825. [CrossRef]
137. Pan, L.C.; Hu, R.Z.; Wang, X.S.; Bi, X.W.; Zhu, J.J.; Li, C. Apatite trace element and halogen compositions as petrogenetic–metallogenic indicators: Examples from four granite plutons in the Sanjiang region, SW China. *Lithos* **2016**, *254*, 118–130. [CrossRef]
138. Richards, J.P. High Sr/Y arc magmas and porphyry Cu $\pm$ Mo $\pm$ Au deposits: Just add water. *Econ. Geol. Bul. Soc. Econ. Geol.* **2011**, *106*, 1075–1081. [CrossRef]
139. Pan, L.-C.; Hu, R.-Z.; Liu, Q.; Li, J.-W.; Li, J.-X. Titanite geochemistry and its indicators for Cu-mineralized magmas: Insight from three dioritic intrusions in the Tongling area of the Lower Yangtze River metallogenic belt. *Ore Geol. Rev.* **2022**, *151*, 105219. [CrossRef]
140. King, P.L.; Sham, T.K.; Gordon, R.A.; Dyar, M.D. Microbeam X-ray analysis of Ce^{3+}/Ce^{4+} in Ti-rich minerals: A case study with titanite (sphene) with implications for multivalent trace element substitution in minerals. *Am. Mineral.* **2013**, *98*, 110–119. [CrossRef]

minerals

MDPI

Article

Comparative Study of Sulfides from Porphyry, Skarn, and Carbonate-Replacement Mineralization at the Recsk Porphyry-Mineralized Complex, Hungary

Máté Biró [1], Johann G. Raith [2], Monika Feichter [2], Máté Hencz [1], Gabriella B. Kiss [1], Attila Virág [3] and Ferenc Molnár [1,*]

[1] Institute of Geography and Earth Sciences, Department of Mineralogy, ELTE Eötvös Loránd University, 1053 Budapest, Hungary; mate.biro@ttk.elte.hu (M.B.); gabriella.b.kiss@ttk.elte.hu (G.B.K.)

[2] Chair of Resource Mineralogy, Department of Applied Geosciences and Geophysics, Montanuniversität Leoben, 8700 Leoben, Austria

[3] Institute of Geography and Earth Sciences, Department of Petrology and Geochemistry, ELTE Eötvös Loránd University, 1053 Budapest, Hungary

* Correspondence: molnar.ferenc@ttk.elte.hu

Abstract: A calc–alkaline dioritic–andesitic–dacitic intrusive–volcanic complex of Early Oligocene (30 Ma) age and its Mesozoic sedimentary basement at Recsk host a well-preserved porphyry–skarn–polymetallic carbonate-replacement–epithermal mineral system. The unique occurrence offers an exceptional possibility to study these related mineralization types at a single locality. This study presents the textural–paragenetic, compositional characteristics, and systematics of sulfide mineral assemblages for the porphyry, skarn, and carbonate-replacement ore types, which are currently situated at a depth of 500–1200 m below the present surface. Detailed petrography combined with EPMA analyses of molybdenite, galena, sphalerite, tetrahedrite-group minerals and Bi-bearing sulfosalts allows for the establishment of characteristic mineral and chemical fingerprints for each mineralization type. Rhenium concentration in molybdenite, occurring as rare disseminations and quartz–carbonate veinlets in altered host rocks in all three mineralization types, shows a decreasing trend towards the more distal mineralization types. High Re contents ($\overline{x}$ = 1.04 wt.%, max. up to 4.47 wt%) are typical for molybdenite from the porphyry mineralization, but Re is not homogeneously distributed, neither within individual molybdenite crystals nor on a mineralization scale. Copper and Se show opposite behavior in molybdenite, both becoming enriched in the more distal mineralization types. Silver, Bi, and Se concentrations increase in galena and tetrahedrite-group minerals, both towards the country rocks, making them the best candidates for vectoring within the whole hydrothermal system. For tetrahedrite-group minerals, Ag, Bi, Se, together with Sb and Zn, are the suitable elements for fingerprinting; all these are significantly enriched in the distal carbonate-replacement mineralization compared to the other, more proximal ore types. Additionally, further trends can be traced within the composition of sulfosalts. Lead-bearing Bi sulfosalts preferentially occur in the polymetallic carbonate-replacement veins, while being under-represented in the skarn and porphyry mineralization. Porphyry mineralization hosts Cu-bearing Bi sulfosalts dominantly, while skarn is characterized by Bi-dominated sulfosalts. Sphalerite, although present in all mineralization types, cannot be used for fingerprinting, vectoring, or thermobarometry based on EPMA measurements only. Trace element contents of sphalerite are low, often below the detection limits of the analyses. This is further complicated by the intense "chalcopyrite disease" occurring throughout the distal mineralization types. All the above-listed major, minor, and trace element ore mineral characteristics enable the characterization of the Recsk ores by mineral geochemical fingerprints, providing a possible vectoring tool in porphyry Cu–(Mo)–Au-mineralized systems.

Keywords: porphyry; sulfide; EPMA; fingerprinting; vectoring; skarn; carbonate replacement

1. Introduction

Porphyry copper deposits provide the majority of the worldwide Cu ore production and are important sources of Mo and Au [1]. The increasing demand for copper caused by the green energy transition [2,3] calls for evolved methods of exploration for this type of mineral system. The porphyry Cu(–Mo–Au)-style mineralization at the subvolcanic level of intrusive–volcanic complexes is a part of a larger-scale mineral system in which skarn, carbonate-replacement, sedimented-hosted distal-disseminated, subepithermal vein-style, and intermediate- and high-sulfidation epithermal ore types may also occur. Preservation of all of those ore types within a single system is rarely seen. The intrusive–volcanic complex of Paleogene age at Recsk provides an excellent opportunity to study porphyry and associated ore systems. This is due to the fact that all of the different mineralization types with common origin at subvolcanic and volcanic levels listed above are known and explored within a small area [4–13].

Mineral trace element fingerprints are widely used to describe the genetic framework and further economic potential of ore deposits [14–17]. Trace element distributions can also be used for mineral exploration purposes, as trace element contents of various mineral phases may change with changing geochemical conditions typical for different mineralizing conditions [18–21]. Sphalerite is widely used for thermobarometry [22,23], molybdenite is well-known for its Re content [1,24–26], and tetrahedrite-group minerals may also provide an effective tool for fingerprinting/vectoring within various mineralization types [13,14,27,28], although galena and sulfosalts are less often used for these purposes in porphyry Cu mineralization.

In this paper, we present results of detailed textural and paragenetic observations combined with electron microprobe (EPMA) data for ore minerals from the porphyry, skarn, and carbonate-replacement mineralization types in the Recsk ore complex. These data interpreted together are appropriate to describe the distribution and zoning of metals and minor elements in ore minerals within the whole mineralized Recsk area, which may provide a further fingerprinting tool applicable for refining ore system models and vectoring towards different types of ores in the diorite porphyry-related mineralized system.

Geology and Mineralization of the Intrusive–Volcanic Complex at the Recsk

The intrusive–volcanic complex of the Paleogene (Early Oligocene) age at Recsk is a part of a metallogenic belt that extends from the western Central Alps at Traversella in Italy eastward as far as Recsk in NE Hungary [10,29–32] along the Insubrian–Periadriatic–Middle-Hungarian Line (Figure 1).

At Recsk, diorite intrusion and associated mineralization systems at subvolcanic levels are not exposed on the surface; they became known from extensive drilling and underground exploration activities that were completed from the late 1950s to early 1990s in the 20th century. Dioritic bodies (stocks and veins) of around 29 Ma age [33] intruded Mesozoic sediments and are overlain by volcanic, volcano–sedimentary, and sedimentary rocks of the Late-Eocene–Early-Oligocene age [4,7,33–36]. The intrusive and volcanic rocks of the Paleogene age show subduction-related, medium-to-high-K calc-alkaline geochemical characteristics and form a single unit within the Paleogene Volcanic Belt of Hungary [10,29,37]. Diorite stocks intruded Jurassic olistostromes, predominantly containing Triassic to Jurassic sedimentary units (limestone, dolomite, chert) and less widespread basaltic rocks [7,36]. The Mesozoic basement is overlain by Late Eocene-Early Oligocene sedimentary rocks (limestone, marl, clay) [33] which are overlain and crosscut by the extrusive and intrusive volcanic units of the ore complex. The Cu–porphyry mineralization is hosted by the diorite stocks and their endoskarn zones along their deep-seated contacts with the Mesozoic sedimentary units (Figure 2), with indicated resources of 781 Mt Cu-ore (0.66% Cu, 0.4% cut-off with 0.005% Mo and 0.28 g/t Au) and 159 Mt Cu-ore (1.16% Cu, 0.01% Mo and 0.17 g/t Au), respectively [38–40].

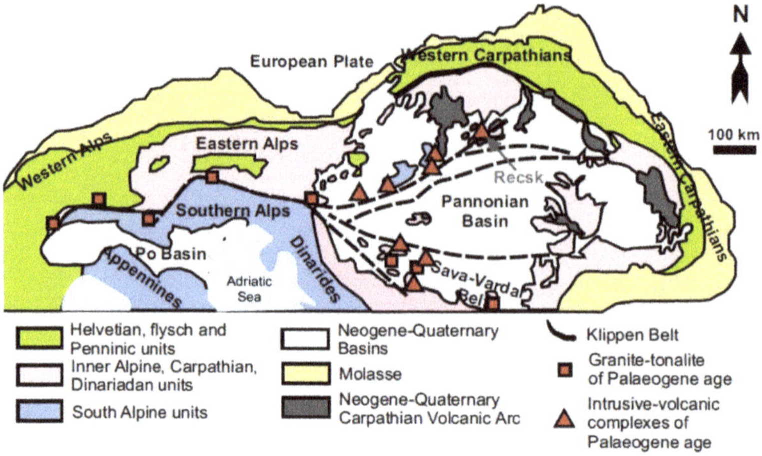

Figure 1. Geological setting of the mineralized Recsk complex along the Insubrian–Periadriatic–Middle-Hungarian line. Note the Paleogene volcanic and intrusive complexes occurring along the tectonic line, their northeastern-most exposure being Recsk. Modified after Molnár 2007.

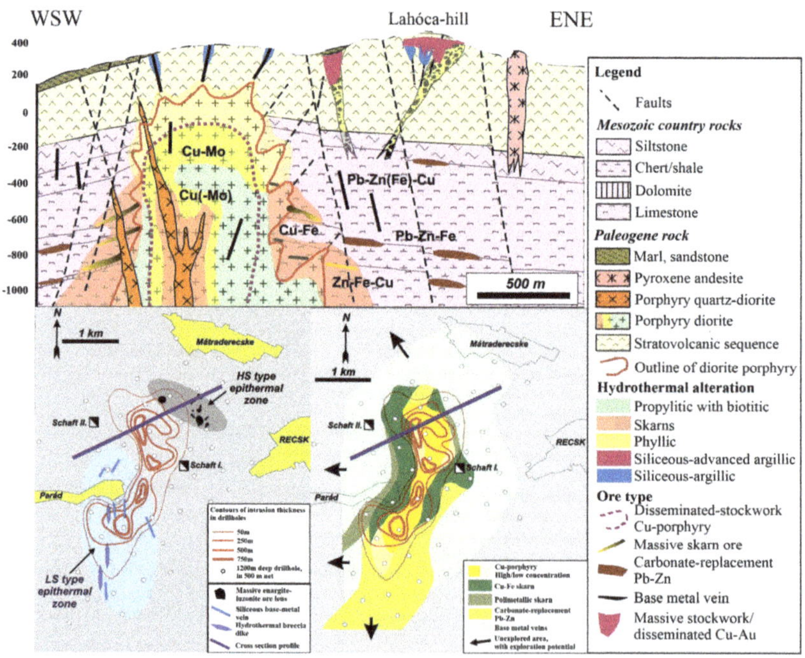

Figure 2. Cross-section and contour maps of the mineralized Recsk complex. Multiple apexes of the complex diorite porphyry can be seen on both the cross-section and the contour maps. Mineralized zones extend as far as 2 km from the porphyry diorite, showing a Cu–Zn–Pb metal zonation trend towards the country rocks. Epithermal zones are situated within the stratovolcanic units, hosted by diatreme breccias and tectonically preformed zones. Based on [7,10,12].

The main porphyry orebody is surrounded by skarns along its contact with the Mesozoic sedimentary rocks (Figure 2), containing horizontally and vertically zoned Cu–

Zn–Pb mineralization with 36 Mt Cu-ore (2.19% Cu) and 11.5 Mt Zn-ore (4.98% Zn). In the zones towards the country rocks, the Triassic–Jurassic tectonically disturbed units host a polymetallic Zn–Pb carbonate-replacement ore with grades as high as 3.15–3.5% Zn and 1.2–2.1% Pb locally [40,41].

The shallow epithermal zones of the Recsk ore complex have been targeted by mining and intense exploration since the early 19th century, but mining of the Cu–Au ore was stopped in the 1970s. In the past 20–30 years, Brownfield mineral exploration has mostly targeted the epithermal Cu–Ag–Au intermediate sulfidation (IS) and Cu–Au high-sulfidation (HS)-type ore zones at shallow depths, above the porphyry stocks (Figure 2) [8,13]. Those exploration programs reported an indicated remaining resource of 32.4 Mt of Au ore with 1.47 g/t of average-grade Au in the HS-type mineralization zones [38].

2. Materials and Methods

2.1. Samples

Based on the results of previous studies on more than 500 samples, a total of 32 samples were selected for the purpose of this study, all of which were accessible from the drill core sample repository of the Department of Mineralogy, Eötvös Loránd University, Budapest (Table 1). The selected samples represent the three deep-seated economically significant mineralization types, i.e., Cu(–Mo–Au) porphyry, Cu, and Zn–Pb skarn and Pb–Zn–(Ag) carbonate replacement mineralization. The selected sample set represents both the rocks from the northernmost part of the mineralization as well as the more southern intrusives and their surroundings. Samples were selected upon mineral composition, preferring those that contain measurable crystals from most of the studied mineral species. Some samples derive from the archive collection of representative drill cores from the mineralization.

Table 1. Drill core samples used for petrography and EPMA studies from the northern intrusive part of the Recsk mineralization. Ore types: PO–Cu–(Mo)–Au-porphyry, SK-skarn, CAR–polymetallic carbonate replacement. Minerals: py—pyrite, cpy—chalcopyrite, fhl—tetrahedrite-group minerals, sph—sphalerite, ga—galena, sphs—sulfosalt, mo—molybdenite, sch—scheelite, va—valleriite, to—tochilinite, mt—magnetite.

Drillcore	Spl. Nr.	Coordinates (WGS 84)			Mineralogy		Mineralization Type
		X	Y	Z	Ore Minerals	Gangue Minerals	
Rm-30	63	47.9349	20.0694	−290.3	py, cpy, fh, ga, sphs	fs, qz, ag,	PO
Rm-30	64	47.9349	20.0694	−400.6	py, ga, sph, sphs, mo	qz, fs, ag, cc, dol	PO-CAR
Rm-34	05	47.9373	20.0615	−602.3	py, sph, ga,cpy, sphs	qz, cc	CAR
Rm-34	06	47.9373	20.0615	−697.3	py, sph, sphs, fhl	qz, cc	CAR
Rm-34	07	47.9373	20.0615	−716.3	py, mo, sch, sph, fhl, sphs	dol, qz,	CAR
Rm-34	42	47.9373	20.0615	−796.3	py, sph, cpy, fhl, sphs, sch, mo	qz, grt	SK
Rm-34	43	47.9373	20.0615	−818.3	py, sphs, sch, cpy	qz	SK
Rm-34	44	47.9373	20.0615	−832.8	py, sphs, sch, cpy	fs, qz, ag	PO
Rm-34	46	47.9373	20.0615	−961.3	py, fhl, sphs	fs, qz, ag	PO
Rm-35	5	47.9359	20.0658	−337.8	py, cpy, sph, ga, sphs	fs, qz, ag	PO
Rm-35	6	47.9359	20.0658	−334.3	py, cpy, sph, ga	fs, qz, ag	PO
Rm-35	7	47.9359	20.0658	−373.8	py, mo, cpy, ga, fhl	fs, qz, ag, cc	PO
Rm-35	8	47.9359	20.0658	−372.8	py, mo, fhl, sph,	fs, qz, ag, cc	PO
Rm-35	13	47.9359	20.0658	−569.3	py, mo, cpy	fs, qz, ag, cc	PO
Rm-35	16	47.9359	20.0658	−677.3	py, mo, cpy	fs, qz, ag, cc	PO
Rm-35	17	47.9359	20.0658	−682.8	py, mo	fs, qz, ag	PO
Rm-35	21	47.9359	20.0658	−782.3	py, mo	fs, qz, ag	PO
Rm-35	23	47.9359	20.0658	−786.3	py, mo, cpy, sphs	qz, grt, px	SK

Table 1. *Cont.*

Drillcore	Spl. Nr.	Coordinates (WGS 84)			Mineralogy		Mineralization Type
		X	Y	Z	Ore Minerals	Gangue Minerals	
Rm-35	24	47.9359	20.0658	−837.3	py, mo, cpy	fs, qz, ag	PO
Rm-45	340	47.9329	20.0709	−839.3	py, cpy, mo,	qz, amf	SK
Rm-61	01	47.9420	20.0625	−664.5	py, sph, cpy, ga,sphs	qz, cc, dol, fl	CAR
Rm-63	03	47.9324	20.0604	−506.3	py, ga, sph, sphs	cc	CAR
Rm-64	01	47.9280	20.06202	−233.0	py, sph, ga, fhl, sphs	qz, cc, dol	CAR
Rm-70	01	47.9203	20.0698	−576.9	py, sph, ga, cpy	cc, px, ag	SK
Rm-70	16	47.9203	20.0698	−313.3	py, cpy, mo, ga	fs, qz, ag	PO
Rm-71	74	47.9172	20.0580	−926.9	py, cpy, ga, sph, sphs	cc, qz, anh	SK
Rm-71	499	47.9172	20.0580	−907.0	py, sph, ga, cpy, sphs	grt, qz, px, cc	SK
Rm-71	02	47.9203	20.0698	−717.9	py, sph, ga, cpy	px, ep, srp, cc	SK
Rm-73	49	47.9186	20.0631	−58.3	py, cpy, mo, sph, ga, sphs	fs, qz, ag	PO
Rm-77	492	47.92569	20.0633	−219.3	py, po, mt, sph, ga, to, va, ala	grt, anh, br	SK
Rm-82	516	47.9021	20.0513	−645.2	py, sph, ga, cpy	anh	SK
Rm-87	01	47.9368	20.0557	−422.3	py, sph, ga, fhl, cpy	qz, cc	CAR

2.2. Methods

Microscopic studies were performed on 30 or 100 µm thick polished thin sections. Optical microscopy was performed with a Zeiss Axioplan polarization microscope equipped with Zeiss Axiocam 208 camera (Carl Zeiss GmBH, Jena, Germany). Scanning electron microscopic (SEM) imaging was implemented using an AMRAY 1830i electron microscope with a PU9800 energy-dispersive detector at the Eötvös Loránd University, Institute of Geography and Earth Sciences, and a Thermo Scientific PhenomXL Desktop SEM, equipped with a CeB_6 electron source and a built-in SDD energy-dispersive spectrometer, provided by AuroScience Ltd (Budapest, Hungary).

Electron microprobe (EPMA) measurements of sulfide minerals were performed on a JEOL JXA-8200 WD/ED-combined microanalyzer at the Eugen F. Stumpfl Electron Microprobe Laboratory at Montanuniversität Leoben. Measurement conditions for all analyses were 20 kV acceleration voltage and 10 nA sample current. Standardization was made with SPI Supplies natural and artificial $CuFeS_2$, ZnS, HgS, PtAs, MoS_2, CuS, PbS, $BiAgSe_2$, Sb_2S_3, SnS_2, PdTe sulfide; V, Ge, Cd, In, and Re pure metals; and $Mn_5Si_5O_{15}$ silicate reference materials. EPMA measurements were performed on sphalerite, galena, tetrahedrite, molybdenite, Bi-sulfosalts, and kesterite. The elements measured in sphalerite were the following: Zn, S, Fe, Mn, Cu, Cd, Hg, Ge, In, V; in galena: Pb, S, Ag, Se, Bi, Cu; in tetrahedrite-group minerals: As, Sb, S, Fe, Zn, Cu, Ag, Te, Se, Bi, Hg, Pb, Cd, Se; in molybdenite: Mo, S, Se, Re, Cu, W; in bismuth sulfides: Bi, S, Pb, Cu, Se, Te, Ag, and Sn; and in kesterite: Zn, V, Fe, Cu, S, Se, In, Cd, Mn. Individual measurements, measurement conditions, times, and detection limits for each element are listed in the electronic appendix.

Data analysis was performed in IMDEX ioGAS and R software environment (version 4.3.1; [42]).

3. Results

3.1. Petrography

3.1.1. Porphyry Cu–(Mo)–Au Ore

The porphyry-copper-type mineralization is mostly hosted by the apical zones of the diorite intrusion, but mineralized zones extend subordinately in the endoskarn rock assemblage as well. The Cu–Mo ore is contained mostly by a quartz-rich stockwork zone in the diorite, while the endoskarns are characterized by lenticular ore bodies with massive-to-disseminated Cu mineralization [5].

The porphyry mineralization is characterized by a 50–150 m thick zone of intensively altered rock assemblage hosting stockwork Cu(–Mo–Au) mineralization. K-metasomatic zones are often overprinted by various degrees of propylitic or phyllic alteration (Figure 2), resulting in the presence of sericite/illite together with quartz, or green sheet-silicate minerals accompanied by epidote. The primary porphyry minerals plagioclase and amphibole are often replaced by quartz–illite–sulfide paragenesis or chlorite–quartz–sulfide–apatite paragenesis. All the mineralized samples are characterized by intense veining. Veins are characterized according to Sillitoe [43]. M-type veins were not observed, but multiple generations of A-, B-, and D-type veins are widespread in the studied samples. These veins are also accompanied by multiple generations of carbonate veins. While A-type veins are characteristically present in the potassic alteration zone, B-type and D-type veins tend to occur within variously altered samples even within the surrounding endoskarns. Carbonate veins with various mineralogy are present throughout the entire mineralized porphyry stock.

The stockwork mineralization is characterized by the mineralogically and genetically diverse veinlets cross-cutting each other, which are the most important carriers of the sulfide ore minerals. As samples come from 63.5 mm halved drill cores, clear cross-cutting relations cannot always be determined, thus multiple generations of the same vein type are not necessarily distinguishable. There are early, exclusively cellular or, more rarely, sugary quartz veins, quartz–sulfide veins, which typically contain pyrite, rarely chalcopyrite, galena, iron oxide, or K-feldspar, in addition to quartz. These are recognized as A-type veins, and by their mineralogy and appearance, at least two different generations can be distinguished. These two types of A-type veins are distinguished by their sulfide mineral contents. There are veins hosting exclusively pyrite, while in other A-type vein generations, other sulfide minerals may also occur, although only in subordinate amounts. A-type veins may occur throughout the mineralized system, their thickness being mostly between 2 and 30 mm.

Multiple generations of B-type veins can be distinguished by their mineralogy. There are mainly quartz–pyrite veins, but quartz–pyrite–chalcopyrite, quartz–pyrite–molybdenite veins and quartz–pyrite–chalcopyrite–molybdenite veins can also be recognized as subclasses of B-type veins (Figure 3e). Sulfide minerals usually occur in the central parts of these veins, but there are veins which show multiple sequences and/or asymmetrical distribution of ore minerals; these latter veins have a banded texture. When B-type veins crosscut mafic porphyry phenocrysts, massive sulfide replacement of the phenocryst usually occurs (Figure 3c,d), forming a "sulfide-bead" laced up on the narrow (1–5 mm) quartz–sulfide vein. B-type veins occur both in the mineralized diorite apexes and the surrounding endoskarns. B-type veins in the endoskarns typically crosscut the original textural features of the rocks such as remaining sedimentary structures, mineral nodules, or massive sulfide lenses. Thickness of B-type veins typically range between 1 and 5 mm.

D-type veins with their typical sericite–illite-rich alteration halos and relatively higher sulfide contents are also abundant. Besides the quartz–sericite–sulfide veins, some pure sulfide (pyrite, chalcopyrite) veins may also be categorized to type-D based on their cross-cutting relations. Besides their dominantly pyrite–chalcopyrite ore mineral assemblage, molybdenite or, rarely, scheelite may also occur within some of these veins. Repeated "bands" of quartz–sulfide parageneses in some veins also suggest multiple re-opening processes. Carbonates also occur as central zones of D-type veins, typically with a calcite–dolomite, pyrite, chalcopyrite, $\pm$molybdenite, galena mineralogy (Figure 3f).

Within the apical and outermost lying parts of the diorite intrusion, but sometimes also within the K-metasomatic zones, multiple generations of carbonate veins can be observed. These are usually thicker (generally 5–40 mm) than the usual B- and D-type veins and banded even within the porphyry mineralization center. These carbonate veins often show a banded texture. The wall rock usually shows sericitic, or propylitic, alteration. The veins generally start with coarse-grained drusy quartz, followed by quartz, pyrite, chalcopyrite, molybdenite, and galena mineral assemblages, followed by the precipitation of carbonates and sericite. Pyrite, molybdenite, and galena tend to occur in the central carbonate part of the veins too.

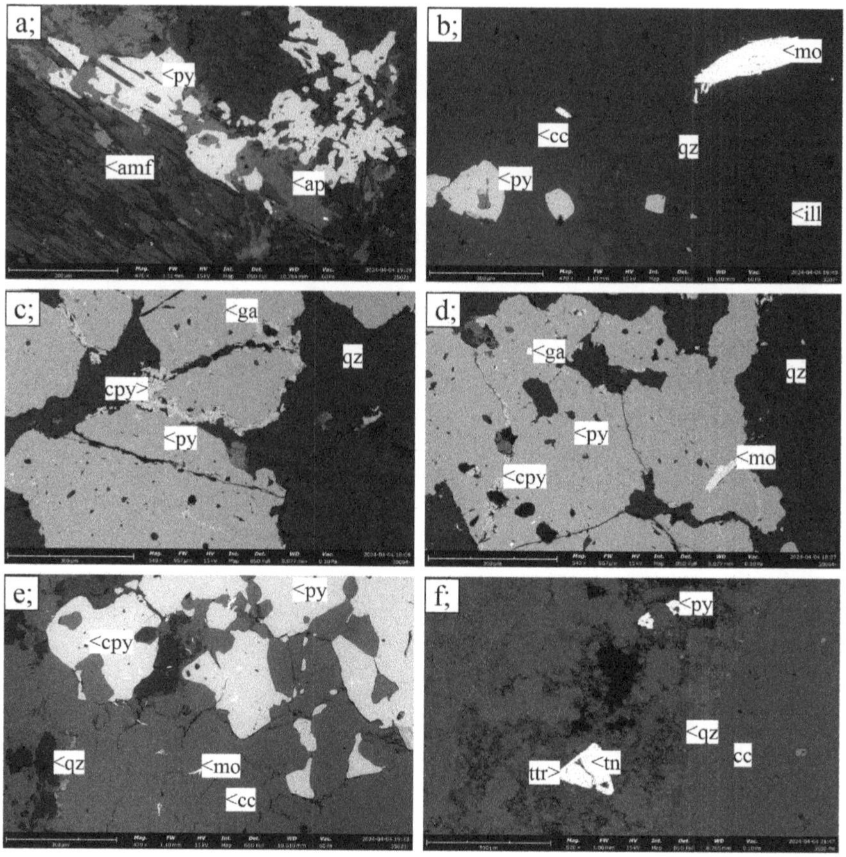

Figure 3. Ore textures on SEM-BSE images from the Cu–(Mo)–Au mineralization: (**a**), replacement of magmatic amphibole grain (amf) by chlorite and skeletal pyrite (py) with apatite (ap) at the rims (Rm35–21); (**b**), euhedral pyrite and subhedral molybdenite (mo) "fish" within carbonate(cc)–quartz(qz)–illite(ill) vein with baddeleyite (Rm35–07); (**c**), anhedral porphyry pyrite replacing glomeroporphyric nodule with chalcopyrite inclusions, crosscut by galena-bearing chalcopyrite veins (Rm30–64); (**d**), sulfide paragenesis replacing tabular porphyry phenocryst. Molybdenite and galena (ga) inclusions in anhedral porphyry pyrite, crosscut by chalcopyrite veins (Rm30–64); (**e**), anhedral pyrite in quartz vein with rounded chalcopyrite (cpy) inclusions associated with thin flakes of molybdenite embedded in quartz (Rm35–16); (**f**), euhedral tennantite (tn) with Sb-enriched tetrahedrite rim (ttr) associated with pyrite in quartz–illite porphyry matrix (Rm35–08).

Carbonate veins are classified into several additional types based on their mineral association: (1) quartz–carbonate–pyrite veins with pyrite being the one and only sulfide, (2) carbonate veins with molybdenite–pyrite–chalcopyrite–galena characteristic mineral association (Figure 3b), (3) quartz–carbonate–pyrite–molybdenite–tennantite veins, and (4) sphalerite–galena–pyrite-bearing carbonate veins containing chalcopyrite and abundant sulfosalts too. In addition, quartz, anhydrite, barite, celestite, clay minerals—typically illite—may appear in any of the above-mentioned carbonate veins. Minor monazite, rare earth carbonates like bastnäsite, apatite, xenotime, rare scheelite and molybdenite, and rheniite also occur as accessory constituents of these carbonate veins.

3.1.2. Skarn Ores

Skarns host the highest grades of Cu and Zn ores within the Recsk ore complex. Skarns occur typically as lenticular or pocket-like bodies with a thickness of 50–200 m, surrounding the mineralized diorite apex. The classification of skarn samples based on mineralogical and textural criteria is more problematic, as the skarn deposits are diverse and several parageneses can be observed even on a hand specimen scale. Skarns at Recsk were classified by their mineralogy as follows [5]: wollastonite marble, garnet–pyroxene skarn, pyroxene (±amphibole) skarn, epidote skarn, phlogopite skarn, anhydrite–serpentine skarn, secondary quartzite metamorphosed rocks. All these rock types also bear signs of hydrothermal overprint by a sulfide paragenesis. The following types were identified among the examined samples: garnet–pyroxene skarn with sulfide veins, pyroxene skarn with massive sulfide lenses, pyroxene–amphibole skarn with replacement—vermicular—fabric, and epidote–anhydrite skarn with replacement quartz–anhydrite–sulfide veins, serpentinite with massive sulfide veins, and contact quartzite with grossular with massive sulfide lenses and B-type veins. Any of these rocks may host Fe, Cu, Zn–Pb, or Cu–Zn–Pb mineralization, i.e., the type of mineralization does not seem to depend on contact metamorphic/metasomatic paragenesis.

Most examined samples belong to the garnet skarn. The size of garnet grains varies between 50 µm and 1 cm. Individual crystals can be euhedral, but they can also be anhedral in veins. Garnet often shows growth-zoning, marked by radially occurring primary 2-phase fluid inclusions. The garnet in these rocks is andradite, less often grossular, showing zonation patterns. As mineral inclusions, diopside, quartz, less often anhydrite, hematite, and sphalerite (Figure 4a) can be found in garnet. The garnet skarns often contain calcite, quartz, pyroxene, epidote–zoisite nodules, tabular anhydrite, and less commonly amphibole, occurring in fibrous clusters.

The pyroxene skarn most often contains massive sulfide ore. Pyroxene has hedenbergite–johannsenite, less often diopside composition, and an average grain size of 200–1000 µm. Pyroxene shows a isometric–hypidiomorphic habit, but it is idiomorphic and tabular, associated with large (5–10 mm) carbonate grains. It often occurs together with 10–200 µm crystals of tremolite having a needle-like or fibrous appearance. The pyroxene in the garnet–pyroxene skarn rocks is not significantly different in composition or form from the pyroxene appearing in other skarn rocks. Pyroxene skarn rocks typically contain mostly Cu-rich ore and, less often, Cu–Mo mineralization, appearing as disseminations of chalcopyrite and molybdenite (Figure 4f,g) associated with massive lenses of pyrite.

Epidote–quartz skarn rocks consist of epidote, pyroxene, and anhydrite. These lenses are often crosscut by a network of 1–10 mm-thick veins containing quartz, carbonate, and anhydrite. Epidote is also present in an epidote–chlorite endoskarn; there, the original texture of the diorite and relics of the magmatic mineral assemblage are preserved. In the epidote skarns, the Zn-, Cu–Zn, Zn–Pb sulfide mineralization is typically found in these veins that sometimes comprise up to 10% of the whole rock. Common ore minerals are pyrite, chalcopyrite, and sphalerite. In some samples, scheelite was observed, often accompanied by molybdenite.

Contact metamorphic quartzite occurs along the contact of the diorite intrusion with siliceous rocks. The massive quartzite lenses are often crosscut by thin garnet-bearing calc-silicate veins. Centimeter-sized massive sulfide lenses and B and D quartz-sulfide veins are common in this rock type. They typically contain a pyrite–chalcopyrite–sulfosalt/telluride–tennantite–scheelite paragenesis (Figure 4c,h).

An unusual anhydrous skarn rock of monticellite–magnetite–anhydrite composition was overprinted by a hydrous retrograde mineral association with brucite–hydrogrossular–serpentine–chlorite–tochilinite–valeriite (Figure 4b,d). This is accompanied by pyrite, pyrrhotite, sphalerite, alabandite, and galena, where galena may host Bi-tellurides and sulfosalts as inclusions (Figure 4e).

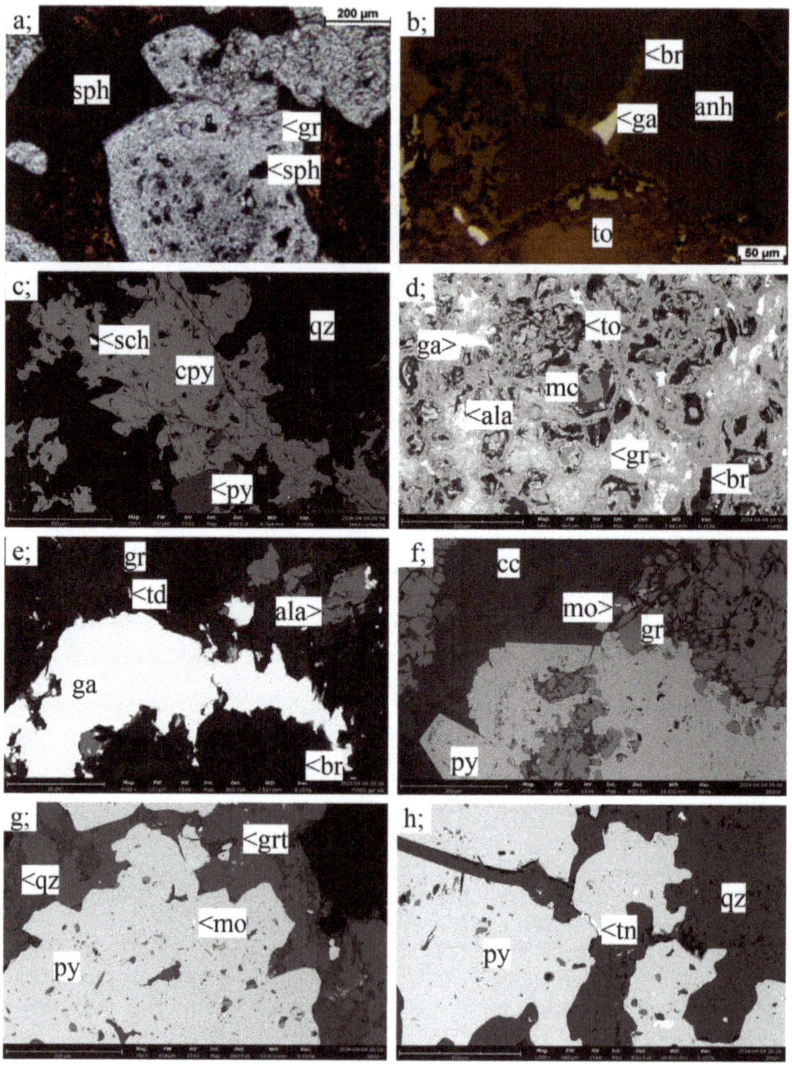

Figure 4. Typical ore textures ((**a**,**b**), polarized light), ((**c**–**h**), SEM-BSE) from the various skarn mineralized rocks: (**a**), transmitted polarized light photograph of sphalerite (sph) inclusions in andradite garnet (gr) surrounded by transparent red sphalerite (Rm71–499); (**b**), reflected polarized light microphotograph of equilibrium-texture anhydrite (anh) being replaced by brucite (br) and galena (ga) along fractures, in pseudomorph equilibrium texture with rust-red tochilinite (to) replacing magnetite (Rm77–492); (**c**), scheelite (sch) inclusion swarm in quartz–chalcopyrite–pyrite vein in garnet–skarn (Rm34–43); (**d**), monticellite (mc) being replaced by brucite–tochilinite assemblage surrounded by xenomorphic, vein-filling hydrogrossular (gr), resorbed anhydrite, and hydrothermal galena, pyrite, and alabandite sulfide grains (Rm77–492); (**e**), Bi-sulfotelluride–tetradymite–(td) grains with anhedral, vein-filling galena associated with sphalerite and alabandite; (**f**), euhedral pyrite crystals with silicate inclusion zones occurring with grossular and calcite (cc), molybdenite (mo) occurs in a subhedral garnet crystal (Rm45–340); (**g**), molybdenite inclusions in subhedral, quartz-inclusion-rich pyrite grains in siliceous matrix (Rm34–42); (**h**), anhedral skarn pyrite with chalcopyrite inclusions crosscut by not zoned tennantite (tn) vein in illite–quartz matrix (Rm34–42).

3.1.3. Carbonate-Replacement Polymetallic Mineralization

Carbonate-replacement (CAR) hydrothermal ores have diverse appearances and occur within variously altered host rocks. This type of mineralization extends as far as 2 km from the porphyry mineralization center. Pyrite-dominated mantos were described as far as 1–1.5 km southwest from the outermost diorite stocks, while few-centimeters-thin carbonate-replacement polymetallic veins occurred up to 2 km from the dioritic bodies in every direction. CAR mineralization is generally represented by altered-recrystallized hydrothermal breccias forming centimeter-sized brecciated veins or a few ten-centimeter-thick breccia pipes in the samples.

The manto ore forms horizontal massive sulfide lenses with a thickness up to few tens of cm. The breccias show a diverse appearance within the mineralized zones. Based on the composition of the clasts, there are exclusively carbonate, carbonate–quartz–clay minerals, as well as exclusively siliceous breccia types, within which the gangue and ore mineral assemblage can show further variability. Among the hydrothermally altered, brecciated sedimentary rocks, the most characteristic common feature is the fragmented and somehow recrystallized bedrock cemented by a carbonate–sulfide ± quartz ± anhydrite assemblage. The shape (roundness and sphericity) and size (from a few 10 μm to >10 cm) of the fragments can vary from sample to sample and within individual samples too. Any rock type within the whole mineralized complex may be cross-cut by carbonate–sulfide ± quartz ± sulfate veins of 0.5–5 cm thickness. These veins are characterized by banded texture, the rims being typically enriched in quartz, their central part being most heavily mineralized. Sulfates within these veins may be anhydrite, celestite, or baryte.

The ore mineral content of the rocks varies from sample to sample, usually ranging from below 1% to 5%, although some samples may contain as much as 50% sulfides connected to veins or the matrix. Higher-ore mineral contents were observed locally in sulfide-rich veins with 10–50% of sulfides, dominantly sphalerite–pyrite, while manto-like, horizontally extensive pyritization accompanied by sphalerite, tetrahedrite-group minerals, kesterite, and, rarely, scheelite and molybdenite (Figure 5a,e,g,h) hosts 15–20% ore minerals on average.

Mantos, hydrothermal breccias, and distal altered sedimentary rocks are commonly crosscut by carbonate or carbonate–sulfide veins, often accompanied by quartz and anhydrite. The carbonate vein network enmeshing the whole mineralized system can be characterized by several textural–paragenetic types. There are exclusively carbonate veins without any sulfides present, as well as carbonate–pyrite veins in which pyrite may host inclusions of galena or tetrahedrite-group minerals—rarely kesterite—, and carbonate–sulfide veins with 1–5% pyrite–sphalerite–galena–chalcopyrite–tetrahedrite–sulfosalt ore mineral assemblage. The latter vein generation represents the highest ore grades throughout the entire CAR-mineralized system. Sulfide-bearing veins usually contain various amounts of euhedral quartz and locally also contain sulfates (celestite, anhydrite), sometimes containing fluorite, bastnäsite, xenotime (Figure 5e,h), monazite, apatite, and pockets of clay minerals (usually illite).

Veins containing only various carbonate minerals occur throughout the altered country rocks. Multiple generations of carbonate veins can be distinguished: there are coarse (1–2 mm) crystalline, exclusively calcite-bearing veins which predate the mineralizing carbonate–sulfide veins based on cross-cutting relations. This vein type is usually thin, 1–5 mm in thickness. There are open fissures and cavities filled with carbonate minerals, where calcite is often accompanied by dolomite and minor bastnäsite; even monazite may be present within those veins. The size of calcite crystals in such veins is usually below 1 mm, although, in open cavities, both calcite and dolomite may form rhombohedral crystals with a size up to 5 mm. These veins can reach a thickness up to 2 cm, their outline being sharp. Both mineralogical and cross-cutting relations suggest that these veins were formed during the post-ore stage. Although sulfides are absent in these carbonate veins, they usually occur in samples with disseminated ore types. There are pure calcite, pure dolomite, or calcite–

dolomite veins, all crosscutting every preexisting mineralized feature. These are usually fine crystalline (below 0.5 mm), and their thickness is usually also maximally 1–2 mm.

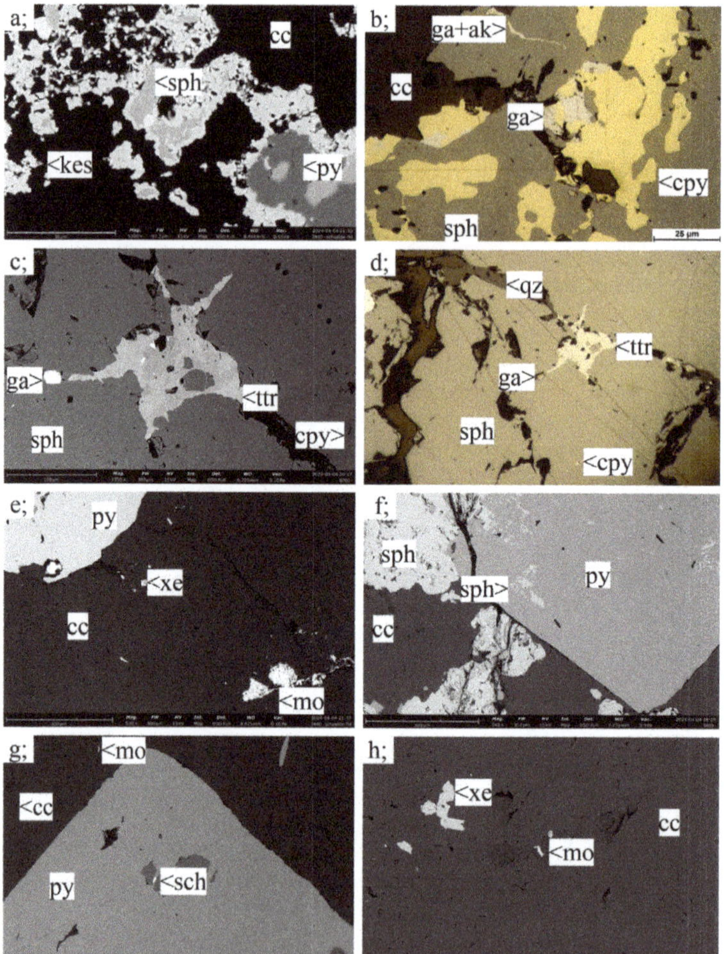

Figure 5. SEM-BSE (**a**,**c**,**e**–**h**) and reflected-light (**b**,**d**) microphotographs of characteristic ore textural features from the CAR mineralization type: (**a**), kesterite–kuramite (kes) rim on resorbed sphalerite (sph) and pyrite (py), accompanied by tetrahedrite, sphalerite showing network-like and granular kesterite disease (Rm34–07); (**b**), chalcopyrite (cpy) replacing both sphalerite and galena (ga) with characteristic dusted disease in sphalerite associated with aikinite (ak) domains replacing galena in vein (Rm61–01); (**c**,**d**), oscillatory zoned tetrahedrite (ttr) patch in sphalerite associated with galena and the same grain in reflected polarized light (Rm87–01); (**e**), microcrystalline molybdenite (mo) patches in massive carbonate (cc) manto with pyrite (Rm34–07); (**f**), unaltered sphalerite (sph>) with poikilitic texture in euhedral pyrite surrounded by subhedral sphalerite (sph) grains showing dusty chalcopyrite disease (Rm34–05); (**g**), scheelite inclusion in euhedral pyrite (Rm34–07); (**h**), molybdenite flakes within illite-bearing vein in carbonate manto associated with idiomorphic xenotime (Rm34–07).

Carbonate–pyrite veins are typical both for mantos, and for country rocks crosscut by massive carbonate–Zn–Pb sulfide veins. There are veins hardly reaching a thickness of 1 mm, with pyrite and calcite being present within them, while other carbonate–pyrite veins can reach a thickness of 1 cm or more. Pyrite within these veins is typically euhedral, accompanied by calcite and dolomite mostly. Accessory minerals in these veins can be bastnäsite, and various minor sulfides found as inclusions in pyrite, such as sulfosalts, sphalerite, galena, and kesterite. In their thickness and their mineral composition, they can be distinguished from veins having more various sulfide mineralogy.

Carbonate–sulfide veins are one order of magnitude thicker within the samples than the previously mentioned groups. Their carbonate mineralogy is dominated by Mn-bearing calcite, dolomite, ankerite, and, rarely, siderite. Drusy quartz with sulfide inclusions is also a common constituent of these veins. Both the pyrite–sphalerite–galena–tetrahedrite-group minerals dominated ore mineral composition, and the various carbonate phases may occur with various crystal sizes between 1 mm and 3 cm. Carbonate–sulfide veins are generally the thickest, reaching thicknesses over 12 cm within the samples, although they are often interlocking and braided. Thus, a 1–4 cm vein is often accompanied by 1–5 mm sized nearly parallel-running veins with similar mineralogy.

Some quartz–carbonate–sulfide veins with accessory anhydrite, celestite, barite within the porphyry apex share the textural, mineralogical, and paragenetic sequential properties of the veins occurring within the carbonate-replacement zones.

3.2. Ore Mineralogy

3.2.1. Porphyry Mineralization

Pyrite is the dominant mineral in the disseminated ore, with various sizes and shapes ranging from 10 μm euhedral to 1 cm subhedral–euhedral crystals. In addition to pyrite, chalcopyrite and molybdenite are locally abundant (Figure 3), primarily as inclusions in pyrite, less often as an independent phase. Pyrrhotite, bornite, sphalerite, galena, tennantite, and various sulfosalts may also occur (Figure 6). Chalcopyrite is never euhedral; it occurs as anhedral 20 μm to 1 cm patches, surrounding pyrite, or as individual sulfide veinlets (Figure 3c–e). Disseminated molybdenite occurring in the diorite is typically 20–50 μm in size, with thin flakes mostly embedded in clay or quartz, replacing primary porphyry components. Disseminated pyrite replaces altered-porphyry grains of any size from 200 μm to 2 cm. The edge and outline of the altered magmatic phenocryst is often contoured by anhedral sulfide mineral nodules (Figure 3a), or the ratio of sulfide replacement increases towards the tips of the porphyritic phenocryst. Zonal sulfide replacement is also common, where the distribution of the fine-grained, typically anhedral–subhedral pyrite and chalcopyrite grains follow the growth zones of the porphyry crystals. Sulfides replacing the phenocrysts are often found alongside dominantly idiomorphic apatite grains zoned with xenotime or monazite inclusions. In the altered matrix of the diorite, which is typically silicified with clay minerals, the enrichment of the sulfides can be associated with a thin, μm-scale crack system that can only be recognized under the microscope. The cracks contain predominantly quartz, pyrite and chalcopyrite. Rarely, disseminated idiomorphic pyrite, molybdenite, or sphalerite grains appear, which cannot be connected to any visible vein or to the position of a previous mineral phase in the matrix. Pyrite with inclusions of galena, sulfosalts, and chalcopyrite, sometimes with inclusions of bornite and kesterite, was the first phase to crystallize. Pyrite rims are often co-crystallized with idiomorphic molybdenite flakes. Chalcopyrite generally occurs as space-filling xenomorphic replacive masses in the porphyry phenocrysts, or as late-stage veins cross-cutting the earlier assemblages. The extent of these latter veins is often limited to the replaced phenocrysts. Aluminum–phosphate–sulfate minerals, apatite, monazite, and hydrothermal zircon of 30 μm size may appear in nests of clay minerals besides the sulfides.

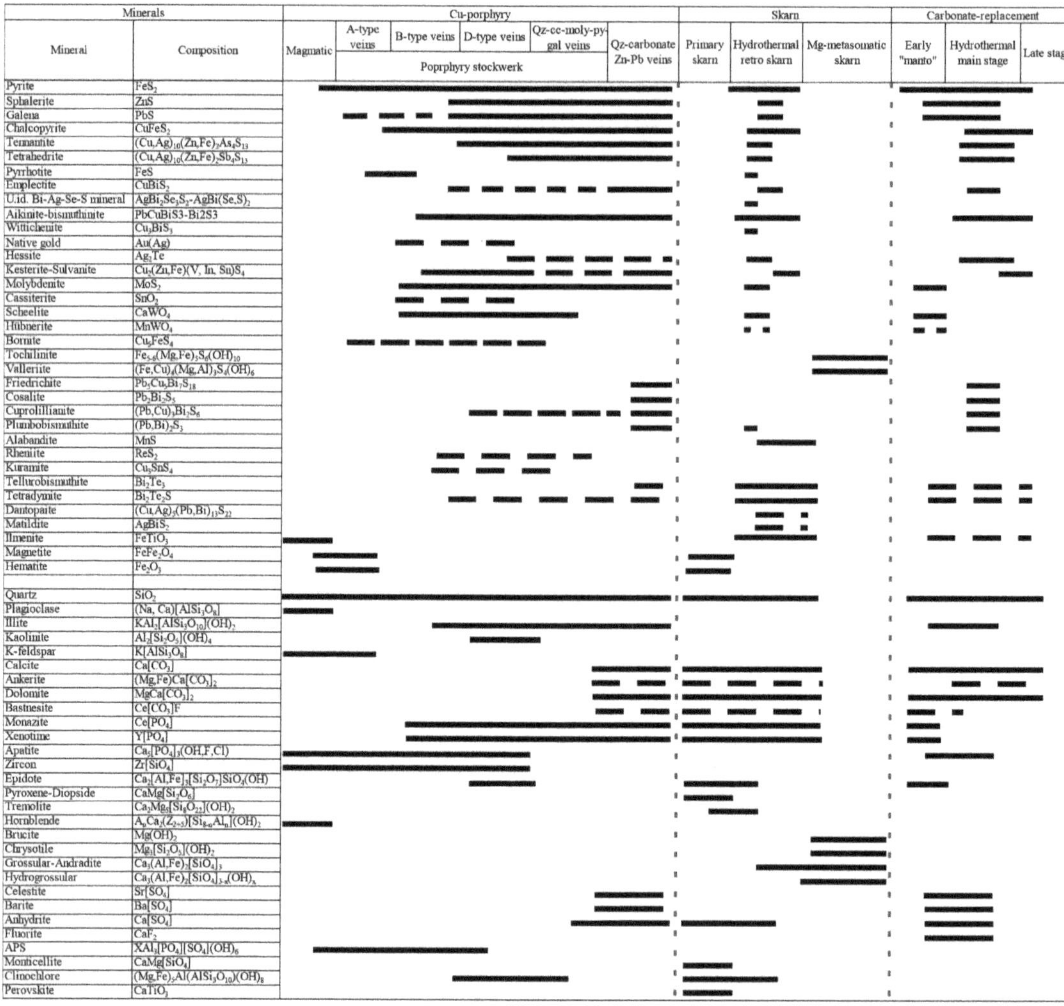

Figure 6. Paragenetic succession table of the most abundant ore and gangue minerals within the studied samples. Copper–porphyry mineralization is listed by textural features, while skarn and carbonate-replacement polymetallic mineralization are listed according to the ore-forming stages. Minerals are given in order of decreasing frequency.

3.2.2. Skarn Mineralization

The main sulfide minerals of the different skarn types are pyrite, chalcopyrite, sphalerite, galena, pyrrhotite, and alabandite (Figure 6). Mainly as inclusions in pyrite or chalcopyrite occur kesterite, tetradymite, tellurobismuthinite, plumbobismuthinite, wittichenite, aikinite, dantopaite, and matildite, as well as an unknown mineral of AgBiSeS composition (Figure 6). Besides these, scheelite, tennantite, molybdenite, valleriite, and tochilinite may occur in some skarn samples, not only as inclusions in pyrite, but as single ore mineral phases, too.

Pyrite is the most abundant ore mineral in the skarns. Various pyrite generations can be distinguished in skarn ores: some idiomorphic, 10–50 μm pyrite crystals occur in the recrystallized carbonate matrix often dispersed, without any inclusions; skarns with originally porphyritic texture may host porphyry-replacement pyrite and chalcopyrite, often accompanied by pyrite and chalcopyrite in B-type veins; pyroxene–garnet skarns host dispersed 0.1–5 mm subhedral pyrite grains within the silicate matrix; while idiomorphic pyrite crystals of 0.1–2 mm size may occur as dispersed crystals, rarely as vermicular veins in pyroxene or pyroxene-amphibole-skarns.

Chalcopyrite generally forms xenomorphic masses within all skarn rock types (Figure 4c). Massive quartz–chalcopyrite lenses occur in pyroxene skarn samples with 10–30% chalcopyrite locally, chalcopyrite with xenomorphic sugary quartz occurring in an 100–1000 μm wide vein network enmeshing the whole pyroxene skarn. Chalcopyrite in these samples locally shows alteration to covellite along the grain boundaries. In Zn–skarn samples chalcopyrite occurs mostly as chalcopyrite disease in sphalerite, often replacing as much as 15% of the original sphalerite grains. Euhedral–subhedral sphalerite and galena tend to occur together. Early sphalerite shows co-precipitation textures with andradite (Figure 4a), while sphalerite is also observed to co-precipitate with pyrite and galena. Sphalerite grains in Zn–Pb skarns are typically 0.1–2 mm in size, occurring in bands typically accompanied by subhedral–euhedral 0.1–5 mm galena and euhedral 0.1–2 mm pyrite. Further ore minerals are of less importance. Alabandite (Figure 4d,e) occurs usually in subhedral 20–200 μm grains, often accompanied by high-Mn high-Fe sphalerite and space-filling galena. Valleriite and tochilinite were only observed in one of the studied samples, accompanied by brucite, serpentine-group minerals and hydrogrossular, replacing magnetite, monticellite, and anhydrite (Figure 4d). Tochilinite and valleriite within this skarn assemblage are often intergrown, forming xenomorphic masses with 20–600 μm pseudomorphs after the pre-existing equilibrium-textured magnetite–monticellite assemblage (Figure 4b). In the original silicate–oxide mineral assemblage of the skarn, hematite and magnetite can be observed as ore minerals and as accessories, as well as accessory allanite, thorianite, rutile, and perovskite. Sphalerite in most of the studied skarn samples shows intense "chalcopyrite disease".

3.2.3. Carbonate-Replacement Mineralization

The most abundant ore mineral throughout the mineralized carbonate host rocks is pyrite with varying size and shape depending on textural position, commonly associated with sphalerite-, galena-, chalcopyrite-, and tetrahedrite-group minerals (Figure 5a,e–g; Figure 6). In addition, accessory molybdenite, scheelite, kesterite, kuramite, aikinite, friedrichite, bismuthinite, telluro-, cupro-, or plumbobismuthinite, tetradymite, hessite, and emplectite occur in the mineralized samples (Figure 6). While molybdenite and scheelite occur as <1% separate phases, kesterite usually occurs connected to sphalerite- or tetrahedrite-group minerals, while the sulfosalts are generally present as 5–20 μm inclusions within pyrite or galena. The ore minerals can either occur in the breccia clasts or in the cementing carbonate vein network.

Pyrite occurs exclusively as 10–500 μm large cubes in the silicified host rock and 100 μm–4 mm large crystals in carbonate veins. Pyrite is generally the first phase to crystallize, often showing growth zonation bands. Inclusions of sphalerite-, galena-, tetrahedrite-group minerals, chalcopyrite, and, rarely, arsenopyrite highlight the zonal patterns visible in pyrite. Pyrite has a typical poikilitic [44] co-precipitation texture with sphalerite occurring in some CAR-mineralized veins (Figure 5f). Sphalerite also occurs as euhedral–subhedral grains, with typically amber, rarely bluish-yellow, or dark-brown color. Early precipitating sphalerite may occur as zonal inclusion swarms in pyrite or euhedral quartz, while sphalerite in carbonate veins is usually intergrown with pyrite and galena, showing polysynthetic twinning. Multiple growth zones of sphalerite can be observed, typically being highlighted by color changes. All samples with sphalerite are characterized by the intense "chalcopyrite disease" (Figure 5b,d) [45]. Chalcopyrite displaces up to 10–15% of the original crystal generally, although the extent of the disease can be different in every specimen. Besides chalcopyrite, kesterite occurring together with tetrahedrite also replace sphalerite (Figure 5a). Sphalerite grain size ranges from 5 μm to 3 cm within different textural positions.

Galena usually is co-genetic with sphalerite, although some galena veins also cut across sphalerite crystals, suggesting a later origin. Galena is typically subhedral, always occurring in the same textural position as sphalerite both in the matrix or in the carbonate veins. The typical crystal size for galena is between 50 μm and 3 mm. In some sphalerite-bearing specimens with chalcopyrite disease in sphalerite, galena shows cellular–granular, ameboid, or less commonly, oriented lamellar replacement texture by aikinite [46] as a quasi "galena disease".

Besides its typical occurrence as "disease in sphalerite", chalcopyrite generally occurs as veins or anhedral patches of 50 μm–10 mm size. Chalcopyrite veining typically shows crosscutting relations with every ore mineral except the late-stage tetrahedrite-group minerals. The tetrahedrite-group minerals occur as 100–500 μm space-filling anhedral masses, veins, or pockets often accompanied by anhedral galena or rarely with hessite and bismuthinite. Kesterite only occurs as rims on diseased sphalerite crystals often accompanied by vanadium-rich sulvanite and tin-rich kuramite-like mineral phases.

3.3. *Compositions of Major Sulfide Minerals in Different Types of Mineralization*

Galena, sphalerite, tetrahedrite-group minerals, molybdenite, and bismuth-bearing sulfosalts all occur in porphyry, skarn, and carbonate-replacement type of ores at Recsk. Thus, comparison of their compositions may indicate their different formation conditions in different parts of the porphyry-, skarn-, and carbonate-replacement-ore-forming system at subvolcanic levels. Chemical compositional data are given in the electronic supplement.

3.3.1. Molybdenite

In molybdenite, the concentrations of Cu, Se, Re, and W are highly variable. Rhenium is the most abundant trace element with a wide range of concentrations. Molybdenite in the porphyry copper ore shows the highest Re contents up to 4.47 wt% ($\overline{x}$ = 1.05 wt% and 2σ = 1.22, n = 47). Molybdenite with the highest Re contents was found in the northernmost apexes of the diorite intrusion where molybdenite is often hosted by D-type quartz veins and carbonate veins with pyrite–molybdenite–galena occurring in the central part of D-type veins. Molybdenite grains within these veins also tend to show heterogeneous distribution of Re. Rhenium contents show no systematic changes throughout most of the grains but oscillatory zoning and lamellar changes in the distribution of Re occasionally also occur. In those latter cases, Re contents tend to change by various layers within one single molybdenite flake. Molybdenite grains with the highest Re contents also host micrometer–nanometer-sized inclusions of rheniite. Molybdenite in the skarn and CAR mineralization shows significantly lower rhenium concentrations compared to the porphyry ore (Figure 6). The Re content of molybdenite is 0.026–1.86 wt% with $\overline{x}$ = 0.267 and 2σ = 0.359 (n = 30) in the skarn ores, and it is always below 0.125 wt% with $\overline{x}$ = 0.047 wt% with 2σ = 0.039 (n = 13) in the CAR mineralization.

The Cu contents of molybdenite show a reverse trend compared to Re. The distal CAR mineralization has the highest Cu contents, while lower values characterize both the skarn and porphyry ores (Figure 7a). In the two latter mineralization types, the Cu values are mostly below the detection limit, while they are always detectable in the CAR mineralization. The Se concentrations in molybdenite show a similar trend as Cu (Figure 7). Selenium is generally low in molybdenite from both the porphyry (b.d.l–0.14 wt% $\overline{x}$ = 0.024 wt% with 2σ = 0.027, n = 47) and skarn (b.d.l–0.034 wt% $\overline{x}$ = 0.01 wt% with 2σ = 0.011, n = 30) ores, though some molybdenite from the porphyry ore have sporadically higher values. The average Se content in molybdenite is highest in the CAR mineralization (up to 0.11 wt% $\overline{x}$ = 0.06 wt% with 2σ = 0.030, n = 13). Tungsten does not show systematic variation among the three mineralization types, but only a few values (n = 12 out of 90) are above the detection limit (180–200 ppm) of the applied analytical setup. These higher values were measured dominantly in molybdenite from the skarn mineralization.

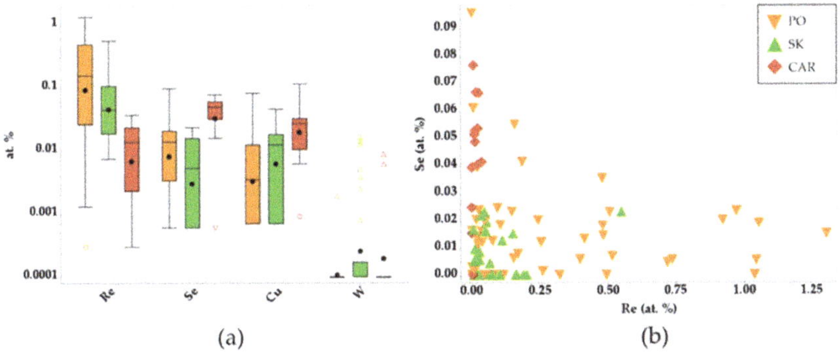

Figure 7. Minor and trace element distributions in molybdenite from the three mineralization types (Porphyry (PO), skarn (SK), and carbonate-replacement (CAR)). (**a**) Boxplots showing the trace element distributions and their interquartile ranges, averages (horizontal lines within the boxes), and medians (black circles) for the three mineralization types. Note that the *y*-axis is logarithmic. While Re shows a decreasing trend from the porphyry towards the carbonate rocks, Cu and Se are both seemingly enriched towards the more distal mineralization. (**b**) Re–Se (atom%) scatterplot showing generally low Se contents coupled with high Re in porphyry, while low Re with relatively higher Se contents in the carbonate-replacement mineralization.

When plotting Re vs. Se data of molybdenite, a grouping of the data according to the mineralization type becomes visible (Figure 7b). The Re contents of molybdenite are lower in the skarn and especially in the CAR mineralization types reaching the highest values in the porphyry molybdenite. In contrast, Se contents are highest in molybdenite from CAR mineralization. Molybdenite with high Se and low Re contents usually is from carbonate veins hosting only pyrite besides molybdenite. Skarn molybdenite has low concentrations of both Se and Re and overlaps with the lower range of porphyry mineralization (Figure 7b).

3.3.2. Galena

Galena is a common ore mineral in all three mineralization types at subvolcanic levels at Recsk. It shows variations in Se, Ag, and Bi contents according to the ore type (Figure 8a,b). Selenium is lower in the porphyry copper ore (b.d.l–1.54 wt%, $\overline{x} = 0.53$ wt% and $2\sigma = 0.1$, n = 39) compared to the skarn (0.23–2.13 wt%, $\overline{x} = 0.97$ wt% and $2\sigma = 0.51$, n = 45), and peripheral CAR mineralization (0.35–4.08 wt%, $\overline{x} = 1.01$ wt% and $2\sigma = 0.95$, n = 34) (Figure 8a). Silver and Bi show a strong positive correlation (Figure 8b) with r = 0.94 for the whole set of samples. The highest Ag and Bi contents are both found in the CAR type. Galena analyses with Ag values above the detection limit show a negative correlation of Se and Ag (r = 0.72). Porphyry mineralization hosts galena with Ag (b.d.l.-0.76 wt%, $\overline{x} = 0.11$ wt% and $2\sigma = 0.19$, n = 39) and Bi (b.d.l–0.75 wt%, $\overline{x} = 0.06$ wt% and $2\sigma = 0.16$, n = 39) contents, and galena from the skarn mineralization shows similar range of Ag (b.d.l–0.43 wt%, $\overline{x} = 0.10$ wt% and $2\sigma = 0.14$, n = 45) and Bi (b.d.l–0.28 wt%, $\overline{x} = 0.05$ wt% and $2\sigma = 0.09$, n = 45) contents. Galena in the CAR mineralization contains the highest amount of both Ag (0.18–0.97 wt%, $\overline{x} = 0.50$ wt% and $2\sigma = 0.24$, n = 34) and Bi (b.d.l–1.28 wt%, $\overline{x} = 0.43$ wt% and $2\sigma = 0.4$, n = 34, respectively). Copper was also analyzed in galena, showing the highest values in the porphyry mineralization, although the data may be affected by chalcopyrite hosting most of the measured galena grains. Thus, Cu data are not further used for deposit type characterization.

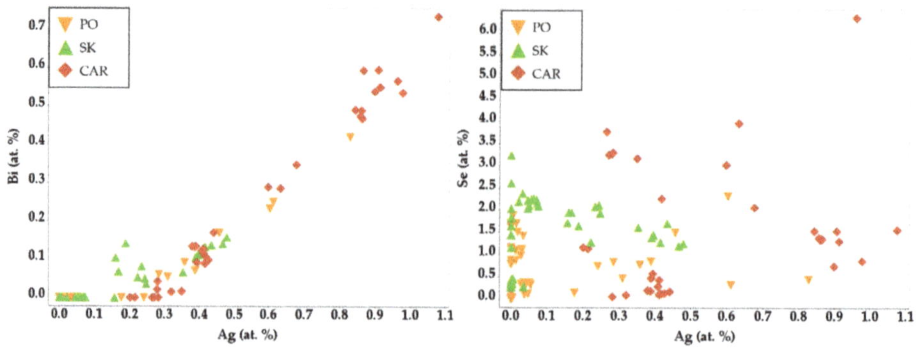

(a) (b)

Figure 8. (**a**): Ag–Bi (atom%) scatterplot for galena. Linear correlation is visible, with increased Ag–Bi contents characteristic to CAR mineralization, while skarn mineralization shows similar trends to porphyry. (**b**): Ag–Se scatterplot for galena shows the lack of correlation between these elements.

For the whole galena dataset above detection limit for Bi (n = 51), the following linear regression can be set for Ag–Bi compositions with a correlation (r = 0.92):

$$\text{Bi(at. \%)} = 0.7512\text{Ag(at. \%)} - 0.1634 \quad (1)$$

If skarn data are excluded—as it shows a different behavior regarding Ag–Se—the remaining porphyry–carbonate-replacement dataset shows a correlation of (r = 0.97) for Bi and Ag. With this dataset (n = 37), the linear regression can be modified as follows:

$$\text{Bi(at. \%)} = 0.8387\text{Ag(at. \%)} - 0.2243 \quad (2)$$

3.3.3. Sphalerite

Sphalerite is characterized by variable Fe, Mn, Cd, and V contents while Cu occurs in sphalerite mostly due to intense "chalcopyrite disease" (Figure 5b) [45]. Thus, high Cu contents are also often associated with similarly high Fe contents, although higher Fe contents are generally observed with Cu being totally absent. All the analyzed sphalerite crystals are generally low (below detection limit) in Ge, In, and Hg. Some exceptions of relatively higher In contents (0.58–0.69 wt%, n = 3) coincide with enrichments of Cu, with no other elements being significantly enriched. All of the above-mentioned In enrichments occur in one CAR sample, showing extensive chalcopyrite disease in all sphalerite grains. Principal component analysis (PCA) of the data (Figure 9.) shows Mn and Fe charging positively on PC2 while being negatively loaded on PC1. Cadmium and Zn show an inverse behavior compared to Fe and Mn, although their characteristics are less distinctly highlighted. Loading positively on both PC1 and PC2, Cu shows a nearly perpendicular trend to all of the aforementioned elements. Hence, while high Mn contents are typical for skarn sphalerite, porphyry sphalerite is Zn–Cd-dominated, while CAR mineralization tends to fall in between, overlapping with the skarn data field.

3.3.4. Kesterite

Besides the chalcopyrite disease, kesterite $Cu_2(Zn,Fe)(V,In,Sn)S_4$ is also found replacing sphalerite with a similar texture (Figure 5a) in the CAR mineralization. Sulfides with Cu(Zn, Fe)(Sn,V)(S,Se) element association occur as inclusions in pyrite together with sphalerite, chalcopyrite, and bornite within the porphyry orebody too. Multiple generations of kesterite have been found as inclusions of pyrite, replacing sphalerite in the CAR mineralization or accompanying tetrahedrite-group minerals (Figure 6), some of them containing significant amounts of V (up to 3.22 wt%, $\overline{x}$ = 0.66 wt% n = 18), and sometimes remarkable amounts of In too (up to 0.63 wt%, $\overline{x}$ = 0.25 wt% n = 18). Both in the porphyry and the

CAR mineralization, crystals of ferrokesterite–stannite occur. Moreover, composition close to kuramite is detected, although within these grains, Fe, Zn, In, and V substitutions also occur, which are known as substitutions within this structure.

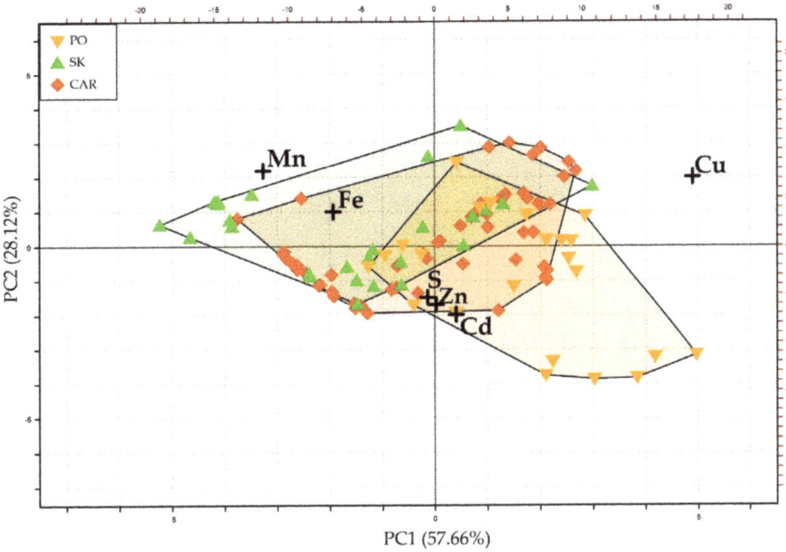

Figure 9. PCA biplot of sphalerite main element compositions from the studied mineralization types. Vector endpoints are indicated by black crosses. Iron and Mn showing inverse behavior compared to Zn and Cd is a result of the well-known substitution mechanisms for these elements, while Cu loading nearly perpendicular to these is an effect of abundant chalcopyrite disease within the sphalerite grains. While porphyry sphalerite can be characterized by relatively higher Cd and Zn contents, skarns are marked by elevated Mn levels. CAR field is falling in between, overlapping strongly with the majority of the skarn data.

3.3.5. Tetrahedrite-Group Minerals

Major and minor element compositions for tetrahedrite-group minerals were also measured from all three deep-seated mineralization types, while data for the epithermal mineralization are available from [13]. It is important to highlight that skarn data come from one single sample, as occurrence of tetrahedrite-group minerals within the skarn mineralization is rather accidental. While the CAR-mineralized zones are dominated by the presence of tetrahedrite, tennantite is dominant in the porphyry and the skarn samples.

Tetrahedrite group minerals in all three mineralization types show strong ($r = 0.72$) correlation between Sb and Ag, while its Fe and Cd contents are negatively correlated ($r = 0.70$). Oscillatory zoning is often observed in the CAR mineralization (Figure 5c), where multiple zones within one single grain were also measured. Due to oscillatory zoning patterns, tetrahedrite- and tennantite-rich compositions may occur next to each other on the single-grain-scale in the CAR mineralization. On the contrary, tennantite is predominant in the porphyry and skarn mineralization; in the latter, zoning is lacking. Oscillatory zoning is caused by changes in As, Sb, Ag, Fe, Zn, and Bi contents, As–Sb being the most influential substitution. While CAR mineralization is characterized by oscillatory zoned tetrahedrite, the mineral grains in the porphyry mineralization are generally homogeneous or can be characterized by one single Sb-enriched rim.

Mineralization types within the porphyry–epithermal system can be distinguished by the Ag + Cu/Fe + Zn (apfu) ratios [13] or Sb/(Sb + As) (apfu) ratios, as well as Fe/(Fe + Zn) (apfu) ratios [27] in the tetrahedrite-group minerals (Figure 10.). Porphyry samples show Sb/(Sb + As) ratios of 0.00014–0.46, i.e., these are tennantite. Sb/(Sb + As) ratios between

0.9 and 0.007 are characteristic for the CAR mineralization type. As mentioned, the wide range of this ratio is caused by oscillatory zoning of this mineral. A single generation of tennantite grains (n = 14) found in a sample from skarn mineralization is characterized by Sb/Sb + As values of 0.0008–0.026.

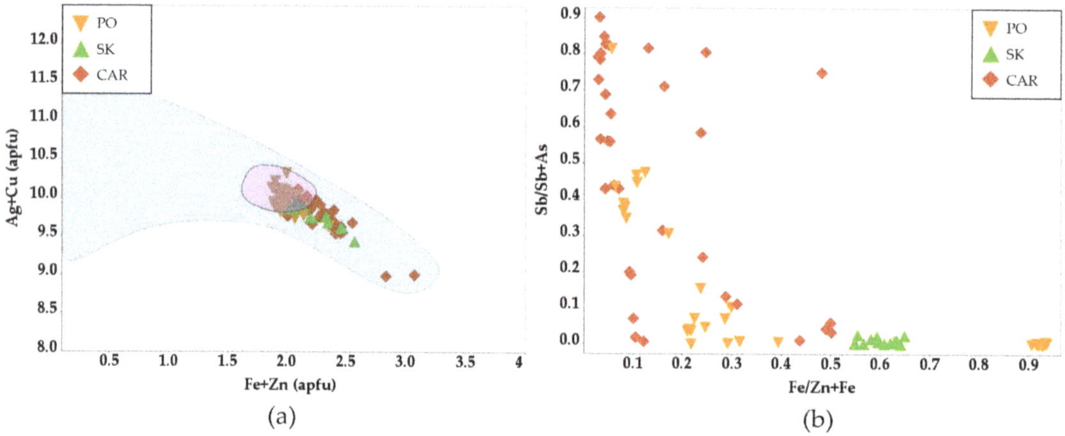

Figure 10. (**a**): Ag + Cu − Fe + Zn scatterplot of tetrahedrite-group minerals showing the data of the deep-seated mineralization types (this study) compared to data fields of all three generations of epithermal high-sulfidation (blue) and intermediate-sulfidation from the Parád area (pink) from [13]. (**b**): Sb/(Sb + As) − Fe/(Fe + Zn) scatterplot of tetrahedrite-group minerals showing the deep-seated mineralization types. Relatively lower Fe contents are characteristic for the CAR mineralization, while relatively higher As contents are characteristic for the porphyry and skarn mineralization types.

Elevated Ag values in the tetrahedrite-group minerals are characteristic for the CAR mineralization type, together with the elevated contents of Sb. Antimony and Bi also show elevated concentration in the CAR mineralization type, although no correlation of Bi with Sb or Ag was found. Fe/(Fe + Zn) in tetrahedrite-group minerals is less distinctive because this ratio shows bimodal distribution in the studied samples, with values between 0.94 and 0.90, as well as 0.40–0.06 for Cu–porphyry, In the skarn, Fe/(Fe + Zn) is 0.64–0.54, and CAR mineralization is richer in Zn (0.50–0.03). Cadmium, occurring in the same crystallographic position as Zn, shows enrichment in the porphyry mineralization, while being constantly lower in the CAR mineralization.

3.3.6. Bismuth Sulfosalts

Bismuth sulfosalts are typical accessory minerals in all ore types at Recsk. The sulfosalts from high-sulfidation and intermediate-sulfidation epithermal mineralization were already described in detail before [8,13], but less so in the deep-seated orebodies addressed in this study. Bismuthinite, galenobismuthinite, and tetradymite occur in all the mineralization types in various amounts. Copper–Pb–Bi ratios allow distinction of different mineral species (Figure 11), such as aikinite–bismuthinite series minerals as well as friedrichite, cosalite, cuprolillianite, wittichenite, and cuprobismuthinite homologous series.

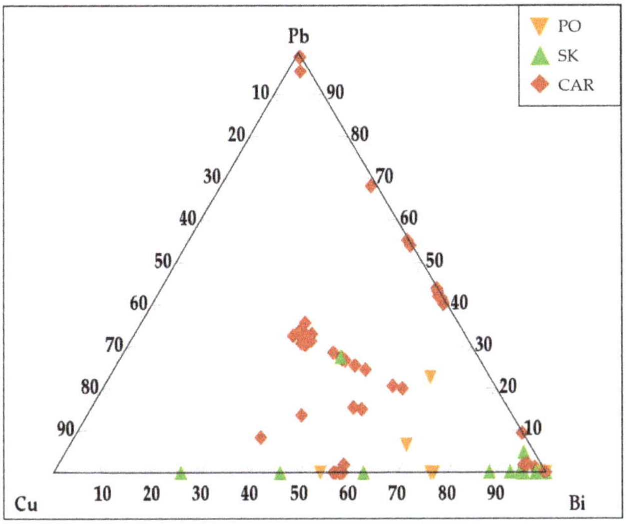

Figure 11. Pb–Cu–Bi ternary plot of the measured Bi sulfosalts. Most sulfosalt grains were found within the CAR mineralization type, most of them being typically Pb-rich, while Pb-bearing sulfosalt phases are less abundant in skarn and porphyry mineralization. PO mineralization is dominated by Cu-bearing sulfosalts, while skarn rocks typically host Pb-poor sulfosalt phases. CAR mineralization hosts a variety of sulfosalts along the aikinite–bismuthinite homologous series and plumbobismuthinite is also abundant.

Identified Ag-bearing Bi sulfosalts are matildite, dantopaite, and hammarite, while an unknown Ag–Bi–Se–S phase could not have been better identified because of its small crystal size. EDS and EPMA measurements for this unknown phase suggest a 3:2 ratio member of the bohdanowitzite–matildite series. Tellurium-bearing phases are similarly abundant in all ore types, and the abundance of Te–Se–S phases show no difference among ore types.

The amounts of identified Bi minerals show some correlation with the type of ore, e.g., aikinite and galenobismuthinite often replace galena in CAR mineralization, while the skarn mineralization is dominated by Pb-poor sulfosalts, like bismuthinite–cuprobismuthinite or tetradymite, and Ag-bearing sulfosalts are also more abundant. Lead-bearing sulfosalt phases are more abundant in the CAR-mineralization type, though emplectite, aikinite–bismuthinite, witthichenite, and hessite occur as accessories within the epithermal, as well as the deep-seated, orebodies.

4. Discussion

Petrography, supported by mineral minor and trace element measurements can serve as a useful tool for mineral exploration in porphyry and related ore deposits. For this purpose, galena, molybdenite, sphalerite, tetrahedrite-group minerals, and sulfosalts can also be discussed regarding stoichiometric, non-stoichiometric, and coupled substitutions documented in the studied mineralization types [16,24,25,27,47–52].

4.1. Molybdenite

In molybdenite, the most notable difference is the change in Re content in the three deep-seated mineralization types (Figure 7). Rhenium gradually decreases from the porphyry mineralization center towards the more distal CAR mineralization, while other elements such as Cu, Se, or W show less distinct characteristics. Detection frequency and quantity of Cu increased from the porphyry stocks towards the distal CAR ore in the carbonate rock. Measurable contents of W were most often analyzed in molybdenite from skarn rocks, which are typically characterized by a low Re content. Higher W contents coincide with occurrences of scheelite within the skarn. It should be emphasized that molybdenite characterized by the highest Re content occurs in carbonate–quartz–sulfide veins of the porphyry mineralization. Moreover, these Re contents are seemingly decreasing with increasing distance from the porphyry mineralization center as already described by Terada et al. [26]. This agrees with the observations of Barton et al. [24], who found high Re content in less felsic diorite intrusions, where the main controls of its precipitation are more oxidative conditions or phase separation. As anhydrite is documented in the porphyry mineralization, relatively oxidized conditions are a realistic assumption for the Recsk porphyry system, while phase separation is also known to occur in the apexes of the mineralization [10,37], where the D-type and carbonate veins are both abundant. The fact that W and Cu occur in molybdenite within mineralization types lacking rhenium also agrees with observations by Barton et al. [24]. These findings altogether indicate that the main source of Re are the subvolcanic or deeper intrusions, and Re can be preferentially enriched in the open fissures of shallow stockwork mineralization simultaneously with the opening up of late veins. Molybdenite from these veins has the highest Re contents (up to 4.47 wt%), and it is associated with rheniite, a rare rhenium mineral.

4.2. Galena

The strong positive correlation of Ag and Bi in galena (Figure 8.) clearly supports the well-established coupled substitution mechanism along the galena–lillianite series [49]. The second significant trace element substituting in galena is Se, which enables continuous S–Se substitution along the galena–clausthalite line [47]. The variable but typically low Cu contents measured in galena raise the possibility of Cu^{+} $+Bi^{3+}$-coupled substitution: $x(Cu^{+} + Ag^{+})$-xBi^{3+} [47,49]. The difference between Cu^{+}–Bi^{3+} and Ag^{+}–Bi^{3+} replacement can be that galena–matildite or galena–lillianite form a nominally continuous series [47,49,53], while aikinite and galena are not known to form an isomorphous series. The aikinite that developed as lamellar or myrmecitic replacement phase in Ag–Bi bearing low Cu galena in CAR mineralization further supports the hypothesis that Cu is less compatible with the crystal structure of galena. The established linear functions for Ag and Bi substitution suggest that for one atom % Ag, approximately 0.6 atom % Bi is incorporated. This means that Sb or As—not measured within this study—may also be incorporated within the galena lattice via coupled substitution, or that other substitutions must be operative [51]. The negative correlation of Ag and Se in skarn galena for values above the lower detection limit for Ag suggests that this galena generation has distinctly different formation conditions, which could reflect the metasomatic processes associated with multi-stage skarn formation.

Galena in porphyry copper and skarn systems is generally characterized by lower Se, Ag, and Bi concentrations compared to the carbonate replacement mineralization. Although the enrichment of all three elements towards the country rocks seems statistically valid, Se shows no correlation with Ag or Bi; thus, its enrichment towards the country rocks is not controlled by the same process causing the enrichment in Bi and Ag. Temperature-dependent changes in galena chemistry are suggested by Craig [53] and Moëlo et al. [49]. The observed patterns in this study underline distance-dependent enrichment processes for Se or Ag and Bi in galena which although Se and Bi–Ag have different enrichment processes. Thus, besides temperature, additional factors must be considered, regarding the enrichment of various elements in galena. Partitioning between sphalerite and galena or further co-existing sulfides like pyrite or chalcopyrite may be present regarding some of the elements.

4.3. Sphalerite

Sphalerite minor and trace element distribution can be affected by the source of the elements and the changes in redox conditions in addition to the changes in temperature [23]. Given that the Ga content of high-temperature hydrothermal sphalerite usually falls below the detection limit of the EPMA method, and Ge as well as In are b.d.l. for more than

70% of the measurements, the sphalerite GGIMF thermobarometer of Frenzel et al. [23] is not applicable for our dataset. Both Fe and Mn occur in very low concentrations in sphalerite from the porphyry copper ore, and both elements show high values in the typically more reduced skarn rocks. In the skarns, this may indicate the subsequent hydrothermal overprinting of the previously formed, high-temperature mineral association [54]. Thus, we suggest that the highest Fe–Mn content measured in sphalerite formed in association with tochilinite, valleriite, alabandite, chalcopyrite, galena, and pyrrhotite during the hydrothermal overprinting/metasomatic replacement of the primary high-temperature magnetite–monticellite skarn mineral assemblage. PCA of sphalerite major element data supports that Fe and Mn behave geochemically similarly, following the well-known Mn and Fe substitution of Zn in sphalerite Frenzel et al. [23]. Cadmium shows opposite behavior to Fe and Mn, which can be interpreted with simple substitution processes. Copper loading nearly perpendicular to the Zn–Cd–Fe–Mn line on the PCA figure (Figure 9) indicates that the enrichment of Cu within sphalerite is caused by an independent process. The interpretation of elevated Cu contents is rather easy in most of the studied sphalerite grains as the whole mineralization is characterized by intense chalcopyrite disease in sphalerite [45]. The appearance of In, V, and Ge is not systematic, although their enrichment is usually associated with some subsequent hydrothermal overprinting of sphalerite. Thus, highest In, V, and Ge contents are associated with chalcopyrite or kesterite "disease", typically in the skarn and carbonate replacement mineralization types. Application of further thermobarometers [22,55] is also limited because of the occurrence of extensive chalcopyrite disease re-equilibrating the Fe content of sphalerite, while Mn and Cd were not measured from galena. To apply modern geothermometers on these sphalerite grains sulfur isotopic ratios of sphalerite and pyrite, fluid inclusion studies, and further trace element analysis must be performed.

4.4. Tetrahedrite-Group Minerals

For tetrahedrite-group minerals, both compositional and textural differences are discriminative. As tetrahedrite-group minerals in skarns are rather rare, the data presented are statistically not representative, as all results originate from one single sample. Tetrahedrite group minerals from the solely examined skarn sample are similarly rich in As as the porphyry mineralization is. They are enriched in Bi, while the concentrations of Se, Ag, and Cd are generally low, below the detection limit. Carbonate-replacement mineralization is dominantly characterized by tetrahedrite with oscillatory growth zoning. However, Sb/(Sb + As) ratios of these grains scatter within a wide range, with the highest Sb contents corresponding to increased Bi, Se and Ag contents. This supports previous observations [56] that increasing As contents are coupled with decreasing Ag contents in tetrahedrite-group minerals. The oscillatory zoning of As-Sb and Fe-Zn as well as the substitution of Cu by Ag may indicate the physico-chemical—most probably temperature—changes during the crystallization of tetrahedrite-group minerals [28]. The observed ratios of Sb/(Sb + As) and Fe/(Fe + Zn) correspond to the data ranges of data presented by Marushchenko et al. [27] from the Baimka Cu–Mo–Au porphyry ore system. Mineral chemical analyses on epithermal tetrahedrite-group minerals were reported by Takács et al. [13]. Their data show that all generations of high sulfidation (HS) epithermal tetrahedrite-group minerals are dominated by Cu and Ag, while Zn and Fe are generally low (Figure 10). HS mineralization is characterized by higher Te and Bi contents as well, while intermediate sulfidation (IS) epithermal tetrahedrite group mineral compositions overlap with the porphyry–skarn–CAR field (Figure 9). IS and CAR data are generally overlapping, though Te is less abundant in the CAR mineralization compared to the IS tetrahedrite-group minerals.

4.5. Sulfosalts

The distribution of sulfosalts with various compositions in the mineralized system is not necessarily conclusive. Lead-rich, or Pb-dominated, sulfosalts were mostly found in the carbonate-replacement mineralization (Figure 11), although some Pb–Cu–Bi sulfosalts are present in late carbonate veins with sphalerite–galena dominant mineral composition of the porphyry mineralization, too. These late veins are therefore petrographically and geochemically linked to the ore forming fluids of the CAR mineralization. In skarn samples, Bi and Cu–Bi sulfosalts and sulfotellurides were found predominantly, while Pb occurs only in galena, not in the form of sulfosalts. The most abundant sulfosalt mineral observed in the deep-seated orebodies was aikinite and transitional compositions within the aikinite–bismuthinite homologous series [49], occurring dominantly in the CAR mineralization type. Occurrence of Pb-bearing sulfosalts mainly in the distal parts of the deep-seated orebodies may be connected to the large scale of distribution of elements in the whole mineral system. The same zonation is also visible in the distribution of ore-forming elements in the skarn mineralization, as well as the zonal occurrence of the different ore types [57]. An alternative explanation for the preferential occurrence of Pb within the carbonate-replacement mineralization can be the partly sedimentary origin of Pb in this mineralization type. Tellurium sulfosalt minerals occur throughout the mineralized system. Tetradymite, hessite, and emplectite occur in all five mineralization types within the mineralized Recsk complex.

4.6. Behavior of Selenum

In the case of molybdenite, galena, and tetrahedrite-group minerals, a major trend was observed of relatively increasing Se concentrations towards the CAR mineralization. Even though this trend is not highlighted by any kind of statistically sound correlation of Se with other elements within the studied mineral grains, the question of temperature or distance-dependent behavior of Se within the whole porphyry hydrothermal system may be raised similarly to the suggestion of Keith et al. [57] for pyrite. The lack of correlation of Se with other elements may be due to the relatively low number of analyses, or further controlling factors, like partitioning of Se between co-precipitating minerals, which were not studied in detail in this study.

5. Conclusions

The magmatic–hydrothermal system at Recsk includes porphyry-style, skarn, and distal carbonate-replacement-type mineralization at subvolcanic levels. The systematic mineralogical and textural analyses of sulfide minerals from the various types of ores revealed that the trace element contents of molybdenite, galena, and tetrahedrite-group minerals can be useful as proximity indicators and for discriminating mineralization types in this porphyry-intrusion-related magmatic–hydrothermal system. The Re content of molybdenite decreases from the central porphyry Cu mineralization towards the more distal skarn and carbonate replacement type of mineralization in the country rocks. In galena and tetrahedrite-group minerals, Ag, Bi, and Se contents increase within the more distal mineralization types. This is also coupled with increasing Sb and Zn contents in the tetrahedrite-group minerals. Thus, Re in molybdenite, and Ag, Bi, and probably Se in galena, as well as Sb, Zn, Ag, and Bi in tetrahedrite-group minerals, may be applied for differentiating mineralization types and as vectors towards the porphyry intrusion within a porphyry–copper-type mineral system.

Compositional variation of sphalerite was found to be of limited use for fingerprinting or vectoring purposes within the Recsk ore complex based on EPMA measurements. In every ore type, anomalous behavior of Mn, Fe, Zn in sphalerite with "chalcopyrite disease" was detected. However, sphalerite crystals that have been co-precipitated with pyrite and without "chalcopyrite disease" in the carbonate replacement type ore show obvious linear correlations between these elements. Kesterite inclusions with similar texture to the "chalcopyrite disease" within sphalerite were also found in the latter type of ore. Any secondary replacement phenomenon occurring in sphalerite (chalcopyrite or kesterite "disease") makes the sphalerite data absolutely inconsistent, limiting its value for fingerprinting, thermobarometry, or vectoring.

Galena replacement by myrmecitic–lamellar aikinite was also observed. The abundant occurrence of these replacement textures within an approximately 2 km wide zone in the distal part of the porphyry system indicates a widespread distal hydrothermal Cu-mobilization event. This could be caused either by remobilization of some central, Cu-rich, mineralized body or the distal presence of primary Cu-bearing ore-forming hydrothermal fluids.

Supplementary Materials: The following supporting information can be downloaded at: https://www.mdpi.com/article/10.3390/min14090956/s1, sum_EPMA_data_minerals.xlsx with all measurement data.

Author Contributions: Conceptualization, M.B. and F.M.; sampling, F.M.; methodology, M.B., M.F., J.G.R. and G.B.K.; software, A.V.; formal analysis, A.V.; investigation, M.B., M.F., M.H. and G.B.K.; resources, J.G.R. and G.B.K.; data curation, A.V. and M.B.; writing—original draft preparation, M.B.; writing—review and editing M.H., J.G.R., G.B.K. and F.M.; visualization, A.V.; supervision, F.M.; project administration, F.M.; funding acquisition, F.M. and J.G.R. All authors have read and agreed to the published version of the manuscript.

Funding: This research was supported by the Central European Exchange Programme for University Studies (CEEPUS) program.

Data Availability Statement: The original contributions presented in the study are included in the article/Supplementary Material, further inquiries can be directed to the corresponding author/s.

Acknowledgments: This research is a part of the PhD thesis of the first author supervised by F. Molnár and T.G. Weiszburg at the Department of Mineralogy, Eötvös Loránd University. The research was supported by a scholarship provided by the Earth Science Doctoral School of the Eötvös Loránd University. The Eugen F. Stumpfl Electron Microprobe Laboratory at Montanuniversität Leoben provided invaluable support to this research in the frame of the Central European Exchange Programme for University Studies (CEEPUS). The work of G. B. Kiss is supported by the János Bolyai Research Scholarship awarded by the Hungarian Academy of Sciences and the ÚNKP-23-5 New National Excellence Program of the Ministry for Culture and Innovation from the source of the National Research, Development, and Innovation Fund. The Auro-Science Consulting Ltd., Hungary also supported the studies by providing access to a desktop scanning electron microscope that was invaluable for high-resolution petrography and selection of mineral phases for electron microprobe analyses.

Conflicts of Interest: The authors declare no conflicts of interest.

References

1. John, D.A.; Taylor, R.D. By-Products of Porphyry Copper and Molybdenum Deposits. In *Rare Earth and Critical Elements in Ore Deposits*; Verplanck, P.L., Hitzman, M.W., Eds.; Reviews in Economic Geology; Society of Economic Geologists: Littleton, CO, USA, 2016; Volume 18, Chapter 7; pp. 137–164. ISBN 9781629490922. [CrossRef]
2. Elshkaki, A.; Graedel, T.E.; Ciacci, L.; Reck, B. Copper Demand, Supply, and Associated Energy Use to 2050. *Glob. Environ. Chang.* **2016**, *39*, 305–315. [CrossRef]
3. Seck, G.S.; Hache, E.; Bonnet, C.; Simoën, M.; Carcanague, S. Copper at the Crossroads: Assessment of the Interactions between Low-Carbon Energy Transition and Supply Limitations. *Resour. Conserv. Recycl.* **2020**, *163*, 105072. [CrossRef] [PubMed]
4. Cseh-Németh, J. Deep-seated base metal ore occurrence of Recsk: Geological pattern of ore accumulation. *Bull. Hung. Geol. Soc.* **1975**, *105*, 692–708. (In Hungarian)
5. Csillag, J. Rocks transformed upon magmatic effect in the Recsk area. *Bull. Hung. Geol. Soc.* **1975**, *105*, 646–671. (In Hungarian)
6. Csongrády, J. Characterization of the deep-seated base metal ore mineralization of Recsk on the basis of ore-microscopic analyses. *Bull. Hung. Geol. Soc.* **1975**, *105*, 672–691. (In Hungarian)
7. Baksa, C. Genetic Aspects of the Recsk Mineralized Complex, Hungary. In *Geology and Metallogeny of Copper Deposits*; Conference Paper; Springer: Berlin/Heidelberg, Germany, 1986; pp. 280–290. [CrossRef]
8. Gatter, I.; Molnár, F.; Földessy, J.; Zelenka, T.; Kiss, J.; Szebényi, G. High-and low-sulfidation epithermal mineralization of the Màtra Mountains, Northeast Hungary. In *Epithermal Mineralization of the Western Carpathians*; Molnár, F., Lexa, J., Hedenquist, J., Eds.; Society of Economic Geologists Guidebook Series; Society of Economic Geologists: Littleton, CO, USA, 1999; Volume 31, pp. 155–179.

9. Zelenka, T.; Szebényi, G. Exploration history of the base metal deposit at Recsk deep levels. In *Mineral Exploration in Hungary during the XXth Century*; Miskolc-Rudabánya; Szakáll, S., Morvai, G., Eds.; 2002; pp. 169–198. Available online: https://library.hungaricana.hu/hu/collection/muze_szak_asva_kozlemenyek/ (accessed on 20 September 2024).
10. Molnár, F. The Cu-Au-Ag-Zn-Pb ore complex at Recsk, Hungary: A uniquely preserved and explored porphyry-skarn-epithermal system in the Palaeogene magmatic belt of the Alp-Carpathian-Dinaride system. In Proceedings of the Ninth Biennial SGA Symposium, Dublin, Ireland, 20–23 August 2007; Volume 1, pp. 153–157.
11. Molnár, F.; Jung, P.; Kupi, L.; Pogány, A.; Vágó, E.; Viktorik, O.; Pécskay, Z. Epithermal Zones of the Porphyry-Skarn-Epithermal Ore Complex at Recsk. In *Recsk and Lahóca—Geology of the Paleogene Ore Complex*; Földessy, J., Hartai, É., Eds.; Publications of the University of Miskolc, Series A, Mining; University of Miskolc: Miskolc, Hungary, 2008; Volume 73, pp. 99–128.
12. Földessy, J.; Hartai, É. *Recsk and Lahóca: Geology of the Paleogene Ore Complex*; Publications of the Univeristy of Miskolc, Series A, Mining; University of Miskolc: Miskolc, Hungary, 2008; Volume 73.
13. Takács, Á.; Molnár, F.; Turi, J.; Mogessie, A.; Menzies, J.C. Ore Mineralogy and Fluid Inclusion Constraints on the Temporal and Spatial Evolution of a High-Sulfidation Epithermal Cu-Au-Ag Deposit in the Recsk Ore Complex, Hungary. *Econ. Geol.* **2017**, *112*, 1461–1481. [CrossRef]
14. Altenberger, F.; Raith, J.G.; Bakker, R.J.; Zarasvandi, A. The Chah-Mesi Epithermal Cu-Pb-Zn-(Ag-Au) Deposit and Its Link to the Meiduk Porphyry Copper Deposit, SE Iran: Evidence from Sulfosalt Chemistry and Fluid Inclusions. *Ore Geol. Rev.* **2022**, *142*, 104732. [CrossRef]
15. Cooke, D.R.; Agnew, P.; Hollings, P.; Baker, M.; Chang, Z.; Wilkinson, J.J.; White, N.C.; Zhang, L.; Thompson, J.; Gemmell, J.B.; et al. Porphyry indicator minerals (PIMS) and porphyry vectoring and fertility tools (PVFTS)–indicators of mineralization styles and recorders of hypogene geochemical dispersion halos. In Proceedings of the Decennial Mineral Exploration Conferences, Toronto, ON, Canada, 22–25 October 2017; pp. 457–470.
16. Cooke, D.R.; Agnew, P.; Hollings, P.; Baker, M.; Chang, Z.; Wilkinson, J.J.; Ahmed, A.; White, N.C.; Zhang, L.; Thompson, J.; et al. Recent Advances in the Application of Mineral Chemistry to Exploration for Porphyry Copper–Gold–Molybdenum Deposits: Detecting the Geochemical Fingerprints and Footprints of Hypogene Mineralization and Alteration. *Geochem. Explor. Environ. Anal.* **2020**, *20*, 176–188. [CrossRef]
17. Sykora, S.; Cooke, D.R.; Meffre, S.; Stephanov, A.S.; Gardner, K.; Scott, R.; Selley, D.; Harris, A.C. Evolution of Pyrite Trace Element Compositions from Porphyry-Style and Epithermal Conditions at the Lihir Gold Deposit: Implications for Ore Genesis and Mineral Processing. *Econ. Geol.* **2018**, *113*, 193–208. [CrossRef]
18. Börner, F.; Keith, M.; Smith, D.J.; Barry, T.L.; Neumann, T.; Klemd, R. Fingerprinting Fluid Evolution by Trace Elements in Epithermal Pyrite, Vatukoula Au-Te Deposit, Fiji. *Ore Geol. Rev.* **2021**, *137*, 104314. [CrossRef]
19. Cave, B.; Lilly, R.; Barovich, K. Textural and Geochemical Analysis of Chalcopyrite, Galena and Sphalerite across the Mount Isa Cu to Pb-Zn Transition: Implications for a Zoned Cu-Pb-Zn System. *Ore Geol. Rev.* **2020**, *124*, 103647. [CrossRef]
20. Cioacă, M.E.; Munteanu, M.; Qi, L.; Costin, G. Trace Element Concentrations in Porphyry Copper Deposits from Metaliferi Mountains, Romania: A Reconnaissance Study. *Ore Geol. Rev.* **2014**, *63*, 22–39. [CrossRef]
21. Keith, M.; Haase, K.M.; Chivas, A.R.; Klemd, R. Phase Separation and Fluid Mixing Revealed by Trace Element Signatures in Pyrite from Porphyry Systems. *Geochim. Cosmochim. Acta* **2022**, *329*, 185–205. [CrossRef]
22. Kullerud, G. The FeS-ZnS system—A geological thermometer. *Nor. Geol. Tiddskrift* **1953**, *32*, 62–147.
23. Frenzel, M.; Hirsch, T.; Gutzmer, J. Gallium, Germanium, Indium, and Other Trace and Minor Elements in Sphalerite as a Function of Deposit Type—A Meta-Analysis. *Ore Geol. Rev.* **2016**, *76*, 52–78. [CrossRef]
24. Barton, I.F.; Rathkopf, C.A.; Barton, M.D. Rhenium in Molybdenite: A Database Approach to Identifying Geochemical Controls on the Distribution of a Critical Element. *Min. Met. Explor.* **2020**, *37*, 21–37. [CrossRef]
25. Ciobanu, C.L.; Cook, N.J.; Kelson, C.R.; Guerin, R.; Kalleske, N.; Danyushevsky, L. Trace Element Heterogeneity in Molybdenite Fingerprints Stages of Mineralization. *Chem. Geol.* **2013**, *347*, 175–189. [CrossRef]
26. Terada, K.; Osaki, S.; Ishihara, S.; Kiba, T. Distribution of Rhenium in Molybdenites from Japan. *Geochem. J.* **1971**, *4*, 123–141. [CrossRef]
27. Marushchenko, L.I.; Baksheev, I.A.; Nagornaya, E.V.; Chitalin, A.F.; Nikolaev, Y.N.; Vlasov, E.A. Compositional Evolution of the Tetrahedrite Solid Solution in Porphyry-Epithermal System: A Case Study of the Baimka Cu-Mo-Au Trend, Chukchi Peninsula, Russia. *Ore Geol. Rev.* **2018**, *103*, 21–37. [CrossRef]
28. Sack, R.O. Fahlore Thermochemistry: Gaps inside the (Cu,Ag)10(Fe,Zn)2(Sb,As)4S13 Cube. *Petrology* **2017**, *25*, 498–515. [CrossRef]
29. Benedek, K. Paleogene Igneous Activity along the Easternmost Segment of the Periadriatic-Balaton Lineament. *Acta Geol. Hung.* **2002**, *45*, 359–371. [CrossRef]
30. Exner, C. *Die Geologische Position der Magmatite des Periadriatischen Lineamentes*; Geologische Bundesanstalt: Vienna, Austria, 1976; pp. 3–64.
31. Földessy, J.; Zelenka, T.; Benedek, K.; Pécskay, Z.; Mádai, F. The Recsk Paleogene magmatism in a regional context. In *Recsk and Lahóca—Geology of the Paleogene Ore Complex*; Földessy, J., Hartai, É., Eds.; Publications of the University of Miskolc, Series A, Mining; University of Miskolc: Miskolc, Hungary, 2008; Volume 73, pp. 7–20.
32. Raith, J.G.; Hutter, F.; Altenberger, F.; Weilbold, J.; Auer, C.; Krause, J.; Berndt, J.; Neinavaie, H. Polymetallic Tungsten Skarn Mineralisation Related to the Periadriatic Intrusion at Lienzer Schlossberg, East Tyrol, Austria. *Austrian J. Earth Sci.* **2024**, *117*, 87–112. [CrossRef]

33. Arató, R.; Dunkl, I.; Takács, Á.; Szebényi, G.; Gerdes, A.; Von Eynatten, H. Thermal Evolution in the Exhumed Basement of a Stratovolcano: Case Study of the Miocene Mátra Volcano, Pannonian Basin. *J. Geol. Soc.* **2018**, *175*, 820–835. [CrossRef]
34. Zelenka, T.; Markó, B. Comparative results of exploratory shaft-sinking and tunnel-driving and exploratory deep drilling at the Recsk ore deposit. *Bull. Hung. Geol. Soc.* **1979**, *109*, 469–477. (In Hungarian)
35. Baksa, C. The genetic framework of the Recsk ore genesis. *Acta Mineroligica-Petrogr. Szeged.* **1983**, *XXVI/I*, 87–97.
36. Kovács, S.; Gecse, Z.; Pelikán, P.; Zelenka, T.; Szebényi, G.; Szabó, I. Upper Triassic conodonts from deep boreholes of the Recsk–Darnó area: New data on the geology of its pre-Cenozoic basement. *Bull. Hung. Geol. Soc.* **2013**, *143*, 29–46. (In Hungarian)
37. Molnár, F.; Gatter, I.; Zelenka, T.; Pécskay, Z.; Bernadett, B. Metallogeny of Paleogene and Neogene Volcanic Belts in Hungary. In Proceedings of the 7th Biannual SGA Meeting, Balkema, Rotterdam, The Netherlands, 24–28 August 2003; Volume 2, pp. 1205–1208.
38. Földessy, J.; Seres-Hartai, É.; Szebényi, G. Distribution of Gold Mineralization in the Recsk Ore Complex, NE Hungary. *Acta Geol. Hung.* **2004**, *47*, 247–258. [CrossRef]
39. Singer, D.A.; Berger, V.I.; Moring, B.C. *Porphyry Copper Deposits of the World: Database and Grade and Tonnage Models*; USGS Reports; USGS: Reston, VA, USA, 2008.
40. Török, K.; Gyuricza, G.; Szeiler, R.B.; Füged, U.; Gál, N.; Gáspár, E.; Gulyás, Á.; Horváth, Z.; Kerékgyártó, T.; Korbély, B.; et al. *Complex Vulnerability and Loadability Assessment Report to Recsk II. Copper Ore Proposed Area for Concession*; Hungarian Office for Mining and Geology: Budapest, Hungary, 2016. Available online: https://mbfsz.gov.hu/sites/default/files/file/2018/07/06/recsk_ii_eng.pdf (accessed on 20 September 2024).
41. Földessy, J.; Szebényi, G. The mineralizations of the the Recsk deeps and Lahóca—Short geological overview. In *Recsk and Lahóca—Geology of the Paleogene Ore Complex*; Földessy, J., Hartai, É., Eds.; Publications of the University of Miskolc, Series A, Mining; University of Miskolc: Miskolc, Hungary, 2008; Volume 73, pp. 85–98.
42. R Core Team. *R: A Language and Environment for Statistical Computing*; R Foundation for Statistical Computing: Vienna, Austria, 2003; Available online: https://www.r-project.org (accessed on 20 September 2024).
43. Sillitoe, R.H. Porphyry Copper Systems. *Econ. Geol.* **2010**, *105*, 3–41. [CrossRef]
44. Ramdohr, P. *Ore Minerals and Their Intergrowths*, 2nd ed.; Pergamon Press: Oxford, UK, 1980.
45. Barton, P.B.; Bethke, P.M. Chalcopyrite Disease in Sphalerite: Pathology and Epidemiology. *Am. Mineral.* **1987**, *72*, 451–467.
46. Foord, E.E.; Shawe, D.R. The Pb-Bi-Ag-Cu-(Hg) chemsitry of galena and some associated sulfosalts: A review and some new data from Colorado, California and Pennsylvania. *Can. Mineral.* **1989**, *27*, 363–382.
47. Blackburn, W.H.; Schwendeman, J.F. Trace Element Substitution in Galena. *Can. Miner.* **1977**, *15*, 365–373.
48. Cook, N.J.; Ciobanu, C.L.; Pring, A.; Skinner, W.; Shimizu, M.; Danyushevsky, L.; Saini-Eidukat, B.; Melcher, F. Trace and Minor Elements in Sphalerite: A LA-ICPMS Study. *Geochim. Cosmochim. Acta* **2009**, *73*, 4761–4791. [CrossRef]
49. Moëlo, Y.; Makovicky, E.; Mozgova, N.N.; Jambor, J.L.; Cook, N.; Pring, A.; Paar, W.; Nickel, E.H.; Graeser, S.; Karup-Møller, S.; et al. Sulfosalt Systematics: A Review. Report of the Sulfosalt Sub-Committee of the IMA Commission on Ore Mineralogy. *Eur. J. Mineral.* **2008**, *20*, 7–62. [CrossRef]
50. Pasava, J.; Svojtka, M.; Veselovský, F.; Ďurišová, J.; Ackerman, L.; Pour, O.; Drábek, M.; Halodová, P.; Haluzová, E. Laser Ablation ICPMS Study of Trace Element Chemistry in Molybdenite Coupled with Scanning Electron Microscopy (SEM)—An Important Tool for Identification of Different Types of Mineralization. *Ore Geol. Rev.* **2016**, *72*, 874–895. [CrossRef]
51. Renock, D.; Becker, U. A First Principles Study of Coupled Substitution in Galena. *Ore Geol. Rev.* **2011**, *42*, 71–83. [CrossRef]
52. George, L.; Cook, N.J.; Cristiana, C.; Wade, B.P. Trace and Minor Elements in Galena: A Reconnaissance LA-ICP-MS Study. *Am. Mineral.* **2015**, *100*, 548–569. [CrossRef]
53. Craig, J.R. Phase Relations and Mineral Assemblages in the Ag-Bi-Pb-S System. *Miner. Depos.* **1967**, *1*, 278–306. [CrossRef]
54. Yin, S.; Wirth, R.; He, H.; Ma, C.; Pan, J.; Xing, J.; Xu, J.; Fu, J.; Zhang, X.N. Replacement of Magnetite by Hematite in Hydrothermal Systems: A Refined Redox-Independent Model. *Earth Planet. Sci. Lett.* **2022**, *577*, 117282. [CrossRef]
55. Bortnikov, N.S.; Dobrovol'skaya, M.G.; Genkin, A.D.; Naumov, V.B.; Shapenko, V.V. Sphalerite-Galena Geothermometers: Distribution of Cadmium, Manganese, and the Fractionation of Sulfur Isotopes. *Econ. Geol.* **1995**, *90*, 155–180. [CrossRef]
56. Ebel, D.S.; Sack, R. Arsenic-Silver Incompatibility in Fahlore. *Mineral. Mag.* **1991**, *55*, 521–528. [CrossRef]
57. Keith, M.; Smith, D.J.; Jenkin, G.R.T.; Holwell, D.A.; Dye, M.D. A Review of Te and Se Systematics in Hydrothermal Pyrite from Precious Metal Deposits: Insights into Ore-Forming Processes. *Ore Geol. Rev.* **2018**, *96*, 269–282. [CrossRef]

minerals

MDPI

Article

Discovery and Exploration of the Luming Porphyry Mo Deposit, Northeastern China: Implications for Regional Prospecting

Bangfei Gao [1,*], Minghua Dong [2], Hui Xie [1], Zhiliang Liu [2], Yihang Li [1] and Tong Zhou [3]

1 China Railway Resources Group Co., Ltd., Beijing 100039, China; xiehui422@163.com (H.X.); lyh181050267@163.com (Y.L.)
2 China Railway Resources Group Survey and Design Co., Ltd., Langfang 065000, China; dmh8311@126.com (M.D.); lzwh.153@163.com (Z.L.)
3 Yichun Luming Mining Co., Ltd., Yichun 152500, China; zhoutong271047553@126.com
* Correspondence: bangfeigao@163.com or gaobangfei3@crecg.com

Abstract: Over the past two decades, significant deposit discoveries were made in Northeastern China, including the super-large Chalukou, Daheishan, and Luming porphyry Mo deposits. The discovery of the Luming deposit was accomplished through verification of stream sediment anomalies, with mineralization closely associated with early Jurassic monzogranite and granite porphyry. Previous studies primarily focused on the mineralization mechanisms of these deposits without adequately addressing the exploration methods and prospecting criteria. This study involved a comprehensive re-evaluation of geological observations, analysis of rock primary halo, gravity and magnetic surveys, and induced polarization surveys conducted during exploration campaigns at the Luming porphyry Mo deposit. The results suggest that hydrothermal breccias play a critical role in controlling the mineralization by forming a central low-grade core within the deposit while the Mo mineralization and hydrothermal alteration exhibit a donut-shaped distribution around it. The primary halo shows a distinct metal zonation moving from a central W-Bi-Mo-(Sb) to a peripheral Cu-Co-Ni and a distal Pb-Zn-Ag-In. The mineralization zone exhibits a low Bouguer gravity anomaly, negative magnetic anomaly, medium to low resistivity, and moderate to high chargeability, indicating the effectiveness of geophysical methods in defining the extent of the ore body. The Luming porphyry Mo deposit and distal skarn-epithermal Pb-Zn mineralization are parts of a porphyry-related magmatic-hydrothermal system. The results of this study offer valuable insights into the genesis of porphyry Mo deposits and their implications for prospecting in the forested region of Northeastern China.

Keywords: stream sediment geochemistry; hydrothermal breccias; low-grade core; primary halo; gravity and magnetic survey; induced polarization; prospecting criteria; Lesser Xing'an Rang–Zhangguangcai Range

1. Introduction

Molybdenum (Mo) is a critical metal that plays a significant role in economic and social development [1]. Porphyry Mo deposits represent the dominant type worldwide, at over 90% [2,3]. Globally, porphyry Mo deposits are primarily found within the Circum-Pacific Metallogenic Belt and the Tethys Metallogenic Belt, which are linked to plate subduction processes [3–7]. China has a significant concentration of Mo deposits, with proven reserves of approximately 5.80 million metric tons accounting for around 38.6% of global reserves [8]. Most Mo deposits in China are located within the eastern Qinling-Dabie Orogenic Belt and the eastern Central Asian Orogenic Belt (i.e., Xing-Meng Metallogenic Belt) [9–12].

Over the past two decades, significant porphyry Mo deposits were discovered in the forested regions of the Lesser Xing'an Range–Zhangguangcai Range. These include super-large Mo deposits at Chalukou [13], Daheishan [11], and Luming [14], as well as large

Mo deposits at Fu'anpu [15], Huojihe [16], Dashihe [17], Jidetun [18], Chang'anpu [19], and numerous other locations (Figure 1). Consequently, this region has emerged as a prominent area for mineral deposit research, with extensive studies conducted in this field [4,6,7,9,20–29]. Previous studies primarily focused on the ore-forming tectonic setting and ore genesis, such as rock geochemistry, geochronology, isotope geochemistry, and fluid circulation [16,30–39]. Limited literature exists discussing mineral prospecting [29,40–45]. As a result, there is currently a lack of comprehensive overview regarding prospecting methodologies for porphyry Mo deposits in forested regions of Northeastern China.

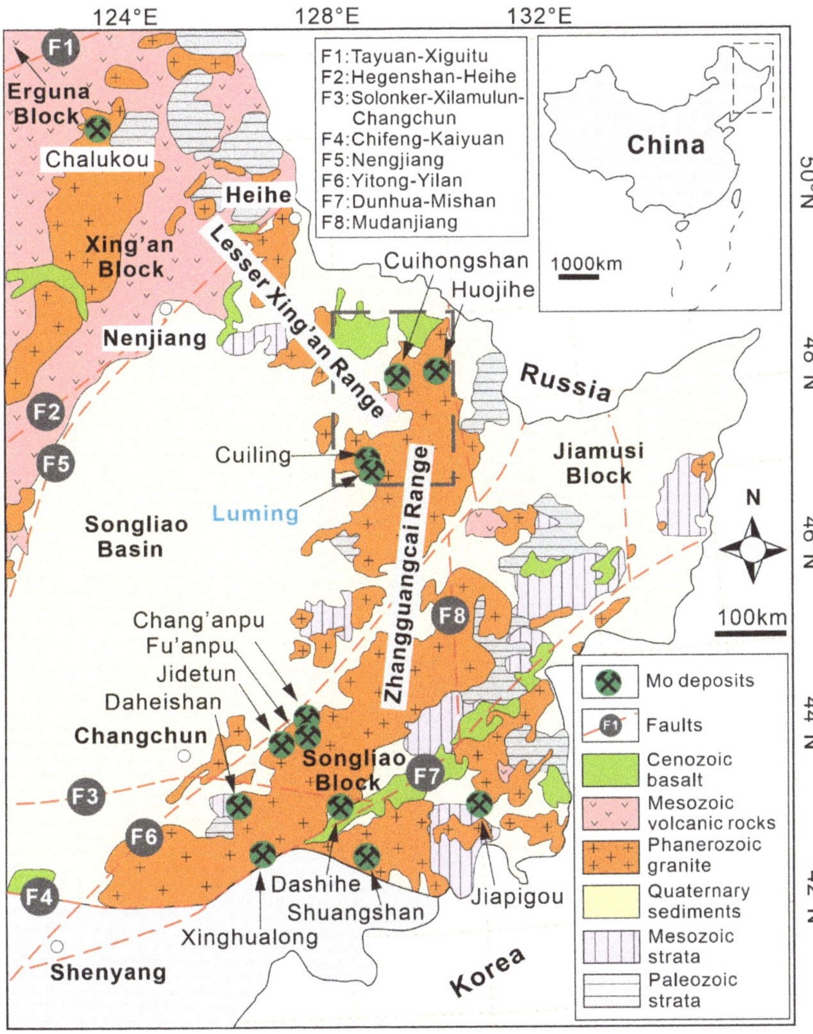

Figure 1. Geological map of the Lesser Xing'an Range–Zhangguangcai Range Metallogenic Belt (LXRZRMB) in Northeastern China and the distribution of Jurassic Mo deposits; the Luming porphyry Mo deposit is located in the northern section of the LXRZRMB [5,16].

The Luming porphyry Mo deposit (Luming PMD) is situated in the northern region of the Lesser Xing'an Range–Zhangguangcai Range Metallogenic Belt (LXRZRMB) (Figure 1). The Luming PMD was initially discovered by stream sediment geochemistry, after which its three-dimensional extent was delineated through drilling exploration. The Luming PMD is estimated to possess indicated and inferred resources totaling 814 million metric tons, with an average Mo grade of 0.092% and a total Mo metal content of 751 thousand metric tons. Furthermore, the ore body exhibits an associated Cu grade averaging 0.015%, corresponding to a Cu metal quantity of 118 thousand metric tons. Notably distinguished by its large-scale operations (processing approximately 15 million metric tons of ores annually), shallow burial conditions (average overburden thickness is about 10 m), and low stripping ratio (approximately 1.0), the Luming PMD demonstrates favorable economic features. Consequently, extensive research has been conducted on the Luming PMD, resulting in the accumulation of substantial fundamental geological data, essential for investigating ore genesis and mineral prospecting [43,46–55].

The purpose of this study is to provide a comprehensive overview of the exploration methods and criteria employed in the Luming PMD, with a particular focus on its discovery history, and a detailed analysis of its geological, geochemical, and geophysical characteristics. The findings derived from this study can offer valuable insights into the genesis of porphyry Mo deposits and their implications for prospecting concealed resources in Northeastern China.

2. Regional Geology and Geophysics

Geologically, the region is situated in the eastern part of the Central Asian Orogenic Belt and at the junction of the Circum-Pacific Orogenic Belt [25,27,56]. Northeastern China can be divided into four blocks from north to south: Erguna, Xing'an, Songliao, and Jiamusi (Figure 1). These blocks are delineated by the Tayuan-Xiguitu Fault, Hegenshan-Heihe Fault, and Mudanjiang Fault [16,27,54]. The combined Erguna, Xing'an, and Songliao blocks are collectively referred to as the Xing'an-Mongolian (Xing-Meng) Orogenic Belt [5,21,28,57]. The tectonic evolution of this region since the Paleozoic era can be divided into three successive stages: the closure of the Palaeo-Asian Ocean, the subduction and closure of the Mongol-Okhotsk Ocean, and subsequent subduction events related to the Palaeo-Pacific Ocean [4,16,19,31,54,58].

The LXRZRMB is situated in the central-eastern part of Northeastern China [6,25,34], spanning across the Xing'an and Songliao Blocks [30,55]. This region is characterized by deep-seated faults that predominantly trend NE, NNE, NNW, and E–W (Figure 1). The distribution of magmatic rocks in the area is controlled by NNE and NE-oriented faults, often intersected by NNW and E–W trending faults [54]. The widespread occurrence of Mesozoic magmatic rocks in this region is believed to be associated with the evolution of the northern Mongol-Okhotsk Ocean and eastern Paleo-Pacific Ocean [24,27,54]. Paleozoic and Mesozoic formations are intermittently distributed within these intrusive rocks. The principal types of mineral deposits in this region are associated with magmatic processes, including porphyry-type, skarn-type, as well as vein-type molybdenum, tungsten, iron, copper, lead-zinc, and gold mineralization [35,36,59]. Porphyry Mo deposits are predominantly located within a NE-trending Jurassic granite belt, as illustrated in Figure 1.

The morphology of the NE-trending tectonic-magmatic belt in this region can be effectively delineated by magnetic anomalies, which exhibit varying intensities ranging from –300 to 500 nT, with a maximum value recorded at 1000 nT (Figure 2).

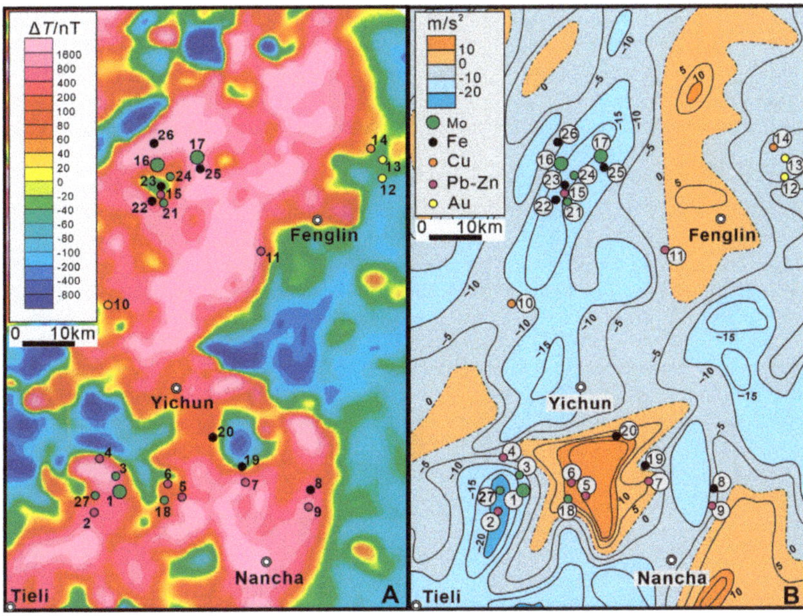

Figure 2. (**A**) Regional aeromagnetic anomaly and distribution of mineral deposits in the Yichun-Tieli area. (**B**) Regional aerogravity anomaly and mineral deposit distribution in the Yichun-Tieli area. The 1:1,000,000 aeromagnetic anomaly data are derived from [60], while the mineral deposit data are sourced from [41,61,62]. The 1:1,000,000 aerogravity anomaly data are based on [63]. 1 = Luming Mo(Cu); 2 = Xulaojiugou Pb-Zn; 3 = Cuiling Mo-Au; 4 = Kunlunqi Pb-Zn; 5 = Qianjin Dongshan Pb-Zn; 6 = Xiling Nanshan Pb-Zn; 7 = Xiaoxilin Pb-Zn; 8 = Fengmao Fe; 9 = Mantoushan Pb-Zn; 10 = Milin Cu-Fe; 11 = Wuxing Pb-Zn; 12 = Pingdingshan Au; 13 = Lianzhushan Au; 14 = Shouhushan Cu-Pb-Zn; 15 = Kubing Pb-Zn; 16 = Cuihongshan Mo-Fe-W-Pb-Zn; 17 = Huojihe Mo; 18 = Qianjin Nanshan Mo; 19 = Xilinshi'er Fe; 20 = Daxilin Fe; 21 = Hongqi Mo; 22 = Kuyuan Fe; 23 = Hongtieshan Fe-Mo; 24 = Yongxu Mo; 25 = Hongqishan Fe; 26 = Cuibei Fe; 27 = 712 Highland (712 HL).

These anomalies display alternating patterns but predominantly show positive characteristics. It is well established that a consistently negative anomaly is typically found in low-grade metamorphic rocks as well as I-type granite belts; conversely, a positive anomaly suggests occurrences related to S-type granites or volcanic belts [64,65]. In this context, Luming PMD is situated within an area characterized by moderate to high intensity levels, with values ranging between 300 and 500 nT. Extensive areas display negative magnetic anomalies on its western, northern, and eastern sides, with intensities ranging from –100 to –200 nT. Most molybdenum and polymetallic deposits in the region are situated within or near the transition zone between positive and negative magnetic anomalies adjacent to the positive anomaly area (Figure 2A).

The aerogravity data exhibit a mirror-image relationship with the aeromagnetic anomaly, indicating predominantly low Bouguer gravity anomalies in the region. The contours of the gravity anomaly exhibit a distribution pattern oriented towards the NE and N–S directions, with values ranging from -20 to 15×10^{-6} m/s^2 (Figure 2B). Low gravity anomalies are primarily associated with igneous rock belts, while high gravity anomalies are attributed to residual sedimentary or metamorphic formations within these igneous rocks [63]. The Luming PMD is situated on a gravity gradient zone that shifts from high to low gravity (Figure 2B). The regional occurrence of both elevated and depressed gravity anomalies facilitates the localization of mineral deposits [64]. For example, the presence of depressed gravity anomalies at Huojihe (points 15–17, 21–26 in Figure 2) and Luming (points 1–4, 27 in Figure 2), and an elevated gravity anomaly at Qianjin-Xiling-XiaoXilin (points 5–7, 18–20 in Figure 2) serve as indicators of numerous mineral occurrences. Regions exhibiting sharp boundaries, curvatures, or distortions in gravitational gradients are often associated with intricate geological structures and intense magmatic-hydrothermal activities, which collectively constitute the most favorable locations for mineralization [64].

3. Discovery History and Exploration Campaigns

The study area is in the densely forested covered region of the Lesser Xing'an Range, which presents significant challenges for mineral exploration [42,44]. The discovery of the Luming PMD originated from the verification of regional stream sediment geochemical anomalies (Figure 3), leading to a shift in exploration focus from Cu-Pb-Zn deposits to Mo deposits. The exploration work can be categorized into several phases, as outlined below.

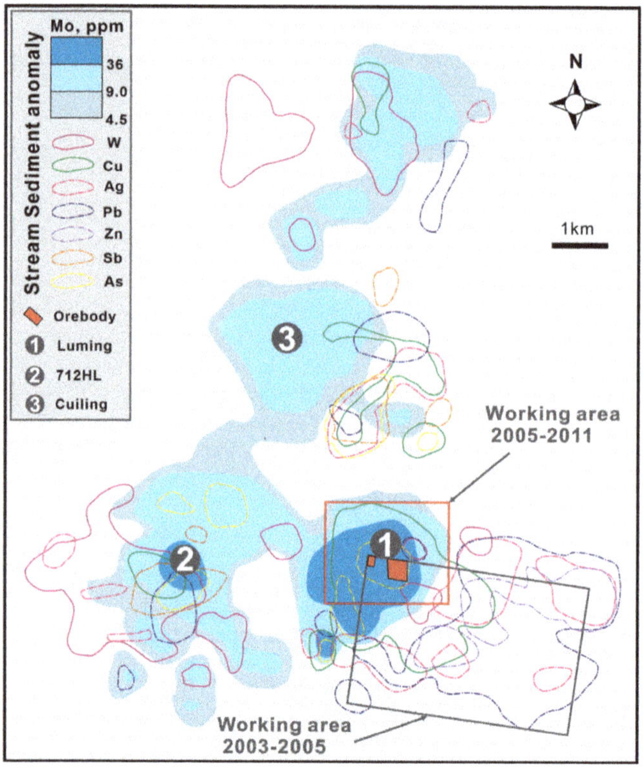

Figure 3. Regional stream sediment anomalies at scale of 1:50,000 show a zoning with W-Mo-Cu and Pb-Zn-Ag-As-Sb, modified after [41,44]. Two Mo ore bodies were discovered at the NW corner of the prospecting area during 2003–2005 (gray rectangle), offering valuable guidance for the following exploration campaigns during 2005–2011(red rectangle).

Between 1968 and 1970, the First Regional Geological Survey Brigade of the Geological and Mineral Resources Bureau of Heilongjiang Province (GMRBHP) conducted a regional geological survey at a scale of 1:200,000. The findings revealed the significant mineralization potential of the late Indosinian-early Yanshanian granites.

In 1987 and 1988, the Geophysical Exploration Brigade of the GMRBHP surveyed streams at a scale of 1:200,000. This study found a multi-element anomaly, including Ag-Cu-Pb-Zn-W-Mo. In 1991, the Second Geological Brigade of the GMRBHP conducted a more detailed stream sediment survey for these anomalies at a scale of 1:50,000. The results exhibited zonation, characterized by the concentration of W-Mo-Cu in the core and the distribution of Pb-Zn-Ag on the periphery [41,44] (Figure 3). Several concentration centers for Mo anomalies were identified at Luming, 712 Highland, and Cuiling.

From 2003 to 2005, the Fifth Geological Survey Institute of the GMRBHP conducted a verification on the concentration center of the 1:50,000 stream sediment anomalies around Luming with the objective of discovering Cu-Pb-Zn deposits [66]. The prospecting work comprised geological mapping, soil geochemistry analysis, and geophysical survey. Nevertheless, no economically significant Cu-Pb-Zn deposits were identified during this initial investigation. The soil geochemistry analysis included the elements Au, Ag, Cu, Pb, Zn, W, Mo, and Bi. Of these elements, only Mo displayed a distinct bimodal distribution in the histogram [44]. Following the verification of surface trenches at the site of the soil geochemical anomaly, a mineralized alteration zone covering approximately 1.5 square kilometers was discovered, delineating two potentially economic Mo ore bodies with an estimated resource of around 20,000 metric tons of Mo [66].

A general exploration was conducted by the Fifth Geological Survey Institute of the GMRBHP during the period 2005–2006 [67]. In March 2008, the technical report was finalized, which estimated an inferred molybdenum resource with a metal content equivalent to 304.2 thousand metric tons.

In 2009, a mineral resource audit and verification was conducted by China Railway Resources Mineral Exploration Co., Ltd. (CRRME), Beijing, China, which involved the drilling of 65 holes at 200 m × 200 m intervals (Figure 4). The estimated mineral resource of that work is 814 million metric tons, with an average Mo grade of 0.092%, containing a total Mo metal of 751 thousand metric tons [68].

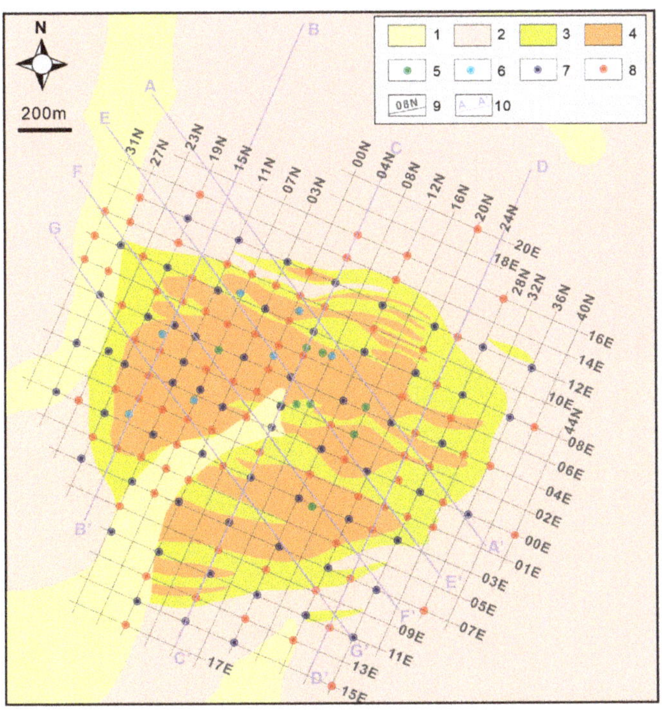

Figure 4. The layout depicts diamond drilling and ground geophysical surveys, with the delineation of ore body outline following drilling exploration, modified after [68]. 1 = Quaternary; 2 = Indosinian-Yanshanian intrusive rocks; 3 = low-grade ores (0.03% ≤ Mo < 0.06%); 4 = economic ores (Mo ≥ 0.06%);

5 = drillings during general exploration; 6 = verification drillings after general exploration; 7 = drillings of resource verification stage; 8 = drillings of complementary exploration stage; 9 = exploration lines and numbers; 10 = locations of geophysical surveys. Gravity and magnetic surveys were conducted at A–A′, B–B′, C–C′, and D–D′, while induced polarization surveys were carried out at A–A′, E–E′, F–F′, and G–G′.

From 2010, CRRME conducted complementary exploration to provide comprehensive geological information for mining design. The spacing of drilling for the primary mining area was increased to 100 × 100 m (Figure 4). Furthermore, ground geophysics and soil geochemistry investigation (to be reported in the following) were also conducted during this phase (in 2011).

4. Deposit Geology

The exploration area of Luming PMD exhibits extensive surface coverage, with a well-developed humus horizon and residual slope deposits that range in thickness from 0.8 to 30.75 m, with an average of 10.36 m [68]. The primary methodology employed in geological mapping is the use of point pits and trenches. In contrast, the drilling information is employed to adjust the projection at the topographic surface of the ore body (Figures 5 and 6).

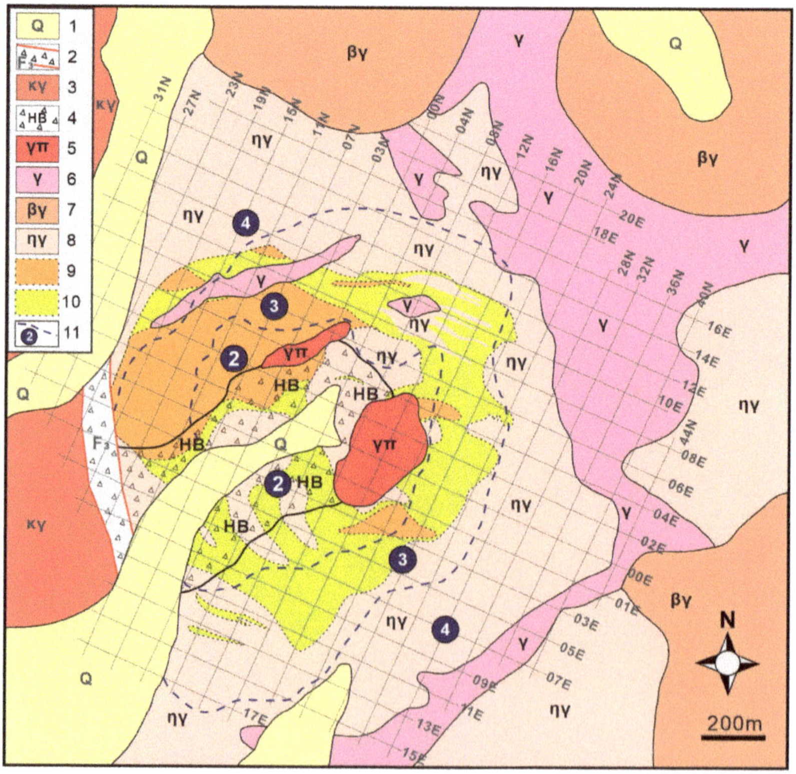

Figure 5. Geological map of the Luming PMD, modified after [68]. 1 = Quaternary; 2 = F3 fault; 3 = K-feldspar granite; 4 = hydrothermal breccias; 5 = granite porphyry; 6 = fine-grained granite; 7 = biotite monzogranite; 8 = monzogranite; 9 = economic Mo ores; 10 = low-grade Mo ores; 11 = alteration zones; ② = silicification-illitization-K-feldsparization; ③ = silicification-illitization-chloritization; ④ = silicification-pyritization-chloritization.

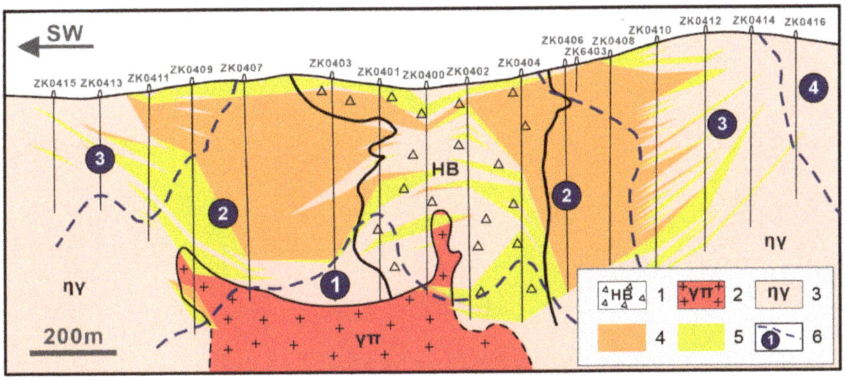

Figure 6. Geological cross section of the 04N exploration line at Luming PMD, modified after [68]. 1 = hydrothermal breccias; 2 = granite porphyry; 3 = monzogranite; 4 = economic Mo ores; 5 = low-grade Mo ores; 6 = alteration zones; ① = silicification-K-feldsparization-biotitization; ② = silicification-illitization-K-feldsparization; ③ = silicification-illitization-chloritization; ④ = silicification-pyritization-chloritization.

4.1. Faults

The principal fault structure within the exploration area is an NNW-trending fault (F3), which was identified through drilling. The dip is towards the northeast, with dip angles of 50° to 60°. The fractured zone has a width ranging from 9 m to 72.3 m. The post-mineralization fault hosts the ore body on its hanging wall, which is primarily composed of monzogranite [68]. The footwall displays no evidence of mineralization or alteration and is predominantly composed of K-feldspar granite. A subsequent NE-trending fault occurred at a later stage and was concealed by Quaternary sediments (Figure 5).

4.2. Intrusive Rocks

The Luming PMD is intruded by a suite of granitic rocks, including monzogranite (ηγ), biotite monzogranite (βγ), fine-grained granite (γ), granite porphyry (γπ), and K-feldspar granite (κγ) [68] (Figure 5). The monzogranite was intruded at an early stage, and subsequent alteration processes occurred predominantly in this area (Figures 5 and 7A). Biotite monzogranite is distributed in the northern part and southeast corner of the exploration area. Jia et al. [67] proposed a regional phase transition relationship between biotite monzogranite and monzogranite. However, direct evidence within the exploration area is lacking. Fine-grained granite is observed to intrude as vein-like bodies into both monzogranite and biotite monzogranite (Figure 5). Field observations indicate at least two stages of vein-like activity for fine-grained granite. The early-stage fine-grained granite underwent silicification and kaolinization alteration (Figure 7B), whereas the late-stage fine-grained granite encloses angular fragments of mineralized monzogranite (Figure 7C). Granite porphyry displays well-developed quartz-feldspar phenocrysts without evident K-feldsparization and biotitization (Figure 7D,E). The drilling results indicate that granite porphyry intruded into the monzogranite (Figure 7D), suggesting that it was emplaced contemporaneously with the fine-grained granites. K-feldspar granite is distributed in the western part of the exploration area (Figure 5), exhibiting no signs of mineralization (Figure 7F,G). This suggests that it was emplaced after mineralization.

In terms of geochronology, the ages of monzogranite and granite porphyry at Luming PMD have been determined to be between 201.1 ± 3.90 Ma and 174 ± 2 Ma [14,16,22,46–48,52], while mineralization of Mo occurred between 182.4 and 176.7 Ma [16,46,47,49,52]. This indicates that the mineralization may be closely associated with the intrusive rocks.

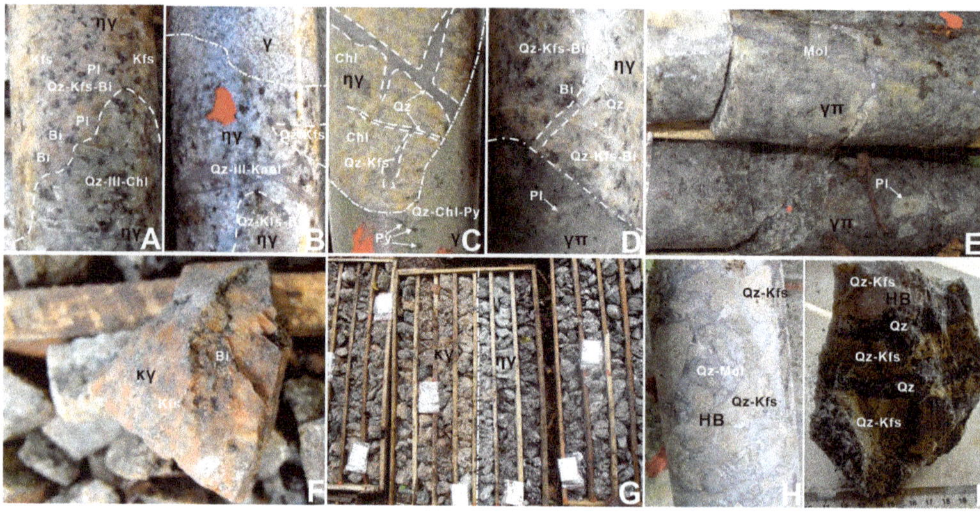

Figure 7. Intrusive rocks and their alteration characteristics of the Luming PMD. (**A**) Monzogranite exhibits the silicification-K-feldsparization-biotitization overlain by silicification-illitization-chloritization. (**B**) Fine-grained granite intruded into the monzogranite, which underwent silicification-kaolinization while enveloping the earlier silicification-K-feldsparization. (**C**) Fine-grained granite encloses breccias of monzogranite. The fine-grained granite experienced silicification-pyrite-chloritization while monzogranite underwent silicification-K-feldsparization-biotitization. Both were superposed by late-stage veinlet silicification. (**D**) Granite porphyry was intruded into monzogranite; the former experienced strong silicification with residual plagioclase phenocrysts, while the latter underwent early silicification-K-feldsparization-biotitization and late veinlet silicification. (**E**) Granite porphyry, characterized by plagioclase phenocrysts, exhibited veinlet Mo mineralization. (**F**) K-feldspar granite underneath F3 fault without any mineralization. (**G**) F3 fault developed at the contact zone between the silicification-illitization monzogranite (grey) and the K-feldspar granite (red). (**H**) Mineralized hydrothermal breccias are composed of clasts of altered monzogranite and a cement matrix comprising silica and molybdenite. (**I**) Barren hydrothermal breccias comprise clasts made up of altered monzogranite within a silica-rich cement matrix. Qz = quartz; Kfs = K-feldspar; Pl = plagioclase; Bi = biotite; ill = illite; Kaol = kaolinite; Chl = chlorite; Py = pyrite; Mol = molybdenite; $\eta\gamma$ = monzogranite; γ = fine-grained granite; $\gamma\pi$ = granite porphyry; $\kappa\gamma$ = K-feldspar granite; HB = hydrothermal breccias.

4.3. Hydrothermal Breccias

The hydrothermal breccias are distributed in a northeasterly direction within the central and western regions of the deposit (Figure 5). In cross section, the hydrothermal breccias exhibit a pipe-like structure above the granite porphyry, gradually transitioning into monzogranite (Figure 6). These breccias are primarily composed of K-feldspathized monzogranite, with a minor proportion of granite porphyry [68]. They exhibit excellent cohesion (Figure 7H,I) and are predominantly composed of siliceous cement. The hydrothermal breccias constitute the primary component of the low-grade core within the ore deposit (Figures 5, 6 and 7I), while the high-grade ores occur in proximity to their contact zone with monzogranite (Figures 5, 6 and 7H). These observations lead to the conclusion that hydrothermal brecciation occurred subsequent to silicification-K-feldsparization but prior to silicification-illitization.

4.4. Ore Body Characteristics

The Luming PMD is defined by a dominant orebody (Figure 8), which is distributed within monzogranite, granite porphyry, and hydrothermal breccias between exploration lines 27N and 28N (Figure 5). The orebody exhibits a donut-shaped configuration on the plane, with an approximate surface exposure length and width of 1200 m and an area of about 1.5 km^2 (Figure 5). The elevation depth of the orebody ranges from 540 m to −220 m. On cross sections, the boundaries converge towards the center, with dip angles ranging from 30° to 70° (Figure 6). The F3 fault cuts across the southwest boundary of the ore body, while K-feldspar granite serves as its footwall. In three-dimensional morphology, this deposit displays a bowl-shaped occurrence (Figure 8), which facilitates reduced stripping ratios during open-pit mining operations. The central portion of the deposit displays a low degree of mineralization, forming a low-grade core within the donut-shaped structure. It is possible to delineate intercalated rock separately within this low-grade core, which dips deeper towards the southeast (Figure 8). From an industrial perspective, the near-surface portion of Luming PMD is primarily composed of oxide ores. The thickness of the oxide zone ranges from 0.8 m to 42.4 m, with an average thickness of 16 m. The oxidation rates of the zone range from 10.7% to 92.9%, with an average of 28% [68]. Below this zone are all sulfide ores, without mixed ore zones.

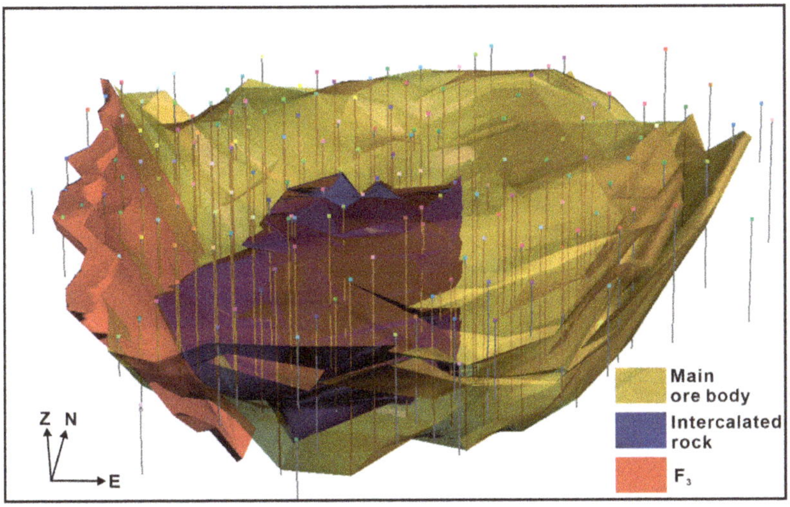

Figure 8. The 3D geological model of the Luming PMD indicates a low strip ratio with a bowl-shaped structure, where intercalated rock can be separately delineated within the low-grade core, which is mainly composed of hydrothermal breccias.

4.5. Ore Compositions and Textures

The economic molybdenum ore is primarily composed of sulfide ores (Figures 9 and 10), with a minor occurrence of oxidized ores near the surface [68]. The primary ore minerals include molybdenite, pyrite, pyrrhotite, ilmenite, chalcopyrite, galena, and sphalerite, accompanied by small amounts of bornite and arsenopyrite. Secondary ore minerals include molybdite, covellite, blue chalcocite, and jarosite. Gangue minerals are primarily composed of quartz, K-feldspar, plagioclase, and biotite, with lesser amounts of illite (sericite), montmorillonite, chlorite, kaolinite, chloritoid, calcite, rutile, titanite, and apatite.

The most common textures observed in ores are euhedral or subhedral crystal textures (Figures 9D,E,H and 10A,I), inclusive texture (Figure 9E,J,K), interstitial texture (Figure 9H,J), metasomatic texture (Figure 9C,D,K,L), cataclastic texture (Figure 10B), and emulsion texture (Figures 9L and 10F). The ore structures mainly include disseminated structure (Figure 10A), veinlet structure (Figures 7E, 9B,C and 10D), stockwork structure (Figure 10E,F), vein structure (Figures 9A and 10G), and brecciated structure (Figures 7H,I and 10A,G). Molybdenite occurs mainly as sheets, plates, or aggregates, with varying degrees of flexibility within fractures or intergranular spaces within the ores. They are distributed either as veinlet-disseminated (Figure 9A,B,D) or as quartz-molybdenite veinlets or stockwork fillings within the ores (Figure 9C,F,G). Most molybdenites coexist with rock-forming minerals such as feldspar and quartz along with illite and chlorite, while a few coexist with sulfide minerals such as pyrite, chalcopyrite, and sphalerite (Figure 9C,F,H,I,J,L). Occasionally, molybdenites also coexist with ilmenite, titanite, pyrrhotite, and galena (Figure 9J,K).

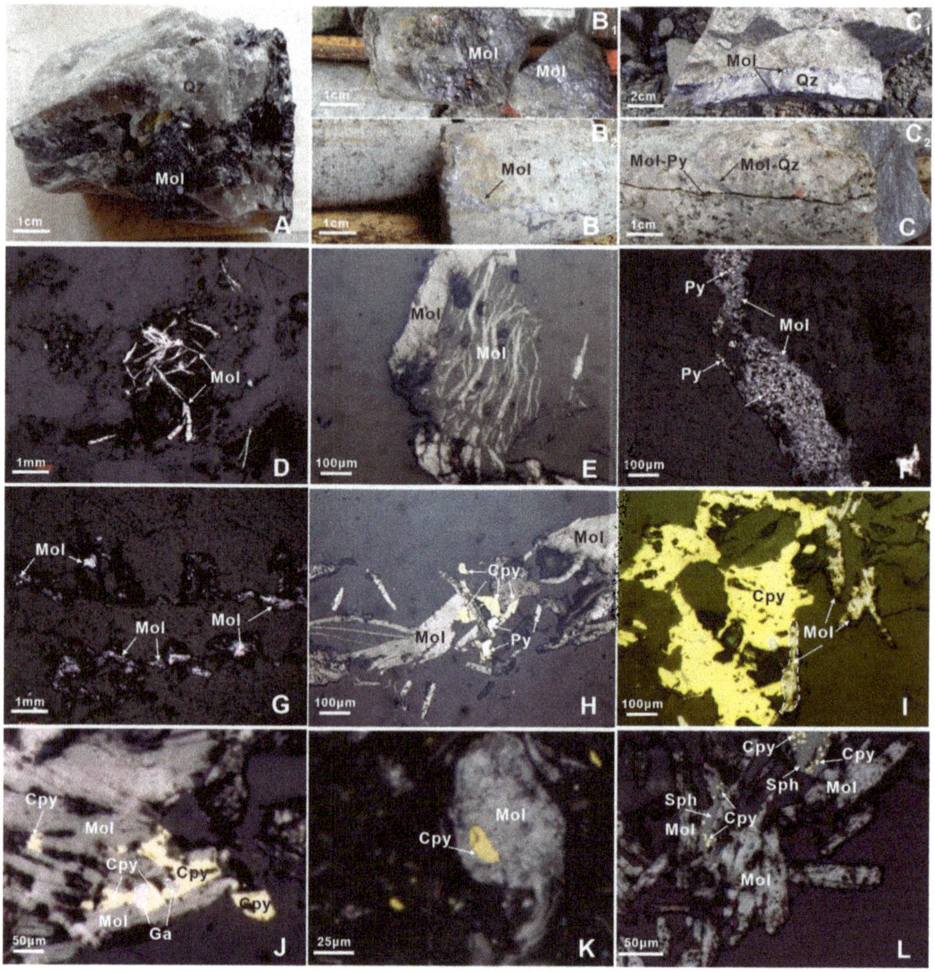

Figure 9. The occurrence of molybdenite at Luming PMD exhibits a multi-phase mineralization. (**A**) Molybdenite is found in clusters within quartz-molybdenite veins. (**B**) Molybdenite fills fractures in the monzogranite (D-type veinlet). (**C**) Molybdenite occurs as quartz-molybdenite veins (early phase, A-type veinlet) and pyrite-molybdenite veins (late phase, D-type veinlet). (**D**) Molybdenite in a radial aggregate. (**E**) Tabular crystals of molybdenite. (**F**) Molybdenite intergrowths with pyrite. (**G**) Molybdenite fills the cracks between mineral grains. (**H**) The paragenesis of molybdenite with pyrite and chalcopyrite. (**I**) Intergrowth of molybdenite and chalcopyrite. (**J**) Late-phase bornite and galena fill the spaces between early-phase molybdenite particles. (**K**) Molybdenite encases early-phase chalcopyrite. (**L**) The late-phase bornite-chalcopyrite solid solution fills the spaces between the early-phase molybdenite particles. (**D–L**) were observed under reflected light; Qz = quartz; Mol = molybdenite; Py = pyrite; Cpy = chalcopyrite; Ga = galena; Sph = sphalerite.

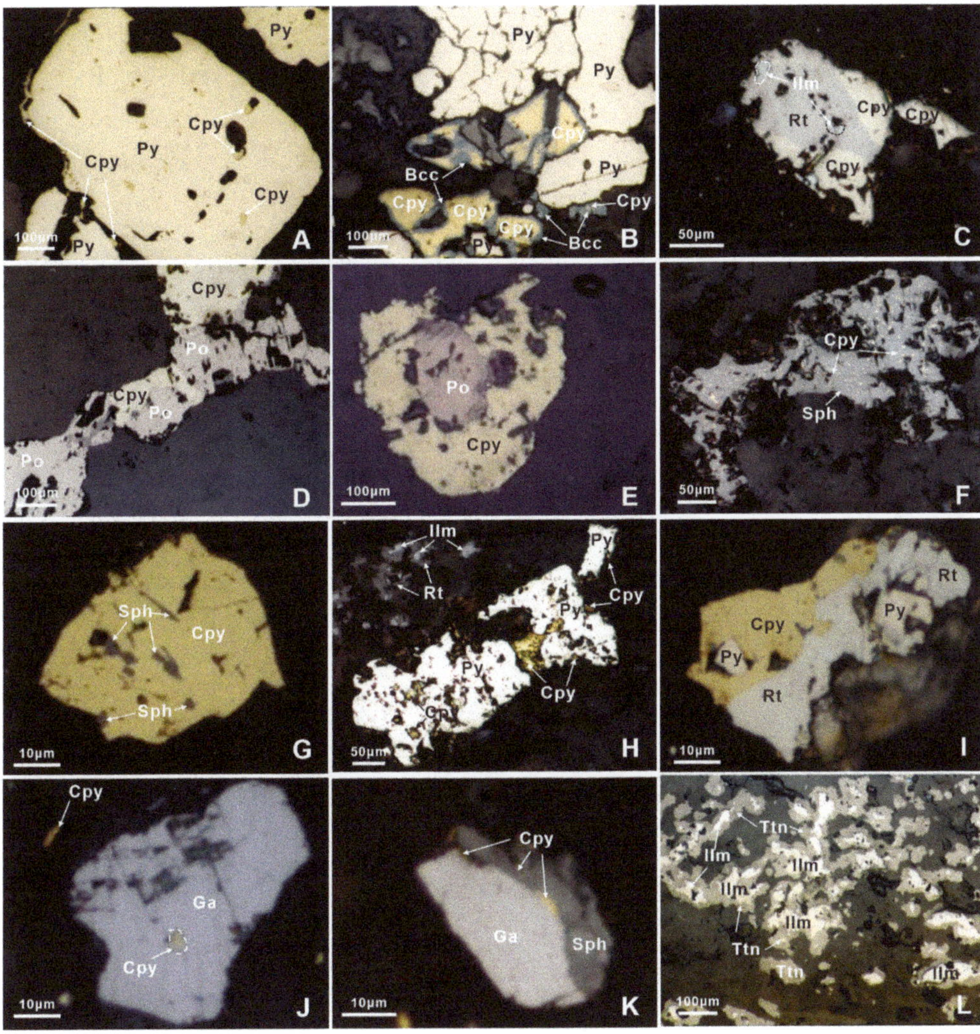

Figure 10. Characteristics of polymetallic mineralization under reflected light at Luming PMD. (**A**) Chalcopyrite (late phase) fills fractures of early pyrite. (**B**) Fractured pyrite was replaced by late-phase chalcopyrite, with oxidized edges forming blue chalcocite. (**C**) Late-phase pyrite replaces rutile encasing early ilmenite. (**D**) Chalcopyrite (late phase) replaces early pyrrhotite, forming a bay contact. (**E**) Chalcopyrite encases early pyrrhotite. (**F**) Chalcopyrite is dispersed in sphalerite. (**G**) Sphalerite is dispersed in chalcopyrite. (**H**) Chalcopyrite fills fractures of early pyrite cutting veinlets formed by rutile and ilmenite. (**I**) Chalcopyrite replaces rutile, both encasing early euhedral pyrite. (**J**) Late galena envelops early chalcopyrite. (**K**) Late galena replaces early sphalerites-chalcopyrite. (**L**) Titanite replaces early ilmenite. Ilm = ilmenite; Rt = rutile; Ttn = titanite; Po = pyrrhotite; Mol = molybdenite; Py = pyrite; Cpy = chalcopyrite; Ga = galena; Sph = sphalerite; Bcc = blue chalcocite.

4.6. Wall Rock Alteration

Hydrothermal activity at Luming results in various alterations, including silicification (Figures 9A–C and 11D–F), illitization (Figures 9A,B and 11C,J), potassic alteration (primarily K-feldsparization followed by biotitization) (Figure 11A–C,I), and chloritization, epidotization, kaolinitization and carbonatization (Figures 7A–C and 11A,G,J). Silicification, potassic alteration, illitization, and chloritization are particularly noteworthy due to their close association with Mo mineralization. They exhibit a wide spatial distribution that includes both disseminations and veinlets.

Figure 11. Alteration characteristics of the Luming PMD, showing a general evolution sequence of K-feldsparization-biotitization→illitization→chloritization→carbonatization, with silicification and pyritization throughout the hydrothermal processes. (**A**) Late calcite veins enclose the early K-feldsparization-biotitization monzogranite. (**B**) Potassic alteration characterized by pervasive K-feldsparization and veinlet biotitization. (**C**) Early silicification-K-feldsparization-biotitization overlain by late silicification-illitization-chloritization. (**D**) Quartz-molybdenite fills the early silicification-K-feldsparization monzogranite, which is cut by late silicification-molybdenite veinlets. (**E**) Early silicification-K-feldsparization ± biotitization overlain by late silicification-molybdenite veinlets shows brecciated structure. (**F**) Late silicification-pyritization intersects early silicification-K-feldsparization. (**G**) Early pyrite-calcite veins enclose early silicification-K-feldsparization. (**H**) Polysynthetic twinning plagioclase is replaced by perthite and quartz. (**I**) Veinlet biotitization. (**J**) Late quartz-calcite vein cut-through altered rocks of silicification-illitization-chloritization. Qz = quartz; Kfs = K-fedlspar; Pl = plagioclase; Bi = biotite; ill = illite; Chl = chlorite; Cc = calcite; Py = pyrite; Mol = molybdenite.

Four alteration zones developed around the orebody: silicification-K-feldsparization-biotitization zone, silicification-illitization-K-feldsparization zone, silicification-illitization-chloritization zone, and silicification-pyritization-chloritization zone [68] (Figures 5 and 6). The central distribution of the silicification-K-feldsparization-biotitization zone is primarily associated with the economic Mo mineralization occurring at its periphery (Figure 6). The silicification-illitization-K-feldsparization zone is characterized by quartz and illite replacement on feldspar, with some residual early-stage K-feldsparization (Figure 11H). The majority of economic ore bodies are found within this alteration zone. The degree of silicification and illitization is relatively limited in the silicification-illitization-chloritization zone. Chloritization is most prevalent at the margins of this alteration zone, where economic ores and low-grade ores are found (Figures 5 and 6). Outside the periphery of the silicification-illitization-chloritization zone, the silicification-pyritization-chloritization zone exhibits slightly different characteristics in different areas [68]. In the northern and northeastern parts of the exploration area, the dominant alteration is extensive high-intensity pyritization and chloritization, which is occasionally accompanied by veinlets of molybdenum mineralization associated with pyritization (Figure 11F). In contrast, in the southern and southeastern parts, carbonatation and pyritization are the dominant processes, occurring primarily in the form of calcite-pyrite veins (Figure 11G,J). The orebody in these areas is composed primarily of low-grade ores, exhibiting a gradual decrease in alteration intensity.

4.7. Mineralization Phases

The development of alteration and mineralization at the Luming PMD is closely related to intrusions and associated hydrothermal activity, which can be divided into several phases (Figures 7 and 9–11). Based on field observations and microscopic examinations, four distinct phases of mineralization were identified: a quartz-pyrite phase; a quartz-molybdenite-chalcopyrite phase; a quartz-polymetallic (molybdenite-chalcopyrite-galena-sphalerite) phase; and a quartz-calcite-pyrite phase (Figure 12).

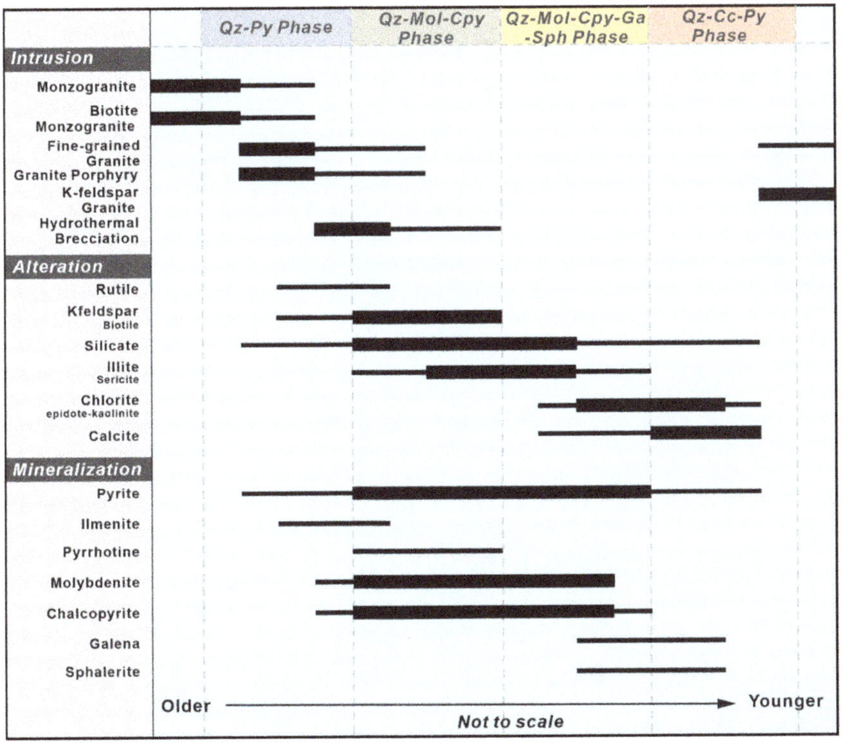

Figure 12. Tentative timeframe for intrusion, brecciation, alteration, and mineralization of the Luming PMD, indicating multi-phase hydrothermal processes. Qz = quartz; Cc = calcite; Py = pyrite; Mol = molybdenite; Cpy = chalcopyrite; Ga = galena; Sph = sphalerite.

5. Material and Methods

In order to enhance comprehension of the factors that influence ore formation and the criteria employed in prospecting, a series of investigations were conducted between 2009 and 2011. These investigations included major and trace elements analysis, rock primary halo geochemistry, and ground geophysical surveys.

Samples for rock geochemistry analysis were obtained from all the drill holes. The sampling technique employed was essentially a whole-hole sampling method, with the exception of a few control holes where no significant alteration or mineralization was observed. The split-center method was employed, with one-half of the sample sent for analysis and the other half retained. Each sample was 2.00 m in length, although it could extend to 3.00 m. Sample preparation and analysis were conducted at the Central Laboratory of the Institute of Geophysical and Geochemical Exploration, Chinese Academy of Geological Sciences. The sample was crushed by a jaw crusher, rollers, and a grinder to a particle size of 100 mesh. Subsequently, the sample was mixed and reduced, with a portion of the material retained at 800–1600 g to 200 mesh by means of rod milling. The sample was then placed in kraft paper bags, while the remainder was stored in bags as analyzed sub-samples for preservation.

The primary halo elements, including Mo, Ag, Bi, Co, Cu, In, Ni, Pb, Sb, W, and Zn, were determined using inductively coupled plasma mass spectrometry (ICP-MS), with high content Mo reanalyzed using inductively coupled plasma optical emission spectrometry (ICP-OES). The test material was initially dried and weighed. It was then decomposed with hydrofluoric acid, nitric acid, and perchloric acid, dissolved with aqua regia, transferred to a polyethylene test tube, fixed, and shaken vigorously. A portion of the solution was analyzed on an ICP-MS instrument (Thermo Fisher Scientific Xseries II, Waltham, MA, USA). The measurement conditions were as follows: nebulizer flow rate (Ar) 0.8–0.9 L/min, residence time 20 ms, scan number 50, number of each mass channel 1, injection pump speed 40 rpm, sampling depth 150–250, and data acquisition mode peak hopping. For samples with more than 250 μg/g Mo, a single determination was conducted using the ICP-OES instrument (Thermo Fisher Scientific IRIS Intrepid II, Waltham, MA, USA). To ensure the reliability and validity of the results, blanks, national standards, and duplicates were included in the test. The sample must undergo retesting with a minimum 95% pass rate. A comprehensive internal quality assurance and quality control (QA/QC) inspection, including duplicates at a ratio of 5%, was conducted on the selected batch. Subsequently, an external QA/QC inspection was conducted, with samples from the analyzed sub-samples sent to the Tianjin Institute of Geology and Mineral Resources for testing at a ratio of 10%. It is imperative that both the internal and external laboratory inspections achieve a success rate of 90%.

A subset of samples from the primary halo investigation was subjected to major and trace element analysis, with the exclusion of weathered or altered rock samples. The major element compositions were analyzed by X-ray fluorescence (XRF) spectrometry, while the trace element compositions were determined using ICP-MS. The major element analysis was conducted in the following steps: (1) The 200-mesh sample was dried in an oven at 105 °C for 2–4 h. (2) Then, a test material was melted with lithium tetraborate and ammonium nitrate, which served as the oxidizing agent, followed by melting with flux at 1150–1250 °C to create glass slides. (3) Finally, the analysis was conducted using a PANalytical Axios PW4400 (Shanghai Spectris Instrument Systems Co., Ltd., Shanghai, China) with a 50 kV X-ray tube, 50 mA current, coarse collimator, and 30 mm field of view. The reference materials were prepared using national standard substances. The relative error for the major element test was within 5%.

The geophysical surveys were conducted by the China Railway Resources Group Survey & Design Co., Ltd. Initially, selected drill core samples were subjected to physical properties measurements. Subsequently, representative profiles were selected for the gravity and magnetic surveys (four profiles: A–A′, B–B′, C–C′, and D–D′) and induced polarization survey (four profiles: A–A′, E–E′, F–F′, and G–G′). The approximate locations of these profiles are depicted in Figure 4.

The gravity survey was conducted using a CG-5 gravimeter with a point spacing of 20 m. Two readings were taken at each point, ensuring that the difference between the two readings was less than 0.005 mGal. After applying zero drift correction and base point network adjustment, the Bouguer gravity anomaly was obtained by normal field correction and Bouguer correction. The potential field was then separated (extending 500 m upwards as the regional field) to obtain the final gravity anomaly values expressed in gravity units (gu; 1 gu = 0.1 mGal).

A magnetic survey was carried out using an EVEIN magnetometer with a point spacing of 20 m. The raw data were subjected to correction and polarization using GeoIPAS V2.0, thereby deriving magnetic anomaly data expressed in nanotesla (nT).

The Induced Polarization Intermediate Gradient survey employed the Phoenix V8 multi-function electric instrument with parameters AB = 3000 m, MN = 40 m, and point spacing of 20 m. The acquired data were then processed and filtered using GeoIPAS V2.0 to derive resistivity values in Ω·m and chargeability expressed as a percentage (%).

6. Results

6.1. Major and Trace Elements

The major and trace element compositions of the monzogranites and granite porphyry in the Luming are presented in Tables S1 and S2. The rocks exhibit relatively high SiO_2 contents, ranging from 65.78% to 77.17%. The $Na_2O + K_2O$ content varies from 5.4% to 9.46%, while the Al_2O_3 content ranges from 11.09% to 15.73%. The samples are enriched in MgO (0.21%–2.325%) but low in TiO_2 (0.14%–0.69%) and P_2O_5 (0.01%–0.22%). On the $Na_2O + K_2O$ versus SiO_2 diagram, the samples fall within the subalkaline granite and quartz monzonite field (Figure 13A). On the K_2O versus SiO_2 diagram, the samples plot predominantly within the high-K calc-alkaline to shoshonitic field (Figure 13B). On the A/NK versus A/CNK diagram, a trend towards peraluminous to metaluminous compositions is observed for the samples (Figure 13C). Most of the samples on the Fe_2O_3/FeO versus SiO_2 plot lie within the porphyry Mo granite field, which aligns with the Endako-type Mo deposits (Figure 13D).

The total rare earth element (ΣREE) concentrations in monzogranites and granite porphyry range from 80.71 to 233.88 ppm and from 68.9 to 162.18 ppm, respectively. On the chondrite-normalized diagram for rare earth elements, all samples exhibit enrichment of light rare earth elements (LREEs) and depletion of heavy rare earth elements (HREEs), indicating moderate differentiation characteristics (Figure 14B). The LREE/HREE ratios range from 6.4 to 19.07, while the $(La/Yb)_N$ ratios range from 6.3 to 28.04. Monzogranites (δEu = 0.37–0.66) exhibit a more pronounced Eu anomaly than granite porphyry (δEu = 0.75–0.93, with the exception of LM-11-B3) (Figure 14B). On the primitive mantle-normalized diagram, the samples exhibit an overall enrichment of large ion lithophile elements, including Rb, Th, U, and Pb, while simultaneously depleting Ba and high field strength elements, including HREE, Nb, and Ta (Figure 14A).

6.2. Primary Halo

The statistical characteristics of 26,900 samples of primary halo data from drilling (27,942 samples specifically for Mo) are presented in Table 1. Coefficients of variation for Mo, Cu, Co, and Ni have comparatively low values ($\leq$200%), while those for other elements range from 247% to 795%. Skewness values indicate varying degrees of right skew for the multi-element data. All kurtosis values are greater than 3, indicating a heavy-tailed characteristic.

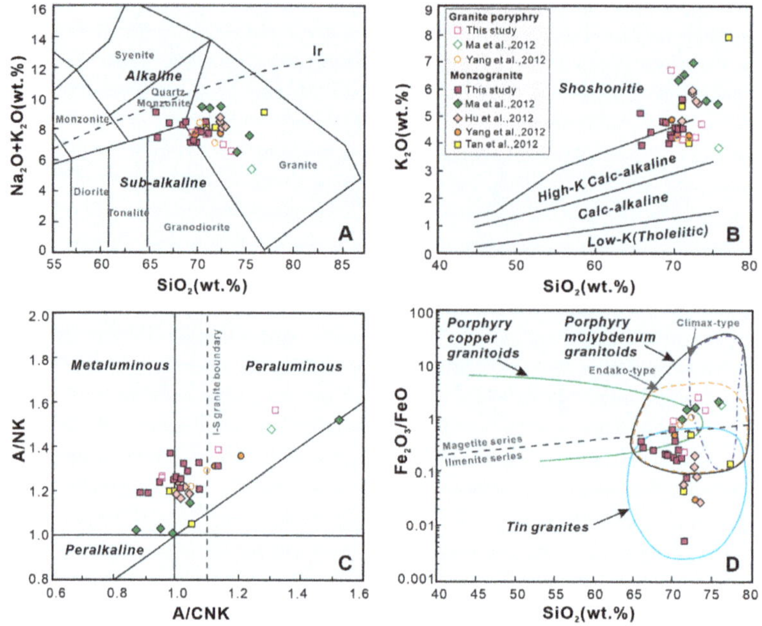

Figure 13. Classification of the Luming monzogranite and granite porphyry samples. (**A**) (Na_2O + K_2O) vs. SiO_2 [69]; (**B**) K_2O vs. SiO_2 [70]; (**C**) A/NK vs. A/CNK [71]; (**D**) Fe_2O_3/FeO vs. SiO_2 diagrams [2]. Major element data are derived from [22,46,47,49].

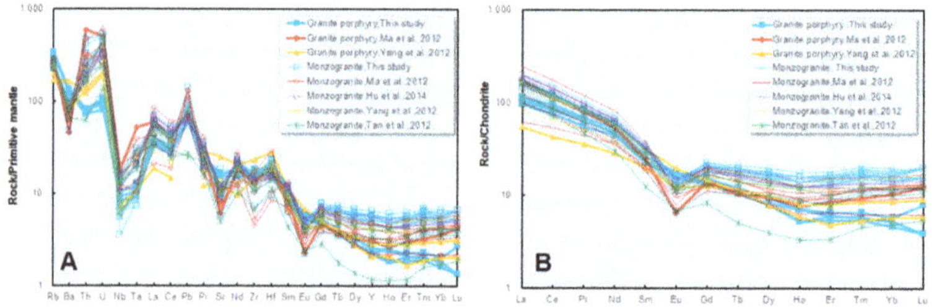

Figure 14. Primitive mantle-normalized trace element (**A**) and chondrite-normalized REE (**B**) spider diagrams for the Luming monzogranite and granite porphyry samples. Data for normalization are from [72]. Trace element data are derived from [22,46,47,49].

Table 1. Statistical characteristics of primary halo at Luming PMD.

	Valid N	Mean ppm	Median ppm	Minimum ppm	Maximum ppm	Std.Dev. ppm	Coef.Var. %	Skewness	Kurtosis
Mo	27,942	516.40	281.41	0.52	24,790.0	758.7	147	7	100
Ag	26,900	307.87	190.99	0.31	82,561.2	1149.8	373	42	2377
Bi	26,900	0.30	0.14	0.01	162.3	1.9	638	48	3187
Co	26,900	7.79	7.86	0.07	272.7	4.3	55	10	548
Cu	26,900	171.46	159.84	1.00	5721.7	121.3	71	9	308
In	26,900	0.04	0.02	0.00	21.9	0.2	612	72	6478
Ni	26,900	3.71	3.78	0.07	978.0	7.7	206	107	12,552
Pb	26,900	24.40	12.71	0.43	18,811.0	194.0	795	62	5010
Sb	26,900	1.02	0.28	0.01	231.8	2.5	248	32	2664
W	26,900	10.99	8.10	0.11	4757.5	31.2	284	132	19,976
Zn	26,900	52.27	29.71	1.99	31,259.7	306.2	586	62	5392

From the perspective of primary halo features along section 00E, the elements W, Bi, and Sb are predominantly enriched in the central area where hydrothermal breccias are distributed (Figure 15A,B,J,L); Mo, Cu, Co, and Ni show a symmetrical distribution on both sides outside the hydrothermal breccias (Figure 15C–F); Pb, Zn, Ag, and In are sparsely distributed outside the Mo mineralization (Figure 15G–I,K).

Overall, the elemental distribution shows a zonation from the center to the outer zone of the deposit as follows: W-Bi-(Sb)→Mo→Cu-Co-Ni→Pb-Zn-Ag-In. The enrichment center of Mo mineralization is located near the contact zone between the silicification-illitization-K-feldsparization zone and the silicification-K-feldsparization-biotitization zone or silicification-illitization-chloritization zone (Figure 15C,L).

6.3. Sample Physical Properties

Physical property measurements on drill core samples are presented in Table 2 and show significant variations in the physical parameters across the different alteration zones. The resistivity (mean values ranging from 22,636 Ω·m to 7211 Ω·m to 4050 Ω·m to 5340 Ω·m), chargeability (ranging from 12.8% to 29.6% to 50.5% to 56.3%), and magnetic susceptibility (ranging from $234.7 \times 10^{-6} \times 4\pi$SI to $7.4 \times 10^{-6} \times 4\pi$SI to $12.5 \times 10^{-6} \times 4\pi$SI to $35.2 \times 10^{-6} \times 4\pi$SI) show a systematic variation from unaltered monzogranite→silicification-K-feldsparization-biotitization zone→silicification-illitization-K-feldsparization zone→silicification-illitization-chloritization zone. Compared to the unaltered monzogranite, the mineralized rock (Mo $\geq$ 300 ppm) exhibits characteristics such as low magnetic susceptibility, low resistivity, relatively high chargeability, and low density (Table 2). These differences in rock physical properties provide a basis for delineating the mineralized rock mass, alteration zones, and unaltered surrounding rocks using gravity and magnetic anomalies and the induced polarization method [43].

6.4. Bouguer Gravity and Magnetic Anomalies

The results of the Bouguer gravity and magnetic anomalies are shown in Figure 16. The observed NE distribution in the low value area of the Bouguer gravity anomalies corresponds well with the extent of mineralization and alteration zones (Figure 16A,C). On the plane, areas below 26 gu show a consistent distribution of hydrothermal breccias and low-grade ores (Figure 16A). Similarly, the B–B′ profile of the Bouguer gravity anomalies shows that the main mineralization (survey points 130–240, Bouguer gravity anomalies = 18.9–31.8 gu, average 24.9 gu) exhibits characteristics of lower gravity anomalies compared to the altered surrounding rocks (survey points 100–130 and 240–300, Bouguer gravity anomalies = 16.9–41.4 gu, average 29.8 gu) (Figure 16C). In unaltered areas, biotite monzogranite and fine-grained granite north of the silicification-pyritization-chloritization zone represent regions of high Bouguer gravity anomalies (>36 gu).

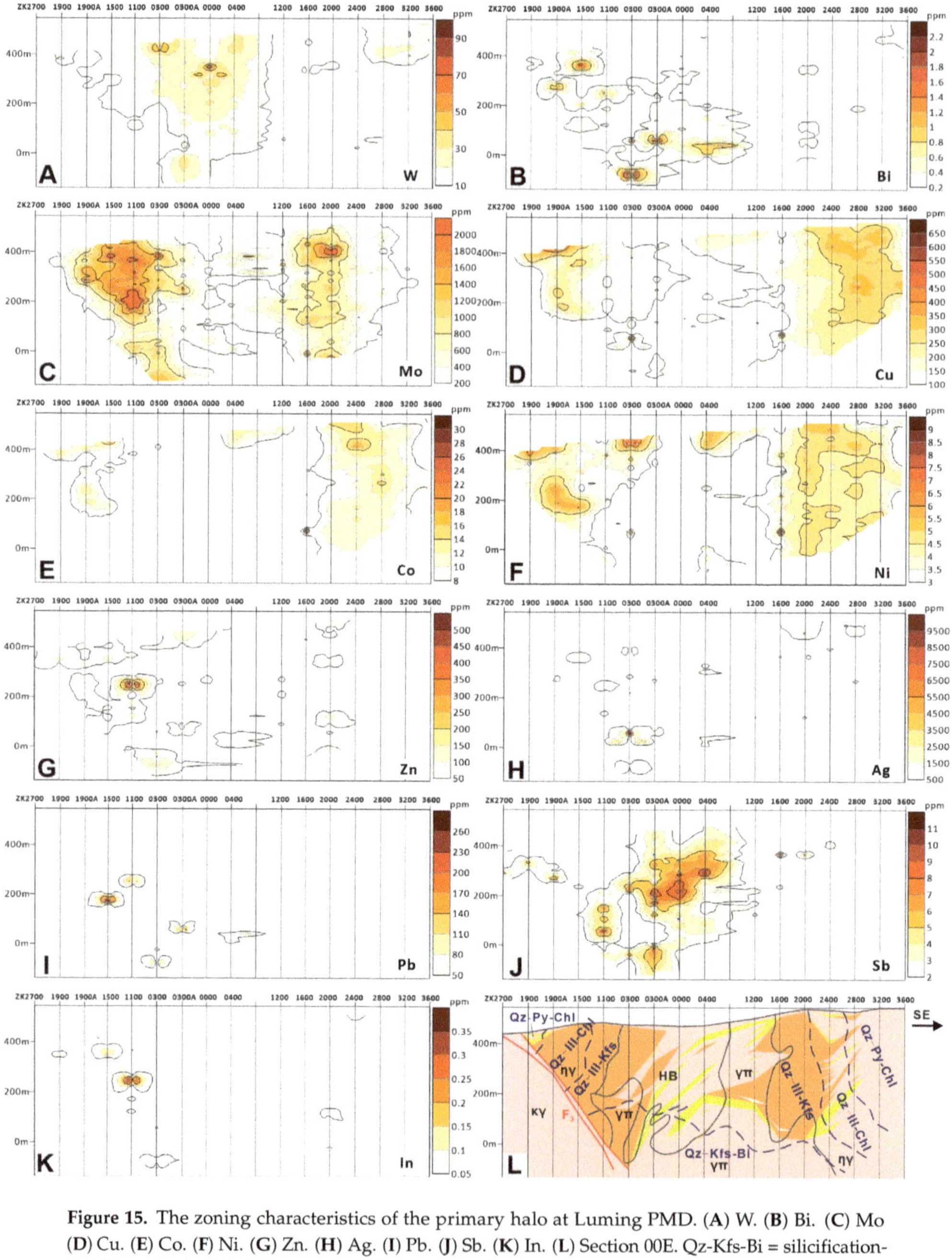

Figure 15. The zoning characteristics of the primary halo at Luming PMD. (**A**) W. (**B**) Bi. (**C**) Mo (**D**) Cu. (**E**) Co. (**F**) Ni. (**G**) Zn. (**H**) Ag. (**I**) Pb. (**J**) Sb. (**K**) In. (**L**) Section 00E. Qz-Kfs-Bi = silicification-K-feldsparization-biotitization; Qz-Ill-Kfs = silicification-illitization-K-feldsparization; Qz-Ill-Chl = silicification-illitization-chloritization; Qz-Py-Chl = silicification-pyritization-chloritization; ηγ = monzogranite; γπ = granite porphyry; HB=hydrothermal breccias; F3 = F3 fault.

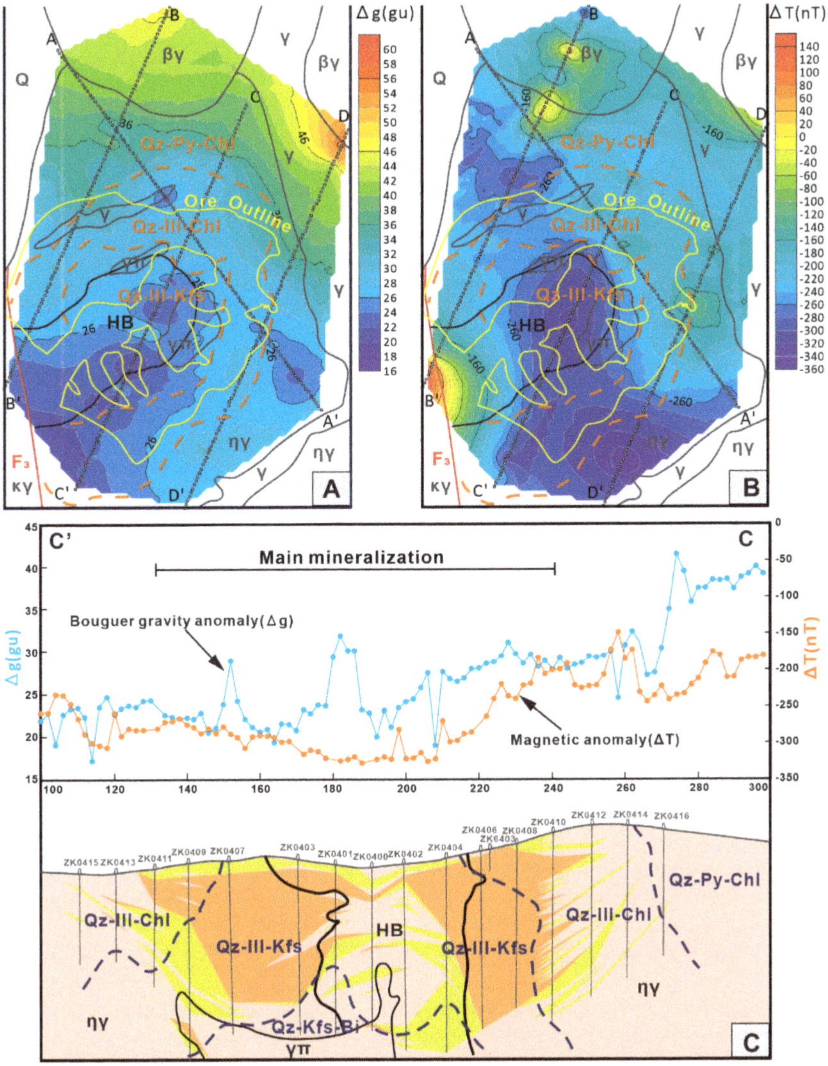

Figure 16. Gravity and magnetic anomalies of the Luming PMD. (**A**) Contour map showing Bouguer gravity anomalies. (**B**) Contour map showing magnetic anomalies. (**C**) Cross section showing gravity and magnetic anomalies along C–C′. Qz-Kfs-Bi = silicification-K-feldsparization-biotitization; Qz-Ill-Kfs = silicification-illitization-K-feldsparization; Qz-Ill-Chl = silicification-illitization-chloritization; Qz-Py-Chl = silicification-pyritization-chloritization; Q = Quaternary; $\eta\gamma$ = monzogranite; $\beta\gamma$ = biotite granite; γ = fine-grained granite; $\gamma\pi$ = granite porphyry; $\kappa\gamma$ = K-feldspar granite; HB = hydrothermal breccias; F3 = F3 fault.

Table 2. Statistics of physical properties of samples at Luming PMD.

Zoning	Lithology	Number of Samples	Mo ppm	Resistivity Ω·m	Chargeability %	Magnetic Susceptibility $10^{-6} \times 4\pi SI$	Density g/cm^3
3	ηr	19	747	5340 (596–19,313)	56.3 (18–169)	35.2 (10–109)	2.54 (2.43–2.63)
2	ηr + GP + HB	58	690	4050 (242–27,683)	50.5 (6.4–204.7)	12.5 (0–74.0)	2.55 (2.40–2.72)
1	HB	10	1015	7211 (949–20,520)	29.6 (9.9–96.2)	7.4 (0–64.0)	2.61 (2.48–2.78)
0 *	ηr	13	103	22,636 (3451–47,698)	12.8 (9.7–18.0)	234.7 (87.0–408.0)	2.58 (2.38–2.72)

* The data set was collected from ZK712-1 in the 712 Highland southwest of the Luming PMD, while other samples were collected from ZK0302 and ZK2001 at the Luming site. Data were summarized as mean (min.–max.) shapes. 0 = unaltered rock; 1 = silicification-K-feldspathization-biotitization zone; 2 = silicification-illitization-K-feldspathization zone; 3 = silicification-illitization-chloritization zone; ηγ = monzogranite; GP = granite porphyry; HB = hydrothermal breccias of monzogranite.

The magnetic survey shows predominantly negative anomalies throughout the deposit, with positive anomalies observed only within the K-feldspar granite distribution area in the southwest corner (Figure 16B). In contrast to the Bouguer gravity anomaly, low magnetic anomalies (≤ -260 nT) generally trend northwest. On a planar scale, the main mineralization is mainly located in regions with magnetic anomalies lower than (-160 nT (Figure 16B). The magnetic anomaly profile (Figure 16C) also shows that the main zone of mineralization coincides with an area of low magnetic anomalies (-394.7~-137.1 nT, averaging -281.3 nT), while altered wall rocks and weakly mineralized areas have relatively higher magnetic anomalies (-306.9~-148.9 nT, averaging -228.3 nT). The hydrothermal breccias have the lowest level of magnetic anomalies along the profile (Figure 16C).

6.5. Resistivity and Chargeability

The results of the Induced Polarization Intermediate Gradient survey are shown in Figure 17. At the surface, the entire deposit has low resistivity (Figure 17A). The main mineralization shows medium to low resistivity, generally below 1000 Ω·m. High resistivity (>1000 Ω·m) is observed in the western area of the deposit where K-feldspar granite is distributed (chargeability less than 3.5%) and in the southeastern area with weaker mineralization but developed pyritization (chargeability greater than 5.5%). The orebody is characterized by moderate to high chargeability, ranging from 3.5% to 8.0% (Figure 17B). High chargeability areas are located within the silicification-pyritization-chloritization zone (chargeability greater than 5.5%). Areas of low chargeability (less than 3.5%) coincide with the central part of the orebody, corresponding to the distribution of hydrothermal breccias (Figure 17B).

The induced polarization profile (Figure 17C) shows that the main mineralization is characterized by moderate to low resistivity (survey points 160–260, resistivity ranging from 264 to 1020 Ω·m, with an average of 655 Ω·m) and moderate to high chargeability (chargeability ranging from 1.4% to 6.1%, with an average of 4.2%). The alteration zone in the northwest region of the deposit is characterized by low resistivity (survey points 90–160, resistivity ranging from 361 to 987 Ω·m, with an average of 565 Ω·m) and high chargeability (chargeability ranging from 4.6% to 9.4%, with an average of approximately 7.0%). The alteration zone in the southeast region is characterized by high resistivity (survey points 260–300, resistivity ranging from 394 to 1600 Ω·m, with an average of 959 Ω·m) and high chargeability (chargeability ranging from 4.9% to 10.7%, with an average of 6.9%).

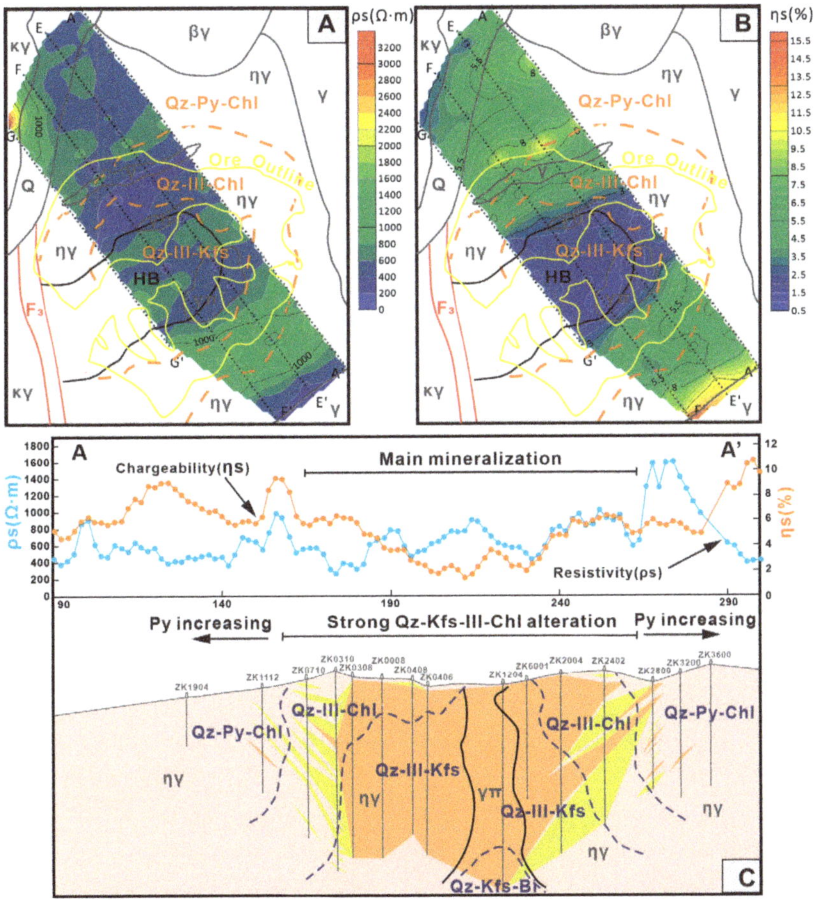

Figure 17. Induced polarization anomalies of the Luming PMD. (**A**) Contour map shows resistivity anomalies. (**B**) Contour map shows chargeability anomalies. (**C**) Profile shows induced polarization anomalies along A–A′. Qz-Kfs-Bi =silicification-K-feldspathization-biotitization; Qz-Ill-Kfs = silicification-illitization-K-feldspathization; Qz-Ill-Chl = silicification-illitization-chloritization; Qz-Py-Chl = silicification-pyritization-chloritization; Q = Quaternary; ηγ = monzogranite; βγ = biotite granite; γ = fine-grained granite; γπ = granite porphyry; κγ = K-feldspar granite; HB = hydrothermal breccias; F3 = F3 fault.

7. Discussion

7.1. *Porphyry-Related Magmatic-Hydrothermal System*

Numerous studies have shown that porphyry Cu-Mo-Au deposits, together with their associated distal skarn and epithermal Pb-Zn-Au-Ag deposits, may be parts of a single porphyry-related magmatic-hydrothermal system [73–78]. Skarn and epithermal mineralization commonly occur within 2–6 km of the periphery of porphyry deposits [77], which has significant implications for exploration criteria and targeting of prospective areas. The Luming PMD and Cuiling porphyry Mo-Au deposits are surrounded by skarn and epithermal Pb-Zn deposits such as Xulaojiugou, Kunlunqi, Qianjin Dongshan, and Xiling Nanshan (Figure 2), indicating the presence of a polymetallic metallogenic system associated with porphyry Mo mineralization. Similar patterns can also be observed at other large porphyry Mo deposits in the LXRZRMB. For example, the Chalukou Mo deposit exhibits vein-type Pb-Zn mineralization in its upper portion adjacent to a deeper porphyry

Mo orebody [13,32,42]; the Cuihongshan and Huojihe porphyry-skarn Mo deposits are surrounded by skarn and epithermal Fe-Pb-Zn deposits [34–36] (Figure 2).

Jin et al. [77] argued that a comprehensive analysis is needed to determine whether porphyry deposits and distal skarn-epithermal deposits belong to the same metallogenic system, considering factors such as their spatiotemporal distribution, mineral paragenesis, and the evolution of fluid and material sources. The comparable compositions and closely related ages of the mineralized porphyries of the Luming PMD (180.7 $\pm$ 1.6 Ma, 181.2 $\pm$ 1.1 Ma) and the Xulaojiu Pb-Zn deposit (179.9 $\pm$ 1.0 Ma) indicate an origin from a common magmatic-hydrothermal system [49]. Furthermore, 1:50,000 scale stream sediment anomalies indicate the presence of central Mo-W-Cu anomalies with peripheral Pb-Zn-Ag distributions [41] (Figure 3), which may exhibit a spatial correlation with this magmatic-hydrothermal system. In the Cuihongshan-Huojihe area, geochronological studies indicate two episodes of magmatic intrusion activity that resulted in the formation of two distinct Mo-polymetallic ore subsystems: the early Cuihongshan porphyry and skarn Mo-Fe-W deposit (207–190 Ma) and the late Huojihe porphyry Mo and skarn Pb-Zn deposit (197–160 Ma) [33–36,62].

7.2. Control of Hydrothermal Breccias on Ore Distribution

The presence of breccias in porphyry deposits is well documented historically [73,79–82]. Lowell [79] reported that of twenty-seven porphyry deposits examined, twenty were found to contain breccias. Within the LXRZRMB, in addition to the Luming deposit, other deposits such as Chalukou [32,42,80], Huojihe [16], and Daheishan [28,30,58] have been identified as containing breccias. Based on their characteristics, Sillitoe [73] classified breccias associated with porphyry deposits into four types: magmatic hydrothermal breccias, phreatic porphyry Cu-level breccias, phreatic epithermal-level breccias, and phreatomagmatic breccias. Although the ore-controlling characteristics vary between the different deposits, most of the breccias host mineralization [79].

The breccias in the Luming PMD consist of K-feldsparization monzogranite and granite porphyry, with silica and sulfide cement, indicating the genesis of magmatic hydrothermal breccias during mineralization (Figure 7H,F). From the 3D geological model (Figure 18), it is clear that the hydrothermal breccias at Luming PMD have significant controls on the ore distribution. They consist of a low-grade core in the center (including the intercalated rock) and high-grade economic mineralization at the periphery. Regionally, a similar configuration of breccias and mineralization has also been observed in the Huojihe Mo deposit [16].

The control mechanism of hydrothermal breccias on mineralization is the fragmentation of porphyries by magma-hydrothermal fluids and volatiles, resulting in the formation of highly permeable zones that facilitate fluid transport and ore accumulation [28,73,79,83]. Therefore, the presence of a central low-grade core consisting of silicification-K-feldsparization hydrothermal breccias at the Luming PMD may indicate a conduit for fluid flow during the early stage of moderate–high temperature magmatic hydrothermal activity [80]. Molybdenum tends to accumulate in the fractured surrounding rock, resulting in a circular distribution pattern on planar surfaces resembling the shape of a donut (Figure 18).

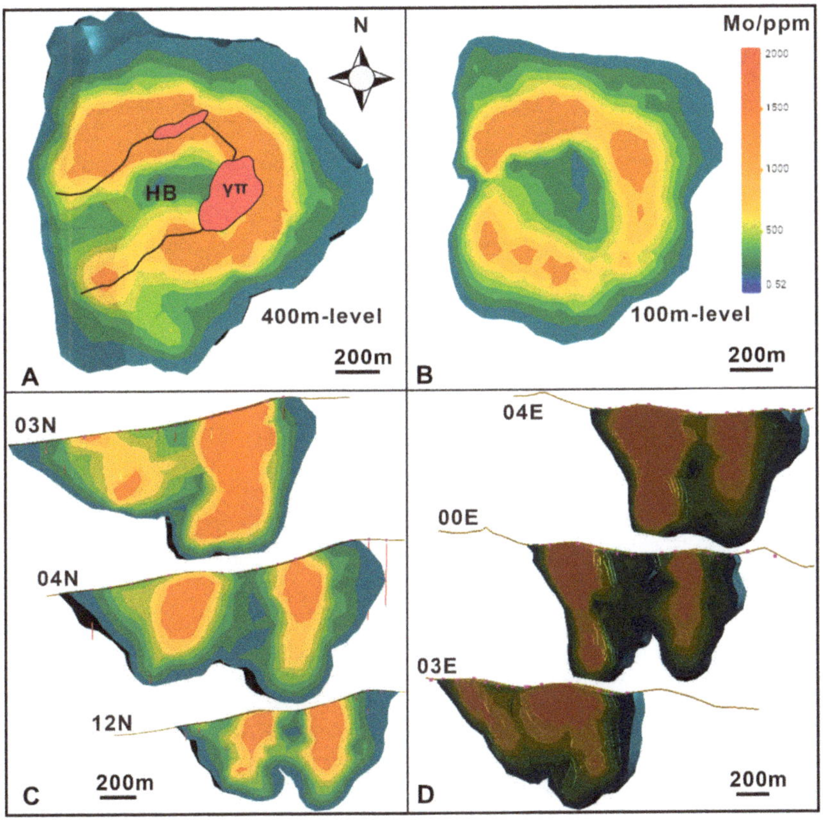

Figure 18. 3D geochemical model of the Luming PMD showing the influence of hydrothermal breccias on the distribution of mineralization. (**A**) Mineralization at 400 m elevation. (**B**) Mineralization at 100 m elevation. (**C**) Mineralization at 03N, 04N, and 12N cross sections. (**D**) Mineralization at 04E, 00E, and 03E cross sections. $\gamma\pi$ = granite porphyry; HB = hydrothermal breccias.

7.3. Primary Halo and Metal Zonation

Understanding primary halo or metal zonation in ore deposits is crucial for the analysis of mineralization mechanisms and the evaluation of the potential of deep-seated ore bodies [81,84–89]. The zoning characteristics of W-Bi-(Sb) →Mo →Cu-Co-Ni →Pb-Zn-Ag-In in the Luming PMD generally conform to the ideal zonation from Cu-Mo to Pb-Zn, as previously observed in other studies [73,76,90]. This can serve as an indicator for targeting the center of porphyry mineralization [81]. However, there is a discrepancy in the distribution of Sb, which is expected to be present along with Pb-Zn-Ag-In outside of Mo mineralization as a front halo. However, its anomaly is clearly located in the central part of the orebody (Figure 15). One potential explanation for this inversion is the possibility of later-stage hydrothermal activity in deeper parts of known ore bodies [87]. Based on the spatial distribution characteristics of existing fine-grained granite, hydrothermal breccias, hydrothermal alterations and ore shells, as well as geophysical anomalies, it can be inferred that ore-forming fluids may have migrated from southwest depths to northeast shallower levels during the mineralization process. This inference is supported by the findings presented in Figures 5, 8, 15, 17 and 18. Further investigation is necessary to ascertain whether the distribution of Sb indicates the presence of an additional superimposed mineralization in the deep southwest side or a genetic connection with the Mo mineralization observed at 712Higland (Figures 2 and 3) on the western side of the Luming PMD.

7.4. Indication of Geophysical Survey

The geophysical properties of rocks are affected by numerous factors, including lithology, type(s)of alteration and mineralization, fracturing, hydraulic conditions, weathering, burial depth, and thickness of surface coverings [91,92]. Therefore, it is necessary to conduct a comprehensive analysis and interpretation from diverse disciplines [93]. The lowest Bouguer gravity values are observed in hydrothermal breccias, indicating the presence of significant fracturing at the center of the Luming PMD (Figure 16A, Table 2). The main mineralization zones at the deposit are closely associated with silicification-illitization, displaying medium-low resistivity characteristics [94]. In contrast, the high pyrite content typically found in the periphery of Cu-Mo mineralization corresponds to the high chargeability anomaly generally observed in the outermost zone of alteration, which is characterized by silicification-pyritization-chloritization (Figure 17B). In general, intense fracturing and fluid flow processes resulted in widespread silicification, illitization, and chloritization overlying K-feldsparization and biotitization. These processes resulted in the destruction of ferromagnetic minerals, resulting in low magnetic susceptibility and negative anomalies at Luming PMD [78,91,93–96] (Figure 16B). Consequently, negative magnetic anomalies serve as an effective indirect technique for the exploration of porphyry deposits [94,97]. This is corroborated by the characteristics of the Duobaoshan Cu deposit [45] and the Chalukou Mo deposit [42], which exhibit characteristics of negative magnetic anomalies within their mineralized or altered zones.

7.5. Exploration Methods and Criteria

In light of the preceding discussion, this paper presents an overview of the exploration criteria and prospecting targets at different scales for Luming-type porphyry Mo deposits.

At the regional scale, prospecting criteria and targets can be defined as follows: (1) tectonic intersection area of the Central Asian Metallogenic Belt and the Circum-Pacific Metallogenic Belt; (2) regions occupied by the NE-trending peraluminous to metaluminous, high-K calc-alkaline to shoshonitic intrusive rocks associated with early Jurassic subduction of the Pacific Plate; (3) anomalous gradient belts in aeromagnetic and aerogravity surveys, which are characterized by moderate–high magnetic anomalies and low Bouguer gravity anomalies; (4) regions with anomalous stream sediment geochemistry exhibiting W-Bi-Cu-Mo anomalies or Pb-Zn-Ag anomalies towards the center; (5) the central area of potential porphyry Mo and polymetallic mineralization systems.

The criteria for exploration and location prediction for ore deposits are as follows: (1) the presence of molybdite in the surface oxidation zone; (2) areas exhibiting typical alteration observed in porphyry deposits, characterized by a central silicification-K-feldsparization-biotitization transitioning to a peripheral silicification-illitization-chloritization and silicification-pyritization-chloritization; (3) the occurrence of hydrothermal breccias; (4) deep or central positions of the Pb-Zn mineralization; (5) primary halo or soil geochemical anomalies indicating W-Bi-Mo, surrounded by Cu-Co-Ni and Pb-Zn-Ag anomalies; (6) areas displaying low Bouguer gravity anomalies and negative magnetic anomalies as revealed through ground geophysical surveys; (7) induced polarization demonstrating moderate–low resistivity and moderate–high chargeability at the center of the target, while featuring low resistance and high chargeability at its periphery.

Further progress could be achieved by application of advanced techniques such as deep-penetration geochemical techniques (such as Geo-gas, [98]), primary halo axial zoning (like CoBA, [88]), three-dimensional geochemical modeling [86], and alteration mapping based on SWIR, pXRF, and indicator minerals [93].

8. Conclusions

(1) The identification of geochemical anomalies in stream sediments represents a pivotal element in the discovery of the Luming PMD. The combination of stream sediment geochemistry with large-scale soil geochemistry and geophysical surveys represents

an effective exploration method. This integrated approach allows the rapid delineation of prospective targets for porphyry Mo deposits.

(2) Hydrothermal breccias play a significant role in controlling mineralization and may act as conduits for ore-forming fluids.

(3) The primary halo exhibits a distinct metal zonation, transitioning from a central W-Bi-Mo-(Sb) to a peripheral Cu-Co-Ni and distal Pb-Zn-Ag-In. The presence of Sb in the tail halo positions may suggest the possibility of deep-seated mineralization.

(4) Gradient belts of aerogravity and aeromagnetic anomalies indicate the locations of the Luming PMD, while the deposit scale geophysical anomalies define the extent of mineralization and alteration. The mineralized rocks exhibit a low Bouguer gravity anomaly, a negative magnetic anomaly, moderate–low resistivity, and moderate–high chargeability.

(5) It is proposed that the porphyry Mo and distal skarn-epithermal Pb-Zn mineralization at Luming may be parts of a single porphyry-related magmatic-hydrothermal system.

Supplementary Materials: The following supporting information can be downloaded at: https://www.mdpi.com/article/10.3390/min14070718/s1, Table S1: Major elements compositions of the monzogranites and granite porphyry of the Luming PMD; Table S2: Trace elements and REE compositions of the monzogranites and granite porphyry of the Luming PMD.

Author Contributions: Conceptualization, B.G.; methodology, B.G. and M.D.; software, Z.L. and H.X.; investigation, B.G., Z.L. and T.Z.; data curation, Y.L. and T.Z.; writing—original draft preparation, B.G., M.D. and H.X.; writing—review and editing, B.G. All authors have read and agreed to the published version of the manuscript.

Funding: This research was jointly funded by China Railway Engineering Corporation's Science & Technology Projects (2011-Key-82) and China Railway Resources Group Co., Ltd.'s Science & Technology Projects (2011-Key-07, 2021-Key-02).

Data Availability Statement: Data are contained within the article.

Acknowledgments: The authors would like to express their gratitude to the anonymous reviewers and the academic editor for their invaluable comments and suggestions, which have significantly enhanced the quality of this manuscript. We would like to express our sincerest gratitude to the dedicated team members from the former China Railway Resources Mineral Exploration Co., Ltd., including Fangliu Dong, Yong Wang, Weiliang Wang, Haijun Yu, Bing Zhang, Xiuqin Chen, Yang Liu, Li Li, Huaixiang Li, Yaoxing Hu, and others for their invaluable contributions during the exploration process of the Luming Mo deposit. Bangfei Gao extends his gratitude to Guofeng Liu, from the Exploration Department of Tenke-Fungurume Mining (D.R. Congo), for his invaluable assistance with the Leapfrog Geo 2023.2 software.

Conflicts of Interest: Authors Bangfei Gao, Hui Xie and Yihang Li are employed by the company China Railway Resources Group Co., Ltd.; Authors Minghua Dong and Zhiliang Liu are employed by the company China Railway Resources Group Survey and Design Co., Ltd.; Author Tong Zhou is employed by the company Yichun Luming Mining Co., Ltd..

References

1. Wang, D.H.; Dai, H.Z.; Liu, S.B.; Wang, C.H.; Yu, Y.; Zhao, Z. Progress in strategic critical minerals exploration and production and proposals for a new round of prospecting in China. *Sci. Technol. Rev.* **2024**, *42*, 7–25. (In Chinese with English abstract) [CrossRef]
2. Sinclair, W.D. Porphyry deposits. In *Mineral Deposits of Canada: A Synthesis of Mzajor Deposit-Types, District Metallogeny, the Evolution of Geological Provinces, and Exploration Methods*; Goodfellow, W.D., Ed.; Geological Association of Canada, Special Publication: St. John's, NL Canada, 2007; Volume 5, pp. 223–243.
3. Li, C.Y.; Liao, R.Q. Formation mechanism and geochemical process of porphyry molybdenum deposits. *Acta Petrol. Sin.* **2020**, *36*, 77–84. (In Chinese with English abstract) [CrossRef]
4. Wu, H.Y.; Zhang, L.C.; Pirajno, F.; Shu, Q.H.; Zhang, M.; Zhu, M.T.; Xiang, P. The Mesozoic Caosiyao giant porphyry Mo deposit in Inner Mongolia, North China and Paleo-Pacific subduction-related magmatism in the northern North China Craton. *J. Asian Earth Sci.* **2016**, *127*, 281–299. [CrossRef]

5. Shu, Q.H.; Chang, Z.S.; Lai, Y.; Zhou, Y.T.; Sun, Y.; Cong, Y. Regional metallogeny of Mo-bearing deposits in Northeastern China, with new Re-Os dates of porphyry Mo deposits in the northern Xilamulun district. *Econ. Geol.* **2016**, *111*, 1783–1798. [CrossRef]
6. Zeng, Q.D.; Guo, W.K.; He, H.Y.; Zhou, L.L.; Cheng, G.H.; Su, F.; Wang, Y.B.; Wang, R.L. He, Ar, and S isotopic compositions and origin of giant porphyry Mo deposits in the Lesser Xing'an Range–Zhangguangcai Range metallogenic belt, Northeast China. *J. Asian Earth Sci.* **2018**, *165*, 228–240. [CrossRef]
7. Yang, J.H.; Xu, L.; Sun, J.F.; Zeng, Q.; Zhao, Y.N.; Wang, H.; Zhu, Y.S. Geodynamics of decratonization and related magmatism and mineralization in the North China Craton. *Sci. China Earth Sci.* **2021**, *64*, 1409–1427. [CrossRef]
8. U.S. Geological Survey. *Mineral Commodity Summaries 2024*; USGS2024; U.S. Geological Survey: Reston, VA, USA, 2024; p. 212. [CrossRef]
9. Zeng, Q.D.; Liu, J.M.; Chu, S.X.; Wang, Y.B.; Sun, Y.; Duan, X.X.; Zhou, L.L. Mesozoic molybdenum deposits in the east Xingmeng orogenic belt, Northeast China: Characteristics and tectonic setting. *Int. Geol. Rev.* **2012**, *54*, 1843–1869. [CrossRef]
10. Fan, Y.; Zhou, T.F.; Zhang, D.Y.; Yuan, F.; Ren, Z.; White, N. Spatial and temporal distribution and metallogical background of the Chinese molybdenum deposits. *Acta Geol. Sin.* **2014**, *88*, 784–804. (In Chinese with English abstract)
11. Han, C.M.; Xiao, W.J.; Windley, B.F.; Zhao, G.C.; Su, B.X.; Ao, S.J.; Zhang, J.E.; Zhang, Z.Y.; Wan, B.; Cui, B.; et al. Re-Os age of molybdenite from the Daheishan Mo deposit in the eastern Central Asian Orogenic Belt, NE China. *Resour. Geol.* **2014**, *64*, 379–386. [CrossRef]
12. Shu, Q.H.; Chiaradia, M. Mesozoic Mo mineralization in northeastern China did not require regional-scale pre-enrichment. *Econ. Geol.* **2021**, *116*, 1227–1237. [CrossRef]
13. Li, Z.Z.; Qin, K.Z.; Li, G.M.; Shunso, I.; Jin, L.Y.; Song, G.X.; Meng, Z.J. Formation of the giant Chalukou porphyry Mo deposit in northern Great Xing'an Range, NE China: Partial melting of the juvenile lower crust in intra-plate extensional environment. *Lithos* **2014**, *202–203*, 138–156. [CrossRef]
14. Shao, J.; Yang, H.Z.; Jia, B.; Peng, M.S. Geological characteristics and ore-forming age of Luming Mo deposit in Heilongjiang Province. *Miner. Depos.* **2012**, *31*, 1301–1310. (In Chinese with English abstract) [CrossRef]
15. Li, L.X.; Song, Q.H.; Wang, D.H.; Wang, C.H.; Qu, W.J.; Wang, Z.G.; Bi, S.Y.; Yu, C. Re-Os isotopic dating of molybdenite from the Fu'anpu molybdenum deposit of Jilin Province and discussion on its metallogenesis. *Rock Miner. Anal.* **2009**, *28*, 283–287. (In Chinese with English abstract) [CrossRef]
16. Hu, X.L.; Yao, S.Z.; Zeng, G.P.; Liu, W.H.; Zhang, Z.J. Multistage magmatism resulting in large-scale mineralization: A case from the Huojihe porphyry Mo deposit in NE China. *Lithos* **2019**, *326–327*, 397–414. [CrossRef]
17. Ju, N.; Ren, Y.S.; Wang, C.; Wang, H.; Zhao, H.L.; Qu, W.J. Ore genesis and molybdenite Re-Os dating of Dashihe molybdenum deposit in Dunhua, Jilin. *Glob. Geol.* **2012**, *31*, 68–76. (In Chinese with English abstract) [CrossRef]
18. Wang, Z.G.; Wang, K.Y.; Wan, D.; Konare, Y.; Yang, T.N.; Liang, Y.H. Metallogenic age and hydrothermal evolution of the Jidetun Mo deposit in central Jilin Province, Northeast China: Evidence from fluid inclusions, isotope systematics, and geochronology. *Ore Geol. Rev.* **2017**, *89*, 731–751. [CrossRef]
19. Zhang, J.; Yang, Y.C.; Han, S.J.; Wutiepu, W. Geochronology, geochemistry, and Lu-Hf Isotopic compositions of monzogranite intrusion from the Chang'anpu Mo deposit, NE China: Implications for tectonic setting and mineralization. *Minerals* **2022**, *12*, 967. [CrossRef]
20. Zeng, Q.D.; Liu, J.M.; Qin, K.Z.; Fan, H.R.; Chu, S.X.; Wang, Y.B.; Zhou, L.L. Types, characteristics, and time-space distribution of molybdenum deposits in China. *Int. Geol. Rev.* **2013**, *55*, 1311–1358. [CrossRef]
21. Ge, W.C.; Wu, F.Y.; Zhou, C.Y.; Zhang, J.H. Porphyry Cu-Mo deposits in the eastern Xing'an-Mongolian Orogenic Belt: Mineralization ages and their geodynamic implications. *Chin. Sci. Bull.* **2007**, *52*, 3416–3427. [CrossRef]
22. Yang, Y.C.; Han, S.J.; Sun, D.Y.; Guo, J.; Zhang, S.J. Geological and geochemical features and geochronology of porphyry molybdenum deposits in the Lesser Xing'an Range-Zhangguangcai Range metallogenic belt. *Acta Petrol. Sin.* **2012**, *28*, 379–390. (In Chinese with English abstract)
23. Zhang, Y.; Sun, J.G.; Chen, Y.J.; Zhao, K.Q.; Gu, A.L. Re-Os and U-Pb geochronology of porphyry Mo deposits in central Jilin Province: Mo ore-forming stages in Northeast China. *Int. Geol. Rev.* **2013**, *55*, 1763–1785. [CrossRef]
24. Zhang, Y.; Sun, J.G.; Xing, S.W.; Zhao, K.Q.; Ma, Y.B.; Zhang, Z.J.; Wang, Y. Ore-forming granites from Jurassic porphyry Mo deposits, east-central Jilin Province, China: Geochemistry, geochronology, and petrogenesis. *Int. Geol. Rev.* **2016**, *58*, 1158–1174. [CrossRef]
25. Wang, Z.G.; Wang, K.Y.; Konare, Y.; Yang, T.N.; Liang, Y.H. Metallogenesis and hydrothermal evolution of Middle Jurassic porphyry Mo deposits in the southern Zhangguangcai Range, NE China: Evidence from fluid inclusions, H-O-S isotopes, and Re-Os geochronology. *Int. Geol. Rev.* **2017**, *59*, 1391–1412. [CrossRef]
26. Shu, S.P.; Li, Q.G.; Liu, S.W.; Chen, Y.J. The origin and significance of the differences in petrogenesis and mineralization of porphyry gold, copper and molybdenum deposits. *Earth Sci. Front.* **2018**, *25*, 237–250. (In Chinese with English abstract) [CrossRef]
27. Qu, P.; Li, N.B.; Niu, H.C.; Shan, Q.; Weng, Q.; Zhao, X.C. Difference in ore-forming magma between the Mesozoic porphyry Cu-Mo and Mo deposits in NE China: Records from apatite and zircon geochemistry. *Ore Geol. Rev.* **2021**, *135*, 104218. [CrossRef]
28. Ju, N.; Zhang, D.; Zhang, G.B.; Zhang, S.; Ren, C.T.; Ren, Y.S.; Wang, H.; Wu, Y.; Liu, X.; Shi, L.; et al. Geology and mineralization of the Daheishan supergiant porphyry molybdenum deposit (1.65 Bt), Jilin, China: A review. *China Geol.* **2023**, *6*, 494–530. [CrossRef]

29. Li, X.G.; Yang, J.F.; Yang, Y.C. Comprehensive geological, physical and chemical characteristics of Xinantun molybdenum deposit in central area of Jilin Province, and prospecting potential analysis. *J. Jilin Univ. (Earth Sci. Ed.)* **2024**, *54*, 1–20. (In Chinese with English abstract) [CrossRef]
30. Zheng, W.; Yu, X.F. Geochronological and geochemical constraints on the petrogenesis and geodynamic setting of the Daheishan porphyry Mo deposit, Northeast China. *Resour. Geol.* **2018**, *68*, 1–21. [CrossRef]
31. Liu, J.; Mao, J.W.; Wu, G.; Wang, F.; Luo, D.F.; Hu, Y.Q. Geochemical signature of the granitoids in the Chalukou giant porphyry Mo deposit in the Heilongjiang Province, NE China. *Ore Geol. Rev.* **2015**, *64*, 35–52. [CrossRef]
32. Zhao, Q.; Zhai, D.; Mathur, R.; Liu, J.; Selby, D.; Williams-Jones, A.E. The giant Chalukou porphyry Mo deposit, Northeast China: The product of a short-Lived, high flux mineralizing event. *Econ. Geol.* **2021**, *116*, 1209–1225. [CrossRef]
33. Hu, X.L.; Ding, Z.J.; He, M.C.; Yao, S.Z.; Zhu, B.P.; Shen, J.; Chen, B. Two epochs of magmatism and metallogeny in the Cuihongshan Fe-polymetallic deposit, Heilongjiang Province, NE China: Constrains from U-Pb and Re-Os geochronology and Lu-Hf isotopes. *J. Geochem. Explor.* **2014**, *143*, 116–126. [CrossRef]
34. Chen, X.; Liu, J.J.; Carranza, E.J.M.; Zhang, D.H.; Collins, A.S.; Yang, B.; Xu, B.W.; Zhai, D.G.; Wang, Y.H.; Wang, J.P.; et al. Geology, geochemistry, and geochronology of the Cuihongshan Fe-polymetallic deposit, Heilongjiang Province, NE China. *Geol. J.* **2018**, *54*, 1254–1278. [CrossRef]
35. Fei, X.H.; Zhang, Z.C.; Cheng, Z.G.; Santosh, M.; Jin, Z.L.; Wen, B.B.; Li, Z.X.; Xu, L.J. Highly differentiated magmas linked with polymetallic mineralization: A case study from the Cuihongshan granitic intrusions, Lesser Xing'an Range, NE China. *Lithos* **2018**, *302–303*, 158–177. [CrossRef]
36. Zhang, Y.; Sun, J.G.; Xing, S.W.; Zhang, Z.J. Geochronology and geochemistry of the Cuihongshan Fe-polymetallic deposit, Northeastern China: Implications for ore genesis and tectonic setting. *Can. J. Earth Sci.* **2018**, *55*, 475–489. [CrossRef]
37. Xing, K.; Shu, Q.H.; Lentz, D.R.; Wang, F.Y. Zircon and apatite geochemical constraints on the formation of the Huojihe porphyry Mo deposit in the Lesser Xing'an Range, NE China. *Am. Mineral.* **2020**, *105*, 382–396. [CrossRef]
38. Xing, K.; Shu, Q.H.; Lentz, D.R. Constraints on the formation of the giant Daheishan porphyry Mo deposit (NE China) from whole-Rock and accessory mineral geochemistry. *J. Petrol.* **2021**, *62*, egab018. [CrossRef]
39. Shu, Q.H.; Chang, Z.S.; Lai, Y.; Hu, X.L.; Wu, H.Y.; Zhang, Y.; Wang, P.; Zhai, D.G.; Zhang, C. Zircon trace elements and magma fertility: Insights from porphyry (-skarn) Mo deposits in NE China. *Miner. Depos.* **2019**, *54*, 645–656. [CrossRef]
40. Zhu, J.H. Role and significance of geochemical exploration in the discovery of the Dayangshugou molybdenum deposit, Liaoning. *Geol. China* **2007**, *34*, 342–346. (In Chinese with English abstract)
41. Han, Z.Z.; Zhao, H.L.; Li, J.J.; Luo, J.; Zhang, W.L.; Li, W.L. Metallogenic prognosis of porphyry style molybdenum mineralization potential in Xing'an area, Tieli city, Heilongjiang province. *Geol. Explor.* **2009**, *45*, 253–259. (In Chinese with English abstract)
42. Meng, Z.J.; Kan, X.S.; Li, X.C.; Wang, J.P.; Zhang, R.Z.; Lv, K.P.; Sun, Z.J.; Shi, Y.J.; Zhang, J.N.; Wang, H.Y.; et al. The discovery and exploration process of the Chalukou giant porphyry molybdenum polymetallic deposit in forest-covered area of Northeastern Da Higgnan Mountains and its geological significance. *Geol. China* **2011**, *38*, 1504–1517. (In Chinese with English abstract)
43. Hu, Y.X.; Zhang, B. Geophysical characteristics and anomaly model of the Luming molybdenum deposit in Heilongjiang Province. *Met. Mine* **2015**, *1*, 82–85. (In Chinese with English abstract)
44. Yin, G.L.; Liang, K.W.; Yang, F.S.; Wen, D.; Wang, Y. A discussion on the geochemical characteristics of stream sediments and ore-prospecting orientation in Weixing-Anquan area of Heilongjiang Province. *Geophys. Geochem. Explor.* **2017**, *41*, 402–409. (In Chinese with English abstract) [CrossRef]
45. Zheng, J.; Yan, H.; Hua, L. The physical characteristics of rock(ore) and prospecting indicator of porphyry Cu-Mo deposits in Duobaoshan ore district. *Chin. J. Eng. Geophys.* **2018**, *15*, 788–797. (In Chinese with English abstract)
46. Tan, H.Y.; Shu, G.L.; Lv, J.C.; Han, R.P.; Zhang, S.K.; Lin, L. LA-ICP-MS zircon U-Pb and molybdenite Re-Os dating for the Luming large-scale molybdenum deposit in Xiao Hinggan Mountains and its geological implication. *J. Jilin Univ. (Earth Sci. Ed.)* **2012**, *42*, 1757–1770. (In Chinese with English abstract)
47. Ma, S.Q.; Chen, J. Geochronology and geochemistry of granite of the Luming molybdenum deposit in Heilongjiang and their geological significance. *Geol. China* **2012**, *39*, 1162–1171. (In Chinese with English abstract)
48. Liu, C.; Deng, J.F.; Luo, Z.H.; Tian, S.P.; Zhang, Y.; Zhong, C.T.; Selby, D.; Zhao, H.D. Post-batholith metallogenesis: Evidence from Luming super large molybdenite deposit in Lesser Xing'an Range. *Acta Petrol. Sin.* **2014**, *30*, 3400–3418. (In Chinese with English abstract)
49. Hu, X.L.; Ding, Z.J.; He, M.C.; Yao, S.Z.; Zhu, B.P.; Shen, J.; Chen, B. A porphyry-skarn metallogenic system in the Lesser Xing'an Range, NE China: Implications from U-Pb and Re-Os geochronology and Sr-Nd-Hf isotopes of the Luming Mo and Xulaojiugou Pb-Zn deposits. *J. Asian Earth Sci.* **2014**, *90*, 88–100. [CrossRef]
50. Sun, Q.L.; Sun, J.G.; Zhao, K.Q.; Tang, C.; Zhang, Y.; Han, S.J.; Yang, F. Re-Os isotopic dating and geological significance of Luming porphyry molybdenum deposit in Heilongjiang. *Glob. Geol.* **2014**, *33*, 418–425. (In Chinese with English abstract)
51. Zhan, C.P.; Wang, E.D.; Bi, Z.W.; Chen, J. Geochemical characteristics of mineralization rock from the Luming molybdenum deposit in Heilongjiang Province. *J. Northeast. Univ. (Nat. Sci.)* **2014**, *35*, 1039–1042. (In Chinese with English abstract) [CrossRef]
52. Cheng, G.H.; Wang, R.L.; Zeng, Q.D.; Guo, Y.P.; Duan, X.X.; Wei, J.J.; Zhang, J.S.; Gao, X.H. Zircon U-Pb ages, Hf isotopes of the granitoids and Re-Os ages of the molybdenites in Luming molybdenum ore area, Heilongjiang Province, and its geological significance. *Acta Petrol. Sin.* **2015**, *31*, 2450–2464. (In Chinese with English abstract)

53. Wang, L.L.; Sun, F.Y.; Xu, J.H.; Wang, K.Y.; Wang, L.; Xu, Q.L. Ore-forming fluid and evolution of Luming molybdenum deposit, Heilongjiang Province. *Acta Petrol. Sin.* **2015**, *31*, 979–990. (In Chinese with English abstract)
54. Chen, L.; Zhang, Y. In situ major-, trace-elements and Sr-Nd isotopic compositions of apatite from the Luming porphyry Mo deposit, NE China: Constraints on the petrogenetic-metallogenic features. *Ore Geol. Rev.* **2018**, *94*, 93–103. [CrossRef]
55. Xi, A.H.; Ge, Y.H.; Liu, J.; Xu, B.W.; Li, B.L.; Zhu, Y.D. Discovery of adakite in Tieli Luming-molybdenum mine, Heilongjiang Province and its geological implications. *Acta Petrol. Sin.* **2018**, *34*, 719–732. (In Chinese with English abstract)
56. Chen, Y.J.; Zhang, C.; Wang, P.; Pirajno, F.; Li, N. The Mo deposits of Northeast China: A powerful indicator of tectonic settings and associated evolutionary trends. *Ore Geol. Rev.* **2017**, *81*, 602–640. [CrossRef]
57. Sun, J.G.; Zhang, Y.; Xing, S.W.; Zhao, K.Q.; Zhang, Z.J.; Bai, L.A.; Ma, Y.B.; Liu, Y.S. Genetic types, ore-forming age and geodynamic setting of endogenic molybdenum deposits in the eastern edge of Xing-Meng orogenic belt. *Acta Petrol. Sin.* **2012**, *28*, 1317–1332. (In Chinese with English abstract)
58. Li, Z.L.; Zeng, Q.D.; Liu, J.M.; Friis, H.; Zhang, Z.L.; Duan, X.X.; Lan, T.G. Geochronology of magmatism and mineralization of the Daheishan giant porphyry molybdenum deposit, JilinProvince, Northeast China: Constraints on ore genesis and implications for geodynamic setting. *Int. Geol. Rev.* **2014**, *56*, 929–953. [CrossRef]
59. Tan, H.Y. Metallogenetic Series and Prospecting Assessment in Lesser Xing'an Range-Zhangguangcai Range Metallogenic Belt of Heilongjiang Province. Ph.D. Thesis, China University of Geosciences (Beijing), Beijing, China, 2013.
60. Xiong, S.Q.; Ding, Y.Y.; Li, Z.K.; Zhang, H.R.; Fan, Z.G.; Guo, Z.H.; Tong, J.; Zhou, D.Q.; Huang, X.H. *Geomagnetic and Geological Structures of the Chinese Mainland: Interpretation of the Chinese Mainland 1:1,000,000 Aeromagnetic Anomaly Map*; Geological Publishing House: Beijing, China, 2016. (In Chinese)
61. Li, Y.C.; Han, Z.Z.; Wu, G.G.; Zhang, D. Metallogenesis and metallogenic differences of the early Mesozoic granites in the Yichun area, Heilongjiang Province. *Geol. Exploation* **2013**, *49*, 28–38. (In Chinese with English abstract)
62. Zeng, G.P.; Hu, X.L.; Yao, S.Z.; Dong, Z.L.; Zhou, M.; Yang, J. Two epochs of Mo mineralization in Cuihongshan Fe-Mo-polymetallic ore field, Heilongjiang Province. *Earth Sci.-J. China Univ. Geosci.* **2022**, *47*, 309–324. (In Chinese with English abstract) [CrossRef]
63. Tian, S.P. Geochemical Characteristics of Ore Rock in the Luming and Culling Mo Deposit, Heilongjiang Province. Master's Thesis, Jilin University, Jilin, China, 2013. (In Chinese with English abstract)
64. Chen, J. Metallogenic Setting and Metallogenesis of Nonferrous-Precious Metals in Lesser Hinggan Mountain, Heilongjiang Province. Ph.D. Thesis, Jilin University, Jilin, China, 2011. (In Chinese with English abstract)
65. Cai, W.J.; Feng, C.Y.; Ni, W.C. Analysis and study of aeromagnetic anomalous characteristics of lead-zinc polymetallic deposits in Yichun area. *Geophys. Geochem. Explor.* **2016**, *40*, 869–875. (In Chinese with English abstract) [CrossRef]
66. Jia, W.L.; Sun, J.S. Non-ferrous metal mineral prospecting report in Luming and Ping'an, Tieli City, Heilongjiang Province. Unpublished Technical Report in Chinese. 2005.
67. Jia, W.L.; Sun, J.S. Geological Exploration Report of Luming Mo Deposit, Tieli City, Heilongjiang Province. Unpublished Technical Report in Chinese. 2006.
68. Wang, Y.; Wang, W.L.; Dong, F.L.; Chen, Y.B.; Gao, B.F.; Chen, X.Q.; Xu, J.L.; Li, L.; Yu, H.J.; Zhang, W.Q.; et al. Verification report on the resource and reserve of Luming Mo deposit, Tieli City, Heilongjiang Province. Unpublished Technical Report in Chinese. 2011.
69. Middlemost, E.A.K. Naming materials in the magma/igneous rock system. *Earth Sci. Rev.* **1994**, *37*, 215–224. [CrossRef]
70. Roberts, M.P.; Clemens, J.D. Origin of high-potassium, calc-alkaline, I-type granitoids. *Geology* **1993**, *21*, 825–828. [CrossRef]
71. Maniar, P.D.; Piccoli, P.M. Tectonic discrimination of granitoids. *Geol. Soc. Am. Bull.* **1989**, *101*, 635–643. [CrossRef]
72. Sun, S.S.; McDonough, W.F. Chemical and isotopic systematics of oceanic basalts: Implications for mantle composition and processes. In *Magmatism in the Ocean Basins*; Saunders, A.D., Norry, M.J., Eds.; Geological Society of London, Special Publications: Oxford, UK, 1989; Volume 42, pp. 313–346. [CrossRef]
73. Sillitoe, R.H. Porphyry copper systems. *Econ. Geol.* **2010**, *105*, 3–41. [CrossRef]
74. Mao, J.W.; Zhang, J.D.; Pirajno, F.; Ishiyama, D.; Su, H.M.; Guo, C.L.; Chen, Y.C. Porphyry Cu-Au-Mo-epithermal Ag-Pb-Zn-distal hydrothermal Au deposits in the Dexing area, Jiangxi province, East China-A linked ore system. *Ore Geol. Rev.* **2011**, *43*, 203–216. [CrossRef]
75. Catchpole, H.; Kouzmanov, K.; Putlitz, B.; Seo, J.H.; Fontboté, L. Zoned base metal mineralization in a porphyry system: Origin and evolution of mineralizing fluids in the Morococha district Peru. *Econ. Geol.* **2015**, *110*, 39–71. [CrossRef]
76. Li, Z.K.; Bi, S.J.; Li, J.W.; Zhang, W.; Cooke, D.R.; Selby, D. Distal Pb-Zn-Ag veins associated with the world-class Donggou porphyry Mo deposit, southern North China Craton. *Ore Geol. Rev.* **2017**, *82*, 232–251. [CrossRef]
77. Jin, L.Y.; Qin, K.Z.; Li, G.M.; Zhao, J.X.; Li, Z.Z. Characteristics, controlling factors and exploration implications of porphyry molybdenum-hydrothermal vein-style lead-zinc-silver metallogenic systems. *Acta Petrol. Sin.* **2020**, *36*, 3813–3839. [CrossRef]
78. Chitalin, A.F.; Baksheev, I.A.; Nikolaev, Y.N.; Nagornaya, E.V.; Khabibullina, Y.N.; Nikolaeva, I.Y.; Kalko, I.A.; Müller, D. Porphyry-epithermal Cu-Mo-Au–Ag mineralization in the Nakhodkaore field, Baimka Trend, Chukotka, Russia: A geological, mineralogical, and geochemical perspective. *Miner. Depos.* **2023**, *58*, 287–306. [CrossRef]
79. Lowell, J.D.; Guilbert, J.M. Lateral and vertical alteration-mineralization zoning in porphyry ore deposits. *Econ. Geol.* **1970**, *65*, 373–408. [CrossRef]

80. Li, Z.Z.; Li, G.M.; Meng, Z.J.; Qin, K.Z.; Song, G.X.; Jin, L.Y.; Kan, X.S.; Wang, J.; Zhang, X.N. Petrofacies classification, characteristics and formation mechanism of breccias in Chalukou giant molybdenum deposit of Da Hinggan Mountains. *Miner. Depos.* **2014**, *33*, 607–624. (In Chinese with English abstract)
81. Qing, M.; Wang, W.C.; Li, W.B.; Zhang, L.J.; Zhang, D.; Han, X.J. Exploration of porphyry Cu±Mo±Au deposits in island arc and continental margin arc environment: Application of geological indexes. *Miner. Depos.* **2019**, *38*, 1223–1244. (In Chinese with English abstract) [CrossRef]
82. Escolme, A.; Cooke, D.R.; Hunt, J.; Berry, R.F.; Maas, R.; Creaser, R.A. The productor a Cu-Au-Mo deposit, Chile: A Mesozoic magmatic-hydrothermal breccia complex with both porphyry and iron oxide Cu-Au affinities. *Econ. Geol.* **2020**, *115*, 543–580. [CrossRef]
83. Landtwing, M.R.; Dillenbeck, E.D.; Heinrich, C.A. Evolution of the breccia-hosted porphyry Cu-Mo-Au deposit at Agua Rica, Argentina: Progressive unroofing of a magmatic hydrothermal system. *Econ. Geol.* **2002**, *97*, 1273–1292. [CrossRef]
84. Carranza, E.J.M.; Sadeghi, M. Primary geochemical characteristics of mineral deposits implications for exploration. *Ore Geol. Rev.* **2012**, *45*, 1–4. [CrossRef]
85. Borden, R.K. Environmental geochemistry of the Bingham Canyon porphyry copper deposit, Utah. *Environ. Geol.* **2003**, *43*, 752–758. [CrossRef]
86. Moradpouri, F.; Ahmadi, S.M.H.; Ghaedrahmati, R.; Barani, K. Geochemical evaluation of the surface erosion for a hidden Cu-Au porphyry deposit using soil samples along with 3D ore-body modeling. *Arab. J. Geosci.* **2022**, *15*, 1164. [CrossRef]
87. Sun, W.T.; Zheng, Y.Y.; Wang, W.; Feng, X.; Zhu, X.S.; Zhang, Z.Y.; Hou, H.X.; Ge, L.S.; Lv, H.Q. Geochemical characteristics of primary halo and prospecting significance of the Qulong porphyry copper-molybdenum deposit in Tibet. *Minerals* **2023**, *13*, 333. [CrossRef]
88. Liu, Y.; Xia, Q.L. Compositional balance analysis: A novel method for tectono-geochemical anomaly identification and blind ore deposit prediction. *Appl. Geochem.* **2024**, *164*, 105939. [CrossRef]
89. Shu, Q.H.; Deng, J.; Chang, Z.S.; Wang, Q.F.; Niu, X.D.; Xing, K.; Sun, X.; Zhang, Z.K.; Zeng, Q.W.; Zhao, H.S.; et al. Skarn zonation of the giant Jiama Cu-Mo-Au deposit in Southern Tibet, SW China. *Econ. Geol.* **2024**, *119*, 1–22. [CrossRef]
90. Stoltnow, M.; Weis, P.; Korges, M. Hydrological controls on base metal precipitation and zoning at the porphyry-epithermal transition constrained by numerical modeling. *Sci. Rep.* **2023**, *13*, 3786. [CrossRef] [PubMed]
91. Clark, D.A. Magnetic effects of hydrothermal alteration in porphyry copper and iron-oxide copper-gold systems: A review. *Tectonophysics* **2014**, *624–625*, 46–65. [CrossRef]
92. Mitchinson, D.E.; Enkin, R.J.; Hart, C.J.R. Linking Porphyry Deposit Geology to Geophysics via Physical Properties: Adding Value to Geoscience BC Geophysical Data. Geoscience BC 2013, Report 2013-14, p. 116. Available online: http://www.geosciencebc.com/reports/gbcr-2013-14/ (accessed on 10 January 2024).
93. Han, Z.H.; Wang, R.; Tong, X.S.; Sun, F.; Li, Y.Y.; Liu, S.Y.; Xue, Q.W. Multi-scale exploration of giant Qulong porphyry deposit in a collisional setting. *Ore Geol. Rev.* **2021**, *139*, 104455. [CrossRef]
94. John, D.A.; Ayuso, R.A.; Barton, M.D.; Blakely, R.J.; Bodnar, R.J.; Dilles, J.H.; Gray, F.; Graybeal, F.T.; Mars, J.C.; McPhee, D.K.; et al. *Porphyry Copper Deposit Model, Chap. B of Mineral Deposit Models for Resource Assessment*; U.S. Geological Survey Scientific Investigations Report 2010-5070-B; USGS Central Mineral and Environmental Resources Science Center: Denver, CO, USA, 2010; 169p.
95. Riveros, K.; Veloso, E.; Campos, E.; Menzies, A.; Véliz, W. Magnetic properties related to hydrothermal alteration processes at the Escondida porphyry copper deposit, Northern Chile. *Miner. Depos.* **2014**, *49*, 693–707. [CrossRef]
96. Howe, B.; Devriese, S.G.R. An empirical geophysical model for porphyry copper deposits in the Laramide Copper Province. In Proceedings of the 4th AEGC: Geoscience-Breaking New Ground-13-18, Brisbane, Australia, 13–18 March 2023.
97. Howe, B.; Kroll, A. The geophysical response of the Tupinda Cu-Au-Mo porphyry prospect, Tabar Islands, Papua New Guinea. *ASEG Ext. Abstr.* **2010**, *2010*, 1–5. [CrossRef]
98. Wang, Q.; Wang, X.; Cheng, Z.; Zhang, B.; Du, Z.; Yan, T.; Yuan, H.; Li, X.; Qiao, Y.; Liu, H. Geogas-carried metal prospecting for concealed ore deposits: A review of case studies in China. *Minerals* **2023**, *13*, 1553. [CrossRef]

MDPI AG
Grosspeteranlage 5
4052 Basel
Switzerland
Tel.: +41 61 683 77 34

Minerals Editorial Office
E-mail: minerals@mdpi.com
www.mdpi.com/journal/minerals

www.ingramcontent.com/pod-product-compliance
Lightning Source LLC
Chambersburg PA
CBHW061505220926
51850CB00004B/171
* 9 7 8 3 7 2 5 8 8 1 8 3 3 *